Studies in Computational Intelligence

Volume 740

Series editor

Janusz Kacprzyk, Polish Academy of Sciences, Warsaw, Poland
e-mail: kacprzyk@ibspan.waw.pl

The series "Studies in Computational Intelligence" (SCI) publishes new developments and advances in the various areas of computational intelligence—quickly and with a high quality. The intent is to cover the theory, applications, and design methods of computational intelligence, as embedded in the fields of engineering, computer science, physics and life sciences, as well as the methodologies behind them. The series contains monographs, lecture notes and edited volumes in computational intelligence spanning the areas of neural networks, connectionist systems, genetic algorithms, evolutionary computation, artificial intelligence, cellular automata, self-organizing systems, soft computing, fuzzy systems, and hybrid intelligent systems. Of particular value to both the contributors and the readership are the short publication timeframe and the world-wide distribution, which enable both wide and rapid dissemination of research output.

More information about this series at http://www.springer.com/series/7092

Khaled Shaalan · Aboul Ella Hassanien
Fahmy Tolba
Editors

Intelligent Natural Language Processing: Trends and Applications

Springer

Editors
Khaled Shaalan
The British University in Dubai
Dubai
United Arab Emirates

Aboul Ella Hassanien
Faculty of Computers and Information Technology
Cairo University
Giza
Egypt

Fahmy Tolba
Faculty of Computers and Information
Ain Shams University
Cairo
Egypt

ISSN 1860-949X ISSN 1860-9503 (electronic)
Studies in Computational Intelligence
ISBN 978-3-319-88371-7 ISBN 978-3-319-67056-0 (eBook)
https://doi.org/10.1007/978-3-319-67056-0

Printed on acid-free paper

This Springer imprint is published by Springer Nature
The registered company is Springer International Publishing AG
The registered company address is: Gewerbestrasse 11, 6330 Cham, Switzerland

Preface

New trends of Natural Language Processing systems are emerging. Applications of these trends have been applied to various domains including education, travel and tourism, health care, among others. Many issues encountered during the development of these applications are resolved by incorporating language technology solutions. The aim of the "Intelligent Natural Language Processing: Trends and Applications" edited book is to bring scientists, researcher scholars, practitioners, and students from academia and industry to present recent and ongoing research activities about the recent advances, techniques, and applications of Natural Language Processing systems and to allow the exchange of new ideas and application experiences.

The edited book contains a selection of papers accepted for presentation and discussion at The Special Track On Intelligent Language Processing: Trends and Applications, International Conference on Advanced Intelligent Systems and Informatics (AISI 2016). Also, an open call for book chapters was targeted to researchers of the Computational Linguistics and Natural Language Processing research Community.

This book is intended to present the state-of-the-art in research on natural language processing, computational linguistics, applied Arabic linguistics and related areas. We accepted 35 submissions. The accepted papers covered the following eleven themes (parts): Character and Speech Recognition, Morphological, Syntactic, and Semantic Processing, Information Extraction, Information Retrieval and Question answering, Text Classification, Text Mining, Text summarization, Sentiment Analysis, Machine translation, Building and Evaluating Linguistic Resources, E-Learning. Each Submission is reviewed by the editorial board. Evaluation criteria include correctness, originality, technical strength, significance, quality of presentation, and interest and relevance to the book scope. Chapters of this book provide a collection of high-quality research works that address broad challenges in both theoretical and application aspects of intelligent natural language processing and its applications.

The volumes of this book are published within Studies in Computational Intelligence Series by Springer which has a high SJR impact. We acknowledge all that contributed to the staging of this edited book (authors, committees, and organizers). We deeply appreciate their involvement and support that was crucial for the success of the "Intelligent Natural Language Processing: Trends and Applications" edited book.

Dubai, United Arab Emirates Khaled Shaalan
Giza, Egypt Aboul Ella Hassanien
Cairo, Egypt Fahmy Tolba

Contents

Part I
Sentiment Analysis

Using Deep Neural Networks for Extracting Sentiment Targets in Arabic Tweets

Ayman El-Kilany, Amr Azzam and Samhaa R. El-Beltagy

Abstract In this paper, we investigate the problem of recognizing entities which are targeted by text sentiment in Arabic tweets. To do so, we train a bidirectional LSTM deep neural network with conditional random fields as a classification layer on top of the network to discover the features of this specific set of entities and extract them from Arabic tweets. We've evaluated the network performance against a baseline method which makes use of a regular named entity recognizer and a sentiment analyzer. The deep neural network has shown a noticeable advantage in extracting sentiment target entities from Arabic tweets.

Keywords Sentiment target recognition · Long short-term memory networks
Recurrent neural networks · Conditional random fields · Arabic

1 Introduction

Sentiment target recognition is the task of identifying the subset of entities in a piece of text which are targeted by the text's overall sentiment. Sentiment target recognition links between sentiment analysis and named entity recognition as it seeks to recognize only entities which are the focus or the subject of a tweet's positive or negative sentiment.

The task we explore is formulated as follows: given some input text annotated with either positive or negative sentiment, identify the entity in that text that is targeted by the given sentiment. For example, in the following text: "So happy that

A. El-Kilany · A. Azzam
Faculty of Computers and Information, Cairo University, Giza 12613, Egypt
e-mail: a.elkilany@fci-cu.edu.eg

A. Azzam
e-mail: a.tarek@fci-cu.edu.eg

S.R. El-Beltagy (✉)
Center for Informatics Science, Nile University, Giza 12588, Egypt
e-mail: samhaa@computer.org

K. Shaalan et al. (eds.), *Intelligent Natural Language Processing: Trends and Applications*, Studies in Computational Intelligence 740,
https://doi.org/10.1007/978-3-319-67056-0_1

Kentucky lost to Tennessee!" there are two entities: 'Kentucky' and 'Tennessee'. The overall sentiment of this tweet is positive and as such the sentiment target recognizer must identify the target of this positive sentiment which is "Tennessee".

Sentiment target recognition in this setting is a specific case of a more general framework which assumes that each entity in the text may have its own sentiment. In this paper, we are proposing a solution to find targets of the overall text sentiment rather than breaking down the text into multiple sentiments with multiple targets for each sentiment. Such a setting allows us to tackle the sentiment target recognition problem as a specific case of named-entity recognition.

In this paper, we propose two models for the sentiment target recognition task. All our experiments are carried out on Arabic text. The first model, is a baseline one which utilizes a named entity recognizer [1, 2] and a sentiment analyzer [3] to recognize the target entities of some given text's sentiment. The second model is a deep neural network for named entity recognition which is provided with target named entities in the training phase in order to identify them during the extraction phase.

The deep neural network which we have used, consisted of a bidirectional LSTM network [4] with a conditional random fields [5] classifier as a layer on top of it. A word2vec model for Arabic words was built in order to represent the word vectors in the network. The deep neural network model has shown a noticeable advantage over the baseline model in detecting sentiment targets over a dataset of Arabic tweets. Figures 1 and 2 depict the components of both baseline and deep neural network models.

The rest of this paper is organized as follows: Sect. 2 provides a background with respect to tasks that are related to sentiment target recognition; Sect. 3 describes the collection and annotation of Arabic tweets; Sect. 4 details the process of building the word embeddings model while Sect. 5 describes both the baseline model and the deep neural model; Sect. 6 presents the evaluation of the proposed model through comparison of its performance against the baseline model while Sect. 7 concludes this paper.

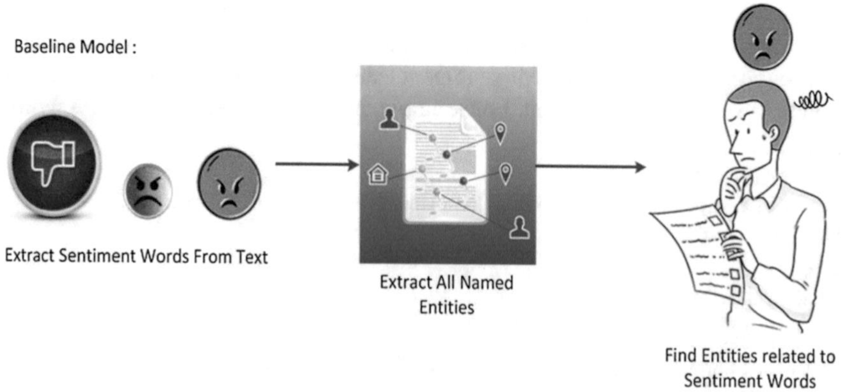
Baseline Model :
Extract Sentiment Words From Text
Extract All Named Entities
Find Entities related to Sentiment Words

Fig. 1 Components of the baseline model

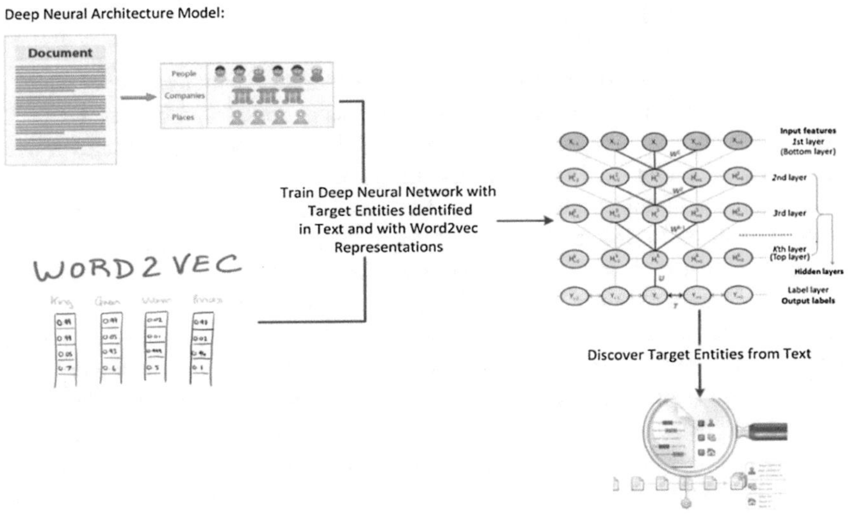

Fig. 2 Components of the deep neural network model

2 Background

The task of sentiment target recognition falls somewhere in between stance detection, targeted sentiment analysis, and named entity recognition. While named entity recognition seeks to find all entities in the text, targeted sentiment analysis seeks to assign each entity or aspect of an entity in some given text, its relevant sentiment which is highly dependent on the context in which it appeared. Stance detection on the other hand, seeks to determine whether the author of some given text, is pro or against some predefined 'target' which can be an entity or an issue. Many models have been developed for each of the aforementioned tasks, but recently the use of deep neural networks models in all of those tasks, has been investigated as detailed here.

Long Short-term Memory Networks (LSTMs) [6] are a special type of recurrent neural networks [7] which operate on sequential data. Given some input as a sequence of vectors, an LSTM network should return a decision about each vector in the sequence. LSTMs are designed to tackle dependencies in long sequences by using a memory-cell to conserve the state of the operations on earlier vectors in the sequence and prevent vectors from forgetting those through multiple iterations. In [8, 4], Bidirectional LSTMs were trained for the tasks of named entity recognition, part of speech tagging and chunking without the inclusion of any handcrafted features. Word embeddings [9] and character embeddings [8] were used as the only input representations for the sentence sequences in the LSTM network. Given the bidirectional theme, each word representation vector was derived from its right context and left context in order to present its state correctly. Recognizing named entities with LSTM network was found to obtain state-of-the-art performance in four different languages [8].

Detecting the polarity of aspects within English text, is another task for which a multitude of systems were developed using deep neural approaches [10]. An example of such as a system is presented in [11] where the authors used a deep convolutional neural network to address the problem.

In [8], a conditional random fields (CRF) classifier was used on top of an LSTM network in order to find the best tagged sequence of words and named entities for an input sequence of words. The CRF classifier was used to model tagging decisions jointly rather than predicating the tag for each word independently. A CRF classifier was also used in [12] for the targeted sentiment analysis task in order to extract each entity contained within some given text and assign it a sentiment value using a set of handcrafted features for both named entities and sentiments.

Recurrent neural networks have also been used for the stance detection task. An example is presented in [13] where a gated recurrent neural network is used to identify the sentiment of some input text given a target. The target word as well as its left context and its right context, are represented clearly in the network in order to obtain a better classification for the sentiment related to this target.

Using a deep neural network to solve similar problems as those presented in [14] has a clear advantage for discovering important features through the network's hidden layers. The automatic discovery of features provides the capability of discovering a variety of hidden information within text without having to worry about feature engineering.

In this work, we use a bidirectional LSTM to learn the features of sentiment targets which are named entities, rather than learning the features of *all* named entities. Tackling such a problem using a regular named entity recognizer would be challenging as the handcrafted features of named entities and sentiment targets aren't expected to be different.

3 Data Collection and Annotation

When we embarked on the task of sentiment target recognition in Arabic tweets, we discovered that no previous work has been carried out in this area. We therefore decided that we need to build our own dataset. To create this dataset, we used a previously collected set of tweets that had 480,000 entries. The dataset was collected by searching the Twitter API using entries from the NileULex sentiment lexicon [15]. Tweets from this dataset were randomly selected, annotated with sentiment using the sentiment analyzer described in [3], and annotated with entities using a modified version of the Named Entity Recognition system described in [1]. Tweets that were annotated as neutral or that had no entities were discarded. A total of 12,897 tweets were generated this way. The number of generated tweets was much more than the target we were aiming to annotate because we knew that some tweets will be ambiguous, while others will have wrong entities. The next step following the collection of this dataset, was revising and correcting the automatically assigned labels, as well as specifying the sentiment target for each tweet. To facilitate this process, a web based, user friendly application was developed. In cases where the assigned sentiment was wrong or where the detected entity was incorrect, the annotator was asked to correct it. The annotator was also asked to specify the target entities for each tweet either by selecting from already identified entities, or by manually entering the target entity.

The annotation process was carried out by a paid expert annotator. To ensure that the resulting annotations would be of high quality, we periodically collected random annotated samples and examined their correctness. The annotator was aware that we were doing this and the agreement was that s/he will only get paid after the annotation process was completed. This process resulted in the annotation of over 5000 target entities.

We created a set of tags similar to those defined by the IOB format [8] which is used for tagging sentiment targets and non-targets. The sentiment target recognition tagging scheme only has three tags: B-target, I-target, O-tag, and aword is tagged with B-target if it is the first word (beginning) of a target entity in a tweet. The I-target tag is used when the target is made up of a sequence of words. While the first word in such a case is tagged with the B-target tag, the remaining words in the sequence are tagged with I-target tag. Any word which is not a target is tagged with the O-tag. This dataset is available by request from any of the authors. It will also be availed online.

4 Building Word Embeddings

One of the goals of this work was to experiment with a bidirectional Long Short-Term Memory (BI-LSTM) neural network. Previous work for text related tasks has shown that the performance of such networks is usually enhanced by the use of word embeddings. Word embeddings [16] refer to a model for feature learning used in NLP tasks to transform input words into a clustered representation of real numbered vectors. The model results in the construction of low dimensional vector representations for all words in a given input text corpus. Words sharing common contexts in that corpus are transformed into vectors that are close to each other in the vector space. As such, it can be stated that word embeddings produce an expressive representation for words as it captures their semantics.

Using word embeddings within deep learning models for NLP tasks such semantic parsing; named entity recognition and sentiment analysis has been shown to result in a great performance boost. These tasks usually use a limited set of labeled instances for training purposes. Word embeddings are usually constructed from huge corpora, so it is very beneficial in these tasks to make use of pre-trained word embeddings in order to have a generalized model with a reliable estimation for the problem parameters.

For the English language, it is quite easy for researchers wishing to experiment with word embeddings to use publicly available pre-trained embeddings. However, for Arabic such embeddings do not exist, so we had to build our own. Training a word embeddings model usually requires a large input corpus. The corpus that we have used consisted of fifty four million Arabic tweets collected over the duration of 4 month (during 2015) by researchers at Nile University. In this work, we have applied two word embeddings generation models to generate a vector for each corpus word.

The first model was built using word2vec [16] which was trained using the skip-n-gram model. Building the model involved a series of experiments which involved using different settings and configurations. The tool used for building the word2vec model was Gensim[1] which is a widely used Python library for topic modeling and other text related tasks. At the end, we used a model which was set to produce a vector of length 100, and whose parameters were set to a window size of 5 and a minimum word frequency cut-off of 5. The advantage of using this method is that it preserves the contextual similarity of words. Moreover, word2vec learns the words order from the context of the large corpus of tweets.

The second applied embeddings model uses character-level embeddings which derives a representation for every character, to create word embeddings. The model that we have used is similar to the one presented in [8]. At the beginning, the character embeddings are randomly initialized. After that character embeddings are constructed by passing each character in every word in the sentiment target recognition task dataset to a bidirectional LSTM from the forward and backward directions. The embedding of each word is generated from the concatenation of the bidirectional LSTM representation of the word characters. The advantage of the character embeddings model is that it allows for building task specific embeddings which are generated from the labeled training data. In addition, it is capable of handling the out-of-vocabulary problem.

5 The Implemented Models

The goal of our work was to experiment with a bidirectional Long Short-Term Memory (BI-LSTM) neural network for addressing the task of sentiment target recognition. However, since this task has not been addressed before for Arabic, we needed to provide another baseline model to which we can compare obtained results. So in this paper, we present two different models for sentiment target recognition in Arabic tweets. The first model is the baseline one, which relies on a windowing method to detect target entities. The second model presents our proposed approach which is a sequence tagging model based on a hybrid approach

[1]https://pypi.python.org/pypi/gensim.

between a bidirectional Long Short-Term Memory (BI-LSTM) neural network and a conditional random fields (CRF) classifier. Both models are detailed in the following subsections.

5.1 *The Baseline Model*

When building the baseline model, we followed a windowing approach which detects sentiment targets based on the distance between each entity in the tweet and sentiment words that appear in that tweet. In this model, input tweets pass through a cleaning step to remove special characters and to normalize the tweet words for consistent representation. The cleaned tweets are provided to a modified version of the named entity recognizer described in [1] for detecting the named entities in each tweet. The tweets are then passed to the El-Beltagy sentiment analyzer for tweet sentiment identification and sentiment words detection [3]. For the latter task, the sentiment analyzer produces two lists; one for the negative words in each tweet and another for the positive ones. The implemented algorithm then selects the target entity based on calculating the Euclidean distance between the generated entities and the words present in the sentiment lists. The followed algorithm details are shown in Fig. 3.

```
Tweet sentiment (tweetSenti): the annotated sentiment for the tweet

Entities (E): the list of entities contained in the tweet

distPos: The accumulated distance between the entity and the list of positive words

distNeg: The accumulated distance between the entity and the list of negative words

posWords: The list of the detected positive words in the tweet

negWords: the list of the detected negative words in the tweet

Boolean isTarget(Entity candEntity)
    distPos = 1000000000;
    distNeg = 1000000000;
    if (posWords.size() > 0)
        distPos = calculate EuclideanDistanceBetweenEntityAndSentimentWords(candEntity, posWords);
    if (negWords.size() > 0)
        distNeg = calculate EuclideanDistanceBetweenEntityAndSentimentWords(candEntity, negWords);

    if ((distPos < distNeg) && tweetSenti.equals("pos"))
            return True;        //Entity is a target
    else if (distNeg < distPos && tweetSenti.equals("neg"))
            return True;        //Entity is a target
    else if ((distPos < distNeg) && (tweetSenti.equals("neg")) && negWords.size() > (1.25 * posWords.size()))
            return True;        //Entity is a target.
    else return False;    //Entity is NOT a target
```

Fig. 3 Algorithm for the Window-based method for detecting sentiment targets

5.2 *The Deep Neural Network Model*

The proposed sentiment target recognition model is based on a hybrid sequence tagging approach that combines a deep neural network (bidirectional-LSTM) with a top layer consisting of a conditional random fields (CRF) classifier. The proposed model follows a similar architecture to ones introduced in [4, 14] for different NLP tasks such as named entity recognition and part of speech tagging. The proposed bidirectional LSTM-CRF model architecture and how it was applied for the sentiment target recognition task is explained in the following subsections.

5.2.1 Bidirectional Long Short-Term Memory Networks (BI-LSTMs)

Recurrent Neural Networks (RNNs) [7] represent a class of neural networks that have the ability of preserving historical information from previous states through loops or iterations. These loops allow a RNN to represent information about sequences as the RNN takes a sequence of vectors (x_1, x_2, …, x_t) as an input and produces another sequence (h_1, h_2, …, h_t) as an output that represents information about the input sequence over time t.

RNNs have been applied to a variety of tasks including speech recognition [17] and image captioning [18]. RNNs have also been used for NLP tasks such as language modeling, named entity recognition (NER) [8], and machine translation [19]. However, RNNs suffer from the problem of long-term dependencies which means that RNNs struggle to remember information for long periods of time and that in practice a RNN is biased to recent input vectors in an input sequence.

An LSTM [6] is a special kind of RNN that is designed to solve the long-term dependency problem through using several gates called the forget gates. Adding these gates gives an LSTM the ability to control the flow of the information to the cell state. The forget gates prevent the back propagated errors from vanishing or exploding points through a sigmoid layer which determines which information to remember and which to forget. In this research, we have used the following implementation of LSTM:

$$i_t = \sigma(W_{xi}x_t + W_{hi}h_{t-1} + W_{ci}c_{t-1} + b_i) \tag{1}$$

$$c_t = (1 - i_t) \odot c_{t-1} + i_t \odot \tanh(W_{xc}x_t + W_{hc}h_{t-1} + b_c). \tag{2}$$

$$o_t = \sigma(W_{xo}x_t + W_{ho}h_{t-1} + W_{co}c_{t-1} + b_o). \tag{3}$$

$$h_t = o_t \odot \tanh(c_t). \tag{4}$$

where σ is the sigmoid function and $\odot$ is dot product.

The bidirectional LSTM network that we have utilized consists of two distinct LSTM networks. The first generates a representation for the left context of the current word ***t*** while the second is responsible for providing the representation for the right context through passing over the tweet in the reverse direction. Generally, the first is called the forward LSTM and the latter is called the backward LSTM. Each word in a bidirectional LSTM is represented through concatenating the output of the forward and backward LSTM. The vector resulting from bidirectional LSTM provides an effective representation for each word, as it takes context into consideration in this representation.

5.2.2 The Conditional Random Fields (CRF) Tagging Model

Sentiment target recognition can be modeled as a binary classification problem where each word/term in a tweet is either classified as a target or not. However, such a model does not consider dependencies between word/terms in a sentence. As natural language enforces grammatical constraints to build interpretable sentences, it is challenging for a classification model to tag the words independently. For the sentiment target recognition task, we take the relationship between words in a sentence into consideration in order to construct a solid interpretation for the state of every word. Towards this end, we've used conditional random fields.

The conditional random fields (CRF) model is a statistical based method which can be applied in machine learning tasks that need structured prediction. A conditional random fields model can tag words in a sentence either as a target or not-while taking into consideration neighboring words that represent the context.

5.2.3 BI-LSTM-CRF

In this work, we have augmented a bidirectional LSTM with a CRF layer in addition to a word embeddings layer to build and train a bidirectional LSTM-CRF sequence tagging neural network for the sentiment target recognition task, where a tag can denote either a target or a non-target word. Given tweet X of size n words (x_1, x_2, …, x_n), each word is represented as an m dimensional vector that is created by the word embeddings layer. Specifically, in our model, the embedding vector for each word in a tweet is obtained from the concatenation of the word2vec embeddings representation of the word and its character-level embeddings representation which yields the weights of the first layer of our model. The bidirectional LSTM layer uses the first layer weights in order to generate the features for the CRF layer. The advantage of using a bidirectional LSTM is its capability to represent the features to the right and left of the word for which the model is trying to predict a tag (right and left contexts). The CRF layer receives these contextual features extracted by the bidirectional LSTM to predict a tag for the current word based on the previous and consequent word tags in the sentence. Figure 4 overviews the interaction between the embeddings layer, the bidirectional LSTM layer and the CRF layer to show an example of words in an Arabic tweet and their predicted tags.

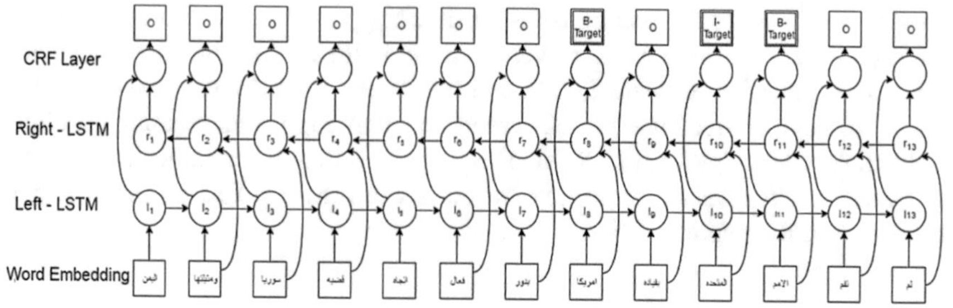

Fig. 4 Overview of BI-LSTM-CRF model layers with an embeddings layer

For the proposed bidirectional LSTM-CRF sentiment target recognition model, we define a set of annotations as follows: T: The number of distinct tags. C: A transition matrix that represents the transition score from tag i to tag j calculated from the training data where C is a square matrix of size T. L: A matrix that represents the output score of the bidirectional LSTM where L is a matrix of size N * T and where each word in the tweet has a score for each distinct tag.

For any input tweet *X* of size *n* words = (x_1, x_2, …, x_n), the network is expected predict a sequence of tags y = (y_1, y_2, …, y_n). A matrix of scores of size $N * T$ is calculated for each word in the tweet. The score function is defined as the summation of both the output of *C* and *L* as follows:

$$S(X, y) = \sum_{i=0}^{N} C_{yi, yi+1} + \sum_{i=1}^{N} L_{i, yi} \tag{5}$$

A softmax function is applied to produce the probability for the tag sequence *y*:

$$P(y|X) = \frac{e^{S(X, y)}}{\sum_{\tilde{y} \in Y_x} e^{s\left(X, \tilde{y}\right)}} \tag{6}$$

As Y_x represents the set of all possible tag sequence predictions for a sentence *X*, the objective function for the sentiment target recognition model during training aims to maximize the log-probability of the correct tag sequence as follows:

$$\log(P(y|X)) = S(X, y) - logadd_{\tilde{y} \in Y_x} S(X, \tilde{y}) \tag{7}$$

During prediction, the output sequence is chosen from the set of all possible sequences that give the maximum score as follows:

$$y_{predicted} = argmax_{\tilde{y} \in Y_x} s\left(X, \tilde{y}\right) \tag{8}$$

6 Performance Evaluation

The objective of our evaluation was to test our hypothesis that using a deep neural network to discover a set of features for a challenging task such as sentiment target recognition in a complex language like Arabic, can outperform a baseline solution for the task.

Towards this goal, training and testing sets were derived from the annotated Arabic tweets dataset previously described in data collection and annotation section. From those, only 3000 were selected. Two thousands tweets were used for building the training set while the rest (1000) were used for testing. Tweets in both the training and testing sets were cleaned by removing all special characters. They were also normalized as follows: "إ" ,"أ", and "آ" characters were replaced with "ا", and "ى" was replaced by "ي", while "ة" was replaced by "ه"

The sentiment target recognition tagging network with a bidirectional LSTM augmented with a CRF layer described in the previous section was built by adapting the implementation and network structure availed by the authors of [8] for tagging named entities. Given the relatively small size of our training set, the network was trained using 100 epochs in order to set the neuron weights.

To analyze the effect of the word2vec vectors on the results, the bidirectional LSTM-CRF network was trained once with the word vectors from the word2vec model concatenated with word vectors from the character embeddings model and once with the word vectors from character embeddings model only. About 94% of words in the training dataset had a vector representation in our set of word2vec embeddings.

Both our model and the previously described baseline method, aim at recognizing sentiment targets in each Arabic tweet in the test set. The precision, recall and F-score metrics were used to compare between the models. An annotated sentiment target is considered to be detected by the model if any of its constituent words is detected. Table 1 shows the scores for baseline method, bidirectional LSTM-CRF with character embeddings only and bidirectional LSTM-CRF with character embeddings and word2vec embeddings.

The results in Table 1 show the clear superiority of the bidirectional LSTM-CRF network over the baseline method with respect to all metrics, especially when using word embeddings. Handcrafting a set of features for sentiment target recognition is a challenging task especially since most targets are simply a special case of a named entity, but through training the bidirectional LSTM network, the network was able to derive features that distinguish between sentiment target entities and other entities.

The results also show that both bidirectional LSTM-CRF networks (with and without word embeddings) have achieved comparable precision, but the network

Table 1 Baseline and BI-LSTM models scores

Model	Precision	Recall	F-score
Baseline method	0.449	0.519	0.4815
BI-LSTM-CRF with character embeddings only	0.732	0.615	0.668
BI-LSTM-CRF with character embeddings and word2vec	0.737	0.714	0.726

with the word2vec model has achieved a much higher recall. This means that the bidirectional LSTM network with the word2vec model has discovered more target entities from the tweets while conserving the same extraction precision. This improvement achieved by the bidirectional LSTM-CRF network with the word2vec model emphasizes the power of word embeddings for improving the performance of NLP tasks.

7 Conclusion

This paper presented a model for detecting sentiment targets in Arabic tweets. The main component of the model is a bidirectional LSTM deep neural network with a conditional random fields (CRF) classification layer, and a word embeddings layer. Results obtained by experimenting with the proposed model, show that the model is capable of addressing the sentiment target recognition problem despite its inherent challenges. The experiments also revealed that the use of a word2vec model enhances the model's results and contributes to the overall effectiveness of the network in extracting sentiment targets. Comparing the proposed model against a baseline method showed a considerable advantage in terms of results.

Acknowledgements We would like to thank Abu Bakr Soliman for building the user interface used for annotating sentiment target data and for collecting the 54 million tweets that we have used for building our word embeddings model. This work was partially supported by ITIDA grant number [PRP]2015.R19.9.

References

1. Zayed, O., El-Beltagy, S.R.: Named entity recognition of persons' names in Arabic tweets. In: Proceedings of Recent Advances in Natural Language Processing (RANLP 2015), Hissar, Bulgaria (2015)
2. Zayed, O., El-Beltagy, S.R.: A hybrid approach for extracting arabic persons' names and resolving their ambiguity from twitter. In: Métais, E. et al. (eds.) Proceedings of 19th International Conference on Application of Natural Language to Information Systems (NLDB2015), NLDB 2015, Lecture Notes in Computer Science (LNCS), Passau, Germany (2015)
3. El-Beltagy, S.R., Khalil, T., Halaby, A., Hammad, M.H.: Combining lexical features and a supervised learning approach for Arabic sentiment analysis. In: CICLing 2016, Konya, Turkey (2016)
4. Huang, Z., Xu, W., Yu, K.: Bidirectional LSTM-CRF models for sequence tagging. arXiv: 1508.01991 (2015)
5. Lafferty, J., McCallum, A., Pereira, F.: Conditional random fields: probabilistic models for segmenting and labeling sequence data. In: Proceedings of the Eighteenth International Conference on Machine Learning, ICML, pp. 282–289 (2001)
6. Hochreiter, S., Schmidhuber, J.: Long short-term memory. Neural Comput. **9**, 1735–1780 (1997)

7. Schuster, M., Paliwal, K.K.: Bidirectional recurrent neural networks. Trans. Sig. Proc. **45**, 2673–2681 (1997)
8. Lample, G., Ballesteros, M., Subramanian, S., Kawakami, K., Dyer, C.: Neural architectures for named entity recognition. In: Proceedings of NAACL-HLT (NAACL 2016) (2016)
9. Mikolov, T., Sutskever, I., Chen, K., Corrado, G., Dean, J.: Distributed representations of words and phrases and their compositionality. In: Proceedings of the 26th International Conference on Neural Information Processing Systems, pp. 3111–3119. Curran Associates Inc., USA (2013)
10. Pontiki, M., Galanis, D., Papageorgiou, H., Androutsopoulos, I., Manandhar, S., AL-Smadi, M., Al-Ayyoub, M., Zhao, Y., Qin, B., De Clercq, O., Hoste, V., Apidianaki, M., Tannier, X., Loukachevitch, N., Kotelnikov, E., Bel, N., Jiménez-Zafra, S.M., Eryi\ugit, G.: SemEval-2016 task 5 : aspect based sentiment analysis. In: Proceedings of the 10th International Workshop on Semantic Evaluation (SemEval-2016), pp. 19–30. Association for Computational Linguistics (2016)
11. Khalil, T., Samhaa R. El-Beltagy: NileTMRG: Deep convolutional neural networks for aspect category and sentiment extraction in SemEval-2016 Task 5. In: Proceedings of the 10th International Workshop on Semantic Evaluation (SemEval 2016), San Diego, California, USA (2016)
12. Mitchell, M., Aguilar, J., Wilson, T., Durme, B. Van: Open domain targeted sentiment. In: Proceedings of the 2013 Conference on Empirical Methods in Natural Language Processing, pp. 1643–1654 (2013)
13. Zhang, M., Zhang, Y., Vo, D.-T.: Gated neural networks for targeted sentiment analysis. In: Proceedings of the Thirtieth AAAI Conference on Artificial Intelligence, Phoenix, Arizona, USA. Association for the Advancement of Artificial Intelligence (2016)
14. Collobert, R., Weston, J., Bottou, L., Karlen, M., Kavukcuoglu, K., Kuksa, P.: Natural language processing (almost) from scratch. J. Mach. Learn. Res. **12**, 2493–2537 (2011)
15. El-Beltagy, S.R.: NileULex: a phrase and word level sentiment Lexicon for Egyptian and modern standard Arabic. In: Proceedings of LREC 2016, Portorož, Slovenia (2016)
16. Mikolov, T., Chen, K., Corrado, G., Dean, J.: Efficient estimation of word representations in vector space. ICLR Work. (2013)
17. Sak, H., Senior, A.W., Beaufays, F.: Long short-term memory based recurrent neural network architectures for large vocabulary speech recognition. CoRR. arXiv:1402.1128 (2014)
18. Mao, J., Xu, W., Yang, Y., Wang, J., Yuille, A.L.: Deep captioning with multimodal recurrent neural networks (m-RNN). CoRR. arXiv:1412.6632 (2014)
19. Zhang, B., Xiong, D., Su, J.: Recurrent neural machine translation. CoRR. arXiv:1607.08725 (2016)

Evaluation and Enrichment of Arabic Sentiment Analysis

Sanjeera Siddiqui, Azza Abdel Monem and Khaled Shaalan

Abstract This paper presents an innovative approach that explores the role of lexicalization for Arabic sentiment analysis. Sentiment Analysis in Arabic is hindered due to lack of resources, language in use with sentiment lexicons, pre-processing of dataset as a must and major concern is repeatedly following same approaches. One of the key solution found to resolve these problems include applying the extension of lexicon to include more words not restricted to Modern Standard Arabic. Secondly, avoiding pre-processing of dataset. Third, and the most important one, is investigating the development of an Arabic Sentiment Analysis system using a novel rule-based approach. This approach uses heuristics rules in a manner that accurately classifies tweets as positive or negative. The manner in which a series of abstraction occurs resulting in an end-to-end mechanism with rule-based chaining approach. For each lexicon, this chain specifically follows a chaining of rules, with appropriate positioning and prioritization of rules. Expensive rules in terms of time and effort thus resulted in outstanding results. The results with end-to-end rule chaining approach achieved 93.9% accuracy when tested on baseline dataset and 85.6% accuracy on OCA, the second dataset. A further comparison with the baseline showed huge increase in accuracy by 23.85%.

Keywords Sentiment analysis · Opinion mining · Rule-based approach
Arabic natural language processing

S. Siddiqui (✉) · K. Shaalan
British University in Dubai, Block 11, 1st and 2nd Floor, Dubai International Academic City, Dubai, UAE
e-mail: faizan.sanjeera@gmail.com

K. Shaalan
e-mail: khaled.shaalan@buid.ac.ae

A.A. Monem
Faculty of Computer and Information Sciences, Ain Shams University, Abbassia, Cairo 11566, Egypt
e-mail: azza_monem@hotmail.com

K. Shaalan
School of Informatics, University of Edinburgh, Edinburgh, UK

K. Shaalan et al. (eds.), *Intelligent Natural Language Processing: Trends and Applications*, Studies in Computational Intelligence 740,
https://doi.org/10.1007/978-3-319-67056-0_2

1 Introduction

The mission of Sentiment Analysis is to perceive the content with suppositions and mastermind them in a way complying with the extremity, which incorporates: negative, positive or nonpartisan. Organization's are taken to huge prestige by living up to the opinions from different people [1, 2]. Subjectivity and Sentiment Analysis characterization are prepared in four measurements: (1) subjectivity arrangement, to estimate on Subjective or Objective, (2) Sentiment Analysis, to anticipate on the extremity that could be negative, positive, or impartial, (3) the level in view of record, sentence, word or expression order, and (4) the methodology that is tailed; it could be standard based, machine learning, or half breed [3]. As stated by Liu [4], Sentiment Analysis is "Sentiment analysis, also called opinion mining, is the field of study that analyzes people's opinions, sentiments, evaluations, appraisals, attitudes, and emotions towards entities such as products, services, organizations, individuals, issues, events, topics, and their attributes."

Arabic is a Semitic dialect, which is distinctive as far as its history, structure, diglossic nature and unpredictability Farghaly and Shaalan [5]. Arabic is broadly talked by more than 300 million individuals. Arabic Natural Language Processing (NLP) is testing and Arabic Sentiment Analysis is not a special case. Arabic is exceptionally inflectional [6, 7] because of the fastens which incorporates relational words and pronouns. Arabic morphology is intricate because of things and verbs bringing about 10,000 root [8]. Arabic morphology has 120 examples. Beesley [8] highlighted the hugeness of 5000 roots for Arabic morphology. No capitalization makes Arabic named substance acknowledgment a troublesome mission [9]. Free order of Arabic Language brings in additional challenges with regards to Sentiment Analysis, as the words in the sentence can be swapped without changing the structure and the meaning. Arabic Sentiment Analysis has been a gigantic center for scientists [10].

The target of this study is to examine procedures that decide negative and positive extremity of the information content. One of the critical result would be to recognize the proposed end-to-end principle binding way to deal with other dictionary based and machine learning-construct approaches in light of the chosen dataset.

The rest of this paper is organized as follows. Related work is covered in Sect. 2, Challenges to perform Arabic Sentiment Analysis is traced in Sect. 3, Data collection is covered in Sect. 4 followed by system implementation in Sect. 5. Section 6 covers evaluation and results, Sect. 7 depicts conclusion.

2 Related Work

To take a shot at Sentiment Analysis, the key parameter is the dataset. Late endeavors by Farra et al. [11] outlined the significance of crowdsourcing as an extremely fruitful technique for commenting on dataset. Sentiment corpora for Arabic brought into existence to prove the methods and perform experiments [12–16].

One of the extreme need to perform sentiment analysis is the availability of corpus. Few attempts are significant in literature illustrating the researchers zest to fill in the barren land of Arabic datasets to perform Arabic Sentiment Analysis. A freely available corpus brought into existence with the aid of manual approach by Al-Kabi et al. [17] covered five areas that is—Religion, Economy, Sport, Technology and Food-Life style, with each of these areas covering 50 subjects.

Al-Kabi et al. [17] collected reviews from Yahoo Maktoob with constraints including a clear discretion in the way the Arabic reviews written, with a mixture of Eygptian, English, MSA and Levantine dialect. Unlike Al-Kabi et al. [17] who have manually compiled the dataset, Farra et al. [18] used crowdsourcing to annotate dataset. Apart from freely available corpus, there has been significant work on different corpora but not shared openly to our knowledge. However some other corpus worked on includes Financial—[19, 20]; News—[21–23].

Other than the corpus, the other main player in the whole process of Arabic Sentiment Analysis is the Lexicon. Lexicons are the most desirable word list playing a key role in defining the text polarity. The lexicons can be available with the inclusion of adjectives. Adjectives are the best way to highlight polarity [24]. Few notable efforts found in literature where lexicons brought to life included, business based lexicons by [25], SentiStrength [26].

Some of the exemplar additions to the Arabic Sentiment Analysis include exploration on different methods in different domains however, piled up with limitations. Likewise, the work by Mountassir et al. [27] on documents with 2,925 reviews in Modern Standard Arabic wherein no experiments performed to judge their proposed method. With the news webpages and social reviews, Al-Kabi et al. [28] worked at sentence level by dealing with colloquial Arabic with a limitation of tiny dataset. However, Elarnaoty et al. [21] worked on document level with only center on news publications.

Bolster Vector Machine classifier accomplished 72.6% precision on twitter dataset of 1000 tweets [29]. The record level assessment investigation utilizing a joined methodology comprising of a dictionary and Machine Learning approach with K-Nearest Neighbors and Maximum Entropy on a blended area corpus involving training, legislative issues and games achieved an F-measure of 80.29% [30]. Shoukry and Refea [29] achieved a precision of 72.6% with the corpus based method. The vocabulary and estimation examination device with an exactness of 70.05% on tweeter dataset also, 63.75% on Yahoo Maktoob dataset [12].

With a mixed approach that is Lexical and Support Vector Machine classifier created 84.01% exactness [31]. 79.90% exactness was shown with Hybrid methodology which involved lexical, entropy and K-closest neighbor [32]. Shoukry and Rafea [33] independently sent two methodologies, one being Support Vector Machine accomplished a precision of 78.80% and other one being Lexical with an exactness of 75.50%.

3 Challenges Arabic Sentiment Analysis

In the first place the challenges with regards to sentiment analysis in Arabic includes a principal that has a quick effect is subjectivity portrayal. In light of the very truth of a trim distinction in a sentence being subjective or objective gets new challenges. Various diverse challenges are found in Sentiment Analysis. If a negative presumption contains a segment from the positive vocabulary, it is general explored as positive limit wherein it is negative. A word found as a section of both positive and negative vocabularies obtains new challenges in the furthest point undertaking. For example, "با قرف" (it makes me sick or disgusting) was found in both negative and positive reviews; finally, joining positive and negative vocabularies. A couple people create rude comments or unmistakable style of commenting using negated sentences in a productive review thusly gets new troubles. For example, "احسن من هيك رئيس وزراء ما في" (There is no Prime Minister better than this one).

As spoke to by Pang and Lee [34], positive words in one zone may hold a substitute furthest point in another space. For example, a word like "بارد" (cold) were found in positive and negative reviews holding specific limit. Distinctive challenges joins, beyond what many would consider possible to 140 characters people winds up using short structures, semantic mistakes in the tweets and specifically usage of slang tongue [31]. Following are few challenges hindering the natural language processing in Arabic.

3.1 Encoding

Windows CP-1256 and Unicode availability, one can read, compose and extract substance written in Arabic. Issues do rise in the midst of the preparation work by some different tasks, as needs be the use of transliteration where the Roman letters and Arabic letters mapping are possible. Subsequently, keeping in mind the end goal to decide this issue, various researchers use transliteration as the pre-processing ready stride [35].

3.2 *Sentiment Analysis Impacted Due to Unavailability of Punctuations*

A great degree smart parts/features shared by Arabic, it gets new troubles. One of the key challenge found in Arabic NLP is the nonattendance of strict and rigid standards with reference to complement (punctuation), get challenges for stick point sentence confines in Modern standard Arabic [36]. Same basic is found in Dialect Arabic, with no fundamentals to manage boundaries of a sentence. No complements (punctuations) are trailed by a huge bit of the all-inclusive community except for development of a full stop toward the end, which even is habitually overlooked. With no highlights, enormous effect is found in most of the NLP assignments including Arabic Sentiment Analysis settling on the decision of right sentences troublesome from a given substance.

3.3 *Excess Resources Required*

Turney [37] highlighted the deficiency of tools and resources in general for Arabic resulting in excess challenges for community doing research in Arabic NLP. The tools and resources forms the assets for Arabic NLP and are found to be the most lacking area when vocabularies and corpus are talked about in sentiment analysis. The vocabularies and corpus forms noteworthy part in Sentiment Analysis.

Because of the absence of culmination in the dictionary for positive and negative thwarts the investigation of Sentiment. Abdalah et al. [12] and Farra et al. [22], expressed regardless of analysts are investing vitality and time creating vocabularies and gathering tweets and audits framing dataset, yet these are not made accessible open for others to utilize and investigate, frustrated the development in Arabic Natural Language Processing with Sentiment Analysis in setting.

Table 1 Lexicons constructed are not complete

Paper	Method used	Lexicon build for domain	Own/translated
[35]	Unsupervised learning	Business reviews	Own
[36]	By translating English lexicons into Arabic	English lexicons for various reviews.	Translated
[35]	By labelling 4000 adjectives as negative, positive and neutral	4000 adjectives words irrespective of any particular domain	Own

Table 2 Corpus for different domains (not openly shared)

Domain	Paper
Financial	[24, 26]
Twitter tweets	[31]
Multi domain	[32, 36]
News corpus	[21, 22]
Web reviews	[29]

Table 1 delineates the vocabulary worked for a portion of the areas which is exceptionally constrained not covering numerous spaces and are not unreservedly accessible. Table 2 portrays the datasets worked for multi space yet are not transparently accessible also are extremely restricted.

3.4 *Sarcastic Tamper*

Wry method for speaking to Arabic suppositions or opinions brings about huge difficulties for the framework to segregate a survey as positive or negative. The snide obstruction causes a gigantic harm to the judgment of general extremity. Due to sarcasm a negative audit could be anticipated as positive and the other way around.

An example sarcastic tweet:

مجرد مشاهد لصعود وهبوط الفك أوه نعم، وأنا أتفق أنه في بعض الأحيان أنا لا أستمع لما تقوله، وأنا "

- Oh yes, I agree that sometimes I don't listen what you say, I just watch your jaw go up and down". Table 3, depicts some of the sarcastic comments which creates the polarity interpretation more cunning task.

Table 3 Sarcastic tamper

Sarcastic comments	
Arabic	English translation
أحيانا أنا لا أعرف إذا كانت اصرخ عليك أو أشفق عليك	Sometimes I don't know whether to shout at you or pity you
أنت مضحك بحيث انك تجعل الجميع يضحكون عليك وليس معك	You are so funny that you are making everyone laugh at you, not with you
الفستان الذي كنت ترتديه يبدو لطيفا ولكن ليس عليك	The dress that you are wearing looks nice, but not on you
أحيانان أحتاج لشيء لا يمكن أن يعطيه أحد إلا أنت وهو أن تغلق فمك	Sometimes I need something that no one can give except you, which is your mouth shut

3.5 *One Word Represents Two Polarities*

In Arabic, addition or deletion of one word turns the sentence into opposite polarity. In the example A: "أنا أحب المدرسة - I like school" and "أنا لا أحب المدرسة - I do not like school". In like manner, a bit much a similar word constantly tends to give a negative significance.

In example B: "لا أحد يصعد الشجرة - No one climbed the tree" contains the word "لا - No" shows up in this sentence however does not pass on any negative assessment. Thus this arrangement of disagreements with single word speaking to two unique assumptions in two varied sentences brings about real concerns.

While fulfilling a negative supposition, one can fulfill "لا - No" as negative estimation by putting the word in negative vocabulary, however neglects to fulfill case B by breaking down the announcement as negative however it is definitely not. Table 4 portrays cases of the words changing sentence extremity. Table 5 delineates the representation for cases on the difficulties that one negative word can convey to the Sentiment Analysis errand in Arabic Natural dialect preparing.

Table 4 One word changing polarity—examples with challenges faced

Positive sentiment word	.indicating positive polarity – لذيذ	happy – سعيد
Positive review	The food is delicious - الطعام لذيذ	أنت سعيد في عملك – You are happy with your work
Addition of one word	ليس Converts the polarity to negative	Addition of لست converts the polarity to negative
Negative review	الطعام ليس لذيذ - The food is not delicious	وأنت لست سعيدا في عملك – You are not happy with your work

Table 5 Challenges to be dealt with

Challenges to tackled both negative and positive polarities	Brings in challenges due to the addition of لذيذ – delicious in positive lexicon, hence the negative sentiment containing the word لذيذ – delicious makes the system identifies the negative sentiment as positive. If "ليس – not" not tackled properly so as to satisfy negative polarity	Brings in challenges due to the addition of "سعيد – happy" in positive lexicon, hence the negative sentiment containing the word "سعيد – happy" makes the system identify the negative sentiment as positive, if "ليس – not" not tackled properly so as to satisfy negative polarity

3.6 *Indifferent Writing Style*

Arabic audits for the most part found on various locales have a reasonable assorted qualities in composing style with Egyptian and Gulf Arabic regularly utilized. A large portion of the Arabic audits are not composed in standard arrangement and doesn't take after Modern Standard Arabic. Individuals compose distinctive vernacular in Arabic which are casual and in this manner includes the utilization of individual elocution and vocabularies [4]. Accordingly, the dictionary building procedure and control based drew nearer are basically tested with this casual writing styles leaving specialists with exhaustive investigation of surveys composed online to perform Sentiment Analysis. Table 6 delineates different written work styles

Table 6 Different writing style resulting in inexactness

Review/comment	Polarity	Negative/positive words	Challenges
انها ليست امرأة سيئة - She is not a bad woman	Positive	Negative words—not and bad	This type of writing style containing negative words but indicating positive polarity makes the job for the system a difficult task to handle. Not and bad due to the very nature of these words indicating their significance in negative reviews, joins the negative lexicon. When the system tries to tackle this positive sentiment, but due to the addition of negative words the polarity system gives as an output is negative but the correct polarity is positive
عليك أن تكون كريما - you have to be generous	Negative	Positive words—generous	This type of writing style with addition of positive words resulting in negative polarity brings in challenges wherein the system matches this with both positive and negative lexicon and indicates the polarity as positive though this sentence is negative, if need to be is not tacked to resolve this challenge
أنت لست انسانن جيدا - You are not a good human being	Negative	Positive word—good negative—not	System struggles to tackle the existence of words like good and not belonging to both positive and negative lexicon. Based on the way the system is designed, the overall polarity goes in favour of either positive or negative. If system indicate this as positive polarity then the answer is wrong, indicating that the classification lacks in handling mixed words in a sentiment
ليس هناك احد حزين في هذا البلد - No one is sad in this country	Positive	Negative words—no sad	This example though include two negative words but fails to satisfy the overall polarity if the system analyse this based on the negative lexicon list and identify this as negative review though the review is positive resulting in inexactness

alongside the difficulties they acquire to the framework to fulfill the negative and positive polarities.

3.7 Free Writing Style

Another key issue found to affect Arabic Natural Language Processing particularly in Sentiment Analysis, wherein individuals don't take after linguistic use, accentuations, and spellings. Individuals are all the time appear to overhaul their status and remark online without breaks, this may bring about an inclination to miss spell or write in rush missing the fundamental structure of composing a legitimate sentence. This makes the Sentiment Analysis prepare more mind boggling. Spelling slip-ups are high in online surveys and tweets [9].

Table 7 delineates couple of cases of free written work style with the spelling botches as greatest test.

3.8 Word Short Forms

Because of the cutoff points on numerous web-based social networking locales case Twitter, with respect to the quantity of words that could be composed, gets new difficulties for performing Natural Language Processing. The way the Arabic tweets are abbreviated by putting the short structures for the words impedes the System to discover the match for the feeling in the Lexicons. Like "غرامة - Fine" is frequently trimmed and composed as "F9" [31].

Table 7 Challenges faced due to spelling mistakes in free writing style

	Lexicon	Review in Arabic	Review in english	Challenges due to the spelling mistakes
Actual sentence	أفضل	أنت أفضل شخص في المبيعات	You're the best sales person	Actual polarity is positive, due to the word misspelled as ضل for أفضل, turns the whole polarity of the review into negative
Spelling mistake	ضل spelled incorrectly for أفضل	أنت ضل شخص في المبيعات	You are lost sales person.	

3.9 Same Word Usage for Both Polarities

At times individuals answer to a remark or audit or tweet through single word or expressions which can bolster both positive and negative survey, subsequently falling in both positive and negative vocabularies. Not just the initial step that is the vocabulary building process gets to be distinctly testing, additionally the framework building turns out to be tremendously testing. Colossal examination is required in this circumstance to deal with such words. Words like "okay, fine, alright, all right, very well – حسنا, "أنا أتفق - I agree", "انا لا اوافق - I do not agree" are frequently observed to bolster assumptions in negative and additionally positive. Building framework understanding on when to counter to these supporting or not supporting words or expressions as negative or positive is a noteworthy hiccup. Table 8 delineates a response to twitter tweets and the difficulties these reactions conveys to the framework to anticipate the extremity of the reaction.

Table 8 Person to person response challenge

	Twitter tweet	Polarity	Challenges
Person-1	ساره دائما تتظاهربأنها جيدة جدا في الطبخ لكنها ليست جيدة في الطبخ. – Sara always pretend to be very good in cooking but she is not a good cook	Negative	This is the challenging phase wherein making the system understand when to give a positive polarity if the word "حق – right" is found in a tweet, as this word belongs to the positive lexicon, but in the current case it supports the negative polarity. Hence illustrating a negative sentiment by supporting a negative statement
Person-2 responding to person-1	انت دائما على حق – You are always right	Supporting a negative opinion—negative polarity	
Person-3	اعتقد انها لا تعرف الف باء الطبخ - I think she does' not know the abc of cooking	Negative comment	Person-3 is negative sentiment. This negative sentiment is not supported by person-4 instead gives a positive response by not agreeing to person-3. Due to the addition of "I do not agree - أنا لا أتفق معك", brings in new challenges as the system will predict this as negative, due to "I do not agree - أنا لا أتفق معك" which falls under negative sentiment and the system will consider this to be a negative review but in actual this is a positive response to a negative review
Person-4 responds to person-3	أنا لا أتفق معك - I do not agree with you	Positive response do not supporting a negative sentiment	

4 Data Collection

The dataset utilized as a part of this paper is Twitter Tweets and film surveys. The dataset is taken from [12, 38]. These datasets are utilized as these are accessible, rich and enough to reached a conclusion, also electronic assets, for example, vocabulary is given.

Abdulah et al. [12] dataset contains 1000 positive tweets and 1000 negative tweets with length ranging from one word to sentences. 7189 words in positive tweets and 9769 words in negative tweets. The tweets were manually collected belong to Modern Standard Arabic and Jordanian Dialect, which covers Levantine language family. The months-long segregation procedure of the tweets was physically led by two human specialists (local speakers of Arabic).

OCA corpus, termed as Opinion Corpus for Arabic, was presented by [38]. This corpus contains total 500 opinions, of which there are 250 positive opinions and 250 negative opinions. The procedure followed by Rushdi et al. [38] included gathering surveys from a few Arabic online journal locales and site pages utilizing a straight-forward bash script for slithering. At that point, they expelled HTML labels and unique characters, and spelling mix-ups were adjusted physically. Next, a preparing of each survey was done, which included tokenizing, evacuating Arabic stop words, and stemming and sifting those tokens whose length was under two characters. In their trials, they have utilized the Arabic stemmer of RapidMiner and the Arabic stop word list. At last, three distinctive N-gram plans were created (unigrams, bigrams, and trigrams) and cross validation was used to assess the corpus for which they have achieved 90.6% accuracy.

The vocabulary utilized as a part of this paper contains the dictionary used by [12], which included opinions, named substances and a few haphazardly set words. In view of the circumspection of reiteration of the words found in negative or positive reviews and their arrangement, they were incorporated into both the rundown that is positive and negative lists.

5 Implementation of Arabic Sentiment Analysis

This paper is a significant extension of [11]. Siddiqui et al. [11] research introduced a system, which contains only two type of rules—"equal to" and "within the text rules", so as to examine whether the tweet is either negative or positive. The rules include a 360° coverage with an improvised segment that is end to end rule chaining principle: (1) in the middle, we termed as "within the text", (2) at the boundary we termed as either "ending with the text" or "beginning with the text", and (3) full coverage, we termed as "equal to the text". Figure 1 delineates the 360° rules coverage. The End-to-End mechanism with rule chaining approach introduced in this paper, includes the chaining of rules based on the positioning of the polarities in the tweets. The key underlying base ground factors which helped us formulate

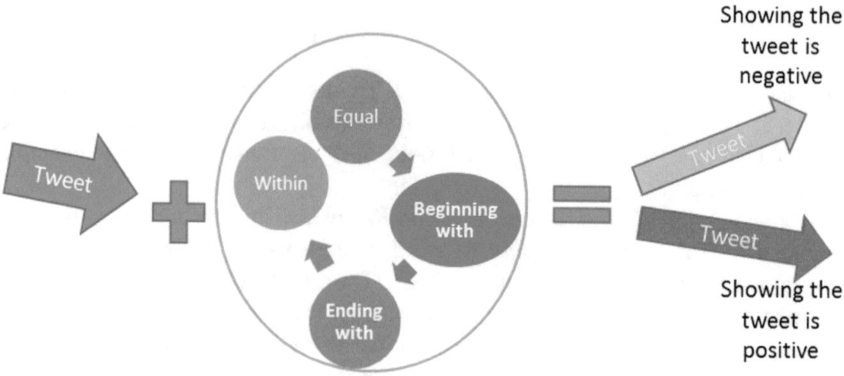

Fig. 1 A 360° coverage of rules to the input tweet

appropriate rules includes analysis of the tweets and the extension of positive and negative lexicons. The analysis of tweets resulted in identifying relations pertaining to words which were either disjoint, intersected or coexisted.

The words which were disjoint that is completely indicating either positive or negative polarity were included in their respective lexicons. The words which intersected that is the ones which were found to be common in both negative and positive reviews were included in positive as well as negative lexicons. The words which coexisted at the same place in the negative and positive reviews, that is the ones which appeared at the beginning or ending were placed in either positive or negative lexicon, based on the highest frequency of the word in the respective reviews.

Rules handling intersection with the end-to-end chaining mechanism: As an example consider the following positive tweet "اوقف القرار للحفاظ على الوطن" (the decision was suspended for protecting the motherland), the word "اوقف" (suspended) appeared in the beginning of this tweet and the same word was found in the negative tweets. Hence, this set of situations was handle with the very use of positioning and chaining of rules. The steps involved to achieve the positioning and chaining of rules includes:

Rules Formation: With the logical discretion of the word "اوقف" (suspended) being seen at the "beginning of" positive tweets and "within the text" for the negative tweets, the rules thus formed were "beginning with" for positive tweets and "within the text" for negative tweets.

With the correct positioning and chaining of rules this problem was resolved. In the current example "beginning with" rule needs to be positioned and chained in an orderly fashioned with the rule "within the text" so as to satisfy both positive and negative reviews. So, the rule "beginning with" was chained with the rule "within the text" for the word "اوقف" (suspended) by positioning the rule "beginning with" first followed by "within the text" rule. Hence, the rules are chained and positioned for the words which are found to repeat themselves in both positive and negative tweets.

Rules handling coexistence: The lexicons which are not seen to repeat themselves in either positive or negative tweets where handled with the rule—"within the text". For example, "الحرامية" (Thieves) is set for search "within the text" "الحرامية" (Thieves), is majorly found within the text in negative review rather than in positive ones. Hence based on the frequency of "الحرامية" (Thieves) the rule is set. Example tweet: "مسؤول عن الحرامية يتفنجرو بمصاري البلد و انا اشحد انا مش" (I am not responsible for that thieves spend lavishly from the funds of the country and I beg).

Rules handling disjoint: The cases wherein the words were not repeated in positive and negative cases and were found to have their significance at the end of the tweet, were handled using "ending with" rule. In the following positive tweet example "من يكذب يكتب عند الله كذابا – اما من يتحرى الصدق يكتب عند الله صادقا تذكر"—(Remember who continues to speak falsehood he is recorded with Allah as a great liar—but who persists in speaking the truth he is recorded with Allah as an honest man).

The word "صادقا" (Honest) appeared at the end of this tweet. Likewise, "صادقا" (Honest) was found to appear at the end in majority of positive reviews. Hence, the "end with" rule is set to search for "صادقا" indicating the system that if the review ends with the word "صادقا" (Honest) then it should be considered as a positive tweet.

Derivation of rules: Table 9 depicts the skeletal examples of derived rules. Conditional search includes two key phases, one includes a condition which checks on the entered tweet mapping with the rules and the second phase is the color coded output which changes its font color to—"Green fill with Green fill text" for negative tweets and "Light red fill with light red fill text" for positive tweets.

Table 9 Derivation of lexicalized rule

Rule type		Derivation of rules
(1) Search within the text	A	If the entered tweet/review contains the word "__" then fill the text with—"Green fill text" for negative tweets
	B	If the entered tweet/review contains the word "__" then fill the text with—"Light fill text" for positive tweets
(2) Ending with the text	A	If the entered tweet/review ends with the word "__" then fill the text with—"Green fill text" for negative tweets
	B	If the entered tweet ends with the word "__" then fill the text with —"Light red fill text" for positive tweets
(3) Beginning with the text	A	If the entered tweet begins with the word "__" then fill the text with—"Green fill text" for negative tweets
	B	If the entered tweet begins with the word "__" then fill the text with—"Light red fill text" for positive tweets
(4) Equal to the text	A	If the Entered Tweet equals to the word "__" Then Fill the text with—"Green fill text" for positive tweets
	B	If the Entered Tweet equals to the word "__" then Fill the text with —"Light red fill text" for negative tweets

With reference to rule 1.A in Table 9 in the primary column, if the word showed up inside of the content in a positive tweet. With reference to manage 2.B or 3.B in this table, if the same word showed up toward the end or toward the starting in the negative tweet, then situating and anchoring of these two rules are finished. For this case the positive principle was situated and fastened underneath the negative guideline. The principle "starting with" or "finishing with" not at all like "inside of the content" search for that word in the first place or end which will tag the tweet as negative. In the event that the situation of guideline is turned around then "inside of the content" as situated and affixed before the tenets "starting" or "consummation" with, when a negative tweet is gone through this, the tweet will be labeled positive. As this is a chained approach for a specific word, this checks its importance at the ending or starting rule, on the off chance that it doesn't have a place with that then it goes through within the text.

End-to-end mechanism with rule chaining approach: After precisely executing the rules taking into account the fulfillment of disjoint word(s) which existed together or basic words in negative and positive surveys, the guideline anchoring was dealt with. End-to-end component with rule chaining approach fulfills a word which has a place with a positive and negative tweet regardless of its real extremity. Consequently, a negative word in a positive audit and a positive word in a negative tweet is fulfilled through apt chaining as examined beneath with the guide of case from negative and positive tweets. Example 1 Rules in use—Rule A and Rule B.

Rule A in Table 9: In the negative tweet "ليست هناك حاجة لحماية وطننا من أيدي المنافقين الأردنيين", there is no need to protect our motherland from hands of hypocrites Jordanian, contains the word "المنافقين" (hypocrites) within the text. Rule B in Table 9: In the positive tweet "اللهم لا تجعلنا من المنافقين" (O Allah! Place us not with the people who are hypocrites) contains the word "المنافقين" (hypocrites) at the last. If Rule A is not chained with Rule B, then only one of the rule will be satisfied. To satisfy both the rules, that is to correctly identify negative and positive polarity, Rule A is chained with Rule B by positioning Rule A Below Rule B, so as to allow the search to first visit the end rule first, then the within the text rule.

Example 2 with explanation on how the system works: For example, the word "تشائم" (pessimism) was found to be part of both positive and negative reviews with a slight variation. In the positive reviews appears only in the middle whereas in the negative review "تشائم" (pessimism) was found to appear at the beginning. Hence the rules thus were created covering "within the text" (refer rule 1.A in Table 9) and "beginning with" (refer rule 2.B in Table 9), but the positioning was varied. By positioning the rule "beginning with" first and then the rule "within the text" helped in satisfying both positive and negative reviews.

As the system checked for the word beginning with "تشائم" (pessimism) in the entered review and if found then the review was tagged as negative. Likewise, the system proceeded with the entered review containing the word "تشائم" (pessimism), if the word "تشائم" (pessimism) was not found at the beginning then the search proceeded further and identified it as positive.

6 Evaluation and Results

To quantify the change also the nature of being trusted and had faith in, assessment plays an essential part. Cross-Validation and accuracy information are regularly used to assess the outcomes in estimation investigation. The exactness measures—Precision, Recall and Accuracy, which are generally being used was conveyed to measure the execution of the instruments utilized as a part of both the analyses. Precision, Recall and Accuracy were utilized to look at the outcomes by [12, 38, 33]. The condition is as per the following:

Precision = TP/(TP + FP)

Recall = TP/(TP + FN)

Accuracy = (TP + TN)/(TP + TN + FP + FN)

where:

TP—True Positive, all the tweets which were characterized accurately as positive

TN—True Negative, all the tweets which were accurately named negative

FP—False Positive, all the tweets which were mistakenly named positive

FN—False Negative, all the tweets which were mistakenly delegated negative

Results: The results fuse the relationship of the impressive number of tests coordinated in this paper. To do the connection, the accuracy of the extensive number of tests are used. Siddiqui et al. [11] system and System 2 (introduced in this paper) were attempted on [12, 38] dataset. Table 10, obviously takes after the outperformance of rules made in Siddiqui et al. [11] with enormous accuracy for Abdulah et al. [12] dataset when contrasted with the results on [38] OCA dataset.

System 1 Versus System 2 Results Comparison: Clear importance in expansion in accuracy in System 2 is seen for Abdulah et al. [12] and OCA dataset. The examination of Siddiqui et al. [11] System 1 and our System 2 doubtlessly answers that the end to end rule chaining improved the performance of sentiment analysis.

Siddiqui et al. [11]-system 1 bound to limits with two norms sort with no attaching and System 2 variation of System 1 with authoritative and reasonable arranging of the standards. System 2 ended up being remarkable for both the datasets. 93.9% accuracy for [12] dataset wherein for OCA dataset 85.6% accuracy was measured. Still the accuracy for our System 2 when tested on Abdulah et al. [12] was high with 8.3% more exactness than OCA dataset. Recall is high for both

Table 10 Comparison of system 1 [11] versus system 2

	System 1		System 2	
	Abdulah et al. [12] dataset	Rushdi et al. [38] OCA dataset	Abdulah et al. [12] dataset	Rushdi et al. [38] OCA dataset
Precision	87.4	50.4	92.5	77.4
Recall	93.3	97.6	95.8	100
Accuracy	89.6	50.1	93.9	85.6

the datasets with 3.3% for Abdulah et al. [12] and 22.6% for Rushdi et al. [38] as very less number of tweets are mistakenly delegated negative.

7 Conclusion

This paper beats the vocabulary building process through the fitting position of words too not barring the basic words found in both the tweets for the vocabularies. The outperformance of rule chaining approach that is System 2 brought about 23.85% in results when contrasted with Abdulah et al. [12] vocabulary based methodology. The incorporation of normal words in light of the examination of tweets in the negative and positive vocabulary list upgraded the general result when contrasted with the gauge dataset. Last yet not the minimum, the situating of principles has a gigantic effect to the rule based methodology as proper situating brought about fulfilling words which were observed to be regular in both negative and positive dictionaries. By and by, the end-to-end rule chaining methodology was the most difficult and costly regarding time and exertion, yet adds to the headways in the cutting edge for Arabic Sentiment Analysis, through the organized set standards and through the right utilization of various principles including "contains content", "equivalent to", "starting with" and "finishing with".

In reality, displayed the recently created assessment investigation framework—System 2, which beat in both arrangements of examinations when contrasted with [12]. System 2 with guidelines reached out to cover all territories was demonstrated to expand the exactness of OCA corpus by 39.8 and 4.3% accuracy for Abdulah et al. [12] when contrasted with System 1's principles of Siddiqui et al. [11]. Thus, starting a wakeup require every one of the scientists to redirect their enthusiasm to lead based methodology. The unmistakable hugeness in results along these lines acquired through the tenets made makes the principle based methodology the most alluring methodology.

References

1. Feldman, R.: Techniques and applications for sentiment analysis. Commun. ACM **56**(4), 82–89 (2013)
2. Taboada, M., Brooke, J., Tofiloski, M., Voll, K., Stede, M.: Lexicon-based methods for sentiment analysis. Comput. Linguist. **37**(2), 9–27 (2011)
3. Korayem, M., Crandall, D., Abdul-Mageed, M.: Subjectivity and sentiment analysis of Arabic: a survey. In: Hassanien, A.E., Salem, A.-B.M., Ramadan, R., Kim, T.-H. (eds.) AMLTA 2012. CCIS, vol. 322, pp. 128–139. Springer, Heidelberg (2012)
4. Albalooshi, N., Mohamed, N., Al-Jaroodi, J.: The challenges of Arabic language use on the internet. In: 2011 International Conference for Internet Technology and Secured Transactions (ICITST), pp. 378–382. IEEE (2011)

5. Farghaly, A., Shaalan, K.: Arabic natural language processing: challenges and solutions. ACM Trans. Asian Lang. Inf. Process. (TALIP) **8**(4), 1–22 (2009). The Association for Computing Machinery (ACM)
6. Hammo, B., Abu-Salem, H., Lytinen, S., Evens, M.: QARAB: a question answering system to support the Arabic language. In: The Proceedings of Workshop on Computational Approaches to Semitic Languages, ACL 2002, Philadelphia, PA, pp. 55–65, July 2002
7. Syiam, M., Fayed, Z., Habib, M.: An intelligent system for Arabic text categorization. Int. J. Intell. Comput. Inf. **6**(1) (2006)
8. Darwish, K.: Building a shallow Arabic morphological analyzer in one day. In: Proceedings of the ACL-2002 Workshop on Computational Approaches to Semitic languages, pp. 1–8. Association for Computational Linguistics, July 2002
9. Khaled, S.: A survey of Arabic named entity recognition and classification. Comput. Linguist. **40**(2), 469–510 (2014)
10. Muhammad, A.-M., Kübler, S., Diab, M.: Samar: a system for subjectivity and sentiment analysis of Arabic social media. In: Proceedings of the 3rd Workshop in Computational Approaches to Subjectivity and Sentiment Analysis, pp. 19–28. Association for Computational Linguistics (2012)
11. Siddiqui, S., Monem, A.A., Shaalan, K.: Sentiment analysis in Arabic. In: Métais, E., Meziane, F., Saraee, M., Sugumaran, V., Vadera, S. (eds.) Proceedings of the 21st International Conference on the Application of Natural Language to Information Systems (NLDB 2016). LNCS. Springer, Heidelberg (2016)
12. Abdulla, N.A., Ahmed, N.A., Shehab, M.A., Al-Ayyoub, M., Al-Kabi, M.N., Al-rifai, S.: Towards improving the lexicon-based approach for Arabic sentiment analysis. Int. J. Inf. Technol. Web Eng. (IJITWE) **9**(3), 55–71 (2014)
13. Abdulla, N., Al-Ayyoub, M., Al-Kabi, M.: An extended analytical study of Arabic sentiments. Int. J. Big Data Intell. **1**(2), 103–113 (2014)
14. Abdulla, N., Ahmed, N., Shehab, M., Al-Ayyoub M.: Arabic sentiment analysis: lexicon-based and corpus-based. In: Proceedings of IEEE Jordan Conference on Applied Electrical Engineering and Computing Technologies, Amman, Jordan, pp. 1–6 (2013)
15. Al Shboul, B., Al-Ayyoub, M., Jararweh, Y.: Multi-way sentiment classification of Arabic reviews. In: Proceedings of the 6th International Conference on Information and Communication Systems, Amman, Jordan (2015)
16. Al-Ayyoub, M., Bani-Essa, S., Alsmadi, I.: Lexicon-based sentiment analysis of Arabic tweets. Int. J. Soc. Netw. Min. **2**(2), 101–114 (2015)
17. Al-Kabi, M.N., Al-Ayyoub, M.A., Alsmadi, I.M., Wahsheh, H.A.: A prototype for a standard Arabic sentiment analysis corpus. Int. Arab J. Inf. Technol. (IAJIT), **13** (2016)
18. Farra, N., McKeown, K., Habash, N.: Annotating targets of opinions in Arabic using crowdsourcing. In: ANLP Workshop 2015, p. 89 (2015)
19. Ahmad, K.: Multi-lingual sentiment analysis of financial news streams. PoS, 001 (2006)
20. Almas, Y., Ahmad, K.: A note on extracting 'sentiments' in financial news in English, Arabic & Urdu. In: The Second Workshop on Computational Approaches to Arabic Script-based Languages, pp. 1–12 (2007)
21. Elarnaoty, M., AbdelRahman, S., Fahmy, A.: A machine learning approach for opinion holder extraction in Arabic language. arXiv:1206.1011 (2012)
22. Farra, N., Challita, E., Assi, R.A., Hajj, H.: Sentence-level and document-level sentiment mining for Arabic texts. In: Feldman, R. (ed.) 2010 IEEE International Conference on Data Mining Workshops (ICDMW), pp. 111 (2013). Techniques and applications for sentiment analysis. Commun. ACM **56**(4), 82–89 (2010)
23. Al-Kabi, M.N., Alsmadi, I.M., Gigieh, A.H., Wahsheh, H.A., Haidar, M.M.: Opinion mining and analysis for Arabic language. Int. J. Adv. Comput. Sci. Appl. (IJACSA) **5**(5), 181–195 (2014)
24. Benamara, F., Cesarano, C., Picariello, A., Recupero, D.R., Subrahmanian, V.S.: Sentiment analysis: adjectives and adverbs are better than adjectives alone. In: ICWSM (2007)
25. El-Halees, A.: Arabic opinion mining using combined classification approach (2011)

26. Elhawary, M., Elfeky, M.: Mining Arabic business reviews. In: 2010 IEEE International Conference on Data Mining Workshops (ICDMW), pp. 1108–1113. IEEE (2010)
27. Mountassir, A., Benbrahim, H., Berrada, I.: An empirical study to address the problem of unbalanced data sets in sentiment classification. In: 2012 IEEE International Conference on Systems, Man, and Cybernetics (SMC), pp. 3298–3303. IEEE (2012)
28. Al-Kabi, M., Gigieh, A., Alsmadi, I., Wahsheh, H., Haidar, M.: An opinion analysis tool for colloquial and standard Arabic. In: The Fourth International Conference on Information and Communication Systems (ICICS 2013), pp. 23–25 (2013)
29. Shoukry, A., Rafea, A.: Sentence-level Arabic sentiment analysis. In: 2012 International Conference on Collaboration Technologies and Systems (CTS), pp. 546–550. IEEE (2012)
30. Alaa, E.-H.: Arabic opinion mining using combined classification approach (2011)
31. Aldayel, H.K., Azmi, A.M.: Arabic tweets sentiment analysis–a hybrid scheme. J. Inf. Sci. (2015). doi:10.1177/0165551515610513
32. Beesley, K.R.: Arabic finite-state morphological analysis and generation. In: Proceedings of the 16th Conference on Computational Linguistics, vol. 1, pp. 89–94. Association for Computational Linguistics, August 1996
33. Shoukry, A., Rafea, A.: Preprocessing Egyptian dialect tweets for sentiment mining. In: The 4th Workshop on Computational Approaches to Arabic Script-Based Languages, pp. 47–56 (2012)
34. Pang, B., Lee, L.: Opinion mining and sentiment analysis. Found. Trends Inf. Retrieval **2**(1–2), 1–135 (2008)
35. ElSahar, H., El-Beltagy, S.R.: Building large Arabic multi-domain resources for sentiment analysis. In: International Conference on Intelligent Text Processing and Computational Linguistics, pp. 23–34. Springer (2015)
36. Diab, M., Hacioglu, K., Jurafsky, D.: Automatic processing of modern standard Arabic text. In: Arabic Computational Morphology, pp. 159–179. Springer, Netherlands (2007)
37. Turney, P.D.: Thumbs up or thumbs down?: semantic orientation applied to unsupervised classification of reviews. In: Proceedings of the Association for Computational Linguistics, pp. 417–424. Association for Computational Linguistics (2002)
38. Rushdi-Saleh, M., Martín-Valdivia, M.T., Ureña-López, L.A., Perea-Ortega, J.M.: OCA: opinion corpus for Arabic. J. Am. Soc. Inf. Sci. Technol. **62**(10), 2045–2054 (2011)

Hotel Arabic-Reviews Dataset Construction for Sentiment Analysis Applications

Ashraf Elnagar, Yasmin S. Khalifa and Anas Einea

Abstract Arabic language suffers from the lack of available large datasets for machine learning and sentiment analysis applications. This work adds to the recently reported large dataset BRAD, which is the largest Book Reviews in Arabic Dataset. In this paper, we introduce HARD (Hotel Arabic-Reviews Dataset), the largest Book Reviews in Arabic Dataset for subjective sentiment analysis and machine language applications. HARD comprises of 490587 hotel reviews collected from the Booking.com website. Each record contains the review text in the Arabic language, the reviewer's rating on a scale of 1 to 10 stars, and other attributes about the hotel/reviewer. We make available the full unbalanced dataset as well as a balanced subset. To examine the datasets, we implement six popular classifiers using Modern Standard Arabic (MSA) as well as Dialectal Arabic (DA). We test the sentiment analyzers for polarity and rating classifications. Furthermore, we implement a polarity lexicon-based sentiment analyzer. The findings confirm the effectiveness of the classifiers and the datasets. Our core contribution is to make this benchmark-dataset available and accessible to the research community on Arabic language.

Keywords Arabic sentiment analysis · Large-scale dataset
Modern Arabic standard and dialects · Lexicon based sentiment analysis

1 Introduction

There has never before been a period in human history where one can easily find an abundance of information, opinions and reviews about any subject. The year 2003 is when social network sites such as Myspace became mainstream; since that time, the number of users on social media has been growing [1]. As such, posts by users on websites such as Twitter are a goldmine of the public's opinions and sentiments

A. Elnagar (✉) · Y.S. Khalifa · A. Einea
Department of Computer Science, University of Sharjah, Sharjah 27272, UAE
e-mail: ashraf@sharjah.ac.ae

K. Shaalan et al. (eds.), *Intelligent Natural Language Processing: Trends and Applications*, Studies in Computational Intelligence 740,
https://doi.org/10.1007/978-3-319-67056-0_3

on any given topic. Similarly, websites—such as Epinions.com—where customers can post their reviews about a variety of products started appearing around 1999. Nowadays, customers have access to numerous review websites such as Yelp, Amazon, TripAdvisor, Booking…etc.

The prevalence of online user reviews has allowed potential customers to read a variety of opinions about a product or service and make an informed decision before purchasing. Studies have shown that online user reviews could significantly improve the sales of "experience goods" such as books, movies, and holiday trips [2].

The question remains, how do organizations take advantage of this wealth of information? The answer lies in machine learning and natural language processing, or more specifically in sentiment analysis. Using sentiment analysis, an organization can gather customer opinions or reviews about various subjects and classify them according to emotion, attitude or opinion as positive, negative or neutral.

Sentiment Analysis is defined as the process of determining the emotion of a given textual input; whether positive, negative or neutral. It is known by many names such as sentiment analysis, opinion mining, opinion extraction, sentiment mining, subjectivity analysis, affect analysis, emotion analysis, review mining…etc. [3].

There are several applications of sentiment analysis; for instance, its use in marketing, healthcare, politics and finance. Marketers can monitor social media and use sentiment analysis to decode customer's opinions on their brands, products and services. As customer's perception of a company/brand is not only influenced by the company's messaging, but also by what their peers (other customers) say, it is thus important for a company to keep track of its reputation online [4]. In the healthcare industry, it can be used to determine the quality of care patients receive at certain health organizations. This can help supplement the information provided by traditional approaches [5]. In the case of politics, there are extensive applications of sentiment analysis ranging from real-time analysis of the public's sentiments towards presidential candidates [6], to predicting election results [3], and identifying the political orientation of news articles [7]. As for the finance sector, blog and news sentiment can be used to study trading strategies, as well as applying sentiment analysis to stocks [3]. The applications do not end there as since the early days of research into sentiment analysis, people have been finding more and more innovative approaches to applying it in a variety of fields and sectors.

As such, the importance of research into sentiment analysis cannot be overstated. This reflects in the increased number of research papers about machine learning in general, and sentiment analysis specifically. Since the year 2000, sentiment analysis has become one of the most active areas of research in the field of Natural Language Processing (NLP) [3]. However, the focus has largely been on the English language; tools, resources and research on other languages, especially Arabic language, are scarce. For example, to build robust classifiers, training on large datasets is a key requirement, which is an important step for a variety of machine learning applications and NLP. A reasonably good number of small-scale, large-scale, and huge datasets are freely available for the research community. However, the majority of such datasets are in the English language. The percentage of datasets that serve the Arabic language is minimal.

The significance of the Arabic language is evident by having it as one of the six official languages of the United Nations. In addition, Arabic came fourth in the top ten Internet languages as reported by the Internet World Stats[1] in June 2016. The latest estimates for Internet users by language. It is reported that there are 168,426,690 Arabic speaking users on the Internet, which represents 4.7% of all the Internet users in the world. In fact, the number of Arabic speaking Internet users has grown 6,602.5% in the last 16 years (2000–2016), which is the highest among all other languages. More details can be found in the source.

Resources that serve sentiment analysis for the Arabic language are scarce. The majority of the available datasets are either small in size or not accessible. Very few-available ones are accessible and reasonable. Namely, recently available large-scale Arabic datasets are LABR [8] and BRAD [9], which consist of 63K and 500K Arabic Book Review, respectively.

In this paper, we introduce the HARD dataset as another huge Arabic dataset of comparable size to similar ones available for the English language. In this paper, we design some sentiment analysis applications, however the dataset can be utilized for other NLP or machine learning applications. The effectiveness of HARD is validated by implementing six widely-known classifiers in addition to a lexicon-based classifier. All seven models are tested with different settings. The results are promising and encouraging. It is our objective to make HARD datasets publicly available and perceived as benchmark in the field of Arabic computing. The inspiration for this work was ignited by the lack of huge-scale Arabic dataset. We anticipate that the research community, working on the Arabic language, would value such contribution.

The paper is organized as follows: Sect. 2 presents the related work. Section 3 introduces some of the challenges in Arabic sentiment analysis. A description of the dataset and tools used in this work are detailed in Sect. 4. The sentiment analysis system is presented in Sect. 5. Performance evaluation of the classifiers is detailed in Sect. 6, and finally, we conclude the work in Sect. 7.

2 Related Work

Research on subjective sentiment analysis has been on the rise for the past 15 years. However, the majority of the reported research work served the English language [3, 10]. Over the past few years, a research momentum has been building up to serve the Arabic language. For example, see [11] for a survey on this fascinating field. The reader may check [12–15] for early research works on Arabic text. However, the datasets used were small-in-size and/or inaccessible. For instance,

[1]http://www.internetworldstats.com/.

Opinion Corpus for Arabic (OCA), [14], has 500 Arabic movie reviews collected from the web. It is a balanced dataset. Another example is AWATIF, which is a multi-genre corpus for Modern Standard Arabic sentiment analysis, [13]. It consists of 2855 sentences of news wire stories, 5342 sentences from Wikipedia talk pages, and 2532 threaded conversations from web forums.

On large available datasets, both [8, 16] used LABR dataset for sentiment analysis. Recently, Elnagar et al. prepared a large-scale Arabic dataset for sentiment analysis, which was made available to the research community. The Large-scale Arabic Book Review dataset (BRAD) is a set of more than 500K book reviews; each with a rating of 1 to 5 stars, [9].

The subarea of lexicon-based sentiment classifiers for Arabic had attracted several researchers. For example, a lexicon-based system for classifying Modern Standard Arabic (MSA) specific to the "news" domain is reported in [12]. The data (sentences) are manually annotated for subjectivity and domain. The work was extended to use Arabic-morphological features. The systems' success rate is reported as 95%. Further, they used a dataset of Arabic social media content and POS tagging. Similar work is reported in [17] using a random graph walk approach while using Naive Bayesian and SVM classifiers. The reported accuracy on the news data and a dataset of tweets is around 80% and 73%, respectively. Similarly, [18] explored the use of SVM and Naive-Bayes to compare the effect of feature selection on classifier performance. First, they collected a total number of 2051 tweets and compiled them in a dataset. These tweets went through some pre-processing steps and were manually labelled as positive, neutral or negative. Then they extracted a number of words and idioms from the corpus and classified them as positive, neutral or negative. As for feature selection, they took two approaches; extracting features using n-grams, and extracting features without n-grams. They also used term frequency–inverse document frequency (TF-IDF) and Binary-Term Occurrence (BTO) as a weighting scheme for the word vector. They compared both the SVM and the Naive-Bayes classifiers and reported that SVM without the n-gram features performed best. However, the Naive-Bayes classifier performed better with n-gram features than without. They conclude that pre-processing is an important step in sentiment analysis and can affect the performance of classifiers.

Another dedicated work on lexicons, [19], compare the performance of two approaches to Arabic sentiment analysis: a corpus based approach, and a lexicon based approach. As for the lexicon based approach, they built it manually. They began with 300 seed words taken from the SentiStrength website. Next, they translated these words from English into Arabic and annotated positive words and negative words. They further extended the lexicon by adding emoticons and synonyms; they gave the synonyms the polarity of the seed words with similar meaning. They ended up with a lexicon consisting of a total of 3479 words (1262 positive, and 2217 negative). The accuracy did not exceed 60%. They concluded that the bigger the lexicon is, the better the results, but only up to a point! In the end they conclude that the corpus based approach using SVM and light stemming gave the highest accuracy and concede that the lexicon based approach performed worse due to a number of possible reasons such as the presence of sarcasm in text.

Along the same line, [20] built an Arabic sentiment lexicon. They gathered a dataset of Arabic tweets and applied some preprocessing steps to clean up the sentence. Their approach was to score each word based on its sentiment value in the lexicon, and then calculate the cumulative score of the sentence by aggregating the scores of all the words in the sentence. The output is the predicted sentiment of the sentence; which is positive, neutral, or negative. The results were satisfactory. A hybrid scheme to predict the sentiment of Arabic tweets is developed in [21]. It is comprised of a lexical based classifier and an SVM classifier. After text pre-processing, they used the lexical classifier to label their data. This labelled data will then be used as training data for the SVM classifier.

Classifiers that are capable of handling MSA as well as DA were also reported. For example, a classifier in [22] is designed for handling both MSA and Egyptian dialectal Arabic tweets using SVM classifier; reported accuracy is 95%. Similar work appeared in [23]. Resources including datasets and pre-processing tools are not fully shared among researchers. For example, cleaning data, handling dialects [12], or negation articles [17, 24]. Therefore, the availability of such resources is a pre-requisite for enabling substantial contributions to Arabic sentiment analysis.

3 Arabic Sentiment Analysis

Sentiment analysis in Arabic faces many unique challenges such as:

- Large variance in informal written Arabic

While MSA has clear grammatical rules, most Arabic speakers do not pay much heed to grammar when writing Arabic informally online i.e. online product/service reviews, tweets, forum posts. E.g. Some may use diacritics, others may not. (e.g. جدا vs. جداً). Moreover, users commonly make spelling mistakes. (e.g. using different versions of the same character (آ أ ا إ)). Furthermore, differing dialects further complicate matters.

- Variety of Dialectal Arabic

Most Arabic speakers do not use Modern Standardized Arabic (MSA) in informal settings. Instead, they may use different dialects which use different words that have the same meaning (e.g. The word for "not" can be مش, مب , مو. Moreover, they may use many slang words.

- Scarcity of Resources.

While there is a growing number of resources available for English language sentiment analysis, other languages, including Arabic, suffer from a lack of resources, support and research. Large Arabic datasets are few in number and their topics are limited. Furthermore, many of the available software used for sentiment analysis either do not support Arabic letters or offer limited support.

4 Hotel Arabic Reviews Dataset (HARD)

4.1 Collection

The dataset is a collection of Arabic reviews compiled from Booking.com; a website that specializes in online accommodation booking. Collected reviews are structured as follows: A rating out of 10 of the accommodation, a title of the review, a positive aspect(s) of the accommodation, a negative aspect(s) of the accommodation, the reviewer's username, and country of residence. Reviews are available in many languages and can be filtered to the user's preference. The data was collected and originally organized in the following columns: Hotel name, rate (reviewer's rating out of 10), user type (family, single, couple), room type, nights (number of nights stayed), review title, positive review, negative review.

4.2 Properties

Initially, the dataset had 981143 reviews. Each review had two parts: positive comments and negative comments. We joined each hotel's positive and negative comments into one review resulting in 492907 reviews. Initially, the dataset contained reviews that had both Arabic and Latin text. After cleaning the reviews and eliminating the Latin text, the total number of reviews is reduced to 373772, which makes up the full unbalanced HARD dataset. However, the balanced subset dataset consists of 94052 reviews made up of 46968 positive and 47084 negative reviews. A negative review is defined as a review that has been given a rating of "1" or "2". Meanwhile a positive review is one where the review has been given a rating of "4" or "5". Neutral reviews with a rating of "3" have been removed from the dataset. The HARD dataset was created from the collected data and consists of the following main attributes:

- Rating—A number from 1 to 5 indicating the extent of the reviewer's satisfaction instead of using the previous scale from 1 to 10.
- Review—The reviewer's opinion written in the MSA or DA.
- Sentiment—Denotes the sentiment of the review with "+1" for a positive review, and "−1" for a negative review.

HARD dataset covers 1858 hotels contributed by 30889 users. The positive reviews constitute 68% of the total number of reviews when compared to the 13% of the negative ones. In addition, 19% of the reviews are "neutral". As expected, the

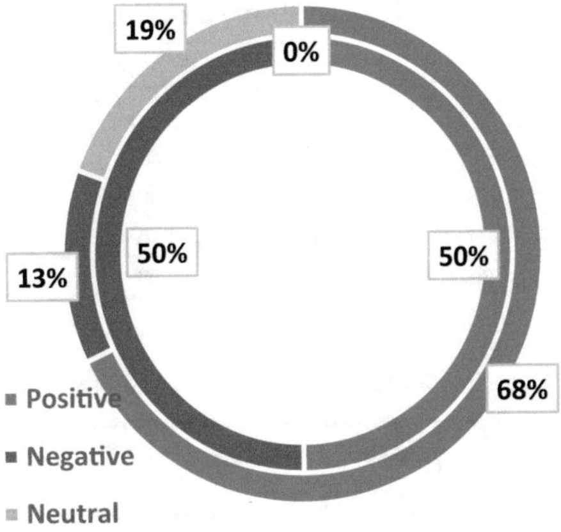

Fig. 1 Reviews distribution in HARD datasets; balanced (inner) and unbalanced (outer)

positive reviews are the majority class. In order to eliminate bias to the majority class, we produced the balanced subset of HARD. Figure 1 shows the distribution of ratings for both the balanced dataset (inner) and the unbalanced dataset (outer).

Figure 2 shows the distribution of reviews per hotel, number of nights stayed, user type, and room type. Table 1 summarizes statistics about the dataset including hotels, users, and reviews. A sample set of Arabic reviews expressed in MSA and

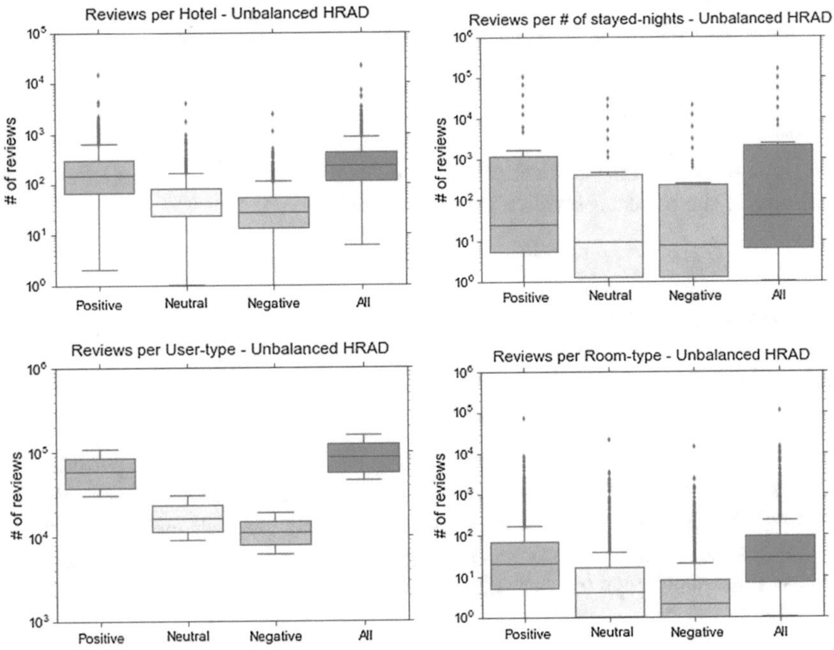

Fig. 2 Box-diagrams of HARD (initially called HRAD) reviews per hotel, number of nights, user, and room type

Table 1 Statistics on HARD dataset

Title	Number	Title	Number
Number of reviews	373,772	Median reviews per hotel	150
Number of hotels	1,858	Min reviews per hotel	3
Avg. reviews per hotel	264	Number of users	30,889
Max reviews per hotel	5,793	Avg. reviews per user	15.8
Median reviews per hotel	150	Number of tokens	8,520,886

Table 2 Sample reviews in MSA and DA

Rating	Arabic Type	Book ID	Review	#
5	MSA	1468	فندق رائع.الفندق مميز بموقعه و تصميمه و ديكورات غرفه. السرير واسع و مريح جدا.لا شئ	1
1	MSA	1468	استغرب تقييم الفندق كخمس نجوم.لا شي.يستحق 2 نجمه	2
2	MSA	1514	ضعيف.كل شي خربان ووسخ	3
1	DA	2125	جيد.موقعه وقريب من الاسواق.السرير **مومويح** سعره **موب** مناسب **عالغرفه**	4
4	DA	1167	جيد.كل شي.النت بفلوس . الفرشه **موب** نظيفه	5
3	DA	1542	مقبول.اعجبني ان سعر مناسب ولكن مافي خصوصية تسمع الي ساكن **جمبك** او في الطابق الي فوقك **بتطرقع ماكو** نوم عدل مع الازعاج.	6

DA is shown in Table 2. Dialectal phrases/words are shown in bold font. The majority of the dialectal reviews are in Gulf dialects.

5 Sentiment Analysis

5.1 Text Pre-processing

The first step is to perform text pre-processing in order to improve the performance of classifiers by transforming the text into a form as consistent as possible. To accomplish this, two phases are implemented. Namely, normalization and removal of stop words. Stemming is not recommended as the reviews include DA. Next, we describe the main steps for each phase.

Normalization

This phase involves the following steps:

1. Removal of punctuation marks from text.
2. Replace each letter Ya'a (ي) with the letter alif-maqsorah (ى).
3. Replace all forms of letter alef-with-hamza (أ، إ، آ) with the normalized form (ا).
4. Removal of kashida and all diacritics.

Stop word removal

The second step is to remove all stop words from the text. Stop words are defined as words that do not add any sentiment value to a sentence; they are usually the most common words in a language. For example, in English, words such as "and, all, also, want, are" can be considered stop words. Stop word lists can either be custom created or acquired from the internet. There is no definite list available; some are more comprehensive and thus more "aggressive" than others. The "right" choice depends on the work needs. Unfortunately, there are few resources available for the Arabic language. For this project, we modified a list of stop words from python.org [25].

Stemming

Stemming may not lead to better results. In fact, the performance of the classifiers deteriorate if stemming is used. For example, using ISRI Arabic Stemmer (available in Python's Natural Language Toolkit's stem package (NLTK)) on the sentence in Table 3 would produce the following poor result (Table 4).

Table 3 Example of text pre-processing

Review	Preprocessing
ضعيف **جداً** .تم الحجز وعندما وصلت فوجئت بان الغرف مشغولة وطلبوا **مني الإنتقال إلى** فندق **أخر**	Original
ضعيف جدا **تم** الحجز **وعندما** وصلت فوجئت **بان** الغرف مشغولة وطلبوا منى الانتقال **الى** فندق اخر	Normalization
ضعيف جدا الحجز وصلت فوجئت الغرف مشغولة وطلبوا منى الانتقال فندق اخر	Stop words

Table 4 Effect of stemming

Review	Preprocessing
ضعيف جدا **الحجز وصلت فوجئت الغرف مشغولة وطلبوا** منى **الانتقال فندق** اخر	Original
ضعف جدا **حجز وصل وجئ غرف شغل طلب** مني **نقل ندق** اخر	Stemming

5.2 Feature Extraction

To assess the performance of classifiers, we used different sets of features. Namely, bag-of-words (BOW) with unigram and bigram terms; BOW with TF-IDF (Token Frequency Inverse Document Frequency) for both unigram and bigram, and a lexicon to limit the number of features of the n-grams. The computation of TF-IDF term is:

$$Tfidf_{word,\,review} = \log(word, review) * \log \frac{\sum reviews}{\sum freq\ of\ words}$$

5.3 Classifiers

We selected six classifiers, which are commonly used in the area of sentiment analysis and reported in [8, 9, 16] as the best classifiers for Arabic sentiment analysis. The selected classifiers are: Logistic Regression (LGR), Passive Aggressive (PAG), Support Vector Machine (SVM), Perceptron (PRN), Random Forest (RFT), and AdaBoost (ABT). In addition, we tested a lexicon based approach. To prepare the dataset for use in these classifiers, it is divided into 3 groups: Training set, validation set and testing set.

6 Experimental Results

In order to show the effectiveness of the proposed dataset, we performed a comprehensive performance evaluation of Arabic sentiment analysis systems using the listed classifiers above. We use the same set of measurement tools used in [9] in order to report the accuracy and f1-measure. We carried out three experiments. The first one is polarity classification, in which each review in the test dataset is classified into either positive (+1) or negative (−1). The second set of experiments is about rating classification, in which each predicted sentiment is classified on a scale of 1 to 5 where 5 is the top score. The third set of experiments is similar to the first one. That is polarity classification but using a lexicon. All experiments were conducted on balanced and unbalanced HARD datasets. The datasets were divided into training and testing subsets using the ratio 80% for training and 20% for testing. We did not use a validation set.

6.1 Bag of Words

As pointed out above, each classifier is tested on unigrams only and unigrams and bigrams together. In order to normalize the words frequencies in the dataset, we repeated the experiment using TF-IDF (Token Frequency Inverse Document Frequency) of the n-grams.

In detail, the performance analysis involves testing the 6 classifiers (LGR, PAG, SVM, RRN, RFT, and ABT) for 2 classification cases, which are polarity classification and rating classification. The parameters used for each study involve:

1. Balanced dataset
   (a) With/without TF-IDF
   (b) Unigrams/bigrams : 1g and 1g + 2g
2. Unbalanced dataset
   (a) With/without TF-IDF
   (b) Unigrams/bigrams : 1g and 1g + 2g

The performance of all classifiers are assessed by two accuracy measures. Namely, f1-measure as well as accuracy. Tables 5 and 6 summarize the result for each classifier on the testing dataset, which reveals the following observations:

- Logistic regression and SVM classifiers outperformed the rest of the classifiers.
- The scores of Logistic regression and SVM classifiers are relatively close. While the logistic regression model outperformed SVM on the testing data sets without

Table 5 Performance evaluation on HARD—polarity classification

Classifier	TFIDF?	Dataset (polarity classification)			
		Balanced		Unbalanced	
		unigram	Uni + bigram	unigram	Uni + bigram
		f1-acc	f1-acc	f1-acc	f1-acc
Logistic regression	No	0.931-0.932	0.944-0.944	0.874-0.962	0.894-0.968
	Yes	0.935-0.935	0.942-0.942	0.871-0.962	0.88-0.965
Passive aggressive	No	0.896-0.896	0.926-0.926	0.823-0.945	0.869-0.96
	Yes	0.885-0.885	0.928-0.929	0.799-0.938	0.871-0.961
SVM	No	0.915-0.916	0.934-0.934	0.854-0.956	0.883-0.965
	Yes	0.927-0.927	0.944-0.944	0.872-0.962	0.897-0.969
Perceptron	No	0.894-0.895	0.929-0.929	0.809-0.939	0.868-0.959
	Yes	0.875-0.875	0.921-0.921	0.776-0.93	0.856-0.955
Random forest	No	0.913-0.84	0.926-0.927	0.776-0.941	0.793-0.945
	Yes	0.908-0.907	0.917-0.917	0.739-0.934	0.768-0.94
AdaBoost	No	0.872-0.875	0.871-0.875	0.719-0.928	0.71-0.926
	Yes	0.876-0.875	0.874-0.876	0.723-0.928	0.701-0.925

Table 6 Performance evaluation on HARD; rating classification

Classifier	TFIDF?	Dataset (rating classification)			
		Balanced		Unbalanced	
		1g	1g + 2g	1g	1g + 2g
		f1-acc	f1-acc	f1-acc	f1-acc
Logistic regression	No	0.677-0.678	0.716-0.717	0.724-0.726	0.749-0.751
	Yes	0.695-0.696	0.726-0.728	0.732-0.735	0.759-0.761
Passive aggressive	No	0.603-0.603	0.672-0.672	0.663-0.665	0.706-0.707
	Yes	0.575-0.575	0.668-0.669	0.641-0.641	0.688-0.689
SVM	No	0.639-0.64	0.691-0.691	0.697-0.699	0.727-0.729
	Yes	0.663-0.664	0.712-0.714	0.709-0.712	0.744-0.745
Perceptron	No	0.587-0.589	0.67-0.67	0.646-0.645	0.693-0.691
	Yes	0.545-0.545	0.653-0.653	0.608-0.607	0.667-0.666
Random forest	No	0.664-0.666	0.686-0.688	0.691-0.696	0.717-0.722
	Yes	0.664-0.666	0.689-0.69	0.69-0.697	0.715-0.72
AdaBoost	No	0.603-0.606	0.648-0.647	0.654-0.663	0.671-0.677
	Yes	0.606-0.608	0.647-0.64	0.658-0.666	0.673-0.68

TF-IDF measure, SVM has higher classification accuracy when utilizing TF-IDF except for the case of unigram terms.

- The depth of random forest (RFT) is limited to 30. The computational time is comparable for the all classifiers without RFT, which took longer time during the training phase.
- For the polarity classification test (Table 5), both the f1-measure and accuracy scores are almost identical for the balanced dataset whereas there is noticeable difference between the 2 measures for the unbalanced dataset.
- For the polarity classification on the balanced test set with unigrams only, the best classifier, which is logistic regression, reported f1-measure-accuracy score of 0.931-0.932 (without TF-IDF) and 0.935-0.935 (with TF-IDF). However, for both unigrams and bigrams, while the best classifier (logistic regression) reported 0.944-0.944 without TF-IDF, SVM reported the best results of 0.944-0.944 for TF-IDF case.
- Similarly, for the polarity classification on the unbalanced test set, the logistic regression produced the highest accuracy for the case without TF-IDF while SVM outperformed the classifiers for the TF-IDF case. See Table 5 for details.
- For the rating classification test (Table 6), the f1-accuracy scores are very close for both the balanced and unbalanced datasets. For the balanced set, the results ranges 0.587-0.589 to 0.677-0.678 for the unigram (without TF-IDF) and from 0.545-0.545 to 0.695-0.696 for the unigram-with TF-IDF case. The best performance is reported for the unbalanced unigram and bigram case. The logistic classifier outperforms the other classifiers in all 4 categories. The performance deteriorates when compared to the polarity experiment, which is justified by the

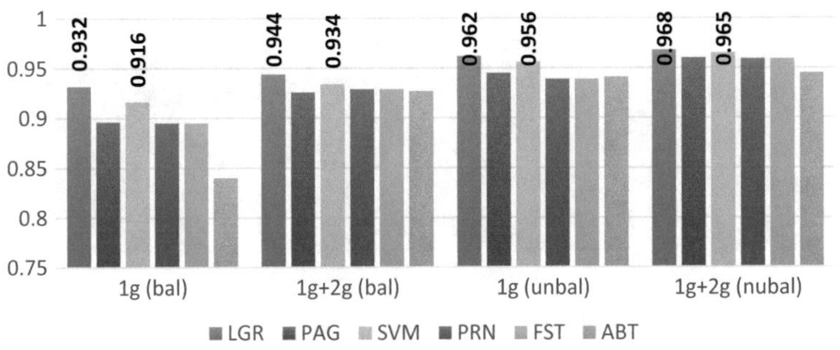

Fig. 3 Accuracy of polarity classification dataset

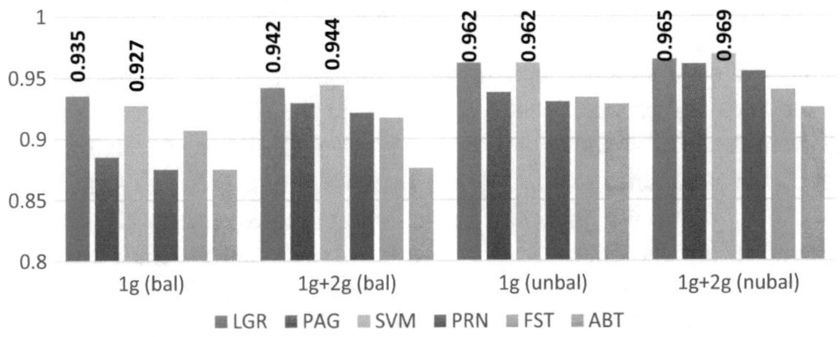

Fig. 4 Accuracy of polarity classification with TF-IDF

relatively large numbers of false positives and false negatives when classifying to 5 classes.

The findings are also depicted in Figs. 3, 4, 5 and 6. Figures 3 and 4 are related to the polarity classification experiment. The figure has 4 groups of 6 sets of bar-charts each. The first 2 groups are for the balanced test subset of HARD dataset. The remaining 2 groups represent the accuracy results on the unbalanced dataset.

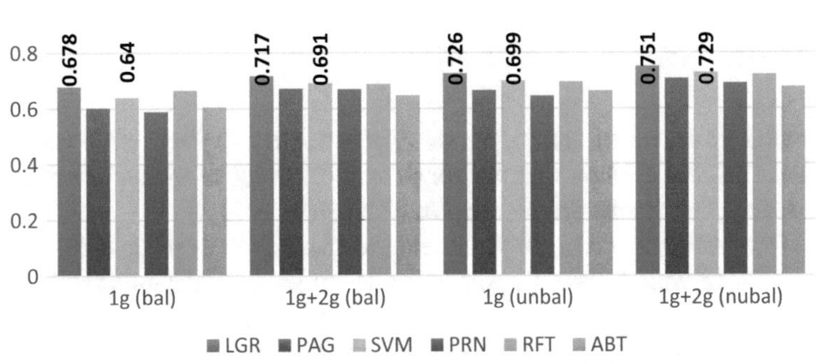

Fig. 5 Accuracy of rating classification without TF-IDF

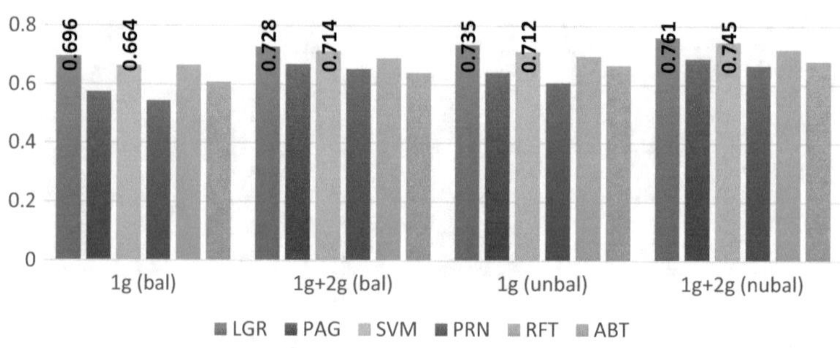

Fig. 6 Accuracy of rating classification with TF-IDF

For each group, we run the classifiers for unigrams (1g), and unigrams + bigrams together (1g + 2g).

For the balanced datasets, consider the first group of bar charts. While Fig. 3 depicts the accuracy scores for all 6 classifiers on the HARD dataset without using TF-IDF measure, Fig. 4 shows the outcome when TF-IDF computation is utilized. On Fig. 3, the accuracies of the logistic regression and SVM classifiers have the highest scores. Figure 4 depicts similar results and improvement rate when TF-IDF is employed in the classifiers, which are a bit higher than the ones in Fig. 3.

For the case of the unbalanced datasets, consider the last 2 groups of bar charts. The accuracies of the logistic regression and SVM classifiers has jumped to 96% and 97%, respectively. Figure 4 depicts similar results and insignificant improvement rate when TF-IDF is employed.

Figures 5 and 6 show the results of the rating classification experiment. As expected, the accuracy would deteriorate as the number of false positives and false negatives would vividly increase as there are 5 possibilities for each sentiment. Considering the balanced dataset, Fig. 5 shows that the logistic regression and SVM classifiers attain the best accuracies in all groups of tests. For the case of the unbalanced datasets, the performance of the classifiers improve. Figure 6 depicts similar results when TF-IDF is employed.

6.2 Lexicon-Based Classification

To further investigate the effectiveness of the HARD dataset, we conducted an experiment using a lexicon that we constructed in order to reduce the number of feature space. The results are reported in Table 7.

A lexicon based approach to sentiment analysis involves using a dictionary of words annotated with the word's polarity to determine a document's sentiment. For any given input text, words are extracted and given a sentiment orientation value (positive or negative) based on the score of the words in the dictionary, then the

Table 7 Performance evaluation on HARD; lexicon-based polarity classification

Classifier	Balanced	Unbalanced
	f1-acc	f1-acc
Logistic regression	0.816-0.794	0.505-0.887
Passive aggressive	0.798-0.772	0.478-0.876
SVM	0.811-0.788	0.495-0.883
Perceptron	0.755-0.743	0.473-0.851
Forest	0.811-0.789	0.519-0.885
AdaBoost	0.796-0.753	0.454-0.88

final score of the text is the aggregation of all the scores of the individual words in the text [26].

The lexicon used for this approach was constructed to better suit the dataset's needs. However, it is mainly accounting for MSA as opposed to DA. The constructed lexicon is a baseline sentiment lexicon extracted from HARD dataset. Simply, we devised an automatic process to generate the most important words by ordering their weights using SVM classifier then selecting the highest weights. We then manually review the list to remove any redundant terms. The resulting lexicon consists of 1453 positive and 3014 negative terms.

To validate the effectiveness of this lexicon, we tested the six classifiers on both the balanced as well as the unbalanced test sets. We used unigrams and bigrams for the experimentation. The feature space is reduced to 4467 features. As a result, the resulting simpler classifiers are computationally faster. For example, the time took for running the random forest classifier was greatly reduced when compared to using the full model. Therefore, more depth trees can be used instead of limiting the depth to 30. Furthermore, other computationally-costly classifiers may be tested.

The reported results in Table 7 and Figs. 7 and 8 show that the accuracy of predicting sentiments using the proposed lexicon vary between 0.743 and 0.794 for the balanced set and between 0.88 and 0.887 for the unbalanced set, which is really inspiring.

Figures 7 and 8 show the results of the polarity classification experiment using the lexicon. These figures show the results of the f1-measure as well as the accuracy

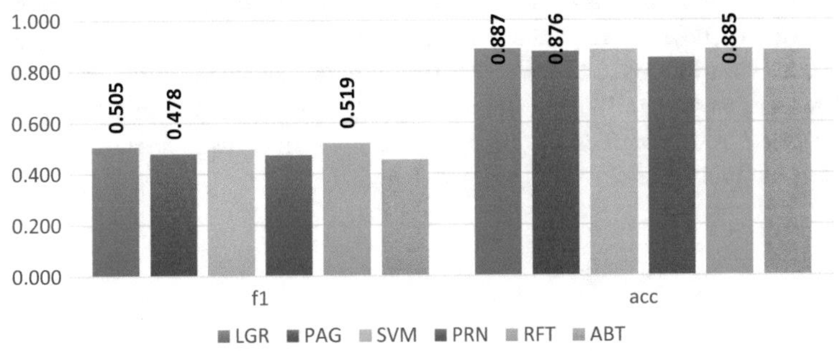

Fig. 7 Polarity Lexicon-based classification—balanced dataset

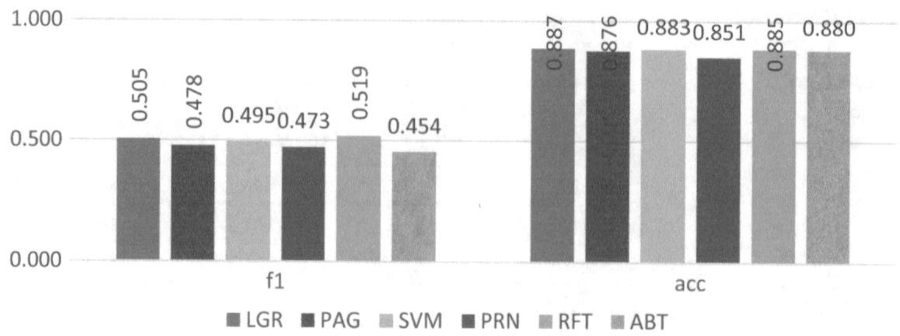

Fig. 8 Polarity Lexicon-based classification—unbalanced dataset

score. The f1-measure scores are much lower as the as the number of false positives and false negatives increase. Both figures show that the performance of the classifiers are comparable and close. However, the logistic regression, SVM, and random forest classifiers attain the best accuracies in both tests.

7 Conclusions

In this work, we introduced HARD, which is a large Arabic dataset for sentiment analysis and machine learning applications. This is a rich dataset of more than 370K reviews expressed in modern standard Arabic and dialectal Arabic. We described the collection, cleaning, construction, and properties of HARD. We intend to make this dataset available to the research community in two forms: the unbalanced complete set as well as the balanced dataset.

To investigate the validity and effectiveness of HARD, we implemented six widely known sentiment classifiers. The classifiers were tested for polarity and rating classification. Logistic regression and SVM classifiers produced the best results. The reported accuracies range from 94 to 97% for polarity classification. However, the accuracies for the rating classification experiment range from 72 to 75%.

In addition, we investigated the use of a constructed lexicon to confirm the effectiveness of HARD. The attained accuracy is 89% for polarity classification.

Our core contribution is to make this benchmark-dataset available and accessible to the research community on Arabic language. We trust that this dataset would make a good resource and the analysis would ignite more research on Arabic sentiment analysis and close-related problems.

References

1. Boyd, D.M., Ellison, N.B.: Social network sites: definition, history, and scholarship. J. Comput-Mediat. Commun. **13**, 210–230 (2008)
2. Gu, B., Law, R., Ye, Q.: The impact of online user reviews on hotel room sale. Int. J. Hosp. Manag. **28**, 180–182 (2009)
3. Liu, B., Sentiment analysis and opinion mining. In Synthesis Lectures on Human Language Technologies, Morgan and Claypool Publishers, pp. 1–133 (2012)
4. Jones, B., Temperley, J., Lima, A.: Corporate reputation in the era of web. J. Mark. Manag. **25** (9–10), 927–939 (2010)
5. Greaves, F., Ramirez-Cano, D., Millett, C., Darzi, A., Donaldson, L.: Harnessing the cloud of patient experience: using social media to detect poor quality healthcare. BMJ Quality & Safety (2013)
6. Wang, H., Can, D., Kazemzadeh, A., Bar, F., Narayanan, S.: A system for real-time Twitter sentiment analysis of 2012 U.S. presidential election cycle. In: Association for Computational Linguistics, Jeju Island, Korea, 2012
7. Park, S., Liu, Y., Kim, J., Ko, M., Song, J.: The politics of comments: predicting political orientation of news stories with commenters' sentiment patterns. In: Proceedings of the Acm 2011 Conference on Computer Supported Cooperative Work, Hangzhou, 2011
8. Nabil, M., Aly, M., Atiya, A.: LABR 2.0: Large scale Arabic sentiment analysis benchmark (2014). arXiv:1411.6718
9. Elnagar, A., Einea, O.: BRAD 1.0: Book Reviews in Arabic Dataset. In: International Workshop for Data Science and Computing (DSC'2016); 13th ACS/IEEE International Conference on Computer Systems and Applications AICCSA 2016, Agadir, 2016
10. Pang, B., Lee, L.: Opinion mining and sentiment analysis. Foundations and Trends in Information Retrieval, pp. 1–135 (2008)
11. Korayem, M., Crandall, D., Abdul-Mageed, M.: Subjectivity and sentiment analysis of Arabic: a survey. In: Advanced Machine Learning Technologies and Applications, Communications in Computer and Information Science series 322, pp. 128–139. Springer (2012)
12. Abdul-Mageed, M., Sandra, K., Diab, M.: SAMAR: a system for subjectivity and sentiment analysis of arabic social media. In: Proceedings of the 3rd Workshop on Computational Approaches to Subjectivity and Sentiment Analysis, Jeju, Republic of Korea, 2012
13. Abdul-Mageed, M., Diab, M.: AWATIF: A multi-genre corpus for Arabic subjectivity and sentiment analysis. In: Proceedings of the 8th International Conference on Language Resources and Evaluation (LREC), stanbul, Turkey, 2012
14. Rushdi-Saleh, M., Mart´ın-Valdivia, M., Urena-L´opez, L., Perea-Ortega, J.: Oca: Opinion corpus for Arabic. J. Am. Soc. Inform. Sci. Technol. 2045–2054 (2011)
15. Elarnaoty, M., AbdelRahman, S., Fahmy, A.: A machine learning approach for opinion Holder extraction Arabic language, *CoRR, abs/1206.1011,* vol. 3, 2012
16. Elnagar, A.: Investigation on sentiment analysis of Arabic book reviews. In: 13th ACS/IEEE International Conference on Computer Systems and Applications AICCSA 2016, Agadir, 2016
17. Mourad, A., Darwish, K.: Subjectivity and sentiment analysis of modern standard Arabic and Arabic microblogs. In: Proceedings of the 4th Workshop on Computational Approaches to Subjectivity, Sentiment and Social Media Analysis (WASSA), Atlanta, Georgia, 2013
18. Li, D., Qiu, R., Al-Rubaiee, H.: Identifying Mubasher software products through sentiment analysis of Arabic tweets. In: Industrial Informatics and Computer Systems, 2016
19. Abdulla, N.A., Ahmed, N.A., Shehab, M.A., Al-Ayyoub, M.: Arabic sentiment analysis: Lexicon-based and Corpus-based. In: 2013 IEEE Jordan Conference on Applied Electrical Engineering and Computing Technologies (AEECT), Amman, 2013
20. Al-Ayyoub, M., Alsmadi, I., Bani Essa, S.: Lexicon-based sentiment analysis of Arabic tweets. Int. J. Soc. Netw. Min. **2**(2), 101–114 (2015)

21. Aldayel, H., Azmi, A.: Arabic tweets sentiment analysis—a hybrid scheme. J. Inform. Sci. **42** (6), 1–16 (2015)
22. Ibrahim, H.S., Abdou, S.M., Gheith, M.: Sentiment analysis for modern standard Arabic and Colloquial. Int. J. Nat. Lang. Comput. **4**(2) (2015)
23. Abdul-Mageed, M., Diab, M.: SANA: a large scale multi-genre, multi-dialect Lexicon for Arabic subjectivity and sentiment analysis. In: The 9th edition of the Language Resources and Evaluation Conference, Reykjavik, Iceland, 2014
24. Shoukry, A., Rafea, A.: Sentence-level Arabic sentiment analysis. In: Collaboration Technologies and Systems (CTS), Denver, CO, USA, 2012
25. Savand, A.: Stop-words 2015.2.23.1: Python Package Index. https://pypi.python.org/pypi/stop-words. Accessed 5 Jan 2017
26. Stede, M., Voll, K., Tofiloski, M., Brooke, J., Taboada, M.: Lexicon-based methods for sentiment analysis. Computational Linguistics, vol. 37, no. 2, p. Computational Linguistics, 2011

Using Twitter to Monitor Political Sentiment for Arabic Slang

Amal Mahmoud and Tarek Elghazaly

Abstract Twitter is one of the most famous applications of social networks that allow users to communicate with each other and share their opinions and feelings in all types of topics: economics, business, science, social, religion, and politics in a very short message of information called Tweets. Users are usually written using colloquial Arabic and include a lot of slang. In this Paper, we studied sentiment analysis of Arabic text retrieved from a twitter focus on presidential elections in Egypt 2012. We are using Naïve Bayes (NB) which is a machine learning algorithm, one time by using N-Gram (unigram and bigram) and another time by using feature selection. The main objective of this paper is to measure the accuracy of each method and determine which method is more accurate for Arabic text classification. The results show that unigram and information gain attribute selection achieves the highest accuracy and the lowest error rate.

Keywords Arabic language · Sentiment analysis · Twitter · Arabic stemmers N-gram · Naïve Bayes · Attribute selection · Weka

1 Introduction

Given the recent political unrest in the Middle East (2011), there has been an increasing interest in harvesting information written in the Arabic language from live online forums, such as twitter. Subjectivity and sentiment analysis aims to determine the attitude of the twitter's user with respect to a topic or the overall

A. Mahmoud (✉) · T. Elghazaly
Computer Science Department, Institute of Statistical Studies and Research (ISSR)
Cairo University, Giza, Egypt
e-mail: amal.mahmoud@pg.cu.edu.eg

T. Elghazaly
e-mail: Tarek.elghazaly@cu.edu.eg

K. Shaalan et al. (eds.), *Intelligent Natural Language Processing: Trends and Applications*, Studies in Computational Intelligence 740,
https://doi.org/10.1007/978-3-319-67056-0_4

contextual polarity of an utterance. Compared to other languages, such as English, research on Arabic text for is sparse. A possible reason for this is the complex morphological, structural and grammatical nature of Arabic [1].

A study prepared and published by Semiocast in 2012 has revealed that Arabic was the fastest growing language on Twitter in 2011, and was the 6th most used language on Twitter in 2012 [2].

A study prepared and published by Tumasjan in 2010 who analyzed 100,000 tweets, in their paper claimed that a mere mention or volume analysis of the tweets related to the election was enough to predict the results. They also said that the people's opinions in the real world are closely related to the tweets' opinions. Till date, this paper has had the most successful attempt in election predictions [3].

Trump and twitter. According to Newsday, "Donald Trump is using social media "especially Twitter "unlike any other 2016 presidential candidate. Or any before. His relentless use of social media to promote himself and bash opponents has been a key element in his bid in the race for the Republican presidential nomination" [4].

Sentiment analysis is the automated mining of attitudes, opinions, and emotions from text, speech, and database sources. Sentiment analysis involves classifying opinions in text into categories like "positive" or "negative" or "neutral".

Many of the researchers in the area of sentiment analysis deal with the English language, but the Arabic language which spoken by more than 300 million people, and is the fastest-growing language on the web there is only a few research.

Arabic has some variants in spelling and typographic forms, Creation of new expressions the usage of which indicates high subjectivity for example "not good فاكس" is used as a negative reference.

The slang intensifiers list, for example, included, "خالص", "موت", "اوى", "جدا" which all corresponds to "a lot", Sarcasm is a form of speech act where a person says something positive while(s) he really means something negative or vice versa [5].

Stemming means finding the root or stem of the given inflected word. It is used in Natural Language processing, Information Retrieval, Text Mining, etc. Mostly, stemming is used to improve the performance for NLP (Natural Language Processing). In this study, we collected political tweets, which were published on Twitter's public message board from March 1st to June 24th, 2012, prior to the Egyptian election, with volume increasing as the election drew nearer. We have annotated 18278 tweets consisting of 11910 positive, 6368 negative to be our training corpus [6].

The organization of this paper is as follows, related work in Sect. 2, proposed methodology, demonstrates the experimentation in Sect. 3, and presents the results and evaluation in Sect. 4, and finally conclusions and future works are given in Sect. 5.

2 Related Work

The study of [7] presents a system to extract business Arabic reviews, and then it analyzed these collected reviews to identify their polarity (positive, negative or neutral). And exhibits the general opinion of the Arab public about different products and services.

The study of [8] Twitter has a great potential of exploring and understanding people, their lives, potential, interests, and opinions. So the study focused on the sentiment analysis of twitter using common machine learning machine learning techniques such as a Naïve Bayes and a Maximum Entropy Model.

The study of [9] uses machine learning classifiers by using both Arabic and English corpora. They employ two machine learning classifiers namely, (SVMs) and (NB) classifiers. The results obtained show that SVMs outperform the NB classifier and also there is not a big difference between using the term frequency (TF) and the term frequency-inverse document frequency (TF-IDF) for weighting methods.

The study of [10] found that light stemming is one of the most superior in morphological analysis.

The study of [11] for light stemming, several variants have been developed. When applying in the AFP_ARB corpus, the authors have found that light stemmer was more effective for cross-language retrieval than a morphological stemmer. They deduce that it is not essential for a stemmer to yield the correct root.

The study of [12, 13] Used character-level trigrams for indexing Arabic documents without prior indexing. It is shown that on average the trigram indexing scheme seems to be a little bit less effective than word-based indexing. No researches have been done to measure the effects of combining word-level unigrams (single words) and bigrams (phrases) as indexing techniques on the efficiency of Arabic text classification.

3 Methodology

The methodology using for Building Machine Learning classifiers consists of 3 steps:

1. Corpus Collection and Preparation
2. Pre-processing
3. Text Classification

3.1 Corpus Collection and Preparation

The total corpus size is 18278 tweets. We have annotated 18278 tweets consisting of 11910 positive, 6368 related to the opinion expressed in Arabic from different

Table 1 Sample of positive and negative tweet

Annotate	English translate	Original tweet
Positive	This man with principles and respectable, we will vote for Abu Alftouh	الراجل ده صاحب مبادئ و محترم صوتنا لـ ابوالفتوح
Negative	Khaled Ali, This man has incredibly bad luck, he will not win	خالد على الراجل ده حظة وحش و مش هيكسب

```
@relation election
@attribute tweet string
@attribute class {"احمد شفيق","محمد مرسى","عمرو موسى", "خالد على"," صباحى حمدين","ابو الفتوح"}
@attribute opinion {0,1}
@data
```

Fig. 1 ARFF file for the election data

domains: “خالد على -(Khaled Ali)”, “عمرو موسى -(Amr Mousa)”, “احمد شفيق-(Ahmed Shafik)”, “محمد مرسى -(Mohammed Morsi)”, “صباحى حمدين -(Hamden Sabahy)”, “"ابو الفتوح" -(Abu Alftouh)”. We define a sentiment as positive or negative opinion, each data instance (Tweet) annotated to positive or negative. Table 1 shows samples of the annotated tweets [14] (Fig. 1).

Then convert the file to ARFF (Fig. 1), which deals with WEKA program. WEKA [15] is a popular suite of machine learning software written in Java. In the Waka tool initial the data set will be loaded. Under Meta classifier, Filtered classifiers were used. The filter used to be StringToWordVector Under this filter IDFTransform, TFTransform, output Word Count, use Stop List was set to True. Stemmer used to be Khoja stemmer and Arabic light stemmer.

3.2 *Pre-processing*

Before being able to properly analyze textual data, the raw twitter data retrieved needed to be preprocessed. Figure 2 displays all the stages of preprocessing the datasets went through before classification.

Tokenization

- Special character Extraction: Every character seems to be non-Arabic has to be removed.

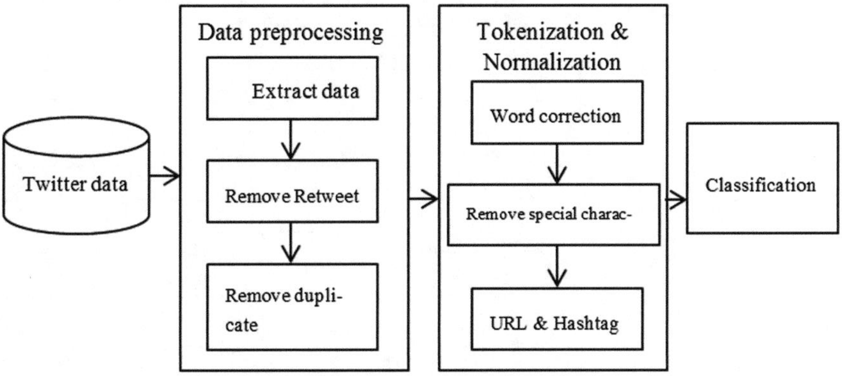

Fig. 2 Preprocessing of data before sentiment classification

- Non-Arabic letter Extraction: This process aims to remove digit, Punctuations, marks, symbol and a special character (e.g. +_, """?). For example, remove a hash letter (#) to capture subjective text in hashtag if available.

  Before: معاااااك يا ريس......
  After: معاك يا ريس
  Translation: with your president.

Normalization:

- Converting text files in UTF-16 encoding.
- Standardizing the Arabic text, this step is to transform some characters in standard form as "آ,أ,إ" to "ا" and "ءي,ءى" to "ئ" and "ؤ" to "و" [16].

Stop words removal:

Words such as prepositions and articles that occur frequently and don't help in discrimination between classes (i.e. Pronouns, prepositions, conjunctions, etc.).

In Arabic words like are "في" (in), "علي" (on), and these words carry no information. And added additional stop words for the Arabic slang. Stop words for the Arabic slang are like (ده, دى, اللى, بس............) (Da, de, Elly, bas…..). These words are usually removed using stop words list.

Stemming

Stemming is the process of reducing words to their stem or root from where morphological information is used to match different variants of words (Table 2).

Arabic Light Stemmer

Stemming reduces words to their stems. Light stemming, in contrast, removes common affixes from words without reducing them to their stems. The main idea of using light stemming is that many word variants do not have similar meanings or semantics, although these word variants are generated from the same root. Thus,

Table 2 Arabic stemmer example

Prefixes + Stem (Root + Pattern) + Suffixes		
Root	لعب	Play
Prefixes	ال	The
Stem	لاعب	Player
Suffixes	ين,ان	Dual
Suffixes	ون	Plural
Suffixes	ة	Feminine
اللاعبان – اللاعبين		The players (dual)
اللاعبون		The players (plural)
اللاعب		Player (masculine)
اللاعبة		The player (feminine)

root extraction algorithms affect the meanings of words. Light stemming aims to enhance feature/keyword reduction while retaining the word meanings. It removes some defined prefixes and suffixes from the word instead of extracting the original root [17, 18].

Khoja stemmer

- The Khoja stemmer follows this procedure [19]:
- Remove definite article (ال).
- Remove inseparable conjunction (و).
- Remove suffixes.
- Remove prefixes.
- Match result against a list of patterns. If a match is found, extract the characters in the pattern representing the root.
- Match the extracted root against a list of known "valid" roots.
- Replace weak letters اوي with و.
- Replace all occurrences of hamza ء،ئ،ؤ with أ.
- Two letter roots are checked to see if they should contain a double character. If so, the character is added to the root.

Term weighting:

Finally Using Term Frequency-Inverse Document Frequency, each document is represented as a weighted Vector of the terms found in the super vector. Every word is given a weight in each document. There are many suggested weighting schemes TF-IDF (*ti*) weight, which is the number of text documents in the corpus in which *ti* occurs at least once and inverse document frequency IDF of the term *ti* defined below [20].

$$\text{IDF}(ti) = \log \frac{D}{DF(ti)} \tag{1}$$

where d is the total number of documents in the dataset. The weight of term *ti* in a document *di* TF.IDF is defined below.

$$\text{TF.IDF}\,(ti, di) = \text{TF}^{*}\,\text{IdF}(ti). \tag{2}$$

In general, it is used as a metric for measuring the importance of a word in a document within a collection, so as to improve the recall and the precision of the retrieved documents.

N-Grams:

N-Gram is a subsequence of N items from a given sequence. It can be conceived as a window of the length of N moves over the text, N-grams could be character level, word-level, or even statement level. N-Grams method is widely used in statistical natural language processing. It is language dependent and works well in case of noisy text [21]. 1-Gram, that is the N-gram of length 1, is called Unigram, 2-Gram is called bigrams, 3-Gram is called Trigram, 4 Gram is called Quadrigram and so on. In our classification method, we used unigrams indexing. For example, the unigrams for the statement are:

"I will elect Khalid Ali, he is not lying - هنتخب خالدعلى مبيكذبش على الاقل"

Are:

"هنتخب", "خالد ", علي ", " مبيكدبش ", "على ","الاقل"

While the bigrams for the same statement are:

"هنتخب خالد"," خالد على", "على مبيكدبش" ",مبيكدبش على", "على الاقل"

Feature Selection:

Feature selection is one of the most important parts in sentiment classification. The tweet dataset has 18278 tweets, even though I only consider the unigrams and the bigrams as features, I still get is too word matrix big for classification and too many features can also cause an over-fitting problem. It is not implementable for a regular computer and quite expensive to do sentiment classification of such a big size. So feature selection is the process of selecting the most important or descriptive set of features/attributes by removing irrelevant and redundant features for reducing the dimensionality of data to improve machine learning performance [22].

For example, if a term "remnant - فلول" occurs in negative class very frequently, then the presence of this term in a new test document indicates that the document belongs to the negative class.

In Feature selection, using an Objective function measure the goodness of subsets of features created by a search algorithm. Three classic models of feature selection are a filter, wrapper, and embedded. A filter model relies on measures about the intrinsic data properties, call these filter methods because they filter out irrelevant features before the induction process occurs.

The present study focuses on filter model, especially Information Gain.

- Filter model

In filter approach, statistical methods such as chi-square, information gain are used to select the top ranked features from a given dataset according to their importance score and remove low ranked ones.

Filter approaches are computationally fast, simple and have the ability to deal with high dimension feature space [23].

Information gain

The Information gain method is frequently employed. Usually, the information gain for each term is computed and the terms with IG less than a pre-determined threshold are removed. In the IG method, the goodness measure of a term for class prediction can be estimated by the presence or absence of that term in a document. Let C be a set of classes in a training dataset, the information gain of a term (t) can be estimated as [24].

$$\mathrm{IG(t)} = \sum_{i=1}^{m} \mathrm{p(ci)} \log p(ci) + p(t) \sum_{i=1}^{m} p(ci|t) \log p(ci|t) + p(\neg t) \sum_{i=1}^{m} p(ci|t) \log p(ci|\neg t). \quad (3)$$

where:

P (c_i) is the probability documents in the dataset belong to a class (c_i);

P (t) and P (¬t) are the probability that a term (t) is in the documents of the dataset or not, and P (c_i|¬t) is the probability that documents in the class (c_i) contain a term (t). m is the number of features.

3.3 Text Classification

The present study focuses on one label binary Classification where each document is assigned one of the two categories positive or negative. Classifiers can be generated in a variety of different ways that in turn also dictates their usage. To analyze sentiment in Arabic text, the process covered in two phases: preprocessing (Phase 1) and attitude prediction (Phase 2). (Figure 3 shows Training and testing a machine learning classifier).

We use the data mining package Weka to perform our tasks of preprocessing and classification [23].

We use 10-fold cross-validation because extensive tests on numerous datasets, with different learning techniques, have shown that 10 is about the right number of folds to get the best estimate of error, and there is also some theoretical evidence that backs this up [15].

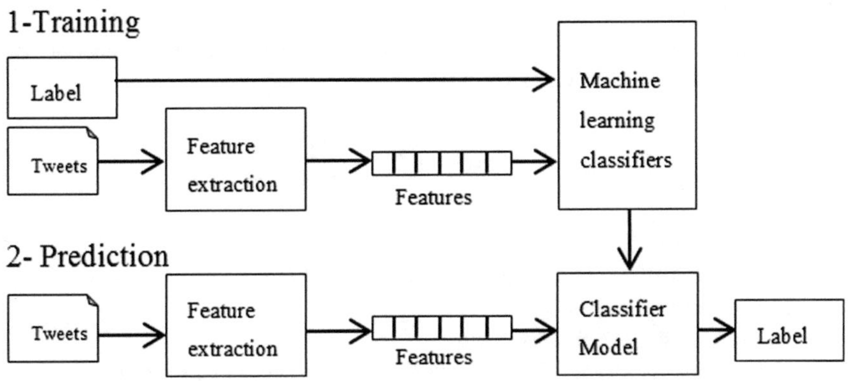

Fig. 3 Training and testing a machine learning classifier

We use Naïve as classification algorithms.

- **Naïve Bayes (NB)**: Naïve Bays (NB) is one of the most common supervised classification methods that can be used to perform text classification (TC). It is computationally very easy and efficient to use. This classifier calculates the conditional probabilities for each class from a training set, in order to predict that a given feature belongs to a particular class [25]. When the NB classifier is applied to the TC problem we use equation [26].

$$\mathrm{p(class|document)} = \frac{P(clas).p(document|class)}{p(document)}. \tag{4}$$

where:

P (class|document): It's the probability that a given document D belongs to a given class C
P (document): The probability of a documentit's a constant that can be ignored
P (class): The probability of a class, it's calculated from the number of documents in the category divided by documents number in all categories
P (document|class): it's the probability of document given class, and documents can be represented by a set of words:

$$\mathrm{P(document|class)} = \prod \mathrm{p_i(word_i|class)}. \tag{5}$$

So:

$$\mathrm{p(class|document)} = \mathrm{p(class).pi(word_i|class)}. \tag{6}$$

where:

p (word$_i$|class) The probability that a given word occurs in all documents of class C, and this can be computed as follows:

$$p(\mathrm{word_i}|\mathrm{class}) = \frac{Tct + \lambda}{NC + V}. \quad (7)$$

where:

- Tct The number of times that the word occurs in that category C.
- Nc The number of words in category C.
- V The size of the vocabulary table.
- λ The positive constant, usually 1, or 0.5 to avoid zero probability.

4 Results and Evaluation

The evaluation is based on two popular machine learning algorithms (NB) using anagram as a feature and using 10-fold cross validation method for testing. The evaluation we used precision, recall, and F-measure to evaluate these approaches (Table 3).

Where TP(c), TN(c), FP(c), and FN(c) defined to denote the number the specific events occurred in the collection for a class c [27].

- TP (true positive) represents the examples that are correctly predicted and marked as positive,
- FP (false positive) indicates the examples that are incorrectly classified as positive,
- TN (true negative) denotes the examples that are correctly predicted as negative,
- FN (false negative) represents the examples that are incorrectly marked as negative.

$$\mathrm{Recall} = \frac{\mathrm{TP}}{\mathrm{TP} + \mathrm{FN}}. \quad (8)$$

Recall rate is a statistical classification function that measures the probability of retrieving relevant examples divided by the total number of the existing examples that are expected to be retrieved.

Table 3 Matrix of predicted and actual class

	Actual		
Predicted		C	¬C
	C	TP(c)	FP(c)
	¬C	FN(c)	TN(c)

$$\text{Precision} = \frac{TP}{TP + FP}. \tag{9}$$

Precision is a statistical classification function that measures the probability of retrieving relevant examples divided by the total number of the retrieved examples.

$$\text{F} - \text{measure} = \frac{2 * \text{Recall} * \text{Precision}}{(Recall + Precision)} \tag{10}$$

The statistical function called F-score (F-measure) estimates the harmonic mean of precision and recall to evaluate the accuracy.

$$\text{Accuracy} = \frac{\text{TP}^{(c)} + \text{TN}^{(c)}}{\text{TP}^{(c)} + \text{FP}^{(c)} + \text{FN}^{(c)} + \text{TN}^{(c)}}. \tag{11}$$

Accuracy is a good measure when classes are distributed uniformly in the collection (Tables 4, 5, 6, 7, 8).

Table 4 Represents the precision, recall, and F-measure for each category and the average values for all categories for the NB classifier using Khoja stemmer, unigram and info again attribute selection

Precision	Recall	F-measure	Class
0.985	0.942	0.963	احمد شفيق
0.929	0.932	0.931	محمد مرسي
0.890	0.917	0.904	عمرو موسي
0.963	0.952	0.957	خالد علي
0.780	0.799	0.789	حمدين صباحي
0.860	0.956	0.906	ابو الفتوح
0.934	0.931	0.932	Weighted avg.

Table 5 Represents the precision, recall, and F-measure for each category and the average values for all categories for the NB classifier using Khoja stemmer, unigram without info again attribute selection

Precision	Recall	F-measure	Class
0.870	0.936	0.902	احمد شفيق
0.926	0.934	0.930	محمد مرسي
0.791	0.961	0.868	عمرو موسي
0.882	0.983	0.930	خالد علي
0.971	0.053	0.101	حمدين صباحي
0.859	0.960	0.907	ابو الفتوح
0.889	0.883	0.857	Weighted avg.

Table 6 Represents the precision, recall, and F-measure for each category and the average values for all categories for the NB classifier using Khoja stemmer, bigrams and info again attribute selection

Precision	Recall	F-measure	Class
0.724	0.934	0.815	احمد شفيق
0.844	0.631	0.722	محمد مرسي
0.808	0.181	0.296	عمرو موسي
0.968	0.961	0.965	خالد علي
0.903	0.363	0.518	حمدين صباحي
0.869	0.956	0.910	ابو الفتوح
0.806	0.790	0.774	Weighted avg.

Table 7 Represents the precision, recall, and F-measure for each category and the average values for all categories for the NB classifier using Khoja stemmer, bigrams without info again attribute selection

Precision	Recall	F-measure	Class
0.734	0.930	0.820	احمد شفيق
0.835	0.653	0.733	محمد مرسي
0.808	0.181	0.296	عمرو موسي
0.968	0.961	0.965	خالد علي
0.903	0.364	0.519	حمدين صباحي
0.870	0.956	0.911	ابو الفتوح
0.808	0.795	0.779	Weighted avg.

Table 8 Evaluating result

Info gain and N-gram	Precision	Recall	F-measure	Accuracy (%)
Info gain-Unigram	0.932	0.931	0.934	93.1268
Without Info gain-Unigram	0.892	0.896	0.889	91.7861
Info gain-bigram	0.808	0.790	0.774	79.0248
Without Info gain-bigram	0.806	0.798	0.783	79.5064

5 Conclusion and Future Work

In this paper, the main goal is to study the effect of preprocessing on Arabic Sentiment analysis specifically on the dialect of Egypt focus on presidential elections in Egypt 2012. Twitter Used as a source of data collection because of the shortness of messages (Tweets) and rich in the vocabulary of slang. The event was determined through some stages; Corpus Collection and Preparation, preprocessing Text Classification.

This study uses NB classifier and our comparison one time by using N-Gram (unigrams and bigrams) and another time by using Infogain attribute selection.

The base of our comparison is the most popular text evaluation measures (F-measure, Recall, and Precision), using Arabic tweets which categorized the Arabic documents into six domains: "خالد على -(Khaled Ali)", "عمرو موسى -(Amr Mousa)", "احمد شفيق-(Ahmed Shafik)", "محمد مرسى(Mohammed Morsi)", "حمدين صباحى -(Hamden Sabahy)", "ابوالفتوح -(Abu Alftouh)". The accuracy of Arabic text categorization has been improved by applying an information gain features selection technique with normalization, stemmer and N-gram. The results are shown that applying text classification with Unigram and Infogain attribute selection achieved the highest classification accuracy of 93%, while it is 79% when info gain and unigram are not used.

We can summarize our paper contributions as follows

1. We experimented with different feature types, to find which one suits best the given problem.
2. We identified several problems and issues regarding Arabic text (coming from Twitter) collection, annotation, and classification.
3. We identified a set of rules that can be followed to facilitate sentiment classification.

The next step for this task is to provide the changes of politician's sentiment at the appropriate time to provide the best temporal data modeling of Twitter data. In addition, it might be possible to compare and validate the performance of the proposed framework for other algorithms such as support vector machine and decision tree.

References

1. Diab, M., Habash, N., Rambow, O., Altantawy, M., Benajiba, Y.: COLABA: Arabic dialect annotation and processing. In: Lrec Workshop on Semitic Language Processing, pp. 66–74 (2010)
2. Farid, D.: Egypt has the largest number of Facebook users in the Arab world: report. Daily News Egypt (2013)
3. Tumasjan, A., Sprenger, T.O., Sandner, P.G., Welpe, I.M.: Predicting elections with twitter: What 140 characters reveal about political sentiment. ICWSM **10**(1), 178–185 (2010)
4. http://www.newsday.com/news/nation/trumpuses-social-media-to-promote-himself-take-down-opponents-1.10835275
5. ElSahar, H., El-Beltagy, S.R.: A fully automated approach for arabic slang lexicon extraction from microblogs. In: International Conference on Intelligent Text Processing and Computational Linguistics, pp. 79–91. Springer (2014)
6. http://r-shief.org
7. Liu, B.: Sentiment analysis and opinion mining. Synth. Lect. Hum. Lang. Technol. **5**(1), 1–167 (2012)
8. Elhawary, M., Elfeky, M.: Mining Arabic business reviews. In: IEEE International Conference on Data Mining Workshops (ICDMW), 2010, pp 1108–1113. IEEE (2010)
9. Parikh, R., Movassate, M.: Sentiment analysis of user-generated twitter updates using various classification techniques. CS224 N Final Report: 1–18 (2009)

10. Rushdi-Saleh, M., Martín-Valdivia, M.T., Ureña-López, L.A., Perea-Ortega, J.M.: OCA: Opinion corpus for Arabic. J. Am. Soc. Inform. Sci. Technol. **62**(10), 2045–2054 (2011)
11. El Kourdi, M., Bensaid, A., Rachidi, T-e.: Automatic Arabic document categorization based on the Naïve Bayes algorithm. In: Proceedings of the Workshop on Computational Approaches to Arabic Script-based Languages, 2004. Association for Computational Linguistics, pp 51–58
12. Larkey, L.S., Ballesteros, L., Connell, M.E.: Improving stemming for Arabic information retrieval: light stemming and co-occurrence analysis. In: Proceedings of the 25th Annual International ACM SIGIR Conference On Research And Development in Information Retrieval, pp. 275–282. ACM (2002)
13. Ibrahim, A., Elghazaly, T.: Arabic text summarization using Rhetorical Structure Theory. In: 2012 8th International Conference on Informatics and Systems (INFOS)
14. Elghazaly, T., Mahmoud, A., Hefny, H.A.: Political sentiment analysis using twitter data. In: Proceedings of the International Conference on Internet of things and Cloud Computing, p. 11. ACM (2016)
15. Witten, I.H., Frank, E., Hall, M.A., Pal, C.J.: Data Mining: Practical Machine Learning Tools And Techniques. Morgan Kaufmann (2016)
16. Ibrahim, A., Elghazaly, T., Gheith, M.: A novel Arabic text summarization model based on rhetorical structure theory and vector space model. Int. J. Comput. Linguist. Nat. Lang. Process. **2**(8), 480–485 (2013)
17. Savoy, J., Rasolofo, Y.: Report on the TREC 11 Experiment: Arabic, Named Page and Topic Distillation Searches. In: TREC (2002)
18. Saad, M.K., Ashour, W.: Arabic morphological tools for text mining. Corpora **18**, 19 (2010)
19. Elghazaly, T.A., Fahmy, A.A.: English/Arabic cross language information retrieval (CLIR) for Arabic OCR-degraded text. Commun. IBIMA **9**(25), 208–218 (2009)
20. Taghva, K., Elkhoury, R., Coombs, J.: Arabic stemming without a root dictionary. In: International Conference on Information Technology: Coding and Computing, 2005 (ITCC 2005). IEEE (2005)
21. Pang, B., Lee, L., Vaithyanathan, S.: Thumbs up?: sentiment classification using machine learning techniques. In: Proceedings of the ACL-02 conference on empirical methods in natural language processing vol. 10. Association for Computational Linguistics, pp 79–86 (2002)
22. Saeys, Y., Inza, I., Larrañaga, P.: A review of feature selection techniques in bioinformatics. Bioinformatics **23**(19), 2507–2517 (2007)
23. Liu, H., Yu, L.: Toward integrating feature selection algorithms for classification and clustering. IEEE Trans. Knowl. Data Eng. **17**(4), 491–502 (2005)
24. Li, Y., Hsu, D.F., Chung, S.M.: Combining multiple feature selection methods for text categorization by using rank-score characteristics. In: 21st International Conference on Tools with Artificial Intelligence, 2009 (ICTAI'09), pp. 508–517. IEEE (2009)
25. Duda, R.O., Hart, P.E., Stork, D.G.: Pattern Classification, vol. 2. Wiley, New York (1973)
26. Alsaleem, S.: Automated Arabic text categorization using SVM and NB. Int. Arab. J. e-Technol. **2**(2), 124–128 (2011)
27. Sokolova, M., Japkowicz, N., Szpakowicz, S.: Beyond accuracy, F-score and ROC: a family of discriminant measures for performance evaluation. In: Australasian Joint Conference on Artificial Intelligence, pp. 1015–1021, 2006. Springer

Estimating Time to Event of Future Events Based on Linguistic Cues on Twitter

Ali Hürriyetoğlu, Nelleke Oostdijk and Antal van den Bosch

Abstract Given a stream of Twitter messages about an event, we investigate the predictive power of features generated from words and temporal expressions in the messages to estimate the time to event (TTE). From labeled training data average TTE values of the predictive features are learned, so that when they occur in an event-related tweet the TTE estimate can be provided for that tweet. We utilize temporal logic rules and a historical context integration function to improve the TTE estimation precision. In experiments on football matches and music concerts we show that the estimates of the method are off by 4 and 10 h in terms of mean absolute error on average, respectively. We find that the type and size of the event affect the estimation quality. An out-of-domain test on music concerts shows that models and hyper-parameters trained and optimized on football matches can be used to estimate the remaining time to concerts. Moreover, mixing in concert events in training improves the precision of the average football event estimate.

Keywords Smart city · Social media analysis · Natural language processing
Time-to-event estimation · Temporal expressions · Skipgrams · Football matches
Music concerts

1 Introduction

Understanding events is crucial for improving life in cities. The more and earlier information we have about events, the better we can handle them. Therefore, we study event time identification from social media in order to make a city smarter in terms of intelligence.

A. Hürriyetoğlu (✉) · N. Oostdijk · A. van den Bosch
Centre for Language Studies, Radboud University,
P.O. Box 9103, 6500 HD Nijmegen, The Netherlands
e-mail: a.hurriyetoglu@let.ru.nl

K. Shaalan et al. (eds.), *Intelligent Natural Language Processing: Trends and Applications*, Studies in Computational Intelligence 740,
https://doi.org/10.1007/978-3-319-67056-0_5

Social media produce data streams that are rich in content. Within the mass of microtexts posted on Twitter,[1] for example, many sub-streams of messages (tweets) can be identified that refer to the same event in the real world. Some of these sub-streams refer to events that are yet to happen, reflecting the joint verbalized anticipation of a set of Twitter users towards these events. In addition to overt markers such as event-specific hashtags in messages on Twitter, much of the information on future events is present in the surface text stream. These come in the form of explicit as well as implicit cues: compare for example 'the match starts in 2 h' with 'the players are on the field; can't wait for kickoff'. Identifying both types of linguistic cues may help disambiguate and pinpoint the starting time of an event, and therefore the remaining time to event (TTE).

We develop a time-to-event estimation method and implement it as an expert system that can process a stream of tweets in order to provide an estimate about the starting time of an event. Our ultimate goal is to provide an estimate for any type of event: football matches, music concerts, labour strikes, floods, etcetera. The reader should consider our study as a first implementation of a general time to event estimation framework. We focus on just two types of events, football matches and music concerts, as it is relatively straightforward to collect gold-standard event dates and times from databases for these types of events. We would like to stress, however, that the method is not restricted in any way to these two types; ultimately it should be applicable to any type of event, also those for which no generic database of event dates and times is available.

Estimating time to event can be defined as the identification of the future start time of an event in real clock time. This estimate is a core component of an alerting system that detects significant events (e.g. on the basis of mention frequency, a subtask not in focus in the present study) that places the event on the agenda and alerts users of the system. This alerting functionality is not only relevant for people interested in attending an event; it may also be relevant in situations requiring decision support to activate others to handle upcoming events, possibly with a commercial, safety, or security goal. A historical example of the latter category was *Project X Haren*,[2] a violent riot on September 21, 2012, in Haren, the Netherlands, organized through social media. This event was abundantly announced on social media, with specific mentions of the date and place. Consequently, a national advisory committee, installed after the event, was asked to make recommendations to handle similar future events. The committee stressed that decision-support alerting systems on social media need to be developed, "where the focus should be on the detection of collective patterns that are remarkable and may require action" [8, p. 31 (our translation)].

In this paper we explore a hybrid rule-based and data-driven method that exploits the explicit mentioning of temporal expressions but also other lexical phrases that implicitly encode time-to-event information, to arrive at accurate and early TTE estimates based on the Twitter stream. At this stage, our focus is only on estimating the

[1]http://www.twitter.com.

[2]http://www.en.wikipedia.org/wiki/Project_X_Haren.

time remaining to a given event. Deferring the fully automatic detection of events to future work, in our current study we identify events in Twitter microposts by event-specific hashtags.

The idea of identifying future event times for selected events gathered (semi-) automatically for subscribers, enriched with features such as personalization and the option to harvest event times both from social media and the traditional news, has been implemented already and is available through services such as Zapaday,[3] Daybees,[4] Songkick,[5] and Recorded Future.[6] To our knowledge, based on the public interfaces of these platforms, these services perform directed crawls of structured information sources, and identify the exact date and time references in posts on these sources through their structure. Some also manually gather relevant event information or collect this through crowdsourcing. However, non-automatic compilation of event data is costly, time consuming, and error prone, while it is also hard to keep the information up to date and ensure its correctness and completeness. Also, when this is the case, users need to know the dedicated platform to learn about a certain type of event. It may not be practical to follow many platforms in order to cover many event types or to discover event types that the user is not aware of.

We aim to develop a method that can estimate the time to any event and that will assist users in discovering new events in a period they are interested in. We automate the TTE estimation part of this task in a flexible manner so that we can generate continuous estimates based on realistic amounts of streaming data over time. Our estimation method offers a solution for the representation of vague terms, by inducing a continuous value for each possible predictive term from training data.

The service offered by our method will be useful only if it generates accurate estimates of the time to event. Preferably, these accurate predictions should come as early as possible. We use a combination of rule-based, temporal and lexical features to create a flexible setting that can be applied to events that may not contain all these types of information. We test our method on football matches and music concerts data, since they contain all of the aforementioned feature types and since event-specific tweets of these types of events as well as gold-standard dates and times of these events can be harvested easily. Any hashtag can be a candidate for detailed analysis with our method.

The remainder of this article is structured as follows. In Sect. 2 we first give an overview of related research before we go on to introduce our time-to-event estimation method in Sect. 3. Next, in Sect. 4, we explain the data used in our experiments, the feature sets we designed, training and test principles, hyper-parameter optimization, and the evaluation method. We then, in Sect. 5, describe the results for football matches and music concerts by measuring the effect of cross-domain parameter and model transfer on the estimation quality. In Sect. 6 we analyze the results and

[3]http://www.zapaday.com.

[4]http://www.daybees.com/.

[5]https://www.songkick.com/.

[6]https://www.recordedfuture.com.

summarize the insights obtained as regards various aspects of the TTE estimation method. Finally, Sect. 7 concludes this paper with a summary of the main findings for each data set and suggestions for future research.

2 Related Research

Using citizen observations about anything happening in city can improve the information collected with traditional methods [32]. Especially event information has the potential to respond to urgent needs for management [27] of a city. Our study aims to contribute to available body of literature in this line with the difference of identifying explicit anticipation and early clues about event start times.

In recent years, there have been numerous studies in the fields of text mining and information retrieval directed at the development of approaches and systems that would make it possible to forecast events. The range of (types of) events targeted is quite broad and varies from predicting manifestations of societal unrest such as nation-wide strikes or uprisings [14, 21, 29], to forecasting the events that may follow a natural disaster [28]. Studies that focus specifically on identifying future events are, for example, [1, 6, 9, 12]. A review of the literature shows that while approaches are similar to the extent that they all attempt to learn automatically from available data, they are quite different as regards the information they employ. For example, [28] attempt to learn from causality pairs (e.g. a flood causes people to flee) in long-ranging news articles to predict the event that is likely to follow the current event. Lee et al. [18] exploit the UP/DOWN/STAY labels in financial reports when trying to predict the movement of the stock market the next day. Redd et al. [30] attempt to calculate the risk of veterans becoming homeless, by analyzing the medical records supplied by the U.S. Department of Veterans Affairs.

Predicting the type of event is one aspect of event forecasting; giving an estimate as to the time when the event is (likely) to take place is another. Many studies, such as the ones referred to above, focus on the event type rather than the event time, that is, they are more generally concerned with future events, but not particularly with predicting the specific date or hour of an event. The same goes for [25] who describe a system for the identification of the period in which an event will occur, such as in the morning or at night. And again, studies such as those by [4, 16] more specifically focus on the type of information that is relevant for predicting event times, while they do not aim to give an exact time. Furthermore, [22] who in their attempt to improve the retrieval of future events extract (candidate) semantic and syntactic patterns with future reference from news articles. Finally, [24] present a method that automatically extracts forward-looking statements from earnings call transcripts in order to support business analysts in predicting the future events of economic relevance.

There are also several studies that are relevant in this context due to their focus on social media, and Twitter in particular. Research by [31] is directed at creating a calendar of automatically detected events. They use explicit date mentions and words typical of a given event. They train on annotated open domain event mentions and use TempEx [19] for the detection of temporal expressions. Temporal expressions that point to certain periods such as 'tonight' and 'this morning' are used by [39] to detect the personal activities at such times. In the same line, [17] show that machine learning methods can differentiate between tweets posted before, during, and after a football match.

Hürriyetoğlu et al. [10] also use tweet streams that are related to football matches and attempt to estimate the time remaining to an event, using local regression over word time series. In a related study, [36] use support vector machines to classify the TTE in automatically discretized categories. The results obtained in the latter two studies are at best about a day off in their predictions. Both studies also investigate the use of temporal expressions but fail to leverage the utility of this information source, presumably because they use limited sets of regular expressions: In each case fewer than 20 expressions were used.

The obvious baseline that we aim to surpass with our method is the detection of explicit time expressions from which the TTE could be inferred directly. Finding explicit temporal expressions can be achieved with rule-based temporal taggers such as the HeidelTime tagger [34], which generally search for a small, fixed set of temporal expressions [15, 31, 33]. As is apparent from studies such as [7, 19, 34], temporal taggers are successful in identifying temporal expressions in written texts as encountered in more traditional genres, such as news articles or official reports. They, in principle, can also be adapted to cope with various languages and genres (cf. [34]). However, the focus is typically on temporal expressions that have a standard form and a straightforward interpretation.

Various studies have shown that, while temporal expressions provide a reliable basis for the identification of future events, resolving the reference of a given temporal expression remains a challenge (cf. [13, 15, 20, 34]). In certain cases temporal expressions may even be obfuscated intentionally [23] by deleting temporal information that can be misused to commit a crime against or invade privacy of a user. Also, temporal taggers for languages other than English are not as successful and widely available as they are for English. Therefore, it may not be the optimal strategy to base TTE estimation exclusively on them.

Detecting temporal expressions in social media text requires a larger degree of flexibility in recognizing the form of a temporal expression and identifying its value than it would in news text. Part of this flexibility may be gained by learning temporal distances from data rather than fixing them at knowledge-based values. Blamey et al. [5] suggest estimating the values of temporal expressions on the basis of their distributions in the context of estimating creation time of photos on online social networks. Hürriyetoğlu [11] develop a method that relaxes and extends both the temporal pattern recognition and the value identification for temporal expressions.

In the present study we use the same estimation method of [11]; we scale up the number of temporal expressions drastically compared to what Heideltime offers and other time-to-event estimation methods have used [10, 31], and also compare this approach to using any other lexical words or word skipgrams in order to give the best possible estimate of the remaining time to an event. Moreover, we implement a flexible feature generation and selection method, and a history function which uses the previous estimates as a context. In our evaluation we also look into the effects of cross-domain parameter and model transfer.

3 Time-to-Event Estimation Method

Our time-to-event (TTE) estimation method consists of training and estimation steps. First, training data is used to identify the predictive features and their values. Second, each tweet is assigned a TTE value based on the predictive features it contains and on the estimates from recent tweets for the same event stored in a fixed time buffer.

The training phase starts with feature extraction and generation. During feature extraction we first detect which of the features defined in lists of rules, temporal expressions and lexical features, occur in a tweet. Then the various features are combined.

The estimation phase consists of two steps. First, an estimation function assigns a TTE estimate for a tweet based on the predictive features that occur in it. Afterwards the estimate is adjusted by a historical function based on a buffer of previous estimations. Historical adjustment restricts consecutive estimates to deviate from each other.

As we aim to develop a method that can be applied to any type of event and any language, our approach is guided by the following considerations:

1. We refrain from using any domain-specific prior knowledge such as 'football matches occur in the afternoon or in the evening' which would hinder the prediction of matches held at unusual times, or more generally the application of trained models to different domains;
2. We avoid the use of any language analysis tools such as part-of-speech taggers or parsers, so that the method can be easily adapted to languages for which such language analysis tools are not available, less well developed, or less suited for social media text;
3. We use basic statistics (e.g. computing the median) and straightforward time series analysis methods to keep the method efficient and scalable;
4. We use the full resolution of time to provide precise estimates. Our approach treats TTE as a continuous value: calculations are made and the results are reported using decimal fractions of an hour.

4 Experimental Set-Up

In this section we describe the experiments carried out using our TTE estimation method. We describe the data sets gathered for the experiments, and then move to describe the three feature sets that are contrasted in the experiments: temporal expressions and word skipgrams that are induced and used in a machine learning setting, and temporal logic rules that are created manually. Next we describe the training and test regimes to which we subject our method. The section concludes with our evaluation method and the description of two baseline systems.

4.1 Data Sets

For our study we collected tweets referring to scheduled Dutch premier league football matches (FM) and music concerts (MC) in the Netherlands. These events trigger many anticipatory references on social media before they happen, containing numerous temporal expressions and other non-temporal implicit lexical clues on when they will happen.

We harvested all tweets from Twiqs.nl, a database of Dutch tweets collected from December 2010 onwards [35].[7] Both for football matches and music concerts we used event-specific hashtags to identify the event, i.e. we used the hashtag that, to the best of our knowledge, was the most distinctive for the event. The hashtags used for FM follow a convention where the first two or three letters of the names of the two teams playing against each other are concatenated, starting with the host team. An example is #ajatwe for a football match in which Ajax is the host, and Twente is the away team. The MC hashtags are mostly concatenations of the first and last name of the artist, or concatenations of the words forming the band name. Although in the latter case for many of the hashtags shorter variants exist, we did not delve into the task of identifying such variants [26, 38] and used the full variants.

The FM dataset was collected by selecting the six best performing teams of the Dutch premier league in 2011 and 2012. We queried all matches in which these teams played against each other in the calendar years 2011 and 2012.[8] The MC dataset contains tweets from concerts that took place in the Netherlands between January 2011 and September 2014. For our definitive datasets we restricted the data to tweets sent within eight days before the event.[9] We decided to refrain from extending the time frame to the point in time when the first tweet that mentions the hashtag was sent, because hashtags may denote a periodic event or a different event that takes place at another time, which may lead to inconsistencies that we do not aim to solve in this study. Most issues having to do with periodicity, ambiguity and inconsistency are absent within the 8-day window, i.e. tweets with a particular event hashtag largely refer to the event that is upcoming within the next eight days.

As noted above, the use of hashtags neither provides complete sets of tweets about the events nor does it ensure that only tweets pertaining to the main event

[7] http://www.twiqs.nl.

[8] Ajax Amsterdam (aja), Feyenoord Rotterdam (fey), PSV Eindhoven (psv), FC Twente (twe), AZ Alkmaar (az), and FC Utrecht (utr).

[9] An analysis of the tweet distribution shows that the 8-day window captures about 98% of all tweets comprises by means of the hashtags that we used.

are included [37]. We observed that some event hashtags from both data sets were used to denote other similar events that were to take place several days before the event we were targeting, such as a cup match instead of a league match between the same teams, or another concert by the same singer. For example, the teams Ajax and Twente played a league and a national cup match within a period of eight days (two consecutive Sundays). In case there was such a conflict, we aimed to estimate the TTE for the relatively bigger event, i.e. in terms of the available Dutch tweet count about it. For #ajatwe, this was the league match. In so far as we were aware of related events taking place within the same 8-day window with comparable tweet counts, we did not include these events in our datasets.

Social media users tend to include various additional hashtags other than the target hashtag in a tweet, including hashtags that may confuse our method. We consider a hashtag to be confusing when it denotes an event different from the event designated by the hashtag creation rule. For football matches, this is the case for example when a user uses #tweaja instead of #ajatwe when referring to a home game for Ajax; for music concerts, we may encounter tweets with a specific hashtag where these tweets do not refer to the targeted event using the hashtag #beyonce for a topic other than a Beyoncé concert. In these cases, the unrelated tweets are not removed, and are used as if they were referring to the main event. We aim for our approach to be resistant to such noise which after all, we find, is present in most social media data.

In the context of the present research we are working under the assumption that the presence of a hashtag can be used as a proxy for the topic addressed in a tweet. In a previous study [11] it was hypothesized that the position of the hashtag may have an effect as regards the topicality of the tweet and investigated whether the position of the hashtag influences the success in estimating the TTE. Hashtags that occur in final position (i.e. they are tweet-final or only followed by one or more other hashtags) are typically metatags and therefore possibly more reliable as topic identifiers than non-final hashtags which behave more like common content words in context. Hürriyetoğlu et al. [11] found that the results for estimating the TTE from tweets in which the hashtag occurs in final position were best between 12 and 0 h, while for tweets where the hashtag occurs in non-final position the best results were obtained between 24 and 13 h before an event. When estimating the TTE of an event that is still more than 24 h away, the best results were obtained using all data, regardless of the position of the hashtag [11]. Since here we aim to give TTE estimates as soon as there is information available about an event, we use all (tweet-final and non-final hashtagged) tweets available within the 8-day window.

We used the simple pattern "rt @" to identify retweets. Retweets repeat the information of the source tweet and occur possibly much later in time than the original post. While including retweets could improve the performance under certain conditions [3], results from a preliminary experiment we carried out using development data show that eliminating retweets yields better or comparable results for the TTE estimation task most of the time. Therefore, we eliminated all retweets in the experiments reported here.

Table 1 Number of events and tweets for the FM and MC data sets (FM = football matches; MC = music concerts). In the 'All' sets all tweets are included, that is, original posts and retweets

	# of Events	# of Tweets			
		Min.	Median	Max.	Total
FM all	60	305	2,632	34,868	262,542
FM without retweets	60	191	1,345	23,976	139,537
MC all	35	15	54.0	1,074	4,363
MC without retweets	32	15	55.5	674	3,479

In Table 1 we present an overview of the datasets that were used in the research reported on in the remainder of this paper. Events that do not have more than 15 tweets were eliminated, reducing the number of events in the MC data set to 32.

Each tweet in our data set has a time stamp of the moment (in seconds) it was posted. Moreover, for each football match and each music concert we know exactly when it took place: the event start times were gathered from the websites of Eredivisie[10] for football matches and Last.fm[11] for the music concerts. This information is used to calculate for each tweet the actual time that remains to the start of the event, as well as to compute the absolute error in estimating the remaining time to event.

We would like to emphasize that the final data set contains all kinds of discussions that do not contribute to predicting time of event directly. The challenge we undertake is to make sense of this mixed and unstructured content for identifying temporal proximity of an event.

4.2 Features

It is obvious to the human beholder that tweets referring to an event exhibit different patterns as the event draws closer. Not only do the temporal expressions that are used change (e.g. from 'next Sunday' to 'tomorrow afternoon'), the level of excitement rises as well, as the following examples show:

122 h before the event: *björn kuipers is door de knvb aangesteld als scheidsrechter voor de wedstrijd fc utrecht psv van komende zondag 14.30 uur* (En: Björn Kuipers has been assigned to be the referee for Sunday's fixture between FC Utrecht and PSV, played at 2.30 PM)

69 h before the event: *zondag thuiswedstrijd nummer drie fc utrecht psv kijk ernaar uit voorbereidingen in volle gang* (En: Sunday the third match of the season in our stadium FC Utrecht vs. PSV, excited, preparations in full swing)

[10]http://www.eredivisie.nl.

[11]http://www.lastfm.com.

27 h before the event: *dick advocaat kan morgenmiddag tegen fc utrecht beschikken over een volledig fitte selectie* (En: Dick Advocaat has a fully healthy selection at his disposal tomorrow afternoon against FC Utrecht)
8 h before the event: *werken hopen dat ik om 2 uur klaar ben want t is weer matchday* (En: working, hope I'm done by 2 PM, because it's matchday again)
3 h before the event: *onderweg naar de galgenwaard voor de wedstrijd fc utrecht feyenoord #utrfey* (En: on my way to Galgenwaard stadium for the football match fc utrecht feyenoord #utrfey)
1 h before the event: *wij zitten er klaar voor #ajafey op de beamer at #loods* (En: we are ready for #ajafey by the data projector at #loods)
0 h before the event: *rt @lighttown1913 zenuwen beginnen toch enorm toe te nemen* (En: RT @lighttown1913 starting to get really nervous)

The temporal order of different types of preparations for the event can be seen clearly in the tweets above. The stream starts with a referee assignment and continues with people expressing their excitement and planning to go to the event, until finally the event starts. Our goal is to learn to estimate the TTE from texts like these, that is, from tweets, along with their time stamps, by using linguistic clues that, explicitly or implicitly, refer to the time when an event is to take place.

In the next sections, we introduce the three types of features that we use to estimate the TTE: temporal expressions, temporal logic rules and word skipgrams.

4.2.1 Temporal Expressions

In the context of this paper temporal expressions are considered to be words or sequences of words that refer to the point in time, the duration, or the frequency of an event. These may be exact, approximate, or right out vague, e.g., 'still three days', 'less than half an hour' and 'tomorrow'.

We created a comprehensive list of temporal expressions that includes among others single words, e.g. adverbs such as *nu* 'now', *zometeen* 'immediately', *straks* 'later on', *vanavond* 'this evening', nouns such as *zondagmiddag* 'Sunday afternoon', and conjunctions such as *voordat* 'before'), but also multi-word temporal expressions such as *komende woensdag* 'next Wednesday. Temporal expressions of the latter type were obtained by means of a set of 615 seed terms and 70 patterns, which generated a total of 15,236 temporal expressions excluding numerals.[12]

A sample of patterns that are used to generate the temporal expressions are listed below. The first and second examples provide temporal expressions without numerals and the third one illustrate how the numbers are included. The square and round brackets denote obligatory and optional items, respectively. The vertical bar is used to separate alternative items. Between curly brackets we give examples of temporal expressions that were derived from that pattern.

[12]Not all temporal expressions generated by the rules will prove to be correct. Since incorrect items are unlikely to occur and therefore are harmless, we refrained from manually checking the resulting set.

1. [in | m.i.v. | miv | met ingang van | na | t/m | tegen | tot en met | van | vanaf | voor] + (de maand) + [month of the year] → {*in de maand Januari* 'in the month of January', *met ingang van Juli* 'as of July'}
2. een + [seconde | minuut | week | maand | jaar | eeuw] + (of wat) + [eerder | later] → {*een minuut eerder* 'one minute earlier', *een week later* 'one week later'}
3. N > 1 + [minuten | seconden | uren | weken | maanden | jaren | eeuwen] + [eerder | eraan voorafgaand | ervoor | erna | geleden | voorafgaand | later] → {*7 minuten erna* 'after seven minutes', *5 uren geleden* '5 h ago', *2 weken later* 'after two weeks'}

The third example introduces N, a numeral that in this case varies between 2 and the maximal value of numerals in all expressions, 120. The reason that $N > 1$ in this example is that it generates expressions with the plural form of 'minutes', 'seconds', etc.

Including numerals in the generation of patterns yields a total of 460,248 expressions expressing a specific number of minutes, hours, days, or time of day. The items on the list that were obtained through generation patterns, which contain numerals, include temporal expressions such as *over 3 dagen* 'in 3 days', *nog 5 minuten* 'another 5 min', but also fixed temporal expressions such as clock times. The patterns handle frequently observed variations in their notation, for example *drie uur* 'three o'clock' may be written in full or as *3:00, 3:00 uur, 3 u, 15.00*, etc. Finally, combinations of a time, weekday or day of the month, e.g., "Monday 21:00", "Tuesday 18 September", and "1 apr 12:00" are included as well. Notwithstanding the substantial number of items included, the list is bound to be incomplete.[13]

We decided to include prepositions only when they occur in larger prepositional phrases, so as to avoid generating too much noise. After all, many prepositions have several uses: they can be used to express time, but also for example location. Compare *voor* in *voor drie uur* 'before three o'clock' and *voor het stadion* 'in front of the stadium'. Moreover, prepositions are easily confused with parts of separable verbs which in Dutch are abundant.

Various items on the list are inherently ambiguous and only in one of their senses can be considered temporal expressions. Examples are *week* 'week' but also 'weak' and *dag* 'day' but also 'goodbye'. For items such as these, we found that the different senses could fairly easily be distinguished whenever the item was immediately preceded by an adjective such as *komende* and *volgende* (both meaning 'next'). For a few highly frequent items this proved impossible. These are words such as *zo* which can be either a temporal adverb ('in a minute'; cf. *zometeen*) or an intensifying adverb ('so'), *dan* 'then' or 'than', and *nog* 'yet' or 'another'. As we have presently no way of distinguishing between the different senses and these items have at best an extremely vague temporal sense, they cannot be expected to contribute to estimating the time

[13]Dates, which denote the complete year, month and day of the month, are presently not covered by our patterns but will be added in future.

to event. Thus, we deciced to discard these in case they do not co-occur with another temporal expression.[14]

For the items on the list no provisions were made for handling any kind of spelling variation, with the single exception of a small group of words (including *'s morgens* 'in the morning', *'s middags* 'in the afternoon' and *'s avonds* 'in the evening') which use in their standard spelling the archaic *'s*, and abbreviations. As many authors of tweets tend to spell these words as *smorgens*, *smiddags* and *savonds* we decided to include these forms.

Despite the large number of pre-generated temporal expressions, the number of temporal expressions actually encountered in the FM data set is only 2,476; in the smaller MC set we find even fewer temporal expressions, viz. 398.

This set also contains skipgrams, i.e. feature sequences that are created via concatenation of neighboring temporal expressions, while ignoring in-between tokens. Consider, for instance, the tweet *Volgende wedstrijd: Tegenstander: Palermo Datum: zondag 2 november Tijdstip: 20:45 Stadion: San Siro* 'Next match: Opponent: Palermo Date: Sunday 2 November Time: 20:45 Stadium: San Siro'. The basic temporal features in this tweet are, in their original order, <Sunday 2 November> and <20:45>. The skipgram generation step will ignore the in-between tokens <Time> and <:>, preserve the feature occurrence order, and result in <Sunday 2 November, 20:45> as a new skipgram feature. From this point onward we will refer to the entire set of basic and skipgram temporal features as **TFeats**.

We compare the performance of our temporal expression detection based on the TFeats list with that of the Heideltime Tagger,[15] which is the only available temporal tagger for Dutch, on a small set of tweets, i.e. 18,607 tweets from the FM data set.[16] There are 10,183 tweets that contain at least one temporal expression detected by the Heideltime tagger or matched by our list. The list and the tagger have identified the same temporal expressions in 5,008 of the tweets. The Heideltime tagger detects 429 expressions that our list does not; vice versa, our list detects 2,131 expressions that Heideltime does not detect. In the latter category are *straks* 'soon', *vanmiddag* 'today', *dalijk* 'immediately', *nog even* 'a bit', *over een uurtje* 'in 1 h', *over een paar uurtjes* 'in a few hours', *nog maar 1 nachtje* 'still only one night'. On the other hand, the Heideltime tagger detects expressions such as *de afgelopen 22 jaar* 'the last 22 years', *2 keer per jaar* '2 times a year', *het jaar 2012* 'the year 2012'. This can easily be explained as our list focuses on temporal expressions that refer to short term future, and not to the past or the long term. Also, the Heideltime tagger recognizes some expressions that we rule out intentionally due to their ambiguous interpretation. This is the case for the aforementioned *dag* 'day', but also for *uur* 'hour', *jan* 'Jan' (name of person or of month), *minuut* 'minute', *volgend* 'next'. In sum, our

[14]Note that *nog* does occur on the list as part of various multi-word expressions. Examples are *nog twee dagen* 'another two days' and *nog 10 min* '10 more minutes'.

[15]We used Heideltime tagger version 1.7 by enabling the interval tagger and configured NEWS type as genre.

[16]This subset is used to optimize the hyper-parameters as well.

list has a higher coverage than Heideltime and focuses only on forward-looking time expressions. We therefore continue working with our TFeats list.

4.2.2 Rules

By using only absolute forward-pointing temporal expressions as features we do not need temporal logic to understand their time-to-event value. These expressions provide the time-to-event directly, e.g. 'in 30 min' indicates that the event will take place in 0.5 h. We therefore introduce rules that make use of temporal logic to define the temporal meaning of TFeats features that has a context dependent time-to-event value. We refer to these features as non-absolute temporal expressions.

Non-absolute, dynamic TTE temporal expressions such as days of the week, and date-bearing temporal expressions such as *18 September* can on the one hand be detected relatively easily, but on the other hand require further computation based on temporal logic using the time the tweet was posted. For example, the TTE value of a future weekday should be calculated according to the referred day and time of the occurrence. Therefore we use the rules list from [11] and extend it. We define rules against the background of:

1. Adjacency: We specify only contiguous relations between words, i.e. without allowing any other word to occur in between;
2. Limited scope: A rule can indicate a period up to 8 days before an event; thus we do not cover temporal expressions such as *nog een maandje* 'another month' and *over 2 jaar* 'in 2 years';
3. Precision: We refrain from writing rules for highly ambiguous and frequent terms such as *nu* 'now' and *morgen* 'tomorrow';
4. Formality: We restrict the rules to canonical (normal) forms. Thus we do not include rules for expressions like *over 10 min* 'in 10 min' and *zondag 18 9* 'Sunday 18 9';
5. Approximation: We round the estimations to fractions of an hour with maximally two decimals; the rule states that *minder dan een halfuur* 'less than half an hour' corresponds to 0.5 h;
6. Partial rules: We do not aim to parse all possible temporal expressions. Although using complex rules and language normalization can increase the coverage and performance, this approach has its limits and will decrease practicality of the method. Therefore, we define rules up to a certain length, which may cause a long temporal expression to be detected partially. A complex rule would in principle recognize the temporal expression "next week Sunday 20:00" as one unit. But we only implement basic rules that will recognize "next week" and "Sunday 20:00" as two different temporal expressions. Having these parts, our method will combine them in way that they can approximate their value as a whole.

As a result we have two set of rules, which we henceforth refer as **RFeats**:

1. Temporal expressions that denote an exact amount of time are interpreted by means of rules that we shall refer to as **Exact rules**. This applies for example to temporal expressions answering to patterns such as "over N {minuut | minuten | kwartier | uur | uren | dag | dagen | week}" (En: in N {minute | minutes | quarter of an hour | hour | hours | day | days | week}. Here the TTE is assumed to be the same as the N minutes, days or whatever is mentioned;
2. A second set of rules, referred to as **Dynamic rules**, is used to calculate the TTE dynamically, using the temporal expression and the tweet's time stamp. These rules apply to instances such as *zondagmiddag om 3 uur* 'Sunday afternoon at 3 p.m.'. Here we assume that this is a future time reference on the basis of the fact that the tweets were posted prior to the event. With temporal expressions that are underspecified in that they do not provide a specific point in time (hour), we postulate a particular time of day as the default for that expression. For example, *vandaag* 'today' is understood as 'today at 3 p.m.', *vanavond* 'this evening' as 'this evening at 8 p.m.' and *morgenochtend* 'tomorrow morning' as 'tomorrow morning at 10 a.m.'.

4.2.3 Word Skipgrams

In contrast to temporal expressions we also generated an open-ended feature set that draws on any word skipgram occurring in our training set. This feature set is crucial in discovering predictive features not covered by the time-related TFeats or RFeats. A generic method may have the potential of discovering expressions already present in TFeats, but with this feature type we expressly aim to capture any lexical expressions that do not contain any explicit start time, yet are predictive of start times. Since this feature type requires only a word list, it can be adapted to any language.

We first compiled a list of regular, correctly spelled Dutch words by combining the *OpenTaal flexievormen and basis-gekeurd* word lists.[17] From this initial list we then removed all stop words, foreign words, and entity names. The stop word list contains 839 entries. These are numerals, prepositions, articles, discourse connectives, interjections, exclamations, single letters, auxiliary verbs and any abbreviations of these. Foreign words were removed in order to avoid spam and unrelated tweets. We list English words which are in the *OpenTaal flexievormen* or *OpenTaal basis-gekeurd* word lists as we come across them.[18] We also used two lists to identify the named entities: *Geonames*[19] for place names in the Netherlands and *OpenTaal basis-*

---

[17]We used the *OpenTaal flexievormen, basis-gekeurd, and basis-ongekeurd word lists from the URL:* http://www.opentaal.org/bestanden/doc_download/18-woordenlijst-v-210g-bronbestanden-.

[18]The foreign word list currently contains 9 entries: *different, indeed, am, ever, field, indeed, more, none, or, wants.*

[19]http://www.geonames.org/.

ongekeurd for other named entities. The final set of words comprises 317,831 entries. The FM and MC data sets contain 17,646 and 2,617 of these entries, respectively.

Next, the words were combined to generate longer skipgram features based on the words that were found to occur in a tweet. For example, given the tweet *goed weekendje voor psv hopen dat het volgend weekend ht weekend wordt #ajapsv bye tukkers* 'a good weekend for psv hoping that next weekend will be the weekend #ajapsv bye tukkers' we obtained the following words in their original order: <goed>, <weekendje>, <hopen>, <volgend>, <weekend>, <weekend>, <tukkers>. From this list of words, we then generated skipgrams up to $n = 7$. Retaining the order, we generated all possible combinations of the selected words. For example, for $n = 2$ the following features were generated[20]:

<goed, weekendje>, <goed, hopen>, <goed, volgend>, <goed, weekend>, <goed, tukkers>, <weekendje, hopen>, <weekendje, volgend>, <weekendje, weekend>, <weekendje, tukkers>, <hopen, volgend>, <hopen, weekend>, <hopen, tukkers>, <volgend, weekend>, <volgend, tukkers>, <weekend, weekend>, <weekend, tukkers>

The feature set arrived at by this feature generation approach is henceforth referred as **WFeats**.

4.3 *Training and Test Regimes*

After extracting the relevant features, our TTE estimation method proceeds with feature selection, feature value assignment, and tweet TTE estimation steps. Features are selected according to their frequency and distribution over time. We then calculate the TTE value of selected features by means of a training function. Finally, the values of the features that co-occur in a single tweet are given to the estimation function, which on the basis of these single estimates generates an aggregated TTE estimation for the tweet.

The training and estimation functions are applied to TFeats and WFeats. The values of RFeats are not trained, but already set to certain values in the rules themselves, as stated earlier.

4.3.1 Feature Selection

Each tweet in our data set has a time stamp for the exact time it was posted. Moreover, for each event we know precisely when it took place. This information is used to calculate for each feature the series of all occurrence times relative to the event start time, hereafter referred as **time series** of a feature. The training starts with selection of features that carry some information regarding the remaining time to an event,

[20]Features that occur multiple times in a tweet are filtered to have just one occurrence of each feature in a tweet.

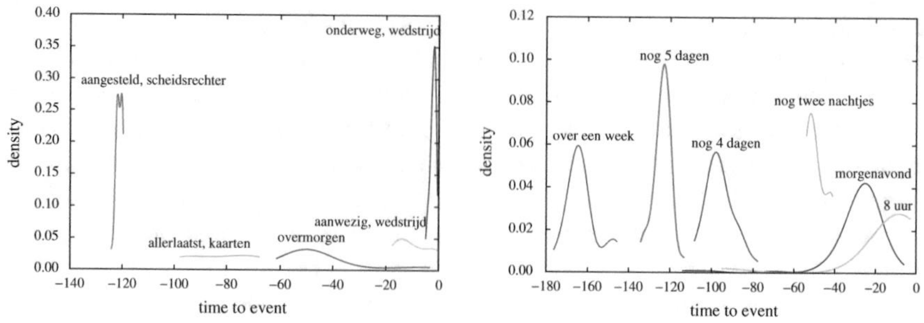

(a) WFeats (b) TFeats

Fig. 1 Temporal distribution of selected features for WFeats and TFeats

based on their frequency and standard deviation of occurrence times relative to an event start time. A feature time series should be longer than one to be taken into account, and should have a standard deviation below a certain threshold for the feature to be considered for the feature value assignment phase. The standard deviation threshold is based on a fixed number of highest quantile regions, a number which is optimized on a development set. Features that are in the highest standard deviation quantile regions are eliminated.

4.3.2 Feature Value Assignment

The value of a feature time series is estimated by a training function. In the current study, the training function is either the mean or the median of the actual TTE values of the selected features encountered in the training data. The proper training function is selected on the basis of its performance on a development set. This method does not need any kind of frequency normalization. We consider this to be an advantage, as now there is no need to take into account daily periodicity or tweet distribution.

Figure 1 visualizes the distribution of TTE values of selected features from both feature sets. The distributions are fitted through kernel density estimation using Gaussian kernels.[21] Kernel density curves visualize the suitedness of a feature: the sharper the curve and the higher its peak (i.e. the higher its kurtosis), the more accurate the feature is for TTE estimation. On the left, Fig. 1a illustrates how peaks in WFeats features may inform about different phases of an event. The features *aangesteld, scheidsrechter* 'appointed, referee', *aanwezig, wedstrijd* 'present, game', and *onderweg, wedstrijd* 'on the road, match' relate to different preparation phases of a football match. The feature *allerlaatst, kaarten* 'latest, tickets' refers to the act of

[21]We used the gaussian_kde method from SciPy v0.14.0 URL: http://docs.scipy.org/doc/scipy/reference/generated/scipy.stats.gaussian_kde.html.

buying tickets; *overmorgen* 'the day after tomorrow', a temporal expression (which is also included in the WFeats set) indicates the temporal distance to the event in terms of days. The curves are either sharply concentrated on particular points in time (e.g. 'on the road, match' refers to travel within hours before the match), or fit a broad array of low-density data points.

In contrast, Fig. 1b displays kernel density curves of selected features from the TFeats set. The features are *over een week* 'over a week', *nog 5 dagen* 'another 5 days', *nog 4 dagen* 'another 4 days', *nog twee nachtjes* 'another two nights', *morgenavond* 'tomorrow evening', and *8 uur* '8 h', and show relatively similar curves. This suggests that these temporal expressions tend to have a similar standard deviation of about a full day. Indeed, the expression *morgenavond* 'tomorrow evening' may be coined in the early morning of the day before up to the night of the day before.

4.3.3 Time-to-Event Estimation for a Tweet

The TTE estimate for a test tweet will be calculated by the estimation function using the TTE estimates of all features observed in a particular tweet.

Since TFeats and RFeats can be sparse as compared to WFeats, and since we want to keep the method modular, we provide estimates for each feature set separately. The TFeats and WFeats that occur in a tweet are evaluated by the estimation function to provide the estimation for each feature set. We use the mean or the median as an estimation function.

To improve the estimate of a current tweet, we use a combination method for the available features in a tweet and a historical context integration function in which we take into account all estimates generated for the same event so far, and take the median of these earlier estimates as a fourth estimate besides those generated on the basis of TFeats, RFeats and WFeats. The combination was generated by using the estimates for a tweet in the following order: (i) if the RFeats generates an estimate, which is the mean of all rule values in a tweet, the TTE estimate for a tweet is that estimate; (ii) else, use the TFeats-based model to generate an estimate; and (iii) finally, if there is not yet an estimate for a tweet, use the WFeats-based model to generate an estimate. The priority order that places rule estimates before TFeats estimates, and TFeats estimates before WFeats estimates, is used in combination with the history function. Our method integrates the history as follows (i) rule estimates only use the history of rule estimates; (ii) TFeats estimates use the history of rule estimates and TFeats estimates; and (iii) WFeats estimates do not make any distinction between the source of previous estimates that enter the history window function calculation. In this manner the precise estimates generated by RFeats are not overwritten by the history of lower-precision estimates, which does happen in the performance-based combination.

4.4 Evaluation and Baselines

The method is evaluated on 51 of the 60 football matches and all 32 music concert events. Nine football matches (15%), selected randomly, are held out as a development set to optimize the hyperparameters of the method.

Experiments on the test data are run in a leave-one-event-out cross-validation setup. In each fold the model is trained on all events except one, and tested on the held-out event; this is repeated for all events.

We report two performance metrics in order to gain different quantified perspectives on the errors of the estimates generated by our systems. The error is represented as the difference between the actual TTE, v_i, and the estimated TTE, e_i. Mean Absolute Error (MAE), given in Eq. 1, represents the average of the absolute value of the estimation errors over test examples $i = 1 \ldots N$. Root Mean Squared Error (RMSE), given in Eq. 2, sums the squared errors; the sum divided by the number of predictions N; the (square) root is then taken to produce the RMSE of the estimations. RMSE penalizes outlier errors more than MAE does.

$$MAE = \frac{1}{N}\sum_{i=1}^{N} |v_i - e_i| \tag{1}$$

$$RMSE = \sqrt{\frac{1}{N}\sum_{i=1}^{N} (v_i - e_i)^2} \tag{2}$$

We computed two straightforward baselines derived from the test set: the mean and median TTE of all tweets. The mean and median are computed by averaging and calculating the mean or the median of the TTE of all training tweets, respectively. The mean baseline estimate is approximately 21 h, while the median baseline is approximately 4 h. Baseline estimations are calculated by assigning every tweet the corresponding baseline as an estimate. For instance, all estimates for the median baseline will be the median of the tweets, which is 4 h.

Although the baselines are simplistic, the distribution of the data make them quite strong. For example, 66% of the tweets in the FM data set occur within 8 h before the start of the football match; the median baseline generates an error of under 4 h for 66% of the tweets.

In addition to error measures, we take the coverage into account as well. The coverage reflects the percentage of the tweets for which an estimate is generated for an event. Evaluating coverage is important, as it reveals the recall of the method. Coverage is not recall (i.e. a match on a tweet does not mean that the estimate is correct) but it sets an upper bound on the percentage of relevant tweets a particular method is able to use. Having a high coverage is crucial in order to be able to handle events that have few tweets, to start generating estimations as early as possible, to apply a trained model to a different domain, and to increase the benefit of using the history of the tweet stream as a context. Thus, we seek a balance between a small error rate and high coverage. Twitter data typically contains many out-of-vocabulary words [2], so it remains a challenge to attain a high coverage.

4.5 Hyperparameter Optimization

The performance of our method depends on three hyperparameters and two types of functions for which we tried to find an optimal combination of settings, by testing on the development set representing nine football matches:

Standard Deviation Threshold—the quantile cut-off point for the highly deviating terms, ranging from not eliminating any feature (0.0), to eliminating the highest standard deviating 40-quantile;
Feature length—maximum number of words or temporal expressions captured in a feature, from 1 to 7;
Window Size—the number of previous estimations used as a context to adjust the estimate of the current tweet, from 0 to the complete history of estimations for an event.

The frequency threshold of features is not a hyperparameter in our experiments. Instead, we only eliminate hapax features occurring once for both WFeats and TFeats.

We want to identify which functions should be used to learn the feature values and perform estimations. Therefore we test the following functions as well.

Training Function—calculates mean or median TTE of the features from all training tweets;
Estimation Function—calculates mean or median value on the basis of all features occurring in a tweet.

Figure 2 shows how the feature length and quantile cut-off point affect the MAE and the coverage on the development data set. The higher the feature length and the higher quantile cut-off point, the lower both the coverage and MAE. We aim for a low MAE and a high coverage.

Long features, which consists of word skipgrams with $n > 1$, are not used on their own but are added to the available feature set. It is perhaps surprising to observe that adding longer features causes coverage to drop. The reason for this is that longer features are less frequent and have smaller standard deviations than shorter features. Shorter features are spread over a long period which makes their standard deviation higher: this causes these short features to be eliminated by the quantile-threshold-based feature selection in favour of the longer features that occur less frequently. Additionally, selecting higher quantile cut-off points eliminates many features, which causes the coverage to decrease.

As a result of our hyperparameter optimization experiments, since the MAE does not get better for features that are longer than 2, a feature length of $n = 2$ is used for

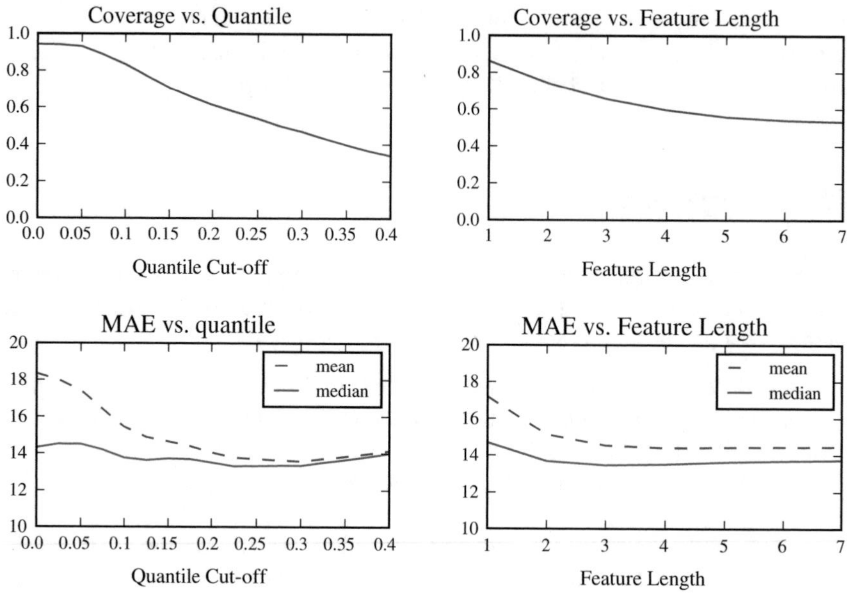

(a) Words (WFeats)

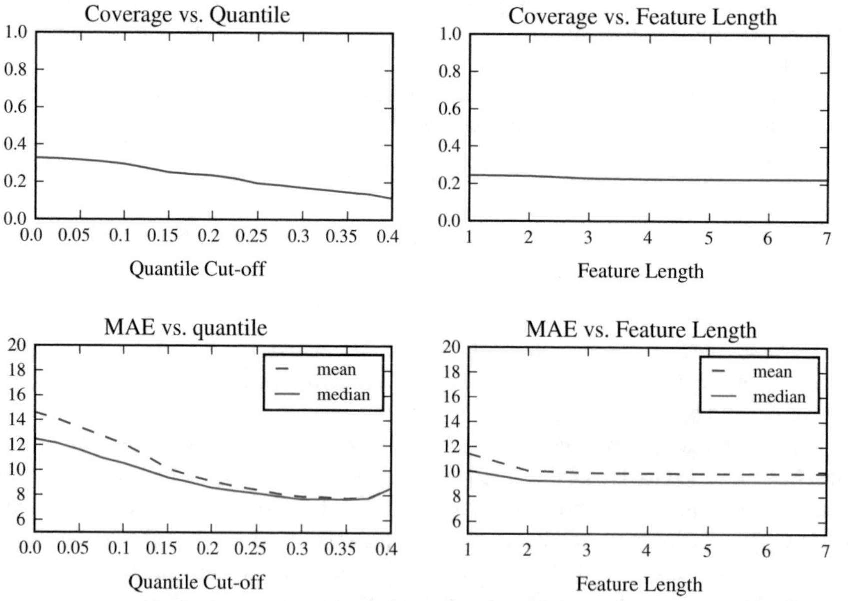

(b) Temporal expressions (TFeats)

Fig. 2 MAE and coverage curves for different quantile cut-off and feature lengths for words (WFeats) and temporal expressions (TFeats)

Table 2 MAE for various training and estimation function combinations on WFeats

	Mean estimation	Median estimation
Mean training	14.11	14.04
Median training	13.15	**13.13**

both WFeats and TFeats. The quantile cut-off point is 0.20 for WFeats since higher quantile cut-off points decrease the coverage; at the same time they do not improve the MAE. For TFeats, the quantile cut-off point is set at 0.25. These parameter values will be used for both optimizing the history window length on the development set and running the actual experiment. Using these parameters yields a reasonable coverage of 80% on the development set.

Table 2 demonstrates that using the median function for both training and estimate generation provides the best result of 13.13 h with WFeats. In this case the median of median TTE values of the selected features in a tweet is taken as the TTE estimation of a tweet.

The window size of the historical context integration function is considered to be a hyperparameter, and is therefore optimized as well. The historical context is included as follows: given a window size $|W|$, for every new estimate the context function will be applied to $W-1$ preceding estimates and the current tweet's estimate to generate the final estimate for the current tweet. In case the number of available estimates is smaller than $W-1$, the historical context function uses only the available ones. The history window function estimate is re-computed with each new tweet.

Figure 3 demonstrates how the window size affects the overall performance for each module and the performance-based combination in Fig. 3a and for a priority-based historical context integration in Fig. 3b. The former figure shows that the history function reinforces the estimation quality. In other words, using the historical context improves overall performance.

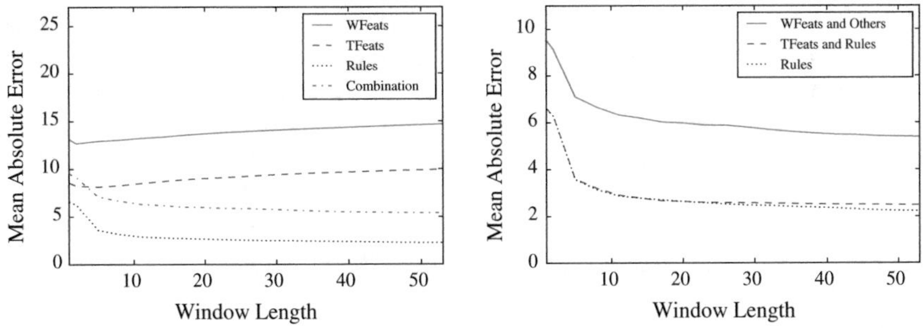

(a) Per feature set and their combination (b) Priority-based history integration

Fig. 3 MAE of estimates and window size of historical context function for **a** various feature set based estimates and **b** the priority-based historical context integration

We aim to pick a window size that improves the estimation accuracy as much as possible while not increasing the MAE for any feature, combination, and integration type. Window size 15 answers to this specification. Thus, we will be using 15 as the size for window length. Furthermore, we will use the priority-based historical context integration, since it provides more precise estimates than the feature type independent historical context integration.

5 Test Results

We test the method by running leave-one-event-out experiments in various domain, feature set and hyperparameter settings. First we perform in-domain experiments, where the system is trained and tested in the scope of a single domain, on the 51 FM events that were not used during the hyperparameter optimization phase, and the full MC data. The in-domain experiment for MC data set is performed using the hyperparameters that were optimized on the development FM data set. We then measure the estimation performance of a model trained on FM data set on the MC data set. Finally we combine the data sets in order to test the contribution of each data set in an open-domain training and testing setting.

Table 3 lists the performance of the method in terms of mean absolute error (MAE), root mean square error (RMSE), and coverage for each experiment. The 'Football Matches' and 'Music Concerts' parts contain the results of the in-domain experiments; the domain union part provides the results for the experiment that uses both the FM and MC data sets to train and test the method. The model transfer part represents results for the experiment in which we use the FM data set for training and the MC data set as a test. Columns denote results obtained with the different feature sets: 'RFeats', 'TFeats', and 'WFeats', priority based historical context integration in 'All', and the two baselines: 'Median Baseline' and 'Mean Baseline'.

For the football matches, RFeats provide the best estimations with 3.42 h MAE and 14.78 h RMSE. RMSE values are higher than MAE values in proportion to the relatively large mistakes the method makes. Although RFeats need to be written manually and their coverage is limited to 27%, their performance shows that they offer rather precise estimates. Without the RFeats, TFeats achieve 7.92 h MAE and 24 h RMSE. This indicates that having a temporal expressions list will provide a reasonable performance as well. WFeats, which do not need any resource other than a lexicon, yield TTE estimates with a MAE of 13.98 and a RMSE of 35.55.

The integration approach, of which the results are listed in the column 'All', succeeds in improving the overall estimation quality while increasing the coverage up to 85% of the test tweets. Comparing these results to the mean and median baseline columns shows that our method outperforms both baselines by a clear margin.

The errors and coverages obtained on music concerts listed in Table 3 show that all features and their combinations lead to higher errors as compared to the FM dataset, roughly doubling their MAE. Notably, WFeats yield a MAE of more than one day. The different distribution of the tweets, also reflected in the higher baseline

Table 3 In domain experiments for football matches (FM), music concerts (MC), domain union and model transfer

	RFeats	TFeats	WFeats	All	Mean baseline	Median baseline
Football matches						
RMSE	14.78	24.39	35.55	21.62	38.34	41.67
MAE	3.42	7.53	13.98	5.95	24.44	18.28
Coverage	0.27	0.32	0.82	0.85	1.00	1.00
Music concerts						
RMSE	25.49	31.37	50.68	38.44	54.07	61.75
MAE	9.59	13.83	26.13	15.80	38.93	34.98
Coverage	0.26	0.27	0.76	0.79	1.00	1.00
Domain union						
RMSE	15.19	24.16	35.76	22.38	38.97	42.43
MAE	3.59	7.63	14.25	6.24	24.95	18.79
Coverage	0.27	0.31	0.82	0.84	1.00	1.00
Model transfer						
RMSE	25.43	36.20	57.65	44.40	57.28	64.81
MAE	9.66	19.62	31.28	19.04	35.51	37.06
Coverage	0.26	0.22	0.74	0.77	1.00	1.00

errors, appears to be causing a higher error range, but still our method performs well under the baseline errors.

The domain union part of Table 3 represents results of an experiment in which we train and test in an open-domain setting. Both the 51 test events from the FM and 32 events from the MC data sets are used in a leave-one-out experiment. These results are comparable to FM in-domain experiment results.

The results of the cross-domain experiment in which the model is trained on FM data set and tested on the MC data set is represented in the 'Model Transfer' part of Table 3. The performance is very close to the performance obtained in the in-domain experiment on the MC data set.

6 Discussion

In this section we take a closer look at some of the results in order to find explanations for some of the differences we find in the results reported in the previous section.

For the in-domain football matches experiment with our combination system, the estimation quality in terms of MAE and RMSE relative to event start time of the integration-based method is represented in the left of Fig. 4a. The MAE of each

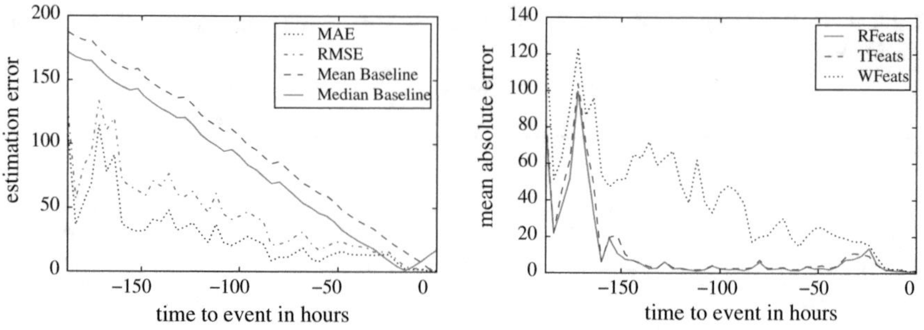

(a) Integration and baselines (b) Feature sets

Fig. 4 Averaged mean estimation error, averaged in 4-h frames, relative to event start time for the in-domain football matches experiment

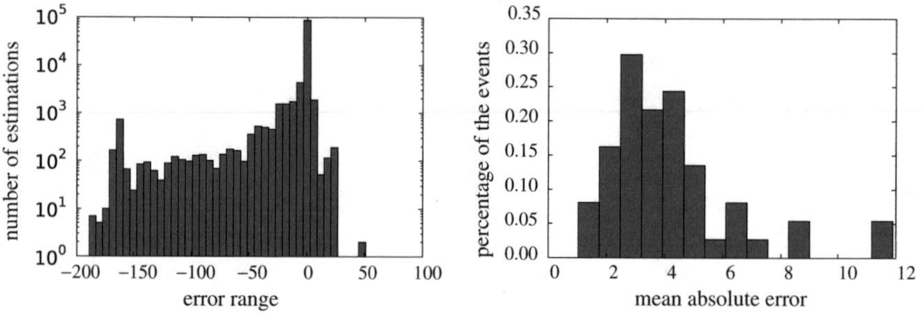

(a) Error Distribution (b) MAE per Event

Fig. 5 Error distribution per estimation and per event for the in-domain football matches experiment

feature type relative to historical context integration based on priority is displayed in the right figure, Fig. 4b.

Figure 4a shows that as the event draws closer, the accuracy of estimations increases. Within approximately 20 h to the event our method estimates TTE nearly flawlessly. This is the region where most of the tweets occur, which means there are enough tweets to learn this pattern. Figure 4b illustrates that all feature sets are relatively more successful in providing TTE estimates as the event start time approaches. RFeats and TFeats provide accurate estimates starting as early as around 150 h before the event. In contrast, Wfeats produce relatively higher errors of 20 or more hours off up to a day before the event.

Error distributions for tweets and events for the in-domain football matches experiment are displayed in Fig. 5. Figure 5a, which has a logarithmic y-axis, illustrates that the majority of the estimations have an error around zero, and that few estimations point to a time after the event as a starting time. Aggregated at the level of events, the per-event estimation performance can be observed in Fig. 5b. The mean absolute error of estimations for an event is mainly under 5 h. There is just one outlier event for which it was not possible to estimate remaining time to it reliably. We observed that the football match #ajatwe, which has an overlapping cup final match in 5 days of the targeted league match, affected the evaluation measures significantly by having the highest number of tweets by 23,976, and having 11.68 and 37.36 h MAE and RMSE, respectively.

Table 4 illustrates some extracted features for each type of feature, and the estimates based on these sets for several tweets. The second and fourth examples draw from WFeats only: *onderweg* 'on the road', *onderweg, wedstrijd* 'on the road, match' and *zitten klaar* 'seated ready', *klaar* 'ready', and *zitten* 'seated' indicate that there is not much time left until the event starts; participants of the event are on their way to the event or ready to watch the game. These features provide a 0.51 and 2.07 h estimate respectively, which are 0.56 and 0.81 h off for the second and fourth examples. The history function subsequently adjusts them to be just 0.17 and 0.49 h, 6 and 29 min, off.

The third example illustrates how RFeats perform. Although the estimate based on *aanstaande zondag* 'next Sunday' is already fairly precise, the history function improves it by 0.24 h. The remaining examples (1, 5, and 6) use TFeats. Although the first example's estimate is off by 3.09 h, the historical context adjusts it to be just 0.28 h off. Finally, the fifth and sixth examples represent the same tweet for different events, i.e. #ajaaz and #psvutr. Since every event provides a different history of tweets, they are adjusted differently.

Repeating the analysis for the second domain, music concerts, Fig. 6 illustrates the estimation quality relative to event start time for the in-domain experiment on music concerts. Figure 6a shows that the mean error of the estimates remains under the baseline error most of the time, though the estimates are not as accurate as with the football matches. Errors do decrease as the event draws closer.

As demonstrated in Fig. 7b the accuracy of TTE estimation vary from one event to another. An analysis of the relation between the size of an event and the method performance (see also Fig. 8) shows that there appears to be a correlation between the number of tweets referring to an event and the accuracy with which our method can estimate a TTE. For events for which we have fewer data, the MAE is generally higher.

The results of the 'Domain union' experiment show that mixing the two event types leads to errors that are comparable to the in-domain experiment for football matches. We take the size of each domain in terms of tweets and the individual domain performances into account to explain this. The proportion of the MC tweets to FM tweets is 0.025. This small amount of additional data is not expected to have a strong impact; results are expected to be, and are, close to the in-domain results. On the other hand, while the results of the domain union are slightly worse than the in-domain experiment on the FM data set, the per-domain results, which were filtered from all results for this experiment, are better than the in-domain experiments for the FM events. The WFeats features for FM event yield 13.85 h MAE and 35.11 h RMSE when music concerts are mixed in, compared to a slightly higher 13.98 h MAE and 35.55 h RMSE for the in-domain experiment on FM. Although the same

Table 4 Sample tweets, extracted features for rules (RFeats), temporal expressions (TFeats) and word skipgrams (WFeats) and estimates for RFeats (REst), TFeats (TEst), WFeats (WEst) and their priority based historical context integration (IEst). Blank cells indicates that the method does not detect relevant features or estimates for this column

	Tweet	RFeats	WFeats	TFeats	TTE	REst	TEst	WEst	IEst
1	nog 4 uurtje tot de klasieker #feyaja volgen via radio want ik ben @ work		radio, volgen radio, uurtje radio, uurtje	nog uurtje, uurtje	4.05		0.86	1.09	4.33
2	we leven er naar toe de 31ste titel wij zitten er klaar voor #tweaja		klaar, zitten klaar, zitten		1.07			0.51	0.90
3	aanstaande zondag naar fc utrecht-afc ajax #utraja #afcajax	aanstaande zondag			112.87	115.37			115.13
4	onderweg naar de galgenwaard voor de wedstrijd fc utrecht feyenoord #utrfey		onderweg, onderweg wedstrijd		2.88			2.07	3.37
5	nu #ajaaz kijken dadelijk #psvutr kijken		kijken dadelijk, dadelijk kijken, kijken kijken, dadelijk, kijken	nu dadelijk, dadelijk	1.85		1.07	0.91	1.88
6	nu #ajaaz kijken dadelijk #psvutr kijken		kijken dadelijk, dadelijk kijken, kijken kijken, dadelijk, kijken	nu dadelijk, dadelijk	1.85		1.07	0.91	1.95

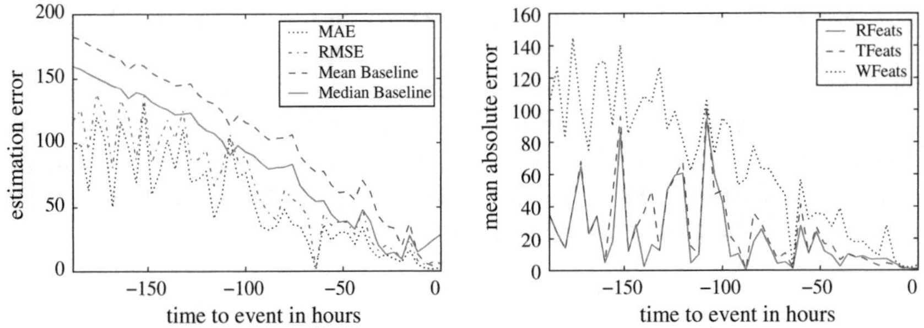

(a) MAE and RMSE for the integration and baselines
(b) MAE for each feature set

Fig. 6 Mean estimation quality relative to event start time for the music concerts

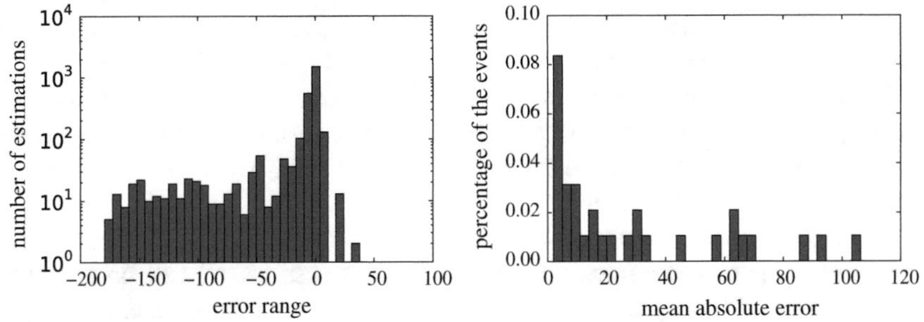

(a) Error Distribution for tweets
(b) MAE per Event

Fig. 7 Error distribution per estimate and per event in the music concerts data

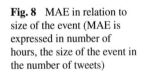

Fig. 8 MAE in relation to size of the event (MAE is expressed in number of hours, the size of the event in the number of tweets)

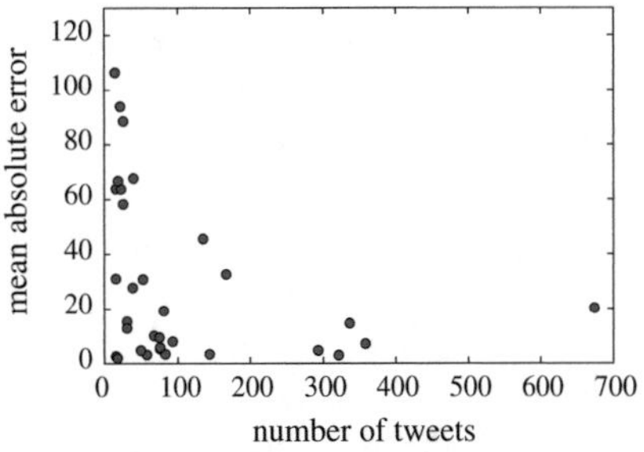

improvement does not hold for the MC events, these results suggests that mixing different domains may contribute to a better overall performance.

Finally, we looked into features that enable the FM-based model to yield precise estimates on the MC data set as well. Successful features relate to sub-events related to large public events that require people to travel, park, and queue: *geparkeerd* 'parked', *inpakken, richting* 'take, direction', *afhalen* 'pick up', *trein, vol* 'train, full', *afwachting* 'anticipation', *langzaam, vol* 'slowly, full', *wachtend* 'waiting', and *rij, wachten* 'queue, waiting'. Moreover features such as *half, uurtje* 'half, hour' prove that WFeats can learn temporal expressions as well. Some example TFeats features learned from FM that were useful with the MC data set are *over 35 minuten* 'in 35 min', *rond 5 uur* 'around 5 o'clock', *nog een paar minuutjes* 'another few minutes', *nog een nachtje, dan* 'one more night, then', and *nog een half uur, dan* 'one more half an hour, then'. These results suggest that the models can be transferred across domains successfully to a fair extent.

7 Conclusions and Future Work

We have presented a time-to-event estimation method that is able to infer the starting time of an event from a stream of tweets automatically by using linguistic cues. It is designed to be general in terms of domain and language, and to generate precise estimates even in cases in which there is not much information about an event.

We test the method by estimating the TTE from single tweets referring to football matches and music concerts in Dutch. We showed that estimates can be as accurate as under 4 and 10 h off, averaged over a time window of 8 days, for football matches and music concerts respectively.

The method provides best results with an ordered procedure that prefers the prediction of temporal logic rules, and then backs off to estimates based on temporal expressions, followed by word skipgram features, of which the individual TTEs are estimated through median training. Comparing the precision of three types of features we find that temporal logic rules and an extensive temporal expression list outperform word skipgrams in generating accurate estimates. On the other hand, word skipgrams demonstrate the potential of covering temporal expressions.

We furthermore showed that integrating historical context based on previous estimates improves the overall estimation quality. Closer analysis of our results also reveals that estimation quality improves as more data are available and the event is getting closer.

We presented results that hint at the possibility that optimized parameters and trained models can be transferred across domains.

Our study has a number of logical extensions in the line of used data, features, and applicability to different domains.

TTE estimation should remain reliable on event data that is noisier than the data collected by using a single hashtag—this could be tested using the output of automatic event detection methods that detect events of any type in the Twitter stream.

The method should also be applied to data that come from different media such as Facebook[22] posts, news articles, official reports, or internet forum posts, to investigate its robustness towards source type.

To handle the high degree of spelling variation on social media, the features included in the study should be more flexible than the current list of temporal expres-

[22]https://www.facebook.com.

sions or a set of word skipgrams filtered by a lexicon. We hardcoded a limited number of frequent spelling variations, where it would be a more generic solution to rely on a more systematic spelling normalization or using all occurring tokens to generate features independent of a lexicon. For example, currently we eliminate the highly ambiguous terms that do not co-occur with a disambiguating term. Eliminating negative patterns like *goede morgen* 'good morning' or 'good tomorrow' can provide more consistent and precise results.

Various functions that were used to select features, assign feature values, and do estimations can be improved in order to identify outliers and perform precise estimates. Analyzing changes in tweet frequencies relative to an event may support the feature selection and estimation phases.

Results obtained in this study contribute to the information available about a city. Even in the case of events are known, the rate of the people anticipating these events will inform the authorities about how they should handle these events.

Acknowledgements This research was supported by the Dutch national programme COMMIT as part of the Infiniti project.

References

1. Baeza Yates, R.: Searching the future. In: In ACM SIGIR Workshop on Mathematical/Formal Methods for Information Retrieval (MF/IR 2005) (2005)
2. Baldwin, T., Cook, P., Lui, M., MacKinlay, A., Wang, L.: How noisy social media text, how diffrnt social media sources. In: Proceedings of the 6th International Joint Conference on Natural Language Processing (IJCNLP 2013), pp. 356–364 (2013)
3. Batista, G.E.A.P.A., Prati, R.C., Monard, M.C.: A study of the behavior of several methods for balancing machine learning training data. SIGKDD Explor. Newsl. **6**(1), 20–29 (2004). doi:10.1145/1007730.1007735. URL http://doi.acm.org/10.1145/1007730.1007735
4. Becker, H., Iter, D., Naaman, M., Gravano, L.: Identifying content for planned events across social media sites. In: Proceedings of the Fifth ACM International Conference on Web Search and Data Mining, WSDM '12, pp. 533–542. ACM, New York, USA (2012). doi:10.1145/2124295.2124360. URL http://doi.acm.org/10.1145/2124295.2124360
5. Blamey, B., Crick, T., Oatley, G.: 'The first day of summer': parsing temporal expressions with distributed semantics. In: Bramer, M., Petridis, M. (eds.) Research and Development in Intelligent Systems XXX, pp. 389–402. Springer International Publishing (2013). doi:10.1007/978-3-319-02621-3_29. http://x.doi.org/10.1007/978-3-319-02621-3_29
6. Briscoe, E., Appling, S., Schlosser, J.: Passive crowd sourcing for technology prediction. In: Agarwal, N., Xu, K., Osgood, N. (eds.) Social Computing, Behavioral-Cultural Modeling, and Prediction. Lecture Notes in Computer Science, vol. 9021, pp. 264–269. Springer International Publishing (2015). doi:10.1007/978-3-319-16268-3_28. http://dx.doi.org/10.1007/978-3-319-16268-3_28
7. Chang, A.X., Manning, C.D.: Sutime: a library for recognizing and normalizing time expressions. In: LREC (2012)
8. Cohen, M.J., Brink, G.J.M., Adang, O.M.J., Dijk, J.A.G.M., Boeschoten, T.: Twee werelden: You only live once. Technical report, Ministerie van Veiligheid en Justitie, The Hague, The Netherlands (2013)
9. Dias, G., Campos, R., Jorge, A.: Future retrieval: what does the future talk about? In: Proceedings SIGIR2011 Workshop on Enriching Information Retrieval (ENIR2011) (2011)

10. Hürriyetoğlu, A., Kunneman, F., van den Bosch, A.: Estimating the time between twitter messages and future events. In: DIR, pp. 20–23 (2013)
11. Hürriyetoğlu, A., Oostdijk, N., van den Bosch, A.: Estimating time to event from tweets using temporal expressions. In: Proceedings of the 5th Workshop on Language Analysis for Social Media (LASM), pp. 8–16. Association for Computational Linguistics, Gothenburg, Sweden (2014). http://www.aclweb.org/anthology/W14-1302
12. Jatowt, A., Au Yeung, C.m.: Extracting collective expectations about the future from large text collections. In: Proceedings of the 20th ACM International Conference on Information and Knowledge Management, CIKM '11, pp. 1259–1264. ACM, New York, USA (2011). doi:10.1145/2063576.2063759. http://doi.acm.org/10.1145/2063576.2063759
13. Jatowt, A., Au Yeung, C.M., Tanaka, K.: Estimating document focus time. In: Proceedings of the 22Nd ACM International Conference on Conference on Information & Knowledge Management, CIKM '13, pp. 2273–2278. ACM, New York, USA (2013). doi:10.1145/2505515.2505655. http://doi.acm.org/10.1145/2505515.2505655
14. Kallus, N.: Predicting crowd behavior with big public data. In: Proceedings of the Companion Publication of the 23rd International Conference on World Wide Web Companion, WWW Companion '14, pp. 625–630. International World Wide Web Conferences Steering Committee, Republic and Canton of Geneva, Switzerland (2014). doi:10.1145/2567948.2579233. http://dx.doi.org/10.1145/2567948.2579233
15. Kanhabua, N., Romano, S., Stewart, A.: Identifying relevant temporal expressions for real-world events. In: Proceedings of The SIGIR 2012 Workshop on Time-aware Information Access, Portland, OR (2012)
16. Kawai, H., Jatowt, A., Tanaka, K., Kunieda, K., Yamada, K.: Chronoseeker: search engine for future and past events. In: Proceedings of the 4th International Conference on Uniquitous Information Management and Communication, ICUIMC '10, pp. 25:1–25:10. ACM, New York, USA (2010). doi:10.1145/2108616.2108647. http://doi.acm.org/10.1145/2108616.2108647
17. Kunneman, F., Van den Bosch, A.: Leveraging unscheduled event prediction through mining scheduled event tweets. In: Roos, N., Winands, M., Uiterwijk, J. (eds.) Proceedings of the 24th Benelux Conference on Artficial Intelligence, pp. 147–154. Maastricht, The Netherlands (2012)
18. Lee, H., Surdeanu, M., MacCartney, B., Jurafsky, D.: On the importance of text analysis for stock price prediction. In: Proceedings of LREC 2014 (2014)
19. Mani, I., Wilson, G.: Robust temporal processing of news. In: Proceedings of the 38th Annual Meeting on Association for Computational Linguistics, ACL '00, pp. 69–76. Association for Computational Linguistics, Stroudsburg, PA, USA (2000). doi:10.3115/1075218.1075228. http://dx.doi.org/10.3115/1075218.1075228
20. Morency, P.: When temporal expressions don't tell time: a pragmatic approach to temporality, argumentation and subjectivity (2006). https://www2.unine.ch/files/content/sites/cognition/files/shared/documents/patrickmorency-thesisproject.pdf
21. Muthiah, S.: Forecasting protests by detecting future time mentions in news and social media. Master's thesis, Virginia Polytechnic Institute and State University (2014). http://vtechworks.lib.vt.edu/handle/10919/25430
22. Nakajima, Y., Ptaszynski, M., Honma, H., Masui, F.: Investigation of future reference expressions in trend information. In: 2014 AAAI Spring Symposium Series, pp. 32–38 (2014). http://www.aaai.org/ocs/index.php/SSS/SSS14/paper/view/7691
23. Nguyen-Son, H.Q., Hoang, A.T., Tran, M.T., Yoshiura, H., Sonehara, N., Echizen, I.: Anonymizing temporal phrases in natural language text to be posted on social networking services. In: Shi, Y.Q., Kim, H.J., Prez-Gonzlez, F. (eds.) Digital-Forensics and Watermarking. Lecture Notes in Computer Science, pp. 437–451. Springer, Heidelberg (2014). doi:10.1007/978-3-662-43886-2_31. http://dx.doi.org/10.1007/978-3-662-43886-2_31
24. Noce, L., Zamberletti, A., Gallo, I., Piccoli, G., Rodriguez, J.: Automatic prediction of future business conditions. In: Przepirkowski, A., Ogrodniczuk, M. (eds.) Advances in Natural Language Processing. Lecture Notes in Computer Science, vol. 8686, pp. 371–383. Springer International Publishing (2014). doi:10.1007/978-3-319-10888-9_37. http://dx.doi.org/10.1007/978-3-319-10888-9_37

25. Noro, T., Inui, T., Takamura, H., Okumura, M.: Time period identification of events in text. In: Proceedings of the 21st International Conference on Computational Linguistics and the 44th annual meeting of the Association for Computational Linguistics, ACL-44, pp. 1153–1160. Association for Computational Linguistics, Stroudsburg, PA, USA (2006). doi:10.3115/1220175.1220320. http://dx.doi.org/10.3115/1220175.1220320
26. Ozdikis, O., Senkul, P., Oguztuzun, H.: Semantic expansion of hashtags for enhanced event detection in twitter. In: Proceedings of the 1st International Workshop on Online Social Systems (2012)
27. Papacharalampous, A.E., Cats, O., Lankhaar, J.W., Daamen, W., Van Lint, H.: Multi-modal data fusion for big events. In: Transportation Research Board 95th Annual Meeting, 16-2267 (2016). https://trid.trb.org/view.aspx?id=1392844
28. Radinsky, K., Davidovich, S., Markovitch, S.: Learning causality for news events prediction. In: Proceedings of the 21st International Conference on World Wide Web, WWW '12, pp. 909–918. ACM, New York, USA (2012). doi:10.1145/2187836.2187958. http://dx.doi.org/10.1145/2187836.2187958
29. Ramakrishnan, N., Butler, P., Muthiah, S., Self, N., Khandpur, R., Saraf, P., Wang, W., Cadena, J., Vullikanti, A., Korkmaz, G., Kuhlman, C.J., Marathe, A., Zhao, L., Hua, T., Chen, F., Lu, C.T., Huang, B., Srinivasan, A., Trinh, K., Getoor, L., Katz, G., Doyle, A., Ackermann, C., Zavorin, I., Ford, J., Summers, K.M., Fayed, Y., Arredondo, J., Gupta, D., Mares, D.: 'beating the news' with embers: forecasting civil unrest using open source indicators. CoRR **abs/1402.7035** (2014)
30. Redd, A., Carter, M., Divita, G., Shen, S., Palmer, M., Samore, M., Gundlapalli, A.V.: Detecting earlier indicators of homelessness in the free text of medical records. Stud. Health Technol. Inform. **202**, 153–156 (2013)
31. Ritter, A., Mausam, Etzioni, O., Clark, S.: Open domain event extraction from twitter. In: Proceedings of the 18th ACM SIGKDD International Conference on Knowledge Discovery and Data Mining, KDD '12, pp. 1104–1112. ACM, New York, USA (2012). doi:10.1145/2339530.2339704. http://dx.doi.org/10.1145/2339530.2339704
32. Roitman, H., Mamou, J., Mehta, S., Satt, A., Subramaniam, L.: Harnessing the crowds for smart city sensing. In: Proceedings of the 1st International Workshop on Multimodal Crowd Sensing, CrowdSens '12, pp. 17–18. ACM, New York, USA (2012). doi:10.1145/2390034.2390043. http://doi.acm.org/10.1145/2390034.2390043
33. Strötgen, J., Alonso, O., Gertz, M.: Identification of top relevant temporal expressions in documents. In: Proceedings of the 2Nd Temporal Web Analytics Workshop, TempWeb '12, pp. 33–40. ACM, New York, USA (2012). doi:10.1145/2169095.2169102. http://doi.acm.org/10.1145/2169095.2169102
34. Strötgen, J., Gertz, M.: Multilingual and cross-domain temporal tagging. Language Resources and Evaluation **47**(2), 269–298 (2013). doi:10.1007/s10579-012-9179-y. http://dx.doi.org/10.1007/s10579-012-9179-y
35. Tjong Kim Sang, E., van den Bosch, A.: Dealing with big data: the case of twitter. Comput. Linguist. Netherlands J **3**, 121–134 (2013)
36. Tops, H., van den Bosch, A., Kunneman, F.: Predicting time-to-event from twitter messages. In: BNAIC 2013 The 24th Benelux Conference on Artificial Intelligence, pp. 207–2014 (2013)
37. Tufekci, Z.: Big questions for social media big data: representativeness, validity and other methodological pitfalls. In: Adar, E., Resnick, P., Choudhury, M.D., Hogan, B., Oh, A. (eds.) Proceedings of the Eighth International Conference on Weblogs and Social Media, ICWSM 2014, Ann Arbor, Michigan, USA, 1–4 June 2014. The AAAI Press (2014). http://www.aaai.org/ocs/index.php/ICWSM/ICWSM14/paper/view/8062
38. Wang, X., Tokarchuk, L., Cuadrado, F., Poslad, S.: Exploiting hashtags for adaptive microblog crawling. In: Proceedings of the 2013 IEEE/ACM International Conference on Advances in Social Networks Analysis and Mining, ASONAM '13, pp. 311–315. ACM, New York, USA (2013). doi:10.1145/2492517.2492624. http://doi.acm.org/10.1145/2492517.2492624
39. Weerkamp, W., De Rijke, M.: Activity prediction: A twitter-based exploration. In: Proceedings of the SIGIR 2012 Workshop on Time-aware Information Access, TAIA-2012 (2012)

Part II
Machine Translation

Automatic Machine Translation for Arabic Tweets

Fatma Mallek, Ngoc Tan Le and Fatiha Sadat

Abstract Twitter is a continuous and unlimited source of data in natural language, which is particularly unstructured, highly noisy and short, making it difficult to deal with traditional approaches to automatic Natural Language Processing (NLP). The current research focus on the implementation of a phrase-based statistical machine translation system for tweets, from a complex and a morphological rich language, Arabic, into English. The first challenge is prepossessing the highly noisy data collected from Twitter, for both the source and target languages. A special attention is given to the pre-processing of Arabic tweets. The second challenge is related to the lack of parallel corpora for Arabic-English tweets. Thus, an out-of-domain corpus was incorporated for training a translation model and an adaptation strategy of a bigger language model for English tweets was used in the training step. Our evaluations confirm that pre-processing tweets of the source and target languages improves the performance of the statistical machine translation system. In addition, using an in-domain data for the language model and the tuning set, showed a better performance of the statistical machine translation system from Arabic to English tweets. An improvement of 4 pt. BLEU was realized.

Keywords Microblogs · Twitter · Statistical machine translation (SMT)
Language model · Parallel corpus · Arabic · English

1 Introduction

Since decades, the Machine Translation (MT) has been the subject of interest of many researches in the domain of Natural Language Processing (NLP). The principal purpose of MT is to translate a natural language into another one. This task seems very simple to a human expert in translation, but not for the computer. For this reason, the research in the domain of MT is ongoing and many machine learning

F. Mallek (✉) · N. Tan Le · F. Sadat
Université du Québec À Montréal, Montreal, Canada
e-mail: mallek.fatma@courrier.uqam.ca

K. Shaalan et al. (eds.), *Intelligent Natural Language Processing: Trends and Applications*, Studies in Computational Intelligence 740,
https://doi.org/10.1007/978-3-319-67056-0_6

methods have been proposed recently. In the literature, we distinguish several types of approaches in MT (1) the example-based approach, (2) the rule-based approach, (3) the statistical approach and finally (4) the hybrid approach. The two first approaches are less and less used due to the considerable human efforts they need to build the rules and the dictionaries. For this reason, current research in the field of MT tends to choose automatic approaches based on machine learning, such as statistical ones.

Statistical Machine Translation (SMT) is one of the most popular approaches, especially the Phrase-based one (PBSMT). It reposes on monolingual and parallel corpora. Parallel corpora consists of a collection of bilingual texts which are generally aligned at the level of the sentences or paragraphs, i.e. texts in the source language with their translations in the target language.

Nowadays, with the emergence of social media sites and the fluidity of digital data, the MT applications are focused more and more on this kind of data [20, 46]. Especially in the recent years, with the political events around the world, it seems very interesting to understand what other people published via the social media in any language. Twitter, is considered as one of the famous social networking platform. It reaches the place of the second most popular social networking site in the world [14]. Registered users can post short messages or tweets, of up to 140 characters at a time. They can also follow messages of other users. According to the study of the company's own blog, the number of active users reached 310 millions per month. Also, about 500 millions of tweets are published by users everyday in more than 40 different languages.[1] In front of this diversity of languages, the company of Twitter has proposed to its users the option to translate their tweets. So, in January 2015, the Bing Translator was integrated in Twitter's platforms. Yet, this option do not reach a good translation performance.[2]

The fluidity and the linguistic characteristics of tweets published continuously has been the subject of several studies in NLP on many topics such as sentiment analysis [31, 38], event detection [8], named entities recognition [29] and machine translation [20, 21, 46]. Tweets are considered as very short texts and are written in non-standard format. Users of Twitter make many orthographic mistakes and often express their ideas in more than one language in the same time. For these reasons, translating the content of social media texts is considered as a very challenging task [9]. This task is more complex when a morphological rich language such as Arabic is involved [17].

Actually the linguistic resources for Arabic social media, such as parallel corpus (Arabic-other language or other language-Arabic) and the NLP tools are very limited. In [31, 38, 42] the authors collected parallel corpora for Arabic-English tweets, but these data collections are not enough to build an efficient machine translation for Arabic-English tweets. Also the data collected in [20] is not open source. For that, in our study, we will train the statistical translation model using a public parallel corpus (Modern Standard Arabic (MSA)/English) and then we will apply adapting strategies for the tweet's contents.

[1] https://about.twitter.com/fr/company.

[2] https://support.twitter.com/articles/20172133.

This research requires to tackle several challenges at the same time and raises several questions:

- What are the linguistic characteristics of the Arabic language used in the social media platforms? Is it closed to the Modern Standard Arabic?
- What are the pre-processing steps for the machine translation applications, when we deal with a morphological rich and complex languages such as Arabic?
- Is it a good choice to build a SMT system for tweets, using an out-of-domain corpus? What is the efficient strategy to adapt the MT system in order to translate noisy and short texts such as tweets?

2 Arabic Language Challenges Within NLP and Social Media

Arabic language is a semitic language that is spoken by more than 300 millions persons, all over the world [18]. There is 28 letters in the Arabic alphabet, and sentences are written from right to left. In the literature, we can distinguish more than one kind of Arabic language: Modern Standard Arabic (MSA) and dialectal Arabic (DA), or colloquial language [40]. MSA is the formal language used in educational and scripted speeches. On the other hand, DA is the daily language of several people in the Arab world that dominantly used on the social media websites [41]. However, the texts on the social media, especially in tweets, are mixed between MSA and DA and with many variations [40]. In the sections below, we detailed the challenges of the MSA language for MT and NLP applications. Afterwards, we focus on the characteristics of the language used on social media.

2.1 Modern Standard Arabic Challenges for MT

The orthographic, morphological and syntactic characteristics of the Arabic language raise many issues and challenges for Machine Translation (MT) applications.

Orthographic Ambiguity: Actually Arabic words can have different semantic and/or syntactic meanings depending on the marks that are added to the consonants. These marks are named Diacritics. For example, the diacritics added to the sentence “كتب الولد” handle to two different sentences with different meanings “كَتَبَ الوَلَدُ” [ktb Alwld] (the boy wrote) or “كُتُبُ الوَلَدِ” [kutubu Alwldi] (the book’s boy) [10]. Such semantic ambiguity in the orthographic representations of sentences deeply affects the quality of the translated Arabic texts [43].

Another orthographic challenge with Arabic is caused by the different spelling ways of some letters. The letter “Hamza” or “Alif” “أ” [A] are differently written.

Also, the letter "Ya" is sometimes written as a dotless "Ya" ('(ى)'). These orthographic ambiguities, lead to poor probabilistic estimations of words in SMT applications [43].

2.1.1 Morphological Complexity

We distinguish two types of sentences in Arabic language: the nominal sentences and the verbal sentences. A nominal sentence is composed by a subject and an attribute, which may be a qualifying adjective, a complement adverb, a complement of an object, etc. For example, the sentence "الطقس جميل" [AlTqs jmyl] (The weather is nice), is a nominal sentence that does not contains a verb (the verb to be (is) in English). It is composed by a subject and an adjective. In the verbal sentences, the verb is presented and it inflects subject (gender, number and person), aspect, voice, or mood.

Actually, Arabic words can belong to one of three categories: verbs, names and particles. Many words, semantically different, can be derived from the same root, by applying different patterns. Hence, the extraction of the lemma from an Arabic word is a challenging task and requires the use of a morphological analyzer [39]. Otherwise, Arabic words are agglutinated words, composed by an inflected word form (base) and attachable clitics. So, a word in Arabic can correspond to multiple English words. This problem complicates the alignment between an Arabic and an English sentence and increases the OOV (out-of-vocabulary) rate in a SMT system [39]. Table 1 shows the derived words from the root عمل [3ml].

In fact, Arabic words are agglutinated words, composed by an inflected word form (base) and attachable clitics. Four categories of clitics were presented in [23]. As the Arabic sentence is written from right to left, the segmentation of an Arabic word can be composed like this (Table 2).

A word in Arabic, can correspond to multiple English words, where the English words are spread at distinct places in the English sentence. This problem complicates the alignment between an Arabic and an English sentences and increases the OOV

Table 1 Example of Arabic words derived from the root "3ml" [13]

Pattern	Arabic word	English Translation
عَمل	فَعل [Eaml]	job
عامِل	فاعِل [EAmil]	worker
عَمَلَ	مَفعل[Eamala]	worked
مَعمل	فَعل [maEml]	workshop
عُمِلَ	فُعِلَ [Eumila]	has been worked

Table 2 Segments of an Arabic word

Enclitic	Suffix	**BASE**	Prefix	Proclitic

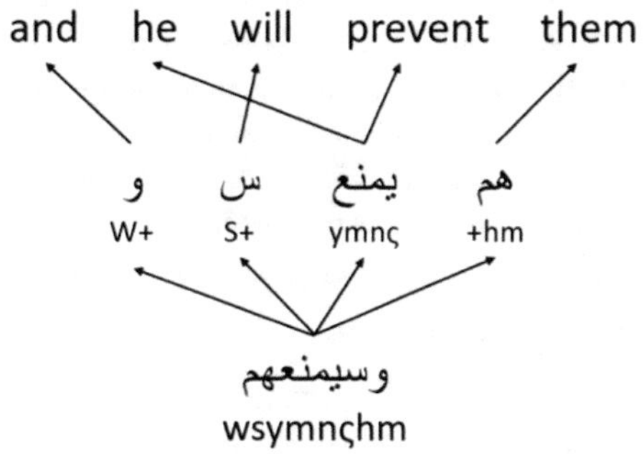

Fig. 1 Illustration of word alignment between a sentence in English and its translation in Arabic [43]

(out-of-vocabulary) rate in a SMT system [39]. The example presented in Fig. 1 illustrates the alignment problem between an agglutinated Arabic word and an English sentence.

As illustrated by the Fig. 1, a single Arabic word can correspond to multiple English words. Hence, to decrease token sparsity and improve the alignment in SMT systems, we lead to the morphological segmentation step of the Arabic text. The segmentation amounts to separating the inflected base from the clitics attached to it. This pre-processing step proves it efficiency in many NLP applications dealing with Arabic language [39]. The segmentation of Arabic words into correct morphemes is a challenging task. The segmentation improves the evaluation scores of the SMT system [39].

2.2 Arabic Language in Microblogs

As presented in the previous section, the complexity of the morphology together with the underspecification of the orthography in Arabic language create a high degree of ambiguity. This ambiguity increases more and more in the social media texts. Nowadays, there are many other issues that appeared with the spread of social media platforms. The texts in microblogs, for example Twitter, are short, noisy and written in a non-standard orthography style.

Indeed, social media users tend to commit spelling defects; They tend to write words in Arabic using the Latin alphabets and numbers. Also, each user transliterates the word in its own way. This phenomenon is called "Arabizi" [1, 3, 7]. For example, the Arabic letter 'ح' [H] is often transliterated by the number '7', the letter 'ق' [q] by the number '9', etc. In tweets, the users transliterate the proper name "احمد" [AHmd] in different ways: *ahmed*, *ahmad*, *ahmd*, *a7mad*, *a7med*, *a7mmd*, *a7md* or *ahmmd*.

Likewise, the proper name "أشرف" [a$rf], is written as *ashraf, ashref, ashrf, shrf, achraf*, or *aschraf* [32].

Moreover, users on social media tend to use more than one language in the text they publish. They often alternate between their dialect and a foreign language. This phenomenon is called *code-switching* and it has been studied in [2]. This problem greatly affects applications that handle only one language.

The use of dialectal Arabic (DA) in social media is a serious problem for Arabic NLP applications. Firstly, the current NLP tools for DA are not able to handle the dialects used in social media texts because they lack strict writing standards. For example, the MSA phrase "لا يلعب" [lA ylEb] can be written in Egyptian dialect in different ways like "مابيلعبش" [mAbylEb$], "مايلعبش" [mAylEb$], "ميلعبش" [mylEb$], "مابلعبش" [mAblE$]. Also they can be transliterated in different ways by many users like *Mayel3absh, mabyelaabsh, mabyel3absh, etc.* [7].

Messages published on Twitter (tweets) are short, noisy and they have a rich structure. It contains different special fields like the *username*, *hashtags*, *retweet*, etc. In fact, *usernames* were the subject of a named entities processing applications for Arabic in [32]. Also *hashtags* were the purpose of an improved machine translation application in [14].

3 Overview of Statistical Machine Translation

Statistical Translation model has been proven in machine translation (MT) applications since 1990. Inspired by the noisy channel model of Claude Shannon [44], Brown and his team at IBM proposed the first Statistical Translation System (SMT) [4, 5]. The authors supposed that any sentence in a source language S, can be translated into another sentence in target language T. Thus for any pair of sentences (s, t), they assign a probability of translation $p(t|s)$. This probability is interpreted as a feature that the machine translation will perform the translation hypothesis $\hat{t}$ in the target language T given a sentence s in the source language S. The problem of SMT aims to maximize the probabilities between the translation model and the language model and choose the best translation hypothesis $\hat{t}$. Hence, applying Bayes' theorem, mathematically, the problem is described as below:

$$p(t|s) = \frac{p(s|t)\, p(t)}{p(s)} \quad (1)$$

Because the probability $p(s)$ is independent to the source sentence t, so the equation (1) is simplified as follow:

$$\hat{t} = argmax_{t \in t^*}\, p(s|t)\, p(t) \quad (2)$$

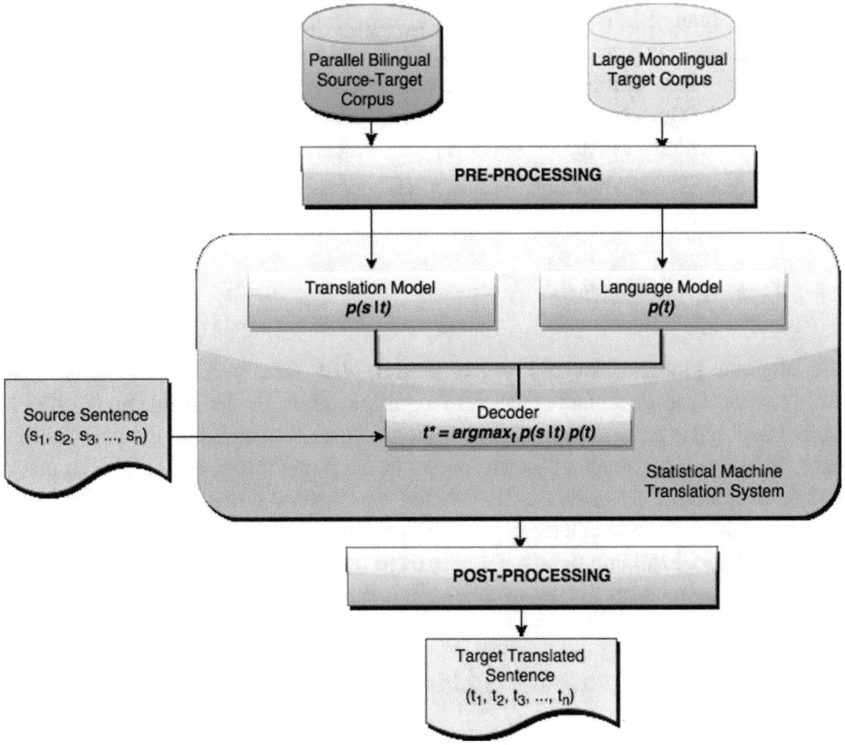

Fig. 2 General architecture of the state of the art of statistical machine translation based system

As described in the Eq. (2), the architecture of a SMT-based system is composed of two important components: the **translation model (TM)** $p(s|t)$ and the **language model (LM)** $p(t)$. In fact, the translation model (TM) contains the list of phrases translations, after the training phase by using a large parallel bilingual corpus and the language model which is built from a monolingual corpus in the target language. The translation model gives the best translation hypothesis according to the source input text while the language model ensures that this hypothesis is syntactically correct for the target text regardless of the source input text.

Moreover, in the architecture of a SMT-based system, there is also a third important component, the **decoder**. It aims to search and to find out the best translation hypothesis $\hat{t}$ among all possibilities proposed by the system. In a non exhaustive list, there are many decoders in the literature such as Pharaoh[3] [25], Portage [22] and Moses[4] [26]. The Fig. 2 describes the general architecture of the state of the art of SMT-based system.

[3]http://www.isi.edu/licensed-sw/pharaoh/.

[4]http://www.statmt.org/moses.

Actually, many SMT-based systems perform translation models either based on words or based on phrases. The first approach is known as the Word-Based Statistical Machine Translation (WBSMT) in which the system is based on an automatic word-to-word alignment [34]. The second approach is known as the Phrases-Based Statistical Machine Translation (PBSMT), in which the system considers the alignment unit as a contiguous sequences of words or segment [27].

3.1 Language Model

The language models (LM) are widely used in NLP for several applications, especially machine translation and speech recognition. Thus, the language model is considered as a basic component in SMT systems. It aims to estimate the probabilities that a phrase or a sequence of words appear in the target language [4]. The language models in the SMT systems are built based on a monolingual corpus in the target language. They allow to calculate the likelihood of the translation hypotheses in the target language. Mathematically, the language model $p(t)$ for the sentence t, which is composed by n words $t = w_1 \ w_2 \ w_3 \ \ldots \ w_n$, is defined as follows:

$$p(t) = \prod_{i=1}^{n} p(w_i | w_1 \ \ldots \ w_{i-1}) \tag{3}$$

To simplify this modeling, we assume that the w_i words of t depend only on the previous $(i - 1)$ words. We can therefore reformulate the Eq. (3) as follows:

$$p(t) = p(w_1)\, p(w_2|w_1)\, p(w_3|w_1 w_2) \ \ldots \ p(w_1|w_1 w_2 \ \ldots \ w_{n-2} w_{n-1}) \tag{4}$$

This language model is known as the n-gram model. It makes predictions based on a fixed size search window, containing n words. Hence, for each word, a probability is calculated by taking into account the $(n - 1)$ words that precede the current word in the target sentence. This probability represents the dependency of each word with respect to the $(n - 1)$ words that precede it, as indicated by the Eq. (4). Statistically, if the sequence of words to be translated does not exist in the language model, its probability will be null.

3.2 Word Alignment

Automatic word alignment remains an important component for all SMT approaches. Given a bilingual sentence pair, the general definition of word alignment refers to any defines set of links between lexical units that are translations of each other. In 1990, the first probabilistic models for machine translation was based on words, i.e. the translation unit that appears in the probability laws is the word [4]. An example of word alignment of a sentence in English and its translation in Arabic was illustrated in Fig. 1.

The first challenge in the word-based SMT models consists of establishing the mapping between words in the source sentence and words in the target sentence. In this modeling problem, a hidden variable a is used to account all the possible pair-wise alignment links between both sentences. This alignment problem is mathematically described, between the source sentence and the target sentence, as below:

$$p(s|t) = \sum_{a} p(s, a|t) \tag{5}$$

$p(s, a|t)$ is generally expressed as:

$$p(s, a|t) = p(J|t) \sum_{j=1}^{J} p(a_j|s_1^{j-1},\ a_1^{j-1}, J, t)\, p(s_j|a_1^{j-1},\ s_1^{j-1}, J, t) \tag{6}$$

where,

J: length of the source sentence s
s_j: word in position j of source sentence s
a_j: hidden alignment of word s_j indicating the position at which s_j aligns in the target sentence t.

The Eq. (6) aims to generate an alignment between a source sentence and a target sentence. Firstly, the length J of the source sentence s is manually chosen, given what the target sentence is known. The choice of the position to link the first source word, given the target sentence and the length of the source sentence, can be made. Then the identity of the first source word is chosen, given the target sentence, the length of the source sentence and the target word linked to the first source position, and so on.

Five models, known as "*IBM models*" from 1 to 5, have been proposed by [4, 5]. They aim to maximize the translation probability $P(t|s)$ of a target sentence given a source sentence.

- *IBM 1*: This model is a simple lexical translation model, which makes use of co-occurrence of word pairs. It assigns only lexicon probabilities.
- *IBM 2*: This model adds absolute reordering probabilities by introducing local dependencies.
- *IBM 3*: This model adds fertility model, which depends only on the source sentence.
- *IBM 4*: This model takes into account the relative reordering probabilities.
- *IBM 5*: This model fixes deficiency. It limits the waste of probabilities mass on impossible situation. It is considered as the non-deficient version of *IBM 4*.

These models did not have much success due to their several disadvantages. They allow only the one-to-many words alignment and the alignment mapping is restricted to source-to-target locations. The IBM models assume that the words are syntactically independent of one another. Hence, the context of the text to be translated is not taken into account, and the models can generate a confusion in the meaning of the sentence to be translated, for example, in case of polysemous words. For example, the word "*livre*" in French, could be translated in English such as a "*book*" or a "*pound*" [10]. For these reasons, most current SMT systems are not based on word-to-word approaches but based on phras-based approaches.

3.3 Translation Model

Most phrases-based translation models are commonly used by the research community in the machine translation domain [27]. A phrase can be a word or a set of words. Actually, a sentence is tokenized and then segmented into many phrases. The one-to-one and one-to-many words alignments are offered by the IBM models while many-to-many words alignments are offered by phrases-based models. Once the alignments are established, a probability score is calculated for all phrases. Each source phrase can have several translation hypotheses in the target language. Then the choice of the candidate phrase is based on the probabilities stored in the *phrases translation table*.

The translation probability $\phi(\bar{s}|\bar{t})$ is defined with score by relative frequency as follows:

$$\phi(\bar{s}|\bar{t}) = \frac{count(\bar{s},\bar{t})}{\sum_{\bar{t}_i} count(\bar{s},\bar{t}_i)} \tag{7}$$

3.4 Decoding

The translation process of a source sentence into a target sentence is known as decoding. This process consists of three following steps: (1) to segment the source sentence into phrases, (2) to translate the source phrases according to the probabilities in the translation model, (3) to reorder the source phrases according to the target language. The decoding process in phrases-based SMT is defined as follows [27]:

$$p(\bar{s}_i|\bar{t}_i) = \prod_{i=1}^{n} \phi(\bar{s}_i|\bar{t}_i)\, d(start_i - end_{i-1} - 1) \tag{8}$$

where:

- $\bar{s}_i$: a set of phrases of the source sentence s
- $\bar{t}_i$: translation of each source phrase $\bar{s}_i$
- $\phi(\bar{s}_i|\bar{t}_i)$: phrase probability according to the translation table
- $d(start_i - end_{i-1} - 1)$: reordering probability of each target phrase $\bar{t}_i$.

The decoding problem is known as a NP-complete problem [24]. To solve this problem, it is very important to reduce the search space, when effective solutions are searched. Wang and Waibel [47] have proposed a stack-based decoder based on the A* search algorithm to find out the best hypothesis. Other researchers used weighted finite states transducers to implement an alignment model [49]. In [12] the authors transformed the decoding problem into a linear programming problem by implementing a beam search algorithm. Also, dynamic programming algorithms with pruning have been implemented in [25, 37]. Koehn et al. [26] have presented the open source Moses decoder which becomes the state-of-the-art used in the SMT research community.

4 Translating Arabic Tweets

The SMT methods depend on the quantity and the quality of the data used for the translation model (TM) and the language model (LM). It's very important that the training set and the test set remain in the same domain (for example the domain of news, medicine, social media, etc.). This gives a more efficient and robust MT system [28].

Although, this is not evident in our work because the parallel corpus for Arabic-English tweets is not available. For this reason, our method is to adapt a state of the art SMT system the most possible, to be able to translate tweets in an efficient way. So, the training data set is an out-of-domain parallel corpus in Modern Standard Arabic (MSA). Inspired by the work in [21], the first strategy is to adapt the system by incorporating a big LM for tweets (an in-domain language model). The second strategy is tuning the SMT with an in-domain data.

4.1 Data Collection

Although, this is not evident in our work as there is no parallel corpus available for Arabic-English tweets. Therefore, our method focuses on the adaptation of a state-of-the-art SMT system to be able to translate tweets in an efficient way. So, the training data set is an out-of-domain parallel corpus in MSA. The first strategy aims to adapt the system by incorporating a big LM for tweets (an in-domain language model). The second strategy uses SMT tuning with an in-domain data [21].

4.2 Data Collection

For the first strategy, we need a large corpus of tweets in English to train the LM, and a small corpus in Arabic for the test set. To do so, we collected tweets via

Table 3 Collected tweets

Language	Number of tweets
Anglais	255 602
Arabe	1 930

the Streaming API of Twitter,[5] using Twitter4j which is an open source library for java [6] [48].

Influenced by the political events in the Arab world, we chose the following list of Arabic keywords to collect tweets in Arabic for the test set: ("الجزيرة" [Aljzyrp], "سوريا" [swryA], "حلب" [Hlb], "بشار" [b$Ar], "ثورة" [vwrp], "الربيع العربي" [AlrbyE AlErby]). Also we adjusted the library to download only tweets written in Arabic.

After that, we used the translations of Arabic keywords to collect English tweets, to generate the LM: (*Syria, Halab, Bachar, revolution, arab spring, AlJazeera*). We collected 255,602 English tweets and 551 Arabic tweets, during four months, from September 2015 to December 2015. The collected tweets are described in the Table 3):

The collected data were very noisy and contained many orthographic and grammar mistakes. Moreover, the tweets follow a standard form and use special terms distinguishing them from other microblogs:

- The *Username* which identified each user and it is preceded by the symbol @ (@UserName).
- The *hashtag* is preceded by the character (#). It highlights the keywords to a specific event. By clicking on a given *hashtag*, tweets that contain the same *hashtag* appear.
- The *retweet* start by (RT; @username). It indicates that the same tweet is republished by another user.
- The URL which is inserted in the end of a tweet.

In this paper, we are interested only in the raw text of the tweet, or the message itself. Thus, we deleted all the specific fields of Twitter listed previously.

Pre-processing English Tweets When we analyzed the collected tweets, we observed that users tend to use abbreviations instead of a whole sentence. For example, the sentence "*just for you*" is written as "*just4u*". Also to express their feelings, users use the stretched words like "*haaaaappy*" instead of "*happy*".

To deal with this problem, we used a dictionary of non-standard words and their standards correspondents. This dictionary was proposed at a shared task organized

[5]https://dev.twitter.com/streaming/public.

[6]http://twitter4j.org/en/index.html.

as part of the W-NUT-2015 (ACL 2015 Workshop on Noisy User-generated Text [W-NUT])[7] [19, 36].

Once the lexical normalization step was done, we obtained a corpus of tweets in English ready for building the LM using the SRILM toolkit [45].

Pre-processing Arabic Tweets This corpus was considered as the reference to evaluate the translation systems. After the pre-processing and the normalization steps, the corpus was translated into English by a human expert.

Evidently, many lexical and orthographic errors were found and were processed. For examples, compound expressions formed by two or more words separated by an underscore such as "الربيع_العربي" [AlrbyE_AlErby], or "تنظيم_الدولة" [tnZym_Al dwlp]. The Stretched words (examples: "تكلنـــــي" [tklmn______y], "دبـــــــي" [db___________y]) and also words with repeated letters (examples: "كبرررررر" [kbrrrrrr], "لايمككنا" [lAymkknnA] had a significant high presence in the collected data set.

The dialectical Arabic, especially the levantine dialect, were frequently used. However, we did not deal with this type of words. Only the lexical and orthographic errors presented before/above have been dealt with in the normalization step.

We applied the state-of-the-art pre-processing steps for Arabic, such the orthographic normalization for the letters "Hamza" [a] and "Ya" [y]. These two letters are inconsistently spelled using different forms. The "Hamza" has different forms like ('أ', 'آ', 'إ'). In this paper we represent it in a single way as bare Alif 'ا' [A]. Also, the letter "Ya" [y] is written with dots or dotless. All the letters 'ي' [y] were normalized by the letter 'ى' [Y], the dotless form of "Ya" [y], to reduce the degree of the spelling ambiguity.

In the next step, we applied a morphological segmentation of Arabic words using the morphological analyzer MADA [15]. This task was deeply studied in many NLP applications for Arabic and it generally improves the state-of-the-art baseline systems in terms of BLEU score [16]. Many segmentation scheme were proposed in literature for Arabic: S1, ON, D1, D2, D3, WA, TB, MR, L1, L2 et EN [16, 39]. In this study, we used the D2 segmentation scheme, which its efficiency for MT applications is well known. In the D2 configuration considered here, four proclitic particles (l+ (to/f or), b+ (by/with), k+ (as/such), and f +(in)) and one conjunction proclitic (w+ (and)) are identified. These are separated from their associated root words.

The normalization of the letters "Hamza" [a] and "Ya" [y] and the segmentation were also applied for Arabic in the parallel corpus.

[7]The dictionary contains 44983 pairs of words, available on-line at: http://noisy-text.github.io/2015/norm-shared-task.html.

Table 4 Details of the parallel corpus

	Size (Mo)	Words	Tokens
Arabic (source)	165,5	16 866 817	18 791 118
English (target)	113,5	18 608 307	19 408 007

4.3 Experiments

Data To build the SMT for Arabic-English tweets, we use the parallel corpus from the United Nation (UN).[8] The training corpus was about 280 mega-octet (Mo). The Arabic side of the parallel corpus contains about 16 million words and 18 million tokens after the segmentation and the normalization steps. The Table 4 presents the detailed statistics for both languages: Arabic and English.

In the experiments, we use three LMs: one LM of tweets (presented in the Sect. 1.4, which is crawled from Twitter), a big LM of tweets used in [6], and a LM in MSA (the target side of the parallel corpus). The LMs are 3-gram LM, generated with the SRILM toolkit [45]. The pre-processing steps, i.e. the normalization of the letters "Hamza" [a], "Ya" [y] and the segmentation, were applied to the Arabic data (LM, parallel data, and test).

For the tuning step, we used a small parallel corpus with around 1000 sentences extracted from the UN corpus. Secondly, we tuned the translation systems using a small in-domain parallel corpus (Arabic-English tweets used in [38]).

Experimental Setup The experiments were carried out using the open source phrase-based SMT system Moses [26], with maximum phrase length of 10.[9] To obtain the word-to-word-alignments during the training between Arabic and English, we used MGIZA [11], a multi-threading version of GIZA++ [34].

Finally, the feature weights were estimated using Moses built-in minimum error rate training procedure (MERT) [33], which uses two different types of tuning sets: MSA and tweets.

Many systems were trained according to defined pre-processing models. We evaluated those systems in terms of BLEU [35], and the out of vocabulary words rate (OOV).

4.4 Results

Table 5 shows different results of the SMT systems for tweets, which was trained using an out-of-domain parallel corpus. To better adapt our system to the domain of social media, especially the tweets, we try to test all the possible combinations

[8]The UN parallel corpus is available at: http://www.un.org/en/documents/ods/.

[9]http://www.statmt.org/moses/?n=FactoredTraining.PrepareTraining.

Table 5 Overview of experiments and results for Arabic-English tweet's MT

Systems	LM	Pre-processing	Tuning	Results	
				BLEU	OOV
1	tweets	–	UN	2.39	28.74
2	tweets	+	UN	6.09	9.95
3	tweets	+	tweets	10.31	7.16
4	UN	–	UN	6.31	24.53
5	UN	+	UN	8.39	12.11
6	UN	+	tweets	8.96	9.61
7	tweets+UN	–	UN	2.49	27.75
8	tweets+UN	+	UN	3.50	11.18
9	tweets+UN	+	tweets	**10.98**	6.89
10	*BIG_tweets*	+	tweets	10.58	7.49

(LM and tuning set). Also, we tested the baseline systems without pre-processing the parallel and the test data. Systems 1, 4 and 7 are considered as baseline systems without any pre-processing strategy (−). All the other experiments (system 2, 3, 5, 6, 8, 9 and 10) were carried out after completing the pre-processing steps (+).

The best system has reached a BLEU score of 10.98. This result remains weaker than the result of a system with training test and dev corpus belong to the same domain, as in [39]. The pre-processing is very important and it has improved the BLEU scores by 1 to 4 points and reduced the OOV rate. As a final step, we tuned the systems using an in-domain corpus, which improved the BLEU scores and the OOV rates. The best combination resulting from these experiments is therefore a system based on two LMs (UN and tweets) with the necessary pre-processing steps for the training, tuning and test corpus.

4.5 *Discussion*

The presented results prove that for the SMT methods, the domain of the training and the test data is very important. Thus, when we combined the LM of MSA and the LM of tweets with an in-domain tuning set, we obtained the best result in term of BLEU score.

However, translation systems were not able to translate many named entities found in the test set like "بوتين" [bwtyn], "درعا" [drEA], "الجزيرة" [Aljzyrp], "تويتر" [twytr], "جوجل" [jwjl], "حماه" [HmAh], "حمص" [HmS], "داعش" [dAE$], "روسيا" [rwsyA].

Also, we noted that DA such as the levantine was used extensively in the collected tweets. As the parallel corpus is in MSA, many levantine expressions were not matched.

For the various works related to MT for microblogs, the BLEU scores are often low and despite the pre-processing performed, the quality of translations of this type of data remains poor. In [30] the BLEU score was 8.75, with out-of-domain parallel corpus. In the same context, [46] obtained 22.57 of BLEU score translating tweets from Spanish to Basque. This result were expected as the two languages are semantically very closed. Also, In [20], the score of the tweet's MT from German to English, was 15.68.

In an advanced comparison study, we tested our translation model with an in domain-data (MSA test set), We used the NIST evaluation set, MT05. The resulted BLEU score was 16.91 and the OOV rate was 3.51. The BLEU outperforms the translation of tweets for the following reasons: the test set is not noisy like tweets, and the parallel corpus and the test corpus are in the same domain.

5 Conclusion and Future Work

In our work, we have explored some of the problems facing (Arabic-English) tweets translation. Thus, we mounted a statistical machine translation using the most popular decoder for segment-based statistical machine translation systems, Moses [26].

Statistical machine translation (SMT) systems are based on a parallel corpus and on a monolingual corpus to train a language model. In our work, the parallel corpus of tweets for the Arabic-English language pair was not available. To overcome this problem, we used several adaptation strategies for the translation systems for tweets. These strategies are based on the use of a big language model for tweets, and a tuning step with a dev in-domain corpus.

Dealing with Arabic language, a morphological rich and complex language, we perform a pre-processing step including the normalisation of the letters "Hamza [a]" and "Ya [y]" and the segmentation of Arabic words. This step was very important to reduce the degree of sparsity and improve the alignment between Arabic and English words.

Also, we carried out the spelling and orthographic mistakes in tweets, by normalising the stretched words, the transliterated expression, etc. These pre-processing steps were very helpful and ameliorate the BLEU score, which reach 10.98.

In the future work, it will be useful to enlarge the parallel corpus by a parallel data, from Twitter or microblogs in general. This will improve the results. Also, dealing with the problem of "arabizi" and "code switching" in tweets is very important, and could improve the quality of the translation system for Arabic tweets.

References

1. Adouane, W., Semmar, N., Johansson, R., Bobicev, V.: Automatic detection of Arabicized Berber and Arabic varieties. VarDial **3**, 63 (2016)
2. Barman, U., Das, A., Wagner, J., Foster, J.: Code mixing: a challenge for language identification in the language of social media. In: First Workshop on Computational Approaches to Code Switching (EMNLP 2014), pp. 13–23. Association for Computational Linguistics (ACL) (2014)
3. Bies, A., Song, Z., Maamouri, M., Grimes, S., Lee, H., Wright, J., Strassel, S., Habash, N., Eskander, R., Rambow, O.: Transliteration of Arabizi into Arabic orthography: developing a parallel annotated Arabizi-Arabic script SMS/chat corpus. In: Workshop on Arabic Natural Langauge Processing (ANLP), pp. 93–103 (2014)
4. Brown, P.F., Cocke, J., Pietra, S.A.D., Pietra, V.J.D., Jelinek, F., Lafferty, J.D., Mercer, R.L., Roossin, P.S.: A statistical approach to machine translation. Comput. Linguist. **16**(2), 79–85 (1990)
5. Brown, P.F., Pietra, V.J.D., Pietra, S.A.D., Mercer, R.L.: The mathematics of statistical machine translation: parameter estimation. Comput. Linguist. **19**(2), 263–311 (1993)
6. Cherry, C., Guo, H.: The unreasonable effectiveness of word representations for twitter named entity recognition. In: HLT-NAACL, pp. 735–745 (2015)
7. Darwish, K.: Arabizi Detection and Conversion to Arabic, pp. 217–224. Association for Computational Linguistics, Doha, Qatar (2014)
8. Dridi, H.E.: Détection d'évènements à partir de Twitter. Ph.D. thesis, Université de Montréal (2015)
9. Farzindar, A., Roche, M.: Les défis de l'analyse des réseaux sociaux pour le traitement automatique des langues. Traitement Automatique des Langues **54**(3), 7–16 (2013)
10. Gahbiche-Braham, S.: Amélioration des systèmes de traduction par analyse linguistique et thématique. Ph.D. thesis, Université Paris Sud (2013)
11. Gao, Q., Vogel, S.: Parallel implementations of word alignment tool. In: Software Engineering, Testing, and Quality Assurance for Natural Language Processing, pp. 49–57. Association for Computational Linguistics (ACL) (2008)
12. Germann, U., Jahr, M., Knight, K., Marcu, D., Yamada, K.: Fast decoding and optimal decoding for machine translation. In: Proceedings of the 39th Annual Meeting on Association for Computational Linguistics, pp. 228–235. Association for Computational Linguistics (ACL) (2001)
13. Ghoul, D.: Outils génériques pour l'étiquetage morphosyntaxique de la langue arabe: segmentation et corpus d'entraînement (2011)
14. Gotti, F., Langlais, P., Farzindar, A.: Translating government agencies tweet feeds: specificities, problems and (a few) solutions. In: The 2013 Conference of the North American Chapter of the Association for Computational Linguistics: Human Language Technologies (NAACL 2013), p. 80 (2013)
15. Habash, N., Rambow, O.: Arabic tokenization, part-of-speech tagging and morphological disambiguation in one fell swoop. In: Proceedings of the 43rd Annual Meeting on Association for Computational Linguistics, pp. 573–580. Association for Computational Linguistics (2005)
16. Habash, N., Sadat, F.: Arabic preprocessing schemes for statistical machine translation. In: Proceedings of the Human Language Technology Conference of the NAACL, Companion Volume: Short Papers, pp. 49–52. Association for Computational Linguistics (ACL) (2006)
17. Habash, N., Sadat, F.: Challenges for Arabic machine translation. In: Abdelhadi Soudi, Ali Farghaly, Günter Neumann, Rabih Zbib (eds.) Natural Language Processing, pp. 73–94. Amsterdam (2012)
18. Habash, N.Y.: Introduction to Arabic natural language processing. Synth. Lect. Hum. Lang. Technol. **3**(1), 1–187 (2010)
19. Han, B., Cook, P., Baldwin, T.: Lexical normalization for social media text. Assoc. Comput. Mach. Trans. Intell. Syst. Technol. (TIST) **4**(1), 5 (2013)

20. Jehl, L., Hieber, F., Riezler, S.: Twitter translation using translation-based cross-lingual retrieval. In: Proceedings of the Seventh Workshop on Statistical Machine Translation, pp. 410–421. Association for Computational Linguistics (ACL) (2012)
21. Jehl, L.E.: Machine Translation for Twitter. Master's thesis, Speech and Language Processing School of Philosophy, Psychology and Language Studies, University of Edinburgh (2010)
22. Johnson, J.H., Sadat, F., Foster, G., Kuhn, R., Simard, M., Joanis, E., Larkin, S.: Portage: with smoothed phrase tables and segment choice models. In: The Workshop on Statistical Machine Translation, pp. 134–137. Association for Computational Linguistics, New York City (2006)
23. Kadri, Y., Nie, J.Y.: Effective stemming for Arabic information retrieval. In: The Challenge of Arabic for Natural Language Processing/Machine Translation NLP/MT, pp. 68–74 (2006)
24. Knight, K., Marcu, D.: Machine translation in the year 2004. In: ICASSP (ed.) ICASSP (5) International Conference on Acoustics, Speech, and Signal Processing, vol. 5, pp. 965–968. Institute of Electrical and Electronics Engineers (IEEE) (2005)
25. Koehn, P.: Pharaoh: a beam search decoder for phrase-based statistical machine translation models. In: Conference of the Association for Machine Translation in the Americas, pp. 115–124. Springer (2004)
26. Koehn, P., Hoang, H., Birch, A., Callison-Burch, C., Federico, M., Bertoldi, N., Cowan, B., Shen, W., Moran, C., Zens, R., et al.: Moses: open source toolkit for statistical machine translation. In: Proceedings of the 45th Annual Meeting of the ACL on Interactive Poster and Demonstration Sessions, pp. 177–180. Association for Computational Linguistics (ACL) (2007)
27. Koehn, P., Och, F.J., Marcu, D.: Statistical phrase-based translation. In: Proceedings of the 2003 Conference of the North American Chapter of the Association for Computational Linguistics on Human Language Technology, vol. 1, pp. 48–54. Association for Computational Linguistics (ACL) (2003)
28. Langlais, P., Gotti, F., Patry, A.: De la chambre des communes à la chambre d'isolement: adaptabilité d'un système de traduction basée sur les segments. In: Les actes de TALN, pp. 217–226 (2006)
29. Le, N.T., Mallek, F., Sadat, F.: UQAM-NTL: named entity recognition in twitter messages. WNUT **2016**, 197 (2016)
30. Ling, W., Xiang, G., Dyer, C., Black, A.W., Trancoso, I.: Microblogs as parallel corpora. Assoc. Comput. Linguist. (ACL) **1**, 176–186 (2013)
31. Mohammad, S.M., Salameh, M., Kiritchenko, S.: How translation alters sentiment. J. Artif. Intell. Res. (JAIR) **55**, 95–130 (2016)
32. Mubarak, H., Abdelali, A.: Arabic to English person name transliteration using Twitter. In: Proceedings of the Tenth International Conference on Language Resources and Evaluation (LREC 2016). European Language Resources Association (ELRA), Slovenia (2016)
33. Och, F.J.: Minimum error rate training in statistical machine translation. In: Proceedings of the 41st Annual Meeting on Association for Computational Linguistics, vol. 1, pp. 160–167. Association for Computational Linguistics (2003)
34. Och, F.J., Ney, H.: A systematic comparison of various statistical alignment models. Comput. Linguist. **29**(1), 19–51 (2003)
35. Papineni, K., Roukos, S., Ward, T., Zhu, W.J.: Bleu: a method for automatic evaluation of machine translation. In: Proceedings of the 40th Annual Meeting on Association for computational linguistics, pp. 311–318. Association for Computational Linguistics (2002)
36. Pennell, D.L., Liu, Y.: Normalization of informal text. Comput. Speech Lang. **28**(1), 256–277 (2014)
37. Quirk, C., Moore, R.: Faster beam-search decoding for phrasal statistical machine translation. Machine Translation Summit XI (2007)
38. Refaee, E., Rieser, V.: Benchmarking machine translated sentiment analysis for Arabic tweets. In: Student Research Workshop (SRW-2015), pp. 71–78 (2015)
39. Sadat, F., Habash, N.: Combination of Arabic preprocessing schemes for statistical machine translation. In: Proceedings of the 21st International Conference on Computational Linguistics and 44th Annual Meeting of the Association for Computational Linguistics (ACL), pp. 1–8. Association for Computational Linguistics, Sydney, July 2006

40. Sadat, F., Kazemi, F., Farzindar, A.: Automatic identification of Arabic language varieties and dialects in social media. In: The 4th International Workshop on Natural Language Processing for Social Media of (SocialNLP 2014) (2014)
41. Sadat, F., Mallek, F., Sellami, R., Boudabous, M.M., Farzindar, A.: Collaboratively constructed linguistic resources for language variants and their exploitation in NLP applications—the case of Tunisian Arabic and the social media. In: Workshop on Lexical and Grammatical Resources for Language Processing, p. 102. Citeseer (2014)
42. Salameh, M., Mohammad, S.M., Kiritchenko, S.: Sentiment after translation: a case-study on Arabic social media posts. In: Human Language Technologies: The 2015 Annual Conference of the North American Chapter of the Association for Computational Linguistics (ACL), pp. 767–777. Association for Computational Linguistics, May 2015
43. Salameh, M.K.: Morphological solutions for Arabic statistical machine translation and sentiment analysis. Ph.D. thesis, University of Alberta (2016)
44. Shannon, C.E.: The Mathematical Theory of Communication. Urbana (1949)
45. Stolcke, A., et al.: Srilm—an extensible language modeling toolkit. In: ICSLP 2, pp. 901–904, Sept 2002
46. Toral, A., Wu, X., Pirinen, T., Qiu, Z., Bicici, E., Du, J.: Dublin City University at the TweetMT 2015 shared task. In: Tweet Translation Workshop at the International Conference of the Spanish Society For Natural Language (SEPLN 2015) (2015)
47. Wang, Y.Y., Waibel, A.: Decoding algorithm in statistical machine translation. In: Proceedings of the Eighth Conference on European Chapter of the Association for Computational Linguistics, pp. 366–372. Association for Computational Linguistics (1997)
48. Yamamoto, Y.: Twitter4J—an open-sourced, mavenized and Google App Engine safe Java library for the Twitter API, released under the BSD license (2009)
49. Zhang, M., Li, H., Kumaran, A., Liu, M.: Report of news 2012 machine transliteration shared task. In: Proceedings of the 4th Named Entity Workshop, pp. 10–20. Association for Computational Linguistics (2012)

Developing a Transfer-Based System for Arabic Dialects Translation

Salwa Hamada and Reham M. Marzouk

Abstract The prominent Arabic Domestic changes have influenced the usage of the Arabic dialects among Arabs communications, which was, previously, limited on daily activities inside their own territories the role of the Modern Standard. Arabic MSA as an official Arabic language started to be diminished, since the Arabic dialects play a greater role than using it during the daily activities. The continuity of using these dialects whether in media or writing may eliminate the dominance of MSA as an official form of Arabic language in the Arab world. Besides, comprehending the Arabic language by non-native speakers, as well as, processing machine translations became a sophisticated process that requires harder effort. Accordingly, a requirement of language processing to interact with the permanent development of the dialects and to flourish the standard Arabic became imperative. Thus, it is planned to built a Hybrid Machine translation system (AlMoFseH) to translate the different Arabic dialects by using the MSA as a pivot. This research is a part of this project which emphasizes on developing a transfer-based system that transfers the Egyptian Arabic dialect EGY used in social media to MSA. For that purpose, a lexical database of 3k words presenting Egyptian Arabic dialect was built. Different texts extracted from Social media were used as a main resource of the database. The system consists of three components: disambiguation of the morphological analysis output using Naive Bayesian learning, a rule based transfer system and a dictionary look up system. The evaluation revealed a high accuracy of the system's performance, since 92.7% of the test data was transferred correctly.

Keywords Machine translation · Lexical database · Transfer system
Naive Bayesian algorithm · Egyptian Arabic dialect

S. Hamada (✉)
Electronics Research Institute ERI, P.O. Box 12611, Giza, Egypt
e-mail: hesalwa@hotmail.com

R.M. Marzouk
Faculty of Arts, Phonetics and Linguistics Department,
Alexandria University, P.O. Box 21526, Alexandria, Egypt
e-mail: marzoukreham@gmail.com

K. Shaalan et al. (eds.), *Intelligent Natural Language Processing: Trends and Applications*, Studies in Computational Intelligence 740,
https://doi.org/10.1007/978-3-319-67056-0_7

1 Introduction

Machine translation MT systems are computer programs that translate from a source language to a target language. The difficulties, that any machine translation confronts, are enlarged by the enlargement of the ambiguity in one or both languages. The Arabic language is one of these languages which the inflectional richness and sparsity of its dialects cause a large scale of ambiguities. Arabic language is classified as a diaglossic language where two forms of the language exist side by side; the formal form is known as Modern standard Language (MSA), while the other form is used in daily communication in each Arabic region and it is called dialect [1]. Both varieties form the linguistic repertoire of MSA written texts without clear boundaries between them [2]. Recently and Due to the permanent growing of the social media texts, Arabic dialects dominated these written texts and became an alternative of the formal MSA form. Thus, the recent studies of linguistics and language technology are directed to study these changes and their effect on the natural language processing. Besides, several sorts of texts were extracted and used as an essential resource for processing these dialects to unify their variations in one comprehensible form for machine translations and the non Arabic native speakers.

Otherwise, the social media user's inclination to improvise during writing enlarged the writing diversity, and added new ambiguities that should be taken into consideration. Accordingly, the need of developing a system that accepts these diversities during the translation of these texts became essential. Thus, developing hybrid system is planned to combine the best achievements of statistical and rules-based paradigms. This system is based on serial combinations of other multiple systems outputs.

This research describes two essential processes needed to achieve the system: the first process is building a lexical database to cover the words that occur frequently and signify the developed Egyptian dialect used in social media. The lexical database is stem based which is divided into lists provided with the required semantic and morpho-syntactic information of the dialectal stems and their equivalents in MSA. The second process describes the way this lexical database is incorporated into the translation process through a transfer system. The task of the transfer system is to normalize the social media texts that contain different varieties to reach to a standard MSA text that can consequently been translated to other language or dialect. This study will elaborate in details the process of transferring a nonstandard text into a standard one. This process requires incorporation of a statistical modeling for classification of the analyzed source text, normalization rules and functional rules to map the surface form of the analyzed texts into the closest form of the lexical database to facilitate the transferring.

Briefly, in this research:

A lexical framework was built to facilitate the selections of the lexical items whether as individual or multiple words. It covers the morphological and phonological distinctions between the Egyptian dialect, (which is taken as a representative dialect of this stage), and MSA in order to produce an underlying form of standard

Arabic sentences that can be generated in a further stage. The lexical items were selected carefully to cover the recent variations of the dialect according to a previous study of the Egyptian corpus ARZ ATB by Marzouk and El KareH (2016).

A transfer system was developed, with the enhancement of the lexical framework, to normalizes social media texts that are composed of a mixture of Arabic dialects and MSA.

Rewrite rules were created to approach the similarity of the morphosyntactic features for both sides, the source dialect and the target MSA.

The objective of the work is to unify the various written forms used in social texts in one standard form that can be comprehended and translated to other languages. The main contribution of this research is that the source of texts are different and significant from the usual texts previously used to present the written forms. Therefore, the results revealed the requirement of handling the new semantic and morphological ambiguities that caused by these texts. The paper is organized as following: Sect. 2 shades lights on the previous studies on dialects transfer and Arabic Machine translation, Sect. 3 describes the main issues that signify the Arabic language and its dialect, Sect. 4 overviews the main machine translation paradigms, Sect. 5 elicits the main modules that are involved in constructing the proposed system, Sect. 6 presents the procedures of building lexical database and the process of collecting the data, Sect. 7 is an evaluation of the systems performance and its results followed by the conclusion and the planned future work.

2 Related Studies

Previous apropos studies on dialect machine translation were limited and most of them were restricted on the normalization of one Arabic dialects words into their equivalents in MSA as a preliminary stage for their translation into English language [3]. Abo Bakr et al. (2008) explained the techniques of transferring Egyptian Arabic dialect into MSA and diacritization of the transferred MSA text. First they used a corpus collected from different pages from WWW to create the Egyptian colloquial to MSA lexicon. Then they depended on Buckwalter Arabic Morphological Analyzer BAMA to segment and analyze the source text. Support Vector Machine SVM multi-Classifier was used for the tokenization and diacritization. Moreover They added Segment type position to indicate the proper order of the segment in the target word or sentence, and new segment type position to move the segment to its proper order. The system's accuracy of converting Egyptian Arabic text into MSA showed that 88% were correct, and the diaritization of the MSA output's accuracy showed that 70% [4]. The main limitation of the work was the unavailability of a TreeBank that represents the Egyptian Arabic dialect. The collected corpus for this system represents a specific genre of the Egyptian Arabic text in social media which is a mixture of EGY and MSA, because written texts in these pages are directed to general communities and educated people. Therefore it may lack some linguistics forms that signify the spoken dialect that transmitted to written texts. Moreover, Buckwalter morphological analyzer was originally designed in order to analyze MSA, therefore the analysis of dialectal data using Buckwalter analyzer in other studies was a cause of reduction in the output's accuracy.

A. Abdel Monem et al. (2008) investigated the usage of the interlingua machine translation approach for morphological and syntactic generation of Arabic texts. They followed rule based grammar generation approach to transform a semantic interlingual representation into Arabic texts. A. Abdel Monem et al. were the first who used the rule based approach from interlingua for morphological and syntactic generation of Arabic text [5]. For the evaluation they used English source sentences of approximately 1900 words and Arabic target sentences of 1600 words. The evaluation achieved a BLEU score with average 0.74, the results of the system performance was confirming the ability of the rule based approach to generate Arabic texts.

Sawaf (2010) developed a hybrid machine translation system to handle Arabic dialects by using a decoding algorithm that normalizes non-standard, spontaneous and dialectal Arabic into MSA. Sawaf's system goes through the following stages: Preprocessing and segmentation modules, Lexical Functional Grammar (LFG) system which incorporates a richly annotated lexicon containing functional and semantic information, functional models to use functional constrains to perform a deeper semantic and syntactic analysis for the source and target language, and Statistical translation models which use the maximum entropy framework [6]. The measured BLEU score of the system reached to 34.6%.

Salloum and Habash (2011) created ELISSA, the Dialectal Arabic DA to MSA Translation System. ELISSA is a ruled based model which relays on an existed morphological analyzer, DA-MSA dictionary and a model to rank and select the generated sentence. It follows certain steps to reach to its target: selection to identifying the word as a dialectal or out-of-vocabulary OOV, translation using classical rule based machine translation flow, morphological analysis ADAM, morphological transfer and morphological generation, and Language Modeling using SRILM for n-best decoding [7].

Other work that concentrated on the Arabic dialects translation, in specific Egyptian and Levantine Arabic was of Zbib et al. (2012), who developed a parallel corpus for the mentioned Arabic dialects using Crowdsourcing. Then they used the data in variant MT experiments. The parallel corpus which consists of 1.5M were classified according the dialects. The resulted dialects were attributed to 4 regions Levantines, Gulf area, Egypt, and Morocco, in addition to MSA. In the next step, The Levantine and Egyptian Arabic texts were translated by non professional translators using Amazon mechanical Turk. Zabib et al. performed a set of experiments to compare system trained using their parallel corpora with other systems trained on larger MSA-English parallel corpora. The experiments objectives were to investigate, first, the effect of the training data size, by examining different sizes of the training set, second, the cross dialect training, by using a training test of one dialect for a translate system of another, third, the validation of independent test data by using test set selected randomly from social media. Zabib et al. concluded that the system trained on the combined Dialectal-MSA data is likely to give the best performance, since informal Arabic data is usually a mixture of Dialectal Arabic and MSA. Also the mismatch between the training and the test data is the main reason beyond the lack of vocabulary coverage [3].

3 Arabic Language Variation

Arabic language is the forth widely spoken language [5]. More than 300 Millions people speak Arabic language [1]. MSA is descended from the Classical Arabic, the language of the Islamic holy book, "Quran". The syntax of MSA is unchanged from the classical Arabic but the changes affect its vocabulary and phraseology [8]. Nowadays, MSA is the written language of Arabic literature, journalism that stands side by side with the spoken regional vernaculars which are known as colloquial Arabic or Arabic dialects [9]. All native speakers learn their dialects as their mother tongue before they begin formal education [8]. These Arabic dialects are distributed along the Arab world from Morrocco in the west to Amman in the east [8]. Each country has its regional dialects but the mentioned dialects in these research are the capital cities dialects for their wide spreading comparing with the other regional dialects. The study of the spoken Arabic language has been dominated by the study of these dialects, but these studies were mostly confronted with negative attitude, as there is a worry that the study of a certain Arabic dialect may affect the supremacy of the study of the Standard Arabic [10]. Although, the Arabic dialects intervention and usage in a wide range of written texts couldn't be resisted for many reasons such as literature purposes in which some novels that talk about certain social and cultural level were preferred to be written using the slang language. As well as the spreading of electronic texts such as in SMS, chatting, and other communication media which became rich sources for the dialects in its written form. The Egyptian Arabic Dialect, and in specific the Cairene (the spoken colloquial Arabic of Egypts capital and the central Delta) is often considered to be the most widely understood dialect throughout the Arabic world [11]. This wide spread intelligibility is a result of the dominance of the Egyptian media in the Arabic world. Besides, unlike most other forms of colloquial Arabic, large resources of Egyptian Arabic can be found in written format in social Media. The difference between Egyptian Arabic dialect and MSA can be limited in certain phonological and morphological exchanges. For instance, Most of MSA nouns are preserved in EGY, but some other nouns have undergone some phonological changes such as:

- Monophthongization, e.g. /Sayf/ in MSA is turned into /Se:f/ in EGY.
- Final hamza deletion and final vowel shortening e.g. /sama:/ in MSA is turned into /sama/ in EGY.
- Atonic shortening e.g. /Sa:ru:x/in MSA is turned into /Saru:x/.
- Compensatory lengthening e.g. /ras/ in MSA is turned into /ra:s/.

Other critical difference between MSA and EGY is the case ending. Cases refer to what in English are called nominative, accusative, and genitive nouns. MSA distinguishes between the three cases by suffixing /u/ for nominative, /a/ for accusative, and /i/ for genitive [12]. However, case ending in EGY are deleted and they are understood by context, suffixes that are used to signify number is an additional concept that distinguishes EGY from MSA. In EGY, the masculine plural suffix /i:n/ is attached to masculine plural nouns, as well as, it is used for some feminine plural nouns beside the feminine suffix /a:t/ (Holes 2004, p. 166).

4 Machine Translation Paradigms

There are two main paradigms of MT: Rules-Based paradigms RBMT, and Statistical paradigms SMT. Hybrid machine translation is an approach to combine the achievements of both paradigms to reach to better results.

4.1 Rule Based MT

Rule based Machine translation uses the linguistic knowledge of the source language and the target language to accomplish the translation, rule based MT covers three main strategies: direct translation which translates word by word or linguistic patterns of the source language SL to others in the target language TL in a single step using bilingual dictionary, transfer system which based on contrastive knowledge to determine the differences between the two languages and it relies on creating rules to overcome these differences, transfer systems involve an analysis of the source text SL to an abstract structure the process that facilitates the transfer a corresponding abstract structure of the target text TL before generating it, Finally, interlingua which is divided into two phases: the analysis phase to encode the input text into interlingua and the generation phase to decode the interlingua into the output text [13, 14], interlingua systems require a transfer step as a part of the translation process [15, 16].

4.2 Statistical MT

It models the probability $P(F/E)$ of any source language F and target language E, the system chooses the translation that maximizes this probability. Initially it worked on the word level but later it is applied on larger chunks of the text [14, 17].

4.3 Hybrid MT

Since each paradigm has its strength and weakness, different approaches arose for combining these systems through hybridization such as:
Hybrid combination: to take one system and improve additional resource to enhance it, e.g. creating rules for a SMT, or vice versa.
Multi engine-Parallel combination: to translate using several independent systems.
Multi pass-Serial combination: to use the output of a system as an input of other system [14, 17].

Transfer systems are considered as a compromise between the ease use of the direct systems and the efficient uses of resources of interlingua systems [15]. The main advantage of the transfer systems is summarized in its ability to use a mediate language, in case of multilingual translation, into and out of which the translation is done [15].

5 (ALMoFseH) Arabic Dialects Machine Translation

The first English to Arabic Machine translation system was built in seventies by Weidner Communications Inc. it was composed of two main stages: the analysis system of the source language, and the generation system of the target language [18]. In this time and for decades after, there were no problem, since all the resourced were written in MSA [1]. After the appearance of the social media, texts which are written using a mixture of MSA and Arabic dialects were augmented, and understanding these texts became imperative. **(ALMoFseH)** project is an attempt to standardize the social media texts by identifying the different dialectal forms in these texts and transfer them into MSA. The project's goal is to develop a hybrid system, based on a multi pass serial combination, to translate the most used Arabic dialects in social media. Selecting these dialects is based on the annual report, that are released by Dubai school of governance and innovation program, to survey the Arabs usage of the social media. Building the system required serial processes, some are designed for the projects purpose and others relied on ready built applications. Hence, the system is planned to be composed of:

- Orthographic normalization to return the words that underwent changes owing to the Phonological Alternation rules into their origins.
- Dialect Identification by using an Automatic classifier to identify the dialectal forms in the text.
- Morphological analysis of the source text that written by the dialect using Egyptian Arabic Morphological analyzer.
- A Transfer system to select the lexical equivalents from dictionary lists. The system is to transfer from the Arabic dialects to MSA and from MSA to Arabic dialect, using machine learning classifier, rules and lexicon lists.
- A Morphological generator for the target dialect.

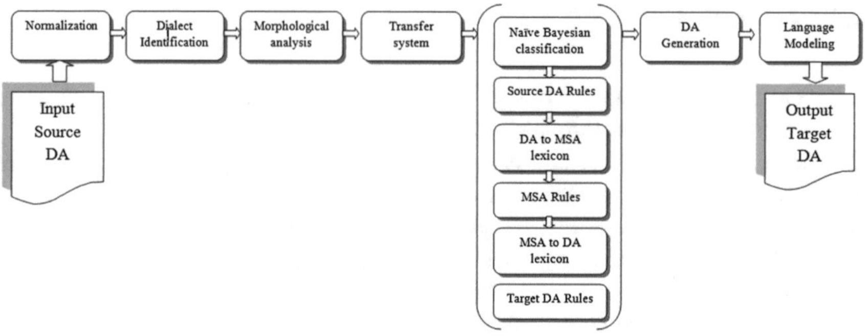

Fig. 1 The diagram shows the processing stages of developing Arabic dialects Machine translation using MSA as a pivot

- A Statistical language Model for the target dialect.

This research concerns with the procedures of creating A transfer system which is considered as the essential module in the Arabic Dialect Machine Translation system (Fig. 1).

6 Methodology

The work in this research is divided into two procedures: first, building a corpus based lexical database. Second, creating a transfer system for the selections from the Egyptian dialect to MSA.

6.1 Building a Lexical Database

A bi-dialectal lexicon is a crucial resource for building the application, since this lexicon is an essential resource to provide morpho-syntactic information such as POS, sub-categorization, tense, case, etc. [19]. Therefore, the objective of this stage is to develop a bi-dialectal lexical database extracted from social media texts such as Whatsapp and discussion forums. The process took different steps:

- Pre-processing of the corpus to clean the input from non-linguistics features
- Selection of most frequent words using a concordance
- Morphological annotating of the selected words
- Generating semantic information for the purpose of translation
- Developing a transfer module to be embedded in the proposed machine translation
- Evaluating the output

Table 1 Table 1 describes the components of the collected Corpus

Corpus componants	Percentage (%)
Non linguistic characters	11
Speech effects	6
Foreign words	18
Arabizi	23
Arabic words	42

A corpus of 250k words were gathered from the Whatsapp messages and the discussion forum. The data was composed of a mixture of Egyptian Arabic words and MSA word, Arabizi (Arabic words written by Latin letters), non-linguistics characters such as emoticons, and sound effects, as well as, the foreign words. The first step in preparing the data was to remove all those nonlinguistic characters. Sound effects were left in their original form since their spreading turned them into consistent standard linguistic forms that render certain meanings. Arabizi were transliterated by native speakers using their own writing style to maintain the realistic variations of the same word. Foreign words were the least among the other characters therefore they were removed. After filtering and manipulating the data the 178k words were listed according to their frequency using the word list in Antconc concordance. The most frequent 3k Egyptian Arabic words were sorted to develop our lexicon.

The developed lexicon is a stem-based lexicon which provides all acceptable stems for the individual word accompanied with their affixes. The lexicon is designed to accept other dialects to be added. Subsequently, the tool can be modified to be able to translate from dialect to other dialects using MSA. In this case, by extending the lexicon to establish a multidialectal lexicon, the input of any dialect D1 can be transferred to any other dialect Dn via MSA as an mediator. The lexicon is divided into 3 lists: the first list includes proclitics and prefixes, the second list includes enclitics and suffixes and the third list includes all possible stems for each word according to the adjacent affixes (Fig. 2).

The stem list is designed to provide all possible morphological and syntactic information for each word stem: category, tense, number, gender, voice, and tense. The list is divided into 3 main sub-lists:

```
( ['>a': 061], ,['prefcat':'1P']['num':'sing'], ['gen':'MASC']),
( ['ti': 062], ,['prefcat':'2P']['num':'sing'], ['gen':'MASC']),
 ( ['ti': 063], ,['prefcat':'2P']['num':'sing'], ['gen':'FEM']),
( ['ti': 064], ,['prefcat':'2P']['num':'PL'], ['gen':'MASC'])
( ['ti': 065], ,['prefcat':'2P']['num':'PL'], ['gen':'FEM']),
( ['yi': 066], ,['prefcat':'3P']['num':'sing'], ['gen':'MASC']),
 ( ['yi': 067], ,['prefcat':'3P']['num':'sing'], ['gen':'FEM']),
( ['yi': 068], ,['prefcat':'3P']['num':'PL'], ['gen':'MASC']),
 ( ['yi': 069], ,['prefcat':'3P']['num':'PL'], ['gen':'FEM']),
( ['yi': 070], ,['prefcat':'3P']['num':'sing'], ['gen':'FEM'])
```

Fig. 2 It shows a part of the prefix list

Non-inflected words list: it includes interrogative pronouns, relative pronouns, personal pronouns, demonstrative pronouns, prepositions, prepositional, adverbs, adverbials, pseudo verbs and non inflected verbs.
Inflected words list: this list contains EGY words that are descended from MSA but underwent orthographic deviations by altering a phoneme or more such as نَائِم and نَايِم, EGY words that have no origins in MSA such as تُورتَة whose equivalent in MSA is كَعكَة, OOV (Out Of Vocabulary) words, and borrowings that are taken from other language for specific purposes, and underwent the Arabic morphological inflection such as pluralization e.g. لَايكَات.
Broken plurals list: it stores the most frequent EGY Broken plurals with MSA broken plural equivalent e.g. فُلُوس and أَموَال, EGY sound masculine/feminine plurals with MSA broken plural equivalents e.g. سِتَّات and نِسَاء and EGY Broken plurals with MSA sound masculine/feminine plural equivalents e.g. أَزَايِز and زُجَاجَات.
3k Egyptian word types, and 3k MSA word types are the total number of the three lists that cover the following features:

- Orthographic variants: words that have several written forms such as بَرضُو ,بَردُو. All variants of each word type were inserted to the stem list.
- Words with no equivalent: some Egyptian words have no equivalent with the same meaning in MSA these word are called interjection and they are inserted to the sentence to express a reaction toward situation such as طَبّ ,بَقَا ,خَلَاص.
- Words with multi-word equivalent: some Egyptian words are translated to MSA using more than one word such as: مَعلِهش whose equivalent is لَا بَأس.
- New entries: lexical items that are now considered as a significant feature of the Egyptian dialect text such as the borrowings e.g. مِشيَّر, and لَايكَات, and other new words such as, فَاكس. Each lexical item in the lexicon is enriched with morpho-syntactic and semantic information and a numerical code (Fig. 3).

6.2 The Transfer System

The process of transferring is the stage that follows the morphological analysis of the source dialect whose output are tokens with their POS. Thus, the target in this stage is to use output that resemble the Egyptian Arabic Morphological analyser CALIMA output [20] to reach to the closest MSA analyzed format of the Buckwalter morphological analysis for the equivalent MSA text [21]. The accomplishment of this process required three main procedures:

```
(['fas~aH':383], ['POS':'PV'],['cat':'PV']),
(['fas~aH':384], ['POS':'PV'],['cat':'V'], ['tense':'past'], ['voice: 'act'], ['SUFF_OBJ':122]),
(['fas~aH':385], ['POS':'PV'],['cat':'V'], ['tense':'past'], ['voice: 'act'], ['SUFF_OBJ':123]),
(['fas~aH':386], ['POS':'PV'],['cat':'V'], ['tense':'past'], ['voice: 'act'], ['SUFF_OBJ':133]),
(['fas~aH':387], ['POS':'PV'],['cat':'V'], ['tense':'past'], ['voice: 'act'], ['SUFF_OBJ':131]),
(['fas~aH':388], ['POS':'PV'],['cat':'V'], ['tense':'past'], ['voice: 'act'], ['SUFF_OBJ':121]),
(['fas~aH':389], ['POS':'PV'],['cat':'V'], ['tense':'past'], ['voice: 'act'], ['SUFF_OBJ':120]),
(['fas~aH':390], ['POS':'PV'],['cat':'V'], ['tense':'past'], ['voice: 'act'], ['SUFF_OBJ':119]),
(['fas~aH':391], ['POS':'PV'],['cat':'V'], ['tense':'past'], ['voice: 'act'], ['SUFF_OBJ':128]),
(['fas~aH':392], ['POS':'PV'],['cat':'V'], ['tense':'past'], ['voice: 'act'], ['SUFF_SUB':111),['SUFF_OBJ':122]),
(['fas~aH':393], ['POS':'PV'],['cat':'V'], ['tense':'past'], ['voice: 'act'], ['SUFF_SUB':111),['SUFF_OBJ':123]),
(['fas~aH':394], ['POS':'PV'],['cat':'V'], ['tense':'past'], ['voice: 'act'], ['SUFF_SUB':111),['SUFF_OBJ':131]),
(['fas~aH':395], ['POS':'PV'],['cat':'V'], ['tense':'past'], ['voice: 'act'], ['SUFF_SUB':111),['SUFF_OBJ':121]),
(['fas~aH':396], ['POS':'PV'],['cat':'V'], ['tense':'past'], ['voice: 'act'], ['SUFF_SUB':111),['SUFF_OBJ':120]),
(['fas~aH':397], ['POS':'PV'],['cat':'V'], ['tense':'past'], ['voice: 'act'], ['SUFF_SUB':111),['SUFF_OBJ':119]),
(['fas~aH':398], ['POS':'PV'],['cat':'V'], ['tense':'past'], ['voice: 'act'], ['SUFF_SUB':111),['SUFF_OBJ':128]),
(['fas~aH':399], ['POS':'PV'],['cat':'V'], ['tense':'past'], ['voice: 'act'], ['SUFF_SUB':111),['SUFF_OBJ':133]),
```

Fig. 3 It shows a part of the EGY stem list of the word /fas aH/ فسّح in the lexicon database

1. A machine learning classifier to select the mentioned token according to the context from the available analyses of the same word.
2. Rewrite rules to normalize the output tokens to suit the lexicon entries to facilitate looking up the tokens from the lists.
3. Looking up the normalized tokens from the lexicon with its all possible equivalents.

6.3 Naive Bayesian Classifier (NB)

Enhancing the system with learning methods to disambiguate the output of the morphological analyzer was urgently required. This stage guarantees the efficiency of the system and the avoidance of further undefined output. For this process, the supervised learning algorithm Naive Bayesian was chosen, since Naive Bayesian is the simplest representative of probabilistic learning methods' [22]. The description of the process of using NB is not the major subject of this research, therefore these lines are a brief overview of the usage of this method to reach to the desired target. According to NB the context in which the ambiguous words appear, is represented by vector of feature variables F = (f1, f2, …, fn), and the sense of these words are represented in a classification variable S = (s1, s2, …, sn). The disambiguation occurs through estimating the maximized sense according to the conditional probability P(w = si/F). Features in NB algorithms are terms as words, collocations, or words assigned by their position in the context [23]. The selected features in the system were:

F1 = a set of individual words,
F2 = a set of part of speech tags,
F3 = a set of words collocations,
F4 = a set of collocations of part of speech tags.

To choose the right sense of the ambiguous word in the given context, the conditional probability of the feature fi and the conditional probability of the sense si were computed using the Maximum Likelihood Estimation as follows:

$$P(s_i) = C(s_i)/N$$
$$P(f_i/w = s_i) = C(f_i, s_i)/C(s_i)$$

where $C(f_i, s_i)$ is the number of occurrence of the feature fi with certain sense si in the training corpus, $C(s_i)$ is the occurrence of this sense si in the corpus training, and N is the total number of the training corpus.

6.4 Rewrite Rules for Dialectal Normalization

After the tokenization and the morphological Analysis of the source dialect, some of the source words appeared differently from their saved forms in the lexicon. Definitely, in this case one of the main pre-processing stages was to normalize these words to assimilate their modified forms in the lexicon. The occurring modifications of these certain words were for the purpose of exceeding the changes between the source words and their equivalents in MSA. We gathered the most common words that fall under these conditions in sub-lists to facilitate writing rules for their normalization. These rules were designed to map the surface form of the word into the closest form to the words stored in the lexicon. These rules were categorized into 3 sorts:
Deletion Rules: for words with affixes that have no equivalent in MSA. Deletion rules were written to delete these affixes during the transfer.

```
Ex: ap → 0 || N[LIST04]_ 0
[HAj+NOUN+ap+NSUFF_SG :$iy>+NOUN+NULL]
```

This rule is designed to delete the suffix /ap/ that are joined to list of words in the source dialect and has no equivalent in MSA, such as /HAjap/ حَاجَة whose equivalent in MSA is /$iy/ شَيْءٌ.
Alternation Rules: these rules are designed to alter certain morphemes with others that differ from their equivalents in the lexicon to match the referring meaning.

```
Ex: [N:1]→ [N:2]|| _[LIST05]
[EalaY+PREP: li+PREP]
```

The previous rule is written to alter the preposition /EalaY/ in the word /Eala/ عَلَشَان to the preposition li to avoid the wrong literal transferring.
Merging rules: these rules were written for individual cases when merging the analyzed morphemes is required to reach to the form existed in the lexicon.

```
Ex: [N:1]→[N:2]|| _ [LIST06]
[fiy+PREP : fiyh+PREP]
```

The word /fiyh/, فِيه is inserted to the lexicon without tokenization to match its equivalent in MSA /hunAka/, هُنَاكَ. Therefor the merging rule's role is to merge the two morphemes to match the lexicon entry. For instance, the Egyptian word /fiyh/ has two entries in the lexicon: the first entry renders the meaning (in it). And the second

```
#RULE 2 "fiyh":

    elif stem == 'fiy' and suf == 'uh':
        print prf+stem+suf
        print prf,'+',prfpos,'+',stem,'+',stpos,'+',suf,'+',sufpos
        prf = 'yu'
        prfpos = 'IV3MS'
        stpos = 'IV_PASS'
        stem = 'jad'
        suf = ''
        sufpos = 'NULL'
        print prf,'+',prfpos,'+',stem,'+',stpos,'+',suf,'+',sufpos
    else:
        for prow in eprefix:
            for enrow in eprefix:
                for prorow in esuffix:
                    for srow in esuffix:
                        for strow in egy_stem:
                            if enrow[x] == enc and enrow[y] == encpos:
                                if prow[x] == prf and   prow[y] == prfpos:
                                    if strow[x] == stem and strow[y] == stpos:
                                        if srow[x] == suf and srow[y] == sufpos:
                                            if prorow[x] == proc and prorow[y] == procpos:

                                                print prf+stem+suf
                                                print prf,'+',prfpos,'+',stem,'+',stpos,'+',suf,'+',sufpos
                                                print enrow[z],'+',prow[z],':',prfpos,'+',strow[z],':',stpos,'+',srow[z],':',sufpos,'+',prorow[z]
                                else:
                                    print prf+stem+suf
                                    print prf,'+',prfpos,'+',stem,'+',stpos,'+',suf,'+',sufpos
```

Fig. 4 It shows one of the merging rules for the word (fiyh)

entry renders the meaning (there is). The rule is designed to cover the second entry. *Splitting Rules*: these rules were created for two purposes: first, to split the merged words in the source dialect, second, to split the words in MSA that stored in the lexicon with their affixes that have no equivalent in the source language.

```
Ex: [N:1] → [N:2]||N in (list06)

(bisml~Ah+NOUN : bi+PREP+Aism+NOUN+All~h+NOUN)
```

This rule is to split the two merged words and normalize them before the process of transferring. The most common merged words in the social media texts were gathered in a database list (list06) with their normalized form and their morphosyntactic information, as shown in the previous rule (Fig. 4).

Lexicon Look Up

Looking up in the systems lexicon follows certain restrictions to cover all the distinctions that distinguish the Egyptian dialect from MSA. As mentioned above, each morpheme in the source dialect lists have its own code number which matches another code number for the equivalent morpheme in the target dialect lists. Looking for the matched morphemes in both dialects were achieved in this stage by using the code numbers. Words in the source language that consists of one morpheme and whose equivalent consists of more than one morpheme has specific codes to facilitate the matching.

```
ending rules

for prow in eprefix:
    for enrow in eprefix:
        for prorow in esuffix:
            for srow in esuffix:
                for strow in egy_stem:
                    if enrow[x] == enc and enrow[y] == encpos:
                        if prow[x] == prf and   prow[y] == prfpos:
                            if strow[x] == stem and strow[y] == stpos:
                                if srow[x] == suf and srow[y] == sufpos:
                                    if prorow[x] == proc and prorow[y] == procpos:
                                        if stpos == 'FV':
                                            case == 'a'
                                            casepos == 'CASE_ACC'
                                            print enrow[z],'+',prow[z],':',prfpos,'+',strow[z],':',stpos,'+',case,':',casepos,'+',srow[z],':',sufpos,'+',prorow[z]
                                            elif stpos == 'IV':
                                                print enrow[z],'+',prow[z],':',prfpos,'+',strow[z],':',stpos,'+',case,':',casepos,'+',srow[z],':',sufpos,'+',prorow[z
                                                elif eprefix == 'NULL':
                                                    if stpos == NOUN:
                                                        if esuffix == 'NULL':
                                                            case == 'AF'
                                                            casepos == 'CASE_ACC'
                                                            print enrow[z],'+',prow[z],':',prfpos,'+',strow[z],':',stpos,'+',case,':',casepos,'+',srow[z],':',sufpos,
```

Fig. 5 Shows a part of the code of the case ending rule

6.5 *Functional Model*

One of the critical distinctions between the EGY and MSA is the case ending. Usually words in EGY dont include the case ending however, ending is understood by context. Therefore, output of the morphological analysis must be manipulated after transferring into MSA by adding the appropriate case ending. 38 rules were created to interpolate the main case endings that occurs persistently with verbs and nouns according to their tense, positions and definiteness. Part of compiling these rules using Python programming language is introduced here (Fig. 5).

7 Evaluation of the System

Evaluations of machine translations and their modules are needed to measure the performance of the systems by revealing haw far the output is accurate, predictable to the real human language. According to the general error metrics, the distance d(t, r) between the produced translation t and the predefined reference translation r is calculated and computed automatically.

An evaluation was conducted to measure the performance of the transfer system. The goal of the evaluation is: first, to measure the lexicon coverage of the Egyptian Arabic words through calculating the word error rate (WER). Second, to measure the accuracy of the output by comparing it to the output of the Buckwalter Morphological analyzer.

The first measurement was a measurement of the lexical database proficiency in the word level, each list (prefix, stem, suffix) in the built lexicon was measured separately. For that purpose, a blind test of 90 texts with total number of words 3000 words were collected from social media forums and SMSs, the data were collected

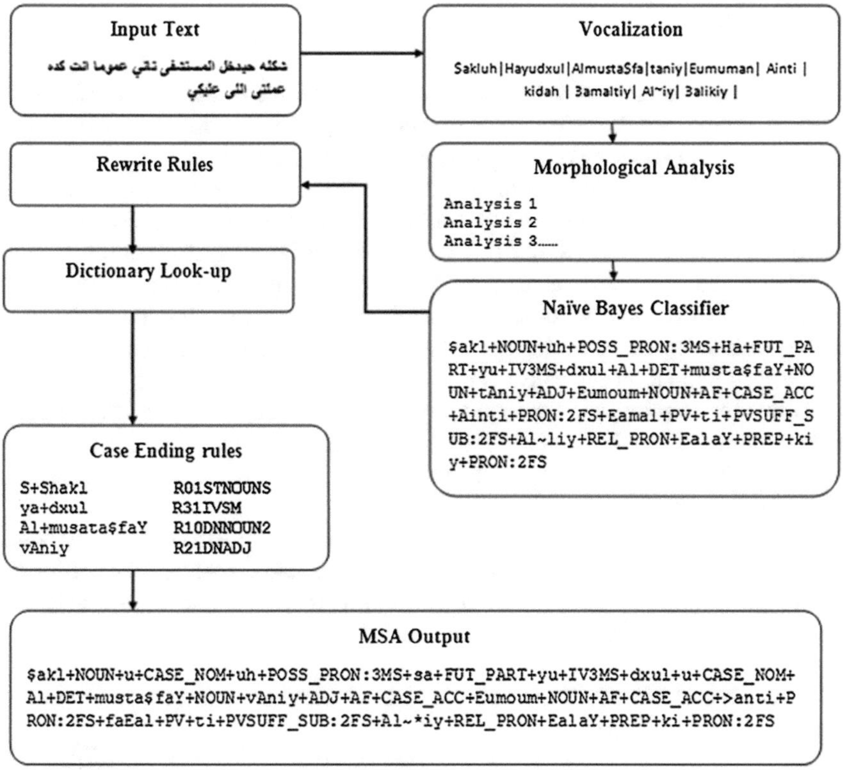

Fig. 6 Shows the process of transferring one of input sentences that are used for the evaluation

from different messages than those used in building the system's lexical database. Then it was analyzed by assistance of the morphological analyzer CALIMA to get the same format of its output. The output of the analysis was entered to the system manually. In this measurement, the Naive Bayes classification were excluded, since the aim of the measurement was to find out the lexicon ability to covers all the possible analyzed forms of the word. Thus, the error rate of each morpheme were calculated according to the following criteria: its existence in the lexicon, its existence in the lexicon with the same meaning according to the context, and the correctness of the equivalent morpheme.

The second measurement was designed to estimate the system's ability to matches the humans translation, and to measure the applicability of the system to provide an output that can act as a source input for a further transfer to other target dialect. Hence, the collected test data were manually translated into MSA by native Egyptian Arabic speakers. Then the MSA texts were morphologically analyzed using the open source of SAMA the last version of Buckwalter morphological analyzer. The output was sorted as database and compared with the systems output to measure the following values: Recall, Precision, F-score (Fig. 6).

8 Results

The results of the first evaluation shows a high coverage of the lexical features of Egyptian Arabic and the efficiency of the rules to facilitate finding the correct equivalent words. The accuracy of the system reached to 92.7%.

Due to the limited time and the number of the researchers who worked in this research, the size of the lexicon wasn't sufficient enough to cover a major number of the Egyptian words. That was the main reason behind the error rate in the evaluation.

Table 2, shows the word error rate (WER): first column presents the percentage of the morphemes that are transferred correctly due to the existence of the morpheme with its correct meaning according to the text, second column presents the morphemes that are not transferred correctly due to the existence of the word but with different meaning, and third column presents the morphemes that are not transferred from EGY to MSA and kept in its source form, due to the inexistence of the morpheme in the lexical database. The average of the correct rate is calculated for all the morphemes (Fig. 7).

The second measurement show that the output of the system could predict most of the analys feature and give an Approximate acceptable analyzed format for the target dialect MSA. Table 3, shows the recall, precision, and f score of the transfer system as a result of the second measurement.

Table 2 Shows the error rate of the lexical database

Feature	Correct (%)	Incorrect (%)	Untransfered (%)
Stem	90.8	2.1	7
Prefixes/Enclitics	95	2.4	2.5
Suffixes/Priclitics	93	7	–

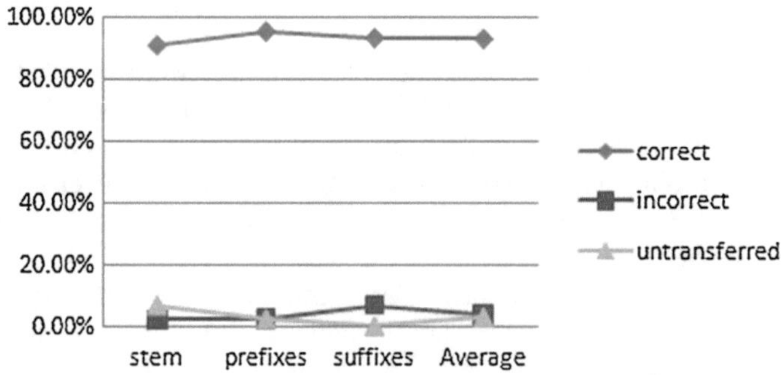

Fig. 7 Shows the error rate of the lexical database

Table 3 Shows the recall, precision, f-score

Feature	Recall (%)	Precision (%)	F-score (%)
Stem	88.1	82	84.8
Prefixes/Enclitics	98	96.8	98
Suffixes/Priclitics	91	89.1	90
Average	90	89	91

9 Conclusion

Arabic dialects Machine translation project (AlMoFseH) is a sophisticated project which demands various sequential processes to be drawn together. Each process should be manipulated separately to reach to satisfying results. This research exposed the paradigm that is used to accomplish one of these processes, and the problems that are confronted during building the transfer system and the procedures to handle them. The results of the primary work shows a high accurate performance of the transfer system. These results are encouraging to expand the work by increasing the database and the required rules before moving to the next stage of the project. For the future work, we would like to investigate the system's capability to transfer the MSA to other dialects by enlarging the lexicon to accept other dialects than the Egyptian dialect to reach to an approximate final phase of the project **(ALMoFseH)**. The project is planned to cover the Egyptian Arabic, Levantine Arabic, and Hijazi dialect.

References

1. Dasigi, P., Diab, M.T.: CODACT: Towards identifying orthographic variants in dialectal Arabic. In: IJCNLP, pp. 318–326 (2011)
2. Ibrahim, Z.: Beyond Lexical Variation in Modern Standard Arabic: Egypt. Cambridge Scholars Publishing, Lebanon and Morocco (2009)
3. Zbib, R., Malchiodi, E., Devlin, J., Stallard, D., Matsoukas, S., Schwartz, R., Makhoul, J., Zaidan, O.F., Callison-Burch, C.: Machine translation of Arabic dialects. In: Proceedings of the 2012 Conference of the North American Chapter of the Association for Computational Linguistics: Human Language Technologies, pp. 49–59. Association for Computational Linguistics (2012)
4. Bakr, H.A., Shaalan, K., Ziedan, I.: A hybrid approach for converting written Egyptian colloquial dialect into diacritized Arabic. In: The 6th International Conference on Informatics and Systems, infos2008. Cairo University (2008)
5. Monem, A.A., Shaalan, K., Rafea, A., Baraka, H.: Generating Arabic text in multilingual speech-to-speech machine translation framework. Mach. Transl. **22**(4), 205–258 (2008)
6. Sawaf, H.: Arabic dialect handling in hybrid machine translation. In: Proceedings of the conference of the Association for Machine Translation in the Americas (AMTA), Denver, Colorado (2010)
7. Salloum, W., Habash, N.: Elissa: a dialectal to standard Arabic machine translation system. In: COLING (demos), pp. 385–392 (2012)

8. Holes, C.: Modern Arabic: Structures, Functions, and Varieties. Georgetown University Press (2004)
9. El-Hassan, S.A.: Educated spoken Arabic in Egypt and the Levant: a critical review of diglossia and related concepts. Archivum Linguisticum Leeds **8**(2), 112–132 (1977)
10. Bani-Khaled, T.A.A.: Standard Arabic and Diglossia: a problem for language education in the Arab World (2014)
11. Hassig, H.: Deriving Cairene Arabic from modern standard Arabic: a framework for using modern standard Arabic text to synthesize Cairene Arabic speech from phonetic transcription (2011)
12. Gadalla, H.A.: Comparative Morphology of Standard and Egyptian Arabic, vol. 5. Lincom Europa Munich (2000)
13. Okpor, M.: Machine translation approaches: issues and challenges. Int. J. Comput. Sci. Issues (IJCSI) **11**(5), 159 (2014)
14. Shilon, R.: Transfer-based machine translation between morphologically-rich and resource-poor languages: the case of Hebrew and Arabic. Ph.D. thesis, Citeseer (2011)
15. Trujillo, A.: Translation Engines: Techniques for Machine Translation. Springer Science & Business Media (2012)
16. Hutchins, W.J., Somers, H.L.: An Introduction to Machine Translation, vol. 362. Academic Press London (1992)
17. Artetxe Zurutuza, M.: Distributional semantics and machine learning for statistical machine translation (2016)
18. Farghaly, A.: Arabic machine translation: a developmental perspective. Int. J. Inf. Commun. Technol. **3**, 3–10 (2010)
19. Tze, L.L.: Multilingual lexicons for machine translation. In: Proceedings of the 11th International Conference on Information Integration and Web-based Applications and Services, pp. 734–738. ACM (2009)
20. Habash, N., Eskander, R., Hawwari, A.: A morphological analyzer for Egyptian Arabic. In: Proceedings of the Twelfth Meeting of the Special Interest Group on Computational Morphology and Phonology, pp. 1–9. Association for Computational Linguistics (2012)
21. Habash, N., Diab, M.T., Rambow, O.: Conventional orthography for dialectal Arabic. In: LREC, pp. 711–718 (2012)
22. Escudero, G., Màrquez, L., Rigau, G.: Machine Learning Techniques for Word Sense Disambiguation (2003)
23. Le, C.A., Shimazu, A.: High WSD accuracy using Naive Bayesian classifier with rich features. Proc. PACLIC **18**, 105–113 (2004)

The Key Challenges for Arabic Machine Translation

Manar Alkhatib and Khaled Shaalan

Abstract Translating the Arabic Language into other languages engenders multiple linguistic problems, as no two languages can match, either in the meaning given to the conforming symbols or in the ways in which such symbols are arranged in phrases and sentences. Lexical, syntactic and semantic problems arise when translating the meaning of Arabic words into English. Machine translation (MT) into morphologically rich languages (MRL) poses many challenges, from handling a complex and rich vocabulary, to designing adequate MT metrics that take morphology into consideration. We present and highlight the key challenges for Arabic language translation into English.

1 Introduction

Natural languages (NLs) are integral to our lives as means by which people communicate and document information. The power of NLs is a reality that should not be taken for granted. To learn more about and to take further advantage of such power, researchers instituted the intense science of computational linguistics. Such science mimics human processing and analysis by means of machines for the purpose of more word-power discoveries from NLs. This led to a still brand-new science called here "Natural Languages Mining" (Abuelyaman 2014).

Machine translation has many challenges, and can be divided into linguistic and cultural categories. Linguistic problems include lexicon, syntax, morphology, text differences, rhetorical differences, and pragmatic factors.

M. Alkhatib (✉) · K. Shaalan
The British University in Dubai, Dubai, UAE
e-mail: Manaralkhatib09@gmail.com

K. Shaalan
e-mail: Khaled.shaalan@buid.ac.ae

K. Shaalan
School of Informatics, Edinburgh, UK

K. Shaalan et al. (eds.), *Intelligent Natural Language Processing: Trends and Applications*, Studies in Computational Intelligence 740,
https://doi.org/10.1007/978-3-319-67056-0_8

Cultural challenges arise for the Arab translator who may find certain phrases in Arabic have no equivalents in English. For example, the term تيمم tayammum, meaning "the Islamic act of dry ablution using a purified sand or dust, which may be performed in place of ritual washing if no clean water is readily available", doesn't have a synonym concept in English.

Arabic has a complex morphology compared to English. Preprocessing the Arabic source by morphological segmentation has been shown to improve the performance of Arabic Machine Translation (Lee 2004; Sadat 2006; Habash 2010) by reducing the size of the source vocabulary and improving the quality of word alignments. The morphological analyzers that cause most segmentors were developed for Modern Standard Arabic (MSA), but the different dialects of Arabic share many of the morphological affixes of MSA, and so it is not unreasonable to expect MSA segmentation to also improve Dialect Arabic to English MT (Zbib et al. 2012).

Quran is a Holy book that teaching Islam, in which, it contains the main principles of Islam and how these principles should be conducted are written. The availability of digitalized translated Quran making the work of finding written knowledge in Quran becomes less complicated, and faster, especially for non-Arabic language familiar or speaker. Machine translations for Quran are available in Internet such as the websites of Islamicity.com and Tafsir.com, and there are more than 100 websites giving access to machine translation for Quran.

Much work has been done on Modern Standard Arabic (MSA) natural language processing (NLP) and machine translation (MT). MSA offers a wealth of resources in terms of morphological analyzers, disambiguation systems, annotated data, and parallel corpora. In contrast, research on dialectal Arabic (DA), the unstandardized spoken varieties of Arabic, is still lacking in NLP in general and in MT in particular (Alkhatib 2016).

The current work on natural language processing of Dialectal Arabic text is somewhat limited, especially machine translation. Earlier studies on Dialectal Arabic MT have focused on normalizing dialectal input words into MSA equivalents before translating to English, and they deal with inputs that contain a limited fraction of dialectal words. (Sawaf 2010) presented a new MT system that is adjusted to handle dialect, spontaneous and noisy text from broadcast transmissions and internet web content. The Author described a novel approach on how to deal with Arabic dialectal data by normalizing the input text to a common form, and then processing that normalized format. He successfully processed normalized source into English using a hybrid MT. By processing the training and the test corpora, his method was able to improve the translation quality.

The Word Sense Disambiguation (WSD) concept is an integral and complex part of natural language processing. The complexity has to be resolved by methods other than human clarification. In the Quran, the verses are written in a particular style, posing a challenge for humanity to dispel any confusion and grasp the intended meaning, as some words and phrases are ambiguous because the component words convey various senses or are polysemous. Problems arise in word sense disambiguation in relation to words that do not have a well-defined meaning and when the sense requires interpretation (Mussa and Tiun 2015).

2 Challenges for Arabic Translation

2.1 Classical Arabic

The Holy Quran text has remained identical and unchanged, since its revelation, over the past 1400 years. The millions of copies of the Quran circulating in the world today match completely, to the level of a single letter. God says in the Holy Quran that he will guard the Quran book: "Surely it is we who have revealed the Exposition, and surely it is we who are its guardians". Translating the Quran has always been problematic and difficult. Many argue that the holy Quran text cannot be mimicked in another language or form. Furthermore, the Quran's words have shades of meanings depending on the context, making an accurate translation even more difficult. Translating the holy Quran requires more wordiness to get the meaning across, which diminishes the beautiful simplicity of the Quranic message.

The various differences between Arabic and English cause many syntactic problems when translating the Holy Quran into English. Verb tense is an obvious syntactic problem that translators usually encounter in translating the Holy Quran. Verb tense means the 'grammatical realization of location in time' and how location in time can be expressed in language (Sadiq 2010). In translating the Holy Quran, the verb tense form should be guided by the overall context as well as by stylistic considerations. In the Holy Quran, there is a transformation from the past tense verb to the imperfect tense verb to achieve an effect, which can pose some problems and challenges in translation. For example

إِذْ جَاءُوكُم مِّن فَوْقِكُمْ وَمِنْ أَسْفَلَ مِنكُمْ وَإِذْ زَاغَتِ الْأَبْصَارُ وَبَلَغَتِ الْقُلُوبُ الْحَنَاجِرَ وَتَظُنُّونَ بِاللَّهِ الظُّنُونَا

(Behold! they came on you from above you and from below you, and behold, the eyes became dim and the hearts gaped up to the throats, and ye imagined various (vain) thoughts about Allah! (Yusuf Ali's Translation 2000) [Surat Al-Aḥzāb 33, verse 10].

The verbs جَاءُوكُم (Ja'ukum, comes against you'), زاغت (zaghat, grew wild) and وبلغت' (wabalaghat', reached) are in the past tense, but the verb وتظنون (think) moves to the present tense. This move is for the purpose of conjuring an important action in the mind as if it were happening in the present. Tenses, in Classical Arabic or in the Holy Quran, cannot be transferred literally. In some cases, they need to move to convey the intended meaning to the target audience (Ali et al. 2012).

The Holy Quran has been interpreted and translated into many languages, including African, Asian, and European languages. The first translation of the Holy Quran was for Surat Al-Fatiha into Persian during the seventh century, by Salman the Persian. Another translation of the Holy Quran was completed in 884 in Alwar (Sindh, India, now Pakistan) under the orders of Abdullah bin Umar bin Abdul Aziz.

2.2 Modern Standard Arabic

A word in Arabic is comprised of morpheme, clitics and affixation, as in the example in Table 1 "وبجلوسهم" (wabajulusihim, and by their sitting). Since there is hardly any difference between complex and compound words in Arabic, this paper uses compound words for both. Cells in the first column are the headers of their respective rows. The first row shows the example of a compound Arabic word. The second breaks down the compound word into its four morphemes.

The third and fourth rows are the transliteration and translation of each morpheme, respectively. For the translation to be tangible, it must be rearranged (permuted), as shown by the arrows in Fig. 1, into the phrase: "and by their sitting." و بـــ جلـــوسهـــم The arrows show the necessary permutation that produces a palpable phrase.

Arabic has different morphological and syntactic perspectives than other languages, which creates a real challenge for Arabic language researchers who wish to take advantage of current language processing technologies, especially to and from English. Moreover, Arabic verbs are indicated explicitly for multiple forms, representing the voice, the time of the action, and the person. These are also deployed with mood (indicative, imperative and interrogative). For nominal forms (nouns, adjectives, proper names), Arabic indicates case (accusative, genitive and nominative), number, gender and definiteness features. Arabic writing is also known for being underspecified for short vowels. When the genus is spiritual or educational, the Arabic text should be fully specified to avoid ambiguity.

From the syntactic standpoint, Arabic is considered as a pro-drop language where the subject of a verb can be implicitly determined in its morphology; the subject is embedded in the verb, unlike in English. For example, the sentence: *She went to the park* can be expressed in Arabic as "ذهبت الى الحديقة" (Dhahabt 'iilaa Alhadiqa, She went to the Park). The subject *She* and the verb *went* are represented in Arabic by the single verb-form "ذهبت" (Thahabat, went) That is, the translated phrase is *She went* to *the park*, with the last part translated as "الحديقة" (Alhadiqa, The Park).

Arabic demonstrates a larger freedom in the order of words within a sentence. It allows permutation of the standard order of components of a sentence—the Subject Verb-Object (SVO), and Verb Subject Object. As an example, the sentence "الطفل أكل الطعام" (Alttifl 'Akl Alttaeam, The child ate the food) can be translated,

Table 1 Compound word

Word	و بـــ جلـــوسهـــم			
Compound	هـــم	جلـــوس	بـــ	و
Transliteration	Himm	Juloos	Bi	Wa
Translation	Their	sitting	By	And

Fig. 1 Translation

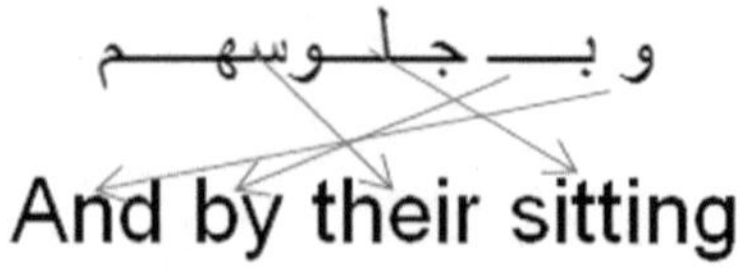

word-by-word, to the English SVO phrase "the child ate the food". The latter may be permuted to the standard Arabic order of a sentence—the VSO form "أكل الطفل الطعام" "('Akl Alttifl Alttaeam, ate the child the food). Both forms preserve the objective of the sentence. Unfortunately, the word by word English translation of the same VSO form is "Ate the child the food." Ironically, most of the online translation programs produce meaningless word by word translations along the lines this one.

2.3 Dialect Arabic

Dialect is the regional, temporal or social variety of a language, distinguished by pronunciation, grammar, or vocabulary; especially a variety of speech differing from the modern standard language or classical language. A dialect is thus related to the culture of its speakers, which varies within a specific community or group of people.

Arabic Dialect poses many challenges for machine translation, especially with the lack of data resources. Since Arabic dialects are much less common in written form than in spoken form, the first challenge is to basically find instances of written Arabic dialects. The regional dialects have been classified into five main groups; Egyptian, Levantine, Gulf, Iraqi and Maghrebin.

3 Machine Translation in Natural Language Processing

3.1 Metaphor Translation

Metaphor is an expression used in everyday life communication to compare between two dissimilar things. It signifies a situation in which the unfamiliar is expressed in terms of the familiar. It is a central concept in literary studies.

Images tend to be universal in languages, as they are basically used to enhance understanding in interaction. Images, especially in speech, economize on time and effort in passing a message to its recipient. Metaphoric expressions are represented by metaphor, simile, and idioms in different languages and contexts.

Metaphor is the key figure of rhetoric, and usually implies a reference to figurative language in general. Therefore, it has always been attended to carefully by linguists, critics and writers. Traditionally, being originally a major aesthetic and rhetorical formulation, it has been analyzed and approached in terms of its constituent components (i.e. image, object, sense, etc.) and types (such as cliché, dead, anthropomorphic, recent, extended, compound, etc. metaphors). However, recently, and in the light of the latest developments of cognitive stylistics, metaphor has received even more attention from a completely different perspective, that of conceptualization and ideologization. Consequently, this change of perspective has its immediate effect on translation theory and practice, which now has to be approached differently with respect to translating metaphor. This paper is an attempt to consider the translation of metaphor from a cognitive stylistic perspective, viewing it primarily as a matter of the conceptualization of topics, objects and people (Alkhatib 2016).

Metaphor is an expression used in everyday life in languages to compare between two dissimilar things. It signifies a situation in which the unfamiliar is expressed in terms of the familiar. In addition, it is a central concept in literary studies. A metaphor is sometimes confused with a simile, especially for translators who may translate metaphor into simile or vice versa. However, it is not too difficult to decide the case of simile because of the correlative existence of simile markers like "as, similar to and like" which are not found in the metaphor (Ahmad Abdel 2014).

Simile refers to something or someone sharing a feature of something or someone else in which a significant commonality is established through one of the simile particles or through the relevant context. The rhetorical analysis of a simile requires the investigation of the two simile ends (طرفي التشبيه). These are the likened-to (المشبه) and the likened (المشبه به) entities. Simile has four components and is divided into four categories. In any simile construction, the likened should be of a higher status, as the characteristic feature is greater than that found in the likened-to. For instance, when we say كلمات كالعسل (words like honey) or وجه كالقمر (her face like the moon), we are comparing (كلمات—Kalemat, words) to (عسل—Asal, honey) in terms of sweetness and (وجه—Wajh, face) to (قمر, Qamar moon) in terms of beauty and brightness. Thus, rhetorically, the likened-to elements are represented by كلمات and وجه and the likened elements are عسل (Asal, honey) and قمر (Qamar, moon). However, the sweetness of honey and the brightness and beauty of the moon cannot be matched and are stronger than the features of the other entities.

Abdul-Raof (2006) stated that simile is realized through the following four components:

a. The likened-to (المشبه): The entity, i.e. a person or thing that is likened to another entity, which is the likened;
b. The likened (المشتبه به): The original entity to which another entity, i.e. the likened-to, is attached;
c. The simile feature: A feature that is common to both the simile ends; and
d. The simile element: The simile particles.

For example: أحمد كالأسد Ahmad Kalasad, Ahmad is like a lion, where:

- The likened to is represented by the noun) أحمد Ahmad);
- The likened is represented by the noun الأسد (Alasad, the lion);
- The simile element is represented by the particle ك (Ka, like); and
- The simile feature is represented by the implicit notion الشجاعة (AlShaja'ah, courage), which is a semantic link that is common to and shared by both nouns الأسد and أحمد.

In Arabic rhetoric, metaphor is referred to as "الاستعارة", which is a form of linguistic allegory and is regarded as the peak of figurative skills in spoken or written discourse. Metaphor is the master figure of speech and is a compressed analogy. Through metaphor, the communicator can turn the cognitive or abstract into a concrete phrase that can be felt, seen, or smelt. Linguistically, الاستعارة is derived from the verb اعار (A'ar, to borrow), i.e. borrowing features from someone or something and applying them to someone or something else.

Rhetorically, however, metaphor is an effective simile whose one end of the two ends, i.e. the likened-to (المشتبه) and the likened (المشتبه به), has been deleted. Metaphor represents a highly elevated effective status in Arabic rhetoric that cannot be attained by effective simile. In metaphor, the relationship between the intrinsic and non-intrinsic signification is established on the similarity between the two significations, i.e. there is a semantic link between the two meanings.

The metaphorical meaning, however, is discernible to the addressee through the lexical clue القرينة available in the speech act. In Arabic, metaphor consists of three major components. As there are different kinds of metaphor, these three components may not all be available in a single metaphor. Abdul-Raof (2006) stated that the three metaphor components are:

1. The borrowed-from: equivalent to the likened element in simile;
2. The lent-to: equivalent to the likened-to in simile; and
3. The borrowed: the borrowed lexical item taken from the borrowed-from and given to the lent-to

For example:

a. زيد أسد (Zaid Asad, Zaid is a lion). (effective simile)
b. رأيت أسدا في المدرسة (Ra'ayt 'asadaan fi Almadrasa, I saw a lion at school). (lion refers to a brave man)

- The lent-to is represented by the noun زيد (Zaid);
- The borrowed-from is represented by the noun أسد (Asad, lion); and
- The semantic feature الشجاعة (Alshaja'a, courage) is shared by and establishes the link between زيد (Zaid) and أسد (Asad) (is the borrowed).
  In example (b), في المدرسة (Fi Almadrasa, at school) is the lexical clue to represent the metaphorical meaning of أسد lion" in this sentence, where lion refers to a brave man.

Although metaphor makes the text more beautiful and charming in the source language text (SLT) through its use of stereotyped words and new images, it can confuse the reader in the target language text (TLT) due to the linguistic and cultural differences between the two languages.

Kuiper and Allan (year) provide a definition about metaphor, as "an easy way to look at metaphor is to see the breaking down of the normal literal selection restrictions that the semantic components of words have in a sentence". When for example, we talk about "نافذة المستقبل", (Nafethat Almustaqbal, a window on the future), we have to ignore some of the semantic components of the word window;

for example, that it is a concrete object, and just take the fact that windows are things that allow us to look outwards from an enclosed space. The metaphor could also be seen out of a window. The metaphor lies in the suppression of some of each word's semantic features.

Metaphor can function as a means of formatting language in order to describe a certain concept, action or object to make it more comprehensive and accurate.

Hashemi (2002) classifies metaphors, i.e. isti'ara (الاستعارة), into three groups:

1. Declarative metaphors (تصريحية, Tasrihiyya): in which only the vehicle is mentioned and the tenor is deleted. In this type of isti'ara, the vehicle is explicitly stated and used to make a comparison between two different concepts that share a feature or a property in order to reveal the senses. A Declarative Metaphor is also considered as a decorative addition to ordinary plain speech. It is also used to achieve aesthetic effects (ibid). For example, in Arabic one might say (وردة, zahra, a rose) "رأيت وردة" *I saw a rose* instead of saying (a beautiful woman) امرأة جميلة, which is the vehicle in a metaphor based on the similarity between a rose and the person in terms of beauty.
2. Cognitive Metaphor (مكنية, Makniya): in which only the tenor is mentioned and the vehicle is deleted. In this type of isti'ara, the vehicle is only implied by mentioning a verb or a noun that always accompanies it. A Cognitive Metaphor is used as a means of formatting language in order to describe a certain concept, action or object to make it more comprehensive and accurate. In this case, it focuses on the denotation rather than the connotation of the metaphor that addresses the receptor in order to highlight its cognitive function.
3. Assimilative Metaphor (تمثيلي , Tamthele): which uses one of the characteristics of a vehicle for tenor. For instance "إذا رأيتَ نُيوبَ اللّيثِ بارِزَةً فَلا تَظُنَّنَّ أَنَّ اللّيثَ يبتَسِمُ" when you see a lion baring his canines, ‹never think he is smiling.

Newmark (1988:105–113) provides another classification of metaphor, divided into six types: dead, cliché, stock, adapted, recent and original.

a. **Dead metaphors**

Dead metaphors are "metaphors where one is hardly conscious of the image, which relate to universal terms of space and time, the main parts of the body, general ecological features and the main human activities." Here the sense of transferred imageno longer exists. Through overuse, the metaphor has lost its figurative value. For example "خلص الوقت" (run out of time).

English words that represent dead metaphors include: "space, field, line, top, bottom, foot, mouth, arm, circle, drop, fall, and rise are particularly used graphically for the language of science to clarify or define.", some other examples are, *I didn't catch his name*, *foot filed*, *top*...etc., and an example in Arabic "عقارب الساعة" which means (hands of the clock). Dead metaphors are not difficult to translate literally; even though they could lose their figurative meaning through extensive popular use. Another example is حقل المعرفة البصرية (field of human knowledge).

b **Cliché metaphors**

Cliché Metaphors are "metaphors that have perhaps temporarily outlived their usefulness, that are used as a substitute for clear thought, often emotively, but without corresponding to the facts of the matter." One example in English would be *at the end of the day*, and an example in Arabic is في نهاية المطاف (Fi nehayat almataf).

c. **Stock or standard metaphors**

Newmark describes this kind of metaphor as "An established metaphor, in an informal context, is an efficient and concise method of covering a physical and/or mental situation both referentially and pragmatically". It has certain emotional warmth, which does not lose its brightness by overuse. These are sometimes difficult to translate since their apparent equivalents may be out of date or now used by a different social class or age group. According to Newmark, a stock metaphor that does not come naturally to you should not be used, which means, if these metaphors are unnatural or senseless in the target language, they should not be used.

d. **Recent metaphors**

Recent metaphors, where an anonymous metaphorical neologism has become something generally used in the source language. It may be a metaphor designating one of a number of 'prototypical' qualities that constantly 'renew' themselves in language. For example: تصفية الخصوم السياسية, (Tasfiyat Alkhosoom Alseyaseyah, head hunting).

e. **Adapted metaphor**

An adapted metaphor is an adaptation of an existing (stock) metaphor. This type of metaphor should be translated by an equivalent adapted metaphor; it may be incomprehensible if it is translated literally, as in "الكرة في ملعبه", (Alkora fi mal'aboh, the ball is in his court).

f. **Original metaphors**

Original metaphors refer to those created or quoted by the Source Language writers in authoritative and expressive texts. These metaphors should be translated literally, whether they are universal, cultural, or obscurely subjective.

3.2 Metaphor in Holy Quran

There are many metaphors in some verses of the Holy Qur'an. These metaphors, that the All-Wise Allah makes, are very effective and advance the understanding of those who read them. Every one of these metaphors and descriptions illustrates the subject in the most effective and the clearest way. Throughout history, critics have rarely defined this word alike (تشبيه tashbeh). The first who is known to have used the term Al-majaz is Abu Ubayda in his book, "Majazal-Quran". However, he did not mean by that Al-majaz is the counterpart of haqiqa and figurative language. He mostly uses the word in the formula: "A, its majaz is B", where A denotes the classical word or phrase and B its "natural" equivalent. In fact, Ubayda was concerned with the first meaning of the term "majaz" which means 'explanatory re-writing' in 'natural' language of idiomatic passages in the Scripture, while the second sense of "majaz" is figurative language, which was developed later. In his Majaz Quran, Ubayda does not define *majaz*, but at the beginning of his work he does give a list of thirty nine cases of deviation from the 'natural' language that can be found in the Qur'an (Alshehab 2015). The following is an instance of the word "آية" Ayah from the Qur'an which is interpreted as a metaphor; a device for presenting a concept. One of the most beautiful metaphors in the Holy Qur'an is the verse:

(اللَّهُ نُورُ السَّمَاوَاتِ وَالْأَرْضِ مَثَلُ نُورِهِ كَمِشْكَاةٍ فِيهَا مِصْبَاحٌ الْمِصْبَاحُ فِي زُجَاجَةٍ الزُّجَاجَةُ كَأَنَّهَا كَوْكَبٌ دُرِّيٌّ يُوقَدُ مِنْ شَجَرَةٍ مُبَارَكَةٍ زَيْتُونَةٍ لَا شَرْقِيَّةٍ وَلَا غَرْبِيَّةٍ يَكَادُ زَيْتُهَا يُضِيءُ وَلَوْ لَمْ تَمْسَسْهُ نَارٌ نُورٌ عَلَى نُورٍ يَهْدِي اللَّهُ لِنُورِهِ مَنْ يَشَاءُ وَيَضْرِبُ اللَّهُ الْأَمْثَالَ لِلنَّاسِ وَاللَّهُ بِكُلِّ شَيْءٍ عَلِيمٌ)

Allah is the Light of the heavens and the earth. The example of His light is like a niche within which is a lamp, the lamp is within glass, the glass as if it were a pearly [white] star lit from [the oil of] a blessed olive tree, neither of the east nor of the west, whose oil would almost glow even if untouched by fire. Light upon light. Allah guides to His light whom He wills. And Allah presents examples for the people, and Allah is Knowing of all things. (Quran: Surah: Al-Noor, Verse 35).

When we use metaphors, it does not mean that we lie; we use metaphor to make the concepts and thoughts sharper and clearer. For instance, in the Holy Quran:

(أَلَمْ تَرَ كَيْفَ ضَرَبَ اللَّهُ مَثَلًا كَلِمَةً طَيِّبَةً كَشَجَرَةٍ طَيِّبَةٍ أَصْلُهَا ثَابِتٌ وَفَرْعُهَا فِي السَّمَاءِ (24) تُؤْتِي أُكُلَهَا كُلَّ حِينٍ بِإِذْنِ رَبِّهَا وَيَضْرِبُ اللَّهُ الْأَمْثَالَ لِلنَّاسِ لَعَلَّهُمْ يَتَذَكَّرُونَ (25) وَمَثَلُ كَلِمَةٍ خَبِيثَةٍ كَشَجَرَةٍ خَبِيثَةٍ اجْتُثَّتْ مِنْ فَوْقِ الْأَرْضِ مَا لَهَا مِنْ قَرَارٍ (26)

"Have you not considered how Allah presents an example: a good word is like a good tree, whose root is firmly fixed and its branches in the sky? (24) It produces its fruit all the time, by permission of its Lord. And Allah presents examples for the people that perhaps they will be reminded. (25) And the example of a bad word is like a bad tree, uprooted from the surface of the earth, not having any stability". [Quran: Surah Ibrahim, Verse 24–26].

The metaphor here; the good word (الكلمة الطيبة) is set in similitude to a good tree (الطيبة الشجرة) that has a firm root and its branches in Heaven (sky) and gives its fruits every now and then by the will of its Lord. On the other hand, the bad word (الكلمة الخبيثة) is likened to a bad tree (الشجرة الخبيثة) which is uprooted from the earth and has no base.

The classical text is a linguistic miracle and was intended to challenge Arabs who are fluent in classic Arabic analogy, and what makes the Qur'an a miracle, is that it is impossible for a human being to compose something like it, as it lies outside the productive capacity of the nature of the Arabic language. The productive capacity of nature, concerning the Arabic language, is that any grammatically sound expression of the Arabic language will always fall within the known Arabic literary forms of prose and poetry. All of the possible combinations of Arabic words, letters and grammatical rules have been exhausted and yet their literary forms with metaphors have not been matched linguistically. The Arabs, who were known to have been Arabic linguists par excellence, failed to successfully challenge the Quran (Mohaghegh 2013).

3.3 Metaphor in Modern Standard Arabic

Metaphor is the process of 'transporting' qualities from one object to another, one person to another, from a thing to a person or animal, etc. When translating a metaphor, it is necessary to start by investigating the concept of metaphor, with the focus on contemporary conceptual approaches of metaphor. There have been rapid and revolutionary changes in communications, computers, and Internet technologies in recent years, along with huge changes in the conceptual studies of metaphor.

A metaphor is a figure of speech that involves a comparison, and a simile is also a figure of speech which involves a comparison. The only difference between them is that in a simile the comparison is explicitly stated, usually by a word such as "like" or "as", while in a metaphor the comparison is implied. Machine translation is much more likely to function correctly for simile than it can for metaphor. For instance, using Google translator:

a. "اشتعل الرأس شيباً" (Eshta'al Alra's Shayban, Flared head Chiba); and
b. "شعره كالثلج " (Sha'aroh Kalthalj, his hair such as snow).

In the second example would help in translation, as it represents a simile, but in the first example the metaphor is implicit and so its translation is much more difficult. Another example is رأيت أسداً في المدرسة (Ra'ayt Asadan fi Almadrasa, I saw a lion in the school), it does not mean that "I saw the lion (the animal), but rather that "I saw a man like a lion in his brave demeanor", here describing the bravery of the man like that of a lion, the king of the forest and the strongest among others.

3.4 Metaphor in Dialect Arabic

Arabic dialects, collectively referred to here as Dialectal Arabic (DA), are the day to day vernaculars spoken in the Arab world. Metaphorical expressions are pervasive in day-to-day speech. The Arabic language is a collection of historically related variants that live side by side with Modern Standard Arabic (MSA). As spoken varieties of Arabic, they differ from MSA on all levels of linguistic representation, from phonology, morphology and lexicon to syntax, semantics, and pragmatic language use. The most extreme differences are at the phonological and morphological levels. We can see the difference in meaning with the use of the word white in metaphorical expressions. For example, the expression in the dialect Arabic

سارة قلبها زي التلج (Sarah Qalbaha zay Althalj, Sara's heart [is] like snow) expresses that Sara is a good person, whereas the expression كدبة بيضة (Kedba Bedhah, a white lie) means a lie that is "honest and harmless". Another example is praise with the word "donkey" in the expression سارة حمارة شغل (Sarah Hemart Shoghol, Sara is a donkey at work) which means "She is a very patient and hard worker". However, describing a person as a donkey in the dialect Arabic is very offensive and has connotations such as foolish or stupid. In dialect metaphors, we usually use the bad words (bad expressions) to express a good adjective and the vice versa.

Dialect Metaphors expressions are day-to-day speech that people use all the time (Biadsy 2009):

- In arguments like "مافيك تدافع عن موقفك" (Mafeek Tedafe'a a;n mauqifak, you cannot defend your position) contain the word "تدافع" (defend); it must be for something like country or building. We consider the person in the argument with us as an opponent and we attack his position. Another example "حكيو ضرب علي الراس", (Haku Dhareb ala Alras, his speech is hitting it on the head), means that he is getting to the heart of the matter.
- Utilizing ideas and peech as food and commodities: "أفكاره مهضومة", (Afkaroh Mahdomeh, his ideas [are] tasty and sweet), means that his ideas are nice and appropriate, while "أفكاره بلا طعمه" (Bla Ta'meh, his ideas [are] without taste) means that they are not useful, or even harmful. Two other examples are "حط ببطنك بطيخة صيفي" (Hoot Bebatnak Batekha Saifi, Eat watermelon), which means 'relax and don't worry', and "طحن الكتب طحن" (Tahan, Elkutob Tahen, he smashed the books), which means that he studied the books thoroughly.
- To express time: "إجا وقت الجد" (Eja Waqt aljad, the time of seriousness has come) means that it is time to work hard and be serious. Other examples of time metaphors are "راح آذار" (March went away), meaning March has ended, and "{الشتا صار على الابواب}," (Alshita sar ala alabwab, winter has reached our doorsteps), which means winter will start soon.
- Times are used as location: "نط التسعين" (Nat Altes'en, He jumped over ninety) means he is over ninety years old, and "العام الي مرق" (Alam Eli Maraq, the year that passed) means the last year, and here describes the year as a person that has walked away.

Dialect metaphors are difficult to understand correctly, unless we are familiar with them and we are from the same culture with the same dialect, as each country (and even each region) has its own metaphor dialect.

4 Named Entity Recognition Translation

The Named Entity Recognition (NER) task consists of determining and classifying proper names within an open-domain text. This Natural Language Processing task is acknowledged to be more difficult for the Arabic language, as it has such a complex morphology. NER has also been confirmed to help in Natural Language

Processing tasks such as Machine Translation, Information Retrieval and Question Answering to obtain a higher performance. NER can also be defined as a task that attempts to determine, extract, and automatically classify proper name entities into predefined classes or types in open-domain text. The importance of named entities is their pervasiveness, which is proven by the high frequency, including occurrence and co-occurrence, of named entities in corpora. Arabic is a language of rich morphology and syntax. The peculiarities and characteristics of the Arabic language pose particular challenges for NER. There has been a growing interest in addressing these challenges to encourage the development of a productive and robust Arabic Named Entity Recognition system (Shaalan 2014). End of editing 18 March (EST) —new editing from here!

The NER task was defined so that it can determine the appropriate names within an open domain text and categorize them as one of the following four classes:

1. Person: person name or family name;
2. Location: name of geographically, and defined location;
3. Organization: corporate, institute, governmental, or other organizational entity; and
4. Miscellaneous: the rest of proper names (vehicles, brand, weapons, etc.).

In the English language the determination of the named entities (NEs) in a text is a quite easy sub-task if we can use capital letters as indicators of where the NEs start and where they end. However, this is only possible when capital letters are also supported in the target language, which is not the case for the Arabic language. The absence of capital letters in the Arabic language is the main difficulty to achieving high performance in NER (Benajiba 2008; Benajiba and Rosso 2007; Shaalan 2014).

To reduce data sparseness in Arabic texts two solutions are possible: (i) Stemming: omitting all of the clitics, prefixes and suffixes that have been added to a lemma to find the needed meaning. This solution is appropriate for tasks such as Information Retrieval and Question Answering because the prepositions, articles and conjunctions are considered as stop words and are not taken into consideration when deciding whether or not a document is relevant for a query. An implementation of this solution is available in Darwish and Magdy (2014); (ii) Word segmentation: separating the different components of a word by a space (blank) character. This solution is more appropriate for NLP tasks that require maintaining the different word morphemes such as Word Sense Disambiguation, Named Entity Recognition, etc.

NER in Dialect Arabic is completely different than it is in MSA. For example, a person name in either DA or MSA could be expressed in DA by more than one form; for example, the name "قمر طارق" (Qamar Tareq) in MSD, can be "امر طارىء" (Amar Tare'a) and "كمر طارك" (Kamar Tarek); the main complication is that the first name is a girl's name, when translated it can be 'moon' and not appear as a Name Entity for a person.

Another issue in NER is the ambiguity between two or more NEs. For example consider the following text: (عيد سعيد عيد مبارك). In this example, the (Eid) is both a person's name and a greeting for Al Eid, thereby giving rise to a conflict situation, where the same NE is tagged as two different NE types. The same in the following names {حصة ، موزة ، شمس ، هند، جمعة} for example "حصة مرحة" (Hesa is a funny girl) and "موزة مهضومة", which means "Mouza is cute", another example is the name "أحمد الفهد الصباح"; these are all person-names and do not refer to an animal or a timing period.

In Machine Translation (MT), NEs require different translation techniques than the rest of the words of a text. The post-editing step is also more expensive when the errors of an MT system are mainly in the translation of NEs. This situation inspired (Babych and Hartley 2003) to conduct a research study in which he tagged a text with an NER system as a pre-processing step of MT. He found achieved a higher accuracy with this new approach which helps the MT system to switch to a different translation technique when a Named Entity (NE) is detected (Othman 2009).

5 Word Sense Disambiguation Translation

The Arabic Language contains several kinds of ambiguity; many words can be in various characteristics based on certain contexts. For example, the word دين has two meaning; the first refers to religion and the second refers to rent money. Such ambiguity can be easily distinguished by a human using common sense, while machine translation cannot distinguish the difference. Instead, MT requires more complex analysis and computation in order to correctly identify the meaning; this process is called Word Sense Disambiguation (WSD) (Mussa 2014); (Hadni 2016).

Word Sense Disambiguation (WSD) is the problem of identifying the sense (meaning) of a word within a specific context. In Natural Language Processing (NLP), WSD is the task of automatically determining the meaning of a word by considering the associated context (Navigli 2009). It is a complicated but crucial task in many areas, such as Topic Detection and Indexing, Information Retrieval, Information Extraction, Machine Translation, Semantic Annotation, Cross-Document Co-Referencing and Web People Search. Given the current explosive growth of online information and content, an efficient and high-quality disambiguation method with high scalability is of vital importance to allow for a better understanding, and consequently, improved exploitation of processed linguistic material (Hadni 2016).

One example of an ambiguous Arabic word is خال (Khal), which can be translated to any of the following three words: "empty", "imagined" or "battalion." Due to the undiacritized and unvowelized Arabic writing system, the three meanings are conflated. Generally, Arabic is loaded with polysemous words. One interesting observation about the Arabic language is its incredible reuse of names of the human body parts. For example, imagining the word رأس 'head' one could think of the neck, nose, eyes, ears, tongue and so on (Abuelyaman et al. 2014).

Apparently, when many researchers translating Quran to English language, several semantic issues have been appear. Such issues poses the ambiguity of words for example ليلاًونهاراً (laylan wanaharan) and يوم الحساب (Yaum Alhesab), which are translated into "day and night" and "judgment day". Such ambiguity has to be omitted by determining the correct sense of the translated word.

In MSA, synonyms are very common, for example the word year has two different synonyms in Arabic for example (سنة sanah, and عام Aam) and both of them are widely used in everyday communication. Despite the issues and complexity of Arabic morphology, this impedes the matching of the Arabic word. The word "year" is written also in two different ways in the Quran سنة sanat, and عام Aam. Both are simple singular forms occur 7 times in the entire Quran, providing one of many examples of word symmetries in the Quran. The words سنة (Sanat) and عام (Aam) are perfect synonyms. This cannot be further away from the miracle why God chose very specific words to be written in His book.

Ambiguity is not limited to Arabic words only, but also to Arabic letters when they affixed to morphemes, lead to ambiguous compound words. Table 2 shows how affixing the letter 'ب' which corresponds to 'b' in English, to an atomic word will turn it into a compound one. This is because, as a prefix, the letter 'ب' takes on any of the following senses: through, in, by, for and at. Table 2 shows only five of the ten possible roles the letter 'ب' plays when prefixed to different words (Abuelyaman et al. 2014).

The ambiguity of letters also appears in the Holy Quran. Twenty-ning surahs of Al-Quran begin with letters, such as Surat Maryam, verse 1 كَهَيَعَصَ "Kaf-Ha-Ya-'Ain-Sad". These letters are one of the miracles of the Qur'an, and none but Allah alone knows their meanings.

Arabic texts without diacritics pose the greatest challenge for WSD, as they increase the number of a word's possible senses and consequently make the disambiguation task much more difficult. For example, the word صوت Sawt without diacritics has 11 senses according to the Arabic WordNet (AWN) (Bouhriz and Benabbou 2016), while the use of diacritics for the same word صوَّتَ Sawata cuts down the number of senses to two. Another example the word مال, which has seven senses in) (Bouhriz and Benabbou 2016):

- Sense 1{فلوس,ثَروة,دَراهم,مَال},
- Sense 2{نقود,مَال},
- Sense 3{تمايل,تَرنح,مَال},
- Sense 4 {انحدر,مَال},

Table 2 Letter ambiguity

Word	Translation	Word‖ بـ	Translation of word‖ بـ
بركة	Blessing	بـــبركة	Through blessing
المدرسة	The school	بالمدرسة	In the school
المال	The money	بـــالمال	By the money
أي	What	بـــأي	For what
الباب	The door	بـــالباب	At the door
القلم	The pen	بـــالقلم	Using the pen

- Sense 5 {مالِ، نزَعَ إلى},
- Sense 6 {أقنع,أمال,مَال},
- Sense 7 {انحرف,انحنى,مال}.

The WSD approach has shown that two words before and after an ambiguous word are sufficient for its disambiguation in almost all languages (Mohamed and Tiun 2015). For the Arabic language, the information extracted from this local context is not always sufficient. To solve this problem, an Arabic WSD system has been proposed that is not only based on the local context, but also on the global context extracted from the full text (Bouhriz and Benabbou 2016). The objective of their approach is to combine the local contextual information with the global one for a better disambiguation using the resource Arabic WordNet (AWN) to select word senses.

All of the WSD approaches make use of words in a sentence to mutually disambiguate each other (Chen et al. 2009; Agire et al. 2009; Ponzetto et al. 2010). The distinction between various approaches lies in the source and type of knowledge made by the lexical units in a sentence. Thus, all of these approaches can be classified into either corpus-based or knowledge-based methods. Corpus-based methods use machine-learning techniques to induce models of word usages from large collections of text examples. Statistical information that may be monolingual or bilingual, raw or sense-tagged is extracted from corpora. Knowledge-based methods instead use external knowledge resources that define explicit sense distinctions for assigning the correct sense of a word in context. (Dagan and Itai 1994); (Gale et al. 1992) used Machine-Readable Dictionaries (MRDs), thesauri, and computational lexicons, such as WordNet (WN). (Dagan and Itai 1994) was the first to resolve lexical ambiguities in one language using statistical data from the monolingual corpus of another language. That approach exploits the differences between the mappings of words to senses in different languages.

6 Conclusion

The paper presents the key the challenges of translating the Arabic language into the English language according to the classical Arabic, Modern Standard Arabic and Dialect Arabic. It also has suggested a line of argument in favors of the conceptualization of Word Sense Disambiguation, Metaphor, and Named Entity Recognition. Up to date, little work has been published on Arabic language translation. Arabic sentences are usually long, the punctuation are not effecting on the text interpretation. Contextual analysis is very important in the Arabic text translation, in order to understand the exact meaning of the word. The absence of diacritization in most of the MSD and completely in Dialect Arabic pose a real challenge in Arabic Natural Language Processing, especially in Machine translation. The Arabic language has many features that are inherently challenging for NLP researchers. The difficulties associated with recognizing the need for full-verbs the likes of "is", and adverbs-of-places—the likes of "there", recognizing the appropriate senses of undiacritized words, and the practice of performing translation at the compound word level are some of the main issues. Classical Arabic is regarded as rhetorical and eloquent because of its stylistic and linguistic manifestations. Translators who are not well-acquainted with this religious discourse cannot succeed in relaying the linguistic, stylistic and cultural aspects in the translated language. Unlike an ordinary text, the classical discourse is featured is noted to be sensitive; its language is euphemistic, indirect, and solicitous of people's feelings. While Dialect can be a crucial element in the process of describing and individualizing characters in literature and therefore should be handled with great care. Dialect phonetic, grammatical and syntactic effect should directly or indirectly be preserved in the target language.

References

Abuelyaman, E., Rahmatallah, L., Mukhtar, W., Elagabani, M.: Machine translation of Arabic language: challenges and keys. In: 2014 5th International Conference on Intelligent Systems, Modelling and Simulation (ISMS), pp. 111–116. IEEE Januray 2004

Abdul-Raof, H.: Arabic rhetoric: A pragmatic analysis. Routledge (2006)

Agire, E., Lacalle, O. L. d., Soroa A.: Knowledge-based WSD and specific domains: performing over supervised WSD. In: Proceedings of the International Joint Conference on Artificial Intelligence 2009, pp. 1501–1506, AAAI Press (2009)

Ahmad Abdel Tawwab Sharaf Eldin.: A Cognitive Metaphorical Analysis of Selected Verses in the Holy Quran. International J. Engl. Linguist **4**(6) (2014)

Ali, A., Brakhw, M.A., Nordin, M.Z.F.B., ShaikIsmail, S.F.: Some linguistic difficulties in translating the holy Quran from Arabic into English. Int. J. Social Sci. Humanity **2**(6), 588 (2012)

Alkhatib, M., Shaalan, K.: Natural language processing for Arabic metaphors: a conceptual approach. In: International Conference on Advanced Intelligent Systems and Informatics, pp. 170–181, Springer International Publishing October (2016)

Alshehab, M.: Two english translations of arabic metaphors in the Holy Qura'n. Browser Download This Paper (2015)

Babych, B., Hartley, A.: Improving machine translation quality with automatic named entity recognition. In: Proceedings of the 7th International EAMT workshop on MT and other Language Technology Tools, Improving MT through other Language Technology Tools: resources and Tools for Building MT, Association for Computational Linguistics, pp. 1–8 April 2003

Benajiba, Y., Rosso, P., Benedíruiz, J.M.: Anersys: an Arabic named entity recognition system based on maximum entropy. In: International Conference on Intelligent Text Processing and Computational Linguistics, pp. 143–153. Springer, Berlin, Heidelberg February 2007

Benajiba, Y., Rosso, P.: Arabic named entity recognition using conditional random fields. In: Proceedings. of Workshop on HLT & NLP within the Arabic World, LREC, vol. **8**, pp. 143–153 May 2008

Biadsy, F., Hirschberg, J., Habash, N.: March. Spoken Arabic dialect identification using phonotactic modeling. In Proceedings of the eacl 2009 workshop on computational approaches to semitic languages (pp. 53–61). Association for Computational Linguistics (2009)

Bouhriz, N., Benabbou, F.: Word sense disambiguation approach for Arabic text. Int. J. Adv. Compt. Sci. Appl. **1**(7), 381–385 (2016)

Chen, P., Ding, W., Bowes, C., Brown, D.: A fully unsupervised word sense disambiguation method using dependency knowledge. In: Proceedings of Human Language Technologies: the 2009 Annual Conference of the North American Chapter of the ACL.pp. 28–36. ACL, Rappel Precision F1-mesure SVM 0,746 0,718 0,732 Naive Bayesien 0,747 0,71 0,782 8 (2009)

Dagan, I., Itai, A.: Word sense disambiguation using a second language monolingual corpus. Comput. linguist. **20**(4), 563–596 (1994)

Darwish, K., Magdy, W.: Arabic information retrieval. Foundations and Trends®. Inf. Retrieval **7**(4), 239–342 (2014)

El Kholy, A., Habash, N.: Techniques for Arabic morphological detokenization and orthographic denormalization. In: LREC 2010 Workshop on Language Resources and Human Language Technology for Semitic Languages, pp. 45–51 (2010)

Gale, W., Church, K., Yarowsky, D.: A method for disambiguating word senses in a large corpus. Comput. Humanit. **26**, 415–439 (1992)

Hadni, M., Ouatik, S.E.A., Lachkar, A.: Word sense disambiguation for Arabic text categorization. Int. Arab. J. Inf. Technol. **13**(1A), 215–222 (2016)

Hashemi, A.: Javaher al-balagha. (H. Erfan, Trans.) Belaghat Publication (2002)

Lee, Y.S.: Morphological analysis for statistical machine translation. In: Proceedings of HLT-NAACL 2004: short Papers, pp. 57–60. Association for Computational Linguistics, May 2004

Mohamed, O.J., Tiun, S.: Word sense disambiguation based on yarowsky approach in english quranic information retrieval system. J. Theor. Appl. Inf. Technol. **82**(1), 163 (2015)

Mohaghegh, A., Dabaghi, A.: A comparative study of figurative language and metaphor in English, Arabic, and Persian with a focus on the role of context in translation of Qur'anic metaphors. TextRoad Publication, **3**(4), pp. 275–282 (2013)

Mussa, S.A.A., Tiun, S.: A novel method of semantic relatedness measurements for word sense disambiguation on english AlQuran. Bulletin Elect. Eng. Inform. vol. 4 (2014)

Mussa, S.A.A., Tiun, S.: Word sense disambiguation on english translation of holy quran. Bulletin Elect. Eng. Inform. **4**(3), 241–247 (2015)

Navigli, R.: Word sense disambiguation: a survey. ACM Comput. Surv. (CSUR) **41**(2), 10 (2009)

Newmark, P.: A textbook of translation (Vol. 66). New York: Prentice hall (1988)

Othman, R.: Trends In Information Retrieval System. IIUM Press (2009)

Ponzetto, S.P., Navigli, R.: Knowledge-rich word sense disambiguation rivaling supervised system. In: Proceedings of the 48th Annual Meeting of the Association for Computational Linguistics, Uppsala, Sweden, pp. 11–16, pp. 1522–1531 July 2010

Sadiq, S.: A Comparative Study of Four English Translations of Surat Ad-Dukhan on the Semantic Level. Cambridge Scholars Publishing (2010)

Sadat, F., Habash, N.: Combination of Arabic preprocessing schemes for statistical machine translation. In: Proceedings of the 21st International Conference on Computational Linguistics and the 44th annual meeting of the Association for Computational Linguistics, pp. 1–8. Association for Computational Linguistics July 2006

Sawaf, H.: Arabic dialect handling in hybrid machine translation. In: Proceedings of the Conference of the Association for machine translation in the Americas (AMTA), Denver, CO November (2010)

Shaalan, K.: A survey of Arabic named entity recognition and classification. Computat. Linguist. **40**(2), 469–510 (2014)

Zbib, R., Malchiodi, E., Devlin, J., Stallard, D., Matsoukas, S., Schwartz, R., Makhoul, J., Zaidan, O.F., Callison-Burch, C.: Machine translation of Arabic dialects. In: Proceedings of the 2012 Conference of the North American chapter of the Association for Computational Linguistics: human language technologies, pp. 49–59. Association for Computational Linguistics June 2012

Zribi, I., Khemakhem, M.E., Belguith, L.H.: Morphological analysis of tunisian dialect. In: IJCNLP pp. 992–996 October 2013

Part III
Information Extraction

Graph-Based Keyword Extraction

Omar Alqaryouti, Hassan Khwileh, Tarek Farouk, Ahmed Nabhan and Khaled Shaalan

Abstract Keyword extraction has gained increasing interest in the era of information explosion. The use of keyword extraction in documents context categorization, indexing and classification has led to the emphasis on graph-based keyword extraction. This research attempts to examine the impact of several factors on the result of using graph-based keyword extraction approach on a scientific dataset. This study applies a new model that processes the Medline scientific abstracts, produces graphs and extracts 3-graphlets and 4-graphlets from those graphs. The focus of the experiment is to come up with a dataset that consists of the keywords and their occurrences in the proposed graphlets patterns for each abstract with its class. Then, apply a supervised Naïve Bayes classifier in order to assign a probability to each word, whether or not it is a keyword, and finally evaluate the performance of the graph-based keyword extraction approach. The model achieved significant results compared to the Term Frequency/Inverse Document Frequency (*TF/IDF*) baseline standard. The experimental results proved the capability of using graphs and graphlet patterns in keyword extraction tasks.

O. Alqaryouti (✉) · H. Khwileh · T. Farouk · K. Shaalan
Faculty of Engineering and IT, The British University in Dubai, Dubai, UAE
e-mail: omar.alqaryouti@gmail.com

H. Khwileh
e-mail: hassan.khwileh@gmail.com

T. Farouk
e-mail: tafarouk@me.com

A. Nabhan
Faculty of Computers and Information, Fayoum University, Fayoum, Egypt
e-mail: ahmed.nabhan@gmail.com

A. Nabhan
Member Technology, Sears Holdings, Hoffman Estates, USA

K. Shaalan
School of Informatics, University of Edinburgh, Edinburgh, UK
e-mail: khaled.shaalan@buid.ac.ae

K. Shaalan et al. (eds.), *Intelligent Natural Language Processing: Trends and Applications*, Studies in Computational Intelligence 740,
https://doi.org/10.1007/978-3-319-67056-0_9

Keywords Graph-based representation · Graph-based methods
Graph patterns extraction · Keyword extraction · Machine learning
Supervised methods

1 Introduction

In the last century, graph theory has gained growing and extensive traction in the explosion of computer networks and internet as a well-studied science in the mathematical field. Graph-based representation of text documents allows powerful and comprehensive methods and algorithms such as random walks as well as frequent subgraph mining. This representation facilitates capturing corresponding features for various Natural Language Processing (NLP) applications. Additionally, graph-based ranking methods were proposed to assist in evaluating the importance of a word in a text document with relevance to its adjacent words [8].

In NLP domain, keyword extraction is considered among the most essential key aspects when it comes to text processing. Readers may take the advantage of keywords as it can help them in deciding whether or not to read a document. As for the website developers, they can use keywords in grouping and categorising the website content and materials by its topics.

As a matter of fact, keyword extraction is said to be an effective method applied to many NLP applications. Through extracting main keywords, one may easily select the relevant document to use for learning the relation among the documents. A study by Gutwin et al. [6], the authors described Keyphind; which basically represents keyphrases and keywords from documents; as the essential building block for Information Retreival (IR) systems. Likewise, Matsuo and and Ishizuka [7] pointed out the significance of keyword extraction techniques for various NLP applications, such as document retrieval, Web page retrieval, document clustering, summarization and text mining. Extracting the proper keywords can assist in easily finding out the documents to read as well as learning how documents are related to each other.

Beliga et al. [2] clarified the way in which the keywords are being assigned through terms of controlled vocabulary or predefined classification. Keywords help in describing the main aspects and concepts discussed in a given context. Keyword assignment in simple terms is the process of identifying few words, phrases and terms that can represent a document. There are various approaches of keyword and key-phrases assignment. Authors and Subject Matter Experts (SME's) manually assign keywords and key-phrases to text documents. Though, the approach remains expensive as compared to other options. This approach is a monotonous and time consuming task apart from being expensive. This could be considered the reason behind the importance of automating the keyword extraction process.

In the literature, several approaches were anticipated by researchers for the keyword extraction process. Beliga et al. [2] proposed three methods for keyword extraction; supervised, unsupervised and semi-supervised. The supervised keyword extraction requires a training set and the use of Machine Learning techniques such as Naïve Bayes and Support Vector Machine. It is domain dependent to the extent that when the domain changes the model will need to be retrained. On the other end of the spectrum, there is the unsupervised methods. At that end there are many statistical-based methods that use frequency based measures, such as *TF/IDF*. These methods are not tied to specific language and does not require training dataset. However, they fare poorly with professional text like health and medical domain; for example, PubMed where a keyword representing medical term might appear rarely.

The following sections will review the work that has been done in the area of keyword extraction in the Related Work Section. Followed by Research Methodology which discusses the concepts of keyword extraction and the proposed model for our research. Based on the approach applied, the experiment will show the essence of proposed approach and its applicability. Then, the Discussion Section reviews the challenges and drawbacks that came across the research project and suggestions for enhancements. Finally, a summary of the impact of using the model for keyword extraction, limitations and future work are discussed in the Conclusion and Future Prospects section.

2 Related Work

In this research project, graph-based supervised methods are being examined and experimented for extracting keywords from text documents. According to the known techniques, keywords get assigned from a list of words that are controlled by authors and librarians. The process of extracting keywords attempts to identify the words from the context that are essential and representative of that particular document. Graph-based techniques in most aspects help in exploiting graph structural features to achieve that objective.

Page et al. [11] presented through their research work PageRank, what is considered a graph-based scoring algorithm. The researchers approach uses random-walks algorithms to score a webpage according to its significance that is driven from its interlinks to each other. Likewise, Mihalcea and Tarau [8] established an adaptation of a similar algorithm in the keyword extraction task. The basis of this adaptation is the central aspect of the natural languages words in a certain text. It is also a key element of narrative connection between each other in the same concept of the links between webpages. The relations in NLP are rich and complicated. The Words in NLP consist of relationships, such as phonological, lexical, morphological, syntactical, and semantical. Graph-based methods in most cases are an outstanding choice towards relation representation. Furthermore, it has the capacity to enrich the relationships by using edges with weights, direction, and other elements. The rich illustration can then get examined and learned via Graph Theory. Apart from that, it has the potential of exploiting throughout the Machine Learning techniques.

In a study developed by Erkan and Radev [5], graph random-walks were used in text summarization process. The sentences were assigned to nodes in the graph while the edges were used to represent the cosine similarity for the connected end nodes.

On the other hand, Rose et al. [14] demonstrated a method that divides the abstract based on keywords and key-phrases candidates through stop words and phrase delimiters. The graph is built from nodes and edges where nodes represent words and edges represents the connection between words. The degree of the vertex and frequency of every word are evaluated, i.e. the edges count linked to the vertex gets calculated. The result of every word is illustrated through the ratio where the word frequency gets divided by its degree. A score gets allocated to the key phrases through the summation of the scores of its words. Then, the key phrases get sorted in descending order according to the scores they achieved. Finally, the main key phrases are derived from the topmost scores. The researchers showed that Rapid Automatic Keyword Extraction (RAKE) does not require Part of Speech (POS) tagging. RAKE produced a single aspect algorithm that creates a low cost as well as a high performance and fast algorithm. Thus, the outcome of RAKE algorithm achieved similar results compared to TextRank.

Palshikar [12] proposed a graph-theoretical concept to identify the significance of a word in a given text. The text was represented as undirected graph with words as vertices and the linkage among the adjacent words as edges labelled by its dissimilarity measure. The researcher used a hybrid approach by adopting various algorithms that use eccentricity, centrality and proximity measures to extract keywords. The word (vertex) centrality in the graph is an indicator to its significance as candidate keyword.

Nabhan and Shaalan [9] presented a method for keyword identification through the use of text graphlet patterns. The proposed experimental methodology showed the competence of the graphlet patterns in keyword identification. The authors confirmed the importance of the set technique stances from a suitable data demonstration that increases the context of texts to span various sentences. In that aspect, it allows the attainment of essential topological features of non-local graph. They defined the graphlets as the small efficiently extracted as well as scored sub-graph patterns. These graphlets show the statistical reliance between the graphlet patterns as well as the words that are labelled as keywords.

3 Research Methodology

3.1 Overview

Ruohonen [15] stated that the graph consists of linked edges and vertices. Thus, a graph is constructed by a set of pairs (V,E) where V signifies the set of vertices and E signifies the set of edges that demonstrates the linkage between two vertices. The degree of the vertex relates to the total count of edges linked to the vertex; where, the vertex is recognized as an end vertex. There are two types of graphs: directed and undirected graphs. Directed graphs have vertices linked through edges

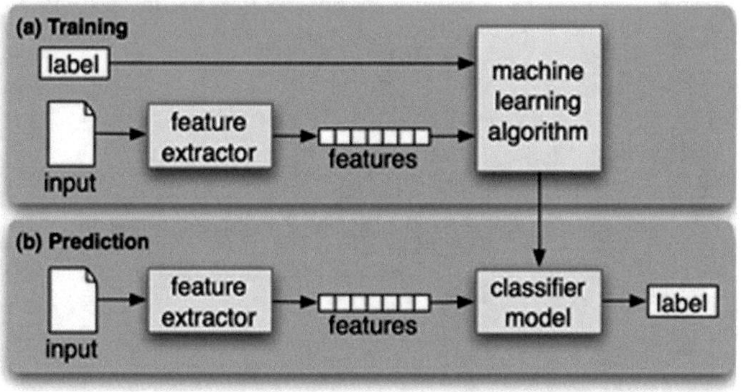

Fig. 1 Supervised classification [3]

with direction. In contrast, the undirected edges do not have the direction between pairs of vertices.

Pržulj [13] defined graphlets as small subgraphs that are linked and non-isomorphic for a large network that allows the capture of local graph or the network topology. A study by DePiero and Krout [4] showed the graph isomorphism and automorphism. These authors claimed that for given graphs G and H with h_k and g_i nodes correspondingly are isomorphic given the existing mapping of:

$$h_k = m(g_i)$$

This mapping maintains all nodes adjacency. Using this concept, the exact application node and the edge must have consistency with this mapping. The condition for subgraph isomorphism is that an isomorphism between the graph (G) and the subgraph (H) should exist. The isomorphism can be verified by the rearrangement of the nodes and edges and then the node-to-node mapping using the adjacency matrix.

As illustrated in Fig. 1, Bird et al. [3] stated that supervised Machine Learning methods adopts the use of training and prediction (testing) data to conclude a model that maps the input features with the anticipated result. Thus, this model is expected to correctly predict the anticipated results based on the given features of new data.

3.2 *The Proposed Methodology*

The proposed methodology (as shown in Fig. 2) is meant to produce keywords using graphs through supervised learning methods. The work on this experiment started by selecting and acquiring a large dataset of abstracts and their associated, manually, predefined keywords. Then, an initial pre-processing took place. Afterwards, graphs were built and sub-graphs were identified. These sub-graphs were used to build a feature vector for each word in the abstract. A Naïve Bayes classifier was then applied on the feature vectors dataset. K-fold cross-validation was used to get precision and

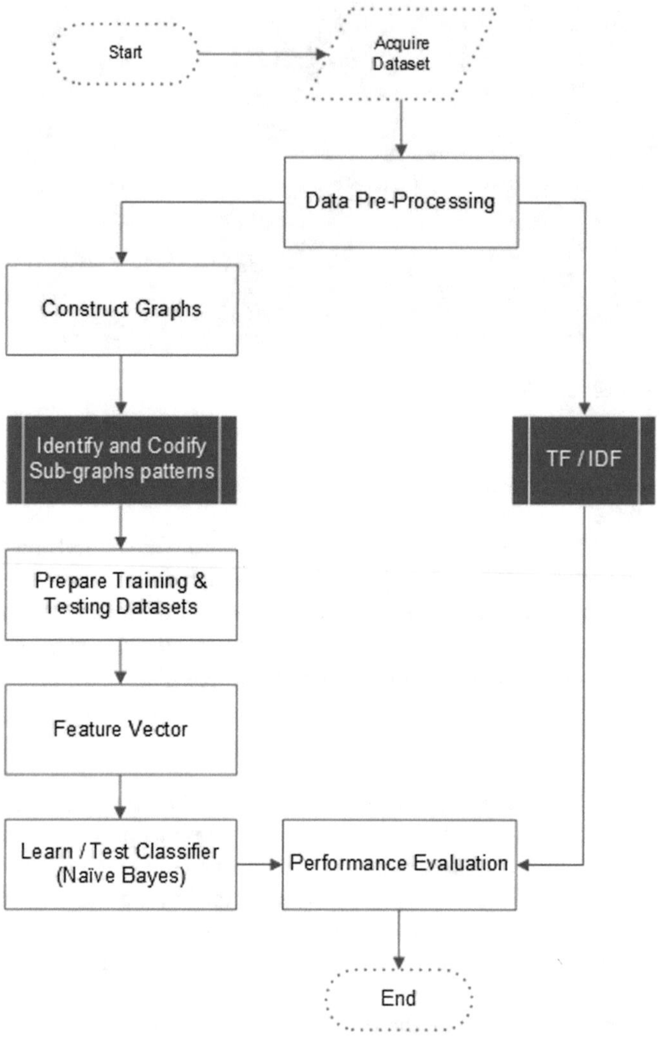

Fig. 2 Research methodology

accuracy averages. The results were then compared to a baseline. The baseline was generated by applying *TF/IDF* on the same dataset.

The process starts by building a dataset from a database of abstracts and their associated preassigned keywords. These abstracts are then pre-processed by taking-out stop words, non-nouns, non-adjectives and words with length less than five characters. Then, abstract graphs are constructed. Each word was represented by a node. The co-occurrence of two words (i.e. adjacent words) was used to connect their respective nodes with an edge. Next, the subgraphs are processed based on the abstract words to identify and codify the sub-graphs patterns. In this step, a novel idea is proposed which is used to identify subgraphs possible patterns using Depth First Search (DFS) by navigating throughout all nodes and extracting all possible 3-graphlet and 4-graphlet patterns. In general, there are three possible 3-graphlet patterns and eleven possible 4-graphlet patterns. A three-digit-code is used to represent the fourteen patterns as follows:

- The first digit is the degree of the word vertex (most left node as shown in Table 1).

$$d(v_i)$$

  where, v_1 is the first vertex.
- The second digit is the sum of degrees of all vertices directly connected to first vertex.

$$\sum_{i=1}^{k} d(v_i)$$

  where, v_i represents the directly connected vertices to the first vertex and k the number of vertices that are directly connected to the first vertex.
- The third digit is the sum of degrees of vertices which are connected to the original word through another word only.

$$\sum_{j=1}^{q} d(v_j)$$

  where,v_j represents the vertices that are indirectly connected to the first vertex and q is the number of vertices that are indirectly connected to the first vertex.

3.3 Dataset

The dataset constitutes of randomly selected abstracts from a scientific dataset that has been obtained from Medline database. PubMed Central is a library of open digital text repository that archives abstracts and references on Medline database and medicine topics. in June 1997 PubMed was published publicly as an open free library. It has more than 26 million records and increasingly about half million records are added yearly. The daily usage of the PubMed is over 3 million searches per day, considered as the world's largest medical library [10].

The same source was also used by Nabhan and Shaalan [9] to process around 10,000 scientific abstracts. Initially, the extracted corpus constituted of around 205,000 abstracts from Medline.

Table 1 The fourteen possible graphlets patterns and its assigned codes. The left most node is the word vertex that is being evaluated

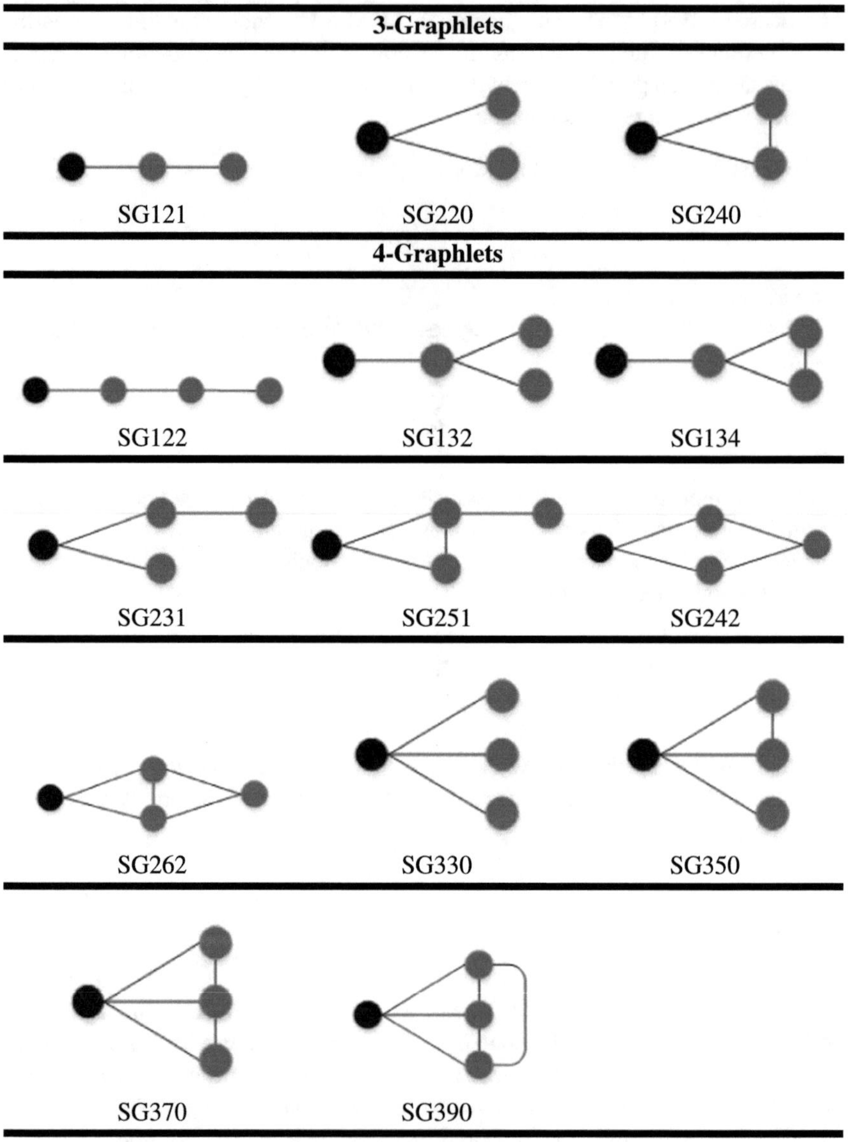

3.4 Data Processing

Each abstract in the PubMed library is accompanied by a keyword list that is provided by either authors or librarians. These pre-assigned keywords were used to train and test the Naïve Bayes classifies. The process starts by filtering-out abstracts that has less than 3 keywords. Also, a keyword was only considered if it was part of the abstract. Afterwards, the abstract was tokenized using NLTK library similar to Bird et al. [3]. Tokenization was followed by a POS tagging process with a filter on the tagged words to exclude all non-nouns and non-adjectives and words with length less than five characters.

In the same way to the TextRank algorithm proposed by Mihalcea and Tarau [8] to mandatory define relationship in which was utilized to have one relationship in this experimental study. Thus, a relationship was constructed to discover the co-occurrence of the words within a predefined range of two adjacent words that identifies the sequences of word pairs that create the un-weighted edges in the abstract graph. An undirected graph will be constructed using the processed words and edges (words pairs) ignoring the pairs order and direction as long as the words are adjacent.

3.5 Learn Classifier and TF/IDF

The proposed model intended to process the produced graphs for the whole corpus as well as identify and codify the sub-graphs possible patterns. An algorithm has been implemented using Python to identify sub-graphs patterns using Depth First Search (DFS) by navigating through all nodes and extracting all possible 3-graphlet and 4-graphlet patterns (see Table 1). The solution uses NLTK, NetworkX, Biopython API's and Rapid Miner. For each word vertex, a mechanism was built for reiteration through the graph to get all possible 3-graphlets and 4-graphlets that the word vertex is part of. This ends up with a new dataset that contains the word, all possible 3-graphlet and 4-graphlets patterns with the count of the number of occurrences for the word in this particular pattern. Moreover, the dataset includes a classification of whether the word was identified as keyword or non-keyword according to the abstract Other Terms (OT's). This training set was used as a data source for the Naïve Bayes classifier. The classifier will assign a probability to each word depending on whether or not it is a keyword. RapidMiner tool was used to implement the supervised classification model of Naïve Bayes (see Figs. 3, 4 and 5).

Gaussian Naïve Bayes has been used in the results evaluation. The continuous values that involve the graphlet patterns counts for each word and associated with a class for the word being a keyword or not being a keyword are distributed according to Gaussian Naïve Bayes. For instance, consider a training dataset that includes continuous attribute. The first step is to cluster the data according to the class and after that compute the mean and variance of the attribute in each class. Assume, for

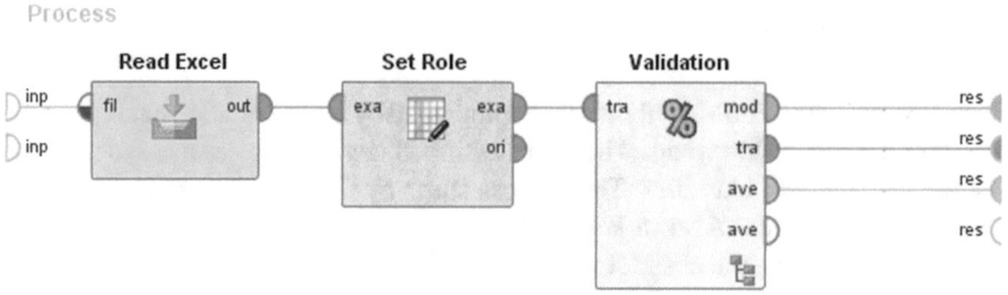

Fig. 3 RapidMiner main process

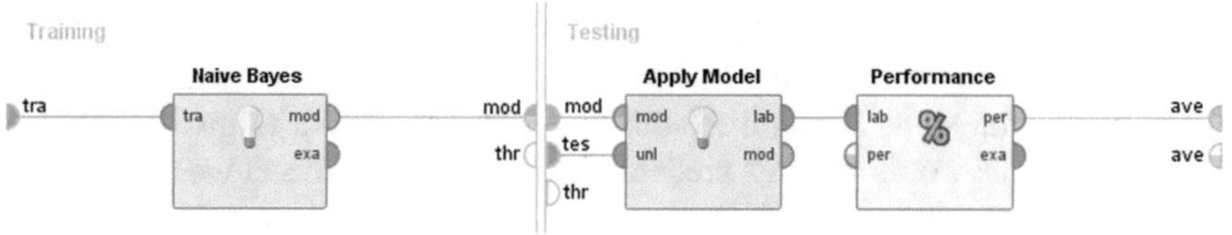

Fig. 4 Naive Bayes subprocess (Validation)

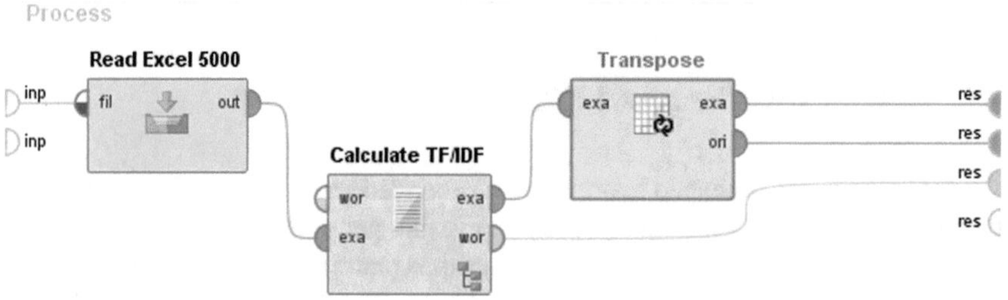

Fig. 5 RapidMiner process to calculate the *TF/IDF* scores

a word from the test dataset, that *v* is the count of graphlet *x*. Then, the probability distribution of *v* given a class *c* can be computed according to the following formula:

$$p(x = v|c) = \frac{1}{\sqrt{2\prod \sigma_c^2}} e^{\frac{(v-\mu_c)^2}{2\sigma_c^2}}$$

where, σ_c^2 is the variance of all counts of the graphlet x in the training dataset that is linked to class c, and μ_c is the mean of all values in *x* linked in class *c*.

In this research study, *TF/IDF* has been used as a baseline standard. Aizawa [1] claimed that in recent information retrieval systems, *TF/IDF* is among the most regular and widely used term weighting techniques. Apart from *TF/IDF* fame, it has been considered as an empirical approach which is a frequency-based approach that considers the words and document frequencies.

The *TF/IDF* is known with its well reasonable performance and efficiency. It consists of three elements: *TF*, *IDF* and *TF/IDF*. The *TF* for a particular word is the number of occurrences for this word W_i in certain document. Whereas, the IDF value can be calculated according to the following formula:

$$idf = log(\frac{N}{N_i})$$

where,

- N represents the entire number of documents in the dataset,
- N_i represents the total number of documents that include the word W_i

Finally, the *TF/IDF* score for a word in a document which is the final product of both *TF* and *IDF* that can be determined using the following formula:

$$tfidf = tf \times idf$$

In order to get higher *TF/IDF* score, there shall be a higher *TF* score for the word in a particular document with a lower document frequency of the same word in the entire dataset.

The *TF/IDF* approach has been used to identify the keywords for each document. As for this research, it has been assumed to identify the top six words as the selection criteria for the candidate keywords. As for abstracts, the authors are usually allowed to define six keywords for their publications. In particular, once the *TF/IDF* scores were calculated for all words in the entire documents collection, the candidate keywords were nominated according to the highest *TF/IDF* scores. A process in RapidMiner has been designed to calculate the *TF/IDF* scores (see Fig. 5). The output of the process has been followed by an automated process in Microsoft Excel 2016 to sort the words to get the top five candidate keywords and then calculate the precision and recall percentages according to the OT values.

3.6 Performance Evaluation

The keywords outcome of RapidMiner process for both learn classifier and *TF/IDF* baseline standard has been used for performance evaluation. These evaluation results for both the proposed model and baseline standard were compared to measure the performance of the proposed model.

The identified keywords have been compared with the annotated keywords that are assigned to each abstract (OT's) to generate the confusion matrix and calculate the Precision (*P*) and Recall (*R*) accuracy measures. The precision percentage indicates the amount of identified keywords that were relevant and its value was calculated according to the following formula:

$$P = \frac{TP}{TP + FP}$$

where,

- *TP* is the True Positive count which illustrates the keywords that were identified using the proposed supervised model or the *TF/IDF* algorithm which were part of the keywords that were assigned to the abstract (OT's).
- *FP* is the False Positive count which is also known as Type I error that illustrates the keywords that were identified using the proposed supervised model or the *TF/IDF* algorithm which were not part of the keywords that were assigned to the abstract (OT's).

Additionally, the Recall percentage value was calculated according to the following formula:

$$R = \frac{TP}{TP + FN}$$

where, *FN* is the False Negative count which is also known as Type II error that illustrates the keywords which were part of the keywords that were assigned to the abstract (OT's) but were not identified using the proposed supervised model or the *TF/IDF* algorithm.

Finally, the F-Measure also known as F-Score represents a combination of Precision and Recall values into a single measurement value. The F-Measure value was computed using the following formula:

$$F - Measure = \frac{2 \times (Precision \times Recall)}{(Precision + Recall)}$$

Table 2 demonstrates the performance evaluation results for both the learned classifier that used the graph-based representation and graphlets frequent pattern identification in addition to the *TF/IDF* baseline standard. The evaluation outcome showed significant results for the proposed model compared to the *TF/IDF*. The results of the proposed model have achieved 76.32% for Precision, 62.88% for Recall and 68.95% for F-Measure. While *TF/IDF* has achieved 64.40% for Precision, 56.48% for Recall and 60.18% for F-Measure.

This section discussed in details the proposed model and how to produce keywords using graphs and graphlet patterns through supervised learning methods. Furthermore, it explained the methodology that was followed and the various work on this experiment throughout the dataset selection, the design and development phase

Table 2 Performance evaluation results

System	Precision (P) (%)	Recall (R) (%)	F-Measure (%)
TF/IDF baseline	64.40	56.48	60.18
Graph representation	76.32	62.88	68.95

and data preparation and validations. Lastly, it illustrated the performance evaluation results.

4 Discussion

This study aims to experiment the impact of using graph-based techniques and subgraphs patterns identification in keyword extraction. A novel efficient method was introduced for exploring significant patterns in word graphs. The dataset used was initially consisted of around 205,000 abstracts. The dataset was filtered to exclude those abstract that does not contain the OT as part of it. Additionally, the abstracts that does not have OT's assigned to them was also excluded. The processed dataset ended up with around 25,000 abstracts after the original dataset was filtered.

The proposed model assumes that the keywords should be part of the documents and does not consider the keywords that are not part of the documents. A better algorithm should not restrict its word space to the given abstract or document but rather should leverage the whole corpus or at least the whole space of keywords of a given corpus. The challenge would be finding strong predictors from a given text that point at the most probable keywords from a pool of keywords from the corpus. These predictors can take the shape of labelled graphlets. Another way could be through embedding the keyword in a graph that connects it to selected words from the abstract. This will give us a database of graphs for each keyword. New abstract can be tested against this database assigning a proximity weight for each keyword in the pool. Then, the candidate keywords can be selected from the most top ones after sorting keywords based on the proximity weight.

5 Conclusion and Future Prospects

Keywords have been proved to be important for document retrieval, text summarization, retrieval of webpages and text mining. Moreover, the keywords help in attracting readers to easily select the relevant topics and documents to read. As stated by Matsuo and Ishizuka [7], the keyword extraction techniques have significant impact on various NLP tasks. The study has experimentally illustrated the significance of graph-based text representation and graphlets patterns approaches in keyword extraction. The results have showed a capability of using graph text representation in extracting keywords compared to the state-of-the-art *TF/IDF*.

There are several areas that can attract researchers for further elaboration in extending the model and the experiment framework with other options that may improve and enhance its outcome. These options may spread to emphasize weighted graphs based on the number of co-occurrences allowing 3-gram and 4-gram, proper treatment of in-line equation and special characters, include edge information like weights and labels [9], measure the impact when extending to use 5-graphlet patterns, cover multi-word keywords/phrases [9], the use of graphs in Text Categorisation [9] and adjoin the extracted keywords that have "and" and "of" in between them which should improve the readability of the key phrases [14]. Also, exploring techniques to mine a corpus for possible keywords for a given abstract rather than restricting the algorithm to words appeared in the given abstract.

Though this experiment used abstracts to automatically extract keywords, it can be equally applied to full-text articles. How effective and efficient is this method on full-text is another good point of future research.

References

1. Aizawa, A.: An information-theoretic perspective of tf-idf measures. Inf. Process. Manag. **39**(1), 45–65 (2003)
2. Beliga, S., Meštrović, A., Martinčić-Ipšić, S.: An overview of graph-based keyword extraction methods and approaches. J. Inf. Organ. Sci. **39**(1), 1–20 (2015)
3. Bird, S., Klein, E., Loper, E.: Natural language processing with Python. O'Reilly Media, Inc. (2009)
4. DePiero, F., Krout, D.: An algorithm using length-r paths to approximate subgraph isomorphism. Pattern Recogn. Lett. **24**(1), 33–46 (2003)
5. Ergan, G., Radev, D.R.: Lexrank: graph-based lexical centrality as salience in text summarization. J. Artif. Intell. Res. **22**, 457–479 (2004)
6. Gutwin, C., Paynter, G., Witten, I., Nevill-Manning, C., Frank, E.: Improving browsing in digital libraries with keyphrase indexes. Decis. Support Syst. **27**(1), 81–104 (1999)
7. Matsuo, Y., Ishizuka, M.: Keyword extraction from a single document using word co-occurrence statistical information. Int. J. Artif. Intell. Tools **13**(01), 157–169 (2004)
8. Mihalcea, R., Tarau, P.: Textrank: bringing order into texts. Assoc. Comput. Linguist. (2004)
9. Nabhan, A.R., Shaalan, K.: Keyword identification using text graphlet patterns. In: International Conference on Applications of Natural Language to Information Systems, pp. 152–161. Springer (2016)
10. Ncbi.nlm.nih.gov. Home-pubmed-ncbi. http://www.ncbi.nlm.nih.gov/pubmed, August (2016)
11. Page, L., Brin, S., Motwani, R., Winograd, T.: The Pagerank Citation Ranking: Bringing Order to the Web (1999)
12. Palshikar, G.K.: Keyword extraction from a single document using centrality measures. In: International Conference on Pattern Recognition and Machine Intelligence, pp. 503–510. Springer (2007)
13. Pržulj, N.: Biological network comparison using graphlet degree distribution. Bioinformatics **23**(2), e177–e183 (2007)
14. Rose, S., Engel, D., Cramer, N., Cowley, W.: Automatic keyword extraction from individual documents. Text Mining, pp. 1–20 (2010)
15. Ruohonen, K.: Graph theory, graafiteoria lecture notes, tut (2013)

CasANER: Arabic Named Entity Recognition Tool

Fatma Ben Mesmia, Kais Haddar, Nathalie Friburger and Denis Maurel

Abstract Actually, the Named Entity Recognition (NER) task is a very innovative research line involving the process of unstructured or semi-structured textual resources to identify the relevant NEs and classify them into predefined categories. Generally, NER task is based on the classification process, which always refers to the previous categorizations. In this context, we propose CasANER, which is a system recognizing and annotating the ANEs. The CasANER elaboration is based on a deep categorization made using a representative Arabic Wikipedia corpus. Moreover, our proposed system is composed of two kinds of transducer cascades, which are the analysis and synthesis transducers. The analysis cascade, which is dedicated to the ANE recognition process, includes the analysis, filtering and generic transduces. However, the synthesis cascade enables to transform the annotation of the recognized ANEs into an annotation respecting the TEI recommendation in order to provide a structured output. The implementation of CasANER is ensured by the linguistic platform Unitex. Then, its evaluation is made using measure values, which show that our proposed system outcomes are satisfactory. Besides, we compare CasANER system with a statistical system recognizing ANEs. The comparison phase proved that the results obtained by our system are as efficient as those of the statistical system in the recognition and annotation of the person's names and organization names.

F.B. Mesmia (✉)
Laboratory MIRACL, Multimedia InfoRmation Systems and Advanced Computing Laboratory, University of Tunis El Manar, Tunis, Tunisia
e-mail: fatmabm@ymail.com

K. Haddar
Laboratory MIRACL, Multimedia InfoRmation Systems and Advanced Computing Laboratory, University of Sfax, Sfax, Tunisia
e-mail: Kais.Haddar@fss.rnu.tn

N. Friburger · D. Maurel
LI, Computer Laboratory, University François Rabelais of Tours, Tours, France
e-mail: nathalie.friburger@univ-tours.fr

D. Maurel
e-mail: denis.maurel@univ-tours.fr

K. Shaalan et al. (eds.), *Intelligent Natural Language Processing: Trends and Applications*, Studies in Computational Intelligence 740,
https://doi.org/10.1007/978-3-319-67056-0_10

Keywords CasANER · Arabic named entity · NER
Transducer cascade · TEI

1 Introduction

Since the MUC (Message Understanding Message conference) conferences, the named entity recognition (NER) has been considered as a sub-task of the Information Extraction (IE) main task. So far, the NER has still been a very innovative research line, which involves processing unstructured or semi-structured textual resources, on one hand, to identify the relevant NEs and on the other hand, to classify them into predefined categories on other hand. The NER task is evolving to realize many objectives, namely enhancing the NLP-application performance. Therefore, recognizing NEs and assigning them to the adequate categories offer an opportunity to create or enrich NE electronic dictionaries and realize the Entity Linking (EL) task. Besides, the NER participates in a large part to enrich documents, in their semantic level that can index Question Answering systems. Furthermore, recognizing and annotating NEs can be a preprocessing to extract SRs between them and increase the search engine efficiency in order to provide relevant responses.

Generally, the NER task is based on the classification process, which always refers to the previous categorizations. However, the NER can encounter many challenges. The first challenge concerns the NE representation that makes its delimitation a complex task. Furthermore, an NER process requires a clear definition to guess the NE limits. The identification phase can suffer from the absence of indicators that may precede an NE. This NE can be similarly enunciated through expressions where the problem of choosing the adequate context arises. Besides, this process may envisage the ANE imbrication, which is very sensitive in its treatment. The second challenge appears after the NE delimitation. This means that the identified NE must be assigned to the adequate category. Nevertheless, the NE categorization is not easy. Moreover, we must determine the best categories to describe the recognized NEs. In addition, the categorization process also contributes to the increase of the granularity level. In fact, an NE can belong to different categories so the ambiguity problem appearing in this case. According to the Arabic NER, the complexity of this language also arises various problems. The main one is its complex morphology due to its agglutinating nature. The complex morphology is highly related to the ambiguity problem. The last challenge is related to the ANE annotation. This annotation must respect a norm that clearly represents the ANE components. Likewise, the ANEs must be described through significant tags having attributes allowing the specification of their categories and sub-categories.

In this context, we exploit the representative Arabic Wikipedia corpora (study and test). We also suggest a deep and detailed ANE categorization to construct a developed hierarchy. In addition, we propose the CasANER system to recognize and annotate the ANEs. Our ANE recognition process is based on (Finite-State) transducers, which couple the recognition and annotation processes. We will exploit generic transducers, which is a new notion improving the recognition transducers. Then, we generate a cascades calling the established transducers in a predefined passage order. The annotation process is based on the TEI recommendation to detail the ANE components.

Our work originality consists in exploiting corpora (study and test) containing articles extracted from Arabic Wikipedia especially form several Arabic countries. In fact, this variety of articles enable us to treat the different styles of the Arabic language, to contemplate the regional writing, such as in the trigger words like "ولاية","محافظة" "مدينة" and "قضاء" to introduce a city name and to envisage the different civilizations, such as the use of various calendars (Syriac, Muslim and Gregorian) to describe dates. Moreover, we refine the categorization, which participates in a large part to improve the realization of other NLP-applications. Similarly, our work originality resides to generate using the TEI recommendation a structured output able to be integrated in the semantic Web and to enrich electronic NE dictionaries.

The present paper is composed of five sections. Section 2 presents the state of the art on NER systems based on different categorizations, approaches and domains. Section 3 consists in describing our linguistic study made to identify the ANE categories and forms from our Arabic Wikipedia corpus and describe the different relationships with the ANEs. Section 4 details our proposed method to elaborate CasANER system to recognize ANEs and to annotate them using the TEI recommendation tags. The implementation ensured by the linguistic platform Unitex and the evaluation are presented in Sect. 5. the linguistic platform Unitex is used in this phase. Finally, we give a conclusion and some perspectives.

2 State of the Art on NER Systems

The establishment of the NER systems is a process composed of three fundamental steps. The first step is related to the NE definition, which facilitates its determination. Then, the second step is based on the first one. In other words, the NE definition enables to guess its category and therefore an NE categorization step arises. Furthermore, the third step concerns the choice of the approach and the associated formalism or techniques, which respond to the future system needs. In the current section, we will present the previous NE categorization and some previous reasearch to elaborate NER systems using the different domains and corpora nature.

2.1 *Previous NE Categorization*

The NER systems revolve around the NE categorization, which is a crucial step for many NLP-applications. It should be noted that the NE categorization is a process aiming to provide adequate NE representation. Furthermore, this process helps elaborate an appropriate hierarchy translating the corpora richness. The NE categorization was proposed for the first time in MUC-6 [20]. In fact, the same MUC-6 categorization was taken up at MUC-7 [13]. However, a new category was added depending on the corpus nature and it appends a slight modification to the last categorization. As well, many conferences and projects adopted the MUC-6 categorization, which is refined in each one based on different corpora related to specific domains and languages. Among these conferences, we quote the two MET (Multilingual Entity Task) conferences that were organized parallel to MUC-6/7. The MET provided a new opportunity to evaluate the NER task progress in Spanish, Japanese and Chinese [25]. Even, the IREX (Information Retrieval and Extraction) is a Japanese project reposing also on the MUC-6 categorization and it allows the adding of the sub-categorization notion [20]. In the CoNLL (Conference on Natural Language Learning), the authors proposed some changes to the MUC categorization by adding a new category called Miscellaneous for the NEs having no determined category. In addition, many evaluation campaigns also adopted the MUC principle, such as the ACE (Automatic Content Extraction) campaigns [15], Ester [18] and annotation models, such as Quaero [19, 5]. The already mentioned categorizations aimed not only at adding new NE categories but also at refining the existent one. At the same time, they set other objectives, such as providing structured and accessible corpora and enhancing the NER task. In the following section, we will present the NER approaches and the associated previous work.

2.2 NER Approaches and Systems

The NER systems are always based on the three main approaches (symbolic, statistical and hybrid) to recognize the NEs and annotate the recognized NEs through a norm, which represents and details their components [8, 33]. Nevertheless, these systems used undefined markup, which responds to their needs and reliability. In what follows, we will present the elaborated NER systems and tools based on the main approaches.

Symbolic approach. The symbolic approach for NER is based on formal local grammars described by hand-crafted rules, which identify the NEs. These hand-crafted rules can be modeled using regular expressions or finite state transducers. In this context, we quote the PERA system developed by [30], which recognizes Person's Names in Arabic scripts. Based on the same PERA functionalities, a NERA system was elaborated by [31]. This system can recognize ANE of 10 categories. In addition, [27] proposed a novel methodology to improve NERA system by identifying its weakness. The improved system is called NERA 2.0, which ameliorates the coverage of the previously misclassified NEs. This enhancement makes the system achieve more reliable results. Using the rule formalism, [2] developed also a system for an Arabic person's name composed of four main rules composing the core system. In [3], the proposed system is dedicated to recognize an event, time and place expressions through a set of general rules: two rules for time expressions and a rule for place expressions. In addition, [12] proposed a tool for both the Part of Speech (PoS) tagging and NER for the Arabic language. According to the NER, the authors elaborate a NE detector, which acts on the text by giving the adequate labels. The elaborated NE detector starts by reading the data set, which are lists containing the person, location and organization names. Then, it splits the input text into words to apply the established rules. Consequently, if the divided words match these rules, then the labels will be assigned, and else the label will be unknown. Moreover, the authors used three labels, which are Person (PERS), Location (LOC) and Organization (ORG). Using the transducer formalism, the authors in [17] proposed a system recognizing ANEs for the sport field. The proposed system is based on syntactic patterns transformed into transducers. Generally, the transducers are called in a specific passage order. This order is known as transducer cascade. Moreover, the transducer cascade is used for the Arabic language to recognize and annotate the ANEs for the Date [9], Person name [10] and Event [11] categories. In fact, the proposed transducer cascades were tested on Arabic Wikipedia corpus and generated using CasSys tool integrated under the linguistic platform Unitex [23].

The statistical approach. The statistical approach takes advantage of ML-algorithms to learn NER decisions from annotated corpora. The current approach requires the availability of large annotated data. Using this approach, we quote the system elaborated in [1] to recognize the ANEs, which can be assigned to 10 identified categories. The elaborated system includes the CRF and bootstrapping pattern recognition. In [34], the authors developed a system to recognize Arabic temporal expressions by exploiting the dashtag–TMP, which is a temporal modifier referring to a point in time or a time span. Furthermore, [26] developed a system recognizing four ANE categories adopted the Artificial Neural Networks (ANN). To improve the NER on microblogs, [14] proposed a system recognizing ANE using the Condition Random Field (CRF) classifier to tag new training set from Tweeter. In [39], the authors proposed a Biomedical NER (Bio-NER) based on a deep neural network architecture. This proposed architecture is composed of multiple layers. Indeed, each layer has abstract features based up on those generated by the lower layers. The elaborated method is absed on a Convolutional Neural Network (CNN) technique. The exploited unlabeled corpora are GENIA corpus and a set of data collected from PUBMED database. The collection from the PUBMED is made through the biopython[1] tool. Based on PUBMED database, [21] proposed a system to extract the biomedical NEs. To realize the proposed system, the authors chose a subset of the relevant document from the PUPMED. Then, they treated the collected corpus through a preprocessing task including the tokenization and the stemming steps. Besides, the preprocessing is a part of the NER phase. In fact, there are other tasks belong to the NER phase, such as the syntactic annotation like the Part of Speech (PoS) tagging and noun phrase chunking. The semantic annotation is

[1]http://biopython.org/wiki/Biopython.

the core of the NER process, which is made in this system using lexical resources and ML based on NLP Model generation using the CRF. Moreover, the system outcome compared to the SVM (Super Vector Machine) algorithm shows that FS-CRF (Feature space based CRF) does better than the SVM.

The hybrid approach. The hybrid approach is the fusion of both the symbolic and statistical approaches, which are complementary. This approach helps to achieve a significant improvement of NER performance. Among the systems based on this approach, we quote: The automatic system developed by [35] to recognize the NEs having the category Event. For the Turkish language, [22] developed a system to recognize the Turkish NEs. For the Arabic language, [32] proposed a system capable of recognizing 11 categories that can represent an ANE. The NER progresses in several languages while it remains a challenge in Indian languages, such as the Assamese. In reality, the Assamese language suffered greatly from the lack of research effort as well as the appropriate textual resources. Besides, the available corpora have a quite small number compared to other languages. For this reason, there are some studied that were carried out to develop systems, which can perform the NER in Assamese texts [37]. In [36], the authors elaborated is the first hybrid system to realize the NER in Assamese. This system recognizes the Assamese NEs and it is capable to treat four categories, such as Person, Location, Organization and Miscellaneous. The authors used three main steps for the NER process. The ML approach is the first step involving both the CRF and HMM techniques. The second step concerns the rule-based approach, which consists in using a set of handcrafted rules. The last step is the gazetteers-based approach, which integrates the NE tagging using lists including location, person and organization names. According to specific domains, [29] elaborated a hybrid model for NER in unstructured biomedical texts. The elaborated model is a framework that has a main task consisting of the identification and classification of the biomedical NEs into five classes namely DNA, RNA, protein, cell-in and cell-type. In fact, the proposed model combines the rule-based and ML approaches applied after a preprocessing step. The rule-based approach is dedicated to the NE identification. However, the NE classification is ensured by the ML approach, especially, the SVM classifier. The authors experimented their model on the data set from GENIA corpus (Medline abstract collection).

In the previous work, the authors adopted no formal definition determining the recognized NE limits. Furthermore, the illustrated NER systems are based on textual resources that are not always exhaustive. In fact, the free resources can be a solution for this difficulty. The identified rules for the NER can cause ambiguity problems if they are applied without a specific order. Therefore, this case requires an adequate formalism to ensure a good NER. Thus, using an annotation standard to detail NE component is necessary to produce structured corpora. In the following section, we will explain our linguistic study to identify ANEs from an Arabic Wikipedia corpus.

3 ANE Identification and Categorization

The objective of our study on the corpus extracted from Arabic Wikipedia is to have a formal and generic system applicable to all domains. This system allows the ANE recognition and treats the various ANE forms appearing in Wikipedia articles with a prediction of those that can be recognized. Generally, the article content is written in Modern Standard Arabic (MSA), Classical Arabic (CA) and Dialectal Arabic (DA) [4]. Anyway, this diversity of the language styles is very motivating and it helps us to collect numerous alternative forms. Indeed, it is not easy to identify an ANE because we must refer to a clear definition that helps determine its limits. Even, we should mention that the ANE categorization is not a trivial task. We know that there are various opinions about which categories should be regarded as an ANE and how the limits of those categories should be. If we want to detect a relevant ANE, then we must analyze all its presented forms while respecting several regional writings. Besides, we should unify the similar forms to eliminate the repeated ones and separate the different forms to avoid ambiguity. Furthermore, we determine an ANE as an expression that can or cannot contain a proper name. The determined ANE can be structured through categories and subcategories. The following sub-section describes our linguistic study to identify ANE categories and forms in the Arabic Wikipedia study corpus.

3.1 *Identification of the ANE Forms and Categories*

Our linguistic study shows that the Arabic Wikipedia study corpus comprises five main ANE categories, which are Date, Person name, Location, Event and Organization. Each category is composed of refined sub-categories.

We identify the "Date" category as a specific part of the numerical expressions. In fact, we find several forms that can describe this category. Moreover, These forms are divided into six sets, which are Period, Century, Year, Date based on Month, full Date or Season followed by a Year. These sets are presented in the following figure associated with the adequate examples.

Figure 1 shows the six sets regrouping the different identified forms, which describe the Date category. For example, we find that period the form can be calculated based on the month, the day, the century and the year. In addition, our study shows that some trigger words can appear in different morphological forms (plural, dual), such as "سنتي/ between" used to calculated period between 2 years.

The "Person name" category represents various forms of an Arabic person name. These different forms are highly associated with the country origin, religion, culture, level of formality and personal preference. Generally, this category contains the following parts, "ism", "kunya", "nasab", "laqab", and "nisba" [33, 11]. Therefore, we remark that a person name form regroups at least one of the already mentioned parts. For example, the ANE "عبد الفتاح أبو غدة/ Abd Al-fattah Abu Ghoda" is composed of an "ism", which is "عبد الفتاح/ Abd Al-fattah" and a laqab, which is

Fig. 1 Forms describing ANE date

"أبو غدة/ Abu Ghoda" expressing a kunya. Otherwise, we classify the trigger words preceding the ANEs in eight classes that are Civilities (الآنسة/ Miss), Profession (المدير/ the director), Peerage function (الأمير/ the prince), Political function (الوزير/ the minister), Religious function (المؤذن/ the muezzin), Sportive function (اللاعب/ the player), Artistic function (الممثل/ the actor) and Military function (الرائد/ the major).

The place names are the most common forms appearing in our study corpus. For this reason, we affect them to the category named "Location". Then, we extend this category to be composed of three sub-categories, which are Absolute, Relative and Geographic Location. We mean by Absolute Location category all the place names defined in its sense by one place, such as the country (تونس/ Tunisia), continent (أفريقيا/ Africa), city (تور/ Tours), delegation (مناخة/ Manakhah) and region (الشبطلية/ Al-Shabtiliyah) names. The Absolute Location forms can be identified through the trigger words and prepositions playing the role of a place indicator. Moreover, we identify the Relative Location sub-category as a place name, which can be quantified by its relationship with an absolute Location as building. We identified 16 sub-categories describing this sub-category. The Relative Location forms are identified through trigger words, which can be a part of the identified ANEs. Grammatically, the trigger words can be defined or undefined as "المدرسة/ the school" and "مدرسة/ school". The richness of our study in terms of Relative location sub-category enables us to refine it into 16 sub-categories. Then, we associate each sub-category with its appropriate forms.

Our identification of the Geographic Location sub-category considers them as specific physical points on earth. In addition, we notice that all the ANE forms related to this sub-category appear with geographic features, which can be a Mountain or a Hydronym. Similarly, the Hydronym forms are divided into two sets where the first set regroups the river or the lake names and the second set is dedicated to the sea names. We illustrate a small hierarchy describing the sub-categories of Relative and Geographic Location.

Figure 2 regroups the sub-categories composing both the Relative and Geographic Location. The forms describing the Relative Location sub-categories have

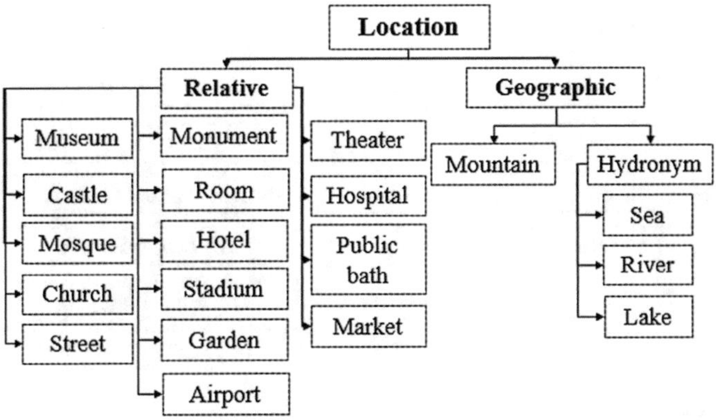

Fig. 2 Sub-categorization of relative and geographic location

different components. These components can be adjectives (الدولي/ the international), such as "المسرح/ الدولي the international theater". It can be noun phrases (العلم الأردني/ the Jordanian flag) in monument names, such as "سارية العلم الأردني/ The Jordanian flagpole". Then, the Relative Location sub-category can contain defined common noun (الثقافة/ culture), such as "شارع الثقافة/ Cultural street". Similarly it can contains the Absolute Location (عمان/ Amman), such as "مطار عمان المدني// Amman civil airport". The Geographic Location forms are also expressed through several trigger words, which help determine their features. For example, the trigger word "جبل/ mountain" is a mountain feature among a set of synonym trigger words, such as {قمة الجبال, تل الجبل, مونتي جبال, جبل}. Notice that some Trigger words are foreign as "مونتي/ mountain" referring to other languages.

Besides, we identified the category Organization and its forms using a set of trigger words allowing not only their identification but also the determination of the organization nature. The organization names can describe an institution, a minister, a university, a society and so on. There are organizations that appear in the acronym form. Nevertheless, we do not treat this case because it is relatively rare in our study corpus.

Figure 3 describes the different sub-categories composing the organization ANEs. The identified forms enable us to treat the agglutination when two common nouns are related by a conjunction, such as "التربية والتعليم/ education and teaching" and analyze the adjectival phrase, such as "التلفزة الوطنية التونسية/ the Tunisian national television", which contains a succession of defined adjectives where "التونسية" is an Arabic adjective expressing a "Nisba".

The last category that we identified is Event, which was well studied in [11]. Therefore, we identify an event as a nominal composition that can have different forms and we classify its forms using the sub-categories. We use the same sub-categories of the ANE event that can be political, cultural and religious.

Figure 4 describes the sub-categories that we used to classify the ANEs having the Event category. Similarly, Event is determined through a set of significant trigger words, which are a part of the identified ANEs. After the ANE form

Fig. 3 Organization sub-categories

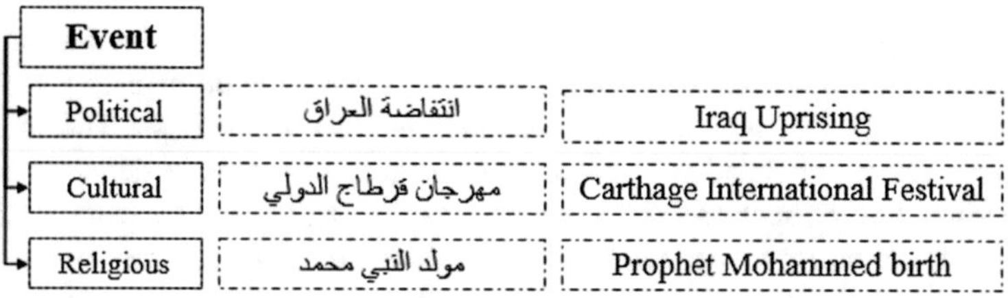

Fig. 4 Event sub-categories

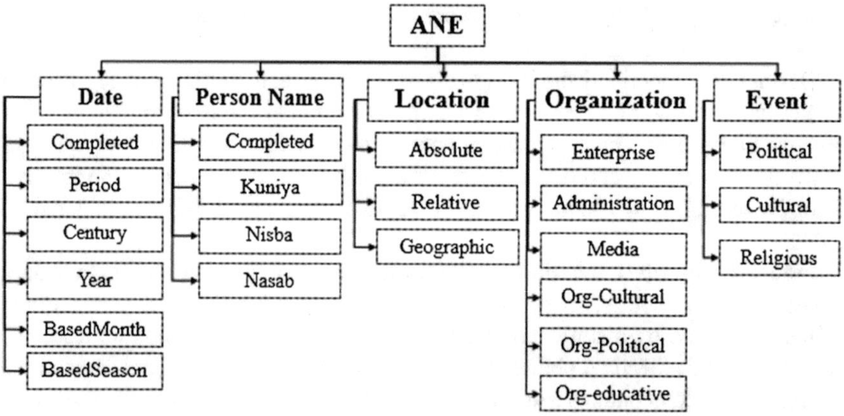

Fig. 5 ANE hierarchy identified from the study corpus

identification, we notice that the ANEs can be assigned to several categories and sub-categories. For this reason, we regroup them in an ANE hierarchy, which helps us to describe our categorization.

Figure 5 describes the ANE hierarchy associated with our study corpus. All the categories are refined to be composed of sub-categories. The illustrated categories are extended to increase the granularity level. This extension depends, in a large part, on the appearance of the ANE forms.

Our linguistic study plays a significant role not only in identifying the ANE forms but also in showing the different relationships that can relate them. In the following sub-section, we describe the different kinds of the relationships between the identified ANEs appearing in our Arabic Wikipedia study corpus.

3.2 *Relationship Between ANEs*

The relationship between the ANEs can be binary (involving two ANEs) or more complex to be an imbrication of ANEs. The ANE imbrication always describes a composition of these ANE without a specific link. However, when a particular link appears through phrases or prepositions, the kind of the relation becomes an SR. In what follows, we will describe the ANE imbrication.

The ANEs having respectively the category Event and Location have a composition relationship with those having the Date category. For example, a relative location name, especially stadium one, can contain a relative date, such as “ملعب 14جانفي/ 14 January stadium”. We notice that the date is composed of a form among those identified during our linguistic study. The imbrication of these ANEs may be surrounded by a left context, such as “ملعب 14جانفي بنابل/ 14 January stadium of Nabeul” and ملعب بالكاف/14 “جانفي January stadium of Kef”. The illustrated left context is in its turn an agglutinated absolute location name. Sometimes, the ANEs Date refer to symbolic events occurred in the past. For example, “يوم 14جانفي / 14th January” is a relative date referring to the event of the Tunisian revolution. Similarly, Event and Location can be a part of the Person Name category, such as “ملعب الطيب المهيري بصفاقس/ El-Taieb Mhiri stadium of Sfax” where “الطيب المهيري/ El-Taieb Mhiri” is a person name of a famous personality. A person name can also be integrated in an organization name, such as “مؤسسة العنود الخيرية/ Anoud Charitable Foundation” and we notice that here the name person is a princess first name. Our study corpus has a rich content, which is very interesting. This richness helps us to value the significant SRs between ANEs. The SRs are always binary relating the ANEs having the same or different categories. The same ANE categories can be related by a synonymy, such as “الثورة التونسية/ Tunisian revolution” and “ثورة الحرية والكرامة/ the freedom and dignity revolution”. The synonyms of ANEs describe an event that can be expressed by an ANE Date, which is “جانفي 14/ 14th January”. Among the SRs linking different categories, we find the meronymy, which expresses an inclusion relation, such as “كمال الملاخ/ Kamal Al-Mallakh” and “القاهرة/ Cairo”. This means that the first ANE is included in the second one. Generally, the linked ANEs surround the relevant expressions or prepositions, which facilitates to guess the SR type. In fact, determining SRs between ANEs is an important challenge that will be studied later. In the following section, we will describe our proposed method to recognize and annotate ANEs.

4 Proposed Method

Our proposed method, which is intended to elaborate CasANER system recognizing and annotating ANEs, is composed of the following steps: the collection of Arabic Wikipedia articles to construct corpora (study and test), the construction of necessary dictionaries and extraction rules and the transducer establishment. We propose the following architecture to describe the different steps.

Figure 6 shows that the CasANER system relies on an important step, which is the resource identification. Furthermore, we construct our extraction rules based on the trigger words, which are identified during the linguistic study. Besides, we convert these established extraction rules into regular expressions and we regroup them into three sets; analysis, filtering and synthesis. Hence, we distinguish two main phases; Analysis and Synthesis.

It should be recalled that a transducer is a graphic representation of the regular expression specification. This establishment is not arbitrary but we fix a principle to create its boxes. Therefore, each transducer contains the collected trigger words, particularly those having common paths. Besides, we always try to separate the overlapped paths because they may create ambiguities. Inside the created transducer, we can call sub-graphs that enables us to reduce the transducer size. According to the annotation, the used tags always encompass the recognized paths. In this context, the analysis and filtering transducers ensure the analysis phase. In fact, the analysis transducers deal with both the recognition and annotation processes. However, we construct the filtering transducers to recognize the ANEs that are not treated by the first analysis. These ANEs are not treated due to the preprocessing, especially, the segmentation step. Otherwise, we try to rectify the recognition paths using variables to organize the output of the filtering transducer. In what follows, we will describe the analysis transducers.

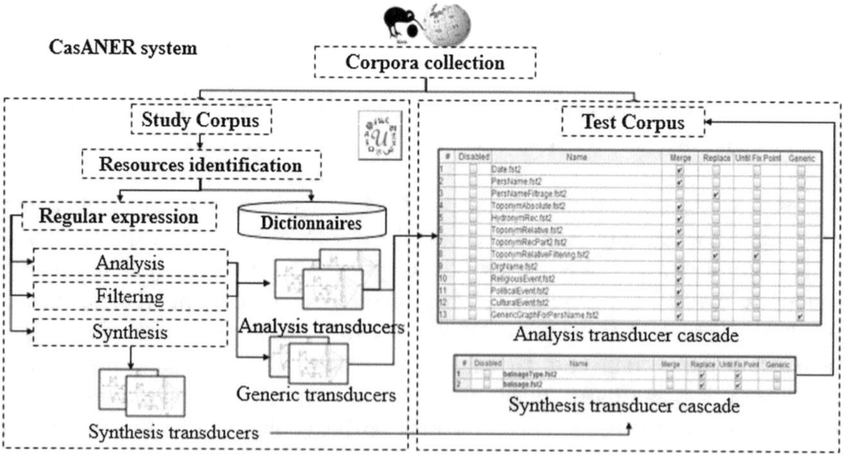

Fig. 6 CasANER architecture

4.1 Analysis Transducer Establishment

The analysis transducers regroup recognition and annotation paths for the identified categories and sub-categories with several kinds of boxes. To analyze the morphological level, we use the morphological mode or filter, which facilitate the agglutination treatment. The morphological filter is used to control the form of a word belonging to the syntactic pattern, which recognizes an ANE. Moreover, the morphological mode is used, for example, to read a conjunction (<CnjCrd>) linked to a first name, like "ومحمد/ and Mohammed". In what follows, we will illustrate some transducers to show their specificity and form.

For the Date category, we create a main transducer, which regroups other transducers recognizing the identified forms (Sect. 3.1). Among these transducers, we illustrate the transducer that recognizes a season followed by a year.

Figure 7 shows the alternative paths to recognize a season followed by a year forms. This transducer resolves the case of the agglutination phenomena. This case is described by two boxes of "<" and ">" to mean that the recognition touches the morphological level. Therefore, the preposition (<Prps>) of the conjunction (<CnjCrd>) will be separated from the box containing <Np + season>. This transducer can recognize the following ANEs "2000 صيف/ summer 2000", "م 1990 ربيع عام/ spring 1990" when the recognition path detects the specific and internal indicators and "فصل الخريف/ the autumn season" if the ANE is detected through a trigger word belonging to the first box.

It is very important to notice that our analysis transducer annotated the recognized ANEs using {} markers. In fact, we do not use them arbitrarily because they make the recognized ANE polylexcial word, which cannot be detected by another transducer. This means that our recognized ANEs are protected. We should also notice that the annotation markers are represented in the node output. Besides, we use the same principle to treat the rest of the categories.

Regarding the Person name category, there are two main transducers. The first one calls the sub-transducers treating this category without trigger words. Nevertheless, there are sub-transducers that need to be preceded by a box storing the sub-graphs containing the trigger words, which are organized based on the

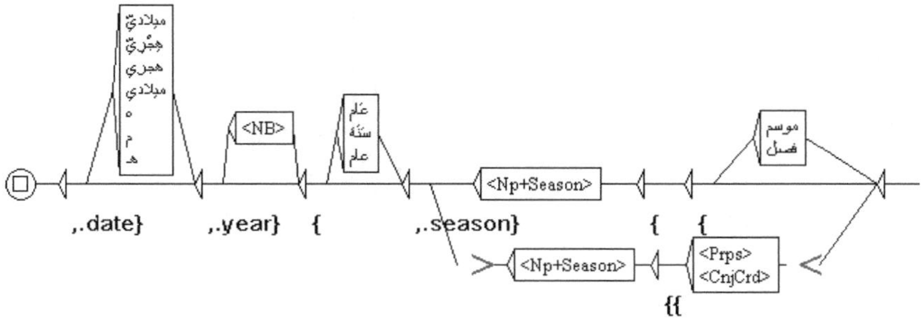

Fig. 7 Transducer recognizing ANE date composed of a season followed by a year

identified classes (Sect. 3.1). In addition, the use of trigger words helps avoid the ambiguity problems related to this category. However, we can decrease this ambiguity using a specific kind of transducer that will be explained later.

Figure 8 describes a path recognizing an identified form, which is composed of a first name, a civility and a last name. The illustrated path can recognize the ANE “بك فؤاد العادلي/ Foued bek Al-Adly”, which contains a civility belonging the list stored in the sub-graph “Civilities”. This transducer does not comprise the final tag “persName” because it will be called in the main transducer.

In the previous transducers, we have not use any element inside the tags because we have treated only the ANE forms related to a category. However, we will use an element named “type” to store the sub-category values. We begin by describing a transducer, which recognizes the Absolute Location sub-category to illustrate the element utility.

Figure 9 describes the recognition of the city names, which will be annotated through the “placeName + type = “city”” tag. It is worth noting that we use the “+” sign just to replace the space, which may cause ambiguity problems during the experimentation phase. Here, the element “type” helps determine the value of the Absolute Location sub-category. Besides, it can take “country”, “continent”, “delegation” and “region” depending on the Absolute Location nature. In the agglutination case, we add the “Prps” tag as output to the boxes containing “Prps” and “CnjCrd” in order to protect them. In fact, this absolute Location can be a part of other ANE. Thus, if we do not separate the ANE and the preposition by tags, then this absolute location will not be recognized.

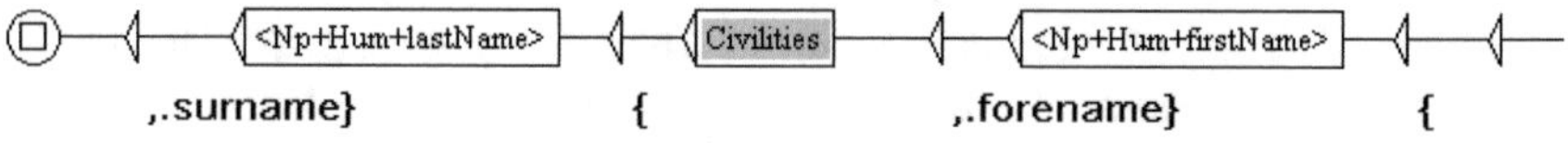

Fig. 8 Transducer recognizing an ANE having person name category

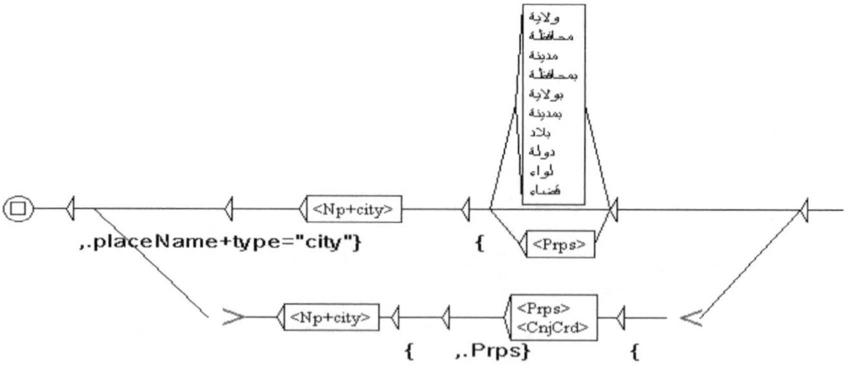

Fig. 9 Transducer recognizing city names

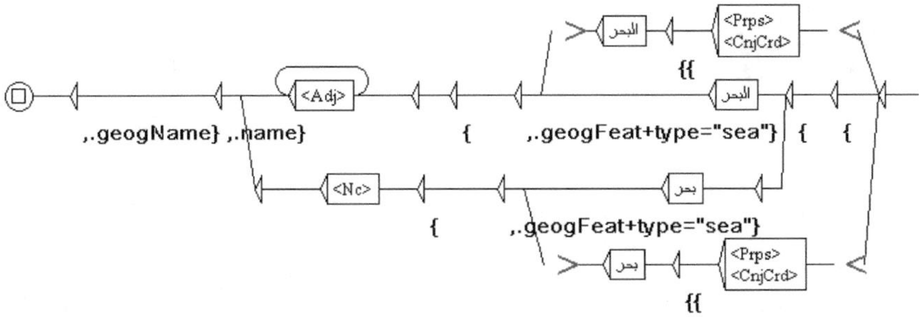

Fig. 10 Transducer recognizing a sea name

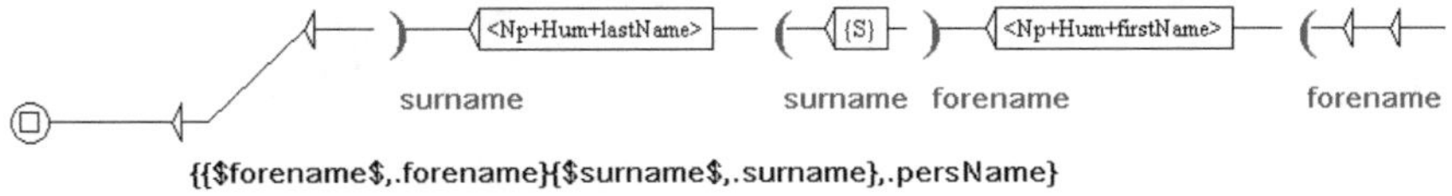

Fig. 11 Filtering transducer to recognize a person name

The "type" element is mainly used to describe a sub-category. Nevertheless, we can use it to describe the value of a geographic feature when we recognize the ANEs having the Geographic and Hydronym sub categories.

Figure 10 shows our manner to annotate the Hydronym sub-category, especially the sea names. At the begging of each path, we consider the trigger words as geographic features associated with a tag called "geogFeat". Then, we surround the rest of the ANEs with a tag entitled "geogName". The illustrated transducer is called in the main one where these paths will be integrated in a global tag "placeName" containing an element type = "Hydronym".

Filtering analysis transducer. The filtering transducers belong to the analysis phase. Their objective is to recognize the ANEs when their components are separated during the segmentation phase. It should be noted that the segmentation is made through a graph provided by the exploited linguistic platform. The filtering transducer is based on the same paths of the analysis one but each box is surrounded by variables to temporary store its value. At the end of the recognition path, we organize the output to obtain an annotated ANE.

Figure 11 describes a transducer treating an exceptional case for the Person name recognition. Here, we have two ANE components separated by {S}, which is a segmentation symbol. The illustrated transducer respects the same annotation principle of the analysis transducer. In addition, we call the used variables (forename and surname) using the $ symbol to retrieve their values.

Generic transducer. Our generic transducers are tagging generalization graphs that aim to locate unrecognized NE occurrences when these NEs appear out of the context with those identified based on specific contexts [28]. Creating a tagging generalization graph consists in building a path that begins with a box having $G

Fig. 12 Tagging generalization graph for the category person name

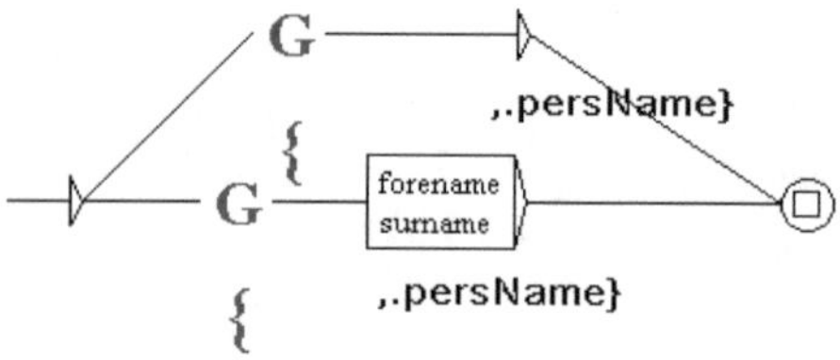

Fig. 13 Tagging generalization graph for the category date

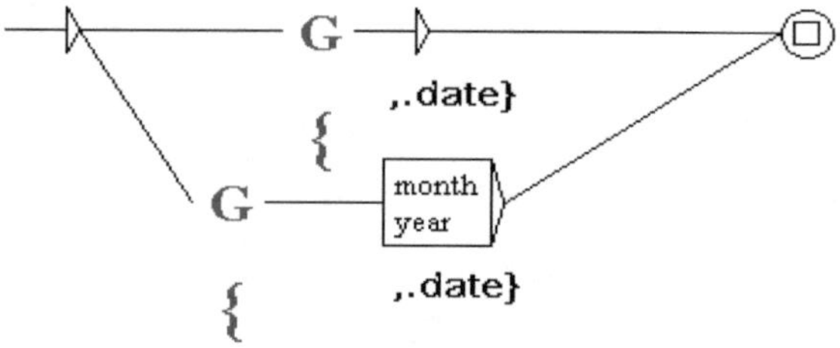

and an opening curly bracket output. Then, the same path has a second box containing the searched categories to find the unrecognized NEs. Checking the generalization mode is necessary when we add this transducer kind to our cascade using the exploited linguistic platform. This enables to consult a text file named "tok_by_alph.txt" of the previous graph placed in the same transducer cascade. In reality, the tagging generalization graph stores in the box all the NEs of the searched category, which are extracted from the "tok_by_alph.txt" file. Finally, the graph recognizes the NEs out of context and attributes the main categories. For example, if the graph recognizes a forename from a full-recognized ANE, then, it will assign it to the "persName" category.

Inside the tagging generalization graph, we can put some restrictions in the second box containing the searched category as an output node. This case is used to treat the recognized ANE components. In fact, we can add new information to the category described in the output node to complete the annotation. The following figure illustrates the restriction created to recognize only forenames and surnames out of the context.

Figure 12 illustrates a tagging generalization graph treating the Person name category. In fact, we duplicate the box containing $G for each path. Moreover, the restriction here is described in the second path.

We also use the restriction principle to recognize the elements composing the Date forms, which helps us to recognize the months and years out of the context (Fig. 13).

The transducers, ensuring the analysis phase, contain an annotation form respecting the tool, which will regroup them into a cascade. However, we organize the annotation syntax as a preprocessing to transform it into the TEI annotation. In what follows, we will describe the principle of this transformation using the synthesis transducers.

4.2 Synthesis Transducer Establishment

The synthesis transducer consists in transforming the annotation made by the analysis phase to the annotation related to the TEI recommendation. It is worth noting that the TEI recommends an international consortium where their goals are the development of a set of standards for the preparation and the exchange of electronic texts [7, 24]. Therefore, we establish our synthesis transducers using the syntax described in [38].

The TEI syntax is defined as follows: an opening tag describing such category, like <persName> and a closing tag </persName> that surround the ANE. The tag <persName> can include an imbrication of the first name, last name and trigger word that can precede a person name that are surrounded respectively by other tags, such as <forename>, <surname> and <roleName>. In the "roleName" tag, it is possible to specify the roleName type, such as military function. Moreover, the addition of the sub-category is made using an element named "type", which can contain different values. Indeed, other categories of an ANE can be presented by the TEI recommendation, such as <orgName> and </orgName> to describe an organization names and <placeName> and </placeName> to describe the Location category, which can have an element "type", such as type = "castle" to annotate castle names, which are Relative Location. To understand the principle of the transformation of the analysis annotation into the TEI recommendation tags, we propose the following architecture (Fig. 14).

After carrying out an analytical study on the annotated files, we have developed two transducers for the synthesis phase. The first one treats the recognized ANEs, which do not have a sub-category. Based on the linguistic study and the refined categorization, we elaborate a second transducer to transform the annotation of the ANEs having an element called "type", including their sub-categories.

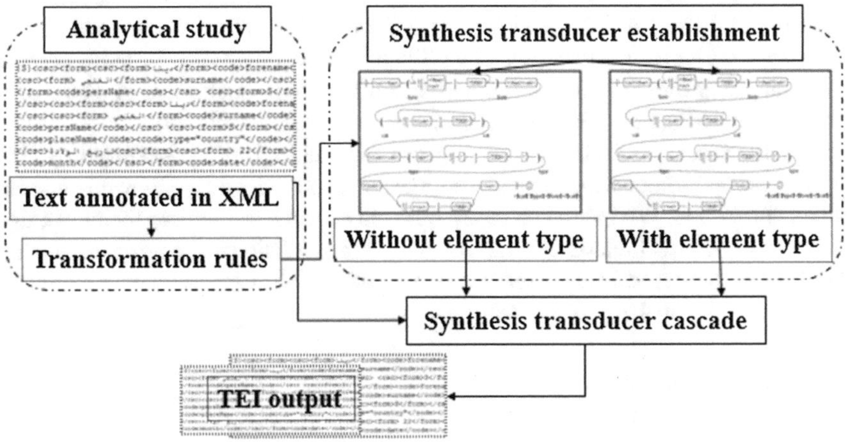

Fig. 14 Principle of transforming annotation

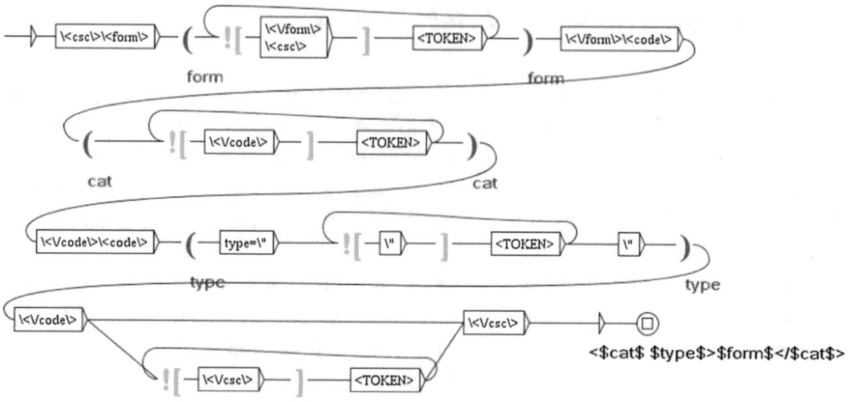

Fig. 15 Transducer transforming the annotation of the analysis phase into TEI

Figure 15 describes the transducer transforming the annotation specific to the analysis phase into TEI when the recognized ANE contains an element "type". In fact, we propose a path taking as input the XML (eXtensible Markup Language) format of the annotated text files generated by the tool creating the transducer cascade. Hence, we use the negative context through ![,] markers and variables using (,) markers, which organize the output annotation.

5 Implementation and Evaluation

Our CasANER system is implemented using the linguistic platform Unitex[2] (version 3.2 alpha), especially, the CasSys tool to generate transducer cascades. However, we experimented our proposed system through a test corpus, which is also collected from Arabic Wikipedia with Kiwix[3] tool. The collected test corpus contains text files for a cumulative of 95 378 tokens. Furthermore, the study corpus, containing text files for a cumulative of 146 000 tokens, enables us to create new dictionaries and update dictionaries available under Unitex platform.

In some cases, we find that Arabic dictionaries under Unitex elaborated by [16] do not respond to our needs in order to make a good recognition. For this reason, we exploit Arabic Wikipedia, precisely our study corpus to improve their coverage or to create new dictionaries. In fact, creating a new dictionary requires respecting the features proposed in a tagset[4] available under Unitex. Indeed, they store different variations that an Arabic entry can have.

[2]http://www-igm.univ-mlv.fr/~unitex/.

[3]http://wiki.kiwix.org/wiki/Main_Page.

[4]The tagset of Arabic Unitex package dictionaries.

Table 1 Dictionary coverage

Dictionary name	Coverage
First name	9 917
Last name	1 991
Arabic adjective	2 938
Toponym	13 757
Common noun	14 976
Direction	14
Season	11
Day	23
Month	48

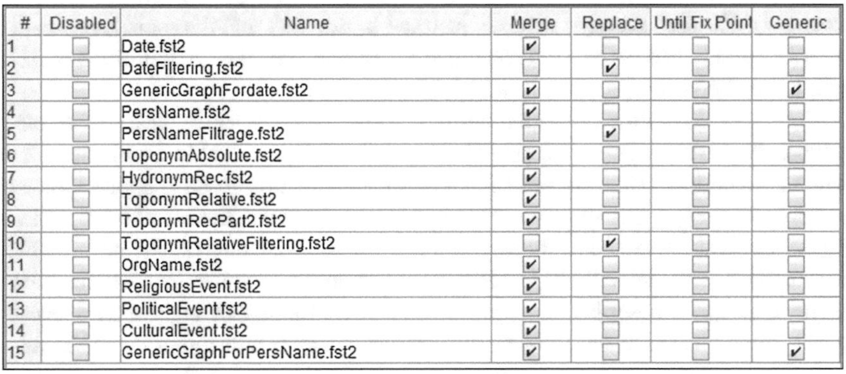

#	Disabled	Name	Merge	Replace	Until Fix Point	Generic
1		Date.fst2	✓			
2		DateFiltering.fst2		✓		
3		GenericGraphFordate.fst2	✓			✓
4		PersName.fst2	✓			
5		PersNameFiltrage.fst2		✓		
6		ToponymAbsolute.fst2	✓			
7		HydronymRec.fst2	✓			
8		ToponymRelative.fst2	✓			
9		ToponymRecPart2.fst2	✓			
10		ToponymRelativeFiltering.fst2		✓		
11		OrgName.fst2	✓			
12		ReligiousEvent.fst2	✓			
13		PoliticalEvent.fst2	✓			
14		CulturalEvent.fst2	✓			
15		GenericGraphForPersName.fst2	✓			✓

Fig. 16 First transducer cascade composing the CasANER system

Table 1 shows the coverage of our dictionaries. In fact, to increase the coverage of the Arabic Adjective and Common noun dictionaries, we improve them using an automatic enrichment based on textual resources.

Our elaborated transducer cascade for the analysis phase calls 15 main graphs in a specific order and respects the adequate mode for each graph. However, we created 178 graphs including analysis, filtering and generic one.

Figure 16 shows the passage order that we have chosen to organize our analysis transducer cascade. In fact, there are graphs that are applied on the text using the mode "Merge". However, the filtering graphs use the "Replace" mode because they use variables at the end of their recognition paths. Therefore, they need to replace the last ANEs with new ones associated with an annotation. The last use mode is Generic for applying the tagging generalization graph.

The generation of the final analysis transducer cascade is not an easy task. In reality, it needs to change the transducer order to fix the best one and test it. During this test task, we detected several kinds of errors. For example, when we treat the category Event we must separate the graphs depending on these sub-categories. Then, if we put cultural event graph before the religious one. We found that the

#	Disabled	Name	Merge	Replace	Until Fix Point	Generic
1	☐	balisageType.fst2	☐	☑	☑	☐
2	☐	balisage.fst2	☐	☑	☑	☐

Fig. 17 Form of the synthesis transducer cascade

ANE "مهرجان عيد الفطر/ Eid Al-fitr festival" was not recognized because it contains the ANE "عيد الفطر/ Eid Al-fitr" which is a religious festival.

The second part composing the CasANER system is the synthesis transducer cascade. This cascade is dedicated to deliver a structure output by transforming CasSys tags to TEI tags. The synthesis cascade also needs a passage order but it is not difficult to choose the best order. Furthermore, it differs from the analysis cascade in the passage mode and in the format of the input file. In fact, the difference of the used passage mode is justified by the fact that the transducers called by this cascade use variables to organize the output. For this reason, the last ANE will be replaced by the same ANE with the new tags.

Figure 17 shows the passage order of our synthesis transducer cascade. In other words, we pass the graph treating the recognized ANEs having an element. The new mode here is "Until fix point", which means that this graph is applied once or re-applied several times until no change occurs in the text. The role of this synthesis transducer is summarized as follows: initially the ANE is annotated using { }, which is a representation specific to CasSys tool. The accolade annotation has its appropriate translation described in XML, which uses significant tags, such as <csc> means the ANE annotated through the cascade, <form> containing the ANE value and <code> detailing the category in which an ANE is assigned. This tag contains other <code> to describe the sub-category of an ANE using the element type. Besides, the TEI recommendation tags replace the already mentioned ones.

Applying our CasANER system, including the two kinds of transducer cascades, provides a structure and normalized corpus (extracted initially from Arabic Wikipedia). This structure output can be used by several NLP-application. However, before its exploitation, we must evaluate the CasANER performance to show its efficiency.

It should be recalled that evaluating our proposed CasANER system is an important process that helps us to prove its reliability. For this reason, we evaluated it in two manners. The first manner is by calculating the measure values and the second phase by applying CasANER on a new corpus annotated by a ML based system. The second evaluation also permits comparing the result of both systems based on different approaches.

The first CasANER evaluation is performed by the precision, recall and F-score measures that are illustrated in Table 2. We should improve the analysis transducer cascade to cover all the ANEs.

Table 2 demonstrates that CasANER shows a precision of 92%, a recall of 91% and an f-score of 91% for ANE. Therefore, we find that the obtained outcomes are very motivating. Notice that the number of ANE detected in error causes the obtained recall value. The Errors presented in this recognition process are due to the

Table 2 CasANER evaluation using measure values

Recall	Precision	F-score
0.91	0.92	0.91

Table 3 CasANER evaluation by categories

	Date	Person name	Event	Location	Organization
Recall	0.78	0.87	0.92	0.95	0.99
Precision	0.81	0.95	0.96	0.94	0.97
F-score	0.79	0.90	0.93	0.94	0.97

fact that the dictionary coverage must be improved. In fact, the performance of the recognition increases if dictionary coverage is enriched. Errors can be caused by the structure of Arabic Wikipedia's articles. For example, there are ANEs having miss-spelled trigger words, such as "اتنفاضة الصدر" instead of "انتفاضة الصدر/ Sadr's uprising" and "محاحة" replacing the word "محافظة/ city" determining city names. Furthermore, the errors can be found in the prepositions, which can play the indicator role to determine the ANE limits, such as "قي" instead of "في/ in".

In order to show the CasANER reliability, we propose an evaluation for each category using also measure values. This decomposition helps us to determine the parts that require an enhancement.

Table 3 proves that our CasANER system excels, especially, in the recognition of the Event, Location and Organization categories. The recognition of the Person name category is interesting because we used three kinds of transducers that are analysis, filter and generalized tagging, which help us to recognize several ANEs. The use of generalized tagging transducer permits to avoid ambiguity problems, for example, we must guess if a word is a first name or surname and it is not an adjective or a common noun. The Event, Location and Organization recognition relies on the important number of the identified extraction rules. However, the year recognition without trigger words and through prepositions cause the obtained measure values for the Date category.

The second evaluation consists in applying our proposed system on ANERcorp.[5] This corpus is freely distributed. [6] collected ANERcorp from several sources to obtain a generalized corpus. The ANERcorp was collected to construct the study and test corpora for the elaborated ANERsys. The ANERSys recognizes and annotates ANE using ML approach. ANERcorp contains more than 150 000 words annotated for the NER task. Each word in this corpus is annotated as one of the tags illustrated in Table 4.

The evaluation admits three steps. The first one consists in deleting tags existing in ANERcorp to recover the initial corpus. The second step is the application of our analysis transducer cascade on the initial ANERcorp to provide an annotated one.

[5]http://users.dsic.upv.es/grupos/nle/?file=kop4.php.

Table 4 Tags used by ANERsys

Tag	Signification
B-PERS	Beginning of the name of a PERSon
I-PERS	Inside of the name of a PERSon
B-LOC	Beginning of the name of a LOCation
I-LOC	Inside of the name of a LOCation
B-ORG	Beginning of the name of an ORGanization
I-ORG	Inside of the name of an ORGanization
B-MISC	Beginning of an NE that does not belong to any previous class called MISCellaneous
I-MSC	Inside of an NE that does not belong to any previous class
O	Annotated word is not an NE (Other)

The recognized ANEs inside the new corpus are annotated using { }, for this reason, it will be the input of the synthesis transducer cascade. We added a new transducer that is dedicated to transform the Location, Person and Organization tags into tags having the following format: <category> and </category> as <LOC> and </LOC> to replace <placeName> and </placeName>. The categories inside the new illustrated tags are the same used by the ANERsys. Thus, the annotated ANERcorp must be adopted to the same format generated by the synthesis transducer cascade. In the third step, the transformation transducer cascade ensures this adoption.

The transformation transducer cascade acts on the ANERCorp to transform the used tags. The first called transducer is dedicated to delete the Inside tags (i,e. I-LOC, I-PERS) and to regroup the annotated words with the beginning of the name as B-LOC in the same line. The second transducer takes the B-LOC, B-PERS, B-ORG and B-MISC and transforms these categories into the new format <Category> and </Category>. For example, the ANE "فرانكفورت/ Frankfurt" was annotated as [فرانكفورت B-LOC] and it is transformed to <LOC> فرانكفورت </LOC>. The last transducer deletes the O tag. All transducers are in mode "replace". The two first transducers are applied in "until fix point" mode in order to treat imbrication inside the ANE. We also applied our synthesis cascade to facilitate the comparison. A new transducer is added to convert the TEI tags related to Location, Person and Organization to the already mentioned format <category> and </Category> tags.

The evaluation before the application of our analysis and synthesis transducer cascade on the ANERCorp is performed by the precision, recall and F-score measures that are illustrated in Table 3. Our measure values are presented to be compared with the ANERsys measure values that were tested on ANERCorp.

Table 5 illustrates the measure values related to our CasANER system and to ANERsys. We can see that the CasANER results are as efficient as ANERsys one in the recognition and annotation of PERS and ORG. However, the ANERsys can recognize and annotate ANE having the category LOC more than our system.

Table 5 Comparison between CasANER and ANERsys using measure values

	ANERSys			CasANER using ANERcorp		
	LOC	PERS	ORG	LOC	PERS	ORG
Recall	0.78	0.41	0.31	0.71	0.81	0.58
Precision	0.82	0.54	0.45	0.66	0.76	0.55
F-score	0.80	0.46	0.36	0.67	0.78	0.63

There are some cases undetected by CasANER, such as abbreviations. In fact, we do not treat this form, which frequently appears in ANERCorp especially in organization names such as <ORG> سي أن أن </ORG> and <ORG> في تي جي </ORG>. In fact, for the category Organization, we recognize it using trigger words. However, we do not treat the famous organization names as "الفيفا/ FIFA" (<ORG> ا لفيفا</ORG>). The NEs written in other languages are not treated by our proposed system, whereas there are several foreign NEs having the category PERS and ORG annotated in ANERCorp as <PERS> Charles I </PERS> and <ORG> El Telegramma Del Rif </ORG>. There are also ANE annotated in error by the ANERsys as "السودان" that was annotated as a person name.

The CasANER application contributes to the enrichement of the ANERCorp. This enrichment was made through, on the one hand, the refinement provided by the use of categories and sub-categories and, on the other hand, through the use of TEI tags. Moreover, our analysis transducer cascade, which is included in CasANER, can resolve the agglutination phenomena untreated in ANERCorp. For example, the ANE "وقاسم العزام/ and Quasim Alaazam" is annotated as follows <PERS> وقاسم العزام </PERS>. However, CasANER annotates this ANE as follows: the conjunction "و/ and" is surrounded by "Prps" tag as <Prps>و </Prps>. This Prps is attached to two sub-tags: <PERS> <forename> قاسم </forename> <surname> العز ام </surname> </PERS>.

6 Conclusion

In the present work, we proposed CasANER system to recognize and annotate ANEs. The system is composed of two kinds of transducer cascades, which are the analysis and the synthesis implemented through the linguistic platform Unitex, especially, using the CasSys tool.

To realize CasANER, we made a deep and detailed ANE categorization in order to develop a category hierarchy. The elaborated hierarchy relies on representative an Arabic Wikipedia corpus containing articles extracted from several Arabic countries. In fact, analysis cascade regroups transducers, which ensure the analysis, filter and generalization tagging phases. The filtering phase is intended to rectify the paths of the analysis transducers in order to have structured recognized ANEs. However, the generalization-tagging phase helps us to improve our system performance. According to the synthesis, this cascade regroups transducers that helps transform the annotation of the recognized ANEs into an annotation respecting the TEI recommendation. This transform enables us to generate structured output corpus that can be used by several NLP-applications. The evaluation using measure values shows that the CasANER proves its reliability because the obtained results are encouraging. Indeed, our system also demonstrates that it provides results more efficient than those of the ANERsys during the comparison process.

In future work, we will exploit our system output to extract the relevant SRs between the recognized and annotated ANEs in order to create electronic ANE dictionary having a convivial interface. In addition, we will improve the CasANER by adding a ML module to enhance its performance. Finally, we will focus on ameliorating our proposed system to realize the EL task based on existing free resources.

References

1. AbdelRahman, S., Elarnaoty, M., Magdy, M., Fahmy, A.: Integrated machine learning techniques for Arabic named entity recognition. Int. J. Comput. Sci. (IJCSI) 27–36 (2010)
2. Aboaoga, M., Aziz, M.J.A.: Arabic person names recognition by using a rule based approach. J. Comput. Sci. 922–927 (2013)
3. Aliane, H., Guendouzi, A., Mokrani, A.: Annotating Events, Time and Place Expressions in Arabic Texts. In: Proceedings of Recent Advances in Natural Language Processing, pp 25–31, Hissar, Bulgaria, 7–13 (2013)
4. Alsayadi, H.A., ElKorany, A.M.: Integrating semantic features for enhancing arabic named entity recognition. Int. J. Adv. Comput. Sci. Appl. (IJACSA) **7**(3), 2016 (2016)
5. Arnulphy, B., and Tannier, X.: Entités nommées événement: guide d'annotation. Notes ET Documents LMSI N: 2013–12 (2013)
6. Benajiba, Y., Rosso, P., Benedíruiz, J.M.: Anersys: An Arabic named entity recognition system based on maximum entropy. In: International Conference on Intelligent Text Processing and Computational Linguistics, pp. 143–153 (2007)
7. Ben Ismail, S., Maraoui, H., Haddar, K., Romary, L.: ALIF editor for generating Arabic normalized lexicons. In: Will Appear in Proceedings of the International Conference on Information and Communication Systems (ICICS) (2017)
8. Ben Mesmia, F., Friburger, N., Haddar, K., Maurel, D.: Construction d'une cascade de transducteurs pour la reconnaissance des dates à partir d'un corpus Wikipédia. Colloque pour les Étudiants Chercheurs en Traitement Automatique du Langage naturel et ses applications, pp 8–11, Sousse, Tunisie (2015)
9. Ben Mesmia, F., Friburger, N., Haddar, K., Maurel, D.: Arabic named entity recognition process using transducer cascade and Arabic wikipedia. In: Proceedings of Recent Advances in Natural Language Processing, pp 48–54, Hissar, Bulgaria (2015)
10. Ben Mesmia, F., Friburger N., Haddar, K., Maurel, D.: Transducer cascade for an automatic recognition of Arabic Named Entities in order to establish links to free resources. In: First International Conference on Arabic Computational Linguistics (ACLing). pp 61–67 (2015)
11. Ben Mesmia, F., Friburger, N., Haddar, K., Maurel, D.: Recognition and TEI annotation of arabic event using transducers. In: Will appear in IEEE proceedings of CiLing'16 (2016)
12. Btoush, M.-H., Alarabeyyat, A., Olab, I.: Rule based approach for Arabic part of speech tagging and name entity recognition. Int. J. Adv. Comput. Sci. Appl. (IJACSA) **7**(6), 331–335 (2016)

13. Chinchor, N.: Overview of MUC-7/MET-2. In Proceedings of the Seventh Message Understanding Conference (MUC-7), p 1–4, Fairfax, VA, USA (1998)
14. Darwish, K., Gao, W.: Simple effective microblog named entity recognition: Arabic as an example. In: LREC, pp 2513–2517 (2014)
15. Doddington, G., Mitchell, A., Przybocki, M., Ramshaw, L., Strassel, S., Weischedel, R.: The automatic content extraction (ACE) program tasks, data, and evaluation. In: Proceedings of the 5th Conference on Language Resources and Evaluation (LREC 2004), pp. 837–840, Lisbon, Portugal (2004)
16. Doumi, N., Lehireche, A., Maurel, D., Ali Cherif, M.: La conception d'un jeu de ressources libres pour le TAL arabe sous Unitex. Paper presented at the TRADETAL2013, Colloque international en Traductologie et TAL, Oran—Algeria, 5–6 may. pp. 5–6 (2013)
17. Fehri, H., Haddar, K., Hamadou, A.B.: Recognition and translation of Arabic named entities with NooJ using a new representation model. In: Constant, M., Maletti, A., Savary, A. (eds.) FSMNLP, 9th International Workshop, pp. 134–142. ACL, Blois, France (2011)
18. Gravier, G., Bonastre, J.F., Galliano, S., Geoffrois, E., Mc Tait, K., Choukri, K.: ESTER, une campagne d'évaluation des systèmes d'indexation d'émissions radiophoniques, Proc. Journées d'Etude sur la Parole (2004)
19. Grouin, C., Rosset, S., Zweignbaum, P., Fort, K., Quintard, L.: Proposal for an extension of traditional named entities: from guidelines to evaluation, an overview. In Proceedings of Linguistic Annotation Workshop, pp. 92–100 (2011)
20. Grishman, R., Sundheim, B.: Message understanding conference—6: a brief history. In: Proceedings of the 16th conference on Computational linguistics (COLING'96), pp 466–471, Copenhagen, Denmark (1996)
21. Kanya, N., Ravi, T.: Named Entity recognition from biomedical text—an information extraction task. ICTACT J. Sort Comput. **06**(04), 1302–1307 (2016)
22. Küçük, D., Yazici, A.: A hybrid named entity recognizer for Turkish. Expert Syst. Appl. **39** (3), 2733–2742 (2012)
23. Maurel, D., Friburger, N., Eshkol, I., Antoine. J.-Y.: Explorer des corpus à l'aide de CasSys. Application au Corpus d'Orléans. G. Willems (ed.). Texte et corpus n°4, Actes des 6es Journées Internationales de Linguistique de Corpus (JLC). pp 189–196 (2013)
24. Maraoui, H., Haddar, K.: Automatisation de l'encodage des lexiques arabes. Colloque pour les Étudiants Chercheurs en Traitement Automatique du Langage naturel et ses applications, pp 74–82, Sousse, Tunisie (2015)
25. Merchant, R., Okurowski, M., Chinchor, N.: The multilingual entity task (MET) overview. In: Proceedings of a workshop on held at Vienna, Virginia. Morristown, NJ, USA. Association for Computational Linguistics. pp 445–447 (1996)
26. Mohammed, N.F., Omar, N.: Arabic named entity recognition using artificial neural network. J. Comput. Sci. 1285–1293 (2012)
27. Oudah, M., Shaalan, K.: NERA 2.0: Improving coverage and performance of rule-based named entity recognition for Arabic. Nat. Lang. Eng. 1–32 (2016)
28. Paumier, S.: UNITEX 3.2 ALPHA. User Manuel. Université Paris-Est Marne-la-Vallée. Date of version, February 23, 2017. 383 p. (2017)
29. Ramesh, D., Sanampudi, S.-K. A Hybrid model for Named Entity Recognition in Biomedical text. Int. J. Sci. Eng. Res. **7**(6), 1164–1166. (2016). ISSN 2229-5518
30. Shaalan, K., Raza, H.: Person named entity recognition for Arabic. In: Proceedings of the 5th Workshop on Important Unresolved Matters, pp. 17–24 (2007)
31. Shaalan, K., Raza, H.: NERA: named entity recognition for Arabic. J. Am. Soc. Inform. Sci. Technol. **60**(9), 1652–1663 (2009)
32. Shaalan, K., Oudah, M.: A hybrid approach to Arabic named entity recognition. J. Inf. Sci. **40** (1), 67–87 (2014)
33. Shaalan, K.: A survey of Arabic named entity recognition and classification. Comput. Linguist. **40**(2), 469–510 (2014)

34. Saleh, I., Tounsi, L., Van Genabith, J.: ZamAn and Raqm: extracting temporal and numerical expressions. In Arabic in Information Retrieval, Lecture Notes in Computer Science, vol. 7097, pp. 562–573 (2011)
35. Serrano, L., Charnois, T., Brunessaux, S., Grilheres, B., Bouzid, M.: Combinaison d'approches pour l'extraction automatique d'événements. In: TALN'2012, volume 2, p: 423–430, Grenoble, France (2012)
36. Sharma, P., Sharma, U., Kalita, J.: Named Entity Recognition in Assamese: A hybrid approach. In: International Conference on Advances in Computing, Communications and Informatics (ICACCI-2016), Jaipur, India (2016)
37. Sharma, P., Sharma, U., Kalita J.: Named entity recognition in assamese. J. Comput. Appl. **142**(8), 1–8 (2016)
38. Text Encoding Initiative Consortium: TEI P5: Guidelines for Electronic Text Encoding and Interchange. Edited by C.M. Sperberg-McQueen and Lou Burnard for the ACH-ALLC-ACL. Version 3.1.0. 1887 p. (2016)
39. Yao, L., Liu, H., Liu, Y., Li, X., Anwar, M.-W.: Biomedical named entity recognition based on deep neutral network. Int. J. Hybrid Inf. Technol. **8**(8), 279–288 (2015)

Semantic Relations Extraction and Ontology Learning from Arabic Texts—A Survey

Aya M. Al-Zoghby, Aya Elshiwi and Ahmed Atwan

Abstract Semantic relations are the building blocks of the Ontologies and any modern knowledge representation system. Extracting semantic relations from the text is one of the most significant and challenging phases in the Ontology learning process. It is essential in all Ontology learning phases starting from building the Ontology from scratch, down to populating and enriching the existing Ontologies. It is challenging, on the other hand, as it requires dealing with natural language text, which represents various challenges especially for syntactically ambiguous languages such as Arabic. In this paper, we present a comprehensive survey of Arabic Semantic Relation Extraction and Arabic Ontology learning research areas. We study Arabic Ontology learning in general while focusing on Arabic Semantic Relation Extraction particularly, as being the most significant, yet challenging task in the Ontology learning process. To the best of our knowledge, this is the first work that addresses the process of Arabic Semantic Relation Extraction from the Ontology learning perspective. We review the conducted researches in both areas. For each research the used technique is illustrated, the limitations and the positive aspects are clarified.

Keywords Knowledge extraction · Arabic semantic relation extraction
Arabic ontology learning · Semantic relation extraction between Arabic NEs
Semantic relation extraction between Arabic ontological concepts

---

A.M. Al-Zoghby (✉) · A. Elshiwi · A. Atwan
Faculty of Computers and Information Systems, Mansoura University,
Mansoura, Egypt
e-mail: aya_el_zoghby@mans.edu.eg

A. Elshiwi
e-mail: Aya.Elshiwi990@gmail.com

A. Atwan
e-mail: Atwan.4@gmail.com

K. Shaalan et al. (eds.), *Intelligent Natural Language Processing: Trends and Applications*, Studies in Computational Intelligence 740,
https://doi.org/10.1007/978-3-319-67056-0_11

1 Introduction

Instant growth in the size of the World Wide Web makes it be the world's biggest repository of data. The amount and the unstructured nature of the majority of this data represent two main issues regarding dealing with the WWW for both humans and machines. As for humans, the amount of data is huge to be processed while the unstructured format cannot be understood by machines. As a result, a challenge in the interoperability between humans and machines showed up. To address this challenge, Tim-Berners Lee invented the Semantic Web as an extension of the current web with the vision that is given a well-defined meaning and provides better cooperation between humans and machines [1].

Ontology can be considered as a gateway to achieve the vision of the semantic web. It is used to represent the data in a way that enables machines to understand its meaning, and allow it to be shared and reused. The most commonly used definition of Ontology is "*Formal, explicit specification of a shared conceptualization*" [2]. The definition is explained in [3] as "*conceptualization refers to an abstract model of some phenomenon in the world by having identified the relevant concepts of that phenomenon. Explicit means that the type of concept used, and constraints on their use are explicitly defined. Formal refers to the fact that the Ontology should be machine-readable. Shared reflects the notion that an Ontology captures consensual knowledge, that is, it is not private of some individual, but accepted by a group*".

Ontology is the "backbone" of the semantic web as described by many researchers. Due to the importance and the full reliance of the semantic web upon Ontology, building it automatically is a significant and, at the same time, a very challenging task. On one hand, it requires dealing with the knowledge acquisition bottleneck [4]. On the other hand, it is affected by the heterogeneity, scalability, uncertainty, and the low quality of web data [5].

Ontologies can be built manually, semi-automatically or automatically. The manual construction of the Ontology has limitations as being expensive, time-consuming, and error-prone [6]. Moreover, it requires specialized domain-experts and Ontology-engineers [7]. To overcome these limitations, a new research area called "Ontology Learning" has emerged aiming to automate or semi-automate the process of building the Ontology [8]. Ontology learning involves the extraction of knowledge through two main tasks: the extraction of concepts (that constitute the Ontology), and the extraction of the semantic relationships among them.

Semantic relations are the building blocks of the Ontologies and any innovative knowledge representation system [9]. Extracting semantic relations from the text is one of the most significant and challenging [4] phases in the Ontology learning process. It is essential in all Ontology learning phases starting from building Ontology from scratch down to populating and enriching existing ones. It is challenging, however, as it requires dealing with natural language texts which represent various challenges especially for syntactically ambiguous languages such as Arabic.

Despite the importance of Arabic language as being one of the six most spoken languages in the world [10], and the obvious growth of its content on the web in the past few years, it has little support in semantic relation extraction in particular and Ontology learning in general. Automatic extraction of semantic relations from Arabic text is not well investigated compared to other languages such as English [11, 12]. While, most of the trials to generate Arabic Ontologies are still done manually [13–16]. Moreover, the nature of the Arabic language added extra challenges in knowledge extraction from Arabic text. As a result, the Arabic language suffers a lack of Ontologies and the semantic web applications in general [17, 18].

In this paper, we study the Arabic Ontology learning in general while focusing on Arabic Semantic Relation Extraction particularly, as being the most significant, yet a challenging task in the Ontology learning process.

To the best of our knowledge, this is the first work that addresses the process of Arabic Semantic Relation Extraction from the Ontology Learning perspective. We reviewed the conducted researches in the area. For each research, the used technique is illustrated, the limitations and the positive aspects are clarified.

The rest of this paper is organized as follows: Sect. 2 discusses the Arabic semantic relation extraction and Ontology learning. Section 3 reviews the Arabic semantic relation extraction trials, while the Arabic Ontology learning literature is represented in Sect. 4. Finally, the paper is concluded in Sect. 5.

2 Arabic Semantic Relation Extraction and Ontology Learning

Arabic is the native language of over 20 countries spoken by approximately 400 million native speakers around the world. It is also the 5th most spoken language in the world [19]. The past few years have witnessed an obvious increase in Arabic content on the web. Therefore, the Arabic language requires a further work in building Ontologies that will exploit the web data, lead to achieve the semantic web vision, and to keep up with the importance of the language. However, the Arabic language has challenges that affect the overall process of Ontology learning in general and the semantic relation extraction as a separated stage of Ontology learning, in particular.

Since learning Ontologies from the natural text requires the usage of NLP techniques, the nature of the Arabic language represents a major challenge. As Arabic is a Semitic language that differs from Latin languages in syntax, morphology, and semantics [20], therefor, the NLP algorithms for other languages cannot be applied directly on Arabic [21]. Also, its morphological analysis is complicated due to its agglutinative, derivational, and inflectional characteristics

[22]. In addition, it is written from right to left, missing capitalization, and it has ambiguities related to typography as the Arabic letters shape changes according to their position in the word [23]. Moreover, the availability of Arabic linguistic resources and tools, that are designed according to the specific features of the Arabic language, represents another challenge. Linguistic resources, such as corpora and dictionaries, are either rare or not free [24, 25]. While, according to [26], there are no corpus analysis tools for the Arabic language. Most of the existing tools of morphological analysis and POS have several limitations, such as being unable to provide full parsing or remove all the ambiguities.

All these challenges are standing as a barrier in the way of learning Ontologies from an Arabic text. Recently, researchers are dedicating extensive work to address the Arabic NLP challenges. Some of these that are related to the highly inflectional and derivational Arabic nature, Part of Speech (POS) tagging, and morphological analysis have been resolved to some extent [27]. But, still there is a lack of Arabic Ontologies and, as a result, a lack of its semantic applications.

In this review, we discussed the semantic relation extraction and Ontology learning both together; as they are highly related in the Arabic language. As we found out, the majority of the work dedicated to extract the semantic relationships was for the purpose of building Ontologies. Moreover, the extraction of those relations was a small part of the research except for some that were mainly dedicated to extract the semantic relationships.

3 Arabic Semantic Relation Extraction

In this section, we reviewed all researches considering the extraction of semantic relations from Arabic text.

In our study, we found out that researches in this area fall into two categories. The first contains the studies considering the extraction of the semantic relations between Arabic NEs. That mainly was for the purpose of using this knowledge in several NLP applications such as: question answering, text mining, and automatic document summarization; see [11, 12, 28–32]. The second category, on the other hand, contains researches considering the extraction of semantic relations between Ontological concepts for the purpose of using them in the construction of Arabic Ontologies, as in [11, 12, 14, 15, 24, 33–42].

From our point of view, we consider that the two categories are complementary. This is because the first category can be used for populating the instance level in an existing Ontology, while the second can be used for constructing the Ontology. Both categories will serve the Arabic Ontologies learning process.

Table 1 Summary of extracting Arabic semantic relation between the NEs

Ref.	Technique
[28]	Rule based
[29]	Rule based
[30]	Machine learning
[32]	Machine learning (Enhanced approach)
[31]	Hybrid (Machine learning and Rule based and Manual)
[11]	Hybrid
[12]	Rule based

3.1 Semantic Relation Extraction Between Arabic Named Entities

The extraction of the semantic relations between NEs hasn't been well investigated in Arabic compared to other languages such as English [32].

As mentioned earlier, there are several challenges related to semantic relation extraction between NEs other than the challenges related to the ambiguity and complexity nature of the Arabic language.

Moreover, the Arabic sentences are long, which might cause the existence of more than one NE in the sentence without being semantically related [30, 31]. Also, the position of semantic relation in the Arabic sentences is hard to be determined as it is not fixed; it may occur before first NE, between the NEs, or after the second NE [30].

Furthermore, the Arabic Semantic relation can be noun, verb, or even preposition. That differs from the English language, in which the relation is usually a verb occurs between the NEs pair [30].

In addition, the is a difficulty in determining the implicit relations that can't be directly specified from the text; it can only be understood from the previous context [28]. Moreover, negative, ambiguous, and multiple words relations between NE pairs, represent extra challenges [28, 31].

The previous trials to extract semantic relations between Arabic NEs succeeded to address most of the stated challenges. Table 1 summarizes the trials of Arabic Semantic Relation Extraction between Named Entities; categorized them into: rule based, machine learning and hybrid approaches.

3.1.1 Rule-Based Approach

The articles [28, 29] proposed a rule based approach for the extraction of semantic relations between Arabic NEs. Both of them used the same technique; by first, identifying a set of linguistic patterns from the training corpus, and then transforming those patterns into rules and transducers using NooJ.[1]

[28] Focused on the extraction of functional relations between the PERS and ORG NEs. A number of journalistic articles and some data from Wikipedia were used as a training corpus. Both NEs and relations were recognized using a collection of manually built dictionaries. For PERS NEs recognition, the *Titles*, *First Names* and *Last Names* dictionaries were used. For ORG NEs recognition, on the other hand, the *Geographical Names*, *Type-Institution* and *Adjective* dictionaries were used. And finally, for relations recognition, the *Functions* dictionary was used. This approach obtained 63%, 78% and 70% in terms of precision, recall and F-measures, respectively.

[29], on the other hand, focused on the extraction of the relations between (PERS-PERS, PERS-LOC, PERS-ORG and ORG-LOC) NEs pairs. For each pair of NEs, several types of relations were identified by the authors. Then, for each type of relations, all the possible syntactic patterns were extracted, and a general pattern for this relation type was built. This approach obtained an F-score of 60%.

The negative aspect of both trials is the manual identification of the relations, which is a tedious time consuming task. This is in addition to the limitations of rule based approach and the necessity to the fully cover of all rules that might satisfy any kind of relations.

3.1.2 Machine Learning Approach

A supervised machine learning method for extracting the semantic relations between Arabic NEs was proposed in [30]. This method was based on the rule mining approach. Its main idea was to extract the position of words surrounding NEs that reflect the semantic relation. In order to apply that idea, two of the previously mentioned challenges of Arabic language were faced. The first was the complex syntax and the length of Arabic sentences, which caused the existence of two NEs in the same sentence without being semantically related. In order to handle this challenge, the authors limited the context by splitting sentences into clauses using Arabic clauses splitter, since each clause is composed of a set of words that contains a subject and a predicate thus, ensuring the existence of a relation between NEs. According to the authors, that solution tackled the problem on average of 80%. The second challenge, on the other hand, was the non-fixed position of the relations in Arabic sentences. In order to handle this challenge, the authors identified the position of words that represent relations in sentences manually. The proposed method consisted of three steps; building training data, automatic rules generation, and selection of the significant rules. In the "first-step", the training dataset was composed by extracting 15 learning features from sentences that contained at least two NEs. Those features were lexical (NEs tag), numerical (number

[1]http://nooj4nlp.net/pages/resources.

of words before, after and between NEs) and semantic (POS of words surrounding NEs). Of these features, 14 were extracted automatically, while the position of the relation was annotated manually by linguistic experts. Each one of the annotated relations was assigned a class that identified its position in the context. In the "second-step", Apriori, tertius, and C4.5 association rules algorithms were applied to automatically extract the highly precision rules. Finally, in the "third-step", the previously generated rules were filtered to select the most significant ones. To evaluate the effectiveness of the proposed method, the authors created an Arabic corpus from electronic and journalistic articles. The method obtained 70%, 53.52% and 60.65% for precision, recall, and f-score, respectively. Authors justified the low recall by the lack of a very efficient Arabic NE recognition tools and the failure of the system to extract the implicit relations between NEs. The main drawback of the system was the need of highly annotated data that was caused from using a supervised machine learning methods. Also, the semantic relations between NEs were extracted fully manually.

In a subsequent enhancement trial to improve the overall coverage of the previous method, authors in [32] used the Genetic Algorithm as a refinement step. The main idea behind using the Genetic Algorithm was to improve the quality of the rules generated from learning algorithms by constructing new rules that are fitter. Crossover and mutation reproduction methods were applied to the selected rules. The results showed that the usage of the Genetic Algorithm increased the precision and recall by 8%.

Another technique to automatically extract relations from the Arabic text was proposed in [12]. The technique was based on the enhancement of Hearst's Algorithm to fit the Arabic language, then integrating it into a four-module framework to extract the relationships. The "preprocessing and feature extraction module" is automatically extracting four different language components; word, POS tag, stem, and phrase. In additionally, three different types of relationships hyponym-hypernym, causality, and hierarchical were manually specified. Using components and relations specified in the previous module, a linguistic expert decided, for each couple of components, the suitable relations. Then, the training set was formed by automatically extracting the matching examples from the text. After that, an enhanced version of Hearst's Algorithm was applied in the "lexical syntactic pattern module". Finally, an evaluation was performed to all the extracted patterns to remove the dirty ones. While, in the "pattern expansion module", the authors expanded the lexical structure of the patterns to include synonyms using Arabic wordNet.[2] Then, enriched the relations with the related concepts. According to the authors, this expansion allowed discovering new relations; however, it caused redundancy as several patterns were covering the same relationship. This problem was handled in the "pattern filtering and aggregation module", where the authors implemented a validation algorithm to filter the redundant patterns that cover the same data instances. To evaluate the efficiency of the framework, it was tested in

[2]http://globalwordnet.org/arabic-wordnet/.

three different datasets: Holy Quran (classical Arabic), Newspapers articles (Modern Standard Arabic), and social blogs (unstructured Arabic text). The overall performance averages of the three datasets in terms of precision, recall and F-measures were 78.57%, 80.71% and 79.54% respectively. According to the authors, when comparing the performance of the proposed technique with the original Hearst's Algorithm, repeated-segments, and co-occurrence based techniques; it achieved the highest performance among all. They, also, studied the effects of different factors such as the type of data in the overall performance of the system. They found that it directly affected the performance; negatively in the classical dataset and positively in Modern Standard Arabic dataset.

3.1.3 Hybrid Approach

So far, authors in [29, 30, 32] were succeeded to label explicit one-word relations between (PERS-LOC, PERS-PERS, PERS-ORG, ORG-LOC, and LOC-LOC) NEs pairs. The authors successfully addressed some of the previously mentioned challenges such as the long Arabic sentences and the non-fixed position of Arabic relations in sentences. For a further enhancement, the authors presented a hybrid approach in [31]. The approach combined the previously discussed machine learning and rule-based methods with a manual technique to extract the semantic relations between the (PER, ORG, LOC) NEs pairs. It presented several enhancement modules for the purpose of addressing further challenges such as multiple words, negative and other complicated relations.

In the first enhancement module, the overall performance of machine learning method was enhanced by partitioning the dataset into verbal and nominal sentences. According to the authors, the partitioning process proved efficiency as it increased the precision and recall by 6.6% and 3%, respectively, compared to the previous machine learning results. It also proved that, for the Arabic language, the position of the relation between NEs depends on phrase structure. In the second module, the challenges of negative and multiple words relations were addressed. This is done by using handcrafted rules proposed by linguistic experts. As for the negative relations, a set of constraints were added to each rule in order to verify the existence of a negative particle that expressed the negation relation. Taking into account that, the position of the negative particles differs according to the sentence type either it is verb or noun. To handle the multiple words relations, on the other hand, the compound words that expressed a relation between NE pairs were collected into a list. Then, using Nooj,[1] all the syntactic grammars were elaborated. The elaborated grammars represented all the classes that the compound word relations belong to, either before first NE, between the two NEs, or after the second NE. Those grammars were then applied to the corpus to extract further multiple words relations. Finally, in the case of implicit and more complicated relations, the authors relied on manual patterns. To evaluate this approach, the authors compared the results of the system against machine learning method in [30] and rule based method in [29] using the same test corpus. The results showed that the system achieved the best improvements and obtained 84.8%, 67.6% and 75.2% in terms of the precision, recall and F-score, respectively. We found that this technique is promising as it succeeded to take advantages of the machine learning, pattern-based approaches, and the manually generated rules in order to handle the complicated relations. Moreover, it is the first work to deal with different types of relations and handle the challenges in a good way with satisfactory results. However, to evaluate its performance in general we can say that handling the complicated sentences manually is not the ideal option due to the nature of the Arabic language, which caused several complicated cases that would require extensive time and effort to be dealt with. Also, the manual identification of relations position in sentences is not practical, as the main challenge is to automatically or semi-automatically extract semantic relations from Arabic sentences.

Another hybrid approach was proposed in [11], the approach mixed the statistical calculus and the linguistic knowledge to extract Arabic semantic relations from a vocalized Hadith corpus. The main idea of this approach was to use the statistical measures to calculate the similarity between terms in order to interpret syntactic relations, then exploiting these relations to infer semantic relations. The authors considered three analysis levels; morphological analysis, syntactic analysis and semantic analysis. In the morphological analysis level, AraMorph[3] analyzer was used to analyze the corpus text and extract tokens, their morpho-syntactic category, and the English translation. While in the syntactic analysis level, one type of noun phrases, which is the prepositional phrases, was extracted from the corpus by applying the grammar rules corresponding to this type of phrases. The first component of the noun phrase was considered as the head, and the second component was the expansion. A syntactic network linking the heads and the expansions with the syntactic relations was generated, and every syntactic relation was represented by the preposition. In the semantic analysis level, a dependency graph, which contained the most common syntactic relations (prepositions) in the corpus, was generated. For each preposition, all the correspondence semantic relations were linked to it. This graph was built based on Arabic grammar books, as authors clarified that the Arabic grammar rules were behind the relationship between the syntactic and semantic relations. By that stage, the syntactic relations were exploited to infer the semantic ones. Finally, the statistical measures were used to solve the ambiguity in the extraction of relations between any couple of terms. By first, calculating the similarity between these terms using contingency table based measures. Then, performing an enrichment of signature of this couple by adding the signature of the nearest couple to it depending on the similarity results. To evaluate this approach, the authors performed experiments in drinks, purification, and fasting domains. They presented two results of experimenting two different definitions of contingency table. The first experiment showed that the success rate of semantic relation extraction in the three domains was weak and didn't exceed 65%. They justified that result by a partial failure in the enrichment operation. So, they

[3]http://www.nongnu.org/aramorph/.

performed the same experiment with the second definition of contingency table. The success rate was improved and reached 75%. Further enrichment had been carried out and it improved the success rate as it reached 97 and 100% in the field of purification. When comparing their results to the co-occurrence approach results, according to the authors, their approach was more effective in some cases while, in other cases the two approaches extracted the same relations. They concluded that the two approaches were complementary.

3.2 Semantic Relation Extraction Between Arabic Ontological Concepts

The extraction of semantic relations between Arabic Ontological concepts is affected by the same challenges and limitations caused by the nature of the Arabic language. In this context, we mean by Ontological concepts; concepts from a seed Ontology, concepts that are extracted from text during the trial to build an Ontology, or a given list of concepts to extract semantic relations between them. Extracting semantic relationships between Ontological concepts was performed as a step in the Ontology learning process. So, we discuss the used techniques in details in a subsequent Sect. (4.2).

4 Arabic Ontology Learning

The effort given to Arabic Ontology Learning research area is very little to keep up with the importance of the Arabic language. We discuss all Arabic researches that either build Ontology from scratch, populate, or enrich an existing Ontology. In this section, we review trials to learn Arabic Ontology from natural text. We categorize the Arabic Ontology learning literature, according to [43–45], into upper Ontology and Domain-specific Ontology. Table 2 summarizes the reviewed trials to build Ontology for the Arabic language.

4.1 Upper Ontology

Upper Ontology, also known as (Top level, Foundational or Universal Ontology), is defined as *"an Ontology which describes very general concepts that are the same across all knowledge domains"* [46]. It provides a foundation to guide the development of other Ontologies. Moreover, it facilitates the process of mapping between them as it is easier to map between two Ontologies that are derived from a standard upper ontology [44]. There are many upper level Ontologies for English

Table 2 Summarizes Arabic ontology learning trials

		Ref.	Learning resource	Ontology learning technique	Domain	Relation extraction technique
Domain ontology	General domains	[41], [42]	Structured	Manual	Arabic verbs derivational ontology	Derivation based
		[15]	Unstructured	Manual	Computer domain	Manual
		[14]	Structured	Manual	Computer domain	Manual
		[34]	Unstructured	Manual	Legal domain	Manual
		[39]	Unstructured	Statistical	Arabic linguistics	Hybrid
	Islamic domain	[58], [13]	Semi-structured	Statistical	Agriculture	–
		[59], [18], [60]	Semi-structured	Linguistics	Wikipedia	–
		[24]	Semi-structured	Hybrid (Manual and Translation)	Food, nutrition and health	Manual
		[33]	Structured	Hybrid (Manual and Statistical)	Arabic linguistics domain	Manual
		[35]	Quran	Manual	Quran ontology	Manual
		[63]	Quran	Hybrid (Manual and Statistical)	Stories of prophets	–
		[36], [37]	Quran	Manual	Islamic knowledge	Manual
		[38]	Quran-Quias-Ijmaa	Manual	Solat (Prayer) ontology	Manual
		[40]	Hadith (Sahih AlBukhary)	Hybrid (Manual and Statistical)	Hadith ontology	–
		[16]	Hadith	A proposed system is suggested to adapt the hybrid approach		
Upper ontology	Arabic WordNet ontology	[48], [49], [50]	NA	NA	NA	NA
	Formal Arabic ontology	[47], [43]	NA	NA	NA	NA

such as BFO, Cyc, DOLCE, SUMO and others. The need to use these Ontologies for different languages including Arabic, lead us to a significant question that is: are the upper level Ontology concepts language dependent or independent? We studied the answer from the two perspectives.

Regarding the application of the language independency point of view for Arabic, using one of the previously mentioned upper Ontologies, requires the integration of some mid-level Ontology that is devoted to the Arabic culture [27]. Also, the translation process should be guaranteed. The application of the language-concept dependency principle, on the other hand, requires building an upper ontology that is specific to the Arabic language. However, there is only one trial to build such Ontology [40, 43], which is still under construction. As there is a challenge to find middle level Ontology for Arabic and the fact that there is no yet standard Arabic Upper Ontology, researchers used the AWN Ontology as an alternative. But, in addition to its practical limitations, mentioned at the following Sect. (4.1.1), it is built based on translation. This presents a big limitation when adopting the language-concept dependency point of view. As, each language has its specific linguistic environment and cultural context [15, 47]. Consequently, this makes it necessary to build the Arabic Ontology that takes into consideration the historical and cultural aspects of the language. In the following two Sects. (4.1.1, 4.1.2) we discuss the two Arabic Upper Ontologies.

4.1.1 Arabic WordNet Ontology

Authors in [48–50] initiated the Arabic WordNet (AWN) project to build a lexical resource for Modern Standard Arabic. AWN was built following the same methodology developed for Princeton WordNet [51] and Euro WordNet [52]. It was constructed by, first, encoding manually the most important concepts to create the core WordNet for Arabic. Then, maximizing the compatibility across other WordNets. A Mapping-based approach was followed to link the Arabic-English corresponding terms. AWN provides a formal semantic framework as it is mapped to SUMO [53] and its associated domain Ontologies. AWN, like all other WordNet projects, was initially constructed as a lexical database, then, it was interpreted and used as a lexical Ontology. Words were collected into sets of synonyms called *synsets* and a number of relations among these *synsets* were recorded. AWN consists of 11,270 synsets and 23,490 Arabic expressions (words & multi words). Interpreting AWN to be used as a lexical ontology is done by formalizing each synset as a concept class. This indeed has led to many practical limitations due to the huge number of concepts (synonym sets) that made it inappropriate for real world applications [54]. Moreover, mapping WordNet to an existing Ontology is a very challenging task [55].

4.1.2 Formal Arabic Ontology

Adopting the language-concept dependency principle point of view, authors in [43, 47] initiated a project to build the first Arabic upper level Ontology. The Ontology was built following the same design as AWN aiming to use it as an alternative. It was constructed following five steps. The first step was, Manual and semi-automatic extraction of Arabic concepts from specialized dictionaries. While, the second step was, reformulation of the concepts manually to strict ontological rules focusing on the intrinsic properties of concepts. Coming to the third step, the generated concepts were mapped automatically with the English WordNet using a smart Algorithm developed by the authors to inherit WordNet semantic relations. Followed by, the fourth step which was, cleaning the inherited semantic relations from WordNet using ONTOClean methodology. Finally, the fifth step was, linking the concepts and the relations with a semantic tree that contains all mother concepts of the Arabic language. This semantic tree was called Arabic Core Ontology; it was built to govern the correctness of the whole Arabic Ontology. It consisted of 10 levels and 400 concepts, and it was built based on DOLCE and SUMO. According to the authors, they have built a logically and semantically well-founded ontology. However, from our point of view, we think that following the same building approach of AWN may lead to the same practical limitations of it.

4.2 Domain Ontology

Domain Ontology, also known as (domain specific Ontology), represents concepts that belong to a specific domain of interest along with relationships interconnecting these concepts [56]. It reduces the conceptual and terminological confusion among users who share electronic documents and various kinds of information that belong to the same domain [40, 57]. Domain Ontology can use upper Ontology as a foundation and extend its concepts, accordingly taking advantage of the semantic richness of the extended concepts and logic that is built into upper Ontology [44]. In this section, we review, to the best of our knowledge, all the conducted trials to build Arabic domain specific Ontology. We categorize domain specific Ontology into two categories, general domains (4.2.1) and Islamic domains (4.2.2).

4.2.1 General Domains

In this section, we review the trials to build Arabic Ontology for general domains such as, computer, legal, linguistics, agriculture, and others. We categorize the trials according to the technique used in the Ontology learning process. In our research, we found out that there are four techniques mainly used in the Arabic Ontology learning process which are manual [14, 15, 34, 41, 42], statistical [13, 39, 58], linguistics [59], and hybrid [24, 33].

Manual Approach

In [41, 42] DEAR-ONTO, a derivational Arabic Meta-Ontology model, was presented. The authors built their hypothesis basing on the fact that Arabic is a "*highly derivational and inflectional language in which morphology plays a significant role*" [22]. The main purpose of that model was to structure the Arabic language into a set of equivalent classes using the derivations and their patterns. To populate the ontology, the authors used a list of selected Arabic verbs and their derivations. The Ontology used verbs as roots, and the derivations formed the equivalent classes. Each equivalent class was represented by a verb and contained all its derived words following derivation and inflection rules of the Arabic language. Then, each equivalent class was modeled as Ontology, and a Meta-Ontology representing the general structures of all those classes was presented. The authors suggested several applications of the model such as, Arabic language development, Arabic language understanding and Arabic morphology analysis.

The work was illustrated as a theoretical stage without practical implementation or evaluation. We found two contradictory opinions regarding that hypothesis. On one hand, authors of [15, 17] criticized it, and proved that, it is imprecise to build Ontology based on the roots. Since, "*85% of Arabic words are derived from tri-lateral roots*" [49] which definitely lead to, the existence of concepts with different meanings sharing the same root and consequently sharing the same class. Moreover, [15] added that there are words in Arabic that have no roots and the model did not handle those cases. On the other hand, authors of [26] supported the hypothesis, by suggesting using the approach in general domain corpora rather than specific domain. As, in general domain corpora there are more frequently similar terms sharing the same root. Moreover, they clarified that the main weakness in this approach was the over-generation of similarity links.

From our point of view, evaluating that hypothesis basically depends on the applications built upon the proposed Ontology model. As, in case of building applications that take advantage of the derivational nature of the Arabic language and exploit its derivation and inflection rules such as applications that are used to understand Arabic language and its rules, the model is valid. In fact, we found the suggested applications by the authors meet this case and as a result we found the hypothesis valid. On the other hand, in case of building applications that require knowledge to be classified and structured correctly such as information retrieval applications and the applications that require studying specific domain knowledge, the hypothesis is invalid. And the model is affected by the limitations mentioned by Al-Safadi [15] and Al-Zoghby [17].

In [14, 15] a "computer" domain Arabic Ontology was constructed. Both researches built the Ontology to use it as a basis for a semantic search engine. In order to enhance the semantic based search results and to solve the traditional search engines related problems, such as, low query precision.

The Ontology presented by Al-Safadi [15] consisted of 110 classes, 78 instances and 48 relations. It was constructed following three steps. In the first step, the Ontology classes were formed by gathering the most relevant concepts in the computer technology domain from users. Then, those translating computer domain English Ontology classes into Arabic and using specialized domain dictionaries in addition to, using domain specific articles. While, in the second step, the instances and properties of the predefined classes were defined. Finally, in the third step, the Ontological relationships were manually defined. In this step the concepts were organized into a hierarchy associated via both taxonomic relations such as, (is-a, part-of and type-of) and non-taxonomic relations as, (produced by (تنتجه شركه), has logo (لها شعار) and use (يستعمل). To evaluate the ontology, the authors imposed an Arabic query on it. The Query was tested using protégé 3.4.4 SPARQL Query and the average precision rate of the experiment was 50%.

While, the Ontology presented in [14] was much simpler. It also was constructed using a set of main concepts in computer domain, such as (computer (الحاسب or العتاد)). Those concepts were associated via inheritance, association, and synonym relations. To evaluate the ontology, the authors performed two queries using a proposed semantic search engine that was built based on the constructed ontology, and "Google", the syntactic search engine. The results of the first query showed that, the semantic search engine returned fewer pages than the syntactic search engine. While, the results of the second query showed that, both search engines returned approximately the same number of pages. According to the authors, the semantic search results were more accurate and specific in both queries. Despite the fact that, the Ontology presented in both trials was initial and it didn't fully cover the computer domain, the trials have a positive aspect in paving the way to build Arabic Ontology based semantic search engine. That provides a better alternative to the keyword based syntactic search engines. While the negative aspects of both trials are, cost, human effort and the time consumed in the manual construction of the Ontology.

In another trial to improve the keyword based search results and to improve Arabic information retrieval in general, [34] presented a simple Ontology to be used as basis of query expansion process in the legal domain. The Ontology was constructed following a top-down strategy, according to the steps mentioned in [7]. Starting with the main concepts in the legal domain, the hierarchy of concepts was constructed. Then, relationships such as, (is-a, and instance-of) were assigned between those concepts. To populate the Ontology, the authors used UN[4] articles in Arabic and a set of selected newspapers articles in the legal domain. In order to improve the precision and recall of the query expansion, the authors associated each concept with its synonyms and derivative set that was selected according to its relevance to the legal domain. To evaluate the efficiency of the generated Ontology in the query expansion process, the authors compared the recall and precision results of an initial query and extended query performed using Arabic engine called Hahooa.[5] The initial query was formulated by a main concept in the legal domain. While the extended query, was formulated by the same concept and all its

[4]http://www.undp.org.

[5]http://www.hahooa.com/nav.php?ver=ar.

synonyms and derivatives after extending it using the generated Ontology. The results showed that, the extended query improved the recall from 115 to 135. while the precision was improved from 2 relevant results out of the first 10 results in case of the initial query, to 7 out of the first 10 results in case of the extended query. Despite the simplicity of the generated Ontology, the results provided by the authors, showed its efficiency in the query expansion process. The main weakness of that trial was the cost of the manual construction of the Ontology and according to the authors, its *"representativeness of the domain"*.

Statistical Approach

Using a statistical approach, authors in [39] constructed Ontology in Arabic Linguistics domain. The Ontology was constructed following a top-down strategy. A seed Ontology was first initialized manually using the general concepts of GOLD. Then, following a three-step process the concepts and relationships between them were extracted and the Ontology was updated. In the first step, a domain corpus was formed and preprocessed. It was formed by 57 documents from books, journal articles, and web documents in Arabic linguistics domain. The documents were selected, prepared manually (by deleting tables, diagrams and graphs), and transformed into plain text. Then a set of preprocessing steps such as normalization, deletion of stop words, and light stemming were performed to prepare corpus for the extraction of ontological elements. While in the second step, the domain concepts were extracted using two techniques; "repeated segments" and "co-occurrence". The "repeated segments" technique considered any term that denotes a concept to consist of four words maximum. The technique extracted all the repeated segments from the corpus after indexing all the words corresponding to their position. The extracted concepts were then filtered to eliminate the unwanted ones using filter of weights and cut filter. The filter of weights filtered concepts according to their total number of occurrences in the corpus compared to a predefined threshold. The cut filter, on the other hand, removed the segments containing certain words, such as verbs, named entities, and numbers. While, the "co-occurrence" technique extracted all the candidate concepts that occur together frequently and at the same time were extracted as repeated segments. Finally, in the third step, the ontological relationships were extracted and the Ontology was updated. The relationships were extracted using two approaches; linguistic markers and hierarchical relations. The linguistic markers approach, used the context between any two candidate concepts to extract elements that identify the relation between them, such as (is-a, and part of). The linguistic markers were organized into categories according to the type of the extracted relation. Hyponym relation category contained (is-a هو هى (هم and meronymy relation category contained (part of تتكون من– تتألف من – تنقسم الى). The hierarchal relations approach, on the other hand, was used as an alternative in case there were no linguistic markers. The approach was responsible for extracting only parent-child or (is-a) relation between two candidate concepts; the first one of them was considered a parent and the second

Table 3 Summarizes the ontology update process using the extracted concepts and relations

		Concept cases		
		Case 1 One concept of the pair was found among the seed ontology concepts and the other one was not	*Case 2* Both concepts of the pair were found among the seed ontology concepts and there was no relation between them	*Case 3* None of pair concepts were found among the seed ontology concepts
Relation extraction	Linguistic Markers Approach	The missed concept of the pair was defined as a new concept and was linked to the other concept with a relation defined by linguistic marker	A New relation between the concept pair was assigned from the linguistic marker	The process does nothing
	Hierarchal Relations Approach	The missed concept was defined as a new son-concept/(father-concept) and was related to the other concept by a subsumption relation "is-a"	A new relation of subsumption "is-a" was assigned between the concept pair	The process does nothing

was the child. It used rules to ensure the existence of those candidate concepts together more frequently also, to ensure the probability of the occurrence of the first concept (parent) before the second (child) is higher than the reverse. After extracting concept pairs and the relationships between them the seed Ontology was updated as illustrated in Table 3. The authors provided results for only the first two steps; the preprocessing and the concept extraction. There was no illustration of the created ontology or the implementation. Therefore, the generate Ontology can't be evaluated.

[13, 58] Used a semi-automatically Ontology learning system to learn a taxonomical Ontology in agriculture domain. The system used a set of semi-structured HTML web documents and a set of seed concepts as input. The domain concepts and taxonomical relationships between them were extracted using two approaches. The first approach utilized the phrases of the HTML documents headings, while the second approach used the hierarchal structure of those headings to extract the taxonomical Ontology.

In the first approach, the Ontology was generated by searching all the headings phrases after extracting them, to find each concept that was considered children of any seed concept. That was done by extracting all the word sequences or what was called by the authors, the N-gram phrases that had one of the seed concepts as their headwords. The concepts extracted from headings were assigned to their parent concepts and the Ontology was formed. While in the second approach, the Ontology was generated by structuring the concepts in a hierarchy based on the heading levels or the HTML structure of the documents. The seed concepts were located at the top level of that hierarchy then considering the concepts at the second level as children of the top level and so on. The two Ontologies generated by both approaches were filtered from fake concepts. Then, both were merged to set the right parent and the right level for each concept that was found in both Ontologies. Resulting in, adjusting the hierarchal structure of the final Ontology.

For evaluating the generated Ontology, the authors followed a Golden Standard Evaluation method that consisted of both lexical and taxonomic evaluation. The generated Ontology was compared to a subset of a golden standard Ontology in agriculture domain called AGROVOC. The best F-score results were 76.5% and 75.66% for lexical and taxonomic evaluation, respectively. The main limitation we found in this work was that the authors didn't clarify how the taxonomical relationships between any two concepts were identified. The main focus of both approaches was to extract and structure concepts which can never be true without defining or extracting the right relations between these concepts. As, in the first approach, the existence of two concepts in any N-gram phrase is not an evidence that there is a relation between them. While in the second approach, depending on the structure of the concepts based on the headings levels is not precise according to the nature of the web.

According to [26], the study didn't explain how to recognize the head of each N-gram, nor how to handle N-grams or how to deal with the syntactic ambiguities. However, the positive aspect of this research is the evaluation of the Ontology as it is the first work in Arabic to evaluate the generated Ontology using Ontology standard evaluation technique. This should encourage other Arabic projects to use such techniques.

Linguistic Approach

In [18, 59, 60], the two earlier publications of the same project, the authors proposed a linguistic-based approach to learn Ontology from Arabic Wikipedia. The approach relied on the semantic field theory, in which concepts were defined using their semantic relations. Applying that theory on Wikipedia, the authors considered each article's title as a concept and its semantic relations were extracted from the list of categories and infoboxes, following a bottom up approach. For each article, the infobox was extracted from the articles text. Each infobox was then parsed to extract (hasFeature), (isRelatedTo) and (hasCategory) relations. The (hasFeature) relation, defined articles features and their values, the (isRelatedTo) relation identified the related Wikipedia articles, and the (hasCategory) relation extracted article's categories. Then, the final Ontology was generated and written as an OWL file.

To evaluate the approach, the authors performed validation testing of the generated ontology and human judgment, from both crowd and domain experts' evaluation. The Ontology passed the validation against violations of OWL rules successfully. As for the human judgment experiment, the authors published an online survey to evaluate a sample of the generated ontology extracted for "Saudi Arabia" article. The overall precision of the experiment was 56%. In the experts' validation experiment, on the other hand, two domain experts evaluated Ontology extracted from 24 Wikipedia articles in the geography domain. The individual precision for each expert was 83.82% and 79.41% respectively, while the average precision of the evaluation according to both experts was 82%. The average precision of the approach for the human judges in the two experiments was 65%.

In fact, we found that the system is very promising. It can be improved by increasing the number of concepts and the ontological relations by extracting them from the Wikipedia text not only from the articles or infoboxes. In fact, the authors suggested future enhancements for the system including this point.

Hybrid Approach

Following a hybrid approach of translation and Manual techniques, authors in [24] Presented an early stage integrated Ontology for food, nutrition, and health domains. It was mainly developed to be used in annotating Arabic textual web resources related to the three domains. At first, a simple Ontology for food and nutrition was built by translating food items, food groups, and nutrition names of the USDA[6] database into Arabic. In addition to inheriting the relationships between food items and nutrition values from the USDA.[6] Then, the health domain was added by defining four classes; diseases, part of the body, body biological function, and people status. To integrate the three domains, the authors created object properties between food, nutrition, and health. Object properties consisted of three types of relation: positive, negative and prevent. Prevent relation was used with disease class only. The authors suggested future enhancements including, the integration of the Ontology with international Ontologies, and the expansion of it with additional concepts and relations extracted from web documents related to food, health and nutrition domains.

Merging the manual and statistical techniques, authors in [33] Presented a prototype of linguistic Ontology that was founded on the Arabic Traditional Grammar (ATG). The Ontology was Extracted following two steps. It was first bootstrapped manually by extracting concepts from Arabic linguistic resources and relating them to the concepts in GOLD. Then, it was enriched by implementing an automatic extraction algorithm to extract new concepts from linguistics text. The text was preprocessed by performing segmentation, light stemming, and stop words elimination before applying the extraction algorithm. The algorithm was based on the repeated segments statistical approach. The newly extracted candidate concepts and their relations were proposed to an expert before being inserted in the ontology. In fact, the authors did not provide any clarification of that algorithm or the enrichment step in general. They only provided two conceptual graphs representing

[6]http://www.ars.usda.gov/Services/docs.htm?docid=8964.

the manually constructed Ontology with both the hierarchal and non-hierarchal relations between concepts.

Uncategorized

In this section, we review researches that don't fall in any of the previously mentioned categories.

In an attempt to implement the Arabic domain Ontology construction process, authors in [26] presented a system called ArabOnto. The system takes N corpora representing different domains as an input. The following five steps briefly summarize how the system works. In the first step, it starts by analyzing documents via performing morphological analysis, disambiguation, and POS tagging using MADA [61]. While in the second step, it generates syntactic trees of all Noun Phrases (NPs) using a syntactic parser developed by the authors. Coming to the third step, an algorithm for morpho-syntactic disambiguation and Domain Relevant Term (DRT) extraction is implemented. Followed by the fourth step, in which, conceptual networks that contain DRTs and their syntactic relations are generated. Finally, in the fifth step, a clustering algorithm was applied to group terms in each network. The groups obtained from different networks were managed in order to merge groups that have many common elements. The authors considered that the obtained structure represents the domain ontology, as terms sharing the same hyperonym (i.e., co-hyperonyms) were clustered.

Authors in [27] built Ontology for semantic based question answering system. The main idea behind that work was to integrate the lexical information extracted from Arabic WordNet (AWN) and the semantic, syntactic and lexical information extracted from Arabic VerbNet (AVN). In order to provide a better representation and to add a semantic dimension to the concepts of the generated Ontology, the authors made that combination between AWN and AVN. The Ontology was constructed following a two-phase process; briefly explained as follows. In the first phase, AWN was used to build the concepts hierarchy by transforming each AWN synset into a concept. Then each concept was assigned a lexicon that contained its lexical information and all the words that were members in its synset. Finally, concepts were categorized according to their type into two nodes; nouns and verbs. In the second phase, AVN was used to extract information related to each concept under verbs node. That was done by first, extracting all the frames related to a specific verb from AVN classes. Each verb had three frames; syntactic, semantic and constraints frames. Those frames were then transformed into sub conceptual Graphs and were related to the verb by syntaxOf, semanticOf, and constraintOf relations. Finally, those conceptual graphs were integrated in the Ontology verb nodes as situations of each verb. In fact, the authors presented in details how each frame was transformed into Conceptual Graph; we are not going to discuss it here since our main focus is on how the Ontology was generated. For more details, we refer to the main article. The authors measured the performance of the question answering system using two approaches; surface and semantic similarity based approaches. The surface similarity based approach measured the similarity between keywords and structure of questions and their candidate passages. While the semantic similarity based approach, on the other hand, used the generated Ontology to build Conceptual Graphs of questions and their candidate passages. Then, it measured the semantic similarity score between those conceptual Graphs. The results showed that the Ontology based semantic approach generally improved the performance of the system. It increased the percentage of the answered questions in general and the correctly answered questions in particular. The percentage of the correctly answered questions increased from 7.39% out of 284 questions in case of surface similarity approach to 16.2% out of 284 questions in case of semantic-based approach. In fact, we found that this research is very promising as, to the best of our knowledge, it is the only research in Arabic that adopted the integration of syntactic and semantic information in order to build semantic based intelligent applications.

4.2.2 Islamic Domain

Islamic Ontology is very important for both Muslims and non-Muslims. As for the non-Muslims, it provides a semantic meaning in order to understand the Islamic Messages as described in Quran and Hadith [37]. As for Muslims, there are 1.7 billion Muslims [62] around the world, most of them do not speak Arabic as their native language. Therefore, it provides them a comprehensive meaning of Quran and Hadith. Arabic plays a significant role in the Islamic scholarship because it is the language of the Holy Quran, and Muslims daily prayers are performed in Arabic [37]. In this section we review trials to build Arabic Islamic domain Ontology. We categorize Islamic Ontology according to the main references that Ontology is built for; to Quran and Hadith.

Quran Ontology

From Al-Quran corpus, authors in [35] built Arabic lexical Ontology called Azhary. The Ontology grouped words into synsets and assigned number of relations between them following the same design as AWN. It contained 26,195 words organized in 13,328 synsets. The Ontology learning system was composed of three modules; word extraction, relation building and Ontology building. The word extraction module, built the seed to start the Ontology. It manually extracted seed words from the Quran corpus. While the relation building module, manually extracted the relations between words from Arabic dictionaries such as (the meaning dictionary (قاموس المعانى), Rich lexicon (معجم الغنى), and Mediator lexicon (المعجم الوسيط). The Ontology contained 7 types of relations; Synonym, antonym, hyponym, holonym, hypernym, meronym and association. The seed words and relationships between them were stored in a table in an excel file. Finally, the Ontology building module, converted the table of words and relations into Ontology. To evaluate the ontology, the authors presented a comparison between Azhary and AWN. According to the authors, Azhary showed a better response time, contained more words, and recorded more word relations. However, we find this comparison is illogical as both Ontologies are different. AWN is a lexical ontology for the whole Arabic language, while *Azhary* was constructed from Quran as the only source. Also, the manual construction of such Ontology takes a tremendous time, cost and effort. In addition, following the same design approach of AWN may lead to the same practical limitation of AWN as we mentioned earlier in Sect. 4.1 (Upper Ontology Section).

[36–38] started another project to build an Islamic Knowledge Ontology. In [36, 37], as a first stage in the project, authors attempt to build Islamic Ontology from Quran text. However, this trial faced some challenges regarding the used approach to extract ontological elements from text. Additionally, the structure of the Quran, which requires another source of Islamic knowledge in order to build a more complete ontology. In [38], as a forward step in the way of Islamic Ontology construction, authors enhanced the Ontology learning approach and also used another knowledge source to build the Ontology. In this trial, Prayers (Solat صلاه) Ontology was presented. It was constructed semi-automatically from Quran as the primary resource, and from Qiyas-analogy and Ijma-consensus as a secondary resource. The Ontology focused on two types of Solat; Obligatory and Sunnah. It was constructed by using the Quran indexes as upper layer TBox for the Ontology. Listing all the important Solat terms and the different types of Solat such as obligatory (Fardhu فرض) and Sunnah (سُنه) to develop the hierarchal taxonomy. In the formation of this taxonomy, the few top-level concepts were associated to the middle level, then all the other classes that can be expanded from Solat were generated. The generated ontology had 48 concepts, 51 relationship properties, and 282 instances. Authors provided a visualization of the generated ontology and we find that it completely covers the two types of solat. We think the approach has proved its efficiency when applied to this small area. The only negative aspect of the research is the cost, time, and effort it requires due to human intervention.

In [63], an attempt to construct Ontology for the holy Quran Chapters (*Surah*) related to stories of the prophets was presented. Al-Quran corpus was used as the knowledge source. The authors followed the same approach as [40] that we illustrated in details in the following section (section "Hadith Ontology").

Hadith Ontology

Authors in [16] suggested a framework for semi-automatic Ontology construction from the Hadith corpus. The presented framework consisted of four modules; documents preprocessing, concept extraction, relation extraction and Ontology edition. It was based on NLP, statistical, and data mining techniques to extract concepts and semantic relations. However, the implementation of this system is still under progress.

Using association rules algorithm, authors in [40] presented a trial to build Ontology for prophetic traditions (Hadith حديث). The trial was considered a first step in the way of building a fully functional Hadith Ontology. The Ontology was constructed from Sahih Al-Bukhari[7] (صحيح البخارى) book as the only knowledge source from the entire hadith collection. The Ontology consisted of two parts; Hadith Metadata Ontology, and Hadith Semantic Ontology. The Hadith Metadata Ontology was built from the structural taxonomy of Sahih Al-bukhary book. By, creating a sub-class-of relationship between a concept and a sub concept based on the structure of the book. For example, each chapter name was considered a concept and all the sections names related to that chapter were considered sub concepts. Hadith semantic Ontology, on the other hand, was built from concepts and semantic relationships extracted from the texts of Hadiths. Concepts were extracted according to the following steps. First, key phrases were extracted using KP-miner[8] after applying preprocessing and tagging operation to the text. Then, key phrases were stemmed to transform all word derivatives to their roots in order to extract the different forms of the same root. Finally, the frequencies of all words that shared the same root were calculated and the words with higher frequencies were defined as concepts.

Relationships were extracted according to the following steps. First, the authors specified certain types of relations such as "kind-of", "Part-of", and "Synonym-of" and assigned tags to the words in the text that represent any type of those relations. Then, Apriori algorithm was applied to extract all the association rules between concepts and the predefined relationships. After that, the rules were selected based on the satisfaction of a condition that; in any rule the higher concept must occur after the relationship. To clarify, when C1 is part of C2; C2 must be the higher concept in the pair. Finally, the confidence of the selected rules was calculated and the rules with higher confidence were extracted. In fact, the authors didn't clarify practically how the higher concepts were identified in case that those concepts weren't included in the first part of the Ontology (Hadith metadata Ontology). The authors presented an illustrative example to extract part of the relationship between concepts from four Hadiths. The authors mentioned that they used OWL to represent the Ontology; they did not provide any representation not even visually of the Ontology.

5 Conclusion

In this review, we studied two highly related topics in Arabic; Arabic Semantic Relation Extraction and Arabic Ontology Learning. Arabic Semantic Relation Extraction is one of the most significant however, least tackled tasks in the Arabic Ontology learning process. As we proceeded in our study, we noticed a gap in this area as we found that most researches considering the extraction of semantic

[7]https://fr.wikipedia.org/wiki/Sahih_al-Bukhari.

[8]http://www.claes.sci.eg/coe_wm/kpminer/.

relationships fall into two categories. The first category, considers the extraction of semantic relations between pairs of NEs for the purpose of using them in several NLP applications. The second category, on the other hand, considers the extraction of semantic relations between Ontological concepts for the purpose of building Arabic Ontologies. We suggest that both categories should be integrated as the first category can be used in the enrichment and population of Ontologies in instance level, while, the second category can be used to build Arabic Ontologies. This integration will certainly enhance the Arabic Ontology learning. Regarding the extraction of Semantic relations between NEs, we found that it is very challenging to fully automate this process due to the nature of Arabic Language and the many odd cases that can only be handled manually. Additionally, research in this area is very little and most of the work was done by the same group of authors. The effort of the authors is well appreciated, but still this area of research in Arabic needs more work in order to be able to compare techniques conducted by other researchers and conducted from different points of view.

Semantic Relation Extraction between Ontological concepts, on the other hand, was discussed as a part of the whole Ontology learning process. We reached a conclusion that, most works extracted it manually or used a set of predefined relations. The main focus was to build the full Ontology neglecting the semantic relation extraction phase.

Arabic Ontology learning is a very critical topic based on which the future of the Arabic semantic web will be determined. Very little trials were directed toward building Arabic Ontologies and these trials are too little to keep up with the significance of the Arabic language. Arabic Ontology learning is facing several obstacles starting from the nature of the Arabic language that makes it very challenging to deal with the Arabic text and the lack of the Arabic linguistic resources and tools. One more obstacle facing the Arabic Ontology construction is the fact that there is no standard upper-level Ontology to work as a foundation to build other Ontologies which causes lack of coherence among the Arabic Ontologies. Regarding the reviewed trials to build domain specific Ontologies, most of these trials are constructed either fully manually or partially manually. Which lead to the limitations of manual techniques represented in time, cost and the simplicity of the generated Ontology. The study also showed an obvious lack in the Islamic Ontologies and the majority of the trials are directed to build a simple domain Ontology. Ontology evaluation and validation approximately not tackled in Arabic researches despite its significance in reflecting the performance of applications. Only few researches performed evaluation of their generated Ontologies. It is necessary for the future development to work on overcoming these limitations and increasing the research work in Arabic Ontology Learning. This subsequently will lead to enhancing the Arabic semantic web.

References

1. Berners-Lee, T.: The semantic web. a new form of web content that is meaningful to computers will unleash a revolution of new possibilities. Sci. Am. **284**(5), 1–5 (2001)
2. Gruber, T.: Toward principles for the design of ontologies used for knowledge sharing. Int. J. Hum. Comput. Stud. **43**(5), 907–928 (1995)
3. Studer, R.: Knowledge engineering: principles and methods. Data Knowl. Eng. **25**(1), 161–197 (1998)
4. Maedche, A.: Ontology learning for the semantic web. Springer Science & Business Media (2002)
5. LEHMANN, J.: An Introduction to Ontology Learning
6. Hazman, M.: A survey of ontology learning approaches. Database **7**(6) (2011)
7. Noy, N.: Ontology development 101: a guide to creating your first ontology (2001)
8. Barforush, A.: Ontology learning: revisted. J. Web Eng. **11**(4), 269–289 (2012)
9. Auger, A.: Pattern-based approaches to semantic relation extraction: a state-of-the-art. Terminology, 1–19 (2008)
10. Wikipedia. Arabic language—wikipedia, the free encyclopedia (2015) [Online; accessed November-2015]
11. Lahbib, W.: A hybrid approach for Arabic semantic relation extraction, pp. 315–320 (2013)
12. Al Zamil, M.: Automatic extraction of ontological relations from Arabic text. J. King Saud Univ. Comput. Inf. Sci. 462–472 (2014)
13. Hazman, M.: Ontology learning from domain specific web documents. Int. J. Metadata Semant. Ontol. **4**(1), 24–33 (2009)
14. Moawad, I.: Ontology-based architecture for an Arabic semantic search engine. In: The Tenth Conference on Language Engineering Organized by Egyptian Society of Language Engineering (ESOLEC'2010) (2010)
15. Al-Safadi, L.: Developing ontology for Arabic blogs retrieval. Int. J. Comput. Appl. **19**(4), 40–45 (2011)
16. Al Arfaj, A.: Towards ontology construction from Arabic texts-a proposed framework. In: IEEE International Conference on Computer and Information Technology (CIT) (2014)
17. Al-Zoghby, A.: Arabic semantic web applications–a survey. J. Em. Technol. Web Intel. **5**(1), 52–69 (2013)
18. Al-Rajebah, N.: Extracting ontologies from Arabic Wikipedia: a linguistic approach. Arabian J. Sci. Eng. **39**(4), 2749–2771 (2014)
19. Wikipedia. List_of_languages_by_total_number_of_speakers, the free encyclopedia (2015) [Online; accessed November-2015]
20. Elkateb, S.: Arabic WordNet and the challenges of Arabic. In: Proceedings of Arabic NLP/MT Conference, London, UK (2006)
21. Al-Khalifa, H.: The Arabic language and the semantic web: challenges and opportunities. In: The 1st International Symposium on Computer and Arabic Language (2007)
22. Attia, M.: Handling Arabic morphological and syntactic ambiguity within the LFG framework with a view to machine translation. Dissertation. University of Manchester (2008)
23. Farghaly, A.: Arabic natural language processing: challenges and solutions. ACM Trans. Asian Lang. Inf. Process. (TALIP) **8**(4), 14 (2009)
24. Albukhitan, S.: Automatic ontology-based annotation of food, nutrition and health Arabic web content. Procedia Comput. Sci. **19**, 461–469 (2013)
25. Soudani, N.: Toward an Arabic ontology for Arabic word sense disambiguation based on normalized dictionaries. In: On the Move to Meaningful Internet Systems: OTM 2014 Workshops. Springer, Berlin, Heidelberg (2014)
26. Bounhas, I.: ArabOnto: experimenting a new distributional approach for building Arabic ontological resources. Int. J. Metadata Semant. Ontol. **6**(2), 8195 (2011)
27. Abouenour, L.: Construction of an ontology for intelligent Arabic QA systems leveraging the conceptual graphs representation. J. Intel. Fuzzy Syst. **27**(6), 2869–2881 (2014)

28. Hamadou, A.: Multilingual extraction of functional relations between Arabic named entities using NooJ platform. In: Nooj 2010 International Conference and Workshop (2010)
29. Ines Boujelben, S.: Rules based approach for semantic relation extraction between Arabic named entities. In: NooJ (2012)
30. Boujelben, I.: Enhancing machine learning results for semantic relation extraction. In: Natural Language Processing and Information Systems, 337–342 (2013)
31. Boujelben, I.: A hybrid method for extracting relations between Arabic named entities. J. King Saud Univ. Comput. Inf. Sci. 425–440 (2014)
32. Boujelben Ines, B.: Genetic algorithm for extracting relations between named entities. In: LTC, pp. 484–488 (2013)
33. Aliane, H.: Al-Khalil: the Arabic linguistic ontology project. In: LREC (2010)
34. Zaidi, S.: A cross-language information retrieval based on an Arabic ontology in the legal domain. In: Proceedings of the International Conference on Signal-Image Technology and Internet-Based Systems (SITIS'05) (2005)
35. Ishkewy, H.: Azhary: an Arabic lexical ontology (2014). arXiv:1411.1999
36. Saad, S.: Islamic knowledge ontology creation. In: International Conference for Internet Technology and Secured Transactions, 2009. ICITST 2009, IEEE (2009)
37. Saad, S.: Towards context-sensitive domain of islamic knowledge ontology extraction. Int. J. Infon. (IJI) **3**(1), 197–206 (2010)
38. Saad, S.: A process for building domain ontology: an experience in developing Solat ontology. In: International Conference on Electrical Engineering and Informatics (ICEEI), 2011. IEEE (2011)
39. Mazari, A.: Automatic construction of ontology from Arabic texts. In: ICWIT (2012)
40. Harrag, F.: Ontology extraction approach for prophetic narration (Hadith) using association rules. Int. J. Islamic Appl. Comput. Sci. Technol. **1**(2), 48–57 (2013)
41. Belkredim, F.: DEAR-ONTO: a derivational Arabic ontology based on verbs. Int. J. Comput. Process. Lang. **21**(03), 279–291 (2008)
42. Belkredim, F.: An ontology based formalism for the Arabic language using verbs and their derivatives. Commun. IBIMA **11**, 44–52 (2009)
43. Jarrar, M.: Building a formal Arabic ontology (invited paper). In: Proceedings of the Experts Meeting on Arabic Ontologies and Semantic Networks. Alecso, Arab League, Tunis (2011)
44. Semy, S.: Toward the use of an upper ontology for US government and US military domains: an evaluation. In: No. MTR-04B0000063. MITRE CORP BEDFORD, MA (2004)
45. Mizoguchi, R.: Part 1: introduction to ontological engineering. New Gener. Comput. **21**(4), 365–384 (2003)
46. Wikipedia. Upper ontology—wikipedia, November (2015)
47. Jarrar, M.: The Arabic ontology. In: Lecture Notes, Knowledge Engineering Course (SCOM7348), Birzeit University, Palestine (2010)
48. Black, W.: Introducing the Arabic WordNet project. In: Proceedings of the Third International WordNet Conference (2006)
49. Elkateb, S.: Building a WordNet for Arabic. In: Proceedings of The fifth international conference on Language Resources and Evaluation (LREC 2006) (2006)
50. Rodríguez, H.: Arabic wordnet: current state and future extensions. In: Proceedings of the Fourth Global WordNet Conference, Szeged, Hungary (2008)
51. Fellbaum (ed.): WordNet—An Electronic Lexical Database. The MIT Press, Cambridge (1998)
52. Vossen, P.: EuroWordNet, A Multilingual Database with Lexical Semantic Networks. Kluwer Academic Publishers, The Netherlands (1999)
53. Pease, A.: The suggested upper merged ontology: a large ontology for the semantic web and its applications. In: Working Notes of the AAAI-2002 Workshop on Ontologies and the Semantic Web, vol. 28 (2002)
54. Wikipedia. Lexical Ontology—wikipedia, the free encyclopedia (2015) [Online; accessed November-2015]

55. Wikipedia. Arabic WordNet Ontology—wikipedia, the free encyclopedia 2015 [Online; accessed November-2015]
56. Wikipedia. Domain Ontology—wikipedia, the free encyclopedia 2015 [Online; accessed November-2015]
57. Navigli, R.: Learning domain ontologies from document warehouses and dedicated web sites. Computational Linguistics, 151–179 (2004)
58. Hazman, M.: Ontology learning from textual web documents. In: 6th International Conference on Informatics and Systems, NLP (2008)
59. Al-Rajebah, N.: Building ontological models from Arabic Wikipedia: a proposed hybrid approach. In: Proceedings of the 12th International Conference on Information Integration and Web-based Applications and Services. ACM (2010)
60. Al-Rajebah, N.: Exploiting Arabic Wikipedia for automatic ontology generation: a proposed approach. In: 2011 International Conference on Semantic Technology and Information Retrieval (STAIR), IEEE (2011)
61. Habash, N., Rambow, O., Roth, R.: MADA + TOKAN: a toolkit for Arabic tokenization, diacritization, morphological disambiguation, POS tagging, stemming and lemmatization. In: Proceedings of the 2nd International Conference on Arabic Language Resources and Tools (MEDAR), vol. 41, p. 62. Cairo, Egypt, April (2009)
62. Wikipedia. Islam_by_country–wikipedia, the free encyclopedia (2015) [Online; accessed November-2015]
63. Harrag, F.: Using association rules for ontology extraction from a Quran corpus

Part IV
Information Retrieval and Question Answering

A New Semantic Distance Measure for the VSM-Based Information Retrieval Systems

Aya M. Al-Zoghby

Abstract One of the main reasons for adopting the Semantic Web technology in search systems is to enhance the performance of the retrieval process. A semantic-based search is characterized by finding the contents that are semantically associated with the concepts of the query rather than those which are exactly matching the query's keywords. There is a growing interest in searching the Arabic content worldwide due to its importance for culture, religion, and economics. However, the Arabic language; across all of its linguistics levels; is morphologically and syntactically rich. This linguistic nature of Arabic makes the effective search of its content be a challenge. In this study, we propose an Arabic semantic-based search approach that is based on the Vector Space Model (VSM). VSM has proved its success, and many studies have been focused on refining its old-style version. Our proposed approach uses the Universal WordNet (UWN) ontology to build a rich index of concepts, **Concept-Space** (CS), which replaces the traditional index of terms, **Term-Space** (TS) and enhances the Semantic VSM capability. As a consequence, we proposed a new incidence indicator to calculate the **Significance Level of a Concept** (SLC) in the document. The new indicator is used to evaluate the performance of the retrieval process semantically instead of the traditional syntactic retrieval that is based on the traditional incidence indicator; **Term Frequency** (TF). This new indicator has motivated us to develop a new formula to calculate the **Semantic Weight of the Concept** (SWC). The SWC is necessary for determining the **Semantic Distance** (SD) of two vectors. As a proof of concept, a prototype is applied on a full dump of the Arabic Wikipedia. Since documents are indexed by their concepts and, hence, classified semantically, we were able to search Arabic documents efficiently. The experimental results regarding the Precision, Recall, and F-measure presented a noticeable improvement in performance.

A.M. Al-Zoghby (✉)
Faculty of Computers and Information Systems, Mansoura University, Mansoura, Egypt
e-mail: aya_el_zoghby@mans.edu.eg

K. Shaalan et al. (eds.), *Intelligent Natural Language Processing: Trends and Applications*, Studies in Computational Intelligence 740,
https://doi.org/10.1007/978-3-319-67056-0_12

Keywords Semantic search systems · Vector space model (VSM) Universal wordnet (UWN) · Concept-space (CS) · Significance level of concept (SLC) · Semantic weight of concept (SWC) · Semantic distance (SD) Arabic language

1 Introduction

The ambiguity of the search query's keywords is one of the main problems that may frustrate the search process efficiency. The use of the terminological variations for the same concept, Synonyms, creates a many-to-one ambiguity. Whereas, the use of the same terminology for different concepts, polysemous, creates a one-to-many ambiguity [1, 2]. The problem becomes more sophisticated with a highly ambiguous language such as Arabic [3, 4]. For example, the optional vowelization in modern Arabic text increases the polysemy of its written words [5, 6].

Traditionally, the search engines are characterized by trading off a high-recall for low-precision. This is caused mainly due to their sensitivity to the query keywords, and the misinterpretation of the synonymous and polysemous terminologies [7]. In other words, not only all relevant pages are retrieved, but also some other irrelevant, which directly affects the Precision. Moreover, the absence of some relevant pages is leading to low Recall. A recommended solution is to use the *semantic search*, which relies on ontological resources for semantic indexing instead of the lexical indexing that are commonly used by traditional search systems. Thus, the *Semantic search* aims to resolve the semantic ambiguity by retrieving the pages referring to the *semantic interpretation*, hence a particular *concept*, of the search query instead of the pages that are just mentioning its keywords [8, 9].

This research proposes an enhanced semantic VSM-based search approach for Arabic information retrieval applications and the like. In the proposed search approach, we built a concept-space which is used to construct the VSM index. This model enabled us to represent the Arabic documents as semantic vectors, in which the most representative concepts are got the highest weights. This representation allows a semantic classification for the search space. Thus, the semantic retrieval abilities, reflected in its Precision and Recall values, can be obtained. The evaluation of the retrieval effectiveness using the concept-space index resulted in a noticeable improvement in terms of the Precision and the Recall as compared to the traditional syntactic term-space baseline.

The rest of the paper is organized as follows: Sect. 2 describes the main aspects of the proposed model. Section 3 represents the architecture of the proposed model and the implementation aspects. A system is implemented, and the experimental results are discussed in details in Sect. 4. Finally, the paper is concluded at the last section. The list of the algorithms developed to implement the proposed system are listed in the article's appendix.

2 The Proposed Approach

This research proposed an enhanced semantic VSM-based approach for Arabic information retrieval applications and the like. The VSM is a conventional information retrieval model that has demonstrated its ability to represent documents in a computer interpretable form [10].

In the proposed model, we built a rich VSM index of concepts, concept-space (CS) that is enabling the representation of the documents as semantic vectors, in which the most relevant concept are given the highest weights. This semantic representation allows a semantic classification of the documents. Thus the semantic search facilities can be obtained. The construction of CS is derived from the semantic relationships obtainable form the UWN. UWN provides a corresponding list of meanings and shows how they are semantically associated [11]. Fortunately, the UWN supports the Arabic language and its dialects as well. As a proof of concept, a system is implemented on the Arabic Wikipedia. The evaluation of the system's semantic retrieval effectiveness is tested in terms of the Precision and Recall. It resulted in noticeable improvements in its performance as compared to the syntactic term-space based systems.

The key contributions of the study, and how it is distinguished from the traditional VSM are highlighted at the next sections.

2.1 A Novel Indexing Approach

In Semantic Web, terms are used to explain concepts[1] and their relations, [8, 9]. Consequently, the concepts sharing some terms in their definition will share many perspectives of their meanings. This can be realized when concepts can be identified and used as a semantic index of the VSM. For performance evaluation purposes, we produced three indices: Term-Space (TS), Semantic Term-Space (STS), and Concept-Space (CS). As specified by Definition 1, each entry of TS considers all inflected forms of the term. For more clarification, see the TS block of Fig. 1.

Each entry of the STS dictionary, on the other hand, is the UWN semantic expansion of the corresponding TS entry. In other words, each term in the TS is represented by its semantic expansions[2]; as specified by Definition 2[3] and the STS block of Fig. 1.

However, the generation of STS index has revealed some drawbacks that need to be addressed. It might produce duplicated entries that are *directly* or *indirectly*

[1]In this paper, whenever the word *term* is used; it refers to a single word. In the VSM, this term is an entry of the TS. Likewise, whenever the word *concept* is used; it refers to a single concept that is defined in terms of the set of related *terms*, and represented by a single entry in the CS.

[2]More precisely, the set of related inflected forms of its semantic expansions.

[3]More details about producing STS index are can be found in [12], and [13].

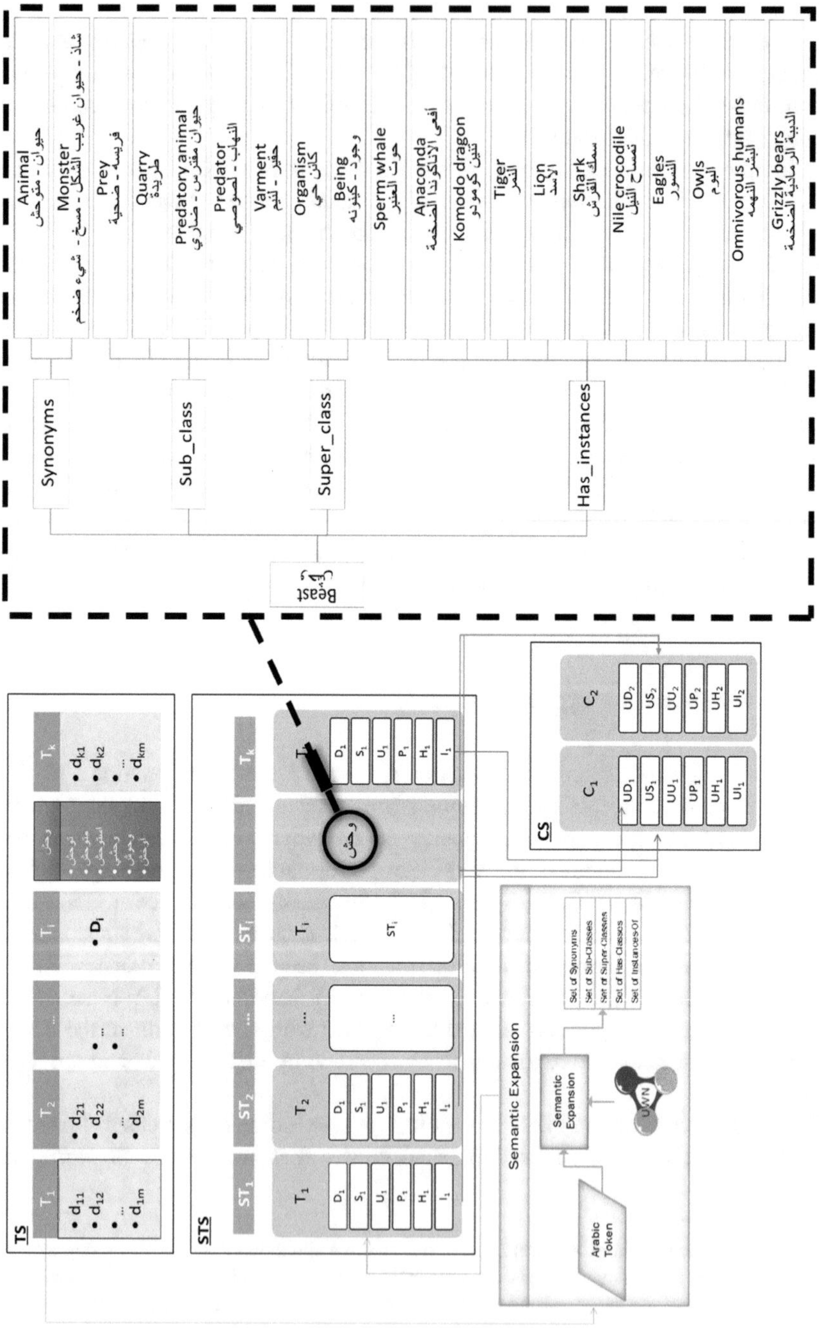

Fig. 1 TS, ST, and CS entries example

linked, see Fig. 2. This duplication causes a kind of semantic redundancy that should be resolved to improve the retrieval performance.

For example, as presented in Fig. 2, the terms:

'Beast[4]/وحش', 'Savage[5]/همجي - بدائي - متوحش', 'Brute[6]/ بهيمة شخص وحشي - بهيمي -', 'Behemoth[7]/ شخص ضخم جدا', and 'Demon[8]/ شيطان - الروح الحارسة - عفريت - شخص ذو قوة -'

are all semantically expanded entries at the STS index. However, there are some direct and indirect redundancies at these expansions. This redundancy is presented at the existence of terms that share their synonyms with other terms or even with the synonyms of other terms. For example, the synonyms '*Beast*' of the terms '*Savage*' and '*Brute*' with the term '*Beast*' itself. Moreover, the shared synonym '*Wildcat*' of both terms '*Savage*' and '*Brute*'. Also, the synonym '*Monster*' of '*Beast*' that is shared by both '*Behemoth*' and '*Demon*'.

To overcome these deficiencies of the STS, we introduced a novel indexing approach to build a CS index that is capable of improving the search performance. As stated at Definition 3, an entry of the CS dictionary is a concept. In our study, the concept is defined by: a main keyword identifying the concept, all of the terms enclosed by the concept's meaning, all of these terms morphological derivations, and all of their UWN semantic expansions. See the CS block of Fig. 1.

The generation and the movement from one indexing type to the next advanced type are depicted at the '*Morphological Indexing*' and '*Semantic and Conceptual Indexing*' phases at the system architecture presented in Fig. 7.

As VSM has proved its capability to represent documents in a computer interpretable form, we tried to improve its performance by replacing its traditional index TS with the new index CS, see Fig. 3. The CS index enabled us to represent documents as semantic vectors, in which the highest weights are assigned to the most representative concept. Therefore, the vector is accurately directed if its highest weight is assigned to the document's fundamental concept, Fig. 3. Thus the semantic search facilities, reflected in its Precision and Recall values, can be gained [4].

2.2 The Significance Level of a Concept (SLC)

In the literature, the performance evaluation of the retrieval capability of indices is usually measured using the following measures: *Document Frequency (df)*, *Term*

---

[4]http://www.lexvo.org/uwn/entity/eng/beast.

[5]http://www.lexvo.org/uwn/entity/eng/savage.

[6]http://www.lexvo.org/uwn/entity/s/n9845589.

[7]http://www.Lexvo.Org/uwn/entity/s/n10128909.

[8]http://www.lexvo.org/uwn/entity/eng/demon.

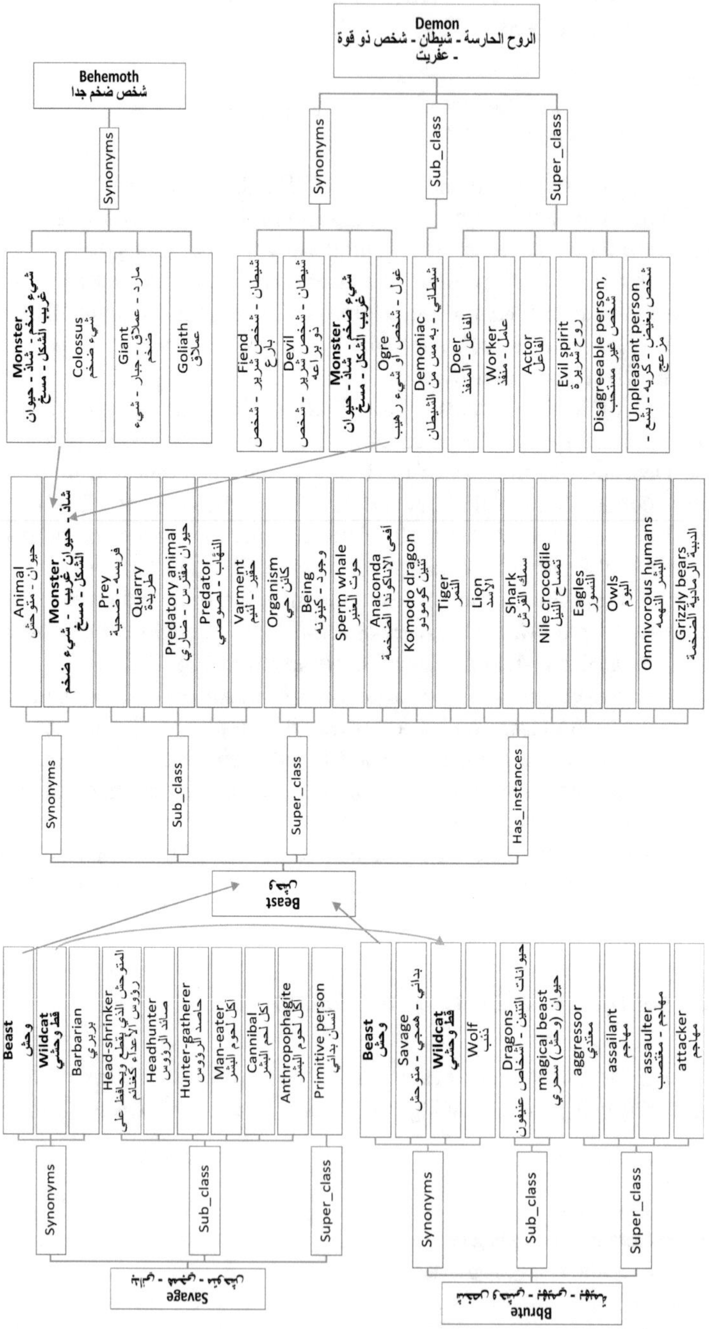

Fig. 2 The semantic expansions redundancy at STS. The semantic expansions presented at the figure are captured from UWN

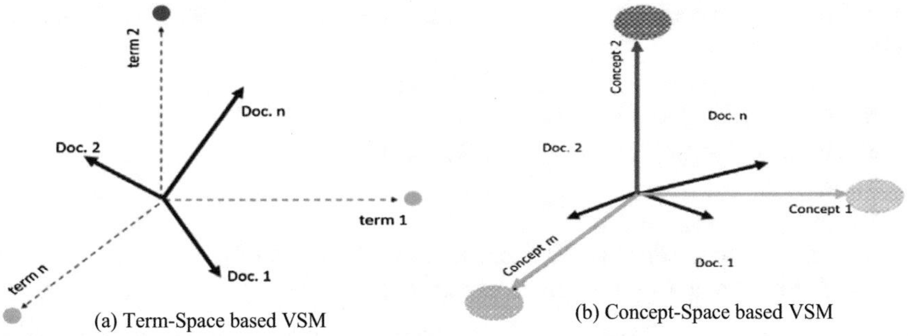

Fig. 3 The enhancement of the traditional VSM to be conceptual VSM

Frequency (tf), and *Weight* (*w*). The bigger the value of the *df, tf, or w* for an index entry, the more relevant documents are found. It is obvious that following the traditional method of calculating the occurrences of the term or its semantic expansions across the document is neither fair nor efficient. It might cause deceptions since documents are considered as relevant basing on the absolute frequency of terms. Therefore, the calculation of the frequency must be controlled by other factors that consider the *relevance degree* instead of the *absolute frequency*. As a matter of fact, the direct increment of *df, tf,* and *w* for each occurrence of the term itself, its semantic expansions, or its conceptualization terms, respectively, may suffer from inaccurate results since there are variations in the relevance levels the expansions. This has motived us to introduce a stage of processing that calculates the *significance level* of the term/concept as a more accurate alternative of the traditional *term frequency* which positively impact the recognition of relevant documents.

In VSM, the weight of the term *t* in document *d* refers to the term's capability of distinguishing the document. Traditionally, the weight *w* of *t* in *d* is defined in terms of its frequency *tf,* which is the number of times that *t* occurs in *d*. However, when the semantic conceptual model is adopted, the equation that calculates the weight will no longer be accurate, and three factors, which are affecting the calculation of the decisive weight, should be taken into consideration.

Definition 1 Term-Space (TS) The TS is defined as the set of all distinct terms belonging to the knowledge source[9] as follows:

$TS = \{T_1, T_2, \ldots, T_i, \ldots, T_k\}$, where:

T_i is a set of inflected forms of Term #i at the TS, i.e. $T_i = \{t_{i1}, t_{i2}, \ldots, t_{ij}, \ldots, t_{im}\}$

k = # terms in TS

m = # inflection forms of Term i

[9]The used knowledge source is the AWDS, which stands for Arabic Wikipedia Documents Space.

Definition 2 Semantic Term-Space (STS) Given TS; the STS is defined as follows:

$STS = \{ ST_1, ST_2, \ldots, ST_i, \ldots, ST_k \}$, where:

$ST_i = Semantic\_Expansion\ (T_i) \cup T_i$[10]

$Semantic\_Expansion\ (T_i) = \{S_i, U_i, P_i, H_i, I_i\}$

Thus,

$ST_i = \{e_{i1}, \ldots, e_{ix,}, \ldots, e_{in}\}$[11]

e_{ix} = the inflectional/semantic expansion #x of the term T_i

n = the total count of T_i sematic expansions

Definition 3 Concept-Space (CS) Given STS, the CS is generated from groups of concepts, each of which is represented by interrelated semantic terms of the STS as follows:

$CS = \{C_1, C_2, \ldots, C_q\}$, where:

$C_i = \{ST_{i1}, ST_{i2}, \ldots, ST_{ij}, \ldots, ST_{ic}\}$[12]

Thus,

$C_i = \{e_{11}, \ldots, e_{1x,}, \ldots, e_{1n}, \ldots, e_{c1}, \ldots, e_{cx,}, \ldots, e_{cn}\}$

q = # concepts extracted from the STS

C_i = the concept #i, which is defined by a set of semantic terms from STS

c = # semantic terms used to define the concept #i

First, the term needs to be matched against all of its **inflected forms** that occur in *d* rather than the input or the normalized form. Therefore, not all matches would have the same weight.

Second, the semantic expansions of terms can be classified into five different types or relations: *Synonyms, SubClasses* and *HasInstances (Specialization expansions)*, and *SuperClasses* and *InstancesOf (Generalization expansions)*. It is evident that the Synonyms expansions are **semantically closer** to the original term than either its generalized or specialized expansions. Moreover, the SubClasses and SuperClasses expansions are worthier than those of HasInstances or InstancesOf, since the former types represent classes that encompass a set of related concepts while the latter types represent instances referring to specifically related elements.

[10]See Ti at Definition 1.

[11]Extracted from the UWN.

Si = {s1,...,sa}, //set of Synonyms

Ui = {u1,...,ub}, //set of Sub-Classes

Pi = {p1,...,pc}, //set of Super-Classes

Hi = {h1,...,hd}, //set of Has-Instances

Ii = {i1,...,ie}, //set of Instances-Of

All of these expansions can be accumulated in the set of all expansions on the term Ti: {ei1, ...ein}.

Each expansion of s, u, p, h, & i, is represented as a pair of the expansion-word and the expansion-confidence as (word, conf.).

[12]Therefore, each concept C is defined in terms of all expansion sets of each ingredient STs. I.e., the accumulation of the subsets {ei1, ..., eix,, ..., ein}.

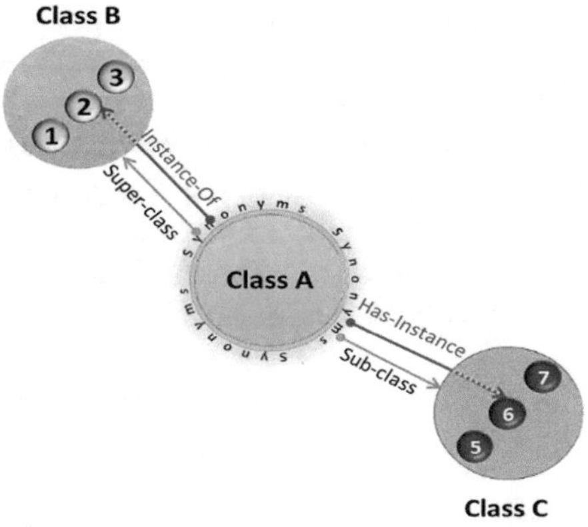

Fig. 4 Synonyms, sub-classes, super-classes, has-instance, and instance-of relationships

This also means that matches with different **semantic relations** also have different **distances** that directly affect the weights.

Third, the available semantic expansions of each semantic relation type have different **confidences** because not all of expansions are relevant to the original term with the same degree or weight. So, an additional confidence factor should be used when calculating the weight.

The factors above impede the term frequency to be used as an accurate *incidence indicator* for the semantic expansions in the semantic search process. Hence, we introduced a new *incidence indicator* of the term *t*, and consequently the concept *c* that is defined in terms of *t*, in document *d*. This indicator is based on the *significance level* of the term in the document instead of its frequency count. The new measurement, *significance level*, is computed in terms of the *association strength* of each expansion of the term. The *association strength* depends on the *distance* of the semantic relation and the *confidence* of the semantic expansion. The distance of the semantic relation is a constant[13] that is heuristically determined by the semantic closeness of the expansion as declared before. The value of the confidence, on the other hand, is directly determined by the UWN specifications.

Figure 4 shows the semantic closeness of the expansion. As depicted in this figure, the set of ***Synonyms*** has an exact matching to the original term. Therefore, we decided to set the value of the multiplying coefficient of synonyms to be 1. On the other hand, the ***Sub/Super classes*** represent some degree of generalizations such that they are not found by the exact match with the original term. However, they are more closely related than ***instances*** since they indicate a general perspective of the meaning. Whereas the instances represent certain items that may share one or more characteristics of the original term's meaning. Therefore, we decided to set the value of the multiplying coefficient of the sub/super classes to be

[13]Multiplying coefficient along the distance scale.

0.75 and that of the has-instances/instances-of classes to be 0.5. More illustrative examples can be found in [14]. Formally, let *AS(e)* denotes the *Association Strength* of the expansion *e*. As presented at Eq. (1), it is defined in terms of the *confidence* of the expansion *e*, and the *distance* of its semantic relation type. The new incidence indicator, which denotes the *Significance Level of a Term ST* at document *d*, is defined by *SLT(ST,d)* as presented at Eq. 2. Where *n* is the number of all expansions *e* of term *ST*.[14] The tf(e_x) factor represents the frequency of expansion e_x instances occurred in document *d*.

$$\mathrm{AS(e)} = \mathrm{confidence(e)} * \mathrm{distance(e)} \tag{1}$$

$$\mathrm{SLT(ST, d)} = \sum_{\mathrm{x}=1}^{\mathrm{n}} \mathrm{AS}(e_x) * \mathrm{tf}(e_x) \tag{2}$$

Let SIDF(ST) denotes the Semantic Inverse Document Frequency of the term ST. It is defined using Eq. 3.[15,16] Thus, the Semantic Weight of term ST in document d is defined by SW(ST,d) presented at Eq. 4. In terms of concepts, the Significance Level of a Concept (SLC) is defined by Eq. 5. Where c is the count of the semantic terms used to define the concept C.

$$\mathrm{SIDF(ST)} = \log \frac{|\mathrm{D}|}{|\{e_x \in \mathrm{d} \mid e_x \in \mathrm{ST}\}|} \tag{3}$$

$$\mathrm{SW(ST, d)} = \mathrm{SLT(ST, d)} * \mathrm{SIDF(ST)} \tag{4}$$

$$\mathrm{SLC(C, d)} = \sum_{\mathrm{i}=1}^{\mathrm{c}} \mathrm{SLT(STi, d)} \tag{5}$$

Finally, let SIDF(C) denotes the Semantic Inverse Document Frequency of the concept C as defined by Eq. 6.[17] The Semantic Weight of each concept in the new Concept Space, SW(C,d), is defined by Eq. 7.

$$\mathrm{SIDF(C)} = \log \frac{|\mathrm{D}|}{|\{e_x \in \mathrm{d} \mid e_x \in \mathrm{C}\}|} \tag{6}$$

$$\mathrm{SW(C, d)} = \mathrm{SLT(C, d)} * \mathrm{SIDF(C)} \tag{7}$$

---

[14]See Definition 2.

[15]D is the entire documents-space.

[16]See ST at Definition 2.

[17]See C at Definition 3.

2.3 Semantic Distance Between Query and CS

As far as the semantic of concepts are considered, we need to accurately match each expanded word in the input query with each concept in CS, see Fig. 5. As stated before, not all matches of each semantic expansion have the same matching consistency.

For example, the first synonyms of the first query word, Qw_1s_1, may match one of the super-classes of the concept x in CS, C_xp_i. The Qw_1s_1 itself may matches one of the synonyms of another concept y, C_ys_j. Thus, as we justified earlier, the second match is stronger than the first. Therefore, for efficiency reasons, each case has to be handled separately. Otherwise, unexpected results will be produced. For instance, weak matches may take the same weights as other stronger ones. To that end, we generated the formulas (8) and (9) for constructing the entries of this sensitive-matching vector of the query.

Let n denotes the count of expansions that are matched between the query and the concepts in the space, and m_x denotes the frequency of each match. The Significance Level of the concept C in query Q is defined by Eq. 8. Where CAS_x is the Association Strength between the concept c and the match m_x, while the $QwAS_x$ is the Association Strength between the query word of the same mach.

Thus, the Semantic Weight of the concept C in the query Q is defined by Eq. 9. Accordingly, the Semantic Distance between query Q and document d_i is calculated

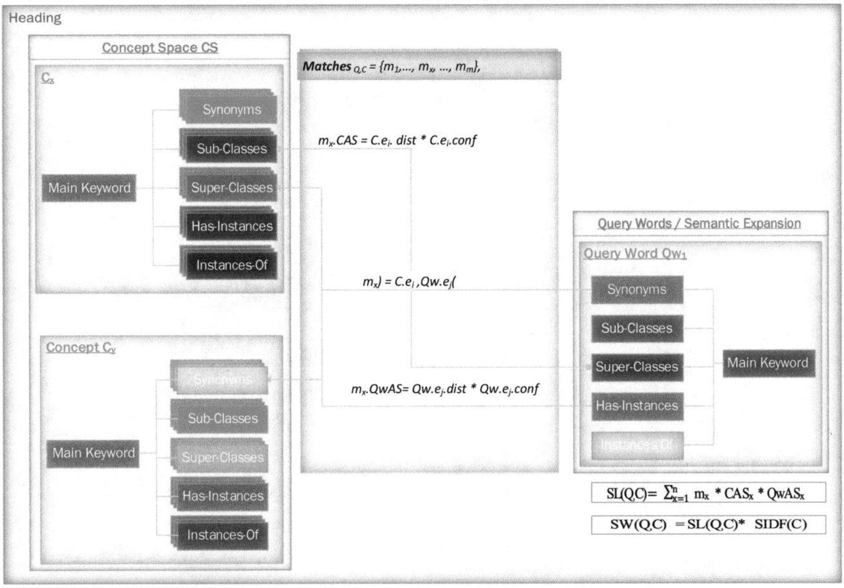

Fig. 5 Query word expansions and Concept's expansions matchings

regarding the semantic weights of the concept C_j in both the query Q and the document d_i by Eq. 10. Where q is the count of concepts in the space.

To sum up, we have described how to calculate weights of concepts in documents to construct the CS index and how to compare the concepts in an input query with these concepts to accurately determine relevant documents.

$$SL(Q, C) = \sum_{x=1}^{n} m_x * CAS_x * QwAS_x \tag{8}$$

$$SW(Q, C) = SL(Q, C) * SIDF(C) \tag{9}$$

$$SDist(d_i, Q) = \frac{\sum_{j=1}^{q} SW(Q, C) * SW(d_i, C)}{\sqrt{\sum_{j=1}^{q} SW(Q, C)^2} * \sqrt{\sum_{j=1}^{q} SW(d_i, C)^2}} \tag{10}$$

3 System Architecture

This section describes the architecture of the proposed system and its components. The overall architecture is portrayed in Fig. 6. Each process is presented in details at the phases of Fig. 7. The generation particulars of the Inflectional, Semantic, and Conceptual vectors are presented more formally by the Algorithms presented at the Appendix.

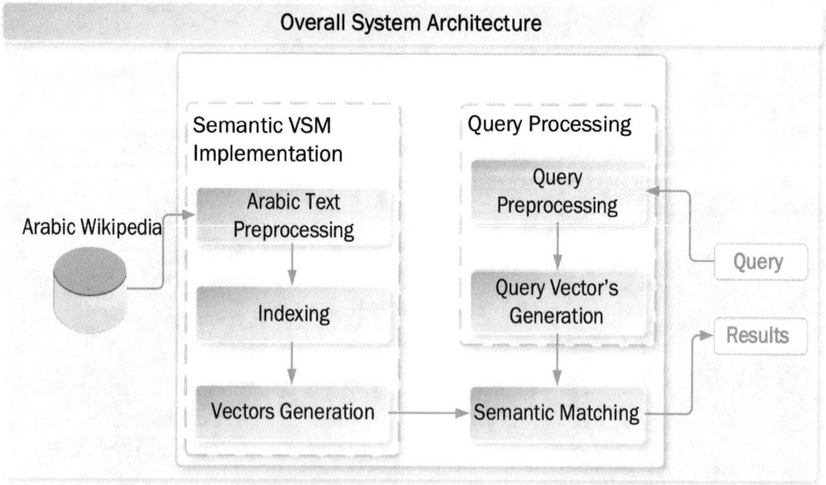

Fig. 6 Overall system architecture

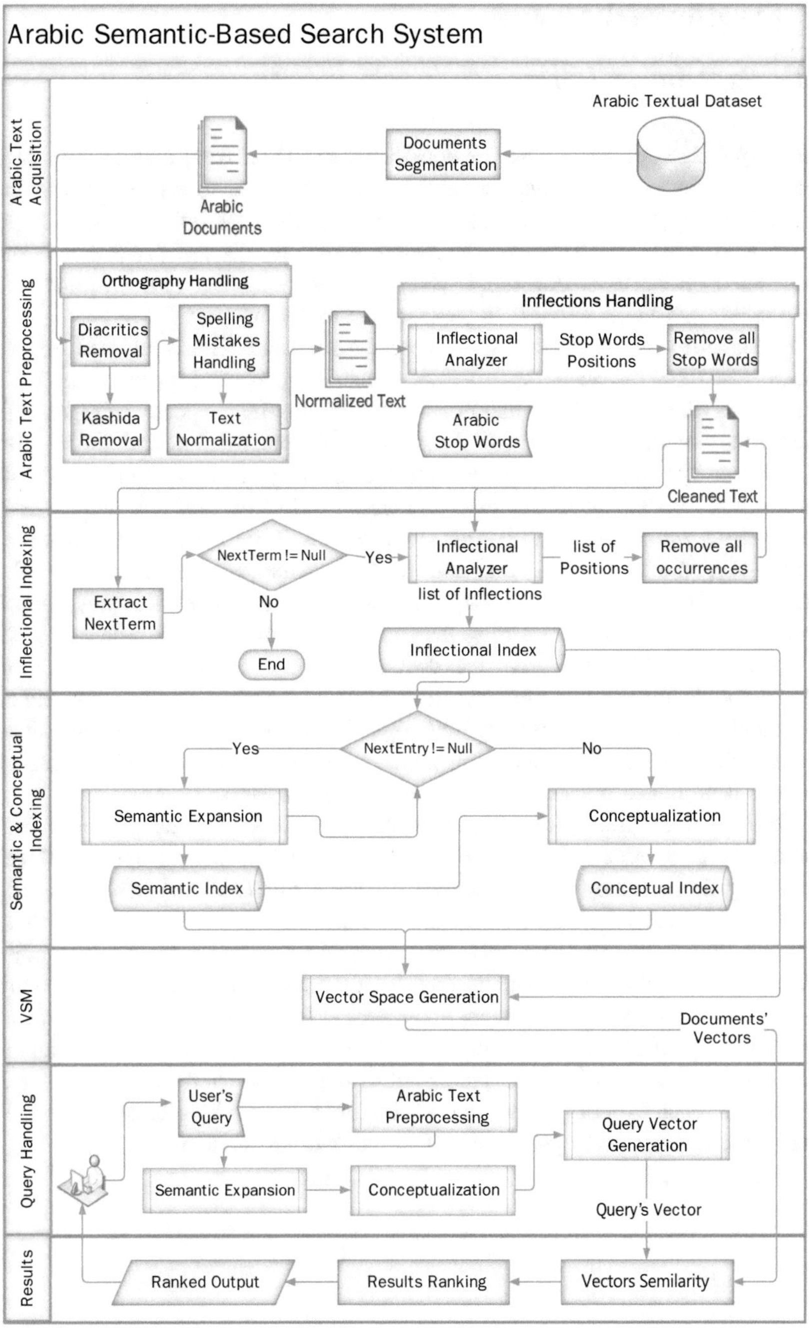

Fig. 7 Detailed system sub-processing

4 Experimental Analysis

4.1 Experimental Setup

The knowledge source is a set of documents extracted from a full dump of the Arabic Wikipedia.[18] We have conducted five experiments to test the effectiveness of the features proposed by the current research. The description of these experiments is presented in Table 1. The experimental results are divided into three interrelated dimensions: document representation, document retrieval, and document ranking. The objective is to measure the following (Fig. 8):

1-The indexing efficiency: when the conceptual indexing is adopted, the documents are retrieved according to their semantic concepts instead of the lexical terms. This dimension of evaluation aims to measure the efficiency of representing documents according to their central concept(s). As presented before, the weight of an index-entry[19] regarding to a document is referring to its capability of distinguishing this document in the space. Therefore, this evaluation is judged by the *Document Frequency (df)* and *Weight (W)* averages. The higher the *df* and *w* for an index-entry, the documents that are more relevant will be retrieved, which increasing the relevance accuracy. Besides the relevance accuracy, another dimension is needed for measuring the capability of the retrieval process itself.

2-The retrieval capability: the documents retrieval method does not only retrieve the documents that are syntactically or even semantically relevant, but also documents which are conceptually (ontologically) relevant. This evaluation aims to measure the impact of the indexing method on the performance of the retrieval process. It is judged by: Precision, Recall, and F-Measure.

However, the retrieved documents do not rank the same. The efficiency of the ranking process affects the overall performance of the search system. Besides measuring the accuracy of the retrieval process, another dimension is needed to measure the accuracy of the document ranking.

3-The accurate ranking: this evaluation aims to measure the accuracy of assigning the corresponding weight that exactly represents the association strength of each index entry with a document in the space. Therefore, the incidence indicator factor directly affects the capability of the accurate ranking. This evaluation is semi-automatic. It calculates the average of distance between the ranking order that results from each experiment and the ranking order judged by a human expert. The smaller the distance average, the closer the rank to the standard.

[18]The Arabic Wikipedia dump is accessed on 29-Aug-2012 from http://dumps.wikimedia.org/arwiki/.

[19]A term or hence a concept.

Table 1 The experiments description and the results summary

Experiment	Indexing	Expanding type	Index size	Incidence indicator	*tf* Average	*df* Average	*w* Average	Ranking-distance average	Experiment description
V1	TS	Inflectional	360486	IF	0.514	2.02	1.02	4.62	Inflectional frequency-based VSM
V2	STS	Semantic		SF	0.704	2.6	2.11	4.4	Semantic frequency-based VSM
V3				SLT			1.65	4.39	Significance-level-of-term based VSM
V4	CS	Conceptual	223502	CF	0.7076	3.8	2.83	4.165	Conceptual frequency-based VSM
V5				SLC			2.32	4.154	Significance-level-of concept based VSM

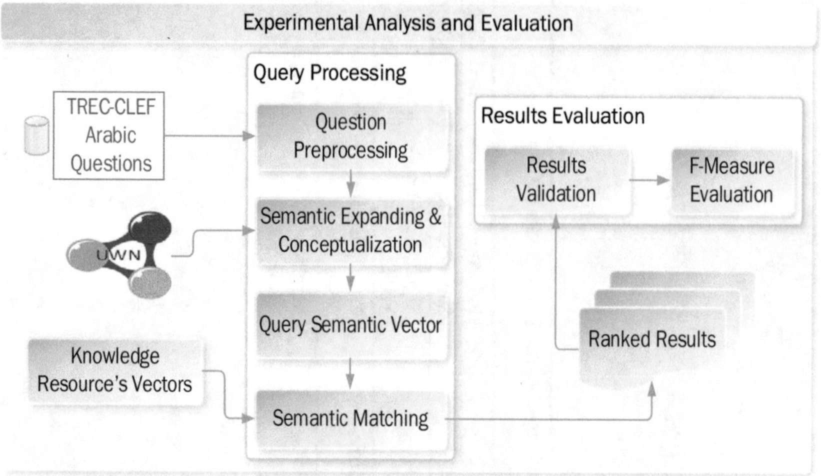

Fig. 8 Experimental analysis and evaluation

4.2 Experiments, Results, and Evaluation

To demonstrate the variations of the retrieval capability when each indexing method is applied, we measured their performance in terms of the values of *Document Frequency (df)*, *Term Frequency (tf)*, and *Weight* (*w*). Table 1 represents the summary of the results presented in the subsequent sections.

4.2.1 The Conceptualization Levels

From the *AWDS*, a terms-space (TS) of 360486 terms is extracted after excluding the Named Entities. The TS is then semantically expanded using the UWN to construct the STS as defined at Definition 2. The two conceptualization levels, presented at Algorithm 3, are then applied on the STS to generate the CS as presented at Definition 3. As a result, the size of the STS is shrunk by 38% to construct the CS version.

This leads to the increment of the representation power of each item in the space since the average of items weights is increased as shown Table 1. Note that the weight average of V3 and V5 are lower than those of V2 and V4, respectively. We are going to explain these results with the discussion of ranking accuracy. However, it is noticeable that the weights of the STS index are higher than those of the inflectional indexing TS. Moreover, the weights of the CS are greater than those of the STS. The important observation is that the obtained results are demonstrating the efficiency of the CS in distinguishing documents according to weights of the corresponding concepts.

4.2.2 The Retrieval Capability

The F-Measure is a very common measurement for calculating the performance of the retrieval systems. It is based on the harmonic mean of the Recall and the Precision scores, which we used to evaluate the retrieval accuracy of the proposed system. The F-Measure is defined by Eq. 11.

$$\mathrm{F-Measure} = 2 * \frac{\mathrm{Precision} * \mathrm{Recall}}{\mathrm{Precision} + \mathrm{Recall}} \tag{11}$$

4.2.3 The Ranking Accuracy

The ranking accuracy of the proposed system is evaluated by measuring the correctness of assigning the weight that precisely characterizes the Association Strength of each index with documents in the space. This is representing how the Incidence Indicator factor affects the ability of the accurate ranking.

The values of the weights averages of the experiments V2 and V4 are greater than those of V3 and V5 respectively. The ranking results show that the V3 experiment is better than V2, while V5 is the best. This is due to the extra error ratios caused by the use of the Sf and CF indicators instead of the SLT and SLC. However, these extra ratios are reduced as a result of using the SLT and SLC indicators at the V3 and V5 experiments, which is directly reflected on the ranking efficiency of these experiments. Still, the experiment V4 gives a better ranking order than V3, since it is based on the conceptual indexing CS, although it suffers from the extra error ratio caused by the CF indicator.

The ranking accuracy is calculated using the Distance Average (DA), Eq. 12, between the experimental order and the standard order delivered by a human specialist.

$$\mathrm{DA(V)} = \frac{\sum_{i=1}^{n} \left| \frac{1}{\mathrm{SRank}_i} - \frac{1}{\mathrm{ERank}_i} \right|}{n} \tag{12}$$

Where n is the count of the retrieved documents as a result of the user's query, while the SRank_i is the standard rank of document i, and the ERank_i is the rank is the experimental rank of document i at experiment V.

The closest ranking order is obtained at the experiment V5, which assures its parameters' capability to rank the retrieved result more accurately.

5 Conclusion and Future Work

This study sheds light on the inaptitude in searching the Arabic Language semantically, which may be attributed to the sophistication of the Arabic language itself. However, this should not stop more effective efforts for achieving the best possible solutions that enable the Arabic Language users getting the benefit from the new electronic technologies.

In an attempt to take a step in that long pathway, we proposed an Arabic semantic search system that based on the Vector Space Model. The VSM is one of the most common information retrieval models for textual documents due to its ability to represent documents in a computer interpretable form. However, as it is syntactically indexed, its sensitivity to keywords reduces its retrieval efficiency. To improve its effectiveness, the proposed system is extracting a concept-space index, using the universal wordnet UWN, to be used as a semantic index of VSM search system. The proposed system enables a conceptual representation of the document space, which in turn permits the semantic classification of them and thus obtaining the semantic search benefits. Moreover, we introduced a new incidence indicator to calculate the significance level of the concept in a document instead of the traditional term frequency. Furthermore, we introduced a new formula for calculating the semantic weight of the concept to be used in determining the semantic distance between two vectors. The system's experimental results showed an enhancement of the F-measure value using the conceptual indexing over that is based on the standard syntactic baseline.

As a future work, we have to solve some problems such as the ambiguity by discriminating the meaning contextually. Also, we may work on refining the processing of the multiword expression expansions. That will improve the results noticeably since. Moreover, the improvement of the Arabic knowledge representation in the UWN will help to overcome its limitations that directly affects the search results. Another open research area is to solve the problems of the Arabic language morphological analysis to prevent the consequent errors occurred in the indexing process, and hence, the construction of the search dictionary. We also may try to use Google Translation API with the UWN to find results for these terms that have results in languages other than Arabic.

Appendix: The Implementation Algorithms

Algorithm 1: V1

```
Input: D
Output: V1 and TS
Begin
  Indexing (D);
  Foreach dj in D
     term = GetNextTerm(dj)[20];
     VSM(term);
  EndFor
  Return V1, TS;
End
Function VSM(t)
   TS.add(t);
   i = TS.size;
   Foreach dj in D
      inflections = Search (t, dj)[21];
      If inflections.size > 0
        TS.inflected_forms[i] = inflections;
        V1.df[i]++;
        V1.tf[i,j] = inflections.size;
      EndIf
   EndFor
   V1.IDF[i] = log(D.size / V1.df[i].size)
   Foreach dj in D
      V1.w[i,j] = V1.IDF[i] * V1.tf[i,j]
   EndFor
End VSM
```

Algorithm 2: V2 and V3

```
Input: D and TS
Output: V2, V3, and STS
Begin
   STS.size = TS.size;
   STS.inflected_forms= TS.inflected_forms;
   Foreach Ti in TS
      STS.semantic_expansions[i] = UWN.Expansion(Ti); // see Def.2
   EndFor
```

[20]Getting the text term in *D* without redundancy.

[21]Search for any inflectional form of *t* in the document d_i using RDI Swift Searcher.

```
    Foreach ST_i in STS
      Foreach e_x in ST_i
        Foreach d_j in D
          semantic_inflections = Search(e_x, d_j)[22];
          If semantic_inflections.size > 0
            If j ∉ df[i]
                V2.df[i].add(j);
                V3.df[i] = V2.df[i];
            EndIf
            V2.tf[i,j]+= semantic_inflections.size;          //SLT(ST_i,d_j)
            AS = e_x.confidence * e_x.distance;               //AS(e)
            V3.tf[i,j]+= AS* semantic_inflections.size;      //SLT(ST_i,d_j)
          EndIf
        EndFor
      EndFor
      V2.IDF[i] = log(D.size / V2.df[i].size);                //SIDF(ST_i)
      V3.IDF[i] = log(D.size / V3.df[i].size);                //SIDF(ST_i)
      Foreach d_j in D
        V2.w[i,j] = V2.IDF[i] * V2.tf[i,j];                   //SW(ST_i,d_j)
        V3.w[i,j] = V3.IDF[i] * V3.tf[i,j];                   //SW(ST_i,d_j)
      EndFor
    EndFor
  Return V2,V3, STS;
End
```

Algorithm 3: V4 and V5

```
Input: V2, V3, and STS
Output: V4, V5, and CS
Begin
    Indexing(S)[23];
```

[22]Search for any inflectional occurrences for the semantic expansion e_x of the term ST in document d_j using RDI Swift Searcher.

[23]S is the set of all Synonyms of all terms in STS.

```
    //1st Conceptualization Phase
    group_ID = 0;
    Foreach ST_i in STS
        x = [STS. inflected_forms[i], STS.semantic_expansions[i].synonyms];
        s = (STS. inflected_forms. Except(STS. inflected_forms[i])) ∪
            (STS.semantic_expansions.synonyms.
                Except(STS.semantic_expansions[i].synonyms));

        relatedTerms = Search(x, s)[24];

        If relatedTerms.size > 0
            group_ID++;
            Foreach r in relatedTerms
              G[group_ID].add(STS[r]);
            EndFor
        EndIf
    EndFor

    //2nd Conceptualization Phase
    Foreach g in G
        concept = g;
        Foreach g' in G. Except(g)
            If g ∩ g' ≠ ∅
              concept ∪= g';
            EndIf
        EndFor
        CS.add(concept);
    EndFor

    //Update V2 and V3 to get V4 and V5.
    Foreach C_i in CS
        Foreach ST_x in C
            V4.df[i] ∪= V2.df[ST_x];                    //SLC(C_i,d_j)
            V4.tf[i,j]+= V2.tf[ST_x,j];                 //SLT(C_i,d_j)
            V5.tf[i,j]+= V3.tf[ST_x,j];                 //SLT(C_i,d_j)
        Endfor
        V5.df[i] = V4.df[i];
        V4.IDF[i] = V5.IDF[i] =  log(D.size / V4.df[i].size);    //SIDF(C_i)
        Foreach d_j in D
            V4.w[i,j] = V4.IDF[i] * V4.tf[i,j];                 //SW(C_i,d_j)
            V5.w[i,j] = V5.IDF[i] * V5.tf[i,j];                 //SW(C_i,d_j)
        EndFor
Return V4,V5,CS;
End
```

[24]Search for any inflectional or Synonyms occurrences for the term ST_i in the set s using RDI Swift Searcher.

References

1. Shaalan, K.: A Survey of Arabic named entity recognition and classification (June 2014)
2. Saleh, L.M.B , Al-Khalifa, H.S.: AraTation: an Arabic semantic annotation tool (2009)
3. Tazit, N.: El Houssine Bouyakhf, Souad Sabri, Abdellah Yousfi. Semantic internet search engine with focus on Arabic language, Karim Bouzouba (2007)
4. Cardoso, J.: Semantic Web Services: Theory, Tools, and Applications. IGI Global, Mar 30 2007
5. Hepp, M., De Leenheer, P., de Moor, A.: Ontology Management: Semantic Web, Semantic Web Services, and Business Applications. Springer, New York; [London] (2008)
6. Kashyap, V., Bussler, C., Moran, M.: The Semantic Web: Semantics for Data and Services on the Web (Data-Centric Systems and Applications). Springer, 15 Aug 2008
7. Panigrahi, S., Biswas, S.: Next generation semantic web and its application, March 2011
8. Unni, M., Baskaran, K.: Overview of approaches to semantic web search, July–December 2011
9. Renteria-Agualimpia, W., López-Pellicer, F.J., Muro-Medrano, P.R., Nogueras-Iso, J., Zarazaga-Soria1, F.J.: Exploring the Advances in Semantic Search (2010)
10. Kassim, J.M., Rahmany, M.: Introduction to Semantic Search Engine. Selangor (2009)
11. Hahash, N.: Introduction to Arabic natural language processing. Association for Computational Linguistics, 30 August 2010
12. Jarrar, M.: Building a Formal Arabic Ontology (Invited Paper), Alecso, Arab League. Tunis, 26–28 April 2011
13. Al-Zoghby, A.M., Shaalan, K.: Conceptual search for Arabic web content. In: Computational Linguistics and Intelligent Text Processing—16th International Conference, vol. 9042, pp. 405–416, 14–20 April 2015
14. Al-Zoghby, A., Ahmed, A., Hamza, T.: Arabic semantic web applications: a survey. J. Emerg. Technol. Web Intell. 52–69 (2013)

An Empirical Study of Documents Information Retrieval Using DWT

Mohamed Yehia Dahab, Mahmoud Kamel and Sara Alnofaie

Abstract Most of the information retrieval models represent documents as bag-of words which takes into account the term frequencies (**tf**) and inverse document frequencies (**idf**). However, most of these models ignore the distance among query terms in the documents (i.e. term proximity). Several researches have appeared in recent years using the term proximity among query terms to increase the efficiency of document retrieval. To solve proximity problem, several researches have implemented tools to specify term proximity at the query formulation level. They rank documents based on the relative positions of the query terms within the documents. They should store all proximity data in the index, affecting on the size of index, which slows down the search process. In the last decade, many researches provided models that use term signal representation to represent a query term, the query is transformed from the time domain into the frequency domain using transformation techniques such as wavelet. Discrete Wavelet Transform **(DWT)**, such as Haar and Daubechies, uses multiple resolutions technique by which different frequencies are analyzed with different resolutions. The advantage of the **DWT** is to consider the spatial information of the query terms within the document rather than using only the count of terms. In this chapter, in order to improve ranking score as well as improve the run-time efficiency to resolve the query, and maintain a reasonable space for the index, two different discrete wavelet transform algorithms have been applied namely: Haar and Daubechies, three different types of spectral analysis based on semantic

M.Y. Dahab (✉) · M. Kamel · S. Alnofaie
Faculty of Computing and Information Technology, Department of Computing Science, King Abdulaziz University, Jeddah 21589, Saudi Arabia
e-mail: mdahab@kau.edu.sa; mohamed.dahab@gmail.com
URL: http://www.kau.edu.sa/Home.aspx

M. Kamel
e-mail: miali@kau.edu.sa

S. Alnofaie
e-mail: salnefaie@kau.edu.sa

K. Shaalan et al. (eds.), *Intelligent Natural Language Processing: Trends and Applications*, Studies in Computational Intelligence 740,
https://doi.org/10.1007/978-3-319-67056-0_13

segmentation are carried out namely: sentence-based segmentation, paragraph-based segmentation and fixed length segmentation; and also different term weighting is performed according to term position. The experiments were constructed using the **T**ext **RE**trieval **C**onference **(TREC)** collection.

Keywords Information retrieval · Term proximity · Term signal
Haar and Daubechies wavelet transform algorithms

1 Introduction

Previous studies on documents information retrieval were based on the bag-of-words concept for documents representation e.g., [1–4], that take into consideration the **tf** and **idf** but the major drawback of this concept is the ignoring of the term proximity. Proximity represents the closeness of the query terms appearing in a document.

For the purpose of improving document retrieval performance, much recent researches have provided new models that consider the proximity among query terms in documents. Those methods use the term positional information to compute the document score. Intuitively, document d_1 ranked higher than documents d_2 when the terms of the query occur closer together in d_1 while the query-terms appear far apart in d_2. Proximity can be seen as a kind of indirect measure of term dependence [5].

Even though the documents ranking has been improved in recent researches (e.g. [6–11]). Nonetheless, it may cause increasing in both time complexity and space complexity because they should make huge number of comparisons and they should store the results of these comparisons. Much recent, many researches use term/word signal representation or time series. The term signal represents a query term in a document and the term signal is transformed from time domain to frequency domain using known transformation techniques such as Fourier transform [12–15]. Cosine transform [16], Wavelet transform [17, 18] etc.

By using the **DWT**, as a filter bank and multi resolution analysis, we are able to analyze a document using a multi-glasses concentration. One glass is to analyze a document as a whole which is similar to vector space method. When using this glass or lens, it is like using a telescope to focus or view the entire document. The glass does not need to be moved at all details (such as the proximity among query terms in specific line or paragraph) of a document. Using another glass, such as an opera glass, which may focuses on the first half of a document and it needs to be moved at all details in this region. A final example of a glass, such as a microscopic glass, which may focuses on the third line of a document and it also needs to be moved at all details in this line.

This chapter presents an information retrieval framework based on **DWT** using both Haar and Daubechies wavelet transform algorithms. The framework achieves the speed of the classical methods such as vector space methods with the benefits of the proximity methods to provide a high quality information retrieval system. The retrieval is performed by the following: (a) locating the appearances of the query terms in each document, (b) transforming the document information into the time scale, and (c) analyzing the time scale with the help of the **DWT**.

Information retrieval systems based on **DWT** has many advantages over classical information retrieval systems but using fixed document segmentation has many disadvantages because of neglecting document semantic representation. The main research hypothesis of this chapter is the necessity of using spectral analysis based on different types of semantic segmentation, namely sentence-based segmentation, paragraph-based segmentation and fixed length segmentation, to improve document score.

This chapter will proceed as follows, Sect. 2 presents the related background, Sect. 3 introduces most important issues of the design and implementation of information retrieval using **DWT** method, Sect. 4 the experiments and results using the dataset Text **RE**trieval **C**onference **(TREC)** collection, and finally the conclusion is shown in Sect. 5.

2 Background

In this section, the major related conceptes are reviewed.

2.1 Term Signal

The notion of term signal was introduced by [14] before being reintroduced by [19] as a word signal. It is term vector representation, which defines the term frequency of appearance in particular bins or segments within a document. The term signal shows how a query term is spread over a document. If a term appearing in a document can be thought of as a signal, then this signal can be transformed using one of the transformation algorithms such as Fourier transform or wavelet transform and compared with other signals. Term signal has representative information of a term in time domain since it describes when the term occurs and its frequency in a certain region. Applying this approach, for each document in the dataset, a set of term signals is computed as a sequence of integers, which shows the term frequency in a specific region of a document. In document *d*, the representation of a query term *t* signal is:

$$s(t,d) = \tilde{f}_{t,d} = [f_{t,1,d}, f_{t,2,d}, \ldots, f_{t,B,d}], \quad (1)$$

where

$f_{t,b,d}$: is frequency of term t in segment b of document d for $1 \leq b \leq B$.

2.2 Weighting Scheme

Zobel and Moffat [20] provided a more general structure of the form AB-CDE-FGH, to allow for the many different scoring functions that had appeared. Their study compared 720 different weighting schemes based on the TREC document set. Their results showed that the BD-ACI-BCA method was the overall best performer. The computed segment components are weighted using a BD-ACI-BCA weighting scheme as follows:

$$\omega_{t,b,d} = \frac{1 + log_e f_{t,b,d}}{(1 - slp) + \frac{slp.W_d}{ave_{d \in D} W_d}}, \quad (2)$$

where

slp: is the slope parameter (usually equals to 0.7),
W_d: is the document vector norm,
$ave_{d \in D} W_d$: is the average document vector norm.

The term signal will be correspond to the weighted term signal as follows:

$$\tilde{\omega}_{t,d} = [\omega_{t,1,d}, \omega_{t,2,d}, ..., \omega_{t,B,d}]. \quad (3)$$

2.3 Document Segmentation

Document segmentation is very important issue in information retrieval specially when using any transformation algorithms such as **DWT** because it determine the acceptance distance between two query terms in a document to be considered in the same region (bin, line, paragraph, etc.). Three suggested segmentation were introduced in [21]. The suggested segmentation includes:

- Sentence-based segmentation.
- Paragraph-based segmentation.
- Fixed length segmentation.

Before diving into details of segmentation, one should understand that if the query terms are found in the same segment/bin the wavelet transform algorithm should gives higher score than if there are scattered in the document. That is why segmentation is a very important issue. The sentence-based segmentation and paragraph-based segmentation were firstly investigated by [21] to give a real meaning of proximity. One of the limitations with wavelet transform algorithm is that it requires exact 2^n where $n \in Z$ number of segments (bins maybe used as a synonym). There is no need for zero padding process in the third item, fixed length segmentation, because the number of segments used is in the form of 2^n where $n \in Z$.

The number of segments when using the fixed length segmentation relies too heavily on the nature of the dataset used. Different length of documents require different size of segments. For instance, datasets such as chats, reviews or microblogs require very small number of bins (4, 8 or 16) [22]. Dataset such as TREC requires larger number of bins (8, 16 or 32) which used in [21] as well as in [23].

Also, another problem relies on the nature of dataset is the definition of *document*. Is the *document* is the whole chat or a single message. While another dataset such as the Holy Quran, *document* maybe a verse, a chapter, a related topic (sequence of verses) or a page [24].

The problem was in the sentence-based segmentation and paragraph-based segmentation because the number of segments is not necessary in the form of 2^n. In this case, zero padding process is needed. Zero padding refers to adding zeros to end of a term signal to increase its length. There is a need to increase the length of term signal if the number of segments used (based on sentence or paragraph) is in between 2^{n-1} and 2^n where $n \in Z$. Much of inaccurate results from the suggested information retrieval model stems from using zero padding process in term signal.

There are two different behaviors regarding zero padding from Haar and Daubechies wavelet transform algorithms:

- The Haars decomposition are discontinuous functions that are not successful in smoothing zero padding.
- Daubechies supports orthogonal wavelets with a preassigned degree of smoothness.

2.4 Wavelet Transform Algorithm

The wavelet transform algorithm has been used in many domains such as image processing, edge detection, data compression, image compression, video compression, audio compression, compression of electroencephalography (**EEG**) signals, compression of electrocardiograph (**ECG**) signals, denoising and filtering, and finally in information retrieval and text mining. Wavelet transform algorithm has the ability to decompose natural signals into scaled and shifted versions of itself [25]. Wavelets are defined by the wavelet function $\psi(t)$ and scaling function $\varphi(t)$ in the time domain. The wavelet function is described as $\psi \in \mathbf{L}^2\mathbb{R}$ (where $\mathbf{L}^2\mathbb{R}$ is the set of functions $f(t)$ which satisfy $\int |f(t)|^2\, dt < \infty$) with a zero average and norm of 1. A wavelet can be scaled and translated by adjusting the parameters s and u, respectively.

$$\psi_{u,s}(t) = \frac{1}{\sqrt{s}}\psi\left(\frac{t-u}{s}\right). \tag{4}$$

The scaling factor keeps the norm equal to one for all s and u. The wavelet transform of $f \in \mathbf{L}^2\mathbb{R}$ at time u and scale s is:

$$Wf(u,s) = \int_{-\infty}^{+\infty} f(t)\frac{1}{\sqrt{s}}\psi^*\left(\frac{t-u}{s}\right)dt, \tag{5}$$

where s and $u \in \mathbf{Z}$ and ψ^* is the complex conjugate of ψ [17].

The scaling function is described as $\varphi_{u,s}(t) \in V_n$. The scaling function satisfies the following:

$$\{0\} \leftarrow \ldots \subset V_{n+1} \subset V_n \subset V_{n-1} \ldots \rightarrow \mathbf{L}^2,$$

where

the set of $\varphi_{u,s}(t)$ for all u is a basis of V_n ($s = 2^n$),
$\bigcup_{n\in\mathbb{Z}} V_n = \mathbf{L}^2(\mathbb{R})$,
$W_n = V_n \cap \overline{V_{n+1}}$

The filter bank tree-structured algorithm can use to compute the **DWT**.The signal $f(t)$ can be transformed using the filter bank tree of **DWT**:

$$\begin{aligned} & f \xrightarrow{DWT^1} A^1 + D^1 \\ & \xrightarrow{DWT^2} A^2 + D^2 + D^1 \\ & \xrightarrow{DWT^3} A^3 + D^3 + D^2 + D^1 \\ & \quad \ldots\ldots \\ & \xrightarrow{DWT^s} A^s + D^s + D^{s-1} + \cdots + D^1 \end{aligned} \tag{6}$$

where A^s correspond to the approximation sub-signal and D^s corresponds to a detail sub-signal in the sth level of transforms of **DWT** of the signal $f(t)$, respectively.

2.4.1 Haar Wavelet Transform

One of the simplest wavelet transforms is Haar transform [26]. The Haar transform is derived from the Haar matrix. It provides the different levels of resolution of a signal. At different resolutions, the positions of the terms are showed by the transformed signal. Every possible wavelet scaled and shifted version is take in Haar transform. Given a term signal $\tilde{f}_{t,d}$ and number of segments N, the wavelet components will be $H_N \tilde{f}_{t,d}$ where H_N is the Haar matrix. When $N = 4$ we have

$$H_4 = \frac{1}{2}\begin{bmatrix} 1 & 1 & 1 & 1 \\ 1 & 1 & -1 & -1 \\ \sqrt{2} & -\sqrt{2} & 0 & 0 \\ 0 & 0 & \sqrt{2} & -\sqrt{2} \end{bmatrix}$$

2.4.2 Daubechies Wavelet Transform

In this research, we are taking the advantage of smoothing the transform of term signal by using Daubechies wavelet transform. There are many types of Daubechies transforms. The simplest one is **Daub4** wavelet. The **Daub4** wavelet transform is defined as the same way as the Haar wavelet transform and differs in how scaling functions and wavelets are defined.

The following code compute the Daubechies wavelet transform (Daub4) using Python. The main function that can be called is *daub4Transform.* The function can be called with one parameter which is the term signal. The length of term signal must equal to 2^n and $n \in Z$. The following is an example of calling the function:

print (daub4Transform([1,2,3,0,0,0,0,0]))

```

def mod_4(i,j):
  if (j==0):
    print "error"
  else:
    value = i
    if (value < 0):
      value = value + abs(j)
    return value

def wrap_4(ival,ilo,ihi):
  jlo = min ( ilo , ihi )
    jhi = max ( ilo , ihi )
  wide = jhi - jlo + 1

  if ( wide == 1 ):
    value = jlo
  else:
    value = jlo + mod_4 ( ival - jlo , wide )
  return value

def daub4Transform ( x ):
  n = len(x)
  Range = [2**i for i in range(2,10)]
  if not n in Range:
            print "the length of term signal must be a power
    of 2."
            print "the length of term signal must be at least
     2^2."
  else:
    y = x
    z = [0*i for i in x]
    m = n
    c = [ 0.4829629131445341E+00, 0.8365163037378079E+00,
    0.2241438680420133E+00, -0.1294095225512603E+00 ]
    while ( 4 <= m ):
      i = 0
      for j in range (0 , m - 1,2):
                        j=j+1
                        j0 = wrap_4 ( j,     1, m )-1
                        j1 = wrap_4 ( j + 1, 1, m )-1
                        j2 = wrap_4 ( j + 2, 1, m )-1
                        j3 = wrap_4 ( j + 3, 1, m )-1
```

```
                    tmp = c[2] * y[j2] + c[3] * y[j3]
                    z[i]       = c[0] * y[j0]+ c[1] * y[j1] + tmp
                    tmp1 = c[1] * y[j2] - c[0] * y[j3]
                    z[i+m/2] = c[3] * y[j0] - c[2] * y[j1] + tmp1
                    i = i + 1
        for j in range (0 , m ):
            y[j] = z[j]

        m =  m// 2

    return y
```

Listing 1.1 Daubechies Wavelet Transform (Daub4) Using Python

The complexity time of matrix multiplication that causes this transformation is $O(N^2)$ for signals of N elements. To enhance the time of this process, one may use the wavelets scaling function and the wavelet function.

3 Design Issues and Implementation of Information Retrieval Using DWT

3.1 *Problems and Design Issues*

Park in [15–17] assumes that documents have fixed length as well as they are unformatted. Using fixed number of bins in all dataset has many disadvantages:

1. It does not distinguish between long documents and short documents. Suppose using of fixed number of bins B such that $B = 2^i$ and $i \in \mathbb{N}$. For any document d, has a number of words $|d|$, the bin b contains number of words $|b| = \frac{|d|}{B}$. That means there are different levels of proximity for different documents. Number of bins in microblogs [22] should be very small in comparison to the number of bins in scientific theses.
2. It does not consider the length of a query. If all terms of a query are found in one bin of a document, that means the document is more related to the query than a document holding the same query terms scattered in more than one bin. To consider this issue, the length of the suggested bin should be $|b| \geq |Q|$.
3. It does not consider the semantic representation of documents. A paragraph, which is a unit of thought that is focused on a single topic, may represent semantically a single bin. Also, a sentence may represent semantically a single bin.
4. Also, it does not consider the weight of matched query terms in a document according to it terms' location. The matched query terms in the first part of a document should have a greater weight more than the matched query terms in the middle and the end in the retrieved documents.

The main hypothesis in this research is to consider the necessity of using different types of segmentation and term weighting beside **DWT** to improve document score. The algorithms and techniques will be described in the following subsections.

3.2 Implementation of the Suggested Model

The suggested model was implemented using python 2.7 and Natural Language Toolkit (**NLTK**) 3.0 that is a leading platform for building Python programs to work with natural language text. It provides easy-to-use interfaces to over 50 corpora and lexical resources such as WordNet, along with a suite of text processing libraries.

3.3 Document Segmentation

Document may be segmented by fixed length, sentences or paragraphs. If we use fixed length the minimum distance between any occurrence of term t_i and term t_j in a document *D* should be considered. Suppose the document contains *C* words, *d* is the maximum distance, *S* represents the number of sentences and *P* represents the number of paragraphs. The main problem is to construct the Haar matrix in these different situations. The following algorithms show how to construct the Haar matrix.

Algorithm 1 Construct The matrix

```
procedure CONSTRUCTMATRIX(D )

    if UsingSentences then
        S ← D.getnumberOfSentences()
        N ← GetMatrixDimension(S)
        Matrix ← GenerateHaarMatrix(N)
    end if
    if UsingParagraphs then
        P ← D.getnumberOfParagraphs()
        N ← GetMatrixDimension(P)
        Matrix ← GenerateHaarMatrix(N)
    end if
    if UsingMaxDistance then
        C ← D.getnumberOfWords()
        Y ← C / d^2
        N ← GetMatrixDimension(Y)
        Matrix ← GenerateHaarMatrix(N)
    end if
    return Matrix
end procedure
```

3.4 Term Weighting

We hypothesize that the significant terms always appear in the front of a document. Formatted document may be segmented by document length, sentences, topics/-subtopics or paragraphs. While unformatted documents may be segmented by document length, sentences or paragraphs. The minimum distance between any occurrence of term t_i and term t_j in a document D should be considered.

Algorithm 2 Get Matrix Dimension

```
procedure GETMATRIXDIMENSION (X)
                              ▷ X is the suggested number of segments
              ▷ the procedure returns the nearest n such that 2^n > X
    power ← 2
    while (power < MaxPower and X > 2^power) do
                                  ▷ MaxPower is set to 13 as default
        power ← power + 1
    end while
    return power
end procedure
```

Algorithm 3 Algorithm for generating Haar matrix

```
procedure GENERATEHAARMATRIX(N)          ▷ to generate N × N matrix
    for i ← 1, N do
        H[1][i] ← 1
    end for
    for i ← 1, N do
        x = log i / log 2
        y = i − 2^x
        for j ← 1, N / 2^(x+1) do
            H[i + 1][j + y × N / 2^x] = 2^(x/2)
            H[i + 1][j + y × N / 2^x + N / 2^(x+1)] = −2^(x/2)
        end for
    end for
    H ← H / √N
    return H
end procedure
```

4 Experiments and Results

The **TREC** dataset is used to evaluate the proposed methods. Exactly, the Associated Press disk 2 and Wall Street Journal disk 2 **(AP2WSJ2)** has been chosen. The **AP2WSJ2** set contains more than 150,000 documents. The selected query set was from 51 to 200 (in TREC 1, 2, and 3). The investigated methods will be compared with the previous high precision methods. The experiments are performed based on a variety of the following segmentation methods:

- Fixed number of segments (8 bins).
- Sentence-based segmentation.
- Paragraph-based segmentation.
- Fixed number of terms in a segment ($d = 3$).
- Fixed number of terms in a segment ($d = 10$).
- Paragraph-based segmentation with position weight.
  These segmentation methods will be compared against another important term proximity measure namely *shortest-substring retrieval* (**SSR**) [6] and *window based bi-gram **BM25** model* [27].

In the experiment, the TREC official evaluation measures have been used, namely the Mean Average Precision (MAP), and Average Precision at K (where $K = \{5, 10, 15, 20\}$).

Table 1 shows that most of the suggested segmentation methods does not have significant improvements. All suggested segmentation methods outperform **SSR** method. However, they did not outperform the fixed bin segmentation method and window based bi-gram **BM25** model (8 bins) while the fixed bin segmentation method outperform window based bi-gram **BM25** model.

Because of using Haar wavelet transform leads to zero padding in the term signal which may affect the accuracy of the results, Daubechies also is used to improve the accuracy of the retreival.

In Paragraph-based segmentation with position weight method, two different additional weight schemes have been used for the first three paragraphs as follows (0.75, 0.50, 0.25) and (0.2, 0.1, 0.05).

Table 2 reveals the results of using the same suggested segmentation methods when using Daubechies (**Daub4**) wavelet transform. Compared with Haar wavelet transform, Daubechies wavelet transform outperforms Haar wavelet transform in most cases and measures. Even the values of results are hardly distinguishable from Haar wavelet transform as in Table 1, but there is a slight improvement of the results.

Table 1 Comparison of Mean Average Precision (MAP), and Average Precision at K (where $K = \{5, 10, 15, 20\}$) for different segmentation methods for **DWT** using Haar Wavelet Transform, SSR and window based bi-gram **BM25** model

Segmentation methods	p@5	p@10	p@15	p@20	MAP	GMAP	Rprecion
Fixed number of segments (8 bins)	0.469	0.439	0.421	0.406	0.2322	0.111	0.270
Sentence-based segmentation	0.431	0.423	0.402	0.381	0.224	0.107	0.262
Paragraph-based segmentation	0.421	0.411	0.388	0.367	0.201	0.092	0.246
Fixed number of terms in a segment (3 terms)	0.399	0.377	0.357	0.340	0.179	0.081	0.225
Fixed number of terms in a segment (10 terms)	0.405	0.392	0.369	0.348	0.183	0.082	0.229
Paragraph-based segmentation with additional position weight (0.75, 0.5, 0.25)	0.421	0.409	0.387	0.370	0.219	0.104	0.258
SSR	0.3718	0.3362	0.3078	0.2856	0.163	–	–
window based bi-gram BM25 model	0.449	0.431	0.418	0.391	0.218	0.103	0.252

Table 2 Comparison of Mean Average Precision (MAP), and Average Precision at K (where $K = \{5, 10, 15, 20\}$) for different segmentation methods for **DWT** using daubechies (**Daub4**) Wavelet Transform, SSR and window based bi-gram **BM25** model

Segmentation methods	p@5	p@10	p@15	p@20	MAP	GMAP	Rprecion
Fixed number of segments (8 bins)	0.464	0.44	0.423	0.407	0.231	0.112	0.272
Sentence-based segmentation	0.429	0.423	0.4	0.382	0.223	0.106	0.263
Paragraph-based segmentation	0.421	0.408	0.388	0.37	0.201	0.092	0.245
Fixed number of terms in a segment (3 terms)	0.392	0.382	0.360	0.342	0.197	0.081	0.235
Fixed number of terms in a segment (10 terms)	0.419	0.393	0.371	0.346	0.184	0.082	0.229
Paragraph-based segmentation with position weight (0.75, 0.5, 0.25)	0.423	0.405	0.389	0.371	0.219	0.103	0.259
Paragraph-based segmentation with position weight (0.2, 0.1, 0.05)	0.428	0.417	0.397	0.397	0.224	0.106	0.262
SSR	0.3718	0.3362	0.3078	0.2856	0.163	–	–
window based bi-gram BM25 model	0.449	0.431	0.418	0.391	0.218	0.103	0.252

Also, there is a slight improvement of the results when using more smoothing, the scheme [0.2, 0.1, 0.05] gives better results than [0.75, 0.50, 0.25, 0.0], for the additional weight schemes for the position of sentences and paragraphs. This confirms the previous findings in the literature as in [28, 29].

5 Conclusion

This chapter has given an account of using **DWT** in information retrieval and the reasons for the reasons for using different document segmentations as well as proposing a variety of segmentation methods to enhance document score and the accuracy of retrieving. All suggested segmentation methods are based on meaning and context. All suggested segmentation methods outperform SSR method while the sentence-based, paragraph-based and sentence-based with position weight outperform window based bi-gram BM25 model. However, they did not outperform the fixed bin segmentation method (8 bins).

This research has argued the using of different wavelet transform algorithms such as Haar and Daubechies and how Daubechies slightly outperforms Haar. One of the most important future work is to find the optimum bin number and to elaborate the additional weight scheme for a given dataset.

References

1. Salton, G., Fox, E.A., Wu, H.: Extended Boolean information retrieval. Commun. ACM **26**, 1022–1036 (1983)
2. Salton, G., Wong, A., Yang, C.-S.: A vector space model for automatic indexing. Commun. ACM **18**, 613–620 (1975)
3. Kang, B., Kim, D., Kim, H.: Fuzzy information retrieval indexed by concept identification. In: International Conference on Text, Speech and Dialogue, pp. 179–186 (2005)
4. Wong, S.K.M., Ziarko, W., Wong, P.C.N.: Generalized vector spaces model in information retrieval. In: Proceedings of the 8th Annual International ACM SIGIR Conference on Research and Development in Information Retrieval, pp. 18–25 (1985)
5. Cummins, R., O'Riordan, C.: Learning in a pairwise term-term proximity framework for information retrieval. In: Proceedings of the 32nd International ACM SIGIR Conference on Research and Development in Information Retrieval, pp. 251–258 (2009)
6. Clarke, C.L.A., Cormack, G.V.: Shortest-substring retrieval and ranking. ACM Trans. Inf. Syst. (TOIS) **18**, 44–78 (2000)
7. Hawking, D., Thistlewaite, P.: Relevance Weighting using Distance Between Term Occurrences (1996)
8. Bhatia, M.P.S., Khalid, A.K.: Contextual proximity based term-weighting for improved web information retrieval. In: International Conference on Knowledge Science, Engineering and Management, pp. 267–278 (2007)
9. Aref, W.G., Barbara, D., Johnson, S., Mehrotra, S.: Efficient processing of proximity queries for large databases. In: Proceedings of the Eleventh International Conference on Data Engineering, 1995, pp. 147–154 (1995)

10. El Mahdaouy, A., Gaussier, E., El Alaoui, S.O.: Exploring term proximity statistic for Arabic information retrieval. 2014 In: Third IEEE International Colloquium in Information Science and Technology (CIST). IEEE (2014)
11. Ye, Z., He, B., Wang, L., Luo, T.: Utilizing term proximity for blog post retrieval. J. Am. Soc. Inf. Sci. Technol. **64**, 2278–2298 (2013)
12. Costa, A., Melucci, M.: An information retrieval model based on discrete fourier transform. In: Information Retrieval Facility Conference, pp. 84–99 (2010)
13. Ramamohanarao, K., Park, L.A.F.: Spectral-based document retrieval. In: Advances in Computer Science-ASIAN 2004. Higher-Level Decision Making, pp. 407–417. Springer (2004)
14. Park, L.A.F., Ramamohanarao, K., Palaniswami, M.: Fourier domain scoring: a novel document ranking method. IEEE Trans. Knowl. Data Eng. **16**, 529–539 (2004)
15. Park, L.A.F., Palaniswami, M., Kotagiri, R.: Internet document filtering using fourier domain scoring. In: European Conference on Principles of Data Mining and Knowledge Discovery, pp. 362–373 (2001)
16. Park, L.A., Palaniswami, M., Ramamohanarao, K.: A novel document ranking method using the discrete cosine transform. IEEE Trans. Pattern Anal. Mach. Intell. **27**, 130–135 (2005)
17. Park, L.A.F., Ramamohanarao, K., Palaniswami, M.: A novel document retrieval method using the discrete wavelet transform. ACM Trans. Inf. Syst. (TOIS) **23**, 267–298 (2005)
18. Arru, G., Feltoni Gurini, D., Gasparetti, F., Micarelli, A., Sansonetti, G.: Signal-based user recommendation on twitter. In: Proceedings of the 22nd International Conference on World Wide Web Steering Committee/ACM, pp. 941–944 (2013)
19. Yang, T., Lee, D.: T3: On mapping text to time series. In: Proceedings of the 3rd Alberto Mendelzon Int'l Workshop on Foundations of Data Management, Arequipa, Peru, May 2009
20. Zobel, J., Moffat, A.: Exploring the similarity space. In: ACM SIGIR Forum, pp. 18–34 (1998)
21. Dahab, M., Kamel, M., Alnofaie, S.: Further Investigations for Documents Information Retrieval Based on DWT. In: International Conference on Advanced Intelligent Systems and Informatics, pp. 3–11. Springer International Publishing (2016)
22. Diwali, A., Kamel, M., Dahab, M: Arabic text-based chat topic classification using discrete Wavelet Transform. Int. J. Comput. Sci. Issues (IJCSI) **12**, 86 (2015)
23. Alnofaie, S., Dahab, M., Kamal, M.: A novel information retrieval approach using query expansion and spectral-based. Int. J. Adv. Comput. Sci. Appl. (IJACSA) **7**(9), 364–373 (2016)
24. Aljaloud, H., Dahab, M., Kamal, M.: Stemmer impact on Quranic mobile information retrieval performance. Int. J. Adv. Comput. Sci. Appl. (IJACSA) **7**(12), 135–139 (2016)
25. Daubechies, I.: Where do wavelets come from? A personal point of view. Proc. IEEE **84**, 510–513 (1996)
26. Haar, A.: Zur theorie der orthogonalen funktionen systeme. Math. Ann. **69**, 331–371 (1910)
27. He, B., Huang, J.X., Zhou, X.: Modeling term proximity for probabilistic information retrieval models. Inf. Sci. **181**(14), 3017–3031 (2011)
28. Hassan, H., Dahab, M., Bahnassy, K., Idrees, A., Gamal, F.: Query answering approach based on document summarization. Int. J. Mod. Eng. Res. (IJMER) **4**(12), 50–55 (2014)
29. Hassan, H., Dahab, M., Bahnassy, K., Idrees, A., Gamal, F.: Arabic documents classification methodastep towards efficient documents summarization. Int. J. Recent Innov. Trends Comput. Commun. **3**(1), 351–359 (2015)

A Review of the State of the Art in Hindi Question Answering Systems

Santosh Kumar Ray, Amir Ahmad and Khaled Shaalan

Abstract Question Answering Systems (QAS) are tools to retrieve precise answers for user questions from a large set of text documents. Researchers from information retrieval and natural language processing community have put tremendous efforts to improve the performance of QASs across several languages. However, Hindi, the fourth most spoken language has not seen a proportional development in the field of question answering to an extent that information seekers accept QASs as a good alternative of search engines. In this chapter, a pipelined architecture for the development of QASs has been explained in the context of English and Hindi languages. This chapter also reviews the developments taking place in Hindi QASs while explaining the challenges faced by researchers in developing Hindi QASs. To encourage and support the new researchers in conducting researches in Hindi QASs, a list of techniques, tools and linguistic resources required to implement the components of a QAS are described in this chapter in a simple and persuasive manner. Finally, the future directions for research in Hindi QASs have been proposed.

Keywords Question answering systems • Tools for hindi information retrieval
Language resources • Architecture • Query expansion • Document retrieval
Answer presentation

S.K. Ray (✉)
Department of IT, Al Khawarizmi International College, Al Ain, UAE
e-mail: santosh.ray@khawarizmi.com

A. Ahmad
College of Information Technology, UAE University, Al Ain, UAE

K. Shaalan
Faculty of Engineering & IT, The British University in Dubai, Dubai, UAE

K. Shaalan et al. (eds.), *Intelligent Natural Language Processing: Trends and Applications*, Studies in Computational Intelligence 740,
https://doi.org/10.1007/978-3-319-67056-0_14

1 Introduction

With the advent of the Internet and the accumulation of huge amount of data in electronic form, there had to be a means for users to retrieve the information they seek from the enormous amount of information. Hence, search engines such as Google and Yahoo were introduced which provided a free platform that would help humans retrieve the required information. However, these search engines assume that if a term in a query was found in a document, then this document could be relevant to that query. The search engines present to the user a list of documents believed to be relevant to his query. They leave it to the user to skim through all the documents to find what he is looking for. In contrast, a Question Answering System (QAS) takes the user question as input, and returns concise and precise answers to the user. So, researchers in QASs investigate approaches and techniques that make it more convenient for users looking for specific and proper information rather than a text file. Thus, a QAS saves the user time, money and the frustration he may have while going through different documents returned by an information retrieval system.

Hindi is the 4th most spoken language in the world. Hindi, a language based on the famous Paninian grammar, is known for its syntactic richness. Nevertheless, research and development in the field of Hindi QASs, compared to efforts in Latin languages, is still in the early stage mainly due to several challenges and language characteristics such as absence of capitalization, free word order etc. To fill this gap and boost up the research work in the field of Hindi question answering, new generation of researchers need to be aware of the state-of-art and know-how of the Hindi QASs. Though some shallow surveys on developments in Hindi QASs have been published, none of them describe the chronological development of QASs, required tools and precise usage of these tools in terms of developments of individual components of QASs. A researcher needs to know how these tools and resources fare in development of QASs. The description of the systems where these tools and resources have been used will provide opportunities to study those systems and organize their research plans accordingly. To the best of our knowledge, the techniques, resources and tools available for designing and developing Hindi QASs and its components have not yet been surveyed extensively, which has motivated us to write this chapter.

The structure of the remaining of the chapter is as follows: Sect. 2 of the chapter provides an architecture of a typical QAS which will work as a roadmap for the development of a Hindi QAS. Section 3 of the chapter reviews the developments taking place in the Hindi QASs. Section 4 introduces the necessary elements of Hindi language and the challenges faced by Hindi QASs. Section 5 provides the details of the tools and resources available for developing Hindi QASs. Finally, Sect. 6 enumerates the possible directions for research in Hindi QASs.

2 A Typical Pipeline Architecture of a Question Answering System

A QAS essentially takes user question as input and presents answers of the question to the user. Accordingly, modern QASs share a number of features and technologies, and the overall designs of the different systems are in most cases quite similar to each other. Most of the QASs follow typical pipeline architecture that divides the question answering process into three distinct phases [1]: question processing, document processing, and answer processing. It should be noted here that none of these phases are mandatory for all QASs. Also, the implementation of these phases may vary to a great extent from one system to another system. In this section, we are providing a generic description of the three phases.

2.1 Question Processing

The question processing phase takes user question as input and applies several processes such as tokenization, stemming, part-of-speech tagging, and query expansion on the question. Thus, the question processing phase can be accomplished by various sub-processes, namely: question classification, derivation of the expected answer type, keyword extraction, and query expansion. Cross-language QASs include an additional process named question translation where the user question is translated into multiple languages [2]. In the rest of this subsection, we shall describe different tasks and subtasks possibly used by different QASs.

2.1.1 Question Classification

Question classification task analyzes the user question and classifies them into one of the predefined classes. The outcome of question classification provides vital information about what to look precisely into documents. Question classification plays a crucial rule in factoid QASs. Moldovoan et al. [3, 4] did a careful survey of the collection of questions in the TREC question collection and identified eight main question patterns; 6 standard Wh-questions, "How" questions, and other questions. Each of these patterns further consists of several sub-patterns. In the following, we provide descriptions of these patterns.[1] These same eight classes can be applied to questions in Hindi also. However, one specific feature of Hindi

[1]Notice that one question can be paraphrased and asked using more than one pattern, getting more than one surface form that share the same meaning. For example the following questions should get the same answer: "During which month tourists visit Kashmir the most?", "What month do tourists visit Kashmir the most?", "Which month do tourists visit Kashmir the most?", and "When do tourists visit Kashmir the most?".

language is worth to mention here. In Hindi, position of question keywords (Wh-questions) may completely alter the meaning of the question. For example, consider the two questions, "क्या आप खाना चाहते हैं? (Do you want to eat?)" and "आप क्या खाना चाहते हैं? (What do you want to eat?)". The second question has been constructed by interchanging the position of first two words of the first question. However, answer of the first question is in Yes/No, but the second question expects name of some food item as answer. Also, though it is grammatically possible to start most of the questions in Hindi with क—question words (counterparts of wh-words in English), in practice, use of क-words at the beginning of question is less frequent compared to that at the middle of sentence, especially if subject (noun) is present in the sentence.

As the focus of this chapter is on Hindi QASs, descriptions of the question patterns are provided first in English, and subsequently equivalent Hindi patterns are described using suitable examples. In all the example questions cited in this section, the Hindi questions are followed by literal English translation with the Hindi word order and then by grammatically correct version of the question.

(i) **Function Word Questions**: These questions contain none of the क—question (non wh-words or how) words. These questions are usually non-factoid questions or explanatory questions. All *Non-Wh-questions (except How questions)* fall under the category of functional word questions.

Example: भारतीय कृषि पर वैश्वीकरण के प्रभाव पर टिप्पणी लिखें। (*Indian agriculture on globalization of effect on comment write*, "Provide comments on the effects of globalization on Indian agriculture.").

(ii) **When Questions**: *"When Questions"* in Hindi contain the keyword "कब (when)" and are temporal in nature. The general pattern in English for "*When Questions*" is "When (do|does|did|AUX) NP [VP] [Complement]?", where AUX, NP, and VP represent auxiliary verb, noun phrase, and verb phrase, respectively. The operator '|' indicates "Boolean OR" operation and 'Complement' can be any combination of words usually playing insignificant roles in the answer type determination. The constituents written inside '[ ]' are optional. The question pattern of "When questions" in Hindi is much different; the keyword कब (when) rarely appears as the first or the last word of the question. Usually, it appears in the middle of the question. Hence a commonly used pattern for when questions is "NP [Complement] when VP [AUX]?"

Example: भारत को अंग्रेजो से आज़ादी कब मिली? (*India Britishers from freedom when got*, "When did India get freedom from the Britishers?").

Like English, there can be a positional reordering of some constituents of the question in Hindi also. For example, the question "भारत को अंग्रेजो से आज़ादी कब मिली? (*India Britishers from freedom when got*?") can be rewritten as "भारत को कब अंग्रेजो से आज़ादी मिली? (*India Britishers when from freedom got?*)". This is true for all types of questions discussed in this subsection.

(iii) **Where Questions**: *"Where Questions"* in Hindi contain the keyword "कहाँ (where)" question word and relate to a location. These may represent natural entities such as mountains, geographical boundaries, man-made locations such as a temple, or some virtual location such as the Internet or a fictional place. The general pattern for "*Where Questions*" in English is "Where (do|does|did|AUX) NP [VP] [Complement]?". The question pattern of "Where questions" in Hindi is much different; the keyword कहाँ (when) rarely comes as the first or last word of the question. Usually it comes in the middle of the question. Hence a commonly used pattern for when questions in Hindi is "NP [Complement] where VP [AUX]?"

Example: श्री सिद्धिविनायक गणपति मंदिर कहाँ है? (*Shree Siddhivinayak Ganapati Temple where is?*, "Where is Shree Siddhivinayak Ganapati Temple?").

(iv) **Which Questions**: The general pattern for *"Which Questions"* in English is "Which NP [do|does|did|AUX] VP [Complement]?". The equivalent questions in Hindi contain the keyword "किस/ कौनसी/ कौनसा (Which)". The expected answer type of such questions varies and is generally decided by the entity type of the first NP following the keyword "किस/ कौनसी/ कौनसा (Which)". The question pattern of "Which questions" in Hindi may or may not contain keywords किस/ कौनसी/ कौनसा at the beginning of question. Hence a commonly used pattern for when questions in Hindi is "Which NP [Complement] [VP] [AUX]?"

Example: किस राज्य की राजधानी अगरतल्ला है? (W*hich state of capital Agartala is,* "Which state's capital is Agartala?") or अगरतल्ला किस राज्य की राजधानी है? (*Agartala which state of capital is,* "Which state's capital is Agartala?").

(v) **Who/Whose/Whom Questions**: Questions falling under Who/Whose/Whom category in English have the general pattern as "(Who|Whose|Whom) [do|does|did|AUX] [VP] [NP] [Complement]?. These questions generally ask about an individual, group of individuals or an organization. The Hindi questions in this categories contain the keywords कौन / किसका /किसकी / किसको /किसने (Who/Whose/Whome). The usually adopted form in Hindi for this type of questions is "NP [Complement] (Who|Whose|Whom) VP [AUX]?"

Example:जय जवान जय किसान का नारा किसने दिया? (*Hail the soldier hail the farmer slogan who gave?*, "Who gave the slogan '*Hail the soldier hail the farmer*'?").

(vi) **Why Questions**: *"Why Questions"* always ask for certain reasons or explanations of some facts or events. The general pattern for "*Why Questions*" in English is "Why [do|does|did|AUX] NP [VP] [NP] [Complement]?". The "Why questions" in Hindi contain the keyword "क्यों". The usually adopted form for this type of question in Hindi is "NP [Complement] Why VP [AUX]?"

Example: सोडियम को मिट्टी के तेल में क्यों रखा जाता है? (*Sodium kerosene oil in why stored is?*, "Why sodium is stored in kerosene oil?").

(vii) **How Question**: *"How Questions"* in English have two types of patterns: "How [do|does|did|AUX] NP VP Complement?" or "How (many|much…) NP Complement?". For the first pattern, Hindi provides a keyword "कैसे", and they usually take the form "NP [Complement] how VP [AUX]?". The expected answer type of this type of questions is a description of some process or event. The second pattern of how questions in Hindi contains the keywords कितना/कितने /कितनी, takes the form "NP How many [Complement]?", and looks for some number as the answer.

Example of the first pattern:

सामान्य लोगों के जीवन का पुनर्निर्माण इतिहासकार कैसे करते हैं

? (*Common people life reconstruction historians how do?*, "How do historians reconstruct the lives of common people?").

Example of the second pattern: भारत में कितने राज्य हैं? (India in h*ow many states?,* "How many states are there in India?").

(viii) **What Questions**: "*What Questions*" are most versatile questions which can ask for virtually anything. *"What Questions"* may have several types of patterns. The most general pattern for "*What Questions*" in English can be written as "What [NP] [do|does|did|AUX] [functional-words] [NP] [VP] Complement?". This type of questions in Hindi contain the keyword "क्या". A commonly used pattern for what type of questions in Hindi is "[NP] [Complement] what [VP] [AUX]?".

Example: कंप्यूटर को हिंदी में क्या कहते हैं? (*Computer Hindi in what say?*, "What do you call Computer in Hindi?").

These question patterns (usually represented by regular expressions or context free grammars) are helpful in predicting the expected answer type for a given question.

2.2 Answer Type Determination

After classifying the user query into one of the eight question classes, a QAS predicts the type of entity expected to be present in the candidate answer sentences. Most of the QASs consider following expected entity types in the answers: Person (व्यक्ति), Location (स्थान), Organization (संस्था), Percentage (प्रतिशत), Date (दिनांक), Time (समय), Duration (अवधि), Measure (माप), and monetary values (मुद्रा). Non-factoid QASs can expect reason (कारण), explanation (व्याख्या) as expected answer types, and return a paragraph to the user query. Table 1 summarizes the question types and corresponding expected answer types.

Table 1 Expected answer type for questions

Question type	Question class	Answer type
Factoid questions	When	Date (दिनांक), Time (समय), Duration (अवधि)
	Where	Location (स्थान)
	Which	Person (व्यक्ति), Location (स्थान), Organization (संस्था), Date (दिनांक)
	Who/Whose/Whom	Person (व्यक्ति), Organization (संस्था)
Non factoid Questions	Why	Reason (कारण)/Explanation (व्याख्या)
Hybrid questions	What	Person (व्यक्ति), Location (स्थान), Organization (संस्था), Date (दिनांक), Number (संख्या), Reason (कारण)/Explanatio (व्याख्या)
	How	Reason (कारण)/Explanatin (व्याख्या),
	How many	Percentage (प्रतिशत), Measure (माप), Monetary value (मुद्रा)
	Function Word	Person (व्यक्ति), Location (स्थान), Organization (संस्था), Date (दिनांक), Number (संख्या), Reason (कारण)/Explanatio (व्याख्या)

2.3 Keyword Extraction

The keyword extraction process starts with tokenizing the user query into keywords. A token is the minimal syntactic unit of a sentence; it can be a word or a group of words. These keywords, if needed, are tagged for part of speech. The keywords can then be stemmed to their roots for finding related words. Usually, a tokenizer is implemented as a preprocessing module of POS tagger or named entity recognizer tasks. A POS tagger takes these tokens as input, and assigns POS category to each token. The keywords are then stemmed to their roots for keyword expansion and passage retrieval. Removal of stopwords can be an optional subtask in keyword extraction process.

2.4 Query Expansion

Query expansion process takes the extracted keywords (both original and stemmed) and adds semantically equivalent words to the question with the help of other linguistic sources such as thesaurus, ontology, treebank etc. Query expansion process helps improving retrieval performance of a QAS by increasing the Recall of the QAS [5]. To understand query expansion, consider the question, भारत को अंग्रेजो से आज़ादी कब मिली ? (When did India get freedom from British?).

This question may fetch only documents which contain the words भारत (India) or आज़ादी (freedom). However, there are several documents which contain the words हिन्दुस्तान, भारतवर्ष, हिन्द (different names of India), स्वाधीनता, स्वतंत्रता, मुक्ति, स्वातंत्र्य (frequently used Hindi words for freedom). If we change the above question to (भारत OR हिन्दुस्तान OR भारतवर्ष OR हिन्द) AND (को अंग्रेजो से) AND (आज़ादी OR स्वाधीनता OR स्वतंत्रता OR मुक्ति OR स्वातंत्र्य) AND कब मिली?, all the documents containing any combination of these words will be retrieved by search engines.

Considering the importance of query expansion process, modern information retrieval engines are using it to reduce the gap between syntax and semantics of the question and the documents. A detailed survey of the literature provides numerous proposals for the query expansion [5]. Query expansion techniques may be broadly classified as manual, automatic, or interactive [6]. In manual query expansion technique, semantically equivalent queries are obtained and compiled manually. Then the semantically equivalent words are added to the original query through logical operators such as AND, OR and NOT. The modified query is fed into search engines to retrieve the relevant documents. In automatic query expansion, the information retrieval system itself is responsible for expanding the initial or subsequent queries based on some methodology. In interactive query expansion, as opposed to manual or automatic query expansion, the retrieval system and the user both are responsible for determining and selecting terms for the query expansion. An interactive retrieval system is first designed to select, retrieve and rank the expansion terms. The user is then presented with the ranked list of terms, and he has to decide which terms are helpful in the expansion of the query.

2.5 *Document Processing*

Document processing typically involves identification of documents relevant to user question, and within the set of relevant documents, identification of the passages most likely to contain the answer to the user question. The accuracy in identification of relevant documents will obviously affect the performance of the answer extraction phase [7]. QASs retrieving documents from locally stored documents implement document retrieval modules. One of the widely adopted techniques for identifying the relevant document is to create an inverted index of the document collection. An inverted index provides list of documents in the document base containing a particular keyword in the user query. For example, if the user puts the query किस राज्य की राजधानी अगरतल्ला है? (Which state's capital is Agartala?); in the simplest form, the inverted index will be a list of the documents containing the words "राज्य" (state), "राजधानी" (capital) and "अगरतल्ला" (Agartala) and these documents will be considered as relevant documents. Some of the QASs use stemming of the keywords to increase the recall of the retrieval while many other systems avoid stemming to avoid compromising the precision of the system [7]. Some other systems use hybrid approaches where they use original keyword as well as stemmed words, but assign less weightage to the stemmed words [8]. On the other hand, web-based QASs usually pass the question keywords and semantically equivalent keywords to one or more search engines such as Google and retrieve the documents with higher ranks [9]. A vast majority of the current information retrieval systems use document retrieval techniques ranging from simple Boolean techniques to sophisticated statistical or NLP based techniques. There is a large variation of document ranking or passage ranking models. Each of these models has its advantages and drawbacks. These models receive the user query and a collection of documents as input and convert them to a non-textual representation. One of the basic document ranking model is Boolean Model. In Boolean model, basic Boolean operators such as AND, OR and, NOT are used for the matching of the query to the document index. Consider the question, कंप्यूटर को हिंदी में क्या कहते हैं? (What do you call Computer in Hindi?), the documents containing both the words "कंप्यूटर" (Computer) and "हिंदी" (Hindi) will be considered more relevant than those containing only one of these words. In this model, the presence or absence of the user query terms in the document is considered, and evaluation of documents only indicates whether they are relevant to the query or not. This set of retrieved documents is presented to the user without giving any consideration to the degree of relevancy. Statistical documents ranking models exploit statistical information about the document such as term frequency, inverse document frequency, document length, etc. to compute the similarity degree of document and the query. Vector Space Model [10] is the most popular model in this category. Probabilistic models provide an intuitive justification for the relevance of matched documents by applying probability theory for ranking documents and uses variant methods for representing the document and the query. One of the well-known of probabilistic models is Inference Model [11], which applies concepts and techniques originating from AI (Artificial Intelligence) without any need to training data sets. Hyperlink based models exploit the hyperlink structures for ranking of documents. These models basically assume that a hyperlink between documents indicates that these documents are on the same topic and one document is recommending some other document. Some of the well-known hyperlink based models are HITS [12], PageRank algorithm [13], WLRank [14] and SALSA algorithm [15]. Finally, Conceptual models [16] work on the principle that there exists some conceptual hierarchy in the documents. These models map the words and phrases in the documents to concepts using the conceptual structures present in the document. Then they extract the concepts of the documents and the query and compare them to compute the degree of similarity.

2.5.1 Passage Retrieval

While identifying relevant documents (or in some other cases after identification of relevant documents), QASs also look for most relevant passages in the documents. Typically a paragraph or a section is selected based on the density or proximity of keywords (or semantically related words) present in them. In this approach, a passage is considered more relevant if it contains a higher number of keywords with minimal distance between the keywords. A review on the keyword density based passage retrieval algorithms and their evaluations can be found in [17]. Another method to retrieve relevant passages is to develop possible answer patterns for the question. To develop the pattern, the question keywords, expanded keywords and expected answer entity obtained from question classification are considered. The passages containing these patterns fully or significantly are considered more relevant. For example, consider the question, **भारत को अंग्रेजो से आज़ादी कब मिली**? (When did India get freedom from British?). The candidate passages should contain sentences like

दिनांक को भारत OR हिन्दुस्तान OR भारतवर्ष OR हिन्द को अंग्रेजो से आज़ादी OR स्वाधीनता OR स्वतंत्रता OR मुक्ति मिली। (On [Date] India got freedom from British.)

भारत OR हिन्दुस्तान OR भारतवर्ष OR हिन्द को अंग्रेजो से दिनांक को आज़ादी OR स्वाधीनता OR स्वतंत्रता OR मुक्ति मिली। (On [Date] India got freedom from British.)

[वर्ष] में भारत OR हिन्दुस्तान OR भारतवर्ष OR हिन्द को अंग्रेजो से आज़ादी OR स्वाधीनता OR स्वतंत्रता OR मुक्ति मिली। (In [Year] India got freedom from British.)

भारत OR हिन्दुस्तान OR भारतवर्ष OR हिन्दको अंग्रेजो से आज़ादी OR स्वाधीनता OR स्वतंत्रता OR मुक्ति [वर्ष] में मिली। (India got freedom from British in [Year].)

[दिनांक] को भारत OR हिन्दुस्तान OR भारतवर्ष OR हिन्द को अंग्रेजो से आज़ादी OR स्वाधीनता OR स्वतंत्रता OR मुक्ति OR स्वातंत्र्य मिली। (On [Date] India got freedom from British.)

भारत OR हिन्दुस्तान OR भारतवर्ष OR हिन्द को अंग्रेजो स[दिनांक] को आज़ादी OR स्वाधीनता OR स्वतंत्रता OR मुक्ति मिली। (On [Date] India got freedom from British.)

[वर्ष] में भारत OR हिन्दुस्तान OR भारतवर्ष OR हिन्द को अंग्रेजो से आज़ादी OR स्वाधीनता OR स्वतंत्रता OR मुक्ति मिली। (In [Year] India got freedom from British).

भारत OR हिन्दुस्तान OR भारतवर्ष OR हिन्द को अंग्रेजो से आज़ादी OR स्वाधीनता OR स्वतंत्रता OR मुक्ति [वर्ष] में मिली। (India got freedom from British in [Year]).

2.6 Answer Extraction

Answer processing is the final phase of a QAS. It consists of small subtasks such as candidate answers identification, answer ranking, and answer formulation. Candidate answers identification requires full parsing of the passage retrieved by passage retrieval phase and comparing it to the expected answer type derived in the question processing phase. This produces a set of candidate answers that are then ranked according to some algorithm or a set of heuristics [18]. These algorithms or heuristics assign weights to candidate answer sentences. Answer sentences with scores lower than a predetermined threshold score are rejected and remaining sentences are ranked according to their scores. The basic strategies employed in answer identification and ranking are to find named entities that match the expected answer type [19], matching syntactic relations from the questions with those from the corpus [20], or attempting to justify the answer using an abductive proof [21]. The answer formulation process restructures the retrieved answer sentences in the user question specific format.

2.6.1 Named Entity Recognition

The Named Entity Recognition (NER) is an important task in answer extraction process of a QAS. The main objective of NER process is to identify the proper names, or temporal and numeric expressions, and classify them under one of the predefined categories such as organization, person, location, date, etc. Thus, the precision of a QAS depends a lot on the correct recognition of named entities. Chu-Carroll et al. [22] investigated the impact of NER on document retrieval precision and observed an improvement of 15.7% in precision of document retrieval when NER was also used.

The approaches to recognize named entities can be broadly classified into two categories: Rule-based approaches and Machine Learning-based approaches. The rule-based approaches [23] rely on handcrafted grammatical rules to recognize named-entities. The rule-based approaches are accurate but more labor intensive. Machine Learning-based approaches, on the other hand, are less time consuming as once developed, trained and tested over a large data set, they adapt themselves according to new patterns or require a little modification. A hybrid approach for NER was recently introduced which combines the machine learning and rule-based approaches together [24, 25]. This has resulted in significant improvement by exploiting the rule-based decisions of named entities as features used by the machine learning classifier.

2.6.2 Answer Scoring and Ranking

A variety of heuristics are used to evaluate whether the candidate entity/sentence/passage is the real answer or not. These heuristics [26] include the frequency and position of occurrence of a given named entity within retrieved passages. Each candidate answer is assigned some score. The top ranked answers are extracted and presented to the user.

2.6.3 Answer Presentation

The last but not the least important issue in the question answering is the presentation of the answers. Different QASs use different approaches to present the answers. Some of the systems present an entity (name, locations, etc.) as an answer to the factoid question along with some additional information [20]. Some other systems present the answer in a sentence/passage form [27] while many other systems present the link to the relevant passage or document along with the candidate answer sentence [28]. Lin et al. [29] showed that users prefer passages over exact phrase answers in a real-world setting because paragraph-sized chunks provide context. Similarly, the number of candidate answer sentences also varies from system to system. Some of the systems present only one answer while other systems present multiple candidate answers.

3 Developments in Hindi Question Answering System

Larkey et al. [30] developed a cross language English-Hindi information retrieval system. They employed several techniques such as normalization, stop-word removal, transliteration, structured query translation, and language modeling using a probabilistic dictionary derived from a parallel corpus in developing this cross language information retrieval system. They tested the system with 15 queries and 41697 Hindi documents from BBC. The reported mean average precision is 0.4298. Some of the challenges posed by Hindi during cross language information retrieval were proprietary encodings of much of the web text, lack of availability of parallel news text, and variability in Unicode encoding.

In the same year another Hindi-English cross language QAS was developed by Sekine and Grishman [31]. This system accepted question in English and analyze the question for expected answer types. The keywords from the questions were translated to Hindi using bilingual dictionary. Then the system searches for answers containing keywords and expected answer type in pre-annotated Hindi newspaper articles. Once the system finds the relevant text in the newspaper containing expected answer type, it translates the answer to English and presents to the users. This system has a web interface designed using Perl-CGI. They collected BBC newspaper article for 6 months to make the corpus. After removing duplicates, the final number of articles in the corpus was 5557. The system was tested with 56 questions. The MRR for the top 5 answers for this system was 0.25 which indicates that cross-language QASs are viable options for question answering.

Shukla et al. [32] developed a restricted domain multilingual QAS. They used Universal Networking Language (UNL) [33] to convert contents of a document in Hindi or English to intermediate language. They analyzed user query to determine its focus and expected answer type. Then an answer template for each question was generated which was again converted to UNL expression. Then the UNL expression for question was matched with UNL expression for documents. The matched answers were finally converted from UNL to natural language. The system provided answers with up to 60% accuracy. However, the authors did not report the details such as number of questions and documents used in testing.

Surve et al. [34] designed another language independent restricted domain QAS named AgroExplorer in 2004. The uniqueness of this system was that instead of doing search on plain text, it first extracts the meaning from the user query using UNL structures and then searches for the extracted meaning in the document base. The document base is created by collecting HTML pages from the web, then parsing and converting these documents in UNL representation. The documents are ranked by matching of UNL graph of user query to the UNL graph of sentences in the documents. Documents have more similarities between query graph and document sentence graphs are given higher ranks. However, this system was tested with a set of only 7 documents in the agricultural domain.

The emphasis on cross-language information retrieval in India can be attributed to the fact that there India is a land of linguistic diversity. Though Hindi is understood by a large section of Indians, it is not only major language of India. According to 2001 census, India has 122 major languages and 1599 other languages. However, not all of these languages are used in academic and administrative communications. In fact, there are 22 schedules languages in India which cover all the states of India. Consider this factor, Government of India initiated a consortium project titled "Development of Cross–Lingual Information Access System" where the users could enter query and retrieve answers in the language of their choice [35].

Kumar et al. [36] developed a QAS for E-Learning Hindi Documents. They classified the question into one of six categories: reasoning questions containing words क्यों (why), क्या (what), वर्णन (explain/describe), कैसे (how); numerical questions containing keyword कितना (how many/how much); time related questions containing keywords कब, जब (when); person and location related questions containing keywords किसने (who), किसको (to whom), कौन (who), कहाँ (where), किधर (which side); questions requiring answers from different passages and containing keywords कौन-कौन (who in plural sense), क्या-क्या (What in plural sense), विभिन्न (different): and miscellaneous questions which do not fit into any of the category. Then stopwords were removed from the question and important keywords were filtered out. The important keywords were stemmed to be used in finding semantically equivalent words for query expansion using a self-constructed small lexical database. The reformulated queries were fed into retrieval engine which used locality based similarity heuristic to select the answer for the given queries. The system was tested with a set of 60 questions whose answers were retrieved from a corpus of Hindi documents related to agriculture and science. According to the authors, the system answered 86.67% of the questions.

Later, Sahu et al. [37] developed a factoid Hindi QAS; Prashnottar, that can answer question questions of type "when", "where", "how many" and "what time". The system uses handcrafted rules to identify question patterns. However, it is not clear how they are extracting answers from document database. The reported accuracy of the system is 68%.

Recently, Nanda et al. [38] propose a Hindi QAS that uses machine learning approach for entity type prediction from the user question. They tested their system over 75 questions. They have not provided any description of the document set. Hence, it is not clear how and from where the system is extracting the answer. The reported accuracy is 90%.

3.1 Developments in Tasks of Question Answering Systems

Cucerzan and Yarowsky [39] developed a language independent model for NER. This model was tested over 5 languages, Hindi being one of them. Among all these language, performance for Hindi was the worst. Later, Li and McCallum [40] developed an NER for Hindi using conditional random fields. The f-value for this model was 71.5. Kumar and Bhattacharyya [41] developed a Hindi NER using Maximum Entropy Model with f-value of 79.7. Saha et al. [42] used a hybrid approach for named entity extraction for Indian languages including Hindi. They used class specific language rules to improve baseline NER based on Maximum Entropy model. They also included some gazetteers and context patterns to improve the performance of the system. The system was trained over half million Hindi words. They reported a precision of 82.76, recall of 53.69, and f-measure of 65.13.

Ekbal and Saha [42] applied simulated annealing based classifier ensemble techniques to POS tagging in Hindi and Bengali. They used, first, the concept of Single Objective Optimization (SOO) for POS tagging, and later developed a method for Multi-objective optimization (MOO). They used Conditional Random Fields and Support vectors for underlying classification. The reported accuracy of POS tagging in Hindi using SOO was 87.67% and 89.88 using MOO. Avinesh and Karthik [43] reported an accuracy of 78.66% for Hindi POS tagging. Ray et al. [44] proposed an algorithm for POS tagger that reduces the number of possible tags for a given sentence by imposing some constraints on the sequence of lexical categories that are possible in a Hindi sentence. Singh et al. [45] used a decision tree based learning algorithm for POS tagging in Hindi. They used a corpora of 15,562 words for training and testing purposes. The reported accuracy of POS tagging is 93.45%.

Akshar et al. [46] developed a parser based on Paninian Grammar formalism to analyse Hindi sentences. This parser based on karaka theory used Integer Programming to analyse simple Hindi sentences.

4 Introduction to Hindi Language and Its Challenges for QASs

Hindi, one of the two official languages of India, is the fourth most-spoken language in the world after Mandarin, Spanish and English. Hindi is written using Devanagari script. The most basic unit of writing Hindi is *Akshara* which can be combination of consonants and vowels. Words are made of aksharas. Words can also be constructed from other words using grammatical constructs called Sandhi and Samaas. Though Hindi is a syntactically rich language, it has certain inherent characteristics that make the computer based processing of the documents in this language, from the information retrieval point of view, a very difficult task. In this section, we are presenting some of these challenges.

- ***No Capitalization***: The factoid QASs require to correctly identify the name of locations, persons and other proper nouns. Identification of proper nouns is done by named entity recognizers which typically exploit the fact that proper nouns in many languages including English are usually started with capital letters. However, Hindi language does not use the capitalization feature to distinguish proper nouns to other word forms such as common nouns, verbs or adjectives. For example, the Hindi proper name "संतोष" [*Santosh*] can be used in a sentence as a first name, or as a common noun.
- ***Lack of uniformity in writing styles:*** In real context, many of translated and transliterated proper nouns tend to be inconsistent. This lack of standardization of the Hindi spelling leads to variants of the same word that are spelled differently but still refers to the same word with the same meaning, creating a many-to-one ambiguity. For example, the word an and (name of a person or happiness) can be spelled as आनंद or आनन्द .
- ***Expressions with multiple words***: It is very common to use same word (or words with similar meaning) consecutively two times in Hindi. For example, the word कौन (who) is used as कौन- कौन in plural sense, धीरे (slow) is used as धीरे-धीरे to emphasize low speed, बहुत (many) सारे (all) are combined together as बहुत सारे (so many). This type of usage of words can be crucial in tokenization process, or it can even negatively affect the performance of cross-language QASs where translation from one language to another language is needed.
- ***Vaalaa morpheme constructs***: The 'vaalaa (वाला)' Hindi morpheme is frequently used in Hindi as suffix to construct new words or to modify the verbs in a sentence. It can take different forms according to gender and number form of the base noun. For example, if we add "vaala" suffix to the word चाय (Tea),

a new word चायवाला (male tea seller) will be formed. However, if we add "vaali" suffix to the word घर (house), a new word घरवाली (wife) will be formed. This can make the automatic word sense disambiguation task more complex.

5 Tools and Resources for Hindi Question Answering

As discussed in Sect. 2, development of a fully functional QAS requires several text processing tasks such as segmentation of user questions and documents in the knowledge base, morphological analysis of question keywords (lemmatization or stemming), determining the part-of-speech (POS) of words, named entity recognition, parsing the question and answers. In order to save their time and energy, researchers can integrate specialized third party open source tools in the main program to perform these tasks. In recent years, a number of tools have been developed for text processing tasks. Many of these tools can be used to implement phases/subtasks of Hindi QASs, and are freely available to the research community. The availability of free tools to the research community will significantly lower down the cost of developing Hindi QAS as compared to tools under license agreements. In this section, we are describing some of the tools and linguistic resources which are freely available and useful in developing components of Hindi QASs.

(a) **Stopwords**: There are certain words in questions which are not useful in question answering once the correct entity type is predicted for the question. These words are called stopwords, and consist of database of most common words that are filtered out prior text processing. Researchers working in the field of Hindi information retrieval have developed their own list of stopwords as and when needed. However, one publicly available list of stopwords can be downloaded from the website.[2]
(b) **Morphological analyzer**: Morphological analysis is an important component of computational linguistic applications. It helps in finding various inflectional and derivational forms of words in a text. As Hindi is a morphologically rich language compared to English, computational linguistic applications such as QASs for Hindi require good morphological analyzers. In order to meet this requirement, a shallow parser was developed at Language Technology Research Centre, IIIT Hyderabad. This parser can be downloaded from the website of LTRC.[3] This parser provides morphological analysis for Hindi sentences and gives the root and other features such as gender, number, tense etc. It also does POS tagging for the sentences.

[2]List of Hindi stopwords, http://members.unine.ch/jacques.savoy/clef/hindiST.txt.

[3]A shallow parser, http://ltrc.iiit.ac.in/showfile.php?filename=downloads/shallow_parser.php.

(c) **Stemmer**: A stemmer conflates morphologically similar words into a single root word. Most of the information retrieval applications use stemmer as one of the most basic components. This helps in reducing the storage size for information retrieval applications as the applications have to store only root words instead of storing several variations of a single word. One of the most popular stemmer used for stemming words in Hindi language was proposed by Ramanathan and Rao [47]. A python implementation of this work is available for public at the website.[4]

(d) **POS Tagger**: POS tagging is an important task in question answering. Classifying the words into various syntactic category helps QASs to parse the questions as well as possible answer sentences. As discussed in Sect. 3.1. Several POS taggers have been developed for tagging Hindi words. However, these POS taggers are not available to public. One publicly available POS tagger for Hindi words can be downloaded from the website.[5] This Hindi POS tagger developed by Reddy and Sharoff [48]. This POS tagger is based on TnT model [49], a popular implementation of the second-order Markov model for POS tagging. The distinctive feature of this tagger is that it does morphological analysis as well as POS Tagging at the same time, and thus mutually benefitting both of the tasks. This Hindi POS tagger supports only Unix based systems.

(e) **Apache openNLP**[6]: Apache OpenNLP is a machine learning based tool for the processing of natural language texts. It can be used for various tasks in QASs such as sentence segmentation, part-of-speech tagging, named entity extraction, parsing, and co-reference resolution. There is no explicit support for any specific natural language from OpenNLP tool. It is a language independent tool which can be used to train models from any language. However, there are some pre-trained models for some tasks in specific languages, These pre-trained models[7] are language dependent and perform well on text in the language of their training only. Because of its language independent nature, OpenNLP has been used for NER [42, 50], for POS tagging and chunking [51] for some of Indian languages including Hindi.

(f) **Ontologies**: Ontologies provide an explicit specification of a conceptualization in a structured knowledge representation formalism that can be used for measuring the similarity between any two fragments of text (a word, sentence, paragraph or document), deriving semantic relations, and finding semantically equivalent words. Some knowledge-based resources are thesaurus, ontology, Wiki, etc. Some ontologies have been constructed in Hindi in various domains such as Grocery [52], health [53, 54], University [55].

(g) **WordNet**: WordNet [56] is a large electronic lexical database of English developed at Princeton University, USA, by a team led by Prof. George Miller

---

[4]A Hindi stemmer, e http://research.variancia.com/hindi_stemmer/.

[5]Hindi POS Tagger, http://sivareddy.in/downloads#hindi_tools.

[6]Apache OpenNLP, http://opennlp.apache.org/download.html.

[7]Pre-trained models for OpenNLP, http://opennlp.sourceforge.net/models-1.5/.

with an aim to create a source of lexical knowledge. WordNet can be downloaded from the Website of Princeton University.[8] It has been used in numerous NLP tasks and applications with a remarkable success, such as POS tagging, Word Sense Disambiguation [57], Text Categorization [58], and Information Extraction [59]. Originally conceived as a full-scale model of human semantic organization, WordNet has become the most used ontological resource for Information Retrieval applications. It has a rich structure connecting its component synonym sets to each other [60]. Semantic relations in WordNet have been extensively used for query expansion [61], building named entity lexical resources [62], and Word Sense Disambiguation [63].

(h) **Hindi Wordnet**[9]: The Hindi Wordnet, like its English counterpart, is a system that provides lexical and semantic relations between different words in Hindi. Hindi Wordnet groups words according to similarity of meaning. For each word there is a synonym set, or synset representing one lexical concept. The current Hindi Wordnet contains 28687 synsets and 63800 unique words. Each entry of Hindi Wordnet describes sysnset (synonyms), gloss (concept) and its position in Ontology. Each synset in the Hindi WordNet is linked with other synsets through the well-known 16 lexical and semantic relations such as hypernymy, hyponymy, meronymy, troponymy, antonymy and entailment. Java APIs have been written to make Hindi WordNet accessible and searchable for Hindi words. A python implementation[10] of Hindi WordNet is publicly available. A broader version of Hindi WordNet called IndoWordnet[11] supports 19 major Indian languages including Hindi and English.

(i) **Hindi Wikipedia**: Wikipedia pages, after its launch in 2001, have been used extensively in English QASs [26, 27]. Hindi Wikipedia was started in 2003 and since then 116,595 pages[12] have been added to it. Hindi Wikipedia API has been used for cross language retrieval [64], query expansion [65].

(j) **DBpedia**: DBpedia is a community based project created to extract structured information from Wikipedia and make it available on the web. DBpedia has localized version in 125 languages, including Hindi. All these versions together describe 38.3 million things while the English version of the DBpedia knowledge base describes 4.58 million things, out of which 4.22 million are classified in a consistent ontology, including 1,445,000 persons, 735,000 places (including 478,000 populated places), 411,000 creative works (including 123,000 music albums, 87,000 films and 19,000 video games), 241,000 organizations (including 58,000 companies and 49,000 educational

---

[8]WordNet, http://wordnet.princeton.edu/wordnet/download/.

[9]Hindi WordNet, http://www.cfilt.iitb.ac.in/wordnet/webhwn/.

[10]Python implementation of Hindi WordNet, http://sivareddy.in/downloads#python-hindi-wordnet.

[11]IndoWordNet, http://www.cfilt.iitb.ac.in/indowordnet/index.jsp.

[12]Hindi Wikipedia, https://hi.wikipedia.org/wiki/विशेष:/Statistics, accessed on January, 25, 2017.

institutions), 251,000 species and 6,000 diseases.[13] Due to its strongly structured information base, DBpedia is a very useful source for question processing task of a QAS [66].

(k) **HindiWac corpus**: HindiWaC corpus[14] contains 65 million tokens crawled from the Hindi Internet and it is tagged [67]. This corpus can be used to design and train various NLP as well as machine learning based algorithms.

(l) **Treebanks**: A treebank is a highly structured corpus which is a linguistic resource that is composed of large collections of manually annotated and verified syntactic analyses of sentences that are carefully and accurately annotated. These annotations are very useful for the development of a variety of applications such as tokenization, POS tagging, morphological disambiguation, base phrase chunking, named entity recognition, and semantic role labeling [68]. Considering the importance of treebanks in Hindi NLP applications, Palmer et al. [69] developed a multi-representational and multi-layered treebank for Hindi and Urdu. The expected number of words in final version of this treebank is 400,000 Hindi words and 200,000 Urdu words.

(m) **Lucene**: Lucene[15] is an open source cross-platform text search engine library written entirely in Java. It can be used to index the documents in the corpus-based QASs. Several QASs have used Lucene in indexing [2] and document analysis phase [58, 70]. Lucene contains several classes[16] to perform analysis of Hindi texts.

(n) **GATE**: GATE[17] is an open source free integrated development environment for performing language processing tasks and developing Information Retrieval/NLP tools. GATE has been used for development of QASs [71], Information Extraction [72], ontology learning [73], corpus annotation [74] and other NLP tasks. GATE provides plugins for processing many non-English languages such as Arabic, Hindi, French, and German.

(o) **QANUS**: QANUS[18] is an open source, Java-based Question Answering framework developed at the National University of Singapore with an aim to assist new researchers in building new QAS quickly, and act as a baseline system for benchmarking the performance of new QASs. QANUS implements the typical pipeline architecture of QASs, and includes modules for NER, POS tagging and question classification. It provides the flexibility to the developers in adding/removing modules so that the newly developed system can be easily trained over different datasets and techniques. A fully functional factoid QAS called QA-SYS [75] has been built using the QANUS framework to

[13]DBPedia, http://wiki.dbpedia.org/about, accessed on January, 25, 2017.

[14]HindiWalC corpus, https://www.sketchengine.co.uk/hindiwac-corpus/.

[15]Lucene, http://lucene.apache.org/core/.

[16]Lucene classes for Hindi, https://lucene.apache.org/core/4_1_0/analyzers-common/org/apache/lucene/analysis/hi/package-summary.html.

[17]GATE, http://gate.ac.uk/.

[18]QANUS, http://www.qanus.com/.

demonstrate the practicality of this framework. QANUS has been used for developing individual components of a QAS such as passage retrieval [76], and it can be extended for non-English languages as demonstrated in [77].

6 Future Scopes

Due to efforts of some selected researchers, there has been some progress in research in Hindi QASs. However, considering the advanced level of work done in other languages such as English and some Asian languages, the progress in Hindi QASs is really very far from satisfactory level. This creates scope for several improvements in Hindi QAS. In this section, we are describing some of these scopes.

(a) **Design of Relevant Resources and Tools**: One of the major impediments in development of high quality Hindi QASs is the lack of availability of freely available NLP/IR tools and integrated development environments for new researchers. For example, in an experiment, it was shown that using the existing POS taggers [43], an accuracy of only 14.7% in named entity tagging could be achieved over Hindi tokens [78]. In order to fill this gap, tools (some of these tools are discussed in the previous section) were developed to accomplish some specific tasks such as POS tagging, named entity recognition, stemming etc. But, as these tools were designed by some researchers to perform very specific tasks in their projects, other researchers either could not avail them or had to borrow and assemble these tools to design QASs. Contrary to their English counterparts such as PowerAnswer [4, 79] and START [80] which utilize the deep NLP techniques namely natural language annotation of the knowledge base, semantic parsing, logic proving, word sense disambiguation and other deep NLP techniques, very few Hindi QASs have attempted to incorporate logical representation, discourse knowledge, and other deep NLP techniques. The consistently good performances of PowerAnswer in TREC and CLEF competitions have demonstrated that deep NLP techniques increase the Precision of the question answering process [81]. Hindi QASs can achieve similar level of efficiency if deep NLP and statistical techniques are tweaked and adopted to the need of Hindi information retrieval.

Open source tools are useful for a large number of researchers due to availability of source codes to the researchers. Some of the QASs in English such as ARANEA [29], QANUS [75] release their source codes to help research community in developing new QASs. These systems can serve as baseline systems for new researchers in order to develop and benchmark the new QASs developed by researchers. A similar practice in Hindi QAS research community will give necessary boost to new researchers to understand the design patterns in better ways.

(b) **Development of Non-factoid QASs**: As most of the questions asked on the Web are factoid questions, it was natural for researchers to focus more on factoid questions, and Hindi question answering research is also not an exception. However, users in many fields such as academic and scientific research, politics, arts, etc. require answers containing several paragraphs. These types of questions are called non-factoid questions and usually start with keywords what and why. For example, consider the question बुजुर्गों की बढ़ती जनसंख्या का क्या राजनीतिक निहितार्थ हैं? (What are the political implications of an increasingly elderly population?). To answer such non-factoid questions more accurately, a system may need to analyze several documents, extract multiple passages, and combine them to present the answers. The biggest challenge in the development of non-factoid QASs is the unavailability of training data and linguistic resources. To overcome this problem, most systems train on a small corpus built manually for the specific system [82] or questions collected from frequently asked questions (FAQs) [83, 84]. As the researches on even factoid Hindi QASs are not at par with the Latin languages, it is not surprising that there is virtually no work reported on Hindi non-factoid QASs. Thus, there is lot of scope for researchers to contribute in the field of non-factoid Hindi QASs.

(c) **Development of Collaborative Question Answering Systems**: Collaborative QASs (also called Community QASs) such as Yahoo answers [85] and Wiki Answers are becoming promising alternatives for information seekers on the web [86]. In collaborative QASs, users provide answers to the questions posed by other users and best answers are selected manually either by the asker or by all the participants by voting. Due to the presence of a large number of internet users, these systems cover a very high volume of questions as well as answers for both factoid and non-factoid questions. Secondly, the processing of these question-answer pairs is also relatively simpler than automated QASs. The only problem with these answers is the quality of answers which, if not controlled or filtered, can be highly irrelevant or even abusing too. Recently, some research has been carried out to rank the answers on collaborative QASs so that the quality of the best answers can be improved [87]. Surprisingly, there is no reported work related to the development of collaborative Hindi QASs in the literature. As the number of Internet users is growing rapidly and crossing 460 million in India, we believe that a collaborative QAS in Hindi will be very effective and helpful for information seekers in Hindi language. This will help users to get more relevant information, especially for non-factoid questions.

(d) **Development and Use of Semantic Web Resources**: The semantic web and ontology have become the key technologies in the development of QASs. The semantic web is a mesh of information linked up in a way that it is easily process-able by machines, on a global scale. Ontology is most widely used method to represent domain-specific conceptual knowledge in order to promote the semantic capability of a QAS. Semantic Web resources and Ontologies have been used extensively for query expansion, and they greatly improve the performance of QASs in answering the questions like “Who wrote ‘The pines of Rome’?” even if the user asks it in a different form. While expanding the query, most of the systems expand the query with words belonging to same POS; however, in several cases the words from different POS, but with equivalent meaning, are more useful. Hence, query expansion phase of a QAS must also include cross-POS semantically related words. Ontologies help in assisting to find the semantically related words from different POS. Thus, the development of computational linguistic applications depends a lot on the availability of the well-developed linguistic corpora such as language dictionary, ontology, or treebank. Therefore, the last decade has witnessed the development of domain specific QASs in all fields of life ranging from education [88] to Medical [89], Tourism [90], and Mobile service consulting [18].

Researches in Hindi QASs are seriously lagging behind in developing semantic web resources and exploiting their richness in development of QASs. There is no open source tool available for designing Hindi semantic web resources. With an exception of Hindi WordNet (HWN), there is not a single ontology resource available on the web for Hindi question answering research community and even HWN has not been used widely. Some researchers attempted to develop ontologies in the field of Grocery [52], health [53, 54], University [55]. Some researchers have developed domain specific QASs in Hindi also [32, 34, 36]. However, none of these QASs used the ontologies available in various domains. This gap stresses the need of development of more domain specific ontological resources in Hindi which should also be exploited in the design and development of Hindi QASs.

(e) **Development of Evaluation Standards and Test Beds**: In Sect. 3 of this chapter, we noted that most of the Hindi QASs are not evaluated properly, which will make it impossible to compare their performances with future improvements and proposals. The set of questions and documents used for evaluation of the QASs are entirely disjoint for different researchers, unlike their English counterparts where the systems are tested over a standard set of questions and document collections compiled by a well-accepted institution such as NSIT. However, a TREC style set of standard questions is needed to be developed and provided to research community so that the performance of the Hindi QASs can be benchmarked.
(f) **Use of Blogs and Social Media Data**: Since the last one decade, people across the world are expressing their views and opinions over the blogs and social media. This has resulted into an explosion of data over blogs and social media across the world and the Hindi language is not an exception. People working in different technical and non-technical fields are providing relevant information on their blogs or social media pages. The processing of information on blogs and social media is not a trivial task due to the relatively large presence of typographical, syntactic and semantic errors [91]. A new set of NLP resources, tools and methods are required for efficient handling of large volume of data. In social media such as Facebook and Twitter, users write their views and

comments using something called code-mixing where phrases and words of one language is embedded into another language. Code-mixing is a serious challenge to conventional QASs which deal contents in only one language. Some researchers [92] have taken up this challenge to develop a full-fledged QAs in code-mixed language. As the first step, they have used Support Vector Machine to build a question classification system that predicts answer type of a question written using code-mixing (Hindi and English). But, the progress in the social media based Hindi QAS is still far from the satisfactory level.

Thus, we can conclude that there is a lot of scope for research in the field of Hindi question Answering. Researchers in Hindi question answering can take inspiration from the developments in QASs in other languages across the globe. In this chapter, we have described some of these developments. This field also requires the development of software tools useful to the research community. The recent trend in the field of question answering is the development of QASs in the form of smartphone-based mobile apps as it happened in the case of True Knowledge which has been turned into mobile application Evi.[19] We expect that the similar mobile applications will be developed for Hindi QASs also in the near future.

References

1. Buscaldi, D., Rosso, P.: Mining knowledge from Wikipedia for the question answering task. In: Proceedings of the 5th International Conference on Language Resources and Evaluation (LREC'06), pp. 727–730 (2011)
2. Dolvera-Lobo, M.-D., Gutiérrez-Artacho, J.: Multilingual question-answering system in biomedical domain on the Web: an evaluation, Lect.e Notes Comput. Sci. **6941**, 83–88 (2011)
3. Moldovan, D., Harabagiu, S., Pasca, M., Mihalcea, R., Girju, R., Goodrum R., Rus, V.: The structure and performance of an open-domain question answering system. In Proceedings of the Conference of the Association for Computational Linguistics (ACL-2000), pp. 563–570 (2000)
4. Moldovan, D., Harabagiu, S., Girju, R., Morarescu, P., Lacatusu, F., Novischi, A., Badulescu, A., Bolohan, O.: LCC tools for question answering. In: Proceedings of the 11th Text REtrieval Conference TREC-2002, NIST, Gaithersburg (2002)
5. Efthimiadis, E.N.: Query expansion. Ann. Rev. Inf. Syst. Technol. **31**, 121–187 (1996)
6. Renals, S., Abberly D.: The THISLSDR system at TREC-9. In: Proceedings of 9th Text Retrieval conference, Gaithersburg, MD (2000)
7. Clarke, C.L.A., Cormack, G.V., Kisman, D.I.E., Lynam, T.R.: Question answering by passage selection (MultiText experiments for TREC-9). In: Voorhees, E., Harman, D. (eds.) Proceedings of the Ninth Text REtrieval Conference (TREC-9, pp. 673–683), NIST Special Publication (2000)
8. Araujo, L., Pérez-Agüera, J.R.: Improving query expansion with stemming terms: a new genetic algorithm approach. In: Proceedings of the 8th European Conference on Evolutionary Computation in Combinatorial Optimization, pp. 182–193 (2008)

[19]True Knowledge, http://www.evi.com/.

9. Li, X., Yang, W.Z.: Research on personalized document retrieval based on user interest model. In: Proceedings of 7th International Conference on, Computer Science & Education, pp. 1771–1773 (2012)
10. Lee, D.L., Chuang, H., Seamons, K.: Document ranking and the vector space model. IEEE Softw. **14**(2), 67–75 (1997)
11. Crestani, F., Lalmas, M., van Rijsbergen, C.J., Campbell, I.: Is this document relevant? Probably. A survey of probabilistic models in information retrieval. ACM Comput. Surv. **30**, 528–552 (1998)
12. Henzinger, Monika, R.: Hyperlink analysis for the web. IEEE Internet Comput. **5**(1), 45–50 (2001)
13. Brin, S., Page, L.: The anatomy of a large-scale hyper-textual web search engine. In: Proceedings of the Seventh International World Wide Web Conference, pp. 107–117, Elsevier Science, New York (1998)
14. Baeza-Yates, R., Davis, E.: Web page ranking using link attributes. In: Proceedings of the 13th International World Wide Web Conference on Alternate Track Papers & Posters, pp. 328–329 (2004)
15. Lempel, R., Moran, S.P.: The stochastic approach for link-structure analysis (SALSA) and the TKC effect. Comput. Netw. Int. J. Comput. Telecommun. Netw. Elsevier North-Holland, New York **33**(1–6), pp 387-401 (2000)
16. Vallet, D., Fernández, M., Castells, P.: An Ontology-based information retrieval model. In: Gómez-Pérez, A., Euzenat, J. (eds.) Proceedings of the 2nd European Semantic Web Conference (ESWC 2005), Heraklion, Greece, Lecture Notes in Computer Science, vol. 3532, pp. 455–470. Springer (2005)
17. Tellex, S., Katz, B., Lin, J., Fernandes, A., Marton, G.: Quantitative evaluation of passage retrieval algorithms for Question Answering. In: Proceedings of the 26th Annual ACM SIGIR International Conference on Research and Development in Information Retrieval (SIGIR 2003), Toronto, Canada (2003)
18. Wang, D.S.: A domain-specific question answering system based on ontology and question templates. In: Proceedings of 11th ACIS International Conference on Software Engineering Artificial Intelligence Networking and Parallel/Distributed Computing (SNPD), pp. 151–156 (2010)
19. AbdelRahman, S., Elarnaoty, M., Magdy, M., Fahmy, A.: Integrated machine learning techniques for Arabic named entity recognition. Int. J. Comput. Sci. Issues **7**(4)(3), 27–36 (2010)
20. Katz, B., Lin, J.: Selectively using relations to improve precision in question answering. In: Proceedings of the EACL 2003 Workshop on Natural Language Processing for Question Answering, Budapest, Hungary, pp. 43–50 (2003)
21. Harabagiu, S.M., Pasca, M.A., Maiorano, S.J.: Experiments with open-domain textual question answering. In: Proceedings of the 18th International Conference on Computational Linguistics, Association for Computational Linguistics, Saarbrucken, Germany, pp. 292–298 (2000)
22. Chu-Carroll, J., Prager, J., Czuba, K., Ferrucci, D., Duboue, P.: Semantic search via XML Fragments: a high-precision approach to IR. In: Proceedings of the Annual International ACM SIGIR Conference on Research and Development on Information Retrieval, Seattle, pp. 445–452 (2006)
23. Shaalan, K.: Rule-based approach in Arabic natural language processing. Special Issue on Advances in Arabic Language Processing, the International Journal on Information and Communication Technologies (IJICT), vol. 3(3), pp 11–19. Serial Publications, New Delhi, India (2010)
24. Shaalan, K., Oudah, M.: A hybrid approach to Arabic named entity recognition. Journal of Information Science (JIS). vol. 40(1), pp. 67–87. SAGE Publications Ltd, UK (2014)
25. Oudah, M., Shaalan, K.: Person name recognition using hybrid approach. In: NLDB 2013, LNCS, vol. 7934, pp. 237–248. Springer, Berlin (2013)

26. Ray, S.K., Singh, S., Joshi, B.P.: Question classification & answer validation—a semantic approach using WordNet and Wikipedia. Pattern Recogn. Lett. **31**(13), 1935–1943 (2010)
27. Cao, Y.G., Liua, F., Simpsonb, P., Antieaua, L., Bennett, A., Cimino, J.J., Ely, J., Yu, H.: AskHERMES: an online question answering system for complex clinical questions. J. Biomed. Inf. **44**(2), pp. 277–288 (2011)
28. Zheng, Z.: AnswerBus question answering system. In: Proceedings of the Second International Conference on Human Language Technology Research, pp. 399–404 (2002)
29. Lin, J.: An exploration of the principles underlying redundancy-based factoid question answering. ACM Trans. Inf. Syst. **27**(2), 1–55 (2007)
30. Larkey, L.S., Connell, M.E., Abduljaleel, N.: "Hindi CLIR in Thirty Days," ACM Transactions on Asian Language Information Processing (TALIP), vol 2, Issue 2, pp. 130–142. ACM, New York, NY, USA, June 2003
31. Sekine, S., Grishman, R.: Hindi-English Cross-Lingual Question-Answering system. ACM Trans. Asian Lang. Inf. Process. (TALIP) **2**(3), 181–192 (2003)
32. Shukla, P., Mukherjee, A., Raina, A.: Towards a language independent encoding of documents: a novel approach to multilingual question answering. In: Proceedings of the 1st International Workshop on Natural Language Understanding and Cognitive Science, NLUCS 2004, pp. 116–125, (2004)
33. Uchida, H.: UNL Beyond machine translation. In: International Symposium on Language in Cyberspace, Seoul, Korea Systems. ICEIS Press (2001)
34. Surve, M., Singh, S., Kagathara, S., Venkatasivaramasastry, K., Dubey, S., Rane, G., Saraswati, J., Badodekar, S., Iyer, A., Almeida, A., Nikam, R., Perez, C.G., Bhattacharyya, P.: AgroExplorer: a meaning based multilingual search engine. International Conference on Digital Libraries (2004)
35. CLIA Consortium: Cross lingual information access system for indian languages. In: Demo/Exhibition of the 3rd International Joint Conference on Natural Language Processing, Hyderabad, India, pp. 973–975 (2008)
36. Kumar, P., Kashyap, S., Mittal, A., Gupta, S.: A query answering system for e-learning Hindi documents. South Asian Language Review, vol. XIII, Nos 1&2, Jan-June, 2003. pp. 69–81 (2003)
37. Sahu, S., Vasnik, N., Roy, D.: Prashnottar: a Hindi question answering system. Int. J. Comput. Sci. Inf. Technol. (IJCSIT) **4**(2) (2012)
38. Nanda, G., Dua, M., Singla, K.: A Hindi question answering system using machine learning approach. In: 2016 International Conference on Computational Techniques in Information and Communication Technologies (ICCTICT) (2016)
39. Cucerzan, S., Yarowsky, D.: Language independent named entity recognition combining morphological and contextual evidence. Proc. Jt. SIGDAT Conf. EMNLP VLC **1999**, 90–99 (1999)
40. Li, W., McCallum, A.: Rapid development of Hindi named entity recognition using conditional random fields and feature induction (Short Paper). In: ACM Transactions on Computational Logic (2004)
41. Kumar, N., Pushpak, B.: Named Entity Recognition in Hindi using MEMM. In Technical Report, IIT Bombay, India (2006)
42. Saha, S.K., Chatterjee, S., Dandapat, S., Sarkar, S., Mitra, P.: A Hybrid Approach for Named Entity Recognition in Indian Languages. In: Proceedings of the IJCNLP-08 Workshop on NER for South and South East Asian Languages, Hyderabad, India, pp. 17–24, January 2008
43. Avinesh, P., Karthik, G.: Part of speech tagging and chunking using conditional random fields and transformation based learning. Proc IJCAI Workshop Shallow Parsing South Asian Lang. India **2007**, 21–24 (2007)
44. Ray, P.R., Harish, V., Basu, A., Sarkar, S.: Part of speech tagging and local word grouping techniques for natural language parsing in Hindi. In: Proceedings of ICON (2003)
45. Singh, S., Gupta, K., Shrivastava, M., Bhattacharyya, P.: Morphological richness offsets resource demand-experiences in constructing a POS Tagger for Hindi. In: Proceedings of the COLING/ACL 2006 Main Conference Poster Sessions, Sydney, pp. 779–786, July 2006

46. Akshar, B., Chaitanya, V., Sangal, R.: NLP A Paninian Perspective. Prentice Hall of India, Delhi (1994)
47. Ramanathan, A., Rao, D.: A lightweight stemmer for Hindi. In: Proceedings of the 10th Conference of the European Chapter of the Association for Computational Linguistics (EACL), on Computational Linguistics for South Asian Languages (Budapest, Apr.) workshop (2003)
48. Reddy, S., Sharoff, S.: Cross language POS Taggers (and other Tools) for Indian languages: an experiment with Kannada using Telugu resources. In: Proceedings of the 5th Workshop on Cross Lingual Information Access (2011)
49. Brants, T.: Tnt: a statistical part-of-speech tagger. In: Proceedings of the Sixth Conference on Applied Natural Language Processing, ANLC'00, Stroudsburg, PA, USA, pp. 224–231. Association for Computational Linguistics (2000)
50. Ekbal, A., Haque, R., Das, A., Poka, V., Bandyopadhyay, S.: Language Independent named entity recognition in Indian languages. In: Proceedings of the IJCNLP-08 Workshop on NER for South and South East Asian Languages, Hyderabad, India, pp. 33–40, Jan 2008
51. Dandapat, S.: Part-of-Speech tagging and chunking with maximum entropy mode. In: Proceedings of SPSAL2007, IJCAI, India, pp. 29–32 (2007)
52. Chaware, S.M., Rao, S.: Ontology approach for cross language information retrieval. Int. J. Comput. Technol Appl. **2**, 379–384 (2011)
53. Bhatt, B., Bhattacharyya, P.: Domain specific ontology extractor for Indian languages. In: Proceedings of 10th Workshop on Asian Language Resources, COLING, Mumbai, pp. 75–84 (2012)
54. Mathur, I., Darbari, H., Joshi, N.: Domain ontology development for communicable diseases. CS & IT-CSCP **3**, 351–360 (2013)
55. Dwivedi, S.K., Kumar, A.: Development of University ontology for aSPOCMS. J. Emerg. Technol. Web Intell. **5**, 213–221 (2013)
56. Miller, G.A.: WordNet: a Lexical database for English. Commun. ACM **38**(11), 39–41 (1995)
57. Segond, F., Schiller, A., Grefenstette, G., Chanod, J.-P.: An experiment in semantic tagging using hidden Markov model tagging. In: Proceedings of the Workshop in Automatic Information Extraction and Building of Lexical Semantic Resources, pp. 78–81 (1997)
58. G´omez-Adorno, H., Pinto, D., Darnes, V.A.: Question Answering System for Reading Comprehension Tests. Pattern Recognition Lecture Notes in Computer Science, vol. 7914, pp. 354–363 (2013)
59. Yue, J., Alan, C., Biermann, W.: The use of lexical semantics in information extraction. In: Proceedings of the Workshop in Automatic Information Extraction and Building of Lexical Semantic Resources, pp. 61–70 (1997)
60. Fellbaum, C.: WordNet(s). In: Brown, K. (ed.) Encyclopedia of Language and Linguistics, 2nd Edn. pp. 665–670. Oxford, Elsevier (2006)
61. Zhiguo, G., Chan, W., Leong, H.U.: Web query expansion by WordNet. Database Expert Syst. Appl. Lect. Notes Comput. Sci. **3588**, 166–175 (2005)
62. Attia, M., Toral, A., Tounsi, L., Monachini, M., van Genabith, J.: An automatically built named entity lexicon for Arabic. In: Proceedings of the International Conference on Language Resources and Evaluation (LREC 2010), Valletta, Malta (2010)
63. Li, X., Szpakowicz, S., Matwin, S.: A WordNet-based algorithm for word sense disambiguation. In: Proceedings of the 14th International Joint Conference on Artificial Intelligence, pp. 1368–1374 (1995)
64. Sharma, V.K., Mittal, N.: Exploiting Wikipedia API for Hindi-english Cross-language Information Retrieval. In: Proceedings of Twelfth International Multi-Conference on Information Processing-2016, 19-21 Aug 2016, Bangalore, India, pp. 434–440 (2016)
65. Barman, U., Lohar, P., Bhaskar, P., Bandyopadhyay, S.: Ad-hoc information retrieval focused on wikipedia based query expansion and entropy based ranking. In: The proceedings of the Forum for Information Retrieval Evaluation (FIRE)—2012. Dec 2012, ISI, Kolkata, India (2012)

66. Adel, T., Okba, T.: DBPedia based factoid question answering system. Int. J. Web Semant. Technol. **4**(3), 23–38 (2013)
67. Kilgarriff, A., Reddy, S., Pomikálek, J., Avinesh, P.V.S.: A Corpus Factory for many languages. In: Proceedings of the Seventh International Conference on Language Resources and Evaluation (LREC'10), 19–21 May 2010. Malta, Valletta (2010)
68. Habash, N., Rambow O., Roth R.: A toolkit for Arabic tokenization, diacritization, morphological, disambiguation, POS tagging, stemming and lemmatization. In: Proceedings of Second International Conference on Arabic Language Resources and Tools, pp. 102–109 (2009)
69. Palmer, M., Bhatt, R., Narasimhan, B., Rambow, O., Misra, D.S., Xia, F.: Hindi syntax: annotating dependency, lexical predicate-argument structure, and phrase structure. In: The Proceedings of the 7th International Conference on Natural Language Processing, ICON-2009, Hyderabad, India, 14–17 Dec 2009
70. Bilotti, M.W., Katz, B., Lin, J.: What works better for question answering: stemming or morphological query expansion? In: Proceedings of Information Retrieval for Question Answering Workshop, at SIGIR (2004)
71. Lopez, V., Victoria, U., Enrico, M., Michele, P.: AquaLog: An ontology-driven question answering system for organizational semantic intranets. J. Web Semant. Elsevier **5**(2), 72–105 (2007)
72. Derczynski, L., Field, C.V., Bøgh, K.S.: DKIE: open source information extraction for Danish. In: Proceedings of the meeting of the European chapter of the Association for Computation Linguistics (EACL), Gothenburg, Sweden (2014)
73. Maynard, D., Bontcheva, K.: Natural language processing. In: Lehmann, J., Voelker, J. (eds.) Perspectives of Ontology Learning. IOS Press (2014)
74. Sabou, M., Bontcheva, K., Derczynski, L., Scharl, A.: Corpus annotation through crowdsourcing: towards best practice guidelines. In: Proceedings of the Language Resources and Evaluation Conference (LREC) (2014)
75. Ng, J.-P., Kan M.-Y.: QANUS: An open source question-answering platform. http://wing.comp.nus.edu.sg/~junping/docs/qanus.pdf (2014). Accessed 1 May 2014
76. Ageev, M., Lagun, D., Agichtein, E.: The answer is at your fingertips: improving passage retrieval for web question answering with search behavior data. In: Proceedings of Conference on Empirical Methods in Natural Language Processing, pp. 1011–1021 (2013)
77. Geirsson, Ó.P.: IceQA: Developing an open source question-answering system. http://www.ru.is/~hrafn/students/IceQA.pdfm (2013)
78. Gali, K., Surana, H., Vaidya, A., Shishtla, P., Sharma, D.M.: Aggregative machine learning and rule based heuristics for named entity recognition. In: Proceedings of the IJCNLP-08 Workshop on NER for South and South East Asian Languages, pp 25–32 (2008)
79. Bowden, M., Olteanu, M., Suriyentrakorn, P., Clark, J., Moldovan, D.: LCC's PowerAnswer at QA@CLEF 2006. In Proceedings of CLEF 2006, pp. 310–317 (2006)
80. Katz, B., Borchardt, G., Felshin, S.: Natural language annotations for question answering. In: Proceedings of the 19th International FLAIRS Conference (FLAIRS 2006), Melbourne Beach, FL (2006)
81. Radev, D.R., Qi, H., Wu, H., Fan, W.: Evaluating web-based question answering systems. In: Proceedings of LREC, Las Palmas, Spain (2002)
82. Higashinaka, R., Isozaki, H.: Corpus-based question answering for why-questions. In: Proceedings of the Third International Joint Conference on Natural Language Processing (IJCNLP), Hyderabad, India, pp. 418–425 (2008)
83. Brill, E., Dumais, S., Banko, M.: An analysis of the AskMSR question answering system. In: Proceedings of the 2002 Conference on Empirical Methods in Natural Language Processing, Pennsylvania, USA, pp. 257–264, 6–7 July 2002
84. Soricut, R., Brill, E.: Automatic question answering using the Web: beyond the factoid. J. Inf. Retr.—Special Issue Web Inf. Retr. **9**(2), 191–206 (2006)

85. Dror, G., Koren, Y., Maarek, Y., Szpektor, I.: I want to answer; who has a question?: Yahoo! answers recommender system. In: Proceedings of the 17th ACM SIGKDD International Conference on Knowledge Discovery and Data Mining, pp. 1109–1117 (2011)
86. Adamic, L.A., Zhang, J., Bakshy, E., Ackerman, M.S.: Knowledge sharing and yahoo answers: everyone knows something. In: Proceedings of WWW '08, pp. 665–674 (2008)
87. Surdeanu, M., Massimiliano, C., Hugo, Z.: Learning to rank answers to non-factoid questions fromweb collections. Assoc. Comput. Linguist. **37**(2), 351–383 (2011)
88. Arai, K., Handayani, A.N.: Question answering system for an effective collaborative learning. Int. J. Adv. Comput. Sci. Appl. **3**(1), 60–64 (2012)
89. Cairns, B.L., Nielsen, R.D., Masanz, J.J., Martin, J.H., Palmer, M.S., Ward, W.H., Savova, G. K.: The MiPACQ clinical question answering system. In: Proceedings of AMIA Annual Symposium, pp. 171–180 (2011)
90. Kongthon, A., Kongyoung, S., Haruechaiyasak, C., Palingoon, P.: A semantic based question answering system for Thailand tourism information. In: Proceedings of the KRAQ11 Workshop, Chiang Mai, Thailand, pp. 38–42 (2011)
91. Baeza-Yates, R., Rello, L.: How bad do you spell?: the lexical quality of social media. In: Proceedings of the Future of the Social Web, WS-11–03 of AAAI Workshops, AAAI (2011)
92. Raghavi, K.C., Chinnakotla, M., Shrivastava, M.: Answer ka type kya he? Learning to classify questions in code-mixed language. In: The Proceedings of the International World Wide Web Conference Committee (IW3C2), pp. 853–858 (2015)

Part V
Text Classification

Machine Learning Implementations in Arabic Text Classification

Mohammed Elarnaoty and Ali Farghaly

Abstract Text categorization denotes the process of assigning to a piece of text a label that describes its thematic information. Although this task has been extensively investigated for different languages, it has not been researched thoroughly with respect to the Arabic language. In this chapter, we summarize the major techniques used for addressing different aspects of the text classification problem. These aspects include problem formalization using vector space model, term weighting, feature reduction, and classification algorithms. We pay special attention to the part of research devoted to text categorization in the Arabic language. We conclude that the effect of language is minimized with respect to this task. Moreover, we list the currently unsolved issues in the text classification context and thereby highlight the active research directions.

Keywords Machine learning · NLP · Supervised classification
Corpus · Text categorization · Bag of words · Stemming

1 Introduction

Inferring a topic from a piece of text has been extensively utilized through the last few decades for addressing a wide range of NLP applications. Basically, the knowledge of text subject is useful for organizing text documents for different purposes like filing patents into patent directories, automated population of web sources into hierarchical catalogues, genre classification or even archiving. Furthermore, many NLP applications such as word sense disambiguation, machine translation, spelling correction, or building semantic linguistic resources utilize the topic inference task as a pre-processing step.

M. Elarnaoty
Computer Science Department, Cairo University, Giza, Egypt

A. Farghaly (✉)
Baaz, Inc, San Francisco, USA
e-mail: alifarghaly@yahoo.com

K. Shaalan et al. (eds.), *Intelligent Natural Language Processing: Trends and Applications*, Studies in Computational Intelligence 740,
https://doi.org/10.1007/978-3-319-67056-0_15

When used as a pre-processing task, topic categorization can be very useful for defining a semantic flag for the document that could be used afterward for filtering or referencing. For instance, a business organization that wants to collect customer reviews about its products would build an opinion mining component over social media. But since they are interested only in opinions about products, they use a topic classifier to filter out user feeds in other categories such as sports or politics.

Another use for topic classification is word sense disambiguation and homograph resolution. Homograph resolution aims at identifying the exact meaning and part of speech of a given word. For example, the word "patient" can be used as a noun to denote a person in need of medical treatment but also it can be used as an adjective to describe a person's level of tolerance. The most discriminating factor that decides the meaning of such ambiguous words is the context. For example, in word sense disambiguation a word like 'jaguar' in a discussion about cars, would refer to a particular brand of cars, whereas if we are talking about mammals, it would refer to particular animal. Similarly, the meaning of 'cloud' in computing is very different from its meaning in weather forecasting discourse. A topic classifier would be extremely helpful for tagging the documents general subject(s). A wide number of NLP applications can thereby make use of this word sense disambiguator such as machine translation systems or misspelling correction.

Inspired by these uses, text categorization has been recognized as one of the earliest fundamental problems in statistical machine learning based NLP. Hundreds of applications and papers were produced to address aspects of this specific problem to the extent that researchers could argue it is already an over studied area of research, at least for some languages such as English. The case can be more controversial for other languages that may still have gaps to fill. Here in this chapter, we will consider the Arabic language in particular.

The Arabic language has its own peculiarities that pose challenges for NLP applications [19, 20]. Text categorization is no exception. First, unlike English, Arabic is an agglutinative language where multiple tokens (word stems, word prefixes, or word suffixes) can be concatenated to form a single written word. A single Arabic word like "أرأيتهم" is equivalent to a number of words in English "Did you see them?". In other words, this single word that can be decomposed to the following four Arabic tokens: "أ" for "did", "رأيـ" for "see", "ت" for "you" and "هم" for "them". It is also a complete sentence with its subject, verb and object. This makes the stemming/tokenization process much harder in Arabic in contrast to easily whitespace separable Latin languages.

Another important property of the Arabic language is the explosion of ambiguity at all levels of the language (morphological, lexical, syntactic and semantic). As will be seen later in the chapter, semantics is the real key to solving text categorization as well as other problems. Unfortunately, Arabic has many sources of ambiguity compared to other languages. One of these sources is the absence of short vowels in almost all written Modern Standard Arabic. The very same letter string can be mapped to multiple different words with different unwritten short vowels where the decision of the correct mapping relies primarily on context. For example, the three-letter Arabic word "ثمن" consists of three letters "th-ث", "m-م"

and "n-ن". However, this three-letter word could be pronounced "thomn" by adding 'o' short vowel after the first letter to mean "eighth". Alternatively, intruding "a" short vowel between each two consonants gives "thaman" which means "price", etc. The Buckwalter Arabic Morphological Analyzer (BAMA) gives 21 different analyses for that word [11].

Normalization adds extra dimensions to the ambiguity problem. With the exception of classic Arabic scripts, most contemporary Arabic writers omit certain diacritic marks such as hamza, shadda, madda and the dots of the terminating yaa letter. Such normalization process causes two distinct words to end up written using the same letters. For instance, "last-آخر" and "retard-أخر"are both normalized by removing the "madda" from the first word and the "hamza" from the second resulting in "اخر" for both words. Note that each of the two words already has vowel ambiguities that add up to this normalization ambiguity. To illustrate more, the first word can map to one of the following two meanings {last, other} based on dropped short vowels whereas the second word maps to the set {retard, was retarded, others, I fell down} summing up to a set of six possible candidates.

There are other sources of ambiguity that we do not mention here since it is less relevant to the text classification problem such as dropped pronouns, lack of capitalization, constituent boundary, anaphoric ambiguity, and structure ambiguities.

In this chapter, we summarize the major techniques used for addressing different aspects of Text Classification (TC) problem. These aspects include problem formalization, term weighting, feature reduction, and classification algorithms. We give special attention to the part of research devoted to text categorization in the Arabic language. For some parts, we include python code snippets to get a better feeling of the process. We also pay attention to answering the question "Which research tasks under Arabic TC umbrella have been already killed, and which tasks have gaps to be filled?"

The chapter is organized as follows: Sect. 2 gives a general formalization of the text classification problem. Section 3 represents the core of this chapter which may be viewed as a tutorial enabling NLP practitioners to go through the task development step by step. We start by data collection and preparation, followed by pre-processing. Term weighting/indexing is then illustrated and finally, we discuss the most important feature reduction methods for this task. Section 4 overviews the commonly used ML algorithms in the literature and how it should be used. Section 5 surveys the topic classifiers developed for Arabic applications. Section 6 tries to answer the question of "what is finished and what is not in the context of Arabic text classification" and highlight open research directions. Finally, we conclude in Sect. 7.

2 Problem Definition

Classification is a generic process that assigns a label to an input based on the input characteristics. A classifier can be as simple as classifying a number as being positive or negative, and also could be as complicated as a physician diagnosing his patient by inspecting his MRI images and ECG signals. Indeed, classification tasks have gone far beyond this in terms of complexity. Modern classifiers rely heavily on statistical learning of models that either mimic each output class (generative model) or alternatively define the boundaries among different classes (discriminative model).

A statistical machine learning classifier uses a collection of input samples to learn its model. The learned model is therefore used to classify future inputs. A perfect classifier that is able to classify all its inputs correctly rarely exists. Classifiers, as well as experts, can miss some examples during the learning phase. They can also fail to learn the discriminating properties that distinguish a given class from another. For hard tasks, different learned classifiers often give uneven results when applied to the same input, which means that some classifiers behave better than others. Hence. there should be a way to evaluate classifiers against each other. The most common way of doing this is F1-measure metric. To apply F1-measure, the trained classifier is tested against a set of unlabelled examples called the testing set. F1-measure defines two metrics for the result of the classifications. These are *precision* and *recall*. Finally F1-measure reports for each individual class these two metrics as well as their harmonic average. For each class, F1-measure defines precision and recall as:

$$\text{Precision} = \frac{\textit{True Positives}}{\textit{True Positives} + \textit{False Positives}}$$

$$\text{recall} = \frac{\textit{True Positives}}{\textit{True Positives} + \textit{False Negatives}}$$

$$\text{F1} - \text{measure} = \frac{2 \times \textit{precision} \times \textit{recall}}{\textit{Precision} + \textit{recall}}$$

Where true examples are those examples classified by the classifier as belonging to the class even if they are not in reality. *T*rue *P*ositives (TP) are those examples classified by the classifier as belonging to the class and they do actually belong to the class. Complementary pieces of information are true negatives, false positives, and false negatives. *T*rue *N*egatives (TN) are those examples classified by the classifier as not belonging to the class and they actually do not. *F*alse *P*ositives (FP) are those examples classified by the classifier as belonging to the class while they are actually not. Finally, *F*alse *N*egatives (FN) are those examples that are incorrectly classified by the classifier as not belonging to the class.

In view of this brief introduction, we move to our problem of interest. In the next two subsections we will precisely define the boundaries of text classification problem in terms of input, output and data representation.

2.1 Problem Scope, Input and Output

Text classification is the task of automatic sorting of a set of documents or text scripts into categories from a predefined set [46]. It can be easily concluded from this definition that this task limits itself to text input in contrast with other classification tasks such as image, voice or signal categorization. Moreover, the set of possible outcomes of the task of concern is known in advance in terms of its cardinality and even its contained elements. This constraint is important because it distinguishes text classification from other NLP tasks such as text clustering, keyword extraction, named entity recognition and extraction, and topic modelling. For instance, if we altered this assumption so that we don't know the previously defined output labels then we end up with a text clustering problem. Even if we added other constraints such as assigning each word in the document to one of the output classes instead of assigning only the document as a single unit then we are no longer solving a text classification problem but rather a topic modeling problem. Therefore, in this chapter, we limit ourselves to solving the text classification problem. An interested reader can consult [9, 36, 41, 55] to learn about other problems.

On the other hand, text classification has a wide range of interpretations depending on the type of text inputs and the labels of output. The input text ranges in length from long documents (articles, reviews) to short micro-blogging text (Tweets). The output is usually defined by the purpose of the application and could be binary or multi-valued.

To put this into perspective, consider the following text classification examples. A text classification task with possible outputs from {positive, negative, neutral} assigned to each document/sentence is known as a sentiment analysis problem. A spam filter, on the other hand, has emails as its input and classifies them into one of two categories {spam, non-spam}. A text classification task that chooses its output class from a set of authors is called author identification or author detection. An essay evaluator, on the other hand, assigns a score (say an integer from 1 to 6) to the essay. In this tutorial, we will limit ourselves to only one task. Other tasks can be easily modelled the same way.

Topic/Text categorization typically defines the subject of a given piece of text within a collection of documents. For example, news stories are typically organized by topics such as politics, sports, and economy. Social media is quite the same. Academic papers are often classified by technical domains and sub-domains. In the next subsection, we formally introduce the problem and discuss how we can represent it in a compact format suitable for machine processing.

2.2 Problem Formalization

The first step in any classification problem is to convert your data set into a suitable form that can be processed by the classification algorithm. In our case, each document represents a single data sample. The most commonly used document representation is *vector space model.*

In vector space model, we define a document as a feature vector of its components (i.e. words). But since we have different documents with different sets of words and the fact that almost all classification algorithms are designed to process feature vectors of equal dimensions, it follows that the constructed feature vector for a document d has to allocate an entry for each possible word in the language. If the word does exist in the document, then its corresponding entry is given a positive value, otherwise, it is given a value of zero. For instance, a document containing the words "أجار - granted asylum", "أخرى - other", and "أخزى - disgrace" but not "أجال – appointed terms" would be represented by the vector constructing the column "Doc1" in the below table of Fig. 1.

Intuitively, we cannot construct an exhausting vector of all words in the language. It will be unnecessarily long, hard to collect and hard to process. Furthermore, we cannot depend on linguistic lexicons for this task. Many words that exist in lexicons are rarely or no longer used in real life. Other modern words are being fabricated from time to time to express emerging concepts. The standard approach to tackling this problem is to use a large collection of documents, typically referred to as 'the training corpus', to construct this vector. All words mentioned in the corpus are used to build a list along with their frequencies. This frequency list is further compressed by removing rare words whose frequencies are less than a pre-known threshold. A common practice also is to construct a set of "stop words" which are functional words in the language that are almost common to all pieces of text and thus do not provide any valuable information about the category of document. Examples of such stop words are "and-و", "he-هو", "in-في" and so on. Stop words list is used then to filter frequency list. In Other words, we have the frequency list reduced in dimension by removing two types of words: rare words, and very common stop words. These basic reduction techniques are not the only ones we can use. We will discuss other effective methods for reducing dimensionality later in Sects. 3.2 and 3.4.

Finally, for each document in the corpus, we construct a feature vector v of n dimensions with each dimension i corresponding to the ith word in the frequency list and n is equal to the number of words in the frequency list. For each word w_i in the words list, we set vi to be equal zero if w_i doesn't exist in the current document, or to a positive value if it does exist. Initially, we can fix this positive value to be one. Later in Sect. 3.3, we will cover a b broader range of values that vi can take. So we have our final representation of a document d_j in vector space model as:

Fig. 1 Bag of words representation for documents

Words	Doc1	Doc2	Doc3	Doc4	Doc5	...
...	...	...	...	...	...	...
أجار	1	1	1	0	1	...
أجال	0	0	1	0	0	...
أخرى	1	1	0	1	1	...
أخزى	1	1	0	1	0	...
...	...	...	...	...	...	...

$$\mathrm{d_j} = \begin{bmatrix} v_1 \\ v_2 \\ \vdots \\ v_i \\ \vdots \\ v_{n-1} \\ v_n \end{bmatrix}$$

This document representation is best known as a *bag of words* (BoW) representation because it expresses the document in terms of its contained words without accounting for their order. The hypothesis here is that order of words doesn't play a significant role in determining the topic of the document. Instead, it is the words itself that make the difference. If we find words like "anatomy-تشريح", "vein-وريد", and "cardiac-قلبي" in a specific document, then this document is probably talking about medicine. It doesn't matter in which order we've seen these words. Even further, we don't have to read the document or understand it so that we recognize it is speaking about medicine. It is just catching "via a single look" the likes of these words in any arbitrary order with a reasonable proportion that is more than enough to judge the document topic. Although simple, the hypothesis is arguably strong enough to carry on a wide range of document classification problems in the literature.

To sum up, we formalize the text classification problem in vector space model as following: Given a set of documents $\{d_1, d_2, \ldots, d_j, \ldots, d_{m-1}, d_m\}^m$, a set of document labels $\{c_1, c_2, \ldots, c_k, \ldots, c_{r-1}, c_r\}^r$, and a set of words $\{w_1, w_2, \ldots, w_i, \ldots, w_{n-1}, w_n\}$, each document is represented as a vector of word frequencies and paired with a class label to form $<v_j, c_j>$ where:

$$\mathrm{v_j} = \begin{bmatrix} v_1^j \\ v_2^j \\ \vdots \\ v_i^j \\ \vdots \\ v_{n-1}^j \\ v_n^j \end{bmatrix}, \text{ and } \mathrm{c^j} \in \{\mathrm{c_1}, \mathrm{c_2}, \ldots, \mathrm{c_k}, \ldots, \mathrm{c}_{r-1}, \mathrm{c_r}\}.$$

And all documents are encoded in an $m \times n$ matrix:

$$\mathrm{V} = \begin{bmatrix} v_1^1 & v_1^2 & \cdots & v_1^j & \cdots & v_1^m \\ v_2^1 & v_2^2 & \cdots & v_2^j & \cdots & v_2^m \\ \vdots & \vdots & \vdots & \vdots & \vdots & \vdots \\ v_i^1 & v_i^2 & \cdots & v_i^j & \cdots & v_i^m \\ \vdots & \vdots & \vdots & \vdots & \vdots & \vdots \\ v_{n-1}^1 & v_{n-1}^2 & \cdots & v_{n-1}^j & \cdots & v_{n-1}^m \\ v_n^1 & v_n^2 & \cdots & v_n^j & \cdots & v_n^m \end{bmatrix},$$

The learning algorithm constructs a model M that is used by the classifier. Classifier receives a new unclassified document d as input and uses M to map it to a class label c $\{c_1, c_2, \ldots, c_k, \ldots, c_{r-1}, c_r\}$.

It is worth noting however that vector space representation is not the only possible modelling for pieces of text. There are other representations that one can use. For example, one can model each word or sentence in a document using a vector of features (such as length, position, some syntactic and semantic features), and hence model the document using the output matrix. Another modelling of the document is simply the document itself. Naturally, a document is a sequence of words in a given order. Sequential labellers such as Hidden Markov Models are an adequate family of classifiers for addressing this input. Although exploited in [32], this family of classifiers is rarely used for document classification and is more suitable for automatic tagging of the words themselves.

3 Text Classification Steps

Text classification goes through a series of steps beginning from data acquisition and ending with applying classification algorithms to test data. Here in this section, we elaborate these steps through five subsections. The entire process flow is depicted in Fig. 2.

The text categorization process starts with document collection. Afterwards, documents are manually annotated. Documents hence are pre-processed by applying stemming, normalization and term indexing. Feature reduction techniques could also be applied. Finally, the documents are split into training and testing parts and are used in the machine learning process.

3.1 Data Selection and Preparation

The first step for building a text classifier is to collect data to be used for learning and testing the classification model. Typically, this data is labelled by humans/experts. For our problem, each document is given a category label (e.g. politics, economy,

Fig. 2 Text classification process flow

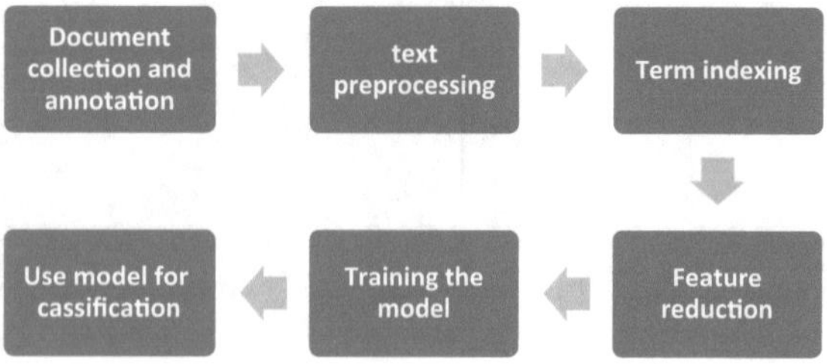

health…etc.). Eventually, as a machine learning task gains more interest from researchers, they begin to compile a standard benchmark datasets to be used for training and testing purpose. Not only, this saves time for future researchers, but also provides a common framework for holding comparisons among different approaches. Although text classification has these kinds of benchmark data, unfortunately, there is no free benchmarking data set for Arabic text classification. As a result, researchers have to compile their own data sets. Further, each publication uses its own tag set (classes) and its own annotation rules. It would be hard then to make any comparisons between different efforts in this case.

So as a starting point, we have to collect our data set, i.e. corpus. This depends basically on our intended application. The corpus could encompass long news articles, shorter emails to filter, or very short social media posts to categorize. Note that this difference in document length is not to be overtaken. As per our discussion in Sect. 2.2, the document has a fixed length representation depending on the final vocabulary size. Nonetheless, while long documents have a considerable number of positive entries in their corresponding vector representation, short text pieces have very rare ones scattered through the vector. In other words, they generate sparse vectors. Experienced readers in machine learning realize how problematic a sparse data can be during the training process. While the details of this problem not within the scope of our article, the reader should keep this piece of information in mind as we will return to it during our discussions later.

After we collect our text units (documents, tweets…etc.), we have to manually assign a label to each document from our defined categories. This setup for the problem is better known as supervised learning, hence, the learning process is guided by the supplied correct answers. Now, with the manually annotated corpus in hand, we can perform some pre-processing operations (Sect. 3.2) on the text before converting each document to its vector space representation (Sect. 2.2).

The final step in corpus preparation is to split the corpus into training and test data. The training data contain the larger portion of the corpus and is used for training the classifier. The testing set is used then to test the performance of the classifier. The test does not guarantee good classification performance due to the overfitting problem. Overfitting occurs when you over-tune your classifier to fit the test set. It could be exemplified by a student preparing for an exam by memorizing the answers to the book exercises rather than understanding the solution approach. A student doing so will do well in the final exam if only the exam included textbook questions. But if he finds other questions he will do poorly, simply because he was not able to generalize the concept and probably will fail. A classifier that is trained on too few data for his model can't learn its model parameters well and will not generalize even if the developers tuned it somehow (usually by feature selection methods) to operate well on the test set.

This problem could be overcome usually by using enough data for the training. However, this will not always be available. Or it may be available but its manual annotation takes very long time. To overcome this, machine learning scholars have introduced the concept of crossvalidation, a process which aims at ensuring that the learned model doesn't overfit. Crossvalidation is done by splitting the data into a

Table 1 Evaluation methods for text classification task

	Microaveraging	Macroaveraging
Precision (π)	$\pi = \frac{\sum_{i=1}^{\lvert c \rvert} TP_i}{\sum_{i=1}^{\lvert c \rvert} TP_i + FP_i}$	$\pi = \frac{\sum_{j=1}^{\lvert c \rvert} \pi_j}{\lvert c \rvert}$
Recall (ρ)	$\rho = \frac{\sum_{i=1}^{\lvert c \rvert} TP_i}{\sum_{i=1}^{\lvert c \rvert} TP_i + FN_i}$	$\rho = \frac{\sum_{j=1}^{\lvert c \rvert} \rho_j}{\lvert c \rvert}$

number of near-size folds k and running k experiments. In each experiment, we set aside one of the k folds for testing and use all the remaining folds for training. A high variance in the results of the experiments refer to overfitting and in that case, you may return to the solution of extending the data size or choosing other less complex training models. A low variance, on the other hand, indicates positively to the problem.

To sum up, researchers usually exploit crossvalidation to make sure their model doesn't overfit the training examples. In order to do this, they compute precision, recall and f-measure for all k runs of the k-fold cross-validation. The average is reported at the end. Moreover, the results may be reported for each category C_i separately, or averaged in a single measure. In the later case, we have either macro-averaging, which is computed on the level of categories or micro-averaging, which is computed on the level of single examples as depicted from Table 1.

Finally, if we are to split our data through different folds, then it may be necessary to make sure that each category examples are uniformly distributed among different folds so that we guarantee that each fold has the same distribution of the entire corpus. In most of the cases, a single random shuffling for the corpus documents would do an acceptable job. But it may be preferable to ensure this property and keep shuffling until it holds.

3.2 *Text Preprocessing*

Text Pre-processing is performed to improve the quality of classification task by excluding worthless pieces of information or importing useful ones. For the topic classification task, excluding useless information is done using stemming and normalization. Importing useful information is done using n-grams.

Stemming. According to the bag of words hypothesis, what matters is the word as a semantic token whereas other morphological information is useless. The two tokens "disease", "diseases" do not count as two distinct words just because they have different suffixes because they share the same stem. Returning a word to its original stem by removing suffixes is denoted as "stemming".

The main advantage of stemming is reducing vector space dimensions, which is the number of words used to represent each document in the vector space model. This reduction is done by aggregating words that have the same stem in the same vector entry. This means that you can have a reduction in vector dimensionality by a factor that varies from one language to another depending on the richness of

```
'''loading Arabic stopwords:'''

stopfile = open(stopwords_path,encoding='utf8')
ar_stop = stopfile.read().split()

'''loading Arabic corpus:'''
ar_corpus = PlaintextCorpusReader(ar_corpus_path, '.*',encoding='cp1256')

''' loading Arabic stemmer:'''
ar_stemmer = snowballstemmer.stemmer('arabic')

''' Reading the corpus'''
arabic_doc_set = [ar_corpus.raw(f) for f in ar_corpus.fileids()]

''' Stemming the corpus'''
tokenizer = RegexpTokenizer(r'\w+')
arabic_texts = [[ar_stemmer.stemWord(tken)
    for tken in tokenizer.tokenize(i.lower()) if not tken in ar_stop]
    for i in arabic_doc_set]

''' Building the dictionary'''
arabic_dictionary = gensim.corpora.Dictionary(arabic_texts)

''' Optionally remove very rare and very common words'''
arabic_dictionary.filter_extremes()

''' Convert documents to bag of words (sparse) representation'''
ar_corpus = [arabic_dictionary.doc2bow(text) for text in arabic_texts]

''' splitting corpus into train and test parts'''
split = int(0.1 * len(arabic_texts))
train = ar_corpus[split:]
test = ar_corpus[:split]
```

Fig. 3 Text classification process flow

language morphology. For example, in the following python code snippet (Fig. 3), we applied Arabic snowball stemmer on an Arabic corpus that contains around 102,000 news articles and we reduced the number of distinct words from 356378 to 119265 tokens (one-third) even before applying normalization. This dimensionality reduction not only results in a significant enhancement in memory and runtime performance but also improves the accuracy of the trained classifier by letting the classifier focus on the truly discriminative features and neglect other irrelevant noisy attachments.

Stemming is considered by the Text Classification community to amplify the performance of the classifiers. However, there is some doubt about the effectiveness of aggressive stemming algorithms such as the Porter algorithm in English [28]. The effect of stemming was investigated also for Semitic languages such as Arabic [25, 43], Aramaic, and Hebrew [22] and is reportedly positive. The vast majority of researchers, as reported by [25], agreed on the effectiveness of stemming on both accuracy and time for the classification problem with very few exceptions, although, there are still other researchers who omitted this pre-processing step and worked with words in their orthographic forms [33].

Normalization. In Sect. 1, we introduced the concept of normalization as a challenge that increases the ambiguity of the Arabic language. In spite of this increased ambiguity, normalization is an inevitable step for many text classification problems. The reason for this is that people use normalization sometimes while writing in Arabic. They also write un-normalized text at other times. In other words, you might find the very same word expressed in various forms with each occurrence contributing to a different entry in the vector model. For instance, a country like "Germany" can be written by different Arabic users as "المانيا" or "ألمانيا". It would be non-sensical if we separated them into two distinct words just because some users omitted the starting "Hamza" while they both refer to the same entity.

N-grams. Stemming is just one form of text pre-processing and other forms were also investigated. For example, a number of researchers had experimented with the replacement of words by character level n-grams [25]. Nonetheless, contradictory results were reported by different researchers. To reach a compromise, we can import only a limited subset of higher order n-grams that occur more frequently than a given threshold in the whole corpus.

Another very important pre-processing step is term weighting, aka indexing as described in the next section.

3.3 Document Indexing and Term Weighting Methods

According to [46] "Document indexing denotes the activity of mapping a document dj into a compact representation of its content that can be directly interpreted (i) by a classifier building algorithm and (ii) by a classifier, once it has been built". From this perspective, vector space representation can be considered a document indexing method. However, document indexing typically refers to more than such mapping. It usually incorporates assigning different weights to distinct words to reflect word's importance within the document. This is referred to as "Term-weighting". A document dj text is typically represented by a vector of term weights $v^j = [v^j_1, v^j_2, \ldots, v^j_n]^T$ where n is the length of the dictionary after stemming, normalization, removing rare words, removing stop words, and adding n-gram phrases (Sect. 3.2) [12].

Term weights may be binary-valued (i.e. $v^j_i \in \{0, 1\}$) associating ones with words that occur at least once in the document or real-valued (i.e. $0 \leq v^j_i \leq 1$). Real valued indexing methods range from simple statistical heuristics to stronger probabilistic techniques. These approaches are based on two empirical observations regarding text [1, 46]:

- *OB1*: The more times a word occurs in a document, the larger its contribution is in characterizing the semantics of a document in which it occurs and the more relevant it is to the topic of the document. This is the intuition of what so called "Term Frequency component, TF".
- *OB2*: The more times the word occurs throughout all documents in the collection, the smaller its contribution is in characterizing the semantics of a document in which it occurs and the more poorly it discriminates between

documents. This is the intuition of what so called “Inverse Document Frequency component, IDF”.

The most important real valued term weighting techniques are [1, 46]:

TF: Based on OB1, the word is weighted only by its frequency within the document [43].

TF-IDF: The tf-idf measure allows us to evaluate the importance of a term to a document. The importance is proportional to the number of times the term appears in the document (from OB1) but is offset by the frequency of the documents in which the term has occurred (OB2).

$$v_i^j = f_i^j \times \log\left(\frac{N}{n_i}\right)$$

where f_i^j denotes the number of times term i occurs in document j, and n_i denotes the document frequency of term i, that is, the number of documents in the data set in which term i occurs. N is the total number of documents [24, 43].

Normalized TF-IDF [TFC]: One shortcoming of tf-idf is that its weighting does not take into account that documents may be of different length. In order for the weights to fall in the [0, 1] interval and for the documents to be of equal length, the tf-idf weights are often normalized using cosine normalization [6, 45], given by:

$$v_i^j = \frac{f_i^j \times \log\left(\frac{N}{n_i}\right)}{\sqrt{\sum_{s=1}^{n}\left[f_s^j \times \log\left(\frac{N}{n_s}\right)\right]^2}}$$

Normalized Logarithmic TF-IDF [LTC]: A slightly different approach uses the logarithm of the word frequency instead of the raw word frequency thus reducing the effects of large differences in frequencies [6].

$$v_j^i = \frac{\log\left(f_i^j + 1\right) \times \log\left(\frac{N}{n_i}\right)}{\sqrt{\sum_{s=1}^{n}\left[\log\left(f_s^j + 1\right) \times \log\left(\frac{N}{n_s}\right)\right]^2}}$$

Darmstadt Indexing: Darmstadt indexing approach is a type of adaptive weighting that enables the system designer to combine a wider range of features for every token. The idea that underlies this indexing approach is to devise and leverage new features based on properties of terms, documents, categories or pairwise relationships among them. Appropriate aggregation functions are applied to these features in order to yield a single value. For example, a weighting function can be decomposed by averaging features that depict location in the document, document length and TF-IDF.

Entropy Weighting: Entropy weighting comes from the well-known information theory, where the entropy is high for equally and near equally distributed random variables and low for irregular ones. Here the random variables are the terms. Terms with high entropies are those that are likely to be present in almost every document, while document specific terms are more likely to have lower entropies. Therefore, entropy weighting scheme replaces term IDF with the negated term entropy. Entropy weighting is said to outperform other indexing schemes by researchers in topic classification community [1].

$$v_j^i = \log\left(f_i^j + 1\right) \times \left(1 + \frac{1}{\log N} \sum_{s=1}^{N} \left[\frac{f_i^s}{n_i} \times \log\left(\frac{f_i^s}{n_i}\right)\right]\right)$$

The same term weighting methods were applied to English as well as Arabic. For Arabic language, LTC and entropy weighting are arguably the preferred indexing methods [6, 25, 33, 43]. However, we must emphasize here that comparisons are not valid unless they are done on the same dataset using very careful analysis. With the absence of a standard benchmark Arabic dataset for the task, we can't assume that the results are absolutely correct, but they still give very significant indications about the robustness of these term weighting techniques. An interesting experiment carried out by [6] investigated mixing different term weighting approaches in a pipeline, but this barely improved the output.

3.4 Feature Reduction

Dimensionality reduction phase is often applied so as to reduce the size of the document representations from T where T is the number of distinct words in the training corpus to a much smaller, predefined number. This has both the effect of reducing overfitting and to make the problem more manageable for the learning method, since many such methods are known not to scale well to high problem sizes [28, 46].

Dimensionality reduction for text categorization problems is tackled either by feature selection techniques, such as mutual information (aka information gain), chi square, and gain ratio or by feature extraction techniques such as latent semantic indexing and term clustering [45].

Feature Selection Methods. The aim of feature-selection methods is the reduction of the dimensionality of the dataset by removing features that are considered irrelevant for the classification. The most popular feature selection methods used for text classification are [1, 25, 28, 38, 40, 46]:

Document Frequency Thresholding: The basic assumption here is that rare words are either non-informative for category prediction or not influential in global performance, and hence are safe to remove. A similar assumption can be made for very common words.

Mutual Information: Mutual information is a symmetric quantity commonly used in statistical modelling of the associations between two random variables. In the text classification context, this criterion is able to provide a precise statistical calculation that could be applied to a very large corpus to produce a table of associations between words and categories [7, 33]. If one considers a two way contingency table of a term t and a category c, then mutual information criterion between t and c is defined to be:

$$\mathrm{I(t, c)} = \log \frac{P(t,c)}{P(t)P(c)}$$

I(t, c) has a natural value of zero if t and s are independent. Using mutual information for feature selection, a word w is retained in the feature vector if maximum_i I(t, c_i) is more than a given threshold. In other words, it is retained when it has high dependency with at least one class. Sometimes the max function is replaced by a weighted average over different categories where term I(t, c_i) is weighted by P(c_i). A weakness of the mutual information is that score is strongly influenced by the marginal probabilities of the terms.

$$I(t) = \max_j I(t, c_j)$$

$$I(t) = \sum_{j=1}^{k} P(c_j) I(t, c_j)$$

Information Gain: Information Gain measure is equivalent to the expectation of Mutual Information quantity. However, we present here a more clever interpretation using entropy. IG of a given word is the number of bits of information obtained for category prediction by knowing the presence or absence of a word at the document. In other words, if we split the corpus into two divisions of the documents containing the word and the documents missing the words (Fig. 4), IG is the difference between the entropy of category distribution before this split and the entropy after.

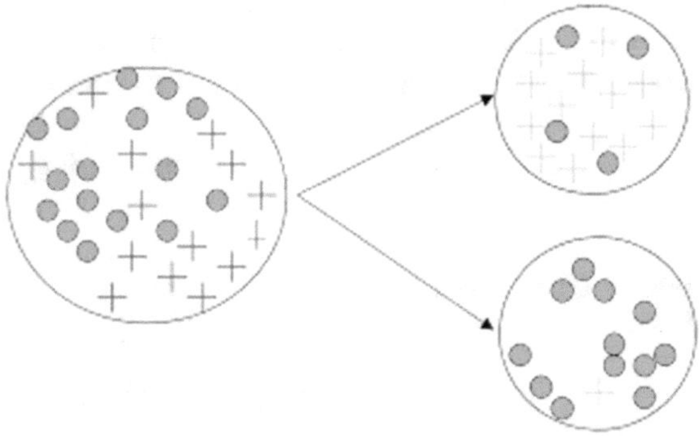

Fig. 4 An example of a word with large information gain. The entire corpus is evenly distributed between two categories. After splitting the corpus by word existence, the documents containing the word (above-right) and the documents missing the word (below-right) have a different class distribution than the original one (left)

This measures exactly the gain in information by considering how the category distribution gets influenced by the intrusion of this specific word. A high information gain for a word is obtained when we use the word to move from, an almost, evenly distributed documents over categories to a split that contains non-uniform distributions of documents over categories (Fig. 4). The information gain of a word is computed as:

$$\mathrm{IG(w)} = \mathrm{entropy}_{\mathrm{before}} - \mathrm{entropy}_{\mathrm{after}}$$

where:

$$\mathrm{entropy}_{\mathrm{before}} = -\sum_{j=1}^{k} P(c_j) \log P(c_j)$$

and

$$\mathrm{entropy}_{\mathrm{after}} = -P(w) \sum_{j=1}^{k} P(c_j|w) \log P(c_j|w) - P(\bar{w}) \sum_{j=1}^{k} P(c_j|\bar{w}) \log P(c_j|\bar{w})$$

Only words with IG larger than a given threshold are retained in the feature vector [6, 33, 35].

Chi-Square Statistics: χ^2 statistic measures the dependency between two distributions, usually depicted in a contingency table (Table 2). For word w and class c_j, χ^2 statistic is given by:

$$\boldsymbol{\chi^2(\mathrm{w}, \mathrm{c_j}) = \frac{\mathrm{N} \times (\mathrm{AD} - \mathrm{CB})^2}{(\mathrm{A}+\mathrm{C}) \times (\mathrm{B}+\mathrm{D}) \times (\mathrm{A}+\mathrm{B}) \times (\mathrm{C}+\mathrm{D})}}$$

Here A is the number of documents from class c_j that contains word w, B is the number of documents from class c_j that does not contain word w, C is the number of documents that contain w but does not belong to class c_j, and D is the number of documents that neither belongs to class c_j nor contains word w. N is still the total number of documents. The final $\chi^2(w)$ is then computed from $\chi^2(w, c_j)$. Like in the mutual information case, two different metrics could be computed for a word w based on the χ^2 statistic; maximum and weighted average [1, 6, 33, 52]:

$$\chi^2(w) = \max_j \chi^2(w, c_j)$$

Table 2 Contingency table for chi-square statistic

Variable	Yes	No	Totals
Var 1: word	A	b	A + D
Var 2: class	C	d	C + D
Total	a + c	b + d	N = A + B + C + D

$$\chi^2(w) = \sum_{j=1}^{k} P(c_j)\chi^2(w, c_j)$$

Odds Ratio: A 'one-sided metric' [54] in the sense that it only discovers features which are positively correlated with a particular class. As a log-odds-ratio for the probability of seeing a feature/term t_k given the class c_i, it is defined [33, 45] as:

$$\frac{P(t_k|c_i)(1 - P(t_k|\bar{c}_i))}{(1 - P(t_k|c_i))P(t_k|\bar{c}_i)}$$

While these are the most popular ones, other term selection methods [45] are also used such as DIA association factor (the conditional probability of class given word) [33], term strength [51], relevancy score, GSS coefficient [6], and NGL coefficient [6]. Table 3 summarizes these methods along with their mathematical formulas.

The text categorization literature also contains a number of methods that depend on wrapper selection methods, where a soft computing method like genetic algorithms, ant colony [37], or swarm optimization [53] is used to filter the features to optimize an objective function. The objective function is usually the error of applying the learned model using the current subset of features. The main problem with these methods is their high computational cost as they need a significant number of train/test cycles to reach their local optimal feature set.

Feature Extraction Methods. Unlike feature selection methods, feature extraction methods do not preserve words but transform the entire feature vector (word set) into another space of lower dimension. That is, the number of features are reduced. However, they are no longer words but a linear combination of them. The two most important feature extraction methods in text categorization literature are term clustering [8] and Latent Semantic Indexing (LSI) [50]. Term clustering methods are the traditional unsupervised machine learning techniques borrowed

Table 3 Feature selection metrics

Metric	Denoted by	Mathematical form
DIA association factor	$z(t_k, c_i)$	$P(c_i \vert t_k)$
Mutual information	$MI(t_k, c_i)$	$\log \frac{P(t,c)}{P(t)P(c)}$
Information gain	$IG(t_k, c_i)$	$\sum_c \sum_t P(t,c).\log \frac{P(t,c)}{P(t)P(c)}$
Chi square	(t_k, c_i)	$\frac{N.(P(t,c)P(\bar{t},\bar{c}) - P(\bar{t},c)P(t,\bar{c}))^2}{P(t)P(\bar{t})P(c)P(\bar{c})}$
Odds ratio	$OR(t_k, c_i)$	$\frac{P(t_k \vert c_i)(1 - P(t_k \vert \bar{c}_i))}{(1 - P(t_k \vert c_i))P(t_k \vert \bar{c}_i)}$
Relevancy score	$RS(t_k, c_i)$	$\log \frac{P(t \vert c) + d}{P(t \vert \bar{c}) + d}$
GSS coefficient	$GSS(t_k, c_i)$	$P(t,c)P(\bar{t},\bar{c}) - P(\bar{t},c)P(t,\bar{c})$
NGL coefficient	$NGL(t_k, c_i)$	$\frac{\sqrt{N}.(P(t,c)P(\bar{t},\bar{c}) - P(\bar{t},c)P(t,\bar{c}))}{\sqrt{P(t)P(\bar{t})P(c)P(\bar{c})}}$

from the area of statistical learning. We can leverage clustering techniques in text categorization by grouping together semantically related words in clusters and hence using the cluster representative(s) to replace all other words in the cluster. This has the impact of shrinking the vector space dimensionality from the size of the dictionary to be as small as the number of cluster representatives. Examples of common clustering methods that could be used are k-means, G-means, and mixture models.

The second method is latent semantic analysis [14], a method which reveals the utility of the singular value decomposition of the document-term matrix [9] in text processing. Given the V matrix from Sect. 2.2 representing document terms matrix, LSI makes use of SVD to construct a low-rank approximation V_k to V for a value of k that is far smaller than the original rank of V. Empirically, k is chosen to be few hundreds. LSI maps each row/column (respectively corresponding to a term/document) to a k-dimensional space; this space is defined by the principal eigenvectors (corresponding to the largest eigenvalues) of VV^T and V^TV. Next, it uses the new k dimensional LSI representation to infer similarities between vectors.

Finally, the new vector is passed to the classification algorithm to get its class using cosine similarity or any other classification algorithm. LSI is reported to have the greatest effect on the results of text categorization over any other method [24, 41, 48].

4 Classification Algorithms

The number of classification algorithms that have been used for topic categorization is overwhelming. These include at the very least, probabilistic methods, regression methods, decision tree and decision rule learners, neural networks, batch and incremental learners of linear classifiers, example-based methods, support vector machines, genetic algorithms, hidden Markov models, and ensemble and boosting methods. We concisely present the most popular of them especially with respect to Arabic topic classification [1, 25, 28, 38, 40, 46]:

Centroid Based classifiers. Centroid based methods are a family of classifiers that relies on the proximity of the observation to the training samples or their representatives. The observation is assigned the label of the closest training sample or alternatively the closest centroid.

The Classic Rocchio algorithm is one of the earliest used centroid methods in the context of information extraction and text classification [44]. In this method, a prototype vector is built for each class c_j and a document vector d is classified by calculating the distance between d and each of the prototype vectors. The distance can be computed using traditional similarity measures [56]. Examples of these measures are the dot product, cosine similarity, and Jaccard similarity. The prototype vector for class c_j is calculated as the mean vector over all training document vectors belonging to class c_j.

Other more productive models besides Rocchio are used for the task. The most popular of them are K Nearest Neighbours, Radial Basis Functions, and Gaussian

Discriminative Analysis. SVM with an RBF Kernel can also be counted as a centroid based classifier. KNN and RBF based models classify an example using the labels of its neighbourhood points. GDA method, on the other hand, models each class as a parameterized Gaussian and assigns the new input sample to the most probable Gaussian of them.

The advantage of these techniques is that they are simple to implement and flexible for text data. They have relatively less computation than other methods in both the learning and classification stages. But they are not quite as discriminative as other techniques. Moreover, it is sometimes hard to choose their hyper-parameters such as k for KNN.

Support vector machines. Support vector machines classifier builds a hyper-plane that maximizes the marginal distance between training examples (points) and then uses this hyper-plane to classify new points. In text classification, the training examples are the labeled documents in vector space representation. The support vector machine (SVM) method has been first introduced in text classification by Joachims [29, 30], and subsequently used in several other text classification methods. In geometrical terms, As argued by Joachims [29], one advantage that SVMs offer for text classification is that dimensionality reduction is usually not needed, as SVMs tend to be fairly robust relative to overfitting and can scale up to considerable dimensionalities. Recent extensive experiments by Brank and colleagues also indicate that feature selection tends to be detrimental to the performance of SVMs [46].

Decision Trees. A decision tree is a flowchart-like structure in which each internal node represents a "test" on an attribute or feature (e.g. whether the current document contains word "opening-افتتاح"), each branch represents the outcome of the test and each leaf node represents a class label (decision taken after computing all attributes). The path from the root to the leaf represents classification rules. Decision tree rules are learned using a number of algorithms like CART, C4.5, C5, and CHAID.

The advantages of decision trees are their robustness to noisy data and their capability to learn disjunctive expressions which seem suitable for document classification. Moreover, decision trees are simple to understand and interpret. A developer can generate the entire tree for the system experts so that they can understand the insights of the classification procedure. The main limitation, however, is its major dependence on greedy heuristic algorithms where decisions are taken locally at each node and cannot guarantee to return the globally optimal tree. Furthermore, the intrusion of irrelevant attributes can affect their performance negatively.

Neural Networks. Neural networks are nonlinear supervised models that are flexible in modelling many real world complex relationships. Neural networks are constructed of an input layer, an output layer and one or more hidden layers. Hidden and output layers compose neurons that linearly combine data from the previous layer and fire their activation function (which is usually nonlinear) on the linear combination. This added complexity in modelling NN output has its pros and cons. The positive thing is their ability to model very complex nonlinear functions especially when the number of layers increases. The main limitation is its hunger for training data. With the increase in the number of input and hidden nodes, the parameters needed for the neural network also increases. This results in over-fitting the data and might be cured only by enlarging the training corpus, with a little support from regularization methods. The neural network is one of the common and early methods in literature [50] which has shown good performance in general. However, compared to its complexity, other methods such as SVM are given preference.

Naive Bayesian: Naive Bayesian is one of the earliest classifiers to be used in text classification and surprisingly is not that naïve in terms of performance. A naive Bayes classifier is a simple probabilistic classifier based on applying Bayes' theorem with strong (naive) independence assumptions. It assumes that the conditional probabilities of the independent variables are statistically independent and hence can decompose the likelihood of data given the class membership as a product of terms, each of which represents one feature. NB is a high bias classifier and can be trained using relatively few training examples. This is probably the most outbalancing property for suggesting NB over other classifiers. It requires, though, a little adaption by using smoothing in order to avoid training data sparseness problem. NB, however, may not be a good choice for problems where the independence assumption doesn't hold, and these are not few in the classification universe.

Graphical Models: Naïve Bayesian is one of the simplest members in a well-known family of classifiers called Probabilistic Graphical Models. PGM comprises Bayesian networks and Markov networks. Of this family, Hidden Markov Models (like HMM, MEMM, and CRF) have gained the most attention for their effectiveness at modelling sequences with latent variables. These classifiers are often referred to as sequential labellers, as they assign labels to an entire sequence of tokens/words instead of classifying a single example. For document modelling, the graph representation of the document is more expressive than the standard bag of words representation and consequently, gives improved classification accuracy. We can also preserve and extract the hidden relationships among terms in the documents. This enables us to go far beyond the categorization task. However, the computational complexity of the graph representation and learning is the main drawback of this approach. Expectation Maximization, Sampling, and Variational Inference are the most popular algorithms used for training these networks.

Voting: Voting algorithms receive as input a classifier and a training set. The voting algorithm, then, trains the classifier multiple times on different versions of the training set. The generated models are then combined to create a final classifier that is used to classify the test set. Voting algorithms fall into two categories: bagging algorithms and boosting algorithms. The main difference between the two types is the way by which the different versions of the training data are generated.

Bagging: Bagging algorithm receives as input a classification algorithm $f(.)$ and a training set T and returns a set of classifiers $f^*(.) = \{f_1(.), \ldots, f_R(.)\}$. A classifier $f_r(.)$ is trained using bootstrap sample T_r of the training set. Each bootstrap sample T_r is constructed by uniformly sampling N random examples from the entire data set with replacement. This means the generated sample, T_r, will be of exactly the same size as the original dataset but some cases may be represented more than once while others may not be represented at all. To classify a new sample d, each classifier $f_r(.)$ from $f^*(.)$ is applied to the sample d resulting in labels $f_1(d), f_2(d), \ldots, f_R(d)$, and the class with the majority of votes is selected for the sample. Al-Radaideh et al. [5] claims the effectiveness of this approach over decision trees and association rules.

Boosting: Just like bagging, boosting selects a training size of N with replacement for each bootstrapping sample. But this time the sampling is no longer uniform. The probability of selecting a given sample for training the classifier f_k is not the same as other samples but rather depends on how often this particular sample was misclassified by the previous k-1 classifiers. Thus boosting aims at devising a new sequence of classifiers that are better able to correctly label the poorly classified examples. AdaBoost is a famous example of boosting classification algorithms. Boosting has proven a powerful intuition for text classification, and the BOOSTEXTER system[1] has reached one of the highest levels of accuracy so far reported in the literature [44, 47, 46].

Other classification algorithms that are worth noting in the literature are generalized linear classifiers like Logistic Regression, and Maximum Entropy (Multinomial Logistic Regression), Multi-Layer Perceptron [24], and Associative Classifiers. According to [25], the majority of the Arabic topic classification community has favored SVM over other classifiers. However, Hmeidi et al. [26], Gharib et al. [21] and Khorsheed and Al-Thubaity [33] have a more interesting finding for Arabic text classification that resembles its peer analysis for English. They have reported that the accuracy of the classification task varies from one algorithm to another depending on the number of used features, the nature and the size of the data. We will come to this point again in Sect. 6 when we state our conclusions.

5 Arabic Text Classification

Text classification has been considered for so many applications in the Arabic NLP community. It is extremely hard to describe all the conducted research. However, we try to cover a subset here that is representative enough for us to build conclusive arguments about the problem. Table 4, in the next page, summarizes the most prominent work in Arabic literature which, as we have said, are by no means intended to be exhaustive.

The Arabic text generally falls into three classes [19, 20]. These are classic Arabic, Modern Standard Arabic (MSA), and dialectic Arabic. Categorized text can also be as long as news articles or as short as tweets or holy book verses. The vast majority of Arabic work, applied the text categorization problem to long news

[1]BOOSTEXTER is available from http://www.cs.princeton.edu/~schapire/boostexter.html.

Table 4 Major Arabic text categorization research

Research	Indexing method	Feature reduction method	Classification algorithm	Notes and findings
Duwairi 06 [15]	Simple bag of words (BoW)	None	**KNN, Naïve Bayesian, similarity based method**	Corpus: 1000 docs Results prefer Naïve Bayesian
Mesleh 08 [37]	TF-IDF	**Ant colony with Chi-square Hybridization**	SVM	Corpus: 1445 docs Using SVM. Ant colony with Chi-square outperformed other popular feature selection methods
Gharib 09 [21]	Normalized TF-IDF	DF along with Information Gain (IG)	**KNN, Naïve Bayesian, SVM Rocchio**	Corpus: 1132 docs Rocchio for small feature sets and SVM for long ones
Zahran 09 [53]	TF_IDF	**Particle Swarm Optimization (PSO)**	RBF	Corpus: 5183 docs PSO outperformed other popular feature selection methods
Harrag 10 [24]	BoW	**SVD**	**NN (RBF kernel and MLP)**	Corpus: 453 docs SVD enhanced performance for large feature sets MLP NN outperforms RBF NN
Saad 10 [44]	**Binary, word count, TF, TF-IDF, normalized TF-IDF**	None	DT (CR4.5), KNN, SVM, NB	Seven corpora with various sizes (293–22429). SVM superiority
Abbas 11 [2]	TF-IDF	MI	**SVM, NN, language model, MI based classifier (TR)**	Corpus: 9000 docs SVM better for binary categorization. Comparable results for multi class categorization
Al-Radaideh 11 [5]	TF-IDF	TF-IDF	**Association rules classifiers**	Corpus: 1008 docs Competing results with classification algorithm readability advantage
AlSaleem 11 [25]	BoW	None	**SVM, Naïve Bayesian**	Corpus: 5121 docs SVM superiority
Alkabi 13 [4]	BoW	None	**DT, KNN, Naïve Bayesian, SVM**	Corpus: 1227 Quran verses Naïve Bayesian outperformed other classifiers
Al-Thubaity 13 [6]	Binary. TF-IDF, Log normalized TF-IDF (LTC)	**Oring/Anding 5 methods (Chi, IG, GSS, NGL, RS)**	Naïve Bayesian	Corpus: 6300 docs Combining feature selection methods has no significant improvements

(continued)

Table 4 (continued)

Research	Indexing method	Feature reduction method	Classification algorithm	Notes and findings
Khorsheed 13 [33]	**TF, TF IDF, LTC, entropy**	**TF, DF, IG, Chi, MI, DIA, NGL, GSS, OddsR, RS**	**DT (C4.5), KNN, Naïve Bayesian, NN, SVM**	**Corpus: 17658 docs (tested different sizes), different feature lengths** Results preferred LTC, GSS, RS, and Naïve Bayesian followed by SVM
Faidi 14 [18]	TF-IDF, added n- grams	None	**DT, Naïve Bayesian, SVM**	Corpus: 795 Hadiths favoured SVM
Haralambous 14 [23]	TF-IDF **Dependency parsing features**	TF-IDF **Dependency parsing features**	Association rules, SVM	Corpus: 6000 docs (balanced) **SVM for large feature sets, Association rules for small ones. Dependency based feature selection outperforms TF-IDF**
Hmeidi 08,15 [26, 27]	TF, TF-IDF	Chi	**Naïve Bayesian, DT (C5.0), decision Table, KNN, SVM**	Corpus: 2700 docs (balanced) SVM outperforms all when increasing number of features
Kechaou 14 [32]	Concatenated text, not BoW. Use n-grams	TF-IDF	**HMM**	Corpus: 1000 docs Proves the validity of HMM for text classification
Elhassan 15 [17]	BoW	None	**Naïve Bayesian, DT (C5.0), KNN, SVM**	Corpus: 750 docs (balanced) SVM outperforms all
Odeh 15 [39]	TF-IDF	TFIDF	**Similarity based approach**	Corpus: 982 docs
Raho 15 [43]	BoW	IG, Chi	**DT, Naïve Bayesian, KNN**	Corpus: 4763 docs KNN gives worst results
Al-Anzi 16 [3]	TF-IDF	**LSI (SVD)**	K-means, self-organizing maps	Corpus: 1000 docs Powerfulness of LSI
Kanan 16 [31]	BoW	None	Naïve Bayesian, random forest, SVM	237,000 docs **Building standard categorization taxonomy** SVM dominance

articles, which falls into the MSA category of Arabic text. Very few projects have addressed the problem on short classical Arabic text, especially on Quraan [4] and Hadith [18].

Table 4 focuses on four types of information for every work. For any given paper, the main objective of the work is bolded. For instance, the second row corresponding to [37] aims basically to investigate the effect of using the ant colony method for feature selection compared to other canonical methods. As you can see from the table, each method has its own focus such as testing weighting methods, classification methods, or the effect of pre-processing. Many of these papers incorporated different stemming methods as a pre-processing task. Since almost all of them agreed with the positive impact of stemming, we omitted this piece of information in the table. Note however, that they noticed the negative impact of aggressive and harsh stemming methods such as root based stemming [23].

Another remarkable feature in the table is the absence of standard benchmark corpus for Arabic topic classification. This is easily determined from the fact that all researchers use different corpora and each of them has compiled his own data collection. This is one of the main problems in Arabic text classification literature. With the absence of such a standard dataset, researchers cannot compare their work against each other and hence cannot validate the significance of their contributions. To come over this, some researchers had to conduct exhaustive comparative studies like in [33]. In order to break this dilemma, Kanan and Fox [31] started to create a standard taxonomy for Arabic news categorization. They used this taxonomy to annotate their enormous corpus, but have yet to make it available.

Granted that this data heterogeneity causes a noticeable diversity in the published results among different researchers, but this doesn't prevent us from extracting some common findings from these studies. For instance, almost all researchers agreed on the superiority of SVM for classifying large datasets with a large amount of features. For smaller data and feature sets, the results tend to prefer other high biased methods such as Naïve Bayesian. Moreover, soft computing methods gave relatively better results than other traditional feature selection methods [37, 53]. These methods, however, are more computationally expensive.

The majority of the listed papers utilize the methods covered in previous methods for filtering data, indexing it, weighting terms, reducing dimensionality and classification with very few exceptions. One interesting exception that is worth noting is the feature selection method used in [23] where authors utilized the dependency parser output to filter out terms. Their assumption is that the most important words usually appear at the top levels of the dependency tree. By selecting only these words, they could retain important words that happen to occur once or twice in a document but that have a clear role in the semantics. Their experiments suggest a significant enhancement by using topic classification results.

6 Directions for Further Research

Topic classification is a mere instance of text classification problems. Even though different text classification problems share the same approaches, these instances, however, are not equally difficult. The reason behind this is that each of these tasks has its own labelling semantics. These semantics define the boundaries among different classes, and how to distinguish between instances of distinct classes. The ease of a classification task relies mainly on the ease of defining these boundaries and ranges from easy tasks such as spam identification to hard tasks such as author recognition. Along this difficulty scale, the topic categorization task is arguably the easy part for long newswires documents, and more difficult for short social media, literature, and holy text. For news text, this claim is supported by the fact that the accuracy of modern news text classification systems rivals that of trained human professionals. The vast majority of recent research aimed at this task obtained a level of accuracy above 80% for Arabic text [2, 4, 15, 16, 21, 32, 33, 43] and 90% for English text [1, 45, 46]. According to the used classifier, the chosen term weighting scheme and the training set size and quality; results are boosted further to surpass 95% and even more.

Considering such reported results, one could conclude that the standard text categorization task is not a challenging task anymore for long text documents. Nevertheless, we can't ignore the fact that researchers constantly keep producing papers that address this problem. This may seem confusing at first glance as new research usually attempts to aim at unsolved or challenging tasks, which is the opposite in our case. This surging interest from researchers in the field can be attributed to a number of factors. Surprisingly, the most important of them is the simplicity of the task itself to the extent that scholars prefer to leverage it to validate their research questions. In other words, many of them don't address topic classification problems intentionally, but, they target other problems such as testing recently devised learning algorithms [29, 30], feature selection or reduction techniques [50], pre-processing stage [43], smoothing or regularization formulas, annotation schemes, and evaluation methods. Defining new specific categorization tasks, creating new linguistic resources, releasing new benchmark datasets and testing on a different language [22] are also obvious research initiatives. For all such tasks, the authors need to empirically validate and argue the significance of their contribution to the literature, and therefore they have to apply it on a well-defined task. From this perspective, topic classification is considered one of the most well-defined tasks in the NLP and machine learning literature. Moreover, the task is very rich in terms of available resources (at least for English), programming libraries and publications. Therefore, it is relatively straightforward to establish the experimental setup and make comparisons with other authors on a truth basis. This explains the popularity of the task from our viewpoint.

Another remarkable note from the results concerns the effect of language. Considering English and Arabic languages for comparison, and granted that we notice a difference in the reported accuracy ranges for both languages (above 80%

for Arabic, and 90% in English) but we can also see that the state of the art results are very competitive for both languages using the same bag of words model. The hypothesis here is that the classification task doesn't actually depend on language. This derives from the rationale that the correct category for a given piece of text depends, in the first place, on the themes itself and not on the literal interpretation of the themes (i.e. words). The themes may be interpreted as different equivalent tokens in different languages. But it is the semantics that defines the latent factor for categorization. Using this reasoning, we can assign the difference in reported accuracy ranges to the level of maturity in the scientific community corresponding to each language. Since English NLP has a more developed literature with a wider community and richer linguistic resources, it takes Arabic researchers more time to reach the level of their peers.

Unlike English, there is no free benchmark dataset for Arabic classification. The effect of this dataset absence is significant. The hypothesis is that the dominant factors in accuracy are the characteristics of the dataset and not the algorithms and in particular the source of the data and methodology of selecting the documents. The use of standard data set would eliminate these factors and enable researchers to make meaningful comparisons between performances of different algorithms [25]. This hypothesis could be justified by the analysis of results reported by different authors. Whereas the SVM classifier is favoured by most of the authors, nonetheless, the reported results for other algorithms are very close. Hmeidi et al. [26] on the other hand have a more interesting finding that is more consistent with both theoretical knowledge and state of practice in the machine learning field. Hmeidi et al. [26], Gharib et al. [21] and Khorsheed and Al-Thubaity [33] have found that the accuracy of the classification task varies from one algorithm to another depending on the number of used features, the nature and the size of the dataset. For instance, a naïve Bayesian classifier is considered a high bias classifier which works fine with small data sets compared to other high variance classifiers. But as the data set size and sparsity increases, naïve Bayesian can hardly compete against other classifiers such as SVM, and neural networks. But as data size increases, these methods could be too slow and naïve Bayesian may come back into use, or alternatively logistic regression.

In summary, we assume that neither the language of classified text nor the classification algorithm has a major impact on the classification accuracy. Even if we have a multi-lingual corpus, we claim it will not make a significant difference given that we trained the classifier on a similar multilingual corpus. This claim, however, needs to be tested empirically.

Another important aspect of this problem is relaxing the bag of words assumption. Wallach [49] argued that the order of words in sentences could affect the topic from which this sentence is generated. For example, the phrases "the department chair couches offers" and "the chair department offers couches" have the same unigram statistics, but are about quite different topics [49]. Moreover, the topic words may evolve over time [10] and these changes need to be depicted by retraining on different corpora every period of time (year or decade for example). In practical applications, the set of categories does change from time to time. For instance, in indexing computer science scientific articles under the ACM classification scheme, one needs to consider that this scheme is revised every five to ten years, to reflect changes in the CS discipline. This means that training documents need to be created for newly introduced categories, and that training documents may have to be removed for categories whose meaning has evolved [46]. These problems, however, trigger a more elegant but expensive way to approach the topic classification task. This is solved by variants of the well accepted probabilistic topic models [9], where considering the order of words is solved, for instance, as in [49], and topic evolution is solved, for instance, as in [10]. Generally speaking, topic models provide a much richer framework for unleashing the semantic structure of multi-topic documents. Nonetheless, topic modeling is beyond the scope of this chapter.

Finally, one more problem that is still open for investigation is the classification of short text such as micro blogging tweets, or Facebook posts. Short text is different from traditional documents in its shortness and sparsity, which hinders the application of conventional machine learning and text mining algorithms. Chen et al. [13] proposed two solutions for this problem. One is to fetch contextual information of a short text to directly add more text; the other is to derive latent topics from an existing large corpus, which are used as features to enrich the representation of short text. Lee et al. [34] have proposed two quite near solutions where for each topic, a document is made from trend definition and a varying number of tweets (30, 100, 300, and 500). From the document text, all tokens with hyperlinks are removed. This document is then assigned a label corresponding to the topic. The website "What's the Trend" provides a regularly updated list of ten most popular topics called "trending topics" from Twitter [34]. They also proposed another solution that classifies tweets using Twitter specific social network information, but this is not within the parameters of our problem. Topic classification for social media streams is now one of the active topics of research in Arabic NLP literature. Interested readers may find that another important research topic that needs to be addressed in Arabic literature is the hierarchical topic classification [42] of both long documents as well as short pieces of text.

7 Conclusion

In this chapter, we introduced the task of text classification for Arabic text. We described basic steps for implementing this task using machine learning techniques. These steps involve text collection and preparation, text pre-processing, term indexing, feature reduction, and classification. We reviewed the traditional approach to carry out these steps in the NLP literature. We also paid special attention to Arabic language related methods. Finally, we tried to analyse what exists and what is missing in the Arabic literature and proposed directions for interested researchers.

References

1. Aas, K., Eikvil, L.: Text categorisation: a survey (1999)
2. Abbas, M., Smaïli, K., Berkani, D.: Evaluation of topic identification methods on Arabic corpora. JDIM **9**(5), 185–192 (2011)
3. Al-Anzi, F.S., AbuZeina, D.: Big data categorization for arabic text using latent semantic indexing and clustering. In: International Conference on Engineering Technologies and Big Data Analytics (ETBDA 2016), pp. 1–4 (2016)
4. Al-Kabi, M.N., Ata, B.M.A., Wahsheh, H.A., Alsmadi, I.M.: A topical classification of Quranic Arabic text. In: Proceedings of the Taibah University International Conference on Advances in Information Technology for the Holy Quran and its Sciences, pp. 22–25, Dec 2013
5. Al-Radaideh, Q.A., Al-Shawakfa, E.M., Ghareb, A.S., Abu-Salem, H.: An approach for Arabic text categorization using association rule mining. Int. J. Comput. Process. Lang. **23** (01), 81–106 (2011)
6. Al-Thubaity, A., Abanumay, N., Al-Jerayyed, S., Alrukban, A., Mannaa, Z.: The effect of combining different feature selection methods on arabic text classification. In: 14th ACIS International Conference on Software Engineering, Artificial Intelligence, Networking and Parallel/Distributed Computing (SNPD), 2013, pp. 211–216. IEEE, July 2013
7. Bali, M., Gore, D.: A survey on text classification with different types of classification methods. Int. J. Innov. Res. Comput. Commun. Eng. **3**, 4888–4894 (2015)
8. Bekkerman, R., El-Yaniv, R., Tishby, N., Winter, Y.: On feature distributional clustering for text categorization. In: Croft, W.B., Harper, D.J., Kraft, D.H., Zobel, J. (eds.) Proceedings of SIGIR-01, 24th ACM International Conference on Research and Development in Information Retrieval, pp. 146–153. ACM Press, New Orleans, New York, US (2001)
9. Blei, D.M.: Probabilistic topic models. Commun. ACM **55**(4), 77–84 (2012)
10. Blei, D.M., Lafferty, J.D.: Dynamic topic models. In: Proceedings of the 23rd International Conference on Machine Learning, pp. 113–120. ACM, June 2006
11. Buckwalter, T.: Issues in Arabic orthography and morphology analysis. In: Farghaly, A., Megerdoomian, K. (eds.) Proceedings of the Workshop on Computational Approaches to Arabic Script-Based Languages, Association for Computational Linguistics, COLING 2004, Geneva, Switzerland 2004, pp. 31–34, Aug 2004
12. Caropreso, M.F., Matwin, S., Sebastiani, F.: A learner-independent evaluation of the usefulness of statistical phrases for automated text categorization. In: Text Databases and Document Management: Theory and Practice, vol. 5478, pp. 78–102 (2001)
13. Chen, M., Jin, X., Shen, D.: Short text classification improved by learning multi-granularity topics. In: IJCAI pp. 1776–1781, July 2011
14. Deerwester, S., Dumais, S., Landauer, T., Furnas, G., Harshman, R.: Indexing by latent semantic analysis. J. Am. Soc. Inform. Sci. **41**(6), 391–407 (1990)
15. Duwairi, R.M.: Machine learning for Arabic text categorization. J. Am. Soc. Inform. Sci. Technol. **57**(8), 1005–1010 (2006)
16. Duwairi, R.M.: Arabic text categorization. Int. Arab J. Inf. Technol. **4**(2), 125–132 (2007)
17. Elhassan, R., Ahmed, M.: Arabic text classification on full word. Int. J. Comput. Sci. Softw. Eng. (IJCSSE) **4**(5), 114–120 (2015)
18. Faidi, K., Ayed, R., Bounhas, I., Elayeb, B.: Comparing Arabic NLP tools for Hadith classification. In: Proceedings of the 2nd International Conference on Islamic Applications in Computer Science and Technologies (IMAN'14) (2014)
19. Farghaly, A.: Statistical and Symbolic Paradigms in Arabic Computational Linguistics, in Arabic Language and Linguistics, pp 31–60. Georgetown University Press (2012)
20. Farghaly, A., Shaalan, K.: Arabic natural language processing: challenges and solutions. ACM Trans. Asian Lang. Inf. Process. (TALIP) **8**(4), 14 (2009)
21. Gharib, T.F., Habib, M.B., Fayed, Z.T.: Arabic text classification using support vector machines. IJ Comput. Appl. **16**(4), 192–199 (2009)

22. HaCohen-Kerner, Y., Boger, Z., Beck, H., Yehudai, E.: Classifying documents' authors to their ethnic group using stems. In: CAINE, pp. 5–11 (2007)
23. Haralambous, Y., Elidrissi, Y., Lenca, P.: Arabic language text classification using dependency syntax-based feature selection (2014). arXiv:1410.4863
24. Harrag, F., Al-Salman, A.M.S., BenMohammed, M.: A comparative study of neural networks architectures on Arabic text categorization using feature extraction. In: 2010 International Conference on Machine and Web Intelligence (ICMWI), pp. 102–107. IEEE (2010)
25. Hijazi, M.M., Zeki, A.M., Ismail, A.R.: Arabic text classification: review study. J. Eng. Appl. Sci. **11**(3), 528–536 (2016)
26. Hmeidi, I., Hawashin, B., El-Qawasmeh, E.: Performance of KNN and SVM classifiers on full word Arabic articles. Adv. Eng. Inf. **22**(1), 106–111 (2008)
27. Hmeidi, I., Al-Ayyoub, M., Abdulla, N.A., Almodawar, A.A., Abooraig, R., Mahyoub, N.A.: Automatic Arabic text categorization: a comprehensive comparative study. J. Inf. Sci. **41**(1), 114–124 (2015)
28. Ikonomakis, M., Kotsiantis, S., Tampakas, V.: Text classification using machine learning techniques. WSEAS Trans. Comput. **4**(8), 966–974 (2005)
29. Joachims, T.: Text categorization with support vector machines: learning with many relevant features. Mach. Learn. ECML-98, 137–142 (1998)
30. Joachims, T.: Transductive inference for text classification using support vector machines. In: ICML, vol. 99, pp. 200–209, June 1999
31. Kanan, T., Fox, E.A.: Automated arabic text classification with P-Stemmer, machine learning, and a tailored news article taxonomy. J. Assoc. Inf. Sci. Technol. (2016)
32. Kechaou, Z., Kanoun, S.: A new-Arabic-text classification system using a hidden Markov model. Int. J. Knowl. Based Intell. Eng. Syst. **18**(4), 201–210 (2014)
33. Khorsheed, M.S., Al-Thubaity, A.O.: Comparative evaluation of text classification techniques using a large diverse Arabic dataset. Lang. Resour. Eval. **47**(2), 513–538 (2013)
34. Lee, K., Palsetia, D., Narayanan, R., Patwary, M.M.A., Agrawal, A., Choudhary, A.: Twitter trending topic classification. In: IEEE 11th International Conference on Data Mining Workshops (ICDMW), pp. 251–258. IEEE, Dec 2011
35. Lewis, D.D., Ringuette, M.: A comparison of two learning algorithms for text categorization. In: Third Annual Symposium on Document Analysis and Information Retrieval, vol. 33 (1994)
36. Mcauliffe, J.D., Blei, D.M.: Supervised topic models. In: Advances in Neural Information Processing Systems, pp. 121–128 (2008)
37. Moh'd Mesleh, A., Kanaan, G.: Support vector machine text classification system: using ant colony optimization based feature subset selection. In: International Conference on Computer Engineering and Systems ICCES, pp. 143–148. IEEE, Nov 2008
38. Nalini, K., Sheela, L.J.: Survey on text classification. Int. J. Innov. Res. Adv. Eng. **1**(6), 412–417 (2014)
39. Odeh, A., Abu-Errub, A., Shambour, Q., Turab, N.: Arabic text categorization algorithm using vector evaluation method (2015). arXiv:1501.01318
40. Patra, A., Singh, D.: A survey report on text classification with different term weighing methods and comparison between classification algorithms. Int. J. Comput. Appl. **75**(7) (2013)
41. Qiang, W., XiaoLong, W., Yi, G.: A study of semi-discrete matrix decomposition for LSI in automated text categorization. In: International Conference on Natural Language Processing, pp. 606–615. Springer, Berlin, March 2004
42. Qiu, X., Huang, X., Liu, Z., Zhou, J.: Hierarchical text classification with latent concepts. In: Proceedings of the 49th Annual Meeting of the Association for Computational Linguistics: Human Language Technologies, short papers-vol. 2, pp. 598–602. Association for Computational Linguistics, June 2011
43. Raho, G., Kanaan, G., Al-Shalabi, R.: Different classification algorithms based on Arabic text classification: feature selection comparative study. Int. J. Adv. Comput. Sci. Appl. **1**(6), 192–195

44. Saad, M.K.: The impact of text preprocessing and term weighting on arabic text classification, Doctoral dissertation, The Islamic University-Gaza (2010)
45. Schapire, R.E., Singer, Y., Singhal, A.: Boosting and Rocchio applied to text filtering. In: Proceedings of the 21st Annual International ACM SIGIR Conference on Research and Development in Information Retrieval, pp. 215–223. ACM (1998)
46. Sebastiani, F.: Machine learning in automated text categorization. ACM Comput. Surv. (CSUR) **34**(1), 1–47 (2002)
47. Sebastiani, F.: Text categorization. In: Encyclopedia of Database Technologies and Applications, pp. 683–687. IGI Global (2005)
48. Sebastiani, F., Sperduti, A., Valdambrini, N.: An improved boosting algorithm and its application to text categorization. In: Proceedings of the Ninth International Conference on Information and Knowledge Management, pp. 78–85. ACM (2000)
49. Wallach, H.M.: Topic modeling: beyond bag-of-words. In: Proceedings of the 23rd International Conference on Machine Learning, pp. 977–984. ACM, June 2006
50. Wiener, E.D., Pedersen, J.O., Weigend, A.S.: A neural network approach to topic spotting. In: Proceedings of SDAIR-95, 4th Annual Symposium on Document Analysis and Information Retrieval, Las Vegas, US, pp. 317–332 (1995)
51. Wilbur, W.J., Sirotkin, K.: The automatic identification of stop words. J. Inf. Sci. **18**(1), 45–55 (1992)
52. Yang, Y., Pedersen, J.O., A comparative study on feature selection in text categorization. In: Fisher, D.H. (ed.) Proceedings of ICML-97, 14th International Conference on Machine Learning, Organ Kaufmann Publishers, San Francisco, Nashville, US, pp. 412–420 (1997)
53. Zahran, B.M., Kanaan, G.: Text Feature Selection using Particle Swarm Optimization Algorithm 1 (2009)
54. Zheng, Z., Wu, X., Srihari, R.: Feature selection for text categorization on imbalanced data. ACM SIGKDD Explor. Newsl. **6**(1), 80–89 (2004)
55. Zhou, S., Li, K., Liu, Y. Text categorization based on topic model. Int. J. Comput. Intell. Syst. **2**(4), 398-409 (2009)
56. Zobel, J., Moffat, A.: Exploring the similarity space. ACM SIGIR Forum, vol. 32, no. 1, pp. 18–34. ACM (1998)

Authorship and Time Attribution of Arabic Texts Using JGAAP

Patrick Juola, Jiří Milička and Petr Zemánek

1 Introduction

One basic task in Natural Language processing is text classification, such as sorting documents by their content. A less well-known variant on this task is classifying documents by inferred metadata, such as the document's (inferred) language, date of composition or authorship. Authorship attribution is a well-studied problem [5, 7, 15, 18, 19, 30, 42], but most of the work done has been in major European languages such as English. Notable exceptions who have studied Arabic, in particular, include [1, 2, 36, 37].

We present a study selected from a new corpus (CLAUDia) containing nearly a half-billion words of Arabic text using a standard authorship analysis tool (JGAAP) to study the effects of author, genre, and time of composition on writing style and by extension on classification. We have selected a subcorpus balanced to permit comparisons between genres as well as between time periods to see how best-performing methods change with genre and time. We also provide an analysis of a larger variety of different feature sets than has previously been done for Arabic.

P. Juola (✉)
Evaluating Variations in Language Laboratory, Duquesne University,
600 Forbes Avenue, Pittsburgh, PA 15282, USA
e-mail: juola@mathcs.duq.edu

J. Milička · P. Zemánek
Faculty of Arts, Institute of Comparative Linguistics, Charles University,
Nám. Jana Palacha 2, Praha 1, 116 38 Prague, Czech Republic
e-mail: jiri@milicka.cz

P. Zemánek
e-mail: petr.zemanek@ff.cuni.cz

K. Shaalan et al. (eds.), *Intelligent Natural Language Processing: Trends and Applications*, Studies in Computational Intelligence 740,
https://doi.org/10.1007/978-3-319-67056-0_16

2 Background

2.1 Authorship Attribution and NLP

Text classification is, of course, a well-understood problem. For example, a patent attorney might want to find all the documents in a database with similar content (and hence relevant) to a pending application [13]. An important emerging application of classification is the attribution of authorship; in other words, classifying documents by the same author into the same category. Perhaps obviously, this cannot easily be done based simply on the semantics of the content (as the same author may write on multiple, unrelated subjects). It is, however, a versatile and useful application. This capacity is obviously of interest to lawyers, not only in resolving plagiarism and copyright infringement cases, but also in validating documents and detecting forgeries [31], establishing identity [24], and even in solving criminal cases such as murder [9, 14]. It can also be used by scholars and journalists interested in resolving long-standing questions in the arts and humanities [6, 10, 33].

In addition to identifying specific authors, similar technology can be used to profile authors, classifying them into broad categories, for example, by gender, age, education level, nationality, native language, or even personality type [3, 26, 29, 38]. Again, the applications are numerous, ranging from police investigations to medical diagnosis [35].

As mentioned above, this type of classification does not depend on traditional content analysis, but instead on the analysis of "style" (hence two alternate names for this subfield: "stylistics" and/or "stylometry"), the way the content is expressed. The basic observation is that human language is an underconstrained system and that there are often many ways to express the same idea. As a simple example, a writer's background may determine her spelling preferences in words such as "honor" (US spelling) or "honour" (Commonwealth spelling). There are dozens, or perhaps thousands of ways to express the idea of BIGGER-THAN-BIG, including "extra large," "giant," "enormous," "elephantine," and so forth. A writer with noted preferences will re-use the same choices frequently and in many different contexts. As McMenamin describes it,

> At any given moment, a writer picks and chooses just those elements of language that will best communicate what he/she wants to say. The writer's "choice" of available alternate forms is often determined by external conditions and then becomes the unconscious result of habitually using one form instead of another. Individuality in writing style results from a given writer's own unique set of habitual linguistic choices. [31]

Coulthard's description is similar:

> The underlying linguistic theory is that all speaker/writers of a given language have their own personal form of that language, technically labeled an idiolect. A speaker/writer's idiolect will manifest itself in distinctive and cumulatively unique rule-governed choices for encoding meaning linguistically in the written and spoken communications they produce. For example, in the case of vocabulary, every speaker/writer has a very large learned and stored set of words built up over many years. Such sets may differ slightly or considerably from the word sets

> that all other speaker/writers have similarly built up, in terms both of stored individual items in their passive vocabulary and, more importantly, in terms of their preferences for selecting and then combining these individual items in the production of texts. [11]

By identifying and detecting the results of these choices, one can, in theory, detect and identify the author that made these choices. There are numerous methods proposed for doing so, but a particularly illustrative example is Binongo's study of the *Oz* books [5]. The backstory is fairly simple: the series was started with L. Frank Baum's publication of *The Wonderful Wizard of Oz* and continued until his death in 1919. After his death, the publishers asked Ruth Plumly Thompson to finish "notes and a fragmentary draft" of what would become *The Royal Book of Oz*, the 15th in the series, and then Thompson herself continued the series until 1939, writing nearly twenty more books. The underlying question is the degree to which this "fragmentary draft" influenced Thompson's writing; indeed, scholars have no evidence that the draft ever existed. Binongo collected frequency statistics on the fifty most frequent function words across the undisputed samples and analyzed them using principal component analysis (PCA). Note that, unlike "honor/honour," both candidate authors used all fifty of these words at relatively high frequencies, but at individually different high frequencies. Reducing these fifty variables down to their first two principal components produced an easily graphable distribution that showed clear visual separation between the two authors. When the *Royal Book* was plotted on the same scale (see Fig. 1) it was shown clearly to lie on Thompson's side of the graph, confirming that "from a statistical standpoint, [the *Royal Book*] is more likely to have been written in Thompson's hand."

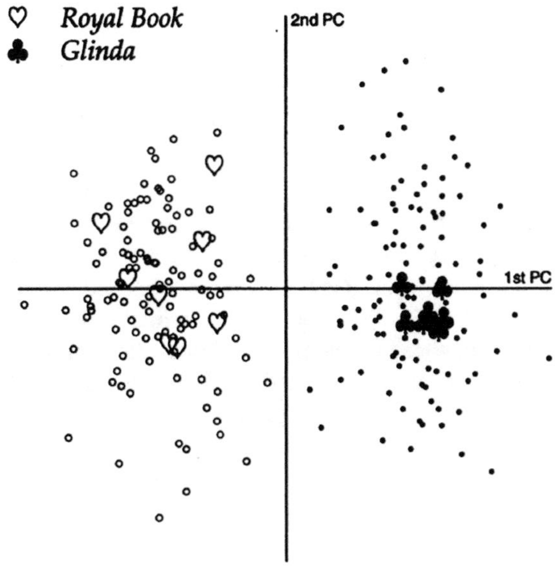

Fig. 1 Visual separation between Thompson (left) and Baum (right). Image from [5], with permission

The specific feature set to use is the subject of much debate and research. One common feature set is the frequency of common words such as articles and prepositions [5, 7, 16]. Because these words tend both to be common and also not to carry strong semantic associations, their frequencies tend to be stable across documents and genres, but these frequencies can also be shown to vary strongly across individuals. Another commonly used feature set is the frequency of common groups of consecutive words (word n-grams) or consecutive characters (character n-grams) [26, 32, 43]. However, Rudman [40] has identified more than a thousand feature sets that have been proposed for this task. Whatever set is used, the features present in a document are automatically identified, gathered into collections of feature representations (such as vector spaces), and then classified using ordinary machine learning algorithms [17, 18, 39] to establish the most likely author.

One issue in much of the research is the question of multiple influence and distractors. For example, plotting a document by an unknown third author on Fig. 1 would almost certainly be nearer to Baum or to Thompson; with only two candidate authors, there are really only two possible decisions. Because gender and language are linked [41], it is at least plausible that a female author would end up on Thompson's side of the graph, while a male author would end up on Baum's. Similarly, both genre and time will create stylistic similarities unrelated to the specific authors; any two romance novels share aspects that may not be present in detective fiction, and of course, any two novels from the 20th century are similar in ways that distinguish them from 17th century plays. (At the extreme, any two works in English will share features, such as the use of the character 'c,' that are not as common in works in Chinese, Greek, or Arabic). This work therefore investigates not merely authorial similarity, but also the comparative effect of authorial similarity, genre similarity, and temporal similarity.

2.2 *Authorship Attribution in Arabic*

As discussed earlier, most research in this area focuses on English or other major European languages, but there are exceptions. Among the first to study Arabic were Abbasi and Chen [1], who applied several different methods to study an ad-hoc corpus of messages from Yahoo groups and obtained surprisingly good accuracy. Ouamor and Sayoud [36, 37] compiled a very small collection of travellers' tales covering ten authors between the early 9th and late 19th century. Unfortunately, the same-author samples appear to have derived from the same work (for example, three samples from Nasser Khasru's (c. 1045) *Book of the Travels*) thus presenting serious issues with potential confounds. Similarly, the time period does not control for language change (imagine comparing *Beowulf* to Charles Dickens' *Oliver Twist*).

Ayadi et al. [4] used data visualization techniques to develop classification trees for a set of only two authors. One of the largest analyses to date is that of Alwajeeh et al. [2], who analyzed a set of five hundred articles in contemporary Arabic by five authors and were able to obtain more than 95% accuracy. However, little work has been done to compare different methods; researchers have focused on methods that work best for English, merely as a proof of concept that they will also work on Arabic.

To that end, we are applying a tool (JGAAP) known to be very good at comparative analyses of different methods [21, 23, 44] in the hopes of determining what types of processing needs to be done differently between Arabic and other, better-studied languages, and, furthermore, to see what differences appear with different genres and time periods.

3 Data

3.1 Corpus Description

CLAUDia (Corpus Linguæ Arabicæ Universalis Diachronicum) is a historical corpus of Arabic covering texts from the 7th to mid-20th centuries. It contains about 2 thousand full text works and roughly 420 million words. The titles are based on edited manuscripts. All the main genres that appeared in the history of Arabic literature are represented in the corpus; however, purely scientific genres, such as mathematics or geology, are present only indirectly in encyclopaedic works, and administrative production (office records, correspondence etc.) are present only as examples of good or bad practice in manuals for clerks. Translations from Greek and other languages (very common in the 9–10 centuries) are not present. Each of the texts has its unambiguous author (which we treat as an unassailable ground truth for this experiment). On the other hand, one should be aware of the fact that many of the titles represent compendia of various types of stories, collections of poetical passages or historical legends, but also sometimes one can find extant passages (many of them verbatim) taken over from earlier texts.

3.2 Selection of Texts

The selection of texts is rather opportunistic one. As the corpus contain texts of various genres which are considered to be crucial to the style and typical topics, the experiment was designed to take into account both time scale and genres.

The 9th and the 11th century were chosen as they are both well-represented in our corpus, as this time period is considered to be the golden age of the Islamic literature.

The following four genres were chosen as they represent the most flourishing part of medieval Arabic literature (and are therefore well-represented as well).

- Adab (Table 1) The Adab genre is probably the least homogeneous one, where we can encounter anything from encyclopaedic writings via some entertainment to rigorous treatises on how a clerk should set up his files and what language

Table 1 List of texts—Adab

Author	Title	Cent.	Length
al-Ǧāḥiẓ	al-Burṣān wa-l-ʕIrǧān	9	38442
	al-Buḫalāʔ	9	49583
ibn Quṭayba ad-Dīnawarī	Kitāb al-Maʕānī al-Kabīr	9	166753
	aš-Šiʕr wa-š-Šuʕarāʔ	9	87681
abū Bakr ibn abī Dunyā	Kitāb al-Muḥtaḍarayn	9	22809
	al-Maṭar wa-r-Raʕad	9	9377
abū al-ʕAbbās al-Mubarrad	al-Kalām fi-l-Luġa wa-l-ʔAdab	9	187662
	al-Fāḍil	9	20819
abū Ḥayyān at-Tawḥīdī	al-ʔImtāʕ wa-l-Muʔānasa	11	114057
	al-Baṣāʔir wa-d-daḫāʔir	11	304648
al-Marzūqī al-ʔIsfahānī	al-ʔAmālī	11	44090
	al-ʔAzmina wa-l-ʔAmkina	11	157523
abū Manṣūr aṯ-Ṯaʕālibī	al-Muntaḥil	11	39007
	Ṯamār al-Qulūb	11	59346
abū ʕAlā al-Maʕarrī	Tafsīr ʔAbyāt al-Maʕānī	11	59346
	al-Fuṣūl wa-l-Ġāyāt	11	83828

he should use. Also the language can be very variable—the same author can use brilliant classical language and some dialectisms. Topics can be so far away from one another that the books by the same authors can differ substantially.

- Fiqh (Table 2) Texts on the Islamic law or to be more precise—human understanding of Islamic law.
- Ḥadīṯ (Table 3) Stories of Muḥammad and his followers complemented with chains of tradents who said the story to each other. The genre also includes studies of these chains of tradents. Both the Fiqh and the Ḥadīṯ genres should be much more unified than the Adab. One can also expect some mutual similarity as the Islamic law is partially dependent on the Hadīṯs. Also typical Hadīṯ texts can be expected to be stylistically close to some Adab texts as they both consist of short narratives. The corpus of those stories is to some extend closed, especially in the later centuries the list of what was considered "sound" or correct story is rather finite.
- Linguistics (Table 4) The Fiqh and Linguistics should be standard scholarly genres, and both can be expected to share some common features.

From the CLAUDia corpus, we selected a set of authors balanced across both all four genres as well as both centuries. We therefore chose eight authors for each genre, four from the 9th and four from the 11th century. Each author was represented twice in our corpus from two different works (e.g., we did not select two chapters from the same work). This produced a set of 64 different documents to be analyzed in various contexts as described below.

Table 2 List of texts—Fiqh

Author	Title	Cent.	Length
ibn ʔIdrīs al-Qurašī	al-ʔUmm (Volume 1)	9	482363
	ʔAḥkām al-Qurʔān	9	47642
al-Harawī al-Baġdādī	aṭ-Ṭuhūr	9	25600
	al-ʔAmwāl	9	124000
Aḥmad ibn Ḥanbal	Risālat aṣ-Ṣalāt	9	7648
	Kitāb al-ʔAšriba	9	9989
al-Buḫārī	Rafʕ al-Yadayni fī ad-Duʕāʔ	9	7227
	al-Qirāʔa ḫalfa al-ʔImām	9	14121
al-Māwardī	al-ʔIqnāʕ	11	28496
	al-ʔAḥkām as-Sulṭānīya	11	84446
ibn Ḥazm al-Qurṭubī	Marātib al-ʔIǧmāʕ	11	38507
	al-ʔAḥkām	11	371321
Aḥmad al-Bayḥaqī	Bayān Ḫaṭaʔ man ʔAḫṭaʔa ʕalā aš-Šāfiʕī	11	19279
	ʔAḥkām al-Qurʔān li-š-Šāfiʕī	11	58789
al-Qāḍī abū Yaʕlā	al-ʕUqda fī ʔUṣūl al-Fiqh	11	185641
	al-ʔAḥkām as-Sulṭānīya	11	74689

Table 3 List of texts—Ḥadīṯ

Author	Title	Cent.	Length
aš-Šaybānī	al-Ǧāmiʕ aṣ-Ṣaġīr (Volume 1)	9	88824
	al-ʔĀṯār	9	14077
ibn Wahb al-Miṣrī	al-Qadr	9	3943
	al-Ǧāmiʕ	9	31386
ibn ʔIdrīs al-Qurašī	as-Sunan	9	39226
	Iḫtilāf al-Ḥadīṯ	9	56291
aṣ-Ṣanʕānī	Muṣannaf ʕAbd ar-Razzāq (Volume 1)	9	277881
	al-ʔAmālī	9	10927
al-ʔAzdī al-Miṣrī	al-Mutawārīn	11	3011
	al-ʔAwhām	11	4505
ar-Rāzī ad-Dimašqī	Fawāʔid Tammām	11	120354
	al-Muqillīn	11	1648
ʔabū Nuʕaym al-ʔIṣbahānī	al-ʔArbaʕūn fī Mahdī	11	5314
	al-ʔArbaʕūn	11	2655
al-Ḫalīl al-Qazwīnī	Fawāʔid ʔAbī Yaʕlā Halīl	11	78911
	al-ʔIršād	11	2963

Table 4 List of texts—Linguistics

Author	Title	Cent.	Length
at-Taymī	Maǧāz al-Qurʔān	9	63288
	al-Ḫiyal	9	22196
Ḥarawī al-Baġdādī	Luġāt al-Qabāʔil al-Wārida fī al-Qurʔān	9	2615
	Ġarīb al-Ḥadīṯ li-Ibn Salām	9	152209
ibn as-Sikkīt	al-Kanz al-Luġawī	9	41561
	Iṣlāḥ al-Manṭiq	9	78345
ibn Quṭayba ad-Dīnawarī	al-Maʕārif	9	93231
	ʔAdab al-Kātib	9	68579
al-Ǧinnī	Kitāb al-Lamʕ fī al-Luġa	11	20405
	al-Mubhiǧ	11	17260
ʔabū Hilāl al-ʕAskarī	Kitāb aṣ-Ṣināʕatayn	11	76820
	al-Furūq al-Luġawīya	11	97321
ibn Fāris	aṣ-Ṣāḥibī	11	38220
	Maqāyīs al-Luġa	11	363736
ʕAbdalqādir al-Ǧurǧānī	Dalāʔil al-ʔIʕǧāz	11	95034
	ʔAsrār al-Balāġa	11	76540

4 Methods

4.1 JGAAP

One well-known tool for this sort of analysis is the Java Graphical Authorship Attribution Program (JGAAP) [19, 25]. This open-source program (available both from github and from `jgaap.com`) provides a large number [21] of interchangeable analysis modules to handle different aspects of the analysis process, allowing for large-scale comparative analyses as part of a search for best practices. [23, 44].

JGAAP approaches the problem via a modular pipelined architecture. The pipeline stages include:

- Document selection—JGAAP supports a wide variety of formats but ultimately converts them to raw UTF text for analysis;
- Canonicalization—Documents are converted to a "canonical" form, and uninformative variations can be eliminated at the option of the analyst. For example, document headlines and page numbers, which are often the product of the editor and not the author, can in theory be removed. This also supports a certain amount of normalization—for example, punctuation (often also a product of an editor) may be stripped out or case variation neutralized to make sure that "The" and "the" are treated as the same word;
- Determination of the event or feature set—The input stream is partitioned into individual non-overlapping "events". (This term is used instead of "feature" to

emphasize that, in text documents, there is an implicit ordering to the features that may or may not be of interest.) At the same time, uninformative events can be eliminated from the event stream. "The fifty most common words" would be an example of such "events," as would "all part-of-speech 3-grams" (assuming the existence of a suitable POS tagger for the documents in question; unfortunately, JGAAP does not currently support Arabic POS tagging) or "all character 4-grams";
- Event culling—the overall set of events can be culled, for example, to make sure that only frequent, only infrequent, only widely distributed, or only events with high information gain are included;
- Statistical inference—The remaining events can be subjected to a variety of inferential statistics, ranging from simple analysis of event distributions through complex pattern-based analysis. JGAAP supports several different distance-based nearest-neighbor analyses (with more than 20 separate distances) as well as other classifiers such as LDA, SVM, and the WEKA suite of classifiers. The results of this inference determine the results (and confidence) in the final report.

Each of these phases is implemented as a generic Java class which can be instantiated by any of several different classes. For example, the event set (EventSet) factory class is defined as EventDriver, which takes in a document and returns the set of events from that document. Specific EventDrivers include words, word *n*-grams, word lengths, parts of speech, characters, character *n*-grams, part of speech tags, word stems, and so forth. Specific AnalysisDrivers include support vector machines, linear discriminant analysis, and so forth. It is not difficult to add a new module, and module selection is performed at run-time through an automatic GUI that searches the codebase for appropriate class files and adds them as options to the various menus.

With combinatorics, JGAAP supports more than one million different analyses and can thus be used for large-scale experiments in search of best practices [27, 28, 44] or to create individual elements for ensemble classification such as mixture-of-experts [20, 22]. We use it here as a testbed to compare different approaches to authorship attribution in Arabic under various temporal and genre conditions as detailed above.

4.2 Canonicizers

The CLAUDia corpus is already tokenized (for our purposes a token is defined as the longest sequence of arabic letters between non arabic letters, digits, whitespaces, or punctuation). Digits are omitted as they were mostly added to the medieval texts by modern editors (as counters of chapters and sections etc.). Vocalization marks are also excluded since they were also mostly added *ex post facto*. For the same reasons, aliphs are unified so that ٱ, ا, إ, أ, and آ are represented as one letter.

We experimented with two (probably unnecessary) canonicizers. JGAAP provides the capacity to unify case, converting all letters to the same case. This case unification is not needed as the Arabic script does not contain case differences as in the Latin script. Similarly, as the CLAUDia corpus is already tokenized (one token per line), JGAAP's capacity for normalizing whitespace should not be necessary, except that the representation of a line break varies from system to system and program to program, thus possibly introducing noise related, not to the contents of the document, but to the editing process for each individual file in the corpus.

4.3 Event Drivers

One finding from prior work [23] is that, while the exact details of the analysis method may not have a strong effect, the choice of event set to analyze is key. For the experiments reported here, we focused on four types of event sets, as follows;

- Character n-grams Character n-grams, groups of n adjacent characters without regard to word boundaries, are a commonly used feature set in many different experiments [26, 32, 43] and are often among the highest-performing methods. Examples of character 4-grams would include the middle ("ndee") of the word "indeed," but would also include the "nth" in the phrase "in the." In English and similar languages, they can capture information at all levels of language, including not only common words, but also morphology (for example, the "psyc" morpheme found in "psychology" "psyche," and "psychiatrist," or the "ing" morpheme found at the end of gerunds), as well as syntactic information gleaned from adjacent words.

  Morphology, of course, is different in Arabic. A key factor in Arabic morphology is its non-concatenative nature; instead, a set of consonants will often define the "root" while intervening vowels and infixes (the so called 'template') produce different specific meanings. An example of this is the k-t-b morpheme, representing (broadly) WRITING, as in *kitāb* BOOK, *kātib* WRITER, *katab-a* HE WROTE, *uktub!* WRITE! (imperative). The interaction of non-concatinative morphology with character n-grams is not obvious, nevertheless, only the combination of the root and the template produces a meaningful word, thus we can assume that the normal n-grams without gaps or anything like that are suitable for handling Arabic as well as other languages. This assumption is in accordance with the results of [2] who found out that even a simple stemmer is rather harmful rather than helpful in this kind of task.
- Word n-grams As with characters, words can also be made into n-grams clusters. The notion of "word" can be linguistically problematic; JGAAP takes a language-agnostic and naive approach and defines them simply as a maximal sequence of non-blank characters; thus "New York City" would be treated as three words," but "bread-and-butter" would be only one.
  The notion of word is even more problematic in Arabic. Not only is the Arabic morphology typologically synthetic (e.g. *yaktubūna* THEY [MASC.] WILL WRITE), but Arabic orthographic words may also contain plethora of clitics (e.g. *fa-sa-*

yaktubūna-hā THEN THEY WILL WRITE IT) and these clitics are not marked in standard written Arabic texts.
Nevertheless, words (and the relationships between them) can be very useful cues both to authorial vocabulary and to syntax. They can even help identify an author's preferred semantics by measuring correspondances between words (for example, what kind of things are described as "good"?)

- Most common words The most common words in a document, as discussed in the Background section, are typically synsemantic words (also known as "function" words) like articles, prepositions and pronouns. However, the information that is expressed in synsemantic words in English is mostly encoded as the part of the autosemantic ("content") word in Arabic (see the example above) or they are at least connected together (*wa-la-hā* AND FOR HER). This does not mean that there are not synsemantic common words at all but their role is, at least if we take the orthographic words as they are, limited. This, in turn, suggests that performance on frequent words differs radically in Arabic from more isolating languages like English.
- Rare words Rare words, words found only once or twice in a document have been proposed as a way to explore the vocabulary richness of an author. Unlike common words, two different authors are unlikely to use exactly the same set of rare words and there are many more rare than common words to chose from. Unfortunately, the notion of "rare" word is not only highly topic-dependent, but also highly-length dependent. A word that appears once in a two thousand word essay is likely to appear dozens of times in a 100,000-word book. Conversely, a word that appears only once in a novel is unlikely to appear at all in a smaller essay, even when the author is the same. Given the size differences present in our corpus, it might be expected to perform poorly in this specific experimental condition.
- Word lengths The idea that some people use longer words than others is perhaps the oldest proposed feature in quantitative authorship studies [12], and although simple analysis using t-tests and other comparisons of central tendency have proven to have low accuracy, better performance can be obtained by taking histograms of word lengths (for example, the sentence "the quick brown fox jumps over the lazy dog" would be represented as zero words of length one or two, four words of length three, two of length four and so forth) and using histogram counts as values in classification space.

4.4 Analysis Methods

A standard IR practice is to embed documents in a high dimensional vector space (using dimensions such as feature frequency) and apply ordinary machine learning algorithms or data visualization techniques to that vector. Binongo [5], for example, used principle component analysis with the assumption that documents by the same author would be "close" in this space. Prior work [34] has shown that simple nearest neighbor methods can accomplish comparable performance to more computationally expensive methods such as support vector machines or decision trees [17, 18, 39]. We therefore applied nearest neighbor classification to this problem. JGAAP supports a wide variety of ways to calculate distances of which normalized cosine distance (defined as 2 minus twice the cosine of the angle between the points as viewed from the origin) [34] is among the best-performing in other studies.

Analysis was performed using leave-one-out-cross-validation (LOOCV), in which each genre corpus was analyzed sixteen times. In each analysis, a single document was "left out," as a nonce testing document. Because each author was represented twice in each corpus, the true author was thus represented once in the remaining training document, while each distractor was represented twice. The overall system thus had a baseline (chance) performance of one in eight, or 0.125.

In addition, each author was classified as 9th or 11th century, with both centuries equally represented. For every analysis, we determined whether the inferred author was of the correct century or not. This allows us to see how much writing style has changed over the approximately 300 years encompassed by the study, and whether the amount of change is similar for all genres.

5 Results

5.1 Character n-grams

As for the best performing n-grams, genres are mutually diverse. The highest precision was reached with 12-grams (6 correctly attributed texts out of 16) for adab, 9-grams for fiqh, 9–10-grams for ḥadīṯ and unigrams for linguistics. The variability can be also observed when we look at the capability of n-grams to determine the century, but the values of n that yield the best results are not the same. The complete results for unigrams to 19-grams for distinct genres are depicted on the Fig. 2.[1] In all cases are the best results are higher than the random guessing (two tailed binomial test at 5% significancy level). The complete results for unigrams to 19-grams for all genres together are depicted on Fig. 3. It is remarkable that the n-grams that are best for determining the century are shorter than the best n-grams for the authorship attribution.

The question is whether the differences between genres are caused by the differences of the typical text length. The length of the text under examination plays less important role than we would expect, for example there are 9 correctly assigned texts within the 32 shortest texts and 7 correctly assigned texts within the 32 shortest texts (for 6-grams). For 13-grams the results are very similar: 8 correctly assigned short texts and 7 correctly assigned long texts. These figures also do not suggest that there is a tendency that the longer n-grams work better for longer texts and vice versa (the

[1] 95 % binomial confidence intervals for single proportion were calculated for each value (Namely the Willson Score interval as described in [45]).

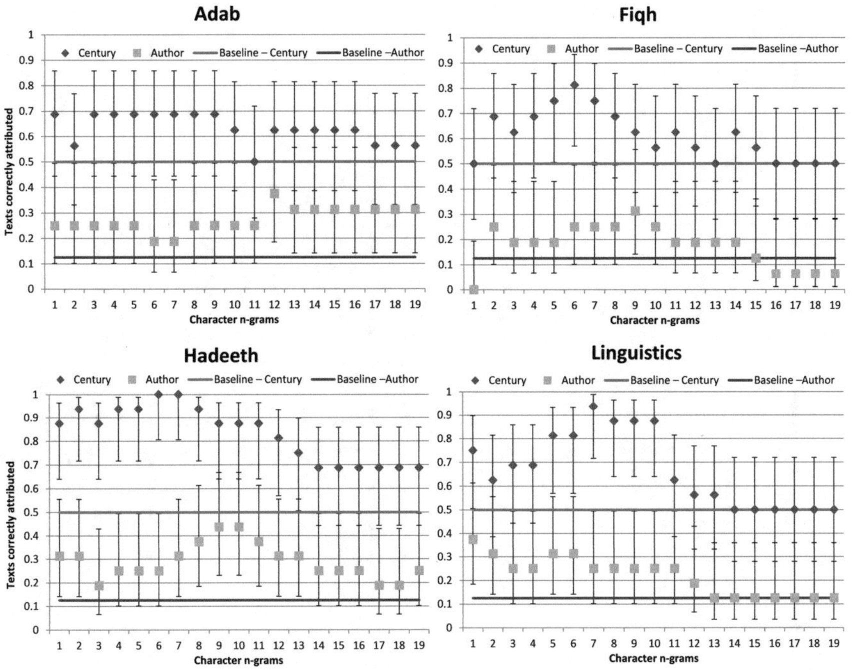

Fig. 2 Relation of the proportion of correctly attributed texts on the character *n*-grams length

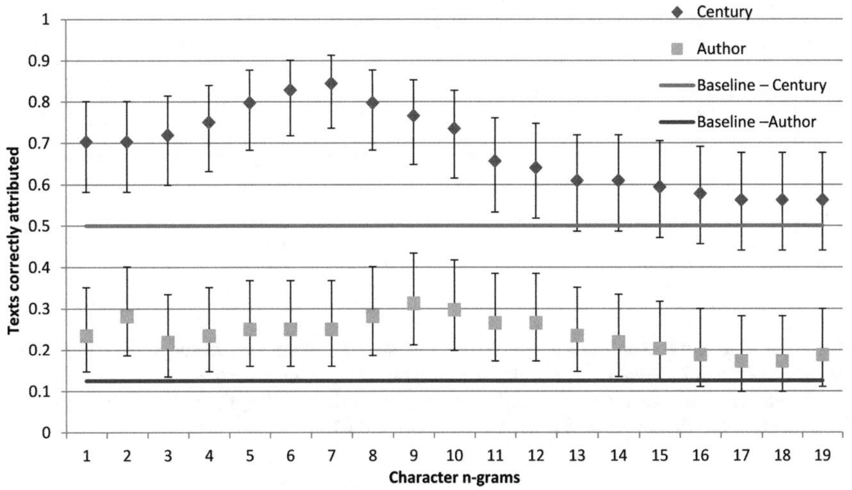

Fig. 3 Relation of the proportion of correctly attributed texts on the character *n*-grams length

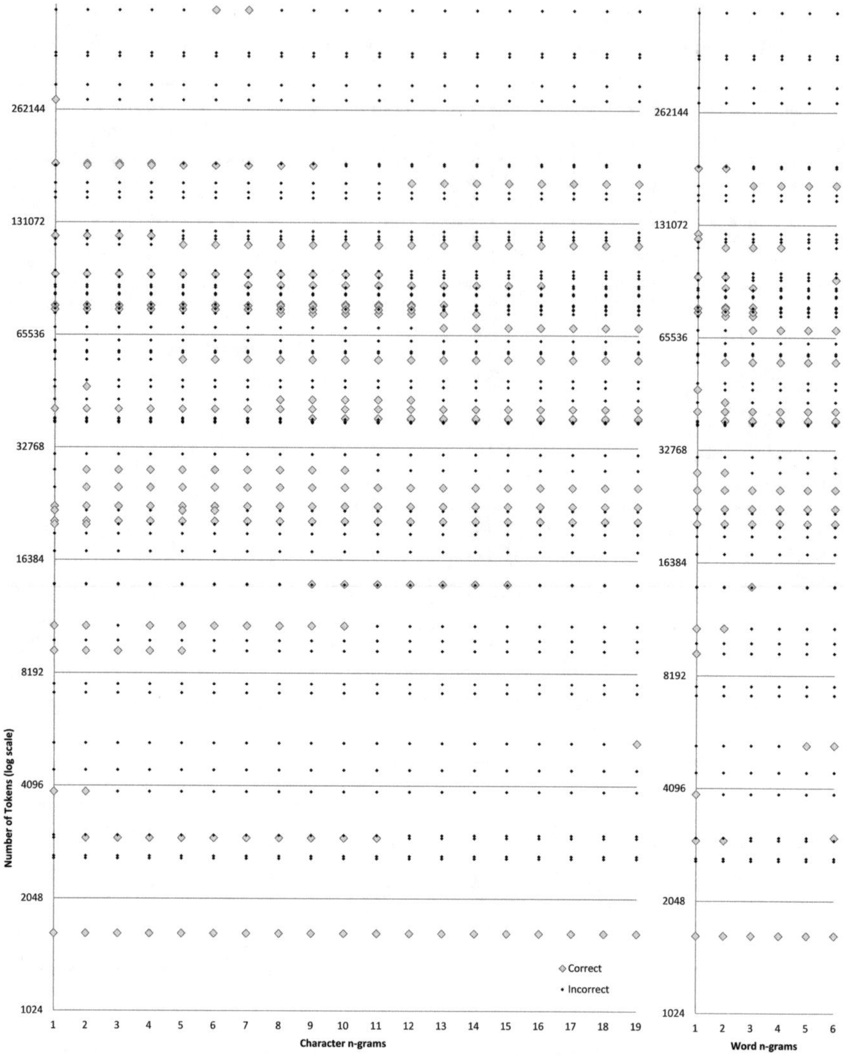

Fig. 4 Each point represents one text—the black ones stand for texts attributed to incorrect author, the light ones stand for the correctly attributed texts. Their position on the y axis stands for their length and the position on the x axis represents the event driver—length of character n-grams on the right chart and the length of the word n-grams on the left

Fisher's test for these values is equal to 1.00). The overall picture can be seen on the Fig. 4 (the left chart).

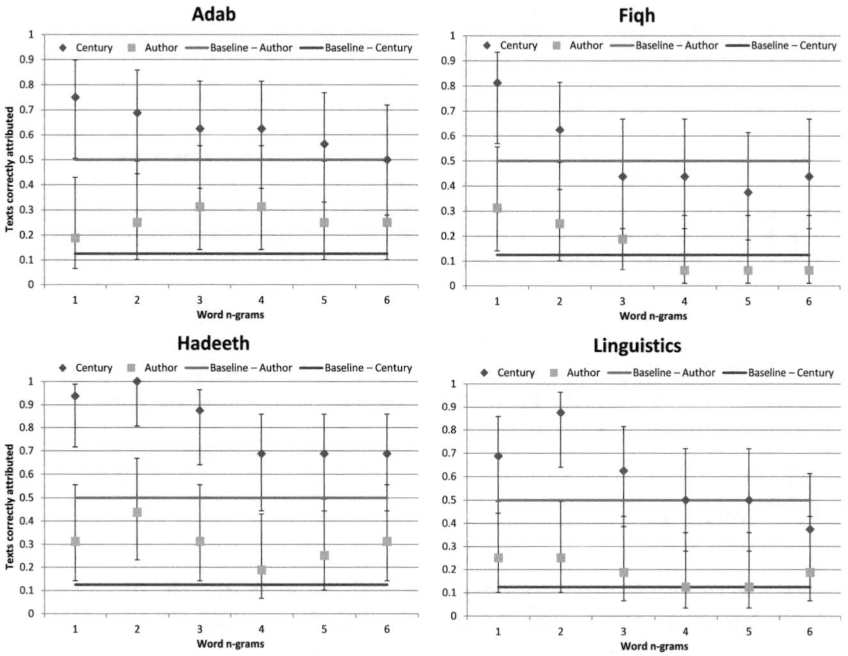

Fig. 5 Left: The relation of the proportion of correctly attributed texts on the word n-grams length. Right: the relation of the number of correctly attributed texts on the number of most common words

5.2 *Word n-grams*

The results for word n-grams correspond to results for character n-grams. Also the errors for word n-grams are similar to the errors for character n-grams, as will be noted in the next section. As the average word length in our sample is about 4 characters, i.e. 5 characters including the whitespace, it is not surprising that word bigrams correspond to character 10-grams, word trigrams to character 15-grams etc. It appears that the longer the n-grams are the closer the similarity is.

Also the effect of text length on the precision that we have failed to observe for character n-grams is irrelevant for the word n-grams (see Fig. 4, the right chart). The difference between genres is depicted on the Fig. 5. The performance of word n-grams for all samples together is represented on the Fig. 6 (the left chart).

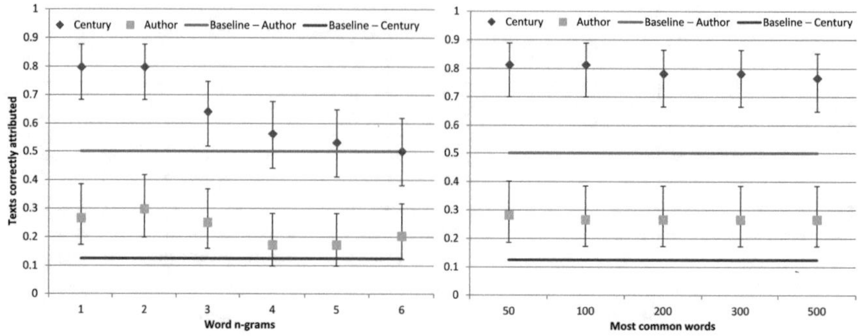

Fig. 6 Left: The relation of the proportion of correctly attributed texts on the word n-grams length. Right: the relation of the proportion of correctly attributed texts on the number of most common words

5.3 *Word Length*

Contrary, the results for the word lengths are vastly different from the character/word n-grams, for details see the next section, especially cf. graphs in the Figs. 9 and 10. At the same time the precision is comparable with the precision of n-grams—see the Fig. 7 (the left chart).

The complementarity of the n-grams and the word lengths means that good results from mixing the two features into one event driver can be expected.

5.4 *Rare Words*

Rare words are the only event-driver that failed to attribute the texts to the right authors statistically better than the random guessing—the results for our sample are

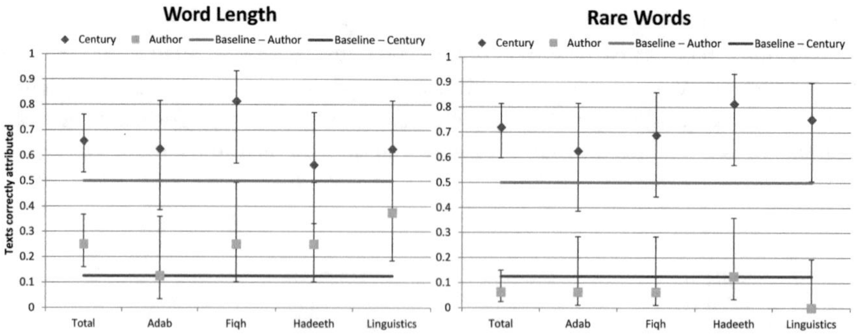

Fig. 7 The proportion of correctly attributed texts in distinct genres according to the word length event driver (on the left) and the rare words (on the right)

mostly even worse than the baseline (see the Fig. 7, the right chart). We attribute the poor performance on this feature set to the size differences within the corpus as discussed in the previous section.

It is remarkable that despite the low precision in the authorship attribution, the results are not that bad when it comes to the time attribution (Fig. 7).

5.5 *Most Common Words*

Most common words are, along with *n*-grams, successful in the authorship attribution and for our samples they are even better in the time attribution whether we compare them genre by genre (Fig. 8) or all together (Fig. 6, the right chart).

The precision of the *n*-grams driven methods was strongly dependent on the *n*-grams length, while in this case the number of most common words does not seem to play an important role as the proportion of correctly attributed texts on the basis of 50 most frequent words does not differ a lot from the resul?ts that are based on the 500 most frequent words (for the direct comparison see Fig. 6).

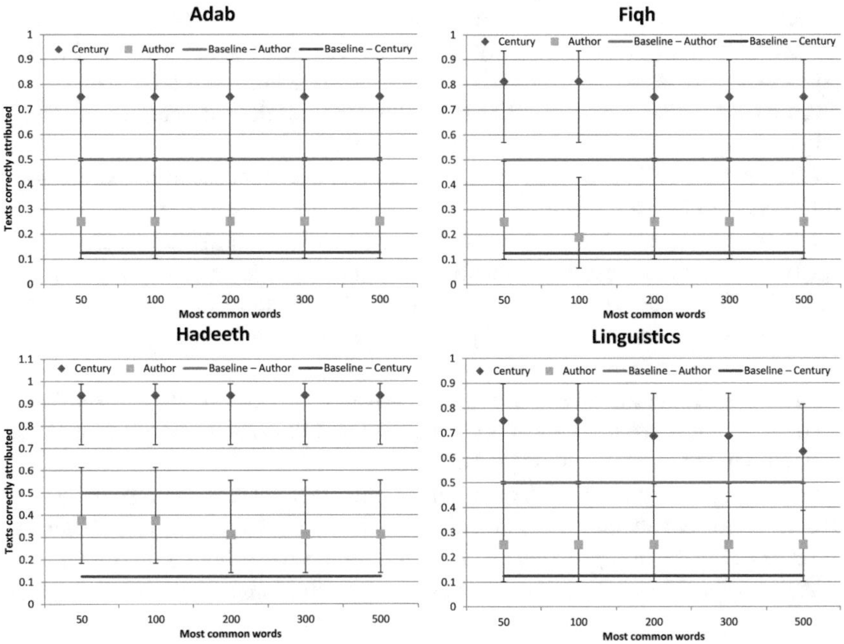

Fig. 8 The relation of the number of correctly attributed texts on the number of most common words

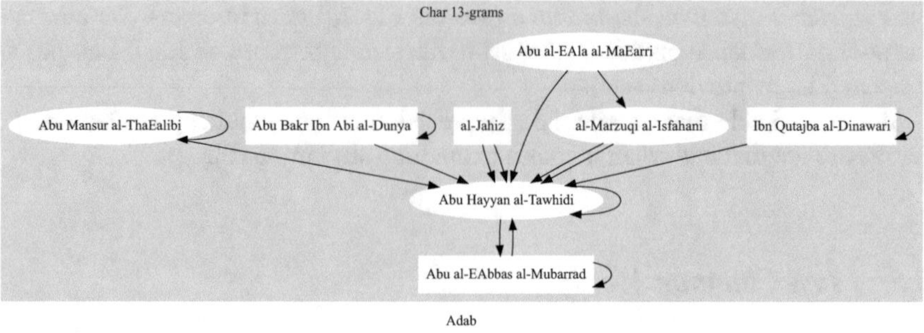

Fig. 9 The graph of the relationships between Adab authors according to the analysis driven by character 13-grams. The graph of the relationships between Adab authors according to the analysis driven by word 3-grams looks identically

6 Analysis of Errors

The results described in the previous section show that there is a high rate of wrongly attributed texts. But are the errors purely random? Or do they reveal the discourse structure of the corpus?

Various event drivers result in similar relationships between authors. For instance the character 13-grams driven analysis of the Adab genre yields exactly the same results as the word 3-grams analysis (Fig. 9).[2] The exceptions are the word length driven analyses that typically differ from the content-oriented analyses (character and word n-grams, most frequent words) (Fig. 10).

Also striking is that the graphs tend to contain one high centrality vertex—the prototypical author of the genre.

The central author of the Adab is abū Ḥayyān at-Tawḥīdī (as can be seen on the Fig. 9); he is not the most famous author—but this impression can be explained by the the fact that we tend to determine the "fame" of the author from our contemporary point of view. Brockelmann [8, 246] categorizes him as an encyclopaedist and his work encloses vast number of topics.

Similarly, the central author of Fiqh is not really well known al-Qāḍī abū Yaʕ lā and only for some event drivers it is al-Buḫārī (Fig. 11)—one of the most influential personalities of the Islamic East [8, 141–142].

In this graph we can see also other remarkable phenomenon: both texts by ibn ʔ Idrīs al-Qurašī are attributed to Aḥmad al-Bayhaqī and vice versa. This means that *Bayān Ḫaṭaʔ man ʔAḫṭaʔa ʕalā aš-Šāfiʕī* by Aḥmad al-Bayhaqī is dissimilar to his *ʔAḥkām al-Qurʔān li-š-Šāfiʕī* and at the same time *al-ʔUmm* by ibn ʔIdrīs al-Qurašī

[2]The vertices in boxes represent authors who lived in the 9th century while the vertices in ellipses represent authors from the 11th century. The edges are directed towards authors who the texts are attributed to—for example the graph in the Fig. 9 shows that both texts by al-Ǧāḥiẓ are attributed to abū Ḥayyān at-Tawḥīdī. In fact the results are typically less similar than this example, for more detailed view please download the complete set of graphs from http://www.milicka.cz/kestazeni/aaa.zip.

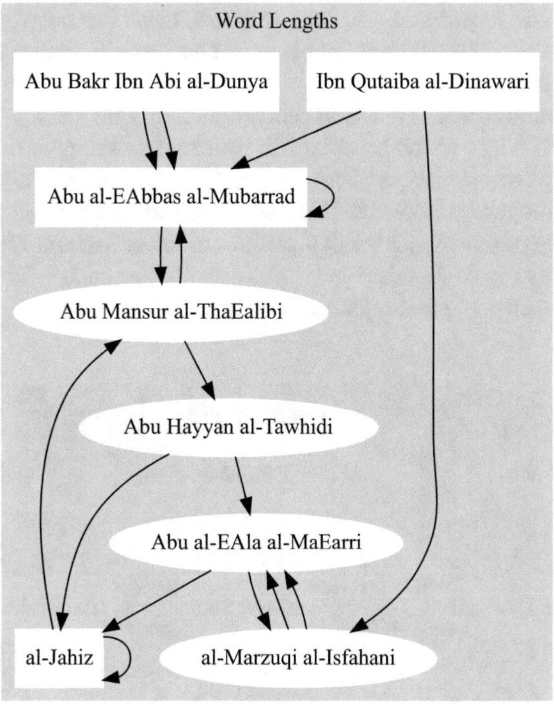

Adab

Fig. 10 The graph of the relationships between Adab authors according to the analysis driven by word lengths

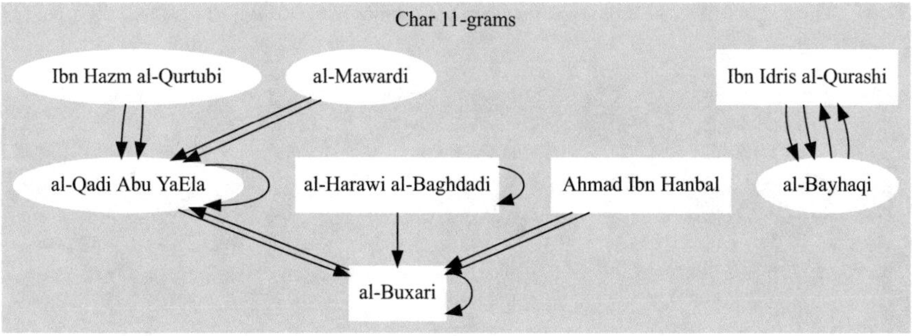

Fiqh

Fig. 11 The graph of the relationships between Fiqh authors according to the analysis driven by caracter 11-grams is dissimilar to his *ʔAḥkām al-Qurʔān*, while *al-ʔUmm* is similar to *Bayān Ḫaṭaʔ man ʔAḫṭaʔa ʕalā aš-Šāfiʕī* and *ʔAḥkām al-Qurʔān* are similar to *ʔAḥkām al-Qurʔān li-š-Šāfiʕī*.

The most central author of the ḥadīṯ genre is ibn ʔIdrīs al-Qurašī. At some point he is also central for Fiqh, which accords with Brockelman's remark that he is viewed as the founder of legal sciences (fiqh), and that he was a mediator between Mālikī and Hanafī theological schools [8, 163–164]. Interesting is that (in both cases) the degree centrality of this author is not that high for lower *n*-grams (Fig. 12) and that it steadilly grows (up to 13 edges out of 16 possible for ḥadīṯ—for both character 19-grams and word 4-, 5- and 6-grams—see Fig. 13).

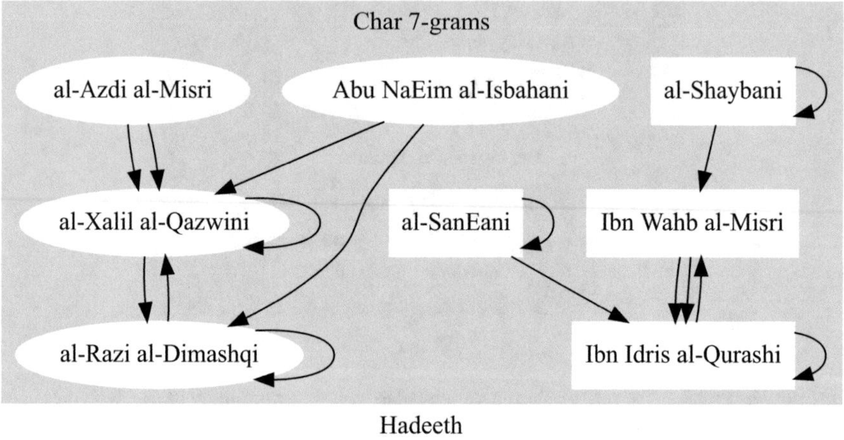

Fig. 12 The graph of the relationships between ḥadīṯ authors according to the analysis driven by character 7-grams

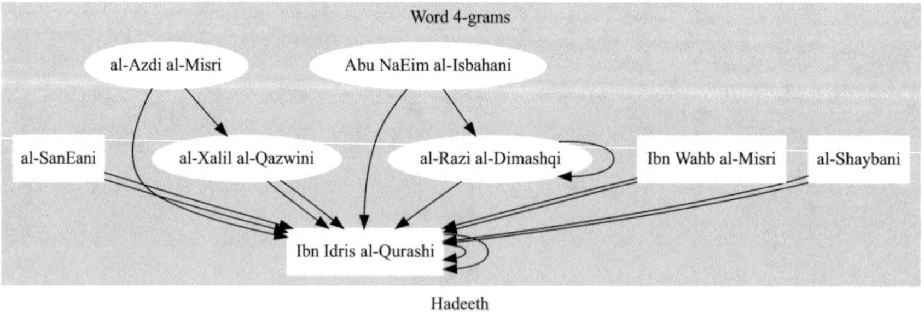

Fig. 13 The graph of the relationships between ḥadīṯ authors according to the analysis driven by word 4-grams

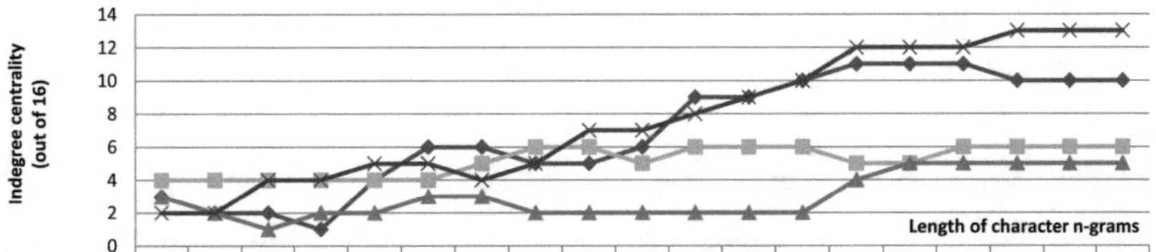

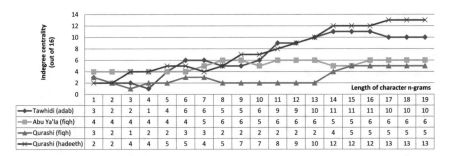

	1	2	3	4	5	6	7	8	9	10	11	12	13	14	15	16	17	18	19
Tawhidi (adab)	3	2	2	1	4	6	6	5	5	6	9	9	10	11	11	11	10	10	10
Abu Ya'la (fiqh)	4	4	4	4	4	4	5	6	6	5	6	6	6	5	5	6	6	6	6
Qurashi (fiqh)	3	2	1	2	2	3	3	2	2	2	2	2	2	4	5	5	5	5	5
Qurashi (hadeeth)	2	2	4	4	5	5	4	5	7	7	8	9	10	12	12	12	13	13	13

Fig. 14 The indegree centrality of the most central authors in adab, fiqh and ḥadīṯ (dependency on the event driver)

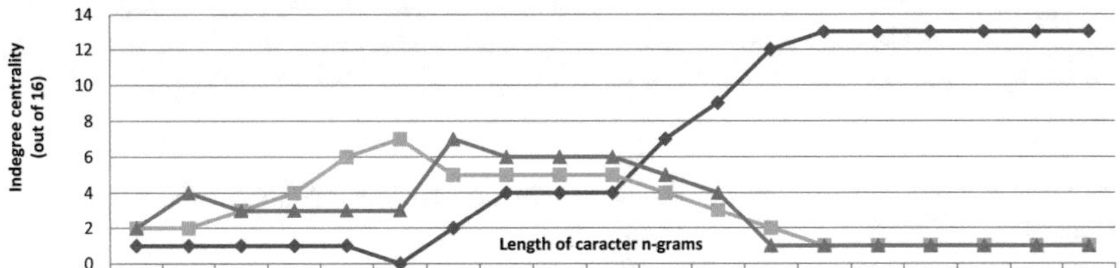

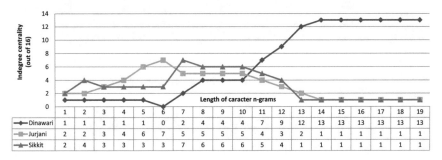

	1	2	3	4	5	6	7	8	9	10	11	12	13	14	15	16	17	18	19
Dinawari	1	1	1	1	1	0	2	4	4	4	7	9	12	13	13	13	13	13	13
Jurjani	2	2	3	4	6	7	5	5	5	5	4	3	2	1	1	1	1	1	1
Sikkit	2	4	3	3	3	3	7	6	6	6	5	4	1	1	1	1	1	1	1

Fig. 15 The indegree centrality of the three most central authors in linguistics (dependency on the event driver)

As we can observe analogous growth of the indegree centrality of the central node in all genres, the phenomena deserves closer look. We have no place to elaborate the topic here but as an illustration see the chart in Fig. 14.

The chart suggests that the longer *n*-grams are nearly useless for authorship attribution as they tend to attribute nearly all texts to one author.

The Linguistic is represented by ibn as-Sikkīt, ʕ Abdalqādir al-Ǧurǧānī and ibn Quṭayba ad-Dīnawarī in the chart (Fig. 15) which is in accordance with the fact that we can observe two centres of the genre as illustrated in the graph on Fig. 16 and that the two centres merge into one (for character 11-grams and longer) (Fig. 17) so that ibn Quṭayba ad-Dīnawarī stays central for higher character *n*-grams and word *n*-grams. The explanation for the bicentrality seems to be self-evident: for both hadīṯ and linguistics the two centers associate authors from the same century.

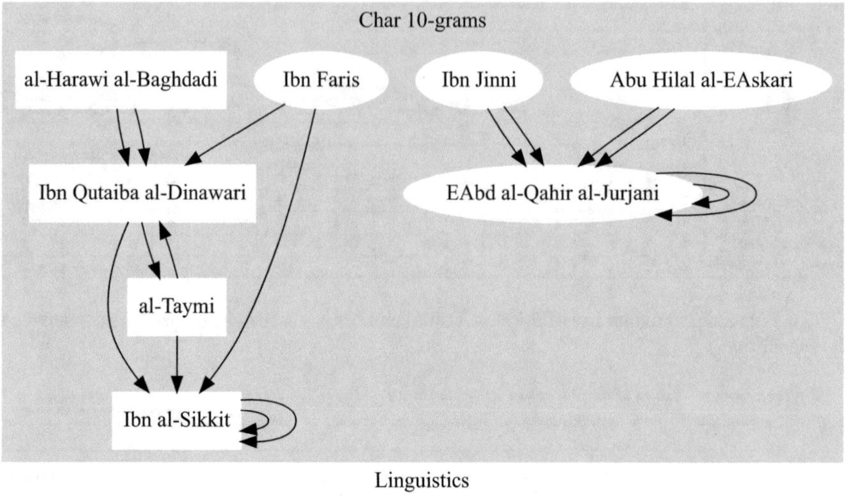

Fig. 16 The graph of the relationships between linguistics authors according to the analysis driven by character 10-grams

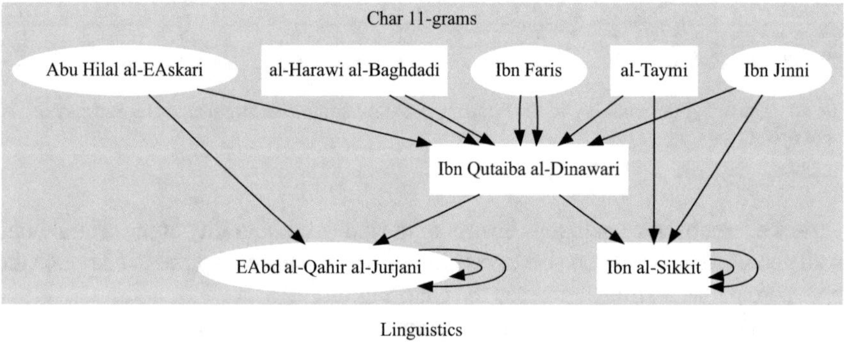

Fig. 17 The graph of the relationships between linguistics authors according to the analysis driven by character 11-grams

7 Future Work and Conclusions

This paper has demonstrated, first, that authorship attribution of Arabic texts using the standard JGAAP framework is possible, and that, by extension, large-scale analysis such as [21, 23, 44] are can and should be performed to determine best practices specifically for Arabic. Similarly, we have demonstrated significant differences in performance between different genres of Arabic writing, suggesting a need for similar experiments to determine genre-specific best practices. The CLAUDia corpus provides a high-quality dataset that should be invaluable for this task.

We have also established, via error analysis, that the attribution graphs tend to contain one high centrality vertex that seems to relate to a key founding or prototypical author. It is not clear if this finding would hold in other languages, and we look forward to investigating this question.

Acknowledgements The research reflected in this article has been supported by the GAČR (Czech Science Foundation), project no. 13-28220S.

References

1. Abbasi, A., Chen, H.: Applying Authorship Analysis to Arabic Web Content, pp. 183–197. Springer, Berlin (2005)
2. Alwajeeh, A., Al-Ayyoub, M., Hmeidi, I.: On authorship authentication of arabic articles. In: 2014 5th International Conference on Information and Communication Systems (ICICS), pp. 1–6. IEEE (2014)
3. Argamon, S., Koppel, M., Pennebaker, J.W., Schler, J.: Automatically profiling the author of an anonymous text. Commun. ACM **52**(2), 119–123 (2009)
4. Ayadi, R., Maraoui, M., Zrigui, M.: Intertextual distance for arabic texts classification. In: International Conference for Internet Technology and Secured Transactions, 2009. ICITST 2009, pp. 1–6. IEEE (2009)
5. Binongo, J.N.G.: Who wrote the 15th book of Oz? an application of multivariate analysis to authorship attribution. Chance **16**(2), 9–17 (2003)
6. Brooks, R., Flyn, C., Rowling, J.K.: The cuckoo in crime novel nest. Sunday Times, 14 July 2013
7. Burrows, J.F.: 'an ocean where each kind...' : statistical analysis and some major determinants of literary style. Comput. Humanit. **23**(4–5), 309–21 (1989)
8. Brockelmann, L.J.C.: History of the Arabic Written Tradition. Volume 1, of Handbook of Oriental Studies. Section 1 the Near and Middle East, lam blg edn. Brill (2017)
9. Chaski, C.E.: Who's at the keyboard: Authorship attribution in digital evidence invesigations. Int. J. Digit. Evid. **4**(1), n/a (2005). Electronic-only journal. http://www.ijde.org (2007). Accessed 31 May 2007
10. Collins, P.: Poe's debut, Hidden in Plain Sight. The New Yorker, Oct 2013
11. Coulthard, M.: On admissible linguistic evidence. J. Law Policy **XXI**(2), 441–466 (2013)
12. de Morgan, A.: Letter to Rev. Heald 18/08/1851. In: De Morgan, S.E. (ed.) Memoirs of Augustus de Morgan by his wife Sophia Elizabeth de Morgan with Selections from his Letters (1851/1882)
13. Fobare-DePonio, D., Koubek, K.: In: Lupu, M., Mayer, K., Tait, J., Trippe, A.J., Alberts, D., Yang, C.B. (eds.) Current Challenges in Patent Information Retrieval. The Information Retrieval Series 29, 1 edn. Springer, Berlin (2011)
14. Grant, T.: Txt 4n6: describing and measuring consistency and distinctiveness in the analysis of SMS text messages. J. Law Policy **XXI**(2), 467–494 (2013)
15. Grieve, J.W.: Quantitative authorship attribution: a history and an evaluation of techniques. Master's thesis, Simon Fraser University, 2005. http://hdl.handle.net/1892/2055 (2007). Accessed 31 May 2007
16. Hoover, D.L.: Delta prime? Lit. Linguist. Comput. **19**(4), 477–495 (2004)
17. Joachims, T.: Learning to Classify Text Using Support Vector Machines. Kluwer (2002)
18. Jockers, M.L., Witten, D.M.: A comparative study of machine learning methods for authorship attribution. LLC **25**(2), 215–23 (2010)
19. Juola, P.: Authorship attribution. Found. Trends Inf. Retr. **1**(3) (2006)

20. Juola, P.: Authorship attribution : what mixture-of-experts says we don't yet know. In: Proceedings of American Association for Corpus Linguistics 2008, Provo, UT USA, Mar 2008
21. Juola, P.: 20,000 ways not to do authorship attribution and a few that work. In: Proceedings of 2009 Biennial Conference of the International Association of Forensic Linguists (IAFL-09), Amsterdam (2009)
22. Juola, P.: Fishing the ocean: lessons from large-scale experiments in styometry. Talk given at School for Advanced Studies, 29 Mar 2011
23. Juola, P.: Large-scale experiments in authorship attribution. Engl. Stud. **93**(3), 275–283 (2012)
24. Juola, P.: Stylometry and immigration: a case study. J. Law Policy **XXI**(2), 287–298 (2013)
25. Juola, P., Noecker Jr., J., Ryan, M., Speer, S.: Jgaap 4.0—a revised authorship attribution tool. In: Proceedings of Digital Humanities 2009, College Park, MD (2009)
26. Juola, P., Noecker Jr., J.I., Stolerman, A., Ryan, M.V., Brennan, P., Greenstadt, R.: Keyboard behavior-based authentication for security. IT Prof. **15**, 8–11 (2013)
27. Juola, P., Vescovi, D.: Empirical evaluation of authorship obfuscation using JGAAP. In: Proceedings of the Third Workshop on Artificial Intelligence and Security, Chicago, IL USA, Oct 2010
28. Juola, P., Vescovi, D.: Authorship attribution for electronic documents. In: Petersen, G., Shenoi, S. (eds) Advances in Digital Forensics VII, International Federal for Information Processing, chapter 9, pp. 115–129. Springer, Boston (2011)
29. Koppel, M., Argamon, S., Shimoni, A.R.: Automatically categorizing written texts by author gender. Lit. Linguist. Comput. **17**(4), 401–412 (2002). https://doi.org/10.1093/llc/17.4.401
30. Koppel, M., Schler, J.N., Argamon, S.: Computational methods in authorship attribution. J. Am. Soc. Inf. Sci. Technol. **60**(1), 9–26 (2009)
31. McMenamin, G.: Declaration of Gerald McMenamin (2011). http://www.scribd.com/doc/67951469/Expert-Report-Gerald-McMenamin
32. Mikros, G.K., Perifanos, K.: Authorship attribution in greek tweets using multilevel author's n-gram profiles. Papers from the 2013 AAAI Spring Symposium Analyzing Microtext, 25–27 Mar 2013, Stanford, CA, pp. 17–23. AAAI Press, Palo Alto, CA (2013)
33. Mosteller, F., Wallace, D.L.: Inference and Disputed Authorship: the Federalist. Wesley, Reading, MA (1964)
34. Noecker Jr., J., Patrick Juola. Cosine distance nearest-neighbor classification for authorship attribution. In: Proceedings of Digital Humanities 2009, College Park, MD, June 2009
35. Noecker Jr., J.I., Juola, P.: Stylometric identification of manic-depressive illness. In: Proceedings of DHCS 2014 (2014)
36. Ouamour, S., Sayoud, H.: Authorship attribution of ancient texts written by ten Arabic travelers using a smo-svm classifier. In: 2012 International Conference on Communications and Information Technology (ICCIT), pp. 44–47. IEEE (2012)
37. Ouamour, S., Sayoud, H.: Authorship attribution of short historical Arabic texts based on lexical features. In: 2013 International Conference on Cyber-Enabled Distributed Computing and Knowledge Discovery (CyberC), pp. 144–147. IEEE (2013)
38. Pennebaker, J.W., King, L.A.: Linguistic styles: language use as an individual difference. J. Personal. Soc. Psychol. **77**, 1296–1312 (1999)
39. Quinlan, J.R.: C4.5: Programs for Machine Learning. Morgan Kauffman (1993)
40. Rudman, J.: On determining a valid text for non-traditional authorship attribution studies: Editing, unediting, and de-editing. In: Proceedings of the Joint International Conference of the Association for Computers and the Humanities and the Association for Literary and Linguistic Computing (ACH/ALLC 2003). GA, May, Athens (2003)
41. Sarawgi, R., Gajulapalli, K., Choi, Y.: Gender attribution: tracing stylometric evidence beyond topic and genre. In: Proceedings of the Fifteenth Conference on Computational Natural Language Learning, CoNLL '11, pp. 78–86. Association for Computational Linguistics, Stroudsburg, PA, USA (2011)
42. Stamatatos, E.: A survey of modern authorship attribution methods. J. Am. Soc. Inf. Sci. Technol. **60**(3), 538–56 (2009)

43. Stamatatos, E.: On the robustness of authorship attribution based on character n-gram features. J. Law Policy **XXI**(2), 420–440 (2013)
44. Vescovi, D.M.: Best practices in authorship attribution of English essays. Master's thesis, Duquesne University (2011)
45. Wallis, S.: Binomial confidence intervals and contingency tests: mathematical fundamentals and the evaluation of alternative methods. J. Quant. Linguist. **20**(3), 178–208 (2013)

Automatic Text Classification Using Neural Network and Statistical Approaches

Tarek ElGhazaly

Abstract Automatic Classification is crucial for text retrieval, knowledge management, and decision making as it converts text from raw data to a real knowledge. In this paper, Text Automatic Classification has been introduced starting from improved methods for the preprocessing algorithms that are common to the different classifiers, then enhanced algorithms for two different classifiers: a Statistical one and a Neural Network based one. The preprocessing algorithm included words features extraction, stop words removal, and enhanced word stemming. For the statistical classifier, weighting techniques have been introduced to enhance the statistical classification concluding that the combination of the Term Frequence X Inverse Document Frequency (TFxIDF) and the Category Frequency (CF) gives the highest classification. For the neural network based classifier, a classification model has been proposed using an artificial neural network trained by the Back propagation learning algorithm. Due to the high dimensionality of the feature space typical for textual data, scalability is poor if the neural network is trained using this high dimensional raw data. In order to improve the scalability of the proposed model, four dimensionality reduction techniques have been proposed to reduce the feature space into an input space of much lower dimension for the neural network classifier. The first three of these techniques are domain dependent term selection methods: the Document Frequency (DF) method, the Category Frequency- Document Frequency (CF-DF) method and the TFxIDF method. The fourth technique is a domain independent feature extraction method based on a statistical multivariate data analysis technique which is the Principal Component Analysis (PCA) an this technique was the best as per the done experiments. The proposed classifiers have been tested through experiments conducted using a subset of the Reuters-21,758 test collection for text classification. Although this paper considered English as the language under research to make use of the standard Reuters-21,758, the proposed model could be used for other languages.

T. ElGhazaly (✉)
Institute of Statistical Studies and Research, Cairo University, Giza, Egypt
e-mail: tarek.elghazaly@cu.edu.eg

K. Shaalan et al. (eds.), *Intelligent Natural Language Processing: Trends and Applications*, Studies in Computational Intelligence 740,
https://doi.org/10.1007/978-3-319-67056-0_17

Keywords Automatic classification · Neural network classifier
Statistical classifier

1 Reviewing the Previous Work

Researches for different knowledge discovery from text ways have been conducted for a long time. This is including automatic classification as introduced by Adams in [1] and Schütze et al. in [2], named entities recognition as described by Shaalan in [3], automatic summarization for English and other languages like Arabic [4–7], and other tools. Although Automatic Classification has been addressed through statistical ways and neural network ways, this paper introduces new techniques for improving both ways and identifies where to use each.

2 Preprocessing Procedures for the Two Proposed Classifiers

2.1 Word Extraction

A text document can be viewed as a long stream of characters. For the purpose of feature identification, it is necessary to convert a text document from a long stream of characters into a stream of words or tokens, where a word or token is defined as a group of characters with collective significance [8]. In information retrieval, this text tokenization process is often called word extraction, word breaking, word segmentation, or lexical analysis.

Depending on the nature language the text document is written in, the word extraction process can involve very different techniques. This process is relatively easy in such languages such as English, where special word-delimiting characters such as spaces and punctuations mark boundaries between words. In this paper, English documents only will be taken into consideration.

2.2 Stop Words Removal

It is well recognized among the information retrieval community, that a set of functional English words (such as 'the', 'a', 'and', 'that', etc.) is useless as indexing terms. Salton and McGill [9] described these words as having very low discrimination value, since they occur in almost every English document, and therefore do not help in distinguishing between documents with contents that are about different topics.

For this reason, these words are not useful in the text categorization task and thus should be removed from the set of words produced by word extraction.

The process of removing the set of non-content-bearing functional words from the set of words being produced by word extraction is called stop words removal, and the functional words being removed are called stop words. In order to remove the stop words, a semi-automatic procedure is followed. This involves first creating a list of stop words to be removed, which is called stop list or a negative dictionary. After this, the set of words produced by word extraction then scanned so that every word appearing in the stop list is removed.

2.3 Word Stemming

In a text document, a word may exist in a number of morphological variants. For example, the word “compute” may also exist in its other morphological variants such as “computing”, “computed”, “computational”, or “computer”. While these morphological variants are different word forms, they represent the same concept. For information retrieval tasks, including text categorization, it is generally desirable to combine those morphological variants of the same word into one canonical form. In information retrieval, the process of combining or fusing together different morphological variants of the same word into a single canonical form is called stemming. The canonical form produced by stemming is called a stem. The module performing is sometimes called a stemmer.

These are various approaches for stemming. The easiest way is to build a translation table with two columns, corresponding to the original word forms and their stems. Stemming is then carried out by looking up the translation table. In order to improve efficiency, indexing methods such as B-tree or hashing could be employed to build an index of table entries. However, one obvious drawback of this approach is that the translation table has to be built manually, which requires a lot of efforts and could be time-consuming. Another problem is that it is very difficult to build a translation table extensive enough to include every possible word that may exist in the documents. Some words, especially that are specialized words in practical domains, are likely to be missed.

Given the drawback of table based stemmer, it s then described to build stemmers which produce stems based on a set of predefined translation rules. This approach is commonly used in a class of stemmers called affix removal stemmer, in whom the stem is formed by removing suffixes and/or prefixes from the original word forms. Affix removal stemmers can be found in various proposals in the literature [10–13]. However, Porter Algorithm is used along with the proposed improvements.

2.4 Improvements

Irregular plurals

For the Porter Algorithm, it did not consider the irregular plurals. It depends only the normal derivation for example the abnormal plurals like: Children → Child, Men → Man, Women → Woman, and Oxen → Ox.

Irregular verbs

Also for the verbs, there are a lot of no-base derivations like: Gone, went → go, Forgotten, forgot → forget. So, and for the algorithm to be more effective, a list of the abnormal derivations is made and added into the algorithm.

2.5 Reuters 21,758 Test Collection for Text Categorization

Reuters 21,758 test collection is a test collection specifically designed for experiments with text categorization systems. Here, the main intention is to give a brief overview of this test collection. Readers who have further interest in the test collection can find a through discussion of the preparation and various characteristics of the collection in paper 8 of [14].

The test documents

As its name suggests, the Reuters-21,578 test collection contains a set of 21,578 documents. These documents are all full text articles distributed via the Reuters newswire during 1998s, which include reports of recent events, long feature stories, quotes on market prices, and corporate earning reports. The articles vary in length from single line text to multiple pages.

The pre-defined categories

There are a total of 689 categories defined in Reuters-21,578 collection. These categories are divided into 5 groups, namely TOPICS, ORGS, EXCHANGES, PLACES and PEOPLE. The TOPICS group corresponds to categories with economic interest; the ORGS group corresponds to categories concerning important regulatory, financial, and political organizations; the EXCHANGES group contains categories about stock and commodity exchanges; the PLACES group contains categories representing different countries of the world; the PEOPLE group consists of categories about important political and economic leaders. These categories were assigned to the set of Reuters articles by two Reuters journalists stationed at Carnigie Group, Inc.

2.6 Term Weighting Techniques

After word extraction, stop words removal and stemming, each text document is transformed into a set of stems corresponding to the set of words appearing in each document. The next step is to find the union of all these sets, such that the union set contains the set of stems corresponding to all of the words appearing in the given set of documents. Duplicates are removed s that each stem is unique within the union set. In information retrieval, this set of stems constitutes the set of indexing terms for the set of documents. This set of indexing terms is called the indexing vocabulary.

In order to set the set of initial features from the indexing vocabulary, measurements must be made for each term in the vocabulary according to the importance of the term of the documents. This involves assigning to each term a weight indicating the relative importance of the term in a document. This process of assigning a weight to each indexing term is commonly known as *term weighting* in information retrieval [15]. With the set of indexing terms in the indexing vocabulary and their corresponding term weights, a document can be represented as the following feature vector:

$$D_j = < w_{j1}, w_{j2}, \ldots, w_{jk} >$$

where K is the number of indexing terms in the vocabulary, or the vocabulary size, and w_{ji} is the term weight of the ith indexing term in the document j.

There are many different approaches for determining the term weights. The simplest method involves a binary term weight. That is, term weight (w_{ji}) is equal to 0 if the ith indexing term is not in documents j and 1 otherwise. In this case, the feature vector representing a document is a bit vector with each bit corresponding to the term weight of each indexing term for the document. This binary representation for documents has been used quite commonly in learned information retrieval tasks [11].

Another approach makes use of the *term occurrence frequencies* information. In this approach, the assumption is that a term that is mentioned more frequently in a document carries more weight in representing the topic of that document, and thus is more important and should be assigned a higher term weight. In this case the term weight (w_{ji}) is a positive integer equal to the term frequency (TF) of the ith indexing term in document j; which is the number of times that the ith indexing term appears in document j.

The problem with the previous approach is that terms with high term frequencies may not always help in distinguishing between documents with different topics, especially when these terms appear frequently in almost all documents in the document set. In order to reflect the *discrimination value* of each term in the term weight, a factor called *inverse document frequency* (IDF) is introduced. A commonly used definition of the IDF is as follows:

$$IDF_i = log\frac{N}{n}$$

where IDF_i is the inverse document frequency of the ith indexing term, N is the number of documents in the document set, and n is the number of documents in which the ith indexing term appears. By this definition, a term that appears in fewer documents will have higher IDF. The assumption behind this definition is that terms that are concentrated in a few documents are more helpful in distinguishing between documents with different topics. For term weighting, the term weight is equal to the product of term frequency (TF) and the inverse document frequency (IDF):

$$W_{ji} = Tf_{ji} \times IDF$$

In this case, terms that appear frequently in a few documents are given higher weights. An extensive survey of the various term-weighting approaches used in text retrieval systems is given in [15].

3 The Proposed Statistical Classifier

The goal of this part is to build an automatic classifier based on statistical clustering. The main issue of this classifier is to be totally automated i.e. no human intervention will occur in any part of this classifier even in choosing the related terms for each category. In the next sections, the different stages for building the automatic statistical classifier will be discussed.

3.1 Converting the Text Documents into a Database

For formatting the test documents as database tables, a SGML parser has been built to parse the Reuters-21,758. its large files were parsed, stop words were removed, words were stemmed and then text of each document was extracted as a text file called Doc_ID.txt. Where Doc_ID is the Document ID in the result database. Then, words of these documents and related categories were extracted and inserted in their related tables in the database.

3.2 The Resulting Database Model

Figure 1 introduces the database model for words and documents after extracting them from the Reuters-21,758 test collection.

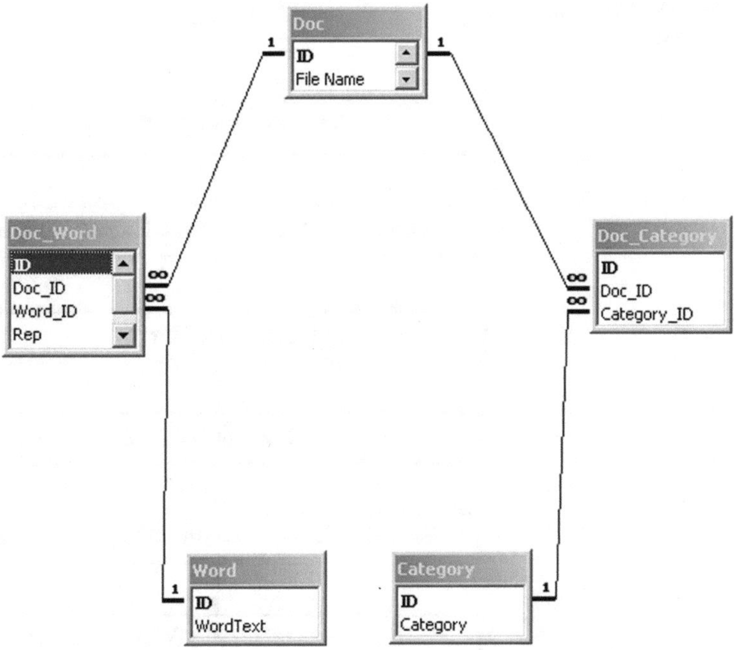

Fig. 1 Database model for the Reuters test set

Figure 2 describes the all the process for building the statistical classifier. For concluding which terms are related to which category, the main goal was to build this relation with an automatic method without human intervention.

So, all the words that are belong to a document are considered with certain categories as these words are related to these categories. This means, if the word w is inside the document d, and the document d is belongs to the category c; then the word w belongs to the category c. Figure 3 describes how the statistical classifier works.

3.3 Weighting Techniques

The weight of each category (from 5 chosen categories) is considered for a document as the sum of repetition of the related words multiplied by their TF_IDF value i.e.:

$$C_k = \sum TF\_IDF\left(w_{ij}\right)\ XRep\left(w_{ij}\right)$$

where Ck is the weight of the category k for the jth document, and wij is the word i which located in the document j and TF_IDF(wij) is its TF_IDF value.

Fig. 2 Building data for a statistical classifier

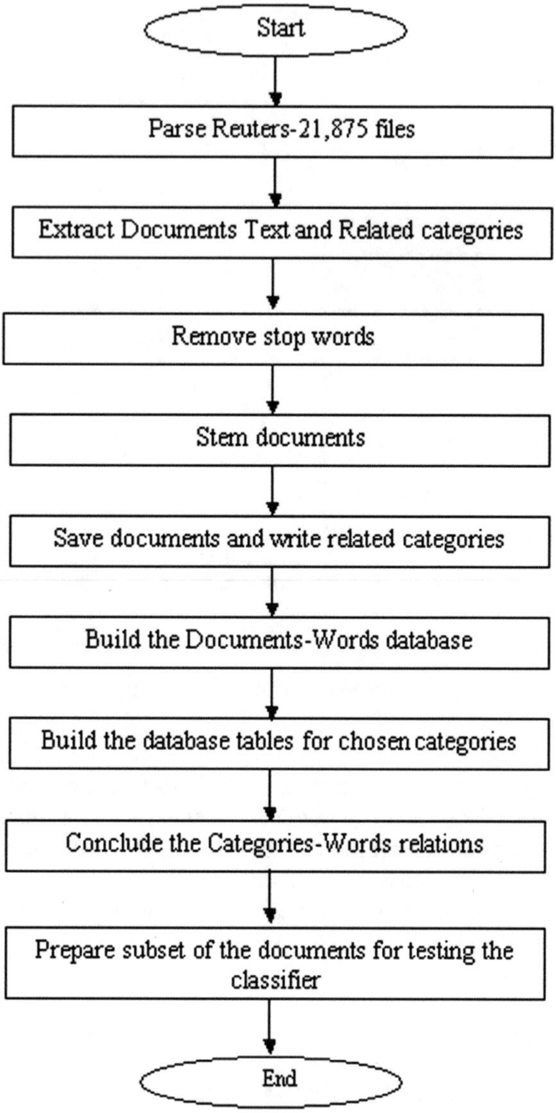

3.4 Improvements for Weighting

However, there will be a problem if there is a word that is located in some documents that are related to more than one category. In this case, there may be a document that will be classified wrongly for a category that it is not belongs to. For example, the word "Network" can be considered for many categories like "Computer", "Communication", and "Mafia". The problem here is that the classifier will

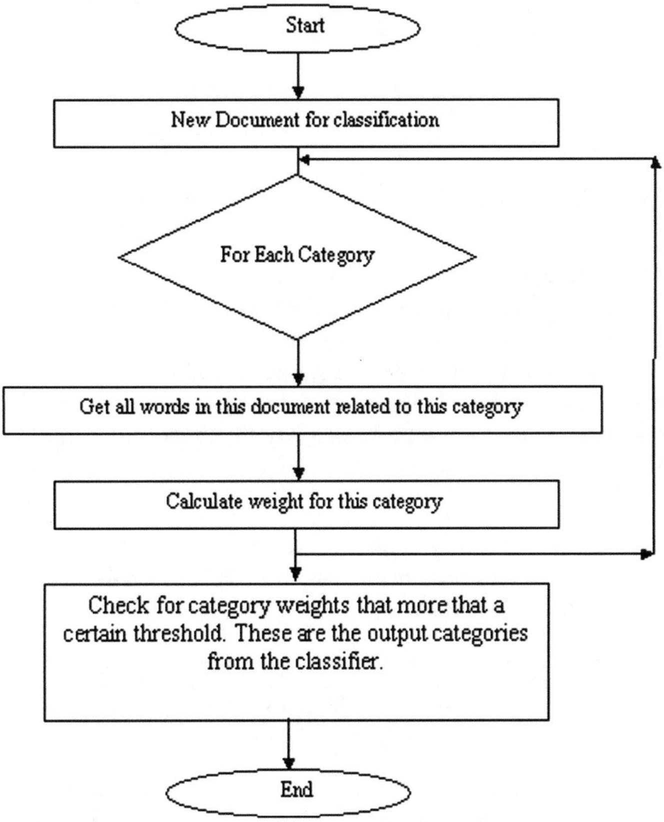

Fig. 3 How the statistical classifier works

classify any document that contains the word "Network" to all its categories including "Mafia" which is not correct in most cases. So, there were two solutions. First, removing all the words that belong to more than one category from the discrimination words. The second solution is to update the above weighting equation as:

$$C_k = \sum \frac{TF\_IDF\left(w_{ij}\right) XRep\left(w_{ij}\right)}{\mathrm{CF}\left(\mathrm{w_{ij}}\right)}$$

where CF(wij) is the number of categories, which this word (wij) belongs to

Table 1 Results of statistical based classifier

Weighting technique	Precision (%)
TF_IDF	5
Words removal (multi-category)	75
TF_IDF and CF	90

3.5 Experimental Details

A Set of 481 documents are used as a subset from the Reuters-21,758 test set. The first 200 documents from them were used for the classifier to relate the words with their categories. Then, different 200 documents were used for the statistical classifier as test documents. The experiments were made with different weighting techniques described in the above point. A threshold has been specified for the categories weights: 1 to consider this category for this document, otherwise 0 is considered.

Precision
Precision measures the accuracy of the retrieval result as indicated by the proportion of the number of categories that are correctly measured to the total no of categories for these documents. It is defined as:

$$\text{Precision} = \frac{\text{Number of categories that are correctly measured}}{\text{Total no of categories for tested documents}}$$

Results
The above results explain clearly that it was not sufficient to use only the TF_IDF method for weighting if the Category-words are extract automatically. Also, it is clear that using the TF_IDF and CF techniques together are very effective for discrimination between categories (Table 1).

4 Neural Network Based Classifier

In this section, a model for text categorization based on a non-linear feed-forward neural network trained by Backpropagation learning rule, will be presented. In the proposed model, the text categorization problem is formulated as a classification problem and the neural network as a classifier created by supervised learning. In order to improve the scalability of the model, various dimensionality reduction techniques will be introduced to reduce the high dimensionality of the raw data before feeding them to the neural network. These techniques in details in this paper will be described.

4.1 Dimensionality Reduction for Text Categorization

The main difficulty in the application of neural network to text categorization is the high dimensionality of the input feature space typical for textual data. This is because each unique term in the vocabulary represents one dimension in the feature space. For a typical document collection, there are thousands or even tens of thousands of unique terms in the vocabulary.

Because of the high dimensionality in the feature space, the feature vectors are not suitable as input to the text classifier since the scalability will be poor. In order to improve the scalability of the text categorization system, dimensionality reduction techniques should be employed to reduce the dimensionality of the feature vectors before they are fed as input to the text classifier. In this paper, four dimensionality reduction techniques applicable to text categorization are studied.

In order to improve scalability of the text classifier, four dimensionality reduction techniques, namely the DF method, the CF-DF method, the TFxIDF method and Principal Component Analysis (PCA) [16], will be introduced to reduce the feature space. The aim of the techniques is to minimize information loss while maximizing reduction in dimensionality.

The DF method

Given a set of training documents together with a specification of which of the pre-defined categories each training document belongs to, the DF method reduces the vocabulary size by term selection based on a local term ranking technique. The categorization information is used for grouping the training documents such that all the documents belonging to the same category are put into the same group.

When there are overlaps between the categories, a document may belong to more than one group. After the documents are grouped, then, groups of the indexing terms can be formed in the vocabulary by putting in a group all terms contained in documents belonging to the same category. This process results in a set of sub-vocabularies corresponding to each category.

In the DF method, terms are ranked based on the document frequency (DF) of each term within a document group. For each document group, the document frequency of a term is defined as the number of documents within that particular group containing the term. By choosing the document frequency as the importance measure, the important terms are being assumed as those appear frequently within a group of documents belonging to the same category.

This is because the set of terms, which are good representatives of the category topics, should be used by most documents belonging to that category. Based on the DF importance measure, terms are ranked separately within each sub-vocabulary. For term selection, a parameter d_ is defined such that within each sub-vocabulary, only the most important d_ terms with the highest ranks are selected. The sets of selected terms from each sub-vocabulary are then merged together to form the reduced feature set. By adjusting the selection parameter d, the dimensionality of the reduced feature vectors can be controlled. A smaller d will result in fewer terms being selected, thus higher reduction in dimensionality.

The CF-DF method

From the discussion of the DF method, terms that appear in most documents within the whole training set are observed that they always have a high within-group document frequency. Even though these frequently occurring terms are of very low discrimination value, and thus not helpful in distinguishing between documents belonging to different categories, hey are likely to be selected by the DF method. The CF-DF method alleviates the problem by considering the discrimination value of a term in the term selection process.

In the CF-DF method, a quantity called category frequency (CF) is introduced. To determine the category frequency of a term, the training documents are grouped according to the categorization information, as in the DF method. For any document group, it could be considered that a term appears in that group if at least one of the documents in that group contains that term. For any term in the vocabulary, the category frequency is equal to the number of groups that the term appears in.

By this definition, terms that are concentrated in a few categories will have a low category frequency, while those that are distributed across a large number of categories will have a high category frequency. The idea is that the discrimination value of a term can be measured as the inverse of its category frequency. In other words, terms that are good discriminators are most likely concentrated in a few categories, and should be considered more important as they are helpful in distinguishing between documents belonging to different categories.

In the CF-DF method, a two-phase process is used for term selection. In the first selection phase, a threshold t is defined on the category frequencies of the terms, such that a term is selected only if its category frequency is lower than the threshold t. In the second selection phase, the DF method is applied for further term selection to produce the reduced feature set.

The TFxIDF method

In the DF method and the CF-DF method, the essential idea is to perform ranking of the terms in the vocabulary based on some importance measure, such that the most important terms can be selected. In both of these methods, the key to minimize information loss as a result of term selection is to define a good importance measure so as to avoid filtering out terms that are useful for the text categorization task. A good measurement of the importance of a term in a document set is the product of the term occurrence frequency (TF) and the inverse document frequency (IDF). The inverse document frequency of the ith term is commonly defined as [17]:

$$IDF_i = log \frac{N}{n}$$

where N is the number of documents in the document set, and n is the number of documents in which the ith term appears. By this definition, a term that appears in fewer documents will have a higher IDF. The assumption behind this definition is that terms that are concentrated in a few documents are more helpful in distinguishing between documents with different topics. In order to examine the effectiveness of this measure for term selection, it is proposed to use the TFxIDF value to measure the importance of a term for term selection. In other words, the terms are ranked according to their TFxIDF values, and a parameter d is set such that only the d terms with the highest TFxIDF values are selected to form the reduced feature set.

Principal Component Analysis

Principal component analysis (PCA) [16] is a statistical technique for dimensionality reduction, which aims at minimizing the loss in variance in the original data. It can be viewed as a domain independent technique for feature extraction, which is applicable to a wide variety of data. This is in contrast with the other three dimensionality reduction techniques, discussed before, which are domain specific feature selection techniques based on feature importance measures defined specifically for textual data.

In order to perform principal component analysis on the set of training documents, the set of feature vectors is represented by an n dimensional random vector (x):

$$x = \langle x_1, x_2, \ldots, x_n \rangle$$

where n is the vocabulary size, and the ith random variable in x(xi) takes on values from the term frequencies of the ith term in the documents. It could be founded now a set of n-dimensional orthogonal unit vectors, u1, u2, …, un, to an orthogonal basis for the n-dimensional feature space. Projections of x(ai) are formed onto the set of unit vectors:

$$a_i = x^T u_i$$

In doing so, a coordinate transformation in the feature space is performed, such that the unit vectors () form the axes of the new coordinate system and transform the original random vector_ into a new random vector a with respect to the new coordinate system: u_1, u_2, …, u_n

$$a = \langle a_1, a_2, \ldots, a_n \rangle$$

In principal component analysis, the choice of the unit vectors (u_1, u_2, …, u_n) is such that the projections (a_i) are uncorrelated with each other. Moreover, if the variance of a_i is denoted by λ_i, for i = 1, 2, …, n, then the following condition is satisfied:

$$\lambda_1 > \lambda_2 > \cdots > \lambda_n$$

In other words, the projections ai contain decreasing variance, these projections ai are called the principal components. It can be shown that the variance (λ1, λ2, …, λn) corresponds to the eigen values of the data covariance matrix R arranged in descending order, and the unit vectors (u1, u2, …, un) are the corresponding eigenvectors of R. In order to reduce the dimensionality of the feature space from n to p where p < n while minimizing the loss in data variance, a reduced feature space is formed by taking the first p dimensions with the largest variance. In this case, the reduced feature vectors of the documents are represented by the p dimensional random vector:

$$a_p = \langle a_1, a_2, \ldots, a_p \rangle$$

4.2 The Proposed Neural Network Based Text Classifier

By means of dimensionality reduction techniques, the set of documents to be categorized is transformed into a set of feature vectors in a relatively low dimensional feature space. This set of reduced feature vectors is then fed to the text classifier as input. In this paper, a 3-layer feed-forward neural network is used as the text classifier.

The neural network employed in this paper is a 3-layer fully connected feed-forward network, which consists of an input layer, a hidden layer, and an output layer. All neurons in the neural network are non-linear units with the sigmoid function as the activation function. In the input layer, the number of input units (T) is equal to the dimensionality of the reduced feature space. In the output layer, the number of output units (m) is equal to the number of pre-defined categories in the particular text categorization task. The number of hidden units in the neural network affects the generalization performance. The choice depends on the size of the training set and the complexity of the classification task the network is trying to learn, and can be found empirically based on the categorization performance.

For classification of the documents, reduced feature vectors representing the documents are fed to the input layer of the neural network classifier as input signals. These input signals are then propagated forward through the neural network so that the output of the neural network is computed in the output layer. As the sigmoid function is used as the activation function in the output units, the output of the neural network classifier is a real-valued classification vector with component values in the range [0, 1]. The classification vector represents a graded classification decision, in which the ith vector component indicates the relevance of the input document to the ith category. If binary classification is desired, a threshold can be set such that a document is considered to be belonging to the ith category only if the ith component of the classification vector is greater than the threshold.

The neural network classifier must be trained before it can be used for text categorization. Training of the neural network classifier is done by the Backpropagation learning rule based on supervised learning [18]. In order to train the neural network, a set of training documents and a specification of the pre-defined categories the documents belong to are required. More precisely, each training example is an input-output pair:

$$T_i = (D_i, C_i)$$

where Di is the reduced feature vector of the ith training document, and Ci is the desired classification vector corresponding to Di. The component values of Ci are determined based on the categorization information provided in the training set. During training, the connection weights of the neural network are initialized to some random values. The training examples in the training set are then presented to the neural network classifier in random order, and the connection weights are adjusted according to the Backpropagation learning rule. This process is repeated until the learning error falls below a pre-defined tolerance level.

4.3 Experimental Details

For performance evaluation, a subset of the documents from the Reuter-21,758 test collection is used for training and testing our text categorization model. A set of 200 documents is used as the training set for the Backpropagation Neural Network based text classifier. For testing, a test set of 200 documents was used. The set of test documents was kept unseen from the system during the training stage. In all experiments, 5 categories are used from the TOPICS group defined in the Reuters test collection.

The documents in the training set were processed by word extraction, stop words removal, and stemming. For word extraction, a word is defined as any consecutive character sequence contained in the character stream of a document, which starts with an alphabet, followed, by any number of alphabets or digits. The end of a word is delimited by non-alphanumeric character. Examples of acceptable words according to this definition include "john", "art", "db2", "b12", and "a300s". The difference between upper case and lower case characters in the words was removed by converting all upper case characters to lower case. After a set of words was extracted, stop words were removed based on a stop list for general English text. The remaining words were then stemmed using the Porter's stemming algorithm [12] with the improvements discussed before. After stemming, the sets of stems are merged from each of the 200 training documents and duplicates are removed. This resulted in a set of 4194 terms in the vocabulary.

To reduce the dimensionality of these high dimensional feature vectors, the Principal Component Analysis (PCA) technique is used, as it is the most effective technique in feature reduction for a Neural Network based Classifier [19].

To create the set of classification vectors corresponding to each training document, the categorization information specified in the test collection was used. This created a set of 200 classification vectors corresponding to each of the 200 training documents. Each classification vector was a 5-dimensional vector of the form:

$$C_j = \langle c_{j,1}, c_{j,2}, \ldots, c_{j,5} \rangle$$

where $c_{j,i}$ was set to 1 if document j belonged to the ith category and was set to 0 if document did not belong to the ith category.

To test the performance of the trained text classifier, the set of reduced feature vectors created from the test documents were fed as input to it, and a set of output classification vectors was obtained. Each output vector was a 5-dimensional vector, with each component being a real number in the range [0, 1]. As binary categorization only was interested in, a threshold at 0.3 was set such that if the ith component of the output vector was greater than 0.3, the decision of the text classifier was to assign the ith category to the corresponding test document. This result was then compared with the categorization information specified for the test documents in the test collection for performing the precision and accuracy.

Precision

Precision measures the accuracy of the retrieval result as indicated by the proportion of the number of categories that are correctly measured to the total no of categories for these documents. It is defined as:

$$\text{Precision} = \frac{\text{Number of categories that are correctly measured}}{\text{Total no of categories for tested documents}}$$

This is the same definition for the precision in the statistical classifier. The same definition is used to be able to compare between the statistical classifier and the Neural Network based classifier.

Results

Figure 4. describes the precision of the neural network classifier according to the input vector i.e. the reduced feature set.

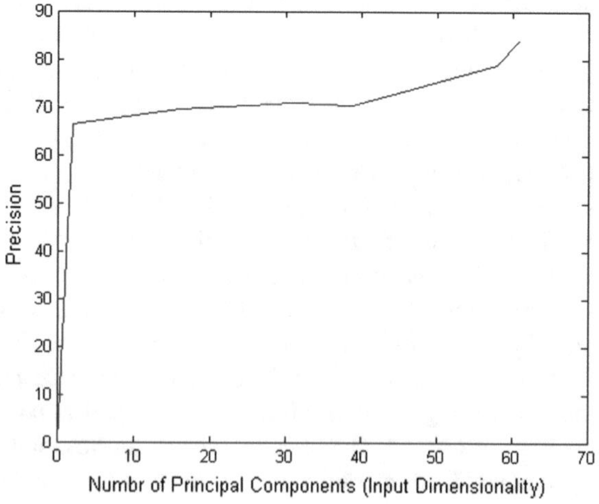

Fig. 4 Precision against number of principal components

It is clear that with feature reduction of 98.55% (from 4194 to only 61 term) the accuracy is acceptable. This means that the neural network based text classifier is very powerful even with a very small training set.

5 Comparison Between Proposed Statistical Classifier and the Neural Network Based Classifier

In this section, a brief comparison between the statistical classifier and the neural network based classifier is introduced. Table 2 introduces this comparison.

It has to be noted that the two classifiers have been tested on the same machine (normal laptop with a single processor). This hardware has limited us from increasing the size of the Neural Network. This is why neural network could be only tested for maximum 61 terms as input. With stronger hardware and increasing the no of terms, increment of the neural network classifier accuracy is expected.

6 Conclusions

In this paper, general procedures for Automatic classification have been introduced and improved. Also, A statistical classifier and a N.N. based classifier have been built and tested. The following points can be concluded:

- Categorization effectiveness and feasibility of the proposed models was tested empirically. Experiments were conducted using the proposed model to categorize real world full text newswire articles contained in the Reuters-21,758 test collection for text categorization. The results showed that the statistical model and the Backpropagation learning in neural networks model were able to give good categorization performance as measured by precision.
- To solve the scalability problem inherent in most applications of neural network techniques to information retrieval, four dimensionality reduction techniques were proposed to reduce the high dimension feature space typical for textual data into a low dimension input space for the neural network. These techniques

Table 2 Comparison between the neural network based classifier and the statistical classifier

Field of comparison	Neural network classifier	Statistical classifier
Reduction techniques	Needed	Not needed
Weighting techniques	Not needed	Needed
Stop words removal	Needed	Needed
Text stemming	Needed	Needed
Training set size	61	4194
Precision	84%	90%

included three domain dependent term selection techniques, namely the DF method, the CF-DF method and the TFxIDF method, and one domain independent statistical data analysis technique called principal component analysis.

- Improvements for the Porter stemming Algorithm have been introduced and tested with the real world full text newswire articles contained in the Reuters-21,758 test collection.
- Also, improvements for the term weighting techniques were introduced and tested to increase the precision of the statistical classifier from 5% using the TFxIDF technique to 75% then 90% after improvements.

References

1. Adams, K., Wranglers, W.: Automatic classification tools transform enterprise documents from bags of words into knowledge resources. Intelligent Enterprise Communities, United States Media (2004)
2. Schütze, H., Hull, D., Pedersen, J.: A comparison of classifiers and document representations for the routing problem. In: 18th Annual International ACM-SIGIR Conference on Research and Development in Information Retrieval, pp. 229–237 (1995)
3. Shaalan, K.: A survey of Arabic named entity recognition and classification. Comput. Linguist. J. **40**, 469–510 (2014)
4. Ibrahim, A., Elghazaly, T.: Arabic text summarization using rhetorical structure theory. In: 2012 8th International Conference on Informatics and Systems (INFOS), pp. NLP-34–NLP-38 (2012)
5. Ibrahim, A., Elghazaly, T.: Rhetorical representation and vector representation in summarizing Arabic text. In: Métais, E., Meziane, F., Saraee, M., Sugumaran, V., Vadera, S. (eds.) Natural Language Processing and Information Systems, NLDB 2013. LNCS, vol. 7934, pp. 421–424. Springer, Heidelberg (2013)
6. Ibrahim, A., Elghazaly, T.: Improve the automatic summarization of Arabic text depending on rhetorical structure theory. In: 12th Mexican International Conference on Artificial Intelligence. IEEE, Mexico City, 2013
7. Ibrahim, A., Elghazaly, T., Gheith, M.: A novel Arabic text summarization model based on rhetorical structure theory and vector space model. IJCLNLP **2**, 480–485 (2013)
8. Frakes, W.B., Baeza-Yates, R.: Information Retrieval Data Structures and Algorithms. P T R Prentice Hall, Englewood Cliffs, New Jersey (1992)
9. Salton, G., McGill, M.J.: Introduction to Modern Information Retrieval. McGraw-Hill, New York (1983)
10. Lovins, J.B.: Development of a stemming algorithm. Mech. Trans. Comput. Linguist. **11**, 22–31 (1968)
11. Salton, G.: Automatic Information Organization and Retrieval. McGraw-Hill, New York (1968)
12. Porter, M.F.: An algorithm for suffix stripping. **14**,130–137 (1980)
13. Paice, C.D.: Another stemmer. SIGIR Forum **24**, 56–61 (1990)
14. Lewis, D.D.: Representation and learning in information retrieval. Ph.D. Thesis, University of Massachusetts, USA (1992)
15. Salton, G., Buckley, C.: Term-weighting approaches in automatic text retrieval. Inf. Process. Manage. **24**, 513–523 (1988)
16. Jolliffe, I.T.: Principal Component Analysis. Springer, New York (1986)

17. Salton. G., Buckley, C.: Term-weighting approaches in automatic text retrieval. Inf. Process. Manage. **24**, 513–523 (1988)
18. Crestani, F.: Learning strategies for an adaptive information retrieval system using neural networks. In: IEEE International Conference on Neural Networks, pp. 244–249 (1993)
19. Lam, S.L., Lee, D.L.: Feature reduction for neural network based text categorization. In: DASFAA-99, 6th IEEE International Conference on Database Advanced Systems for Advanced Application, pp. 195–202. IEEE Computer Society Press, Los Alamitos (1999)

Part VI
Text Mining

Using Text Mining Techniques for Extracting Information from Research Articles

Said A. Salloum, Mostafa Al-Emran, Azza Abdel Monem and Khaled Shaalan

Abstract Nowadays, research in text mining has become one of the widespread fields in analyzing natural language documents. The present study demonstrates a comprehensive overview about text mining and its current research status. As indicated in the literature, there is a limitation in addressing Information Extraction from research articles using Data Mining techniques. The synergy between them helps to discover different interesting text patterns in the retrieved articles. In our study, we collected, and textually analyzed through various text mining techniques, three hundred refereed journal articles in the field of mobile learning from six scientific databases, namely: Springer, Wiley, Science Direct, SAGE, IEEE, and Cambridge. The selection of the collected articles was based on the criteria that all these articles should incorporate mobile learning as the main component in the higher educational context. Experimental results indicated that Springer database represents the main source for research articles in the field of mobile education for the medical domain. Moreover, results where the similarity among topics could not be detected were due to either their interrelations or ambiguity in their meaning. Furthermore, findings showed that there was a booming increase in the number of

S.A. Salloum (✉) · K. Shaalan
Faculty of Engineering & IT, The British University in Dubai, Dubai, UAE
e-mail: ssalloum@uof.ac.ae

K. Shaalan
e-mail: Khaled.shaalan@buid.ac.ae

S.A. Salloum
University of Fujairah, Fujairah, UAE

M. Al-Emran
Faculty of Computer Systems and Software Engineering, Universiti Malaysia Pahang, Gambang, Malaysia
e-mail: malemran@buc.edu.om

M. Al-Emran
Al Buraimi University College, Buraimi, Oman

A.A. Monem
Faculty of Computer & Information Sciences, Ain Shams University, Cairo, Egypt
e-mail: azza_monem@hotmail.com

K. Shaalan et al. (eds.), *Intelligent Natural Language Processing: Trends and Applications*, Studies in Computational Intelligence 740,
https://doi.org/10.1007/978-3-319-67056-0_18

published articles during the years 2015 through 2016. In addition, other implications and future perspectives are presented in the study.

Keywords Text mining · Information extraction · Topic identification Scientific databases · Mobile learning · Higher education

1 Introduction

Nowadays, almost all of the existing information in different institutions (e.g. government, business, industry, and others) is preserved in electronic documents in which it contains semi-structured data. In these documents, the "abstract" is an example of unstructured text component. Whereas, examples of structured fields in a document are: author's name, publication date, title, and category [1]. A study by [2] stated that text mining has become one of the trendy fields that has been incorporated in several research fields such as computational linguistics, Information Retrieval (IR) and data mining. Text mining is different from data mining [3]. Data mining is focused on discovering interesting patterns from large databases rather than textual information [4]. Information recovery methodologies like text indexing techniques have been developed for handling unstructured documents. In conventional researches, it is assumed that a user mostly searches for known terms, which have been previously used or written by someone else. The main problem is that the search results are not relevant to the user's requirements. One solution is to use text mining in order to find out relevant information, which is not indicated explicitly nor written down so far. The procedure of text mining begins with gathering documents through different resources. A particular document would be recovered through text mining instrument and by checking its format and character sets; it will be pre-processed by this instrument. The document would then pass through a text analysis stage. Text analysis includes semantic analysis intended to obtain high-quality information through text. Different text analysis methods are available. Different methods can be used based on the organization's objective. In some cases, text analysis methods are repeated until information is extracted. The outcomes can be stored in a management information system that provides a large amount of significant information for the user of that system.

Text mining intends to detect the information that was not recognized before through extracting it automatically from various text-based sources. Structured data can be handled through data mining tools while unstructured or semi-structured datasets like full-text documents, emails, and HTML files can be handled through text mining. Typically, the information will be kept in a natural form known as text. Text mining is not similar to web mining. When something is explored on the web by the user, it means that it is previously known and it was written by someone else [5]. For example, in E-commerce, a major issue with web mining is buying all the materials which are not relevant to the user's search and it will not show unknown (hidden or implicit) information, while the major objective of text mining is to find out the unknown information [6]; something that is not recognized by anyone.

Data is the basic kind of information, which is required to be organized and mined for the knowledge generation. Discovering patterns and trends from huge data is a significant challenge. Finding out the unknown trends and patterns from databases properly is a major objective of data mining. It is a method where data pre-processing is necessary before applying any other method. Many approaches like clustering, classification, and decision trees are involved in data mining. All the textual based information is stored by electronic means, either on a client's personal computers or on a web server. Due to the increasing growth in hardware storage devices, any computer or laptop has the ability to store an enormous amount of data. Creating new information can be simple while finding out relevant information from a huge amount of data is challenging. In order to extract the relevant information, knowledge, or patterns from various sources that are in unstructured form, text mining technique can be employed. The common structure of text mining involves two consecutive stages: text refining and knowledge distillation. In text refining, free-form text documents are converted into an intermediate form, whereas in knowledge distillation, patterns or knowledge are derived from intermediate form. Intermediate form (IF) can be either semi-structured like the theoretical graph illustration or structured like the relational data illustration. IF can be either a document-based where every entity symbolizes document, it can be a concept-based where every unit symbolizes an object or a concept of interest in a particular area.

Various research areas, techniques, and models are involved in different research domains. The hottest topics of the research domains are the primary focus of many research papers. The research results of a particular domain may influence other research domains since some research domains may have similar topics. These research topics always discuss such a promising research area that is worth studying. Therefore, the trend of cross domain is determined in this research. The longitudinal trends of academic articles in Mobile Leaning (ML) were explored in this research with the help of text mining methods. We recovered and examined (300) refereed journal articles and conference proceedings from various authentic databases.

The primary goals of this research are (1) Using text mining techniques for identifying the topics of a scientific text related to ML research and developing a hierarchical and evolutionary connection among these topics. (2) Using visualization tools for presenting both the topics and the association among them as a convenient way to help users to determine relevant topics.

This paper is categorized as follows: Sect. 2 provides an inclusive background concerning in the text mining field. Other related studies are addressed by Sect. 3. Research methodology is presented in Sect. 4. The results are demonstrated in Sect. 5. Conclusion and future perspectives are presented in Sect. 6.

2 Background on Text Mining and Information Extraction

2.1 Text Mining

The development in the fields of web, digital libraries, technical documentation, medical data has made it easier to access a larger amount of a textual documents, which come together to develop useful data resources [7]. Therefore, it makes text mining (TM) or the knowledge discovery from textual databases a challenging task owing to meet the standards of the depth of natural language which is employed by most of the available documents. The available textual information in the form of databases and online sources [7–9] raises a question about who is responsible for keeping a check on the data and analyzing it? Keeping in view the pertaining condition, it is not possible to analyze and effectively extract the useful information manually. There is a need to employ software solutions which may employ automatic tools for analyzing a considerable amount of textual material, extract relevant data, analyze relevant data, and organize relevant information. Owing to the increasing demands to obtain knowledge from a large number of textual documents accessible on the web, text mining is gaining a significant importance in research [10, 11]. Generally, text mining and data mining are considered similar to one another, with a perception that same techniques may be employed in both concepts to mine text [4, 12, 13], and [3]. However, both are different in a sense that data mining involves structured data, while text deals with certain features and is relatively unstructured and usually require preprocessing. Furthermore, text mining is an interrelated field with Natural Language Processing (NLP). NLP is one of the hot topics that is concerned with the interrelation among the huge amount of unstructured available text [14], besides the analysis and interpretation of human-being languages [15, 16].

2.2 Information Extraction

An initiation point for computers to evaluate unstructured manuscripts is to use Information Extraction (IE). IE software recognizes key phrases and relationships included in the manuscript. This is performed through finding the predefined arrangements in a text; this technique is called pattern matching. Regular language text documents consist of information that cannot be utilized for mining. IE agrees with the documentation, choosing appropriate articles, and the association among them to make them more available for added guidance [17, 18]. Contrary to Information Retrieval, which deals with how to recognize relevant documents from a document collection, IE yields structured information prepared for post-processing, which is essential to various applications of Web mining and searching instruments [19]. IE deals with discovering and extracting important information from natural language texts [18]. It consists of separating appropriate text parts, extracting the offered data in such parts, and transforming the data into the functional form. Fractional extraction from domain-particular texts is currently possible; though complete IE from the random text is still a continuing study target [20].

2.3 Extracting Knowledge from Text

Under most of the conditions, only specific data is obtained from the information extracted from unstructured text instead of abstract knowledge. In such a case, it is required to employ a text mining task along with additional techniques to mine knowledge from the data in hand [21, 22]. DiscoTEX (Discovery from Text EXtraction) is one of the major approaches employed for text mining. It involves using IE first to gather structured data from unstructured text, followed by employing traditional Knowledge Discovery from Database (KDD) tools to discover knowledge from this data. This framework for text mining was presented by [21]. In this method, the learned IE system is used to convert unstructured text into more structured data. This data is then subjected to mining to develop meaningful relationships. In a case that the information extracted from a corpus of documents is in the form of abstract knowledge instead of concrete data, IE tends to serve as the "discovering knowledge" from text. Discovery of knowledge by extracting information, such as key-phrases or keywords extraction from the text may be used for other text mining tasks, i.e. classification, clustering, summarization, and topic detection [23].

2.4 Text Mining Methods and Techniques

Text mining is usually employed to obtain quick results [24]; it has been subjected research under a number of application areas. On the basis of respective areas of application, text mining can be categorized as text categorization, text clustering, association rule extraction, and text visualization. They are discussed in the following sub-sections.

Text Clustering

Text clustering is based on the Cluster hypothesis which proposes that relevant documents must have more similarities with one another than the non-relevant ones [25]. The Clustering technique is a trust-worthy technique that is generally employed for analyzing larger amounts of data like data mining. It has been proven that text clustering is one of the most effective tools used for text theme analysis [26]. Moreover, it facilitates the method of topic analysis in which named entities having concurrent occurrence are grouped together, followed by subjecting them to the clustering process in such a way that frequent item are placed in sets by applying the hyper graph-based method [27]. Each set of named entities is represented by a cluster that is related to one of the ongoing topics in the corpus. The process of topic tracking within dynamic text data has gained the interest from the researchers who are working on the subject of text clustering in the digital field. Various methods and algorithms based on unsupervised document management are included in the process of document clustering. In the clustering process, the numbers, properties, and associations of the grouped sets are initially unknown. The grouping of documents is performed by categorizing them into a particular category such as medical, financial, and/or legal [28].

Association Rule Extraction

A study by [29] argued that the method of association rule mining (ARM) is employed to identify relationships within a larger group of variables in a dataset. The ARM identifies the variable-value combinations which tend to occur frequently. The method of ARM in data mining also known as knowledge discovery in databases; that is similar to the correlation analysis that finds out the relationships between two variables. Wong et al. [30] provided that the Association Rules for Text Mining are majorly concerned to explore the relationships between various topics or factual notions employed for characterizing a corpus. They intend to discover key association rules relative to a corpus in such a way that the occurrence of certain topics in an article may correspond to the occurrence of another topic as well.

K-Means Algorithms

The k-mean approach divides the data set into k clusters, where every cluster is subjected to be represented by the mean of points; called the centroid. A two-step repetitive process is employed for the application of the algorithm: (1) Assigning every point to the nearest centroid. (2) Evaluating the centroids for a recently developed group. The process is ended when the cluster centroid comes to a constant value. The k-mean algorithm has an extensive application owing to its direct parallelization. Furthermore, the order of respective data does not affect the k-mean algorithm which attributes the numerical characteristics to it. It is required to mention the maximum value of k at the beginning of the process. The representation of the cluster is made by the k-medoid algorithm that chooses the object adjoining the center of the cluster. Though, the selection of the k objects is done randomly in the algorithm. The selected objects help to determine the distance. A cluster is formed on the basis of the nearest object to k, whereas the other objects acquire the position of k recursively till the required quality of the cluster is achieved [28].

Information Visualization

Information visualization puts great textual bases in a visual hierarchy or plan and offers browsing abilities as well as general searching. This technique offers improved and quicker comprehensive knowledge, which assists us to mine enormous accumulation documents. The operators can distinguish the colors, associations, and gaps. The assortment of documents can be demonstrated as a structured layout utilizing indexing or vector space model.

Word Cloud

Jayashankar and Sridaran [31] defined word clouds or tag clouds as the visual representation of words for a certain written content structured as per its frequency. Word cloud is among the most frequently used method to present text data in a graphical manner; making it helpful for analyzing various forms of text data such as essays and short answers or written opinions to a survey or questionnaire [32]. Word cloud tends to serve as a preliminary stage for in-depth analysis of certain text material [33, 34]. For example, word cloud assists in finding the relevancy between given text and the required information. Nonetheless, the method has certain drawbacks as well. One of the major drawbacks that is it does not consider the linguistic knowledge about the words and their respective link to the given subject while providing a purely statistical summary to the segregated words. As a result, in most systems, the word clouds are often employed in a statistical manner for summarizing text, providing very little or no means for correlating the data. It is perceived that this could be one of the most influencing paradigms of visualization for most of the analysis conditions. Thus, in this paper, we have employed the use of word clouds as the central method to text analysis.

3 Related Work

Many research works contributed to the field of IE through the use of various techniques. The primary focus of these researches was to determine how different text mining procedures can be utilized as the structured data sets exist in the text document format. This part begins with defining the topic of the research, evaluating previous researches, and then major techniques are applied using information extraction and text mining. In order to determine the topic of each research area and to develop an evolutionary and hierarchical connection between these topics, [35] used the method of text mining. Topics are presented through visualization tools. Moreover, these tools are used in order to show the connection between these topics and to offer interactive functions so that users can effectively find the cross-domain topics and know the trends of cross-domain research.

Moloshnikov et al. [36] developed an algorithm for finding documents on a particular topic depending on a selected reference collection of documents. In addition, the context-semantic graph for visualization themes in search results was also developed. The algorithm depends on the incorporation of a group of entropic, probabilistic and semantic developers for mining of weighted keywords and set of words that explain the specified topic. Results indicated that the average precision is 99% and the recall is 84%. A unique technique was also created for making graphs on the basis of the algorithm, can remove key phrases with weights. It offers the opportunity to show an arrangement of sub-topics in huge sets of documents in compact graph form.

In order to offer a reference for additional researches of other researchers, [37] discussed the research status of text mining technology when it was used in the biomedical field that covers 10 years. Biomedical text mining literature incorporated in SCI from 2004 to 2013 were recovered, filtered and then examined from the viewpoint of research institutions, yearly changes, research areas, local distribution, journals sources, and keywords. A prominent increase in the amount of worldwide biomedical text mining literature is observed. Among this global literature, a huge percentage is taken up by literature related to named entity recognition, entity relation extraction, text categorization, text clustering, abbreviations extraction, and co-occurrence analysis. Studies carried out in USA and UK are considered to be present in the primary position.

In order to extract inter-language clusters through multilingual documents depending on Closed Concepts Mining and vector model, a new statistical approach was suggested by [38]. Formal Concept Analysis methods are used for mining Closed Concepts from similar corpora and later these Closed Concepts and vector models are utilized in the clustering and arrangement of multilingual documents. An experimental assessment is carried out over a set of French-English bilingual documents of CLEF's 2003. With a notable comparability score and in order to remove the bilingual classes of documents, results revealed that the interaction between vector model and Formal Concept Analysis is very useful.

Santosh [39] suggested the graph mining-based document content (i.e. text fields) exploitation. That is, the query generated the graph depending on the users' requirements. This is an easy and effective graph mining method to extract similar patterns through the documents and changed the query graph into model graphs which are utilized when the users are not present. An intelligent solution for document information exploitation has been created. This is characterized by simplicity, ease of use, accuracy, ease of development, and flexibility. In order to understand graph models, it does not need a huge collection of document images. Moreover, since model learning consumes less than 10 s for an input pattern per class on average, changes, amendments, and replacements can be done in the input patterns. Information exploitation average performance is shown to have 86.64% as Precision, and 90.80% as a Recall. However, the suggested technique failed to offer inclusive and accurate solutions for the patterns that have a huge collection of fields in a zigzags arrangement due to the query graph intricacy.

Sirsat et al. [23] proposed two techniques for mining text through online sources. The first technique dealt with the knowledge that is required to be shown directly in the documents that need to be mined. Text mining and IE are considered as the only effective tools for performing that technique. The second one concerned with the documents that hold an actual data in unstructured format instead of nonfigurative knowledge. IE can help to change the unstructured data presented in the document corpus into structured one. In order to discover nonfigurative patterns in the extracted data, data mining algorithms and techniques can be used.

Song and Kim [40] presented the first attempt to apply text mining approaches to a huge collection of full-text articles for discovering the knowledge structure of the area. Instead of depending on the citation data presented in Web of Science, PubMed Central full-text articles have been used for bibliometric examination. Above all, this assisted the creation of text mining routines in order to develop a custom-made citation database following the full-text mining. Findings showed that most of the documents that were published in bioinformatics area were not cited by others. Additionally, a constant and linear rise has been observed in the amount of publications across publication years. Results revealed that the majority of the retrieved studies were inspired by USA-based institutes followed by European institutes. Results reported that the major primary focus of the important topics was on biological factors. However, according to PageRank, the top 10 articles were highly concerned with the computational factors.

In order to facilitate the accurate extraction of text from PDF files of research articles that can be utilized in text mining applications, a "Layout-Aware PDF Text Extraction" (LA-PDF Text) system was presented by [41]. Text blocks are mined from PDF-formatted fill-text research articles under this system and then the system categorizes them into logical units depending on rules that typify particular sections. Only the textual content of the research articles is focused in the LA-PDF Text system. This system serves as a basis for new experiments into more developed extraction methods dealing with multi-modal content like images and graphs. The system goes through three phases: (1) Identifying contiguous text blocks with the help of spatial layout processing in order to discover blocks of contiguous text, (2) Categorization of text blocks into metaphorical categories with the help of a rule-based method, and (3) Joining categorized text blocks together by arranging them accurately which results in the extraction of text from section-wise grouped blocks. An evaluation of the accuracy of the block discovery algorithm used in step 2 was performed. It was also shown that the system can identify and classify them into metaphorical categories with Recall = 0.89%, Precision = 0.96%, and F = 0.91%. Moreover, the accuracy of the text mined with the help of LA-PDF Text is compared to the text from an Open Access subset of PubMed Central. This accuracy is then compared with the text that was mined using the PDF2Text system. These are the two frequently used techniques to extract text from PDF.

Mooney and Bunescu [42] described two techniques for using the natural language information extraction for text mining. First, general knowledge can be mined directly from the text. A project where a knowledge base of 6580 human protein interactions was extracted by mining around 750,000 Medline abstracts in which reconsidered as an example of this technique. Second, structured data can be mined through text documents or web pages. In order to find out patterns in the mined data, traditional KDD methods can be applied. The performed work on the DiscoTEX system and its application to Amazon book descriptions, computer science job postings, and resumes were considered as an example of this technique. In order to discover units and relations in text, research in IE keeps on creating more efficient algorithms. Valuable and significant knowledge can be mined effectively from the constantly developing body of electronic documents and web pages by using modern approaches in human language technology and computational linguistics, and linking them with the modern techniques used in machine learning and conventional data mining techniques. IE deals with determining a particular set of relevant items through natural language documents.

In order to discover topics that recur in articles of text corpus, another method TopCat (Topic Categories) was proposed by [22]. IE was used by this technique in order to discover named entities in individual articles and to characterize them as a collection of items of an article. Therefore, through recognition of frequent item sets which commonly occurred with named entities, the issues in data mining or database context were studied. Association rule data mining technique is used by TopCat to discover these frequent item sets. By using a hypergraph splitting technique, TopCat further clusters the named entities which discovers a collection of frequent item sets with significant overlie. In order to discover documents regarding the topic, IR technique was used. Different technologies like IE for named entity extraction, association rule data mining, clustering of association rules, IR techniques were used in this method. TopCat discovers topics that have a logical accuracy with reasonable identifiers. Callan and Mitamura [43] presented a new technique for named entity detection, called KENE. In order to understand the extraction rules, the knowledge-based technique is used in it. Generate-and-test approach is used for named entity extraction from structured documents.

We can observe from the surveyed literature that there is a limitation in addressing the issue of IE from research articles using data mining techniques. The synergy between these approaches (i.e. IE with data mining techniques) helps to discover different interesting text patterns in the retrieved articles. This approach could be applied to a variety of research topics, where in each topic can generate a wide range of knowledge patterns. Mobile learning (M-learning) has become one of the trendy fields in the higher education [44–50]. In accordance to the existing literature, we can perceive that IE and data mining techniques were never applied to the M-learning field. This creates a need for collecting several research articles in the field of M-learning from different scientific databases and applies the synergic approach on them. Additionally, we are trying to respond to the following research questions:

RQ1: What are the most frequent keywords in the collected articles?
RQ2: What are the most frequent terms among the collected articles?
RQ3: What are the most common topics among the collected articles?
RQ4: How are the articles interrelated to each other?
RQ5: How are the articles distributed in terms of publication year?

4 Research Methodology

4.1 Text Mining Processing Framework

We have developed our customized framework which is inspired by the designed framework proposed by [51], see Fig. 1. Three steps are included in text mining: text pre-processing, text mining operations, and post processing. Text pre-processing involves the following tasks: data selection, classification, feature extraction and text

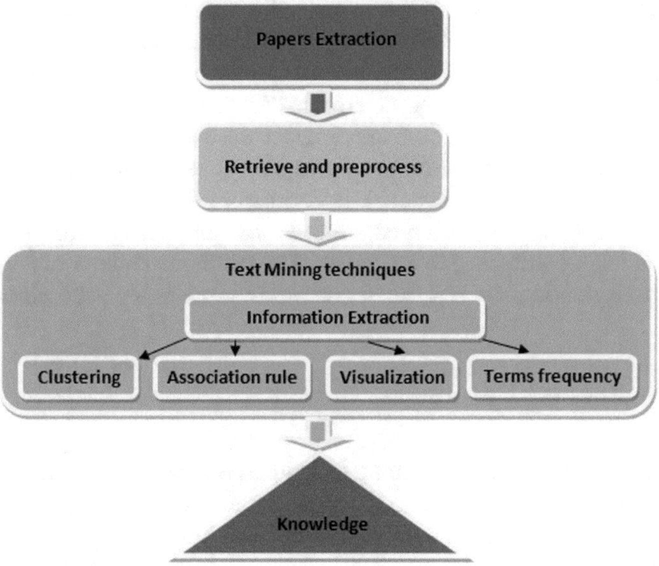

Fig. 1 Text mining processing framework

normalization, i.e. transforming the documents into an intermediate form for ensuring compatibility for various mining tools. The second step deals with different text mining techniques like clustering, association rule detection, visualization, and terms frequency. During the third step, alterations and changes are made on the data (i.e. research articles) through text mining functions like evaluation and choice of knowledge, analysis and visualization of knowledge. The main aim of this study is to extract interesting information from the collected articles using the text mining techniques.

4.2 Data Collection and Pre-processing

The research articles were collected from six scientific databases, namely: Springer, Wiley, Science Direct, SAGE, IEEE, and Cambridge. The search term used for data collection is simply "Mobile Learning in higher education". Based on that, 300 research articles in the field of mobile learning were collected. These articles are categorized into six folders, where each folder represents the database where these articles were retrieved.

The presence of the linguistic noise is a common problem in the content of the extracted articles and we have dealt with. Then, the cleaned data are uploaded into RapidMiner tool while the misplaced and unnecessary data have been removed from the dataset. In order to improve the performance and data quality, all the irrelevant characteristics are debarred while the data is being uploaded into

RapidMiner tool. The major steps involve the separation of the document into tokens; this task is called Tokenization [52]. The next step is concerned with the transformation process of all the characters where each document title is created in a lower case. Stop words filtering is involved in the third step, where English is filtered through this operator. A single English word is required to be signified by each token. All tokens that were similar to stop words were eradicated from the provided document by an operator. The document must have only one stop word per line. The last step is concerned with the text processing phase that involves filtering the tokens according to the length. The minimum number of the characters that the token should have is 4, while the maximum number is 25 characters.

5 Experimental Results

The application of various text mining techniques on the collected articles presents different results and suggestions. In the present study, we are trying to apply almost all of the text mining techniques that were mentioned in the literature on the collected articles. Nevertheless, these techniques have not been applied to the research articles concerning mobile learning in higher education; the reason that makes this study is unique and adds a value to the research community.

Q1: What are the most frequent keywords in the collected articles?

As per the study of [31], we used the cloud technique in order to answer the above research question. As shown in Fig. 2, we can notice that "Learning" is the most keyword that was mentioned across all the collected articles. The second highest frequent words are "Patients" and "Students" respectively. The increasing number of the words (learning and students) could be attributed to the fact that learning and students form the core of the higher educational processes. In addition, the appearance of the word (Patients) in many articles shows that most of these articles are focusing on mobile learning in medical education. Table 1 shows the distribution of the top 5 most frequent terms that were mentioned in the collected articles across each database.

Fig. 2 Words cloud across all databases

Table 1 Words cloud terms distribution across all databases

Scientific database	Term	Frequency
Cambridge	Learning	1165
	Education	854
	University	847
	Students	831
	Higher	490
IEEE	Education	793
	Students	777
	Learning	579
	Engineering	417
	Higher	256
Science Direct	Learn	2043
	Use	837
	Mobile	701
	Education	584
	Student	551
Springer	Patients	9046
	Care	6458
	Learning	5423
	Medical	5180
	Health	4086
SAGE	Learning	2033
	Students	1719
	Mobile	1611
	Education	1112
	University	921
Wiley	Learning	7556
	Patients	6392
	Students	3266
	Care	3212
	Mobile	3193

Moreover, we have applied the word frequency technique on the text in the collected articles. As per (Fig. 3), we can notice that the most frequent linked words among all the articles are: "Learning" followed by "Patients", "Students", "Education", "Care", "Mobile", "Study", "University", "Medical", and "Clinical" respectively. These results indicate that the most frequent linked words are focused on studies targeting mobile learning in medical education. These results match the above mentioned results in terms of the word cloud. Springer database represents the most source that contains these words followed by Wiley and Science Direct respectively. The results reveal that the words (patients, care, medical, and clinical) were highly mentioned in Springer database. That is, researchers who are specialized in mobile medical education should benefit from these results as it shows

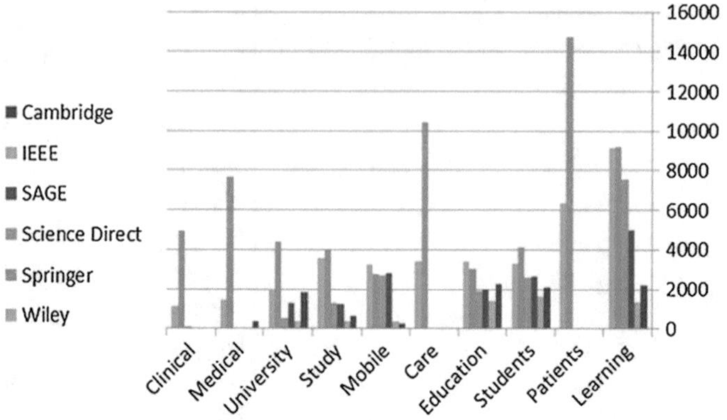

Fig. 3 Word frequency distribution across all databases

them that Springer is the top among other databases for collecting research articles in that field.

For further investigation, the above most frequent words were analyzed and distributed among all the Scientific Databases in order to represent each word separately. As per (Fig. 4), the word "Learning" was frequently mentioned by Springer database followed by Wiley, Science Direct, SAGE, IEEE, and Cambridge, respectively.

According to (Fig. 5), the word "Patients" was frequently used by Springer database followed by Wiley. This indicates that almost all of the mobile medical education articles are published under springer and Wiley databases.

As per (Fig. 6), the word "Students" was frequently utilized by Springer database followed by Wiley, SAGE, Science Direct, Cambridge, and IEEE, respectively.

According to (Fig. 7), the word "Education" was frequently reported by Wiley followed by Springer, Cambridge, SAGE, Science Direct, and IEEE, respectively.

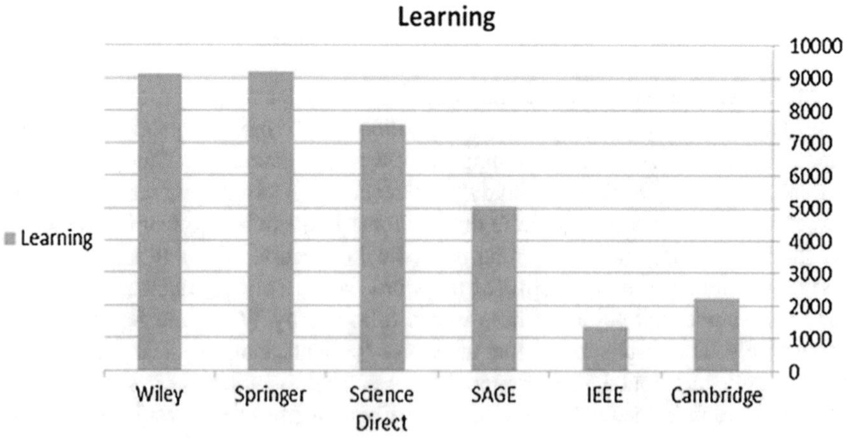

Fig. 4 The distribution of the word "Learning" among all sources

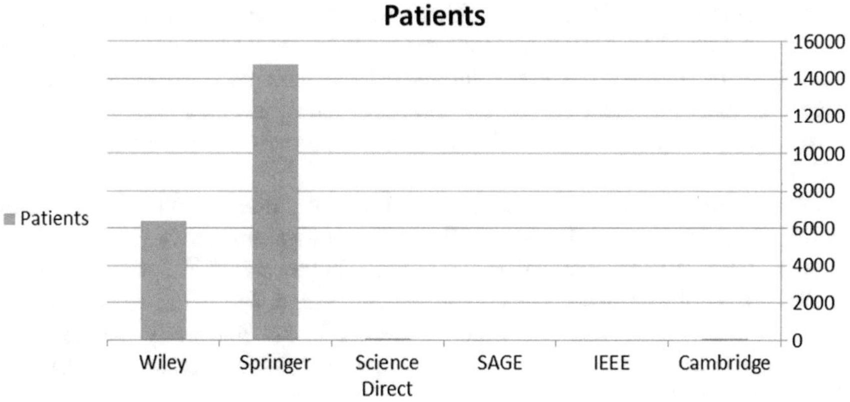

Fig. 5 The distribution of the word "Patients" among all sources

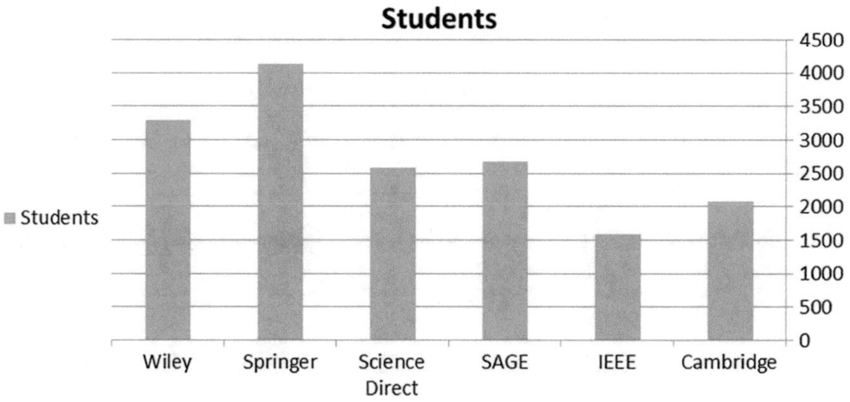

Fig. 6 The distribution of the word "Students" among all sources

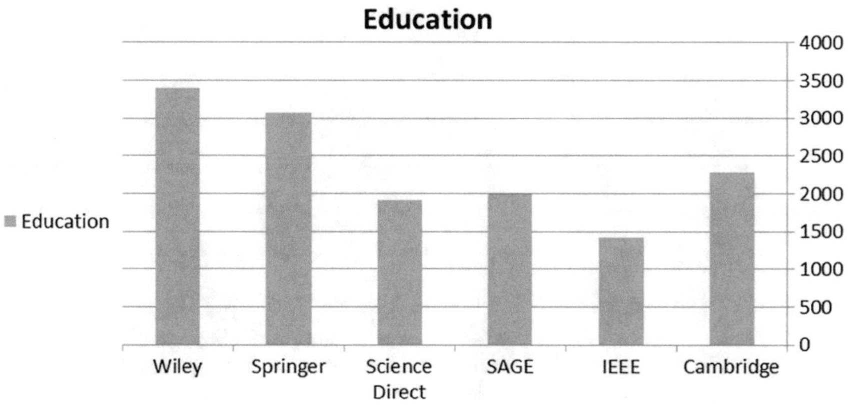

Fig. 7 The distribution of the word "Education" among all sources

According to (Fig. 8), the word “Care” was frequently mentioned by Springer database followed by Wiley. On the other side, other databases don’t show the occurrence of this word. These results assist the researchers of mobile medical education that their field research articles are mainly available in Springer and Wiley, while other databases don’t have enough journals that accommodate these articles.

As per (Fig. 9), the word “Mobile” was frequently used by Wiley followed by SAGE, Springer, Science Direct, IEEE and Cambridge, respectively.

As per (Fig. 10), the word “study” was frequently occurred in Springer database followed by Wiley, Science Direct, SAGE, Cambridge, and IEEE, respectively.

According to (Fig. 11), the word “University” was frequently mentioned in Springer database followed by Wiley, Cambridge, SAGE, Science Direct, and IEEE, respectively.

As per (Fig. 12), the word “Medical” was frequently used by Springer database followed by Wiley, Cambridge, and Science Direct respectively.

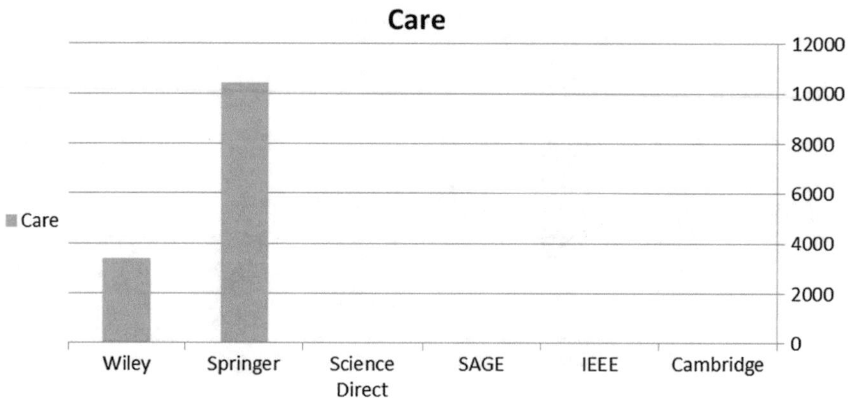

Fig. 8 The distribution of the word “Care” among all sources

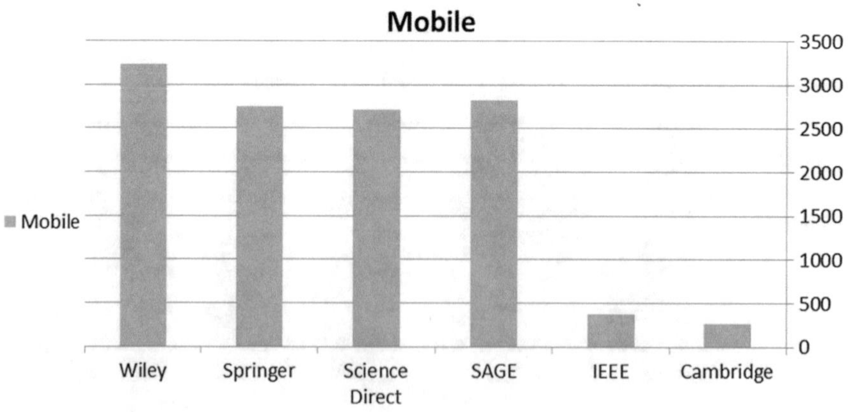

Fig. 9 The distribution of the word “Mobile” among all sources

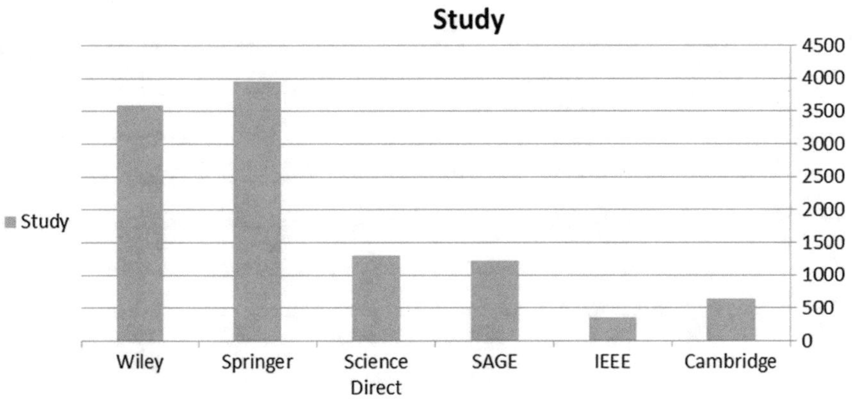

Fig. 10 The distribution of the word "Study" among all sources

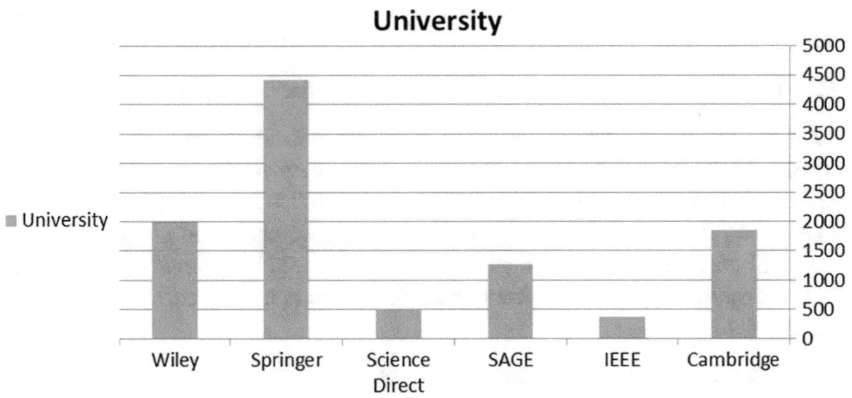

Fig. 11 The distribution of the word "University" among all sources

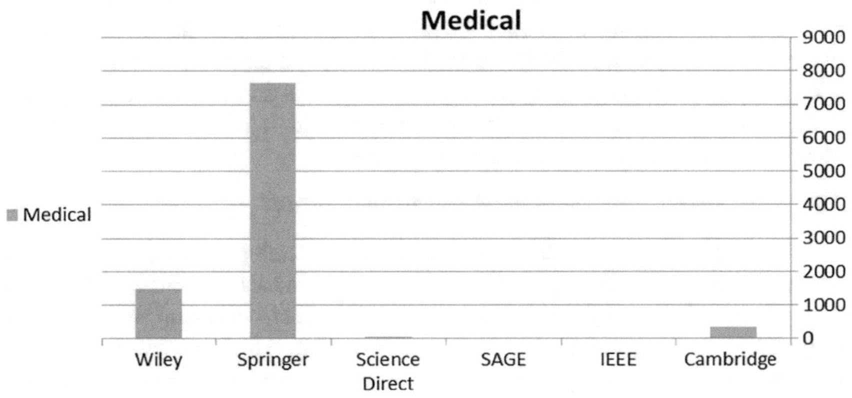

Fig. 12 The distribution of the word "Medical" among all sources

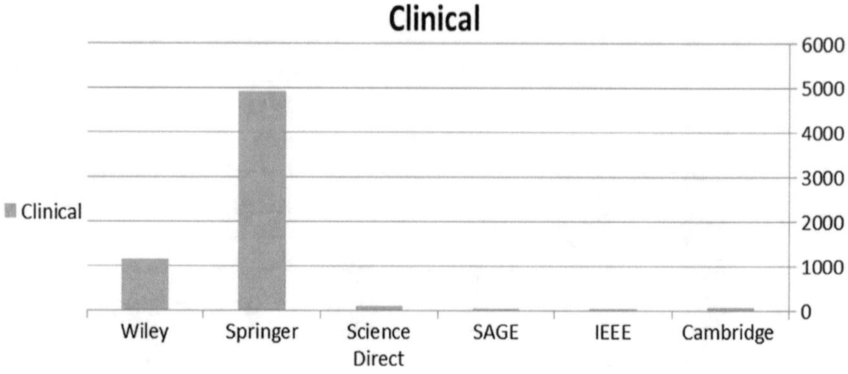

Fig. 13 The distribution of the word "Clinical" among all sources

According to (Fig. 13), the word "Clinical" was frequently occurred in Springer database followed by Wiley, Science Direct, Cambridge, IEEE, and SAGE, respectively.

Q2: What are the most frequent terms among the collected articles?

As per the study of [22], the method of the association rule is employed to identify and visualize the terms that have strong connections to each other. The most connected terms are termed as being strongly related to each other. According to (Fig. 14), the term "Education" is shown as being central to the tree structure having all the relevant words connected to it. This could be referred to the fact that the text acquired from the collected research articles is mainly concentrated on the learning field.

Q3: What are the most common topics among the collected articles?

As per the study of [25], we performed the similarity measure on the collected articles in order to identify the topics that are highly similar to each other. Figure 15 shows the similarity relationships among all the articles. As we can observe from the figure, it is very difficult to track the relationships among all the depicted topics. This could be attributed to the fact that all the collected articles are in one research field (i.e. mobile learning in higher education). To this end, the similarity operator could not detect a clear similarity among the topics since all these topics are interrelated and similar in meaning to each other.

Q4: How are the articles interrelated to each other?

According to [53] and [28], we applied the clustering technique in order to answer the above research question. We used the k-means algorithm through the use of different k values. By examining different k values, we end up with ($k = 6$) as it represents the most reasonable value for answering the above question. As per (Fig. 16), there are six clusters. Cluster 0 contains 3 items (i.e. 3 articles), cluster 1 includes 2 items, cluster 2 contains 3 items, cluster 3 includes 5 items, cluster 4

AssociationRules

```
Association Rules
[university] --> [learning] (confidence: 0.964)
[education] --> [learning] (confidence: 0.964)
[education] --> [students] (confidence: 0.964)
[learning] --> [university] (confidence: 0.990)
[learning] --> [education] (confidence: 0.990)
[university] --> [education] (confidence: 0.990)
[education] --> [university] (confidence: 0.990)
[students] --> [education] (confidence: 1.000)
```

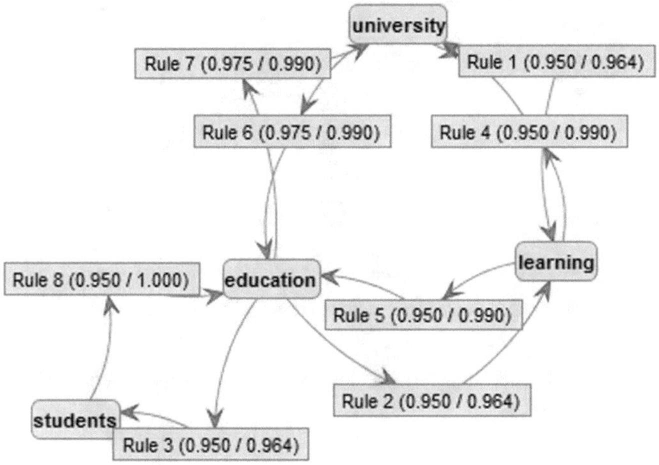

Fig. 14 Concept link diagram

includes 285 items, and cluster 5 includes 2 items. To that end, almost all of the articles (N = 285) are accumulated in cluster 4; this indicates that these articles are discussing the main studied topic (i.e. mobile learning in higher education). On the other side, by further investigating the remaining articles (N = 15) that are accumulated in the other clusters; it has been found that these articles are discussing other topics in learning and education rather than the studied topic. In addition, the reason that brought these articles in the search results when we collect them is that these articles include the word "mobile" as just a cited term in the text.

Q5: How are the articles distributed in terms of publication year?

Figure 17 shows the distribution of the collected research papers across their years of publication. The collected papers (N = 300) were published between years 1990 through 2016. It is clearly demonstrated that the mobile learning field was not common in the early 1990s. In 2009, there is a noticeable increase in the number of published articles as mobile learning becomes very popular during that period.

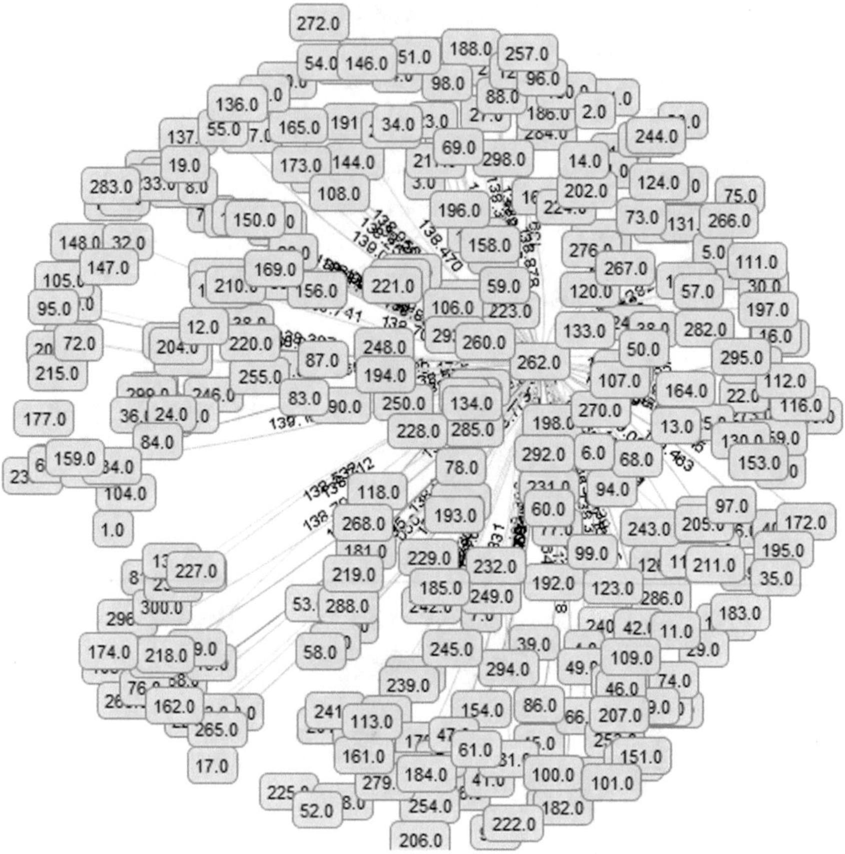

Fig. 15 Similarity diagram

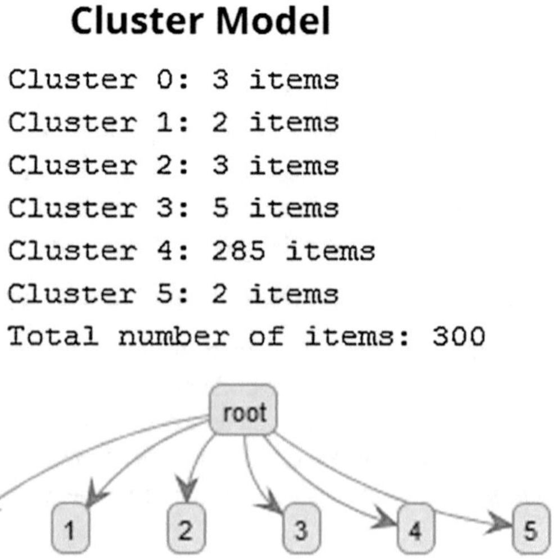

Fig. 16 Cluster model

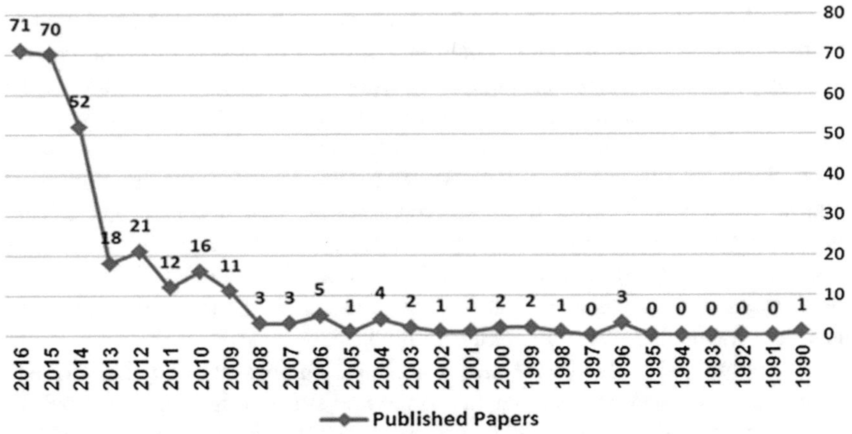

Fig. 17 Distribution of research articles in terms of publication year

Furthermore, during the years 2015 through 2016, mobile learning witnessed an enormous attraction from a lot of scholars who published many articles that contribute to the evolvement of mobile learning.

6 Conclusion

The present study demonstrates a comprehensive overview about text mining and its current research status. According to the surveyed literature, there is a limitation in discussing the issue of information extraction from research articles using data mining techniques. The synergy between information extraction and data mining techniques helps to discover different interesting text patterns in the retrieved articles. This approach could be applied to a variety of research topics, where in each topic it can generate a wide range of knowledge patterns. Mobile learning has become one of the trendy fields in the higher education. Accordingly, we can perceive that information extraction and data mining techniques were never applied to the mobile learning field. This creates a need for collecting a dataset that consists of several research articles in the field of mobile learning from different scientific databases, and applying the proposed approach on them.

Three hundred refereed journal articles from six scientific databases were collected, and textually analyzed through text mining techniques. The six databases are Science Direct, IEEE, Wiley, Cambridge, SAGE, and Springer. The selection of the collected articles was based on the criteria that all these articles should incorporate mobile learning as the main component in the higher educational context. In the present study, text clustering, association rule, word cloud, and word frequency are the main tasks used for text analysis.

By applying the word cloud and the word frequency techniques, results indicated that "Learning" is the most frequent keyword across all the collected articles;

followed by "Patients" and "Students", respectively. The increasing number of the words: "learning" and "students" could be attributed to the fact that learning and students form the core of the higher educational processes. In addition, results revealed that the words: "patients", "care", "medical", and "clinical", were frequently mentioned in Springer database. These results indicate that the most frequent linked words are those focused on studies targeting mobile learning in medical education. Springer database represents the richest source that contains these words followed by Wiley and Science Direct, respectively. That is, researchers who are specialized in mobile medical education should benefit from these results as it shows them that Springer database is the topmost among other databases for finding research articles in this field.

By applying the association rule technique, findings showed that the term "Education" is shown as being central to the tree structure having all the relevant words connected to it. This could be referred to the fact that the text acquired from the collected research articles is mainly concentrated on the learning field. In addition, we performed the similarity measure on the collected articles in order to identify the topics that are highly similar to each other. Results revealed that the similarity operator could not detect a clear similarity among some topics the reason is that these topics are interrelated and similar in meaning to each other (i.e. all the articles are discussing the topic of mobile learning in higher education).

By applying the clustering technique, we used the k-means algorithm through the use of different k values. Results indicated that there were six clusters. Almost all of the articles (N = 285) were accumulated in one cluster; this indicates that these articles are discussing the main studied topic (i.e. mobile learning in higher education). On the other side, by further investigating the remaining articles (N = 15) that are accumulated in the other clusters; it has been found that these articles are discussing other topics in learning and education rather than the studied topic. By distributing the collected research papers across their years of publication, findings showed that there was a booming increase in the number of published articles during the years 2015 through 2016. This could be referred to the reason that mobile learning has witnessed in these years an enormous attraction from a lot of scholars who published many articles that contribute to the evolvement of mobile learning.

As a future work, we are interested in collecting articles from various research topics, i.e. not to focus on one area. This will help us to find more interesting patterns in these articles and how such articles are distributed among the targeted databases. In addition, this will allow the similarity operator to work properly and to draw a clear relationship among the articles.

References

1. Gaikwad, S.V., Chaugule, A., Patil, P.: Text mining methods and techniques. Int. J. Comput. Appl. **85**(17) (2014)
2. Salloum, S.A., Al-Emran, M., Monem, A.A., Shaalan, K.: A Survey of text mining in social media: facebook and twitter perspectives. Adv. Sci. Technol. Eng. Syst. J. (2017)

3. Navathe, S.B., Ramez, E.: Data warehousing and data mining. Fundam. Database Syst., 841–872 (2000)
4. Gupta, V., Lehal, G.S.: A survey of text mining techniques and applications. J. Emerg. Technol. Web Intell. **1**(1), 60–76 (2009)
5. Gupta, S., Kaiser, G.E., Grimm, P., Chiang, M.F., Starren, J.: Automating content extraction of html documents. World Wide Web **8**(2), 179–224 (2005)
6. Hassani, H., Huang, X., Silva, E.S., Ghodsi, M.: A review of data mining applications in crime. Statistical Anal. Data Min.: ASA Data Sci. J. **9**(3), 139–154 (2016)
7. Feldman, R., Dagan, I.: Knowledge discovery in textual databases (KDT). KDD **95**, 112–117 (1995)
8. Tan, A.H.: Text mining: The state of the art and the challenges. In: Proceedings of the PAKDD 1999 Workshop on Knowledge Disocovery from Advanced Databases, vol. 8, pp. 65–70 (1999)
9. Hearst, M.A.: Untangling text data mining. In: Proceedings of the 37th Annual Meeting of the Association for Computational Linguistics on Computational Linguistics, pp. 3–10. Association for Computational Linguistics (1999)
10. Rajman, M., Besançon, R.: Text mining: natural language techniques and text mining applications. In: Data Mining and Reverse Engineering, pp. 50–64. Springer, US (1998)
11. Mahgoub, H., Rösner, D., Ismail, N., Torkey, F.: A text mining technique using association rules extraction. Int. J. Computat. Intell. **4**(1), 21–28 (2008)
12. Akilan, A.: Text mining: challenges and future directions. In: 2015 2nd International Conference on Electronics and Communication Systems (ICECS), pp. 1679–1684. IEEE (2015)
13. Sukanya, M., Biruntha, S.: Techniques on text mining. In: 2012 IEEE International Conference on Advanced Communication Control and Computing Technologies (ICACCCT), pp. 269–271. IEEE (2012)
14. Salloum, S.A., Al-Emran, M., Shaalan, K.: A Survey of lexical functional grammar in the Arabic context. Int. J. Com. Net. Tech. **4**(3) (2016)
15. Al Emran, M., Shaalan, K.: A survey of intelligent language tutoring systems. In: 2014 International Conference on Advances in Computing, Communications and Informatics ICACCI, pp. 393–399. IEEE (2014a)
16. Al-Emran, M., Zaza, S., Shaalan, K.: Parsing modern standard Arabic using Treebank resources. In: 2015 International Conference on Information and Communication Technology Research (ICTRC), pp. 80–83. IEEE (2015)
17. Pazienza, M.T. (Ed.): Information extraction: Towards scalable, adaptable systems. Springer (2003)
18. Cowie, J., Lehnert, W.: Information extraction. Commun. ACM **39**(1), 80–91 (1996)
19. Velasco-Elizondo, P., Marín-Piña, R., Vazquez-Reyes, S., Mora-Soto, A., Mejia, J.: Knowledge representation and information extraction for analysing architectural patterns. Sci. Comput. Program. **121**, 176–189 (2016)
20. Hsu, J.Y.J., Yih, W.T.: Template-based information mining from HTML documents. In: AAAI/IAAI, pp. 256–262 (1997)
21. Mooney, R.J., Nahm, U.Y.: Text mining with information extraction, multilingualism and electronic language management. In: Proceedings 4th International MIDP Colloquium, pp. 141–160 (2003)
22. Clifton, C., Cooley, R., Rennie, J.: TopCat: data mining for topic identification in a text corpus. IEEE Trans. Knowl. Data Eng. **16**(8), 949–964 (2004)
23. Sirsat, S.R., Chavan, D.V., Deshpande, D.S.P.: Mining knowledge from text repositories using information extraction: A review. Sadhana **39**(1), 53–62 (2014)
24. Madani, F.: Technology Mining bibliometrics analysis: applying network analysis and cluster analysis. Scientometrics **105**(1), 323–335 (2015)
25. Huang, A.: Similarity measures for text document clustering. In: Proceedings of the sixth New Zealand Computer Science Research Student Conference (NZCSRSC2008), Christchurch, New Zealand, pp. 49–56 (2008)

26. Clifton, C., Cooley, R.: TopCat: Data mining for topic identification in a text corpus. In: European Conference on Principles of Data Mining and Knowledge Discovery, pp. 174–183. Springer, Heidelberg (1999)
27. Han, E.H., Karypis, G., Kumar, V., Mobasher, B.: Clustering based on association rule hypergraphs. In: DMKD (1997)
28. Irfan, R., King, C.K., Grages, D., Ewen, S., Khan, S.U., Madani, S.A., … & Tziritas, N.: A survey on text mining in social networks. Knowl. Eng. Rev. **30**(2), 157–170 (2015)
29. Goh, D.H., Ang, R.P.: An introduction to association rule mining: An application in counseling and help-seeking behavior of adolescents. Behav. Res. Methods **39**(2), 259–266 (2007)
30. Wong, P.C., Whitney, P., Thomas, J.: Visualizing association rules for text mining. In: 1999 IEEE Symposium on Information Visualization, 1999. (Info Vis' 99) Proceedings, pp. 120–123. IEEE (1999)
31. Jayashankar, S., Sridaran, R.: Superlative model using word cloud for short answers evaluation in eLearning. Educ. Inf. Technol., 1–20 (2016)
32. DePaolo, C.A., Wilkinson, K.: Get your head into the clouds: using word clouds for analyzing qualitative assessment data. TechTrends **58**(3), 38–44 (2014)
33. Sinclair, J., Cardew-Hall, M.: The folksonomy tag cloud: when is it useful? J. Inf. Sci. **34**(1), 15–29 (2008)
34. Viegas, F.B., Wattenberg, M., Van Ham, F., Kriss, J., McKeon, M.: Manyeyes: a site for visualization at internet scale. IEEE Trans. Vis. Comput. Graphics **13**(6), 1121–1128 (2007)
35. Jiang, X., Zhang, J.: A text visualization method for cross-domain research topic mining. J. Vis., 1–16
36. Moloshnikov, I.A., Sboev, A.G., Rybka, R.B., Gydovskikh, D.V.: An algorithm of finding thematically similar documents with creating context-semantic graph based on probabilistic-entropy approach. Proc. Comput. Sci. **66**, 297–306 (2015)
37. Zhai, X., Li, Z., Gao, K., Huang, Y., Lin, L., Wang, L.: Research status and trend analysis of global biomedical text mining studies in recent 10 years. Scientometrics **105**(1), 509–523 (2015)
38. Chebel, M., Latiri, C., Gaussier, E.: Extraction of interlingual documents clusters based on closed concepts mining. Proc. Comput. Sci. **60**, 537–546 (2015)
39. Santosh, K.C.: g-DICE: graph mining-based document information content exploitation. Int. J. Doc. Anal. Recogn. (IJDAR) **18**(4), 337–355 (2015)
40. Song, M., Kim, S.Y.: Detecting the knowledge structure of bioinformatics by mining full-text collections. Scientometrics **96**(1), 183–201 (2013)
41. Ramakrishnan, C., Patnia, A., Hovy, E., Burns, G.A.: Layout-aware text extraction from full-text PDF of scientific articles. Source Code Biol. Med. **7**(1), 1 (2012)
42. Mooney, R.J., Bunescu, R.: Mining knowledge from text using information extraction. ACM SIGKDD Explor. Newsl. **7**(1), 3–10 (2005)
43. Callan, J., Mitamura, T.: Knowledge-based extraction of named entities. In: Proceedings of the Eleventh International Conference on Information and Knowledge Management, pp. 532–537. ACM (2002)
44. Al-Emran, M.N.H.: Investigating Students' and Faculty members' Attitudes Towards the Use of Mobile Learning in Higher Educational Environments at the Gulf Region (2014)
45. Al Emran, M., Shaalan, K.: E-podium Technology: A medium of managing Knowledge at Al Buraimi University College via M-learning. In: BCS International IT Conference (2014)
46. Al-Emran, M., Shaalan, K.: Attitudes towards the use of mobile learning: a case study from the gulf region. Int. J. Interact. Mobile Technol. (iJIM) **9**(3), 75–78 (2015)
47. Al-Emran, M., Shaalan, K.: Learners and educators attitudes towards mobile learning in higher education: State of the art. In: 2015 International Conference on Advances in Computing, Communications and Informatics (ICACCI), pp. 907–913. IEEE (2015)
48. Al-Emran, M., Elsherif, H.M., Shaalan, K.: Investigating attitudes towards the use of mobile learning in higher education. Comput. Human Behav. **56**, 93–102 (2016)

49. Al-Emran, M., Malik, S.I.: The Impact of Google Apps at Work: Higher Educational Perspective. Int. J. Interact. Mobile Technologies (iJIM) **10**(4), 85–88 (2016)
50. Al-Emran, M., Shaalan, K.: Academics' awareness towards mobile learning in Oman. Int. J. Com. Dig. Sys. **6**(1) (2017)
51. Zhang, Y., Chen, M., Liu, L.: A review on text mining. In: 2015 6th IEEE International Conference on Software Engineering and Service Science (ICSESS), pp. 681–685. IEEE (2015)
52. Verma, T., Renu, R., Gaur, D.: Tokenization and Filtering Process in Rapid Miner. Int. J. Appl. Inf. Syst. **7**(2), 16–18 (2014)
53. Zaza, S., Al-Emran, M.: Mining and exploration of credit cards data in UAE. In: 2015 Fifth International Conference on e-Learning (econf), pp. 275–279. IEEE (2015)

Text Mining and Analytics: A Case Study from News Channels Posts on Facebook

Chaker Mhamdi, Mostafa Al-Emran and Said A. Salloum

Abstract Nowadays, social media has swiftly altered the media landscape resulting in a competitive environment of news creation and dissemination. Sharing news through social media websites is almost provided in a textual format. The nature of the disseminated text is considered as unstructured text. Text mining techniques play a significant role in transforming the unstructured text into informative knowledge with various interesting patterns. Due to the lack of literature on textual analysis of news channels' in social media, the current study seeks to explore this genre of new media discourse through analyzing news channels online textual data and transforming its quantifiable information into constructive knowledge. Accordingly, this study applies various text mining techniques on this under-researched context aiming at extracting knowledge from unstructured textual data. To this end, three news channels have been selected, namely Fox News, CNN, and ABC News. Data has been collected from the Facebook pages of these three news channels through Facepager tool which was then processed using RapidMiner tool. Findings indicated that USA elections news received the highest coverage among others in these channels. Moreover, results revealed that the most frequent shared posts regarding the USA elections were tackled by the CNN followed by ABC News, and Fox News, respectively. Additionally, results revealed a significant relationship between ABC News and CNN in covering similar topics.

C. Mhamdi (✉)
Manouba University, Manouba, Tunisia
e-mail: shaker@buc.edu.om

C. Mhamdi · M. Al-Emran
Al Buraimi University College, Buraimi, Oman
e-mail: malemran@buc.edu.om

M. Al-Emran
Universiti Malaysia Pahang, Gambang, Malaysia

S.A. Salloum
The British University in Dubai, Dubai, UAE
e-mail: ssalloum@uof.ac.ae

S.A. Salloum
University of Fujairah, Fujairah, UAE

K. Shaalan et al. (eds.), *Intelligent Natural Language Processing: Trends and Applications*, Studies in Computational Intelligence 740,
https://doi.org/10.1007/978-3-319-67056-0_19

Keywords Text mining · Social media · News channels · Facebook

1 Introduction

In today's globalized world, media is no longer the same as it has previously been in the pre-technology era. For long, viewers, listeners, and readers had to wait for their morning papers, listened to morning radio news leads or assembled to watch news channels in order to follow events around the world. Currently, a growing number of audiences are constantly going online seeking instant breaking news [1]. In the new and highly competitive environment of news creation and dissemination, social media is swiftly altering the media landscape. Today, with the swift pace of globalization affecting all facets of life and revolutionizing the information technologies, the context of news creation and dissemination has been radically transformed. Nowadays, all internet users can easily connect and access a global platform where news is freely available widely shared and easily disseminated. Seeking to define social media, [2] states that "the term 'social media' refers to the wide range of internet-based and mobile services that allow users to participate in online exchanges, contribute user-created content, or join online communities" (p. 1). It is widely argued that "the great wave of web innovation since Google in 1998 has been in social media. Social media is about networking and communicating through text, video, blogs, pictures, and status updates on sites such as Facebook, MySpace, Linkedin or microblogs such as Twitter" [3] (p. 3). Social media is undoubtedly altering the traditional norms of journalism and media practices. A study by [1] argues that "news media is increasingly a forum for information and debate with a non-linear flow of information and open-sourced journalism" (p. 272). In fact, social media has gained such a paramount importance in the realm of journalism due to its interactive aspect and significant role in communication and breaking news.

The dissemination of news through social media websites is almost provided in a textual format. The nature of the disseminated text is considered as unstructured text. Text mining is perceived as the process of extracting indefinite and practical patterns or knowledge from a collection of vast and unstructured data or corpus [4–6]. Scholars of [7] stated that text mining has become one of the trendy fields that has been incorporated in several research fields such as computational linguistics, Information Retrieval (IR) and data mining. It is a branch of data mining where the process involves extorting constructive knowledge starting from unstructured text data. Text mining seeks to increase the efficiency of analyzing texts through the introduction of a number of techniques such as topic detection and tracking, keyword extraction, sentiment analysis, document clustering, and automatic document summarization.

A study by [8] asserts that in this era of sophisticated information technologies and social media, sharing has become an essential element of news dissemination. Consequently, this has led to a mounting integration of User Generated Content

(UGC) into professional news feeds [1]. This multi-media feature has made current media discourse, namely news creation and consumption, a feasible area of research. Researching language processing in its diverse associations with media enables researchers to explore language use that fosters disseminating information in a certain society. Additionally, text mining is an interrelated field with Natural Language Processing (NLP). NLP is concerned about the mutual relations among the huge amount of unstructured existing text [9]. Furthermore, human-being languages are analyzed and interpreted through various NLP techniques [10, 11].

The other sections of the paper are as follows: Sect. 2 provides a detailed literature review about social media and text mining. The methodology is presented in Sect. 3. Section 4 demonstrates the results, while the conclusion and future work are described in Sect. 5.

2 Literature Review

Various researches reveal that social media has gradually been attracting an interestingly rising number of active readers and writers [1, 12, 13]. The interactive aspect of all types of social media lies behind its attractiveness and worldwide fame among internet users. Many studies revealed that social media activists have succeeded to foster a decentralized flow of information and challenge the traditional hierarchical monopoly [1, 14, 15].

Notably, several studies point out that an enormous move is taking place in the media industry where news channels made no exception. Because of the universal reach of social media, "mass media is passé, today it is all about personal media" [3] (p. 9). Additionally, it is evident that today social networking websites have become instant news sources. These days, audiences enjoy direct access to a various online news sources where everyone can decide on his own what to read. A study by [16] advocates that "the public is increasingly seeking its news not from mainstream television networks or ink-on-dead-trees but from grazing online. When we go online, each of us is our own editor, our own gatekeeper. We select the kind of news and opinions that we care the most about" (p. 6).

Seeking to understand how social networking websites provide opportunities to establish an interactive relation between people resulting in a mutual sharing of knowledge, [17] investigated web-based applications and reviewed text mining techniques to explore textual patterns of social web. Through a detailed survey, the authors aimed at providing a comprehensive understanding of various text mining techniques and their appliance in social networks. The study presented the recently developed advancement in the area of intelligent text analysis covering two main approaches to text mining, namely classification and clustering.

Building on previous multimedia research, [18] introduced an approach of "social multimedia" applications which are based on a variety of successful applications that are rooted in mining multimedia content analysis in the context of social multimedia. The study focused on two web-based sharing services that have remarkably made a rising amount of online multimedia content, namely Flickr and Youtube. The researcher concluded that nowadays we live in an ever-changing time that he called the "age of social multimedia" which calls for new challenges which can leverage the new trends.

Focusing the so-called "Arab Spring", [19] explored Facebook aiming at extracting useful information about users' sentiments during this important phase of history. To this end, the researchers used a methodology based on Support Vector Machine (SVM) and Naïve Bayes. Additionally, the study constructed a sentiment lexicon derived from the emoticons, interjections and acronyms which were extracted from the users' statuses updates. Though this study reached insightful findings about Tunisian Facebook users' sentiments and reactions during one of their historical moments, Tunisian revolution of January 2011, it posed some weaknesses related to tracking users' changing sentiments on a particular topic. The study ignored the factor of time dependency in its analysis and discussion which affected the findings partially. It would have been more interesting if the investigation included the temporal feature in its analysis.

Although the use of social media websites has gradually been towering, little research has been conducted to explore the enormous data posted by customers which can be useful for companies' benefits. Trying to tackle this gap, [20] sought to present a case study in order to reveal how social media data analysis can be of great importance to business decision makers and operations management research and practice. Selecting Facebook as a source, data was collected through accessing and downloading SAMSUNG Mobile Facebook page. Data which represented the corpus of this study consisted of 128,371 comments from June 10th to September 10th 2013 and was captured using NCapture for NVivo 10.

Analyzing only the comments in English language, the researchers suggested a structured approach to analyze social media data as well as a statistical cluster analysis to recognize inter-connections between significant factors. Scholars of [20] contributed to the existing knowledge through outlining a clear straight-forward approach to derive quantifiable data from social media for further analysis. The outcome of such quantification can be used in surveys, questionnaires, interviews and design of decision-making systems. However, the study overlooked the ever-changing pattern and dynamics of Facebook users and the challenge of real-time basis.

In this same line of thoughts, [21] applied visual data mining techniques on the Yahoo! Answers to tackle health care consumers' terms use behavior. The investigators collected and analyzed data related to diabetes in the health category of Yahoo! Answers over a period of three months. After processing, validating and classifying the terms from the data, the researchers utilized Multi-dimensional Scaling and Social Network Analysis visualization methods to explore the relationships of terms from related categories. The findings of such an investigation are beneficial to the health sector whether professionals or health consumers.

Since social networking websites offer vast venues for people to express and share their opinions, emotions and experiences, students made no exception. Accordingly, this area was a fertile field of research for a number of researchers.

Researchers of [22] studied students' informal conversations on social media focusing on their feelings, opinions and main concerns about their learning experience. The researcher investigated a sample of 25.000 engineering students' tweets related to their college life. The findings of the study revealed that engineering students encounter various problems such as lack of social engagement, study load, and sleep deprivation.

Moreover, the researchers employed a second level of analysis where a multi-label classification algorithm was utilized to categorize tweets mirroring students' challenges and problems. The same algorithm was then used to detect and stream tweets according to the geo-location of Purdue University. The originality of this work stemmed from its methodology and findings that reveal how social media can offer insightful conclusions about students' experiences. It is useful to researchers and students interested in educational data mining, learning analytics, and learning technologies. It overcame the pitfalls of manual qualitative analysis and user-generated textual content of large scale computational analysis.

Exploring the same context of educational data mining, [23] studied the massive open online courses (MOOCs) and their intersection with social media tools. The study employed three main approaches where it first calculated the significant descriptive statistics of tweets pertaining to MOOC. The second approach was to examine how to assess users' sentiment toward MOOC based on related tweets. Third, the authors scrutinized the positive and negative re-tweets associated with MOOCs to identify influencers. The three approaches together adopted in this study enabled the researchers to overcome previous limitations especially those pertaining to the temporal factor, users' perceptions and sentimental messages, and influencers' identification. The findings of this research provide a thorough understanding of MOOC trends.

A further examination of social media data mining in the context of university students is that of [24] where the focus was on extracting knowledge from university students' data floating on social networking websites. Using the data mining techniques K-means to extract useful information related to the educational sector, the author conducted a questionnaire for university students from different disciplines where responses were analyzed via data mining model. The findings of the study identified the most frequent types of data flow on social media, namely Facebook, Orkut, and Twitter among university students.

Another area which has recently been receiving attention is that of online User Generated Content (UGC). This recent attention stems from its extensive applications from academicians and industry stakeholders [25]. In their pilot study, these latter authors proposed a new Big Data processing approach to scrutinize a niche subset of user-generated popular culture content on one of the most renowned social networks operating in Chinese language known as *Douban*. Based on huge data samples, the researchers discussed how heterogeneous features from samples can be manipulated in order to smooth the process of analyzing online user-generated content such as review comments, users' profiles, and film details on *Douban*.

The study presented a novel framework and demonstrated its applicability and flexibility which enable researchers to extract useful information from social media data and turn it into relevant knowledge. The authors recommended that this knowledge itself could in turn be extended and used to feed producers and decision makers of digital media content especially films and television shows.

In spite of the substantial research on social media data mining and extracting relevant information from various sectors and turning it into useful knowledge, an important area seems to be overlooked. In fact, so far, no research has particularly addressed textual analysis of news channels' social networking websites as well as their User Generated Content. Actually, this genre of new media discourse has increasingly been witnessing a notable worldwide expansion following the technological boom in our postmodern information age.

This has made the news channels' online sites, their social networks WebPages and newly-introduced User Generated Content a rich field of research for analysing online textual data and transforming the quantifiable information into constructive knowledge. Accordingly, this paper presents the authors' attempt to bridge this gap through building on existing researches and applying various text mining techniques on a new under-researched context. Seeking to address this research gap, the authors seek to answer the following research questions:

RQ1: What are the most frequent linked words that are posted on news channels' Facebook pages?
RQ2: To what extent do the news channels share posts on Facebook?
RQ3: To what extent do the news channels cover similar topics?

3 Research Methodology

The collected data has been retrieved from three news channels pages on Facebook, namely Fox News, CNN, and ABC News. Facepager tool is used for the process of the data collection. Facepager stores the extracted data in a local database and then exports these data into a CSV form. The present study analyzes the posts that the aforementioned news channels post on their Facebook pages. The total number of the collected posts is 39,003. According to [26], data has been filtered through the elimination of all missing data. Then, the clean data version is uploaded into RapidMiner tool for text analysis. All irrelevant variables were excluded from the analysis process in order to leverage the data quality and performance.

During the pre-processing stage, we found that the collected data contains some special characters and empty cells. According to [27], we eliminate such kind of data. According to [28], we separate the documents into tokens along with the representation of each word; this process called "Tokenization". Afterwards, we transform all the characters into a lower case form and store them in a separate document. The next step involves filtering the stop-words. Based on that, the operator eliminates all tokens that match the corresponding stop-word. Figure 1 illustrates the aforementioned process.

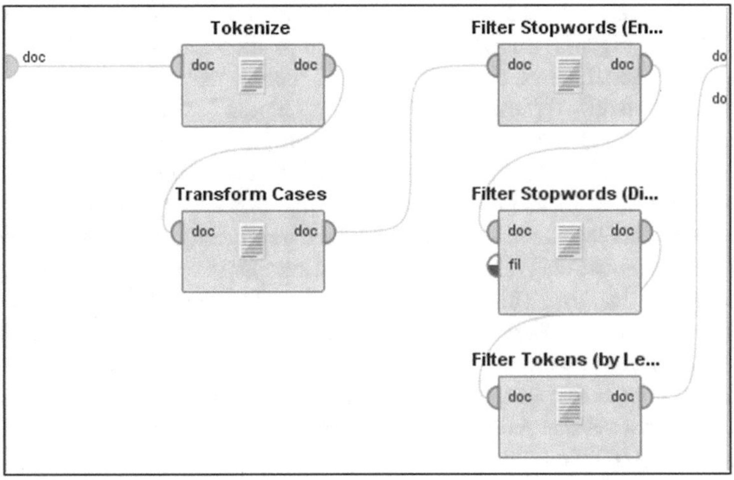

Fig. 1 Document processing

4 Results

Q1: What are the most frequent linked words that are posted on news channels' Facebook pages?

According to [29], we used the word cloud technique in order to answer the above research question. Figure 2 depicts the most frequent linked keywords across the three targeted news channels. We can observe that the most frequent keyword is "Trump" followed by "Donald", "President", and "Elect", respectively. This is an indicator that the targeted news channels are focusing on the political issues above other topics. USA elections news takes the first place among others in these channels.

Fig. 2 Word cloud for the most frequent keywords

For further investigation, we applied the word frequency technique on the collected data. Figure 3 shows the distribution of the most frequent linked words across the targeted news channels. We can notice that the top 10 keywords, "Trump", "Donald", "President", "Elect", "Clinton", "Obama", "Campaign", "World", "Breaking", and "America", refer to the USA elections. The interrelation among these words reveals that the world breaking news is concerned with the USA elections with more focus on the situations of the previous president (i.e. Obama), the candidate for elections (i.e. Clinton), and the new president (i.e. Donald Trump). It is clearly depicted that "Trump" is the most frequent keyword that was reported by these news channels. This is followed by "Donald", "President", "Clinton", and "Elect", respectively. The results of the word frequency technique are almost similar to that of the word cloud. That is, whatever the techniques used; the results confirm the consistency of the collected data. Moreover, Fig. 3 illustrates that Fox News is the channel that most frequently posts news concerning the USA elections followed by "ABC News" and "CNN" respectively.

In addition, the most frequent keywords that were illustrated in (Fig. 3) were analyzed for further investigation. As per (Fig. 4), the word "Trump" was frequently mentioned by "Fox News" channel followed by "ABC News" and "CNN", respectively.

As per (Fig. 5), the word "Donald" was frequently mentioned by "Fox News" channel followed by "ABC News" and "CNN", respectively.

As per (Fig. 6), the word "President" was frequently mentioned by "Fox News" channel followed by "ABC News" and "CNN", respectively.

As per (Fig. 7), the word "Elect" was frequently mentioned by "Fox News" channel followed by "ABC News" and "CNN", respectively.

As per (Fig. 8), the word "Clinton" was frequently mentioned by "Fox News" channel followed by "CNN" and "ABC News", respectively.

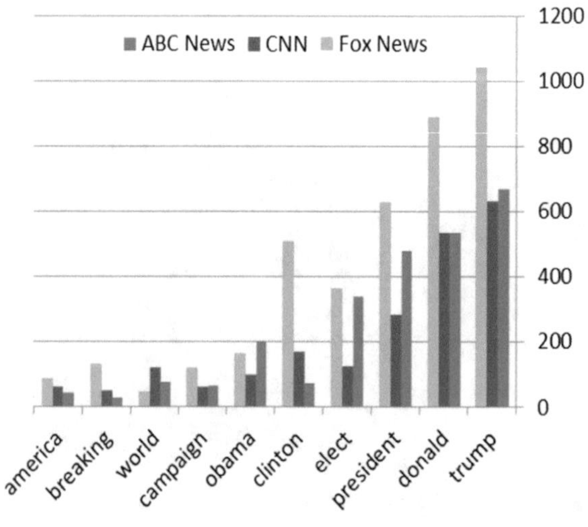

Fig. 3 Word frequency distribution among the news channels

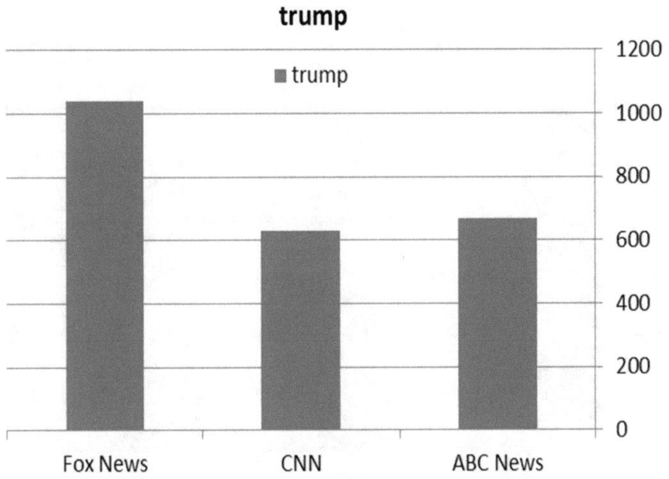

Fig. 4 The frequency distribution of the word "Trump" across all news channels

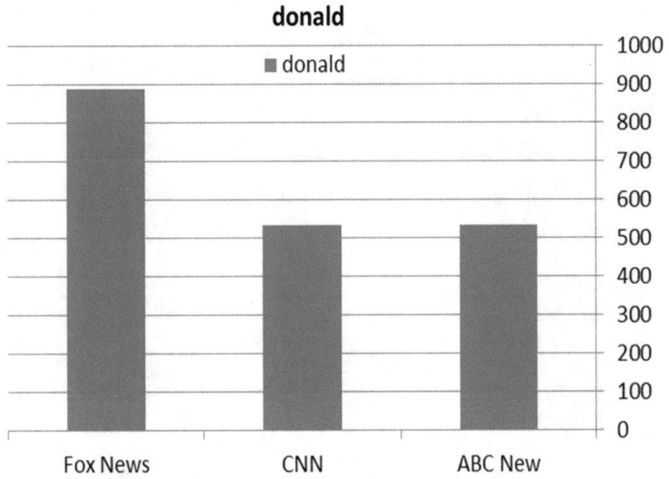

Fig. 5 The frequency distribution of the word "Donald" across all news channels

As per (Fig. 9), the word "Obama" was frequently mentioned by "ABC News" channel followed by "Fox News" and "CNN", respectively.

As per (Fig. 10), the word "Campaign" was frequently mentioned by "Fox News" channel followed by "ABC News" and "CNN", respectively.

As per (Fig. 11), the word "World" was frequently mentioned by "CNN" channel followed by "ABC News" and "Fox News", respectively.

As per (Fig. 12), the word "Breaking" was frequently mentioned by "Fox News" channel followed by "CNN" and "ABC News", respectively.

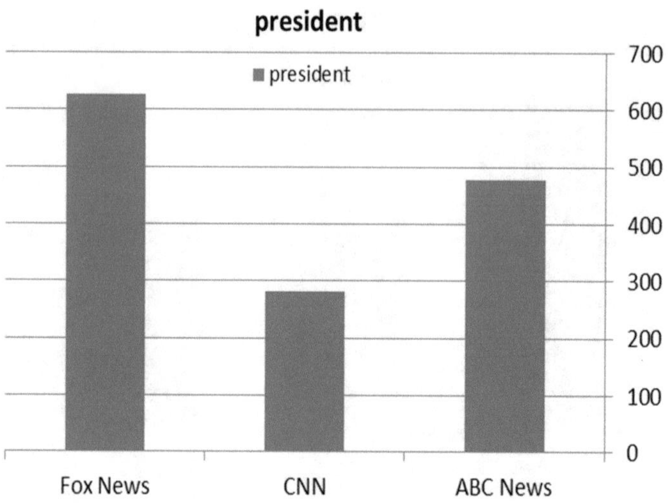

Fig. 6 The frequency distribution of the word "President" across all news channels

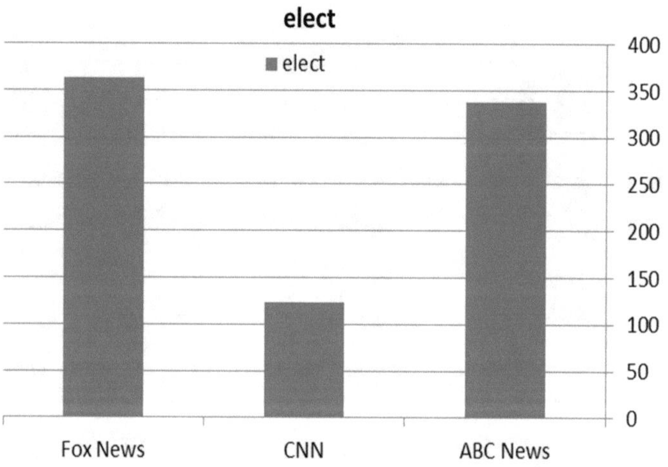

Fig. 7 The frequency distribution of the word "Elect" across all news channels

As per (Fig. 13), the word "America" was frequently mentioned by "Fox News" channel followed by "CNN" and "ABC News", respectively.

Q2: To what extent do the news channels share posts on Facebook?

The word frequency technique was performed on the collected data for the purpose of determining the most frequent shared posts across the targeted news channels and identifying the channel that most frequently shares posts on Facebook. As per (Fig. 14), results revealed that the most frequent shared posts regarding the

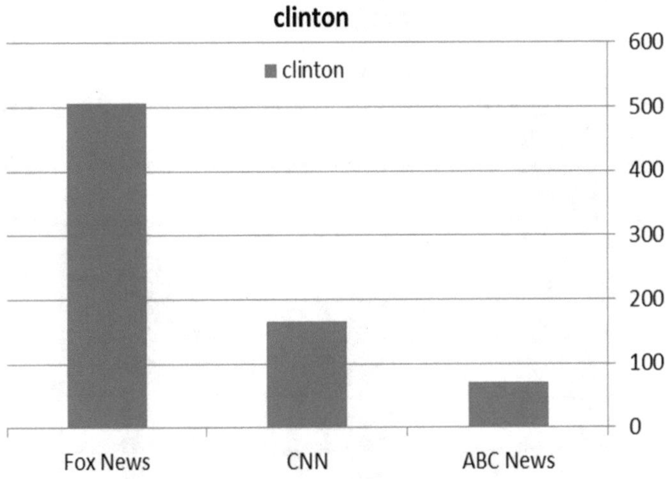

Fig. 8 The frequency distribution of the word "Clinton" across all news channels

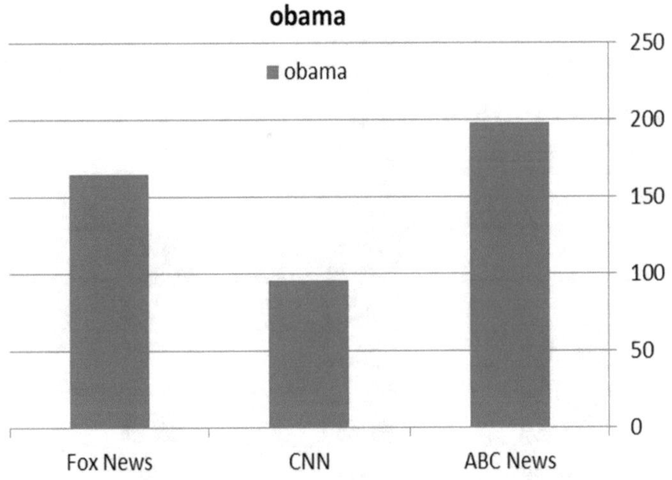

Fig. 9 The frequency distribution of the word "Obama" across all news channels

USA elections were tackled by CNN followed by ABC News and Fox News, respectively. This indicates that the shared posts were highly interesting, which was the reason behind sharing them by many people.

Additionally, results indicated, as per (Fig. 15), that the most frequent news channel that shared posts on Facebook regarding the USA elections is Fox News followed by CNN and ABC News, respectively. That is, Fox News channel was the most interested channel in sharing news regarding the USA elections followed by CNN and ABC News, respectively.

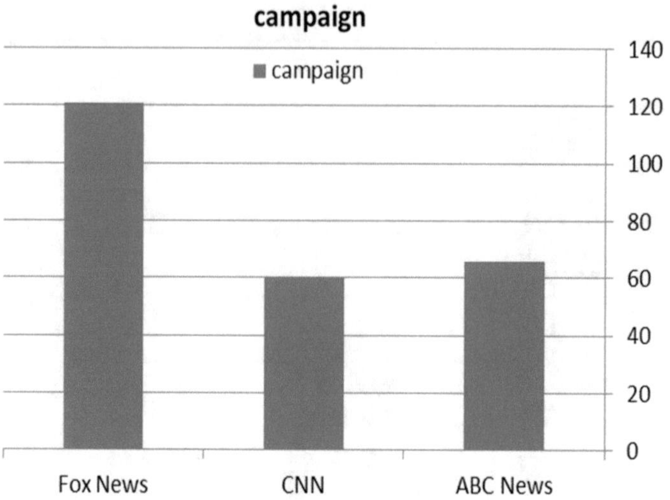

Fig. 10 The frequency distribution of the word "Campaign" across all news channels

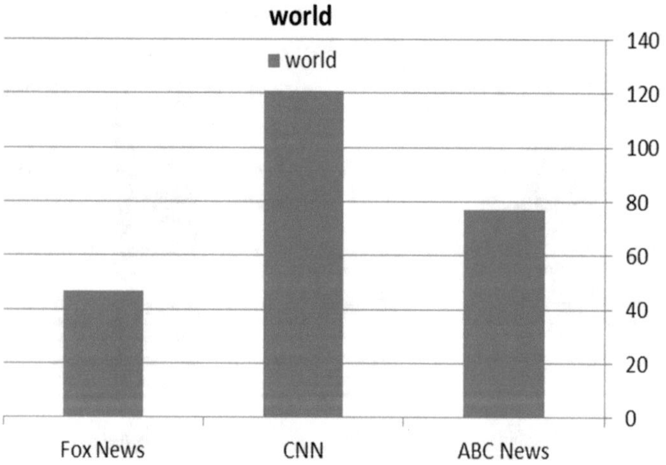

Fig. 11 The frequency distribution of the word "World" across all news channels

Q3: To what extent do the news channels cover similar topics?

According to [30], we applied the similarity operator on the collected data for the purpose of determining the topics that are highly similar to each other. According to (Fig. 16), the value (1.0) stands for (ABC News), (2.0) denotes (CNN), and (3.0) represents the (Fox News) channels. We can perceive that there is a significant relationship between ABC News and CNN in covering similar topics. That is, 38% of the posted topics on ABC News and CNN Facebook pages are similar to each other.

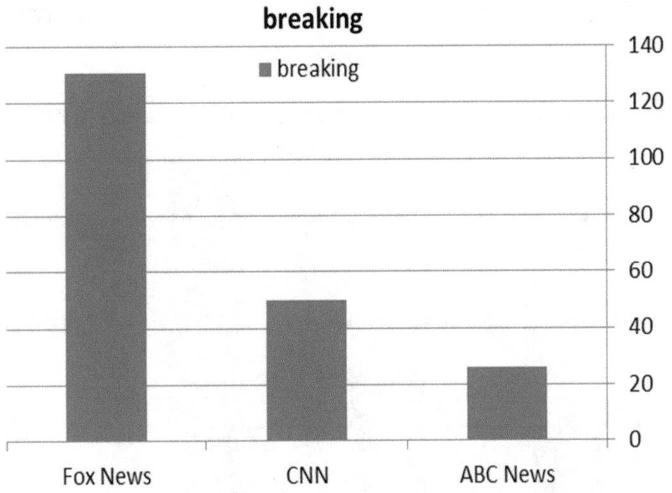

Fig. 12 The frequency distribution of the word "Breaking" across all news channels

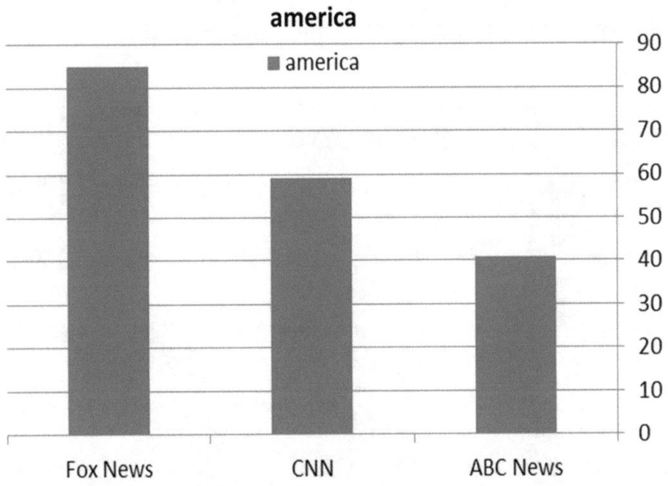

Fig. 13 The frequency distribution of the word "America" across all news channels

Moreover, the second highest percentage (32%) of the similar posted topics is between ABC News and Fox News. In addition, the similar posted topics between CNN and Fox News represents the lowest percentage with (30%).

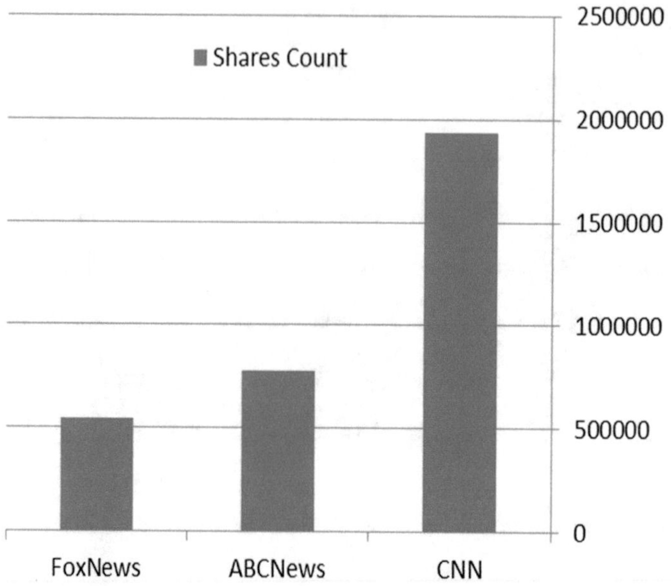

Fig. 14 The most freuqent shared posts across news channels

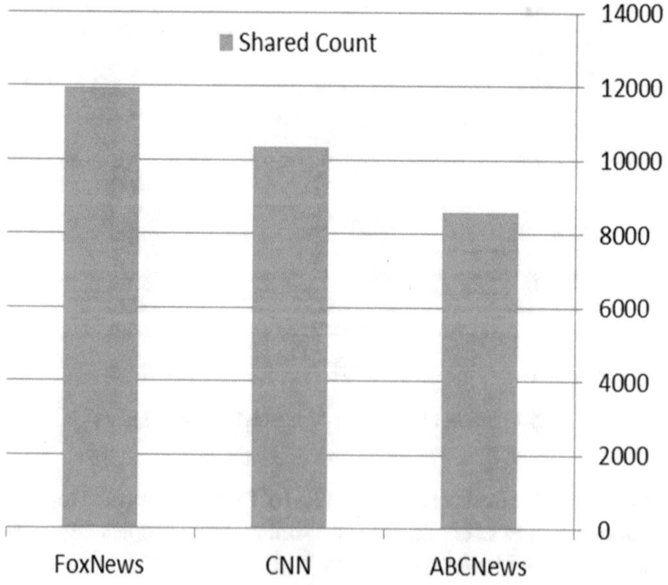

Fig. 15 The most freuqent news channel that shares posts on Facebook

Fig. 16 Similarity among news channels

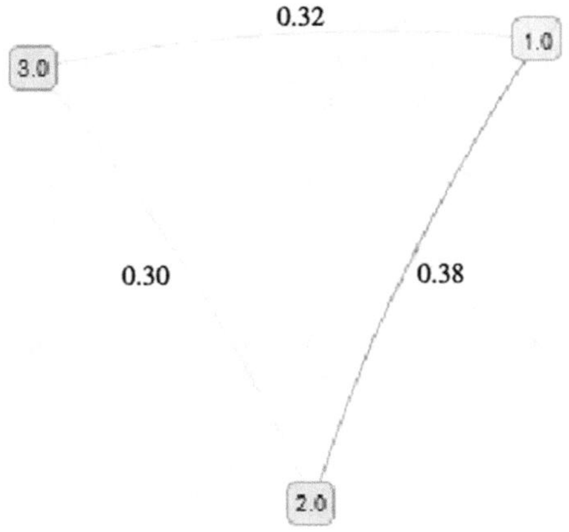

5 Conclusion

In today's globalized world, media is no longer the same as it has previously been in the pre-technology era. Currently, a growing number of audiences are constantly going online seeking instant breaking news [1, 32]. In the new and highly competitive environment of news creation and dissemination, social media is swiftly altering the media landscape. The dissemination of news through social media websites is nearly provided in a textual format. The nature of the disseminated text is considered as unstructured text. Text mining techniques play a vital role in transforming the unstructured text into informative knowledge with various interesting patterns. According to the literature, no research has particularly addressed textual analysis of news channels' social networking websites as well as their User Generated Content. Accordingly, this paper contributed to the existing body of literature to bridge this gap through building on existing researches and applying various text mining techniques on a new under-researched context.

The collected data has been retrieved from three news channels pages on Facebook, namely Fox News, CNN, and ABC News. Facepager tool is used for the process of the data collection. The current study seeks to analyze the posts that these news channels post on their Facebook pages. The same procedures were followed by [31]. The total number of the collected posts is 39,003.

By applying the word cloud method, findings pointed out that the most frequent keyword is "Trump" followed by "Donald", "President", and "Elect" respectively. This indicated that the targeted news channels are focusing on the political issues over other issues. USA elections news takes the first place among others in these channels. By applying the word frequency technique, results showed that the top 10 keywords, "Trump", "Donald", "President", "Elect", "Clinton", "Obama", "Campaign", "World", "Breaking", and "America", refer to the USA elections. The interrelation among these words reveals that the world breaking news is concerned with the USA elections with more focus on the situations of the previous president (i.e. Obama), the candidate for elections (i.e. Clinton), and the new president (i.e. Donald Trump). Moreover, results revealed that the most frequent shared posts regarding the USA elections were tackled by the CNN followed by ABC News and Fox News, respectively. This indicates that the shared posts were highly interesting; the reason that causes many people to share them. Additionally, results indicated that the most frequent news channel that shared posts on Facebook regarding the USA elections is Fox News followed by CNN and ABC News, respectively. That is, the Fox News channel was the most interested channel in sharing news regarding the USA elections followed by CNN and ABC News respectively. We applied the similarity operator on the collected data for the purpose of determining the topics that are highly similar to each other. Results indicated that there is a significant relationship between ABC News and CNN in covering similar topics. That is, 38% of the posted topics on ABC News and CNN Facebook pages are similar to each other.

As a future work, we are interested in studying the human's sentiments through their posts, posts' share, and comments on social media. Moreover, we are planning to collect data from other news channels for drawing more interesting patterns. Furthermore, text mining researchers may focus on analyzing the comments of the most shared posts. This will add more value to the current research results.

References

1. Mhamdi, C.: Transgressing media boundaries: news creation and dissemination in a globalized world. Mediterr. J. Soc. Sci. **7**(5), 272 (2016)
2. Dewing, M: Social media: an introduction. In: Library of Parliament Publication no. 2010-03-E, pp. 1–5 (2012)
3. Alejandro, J.: Journalism in the age of social media. In: Reuters Institute for the Study of Journalism, pp. 1–47 (2010)
4. Tan, A.H.: Text mining: the state of the art and the challenges. In: Proceedings of the PAKDD 1999 Workshop on Knowledge Disocovery from Advanced Databases, vol. 8, pp. 65–70 (1999)
5. Hearst, M.A.: Untangling text data mining. In: Proceedings of the 37th Annual Meeting of the Association for Computational Linguistics on Computational Linguistics, pp. 3–10. Association for Computational Linguistics (1999)
6. Feldman, R., Dagan, I.: Knowledge discovery in textual databases (KDT). In: KDD, vol. 95, pp. 112–117 (1995)
7. Salloum, S.A., Al-Emran, M., Monem, A.A., Shaalan, K.: A survey of text mining in social media: facebook and twitter perspectives. Adv. Sci. Technol. Eng. Syst. J. (2017)
8. Hermida, A., Fletcher, F., Korell, D., Logan, D.: Share, like, recommend. Journal. Stud. **13**(5–6), 819–831 (2012)
9. Salloum, S.A., Al-Emran, M., Shaalan, K.: A survey of lexical functional grammar in the arabic context. Int. J. Com. Net. Tech. **4**(3) (2016)
10. Al Emran, M., Shaalan, K.: A survey of intelligent language tutoring systems. In: Advances in Computing, Communications and Informatics (ICACCI, 2014 International Conference on, pp. 393–399. IEEE (2014)

11. Al-Emran, M., Zaza, S., Shaalan, K.: Parsing modern standard Arabic using Treebank resources. In: 2015 International Conference on Information and Communication Technology Research (ICTRC), pp. 80–83. IEEE (2015)
12. Comscore Media Matrix. (2008). Huffington Post and Politico leadwave of explosive growth at independent political blogs and news sites this election season. Retrieved from http://www.comscore.com/press/release.asp?press=2525
13. Sifry, D.: State of the blogosphere. Retrieved from Technorati (2008). http://technorati.com/blogging/state-of-the-blogosphere/
14. Rosen, J.: The people formerly called the audience (2006). Retrieved from http://jounalism.nyu.edu/pubzone/weblogs/pressthink/2006/06/27/ppl_frmr.html
15. Weinberger, D.: Everything is miscellaneous. Henry Holt, New York (2008)
16. Kristof, N.: The daily me. The New York Times. March 19, 2009
17. Irfan, R., King, C.K., Grages, D., Ewen, S., Khan, S.U., Madani, S.A., … & Tziritas, N.: A survey on text mining in social networks. Knowl. Eng. Rev. **30**(2), 157–170 (2015)
18. Naaman, M.: Social multimedia: highlighting opportunities for search and mining of multimedia data in social media applications. Multimed. Tools Appl. **56**(1), 9–34 (2012)
19. Hamouda, S.B., Akaichi, J.: Social networks' text mining for sentiment classification: The case of Facebook'statuses updates in the 'Arabic Spring'era. Int. J. Appl. Innov. Eng. Manag. **2**(5), 470–478 (2013)
20. Chan, H.K., Lacka, E., Yee, R.W., Lim, M.K.: A case study on mining social media data. In: 2014 IEEE International Conference on Industrial Engineering and Engineering Management, pp. 593–596. IEEE (2014)
21. Zhang, J., Zhao, Y.: Visual data mining in a Q&A based social media website. In: Library and Information Sciences, pp. 41–55. Springer, Heidelberg (2014)
22. Chen, X., Vorvoreanu, M., Madhavan, K.: Mining social media data for understanding students' learning experiences. IEEE Trans. Learn. Technol. **7**(3), 246–259 (2014)
23. Shen, C.W., Kuo, C.J.: Learning in massive open online courses: Evidence from social media mining. Comput. Human Behav. **51**, 568–577 (2015)
24. Singh, A.: Mining of social media data of university students. Educ. Inf. Technol., 1–12 (2016)
25. Yang, J., Yecies, B.: Mining Chinese social media UGC: a big-data framework for analyzing Douban movie reviews. J. Big Data **3**(1), 1 (2016)
26. Zaza, S., Al-Emran, M.: Mining and exploration of credit cards data in UAE. In: 2015 Fifth International Conference on e-Learning (econf), pp. 275–279. IEEE (2015)
27. Atia, S., Shaalan, K.: Increasing the accuracy of opinion mining in Arabic. In: 2015 First International Conference on Arabic Computational Linguistics (ACLing), pp. 106–113. IEEE (2015)
28. Verma, T., Renu, R., Gaur, D.: Tokenization and Filtering Process in Rapid Miner. Int. J. Appl. Inf. Syst. **7**(2), 16–18 (2014)
29. Jayashankar, S., Sridaran, R.: Superlative model using word cloud for short answers evaluation in eLearning. Educ. Inf. Technol., 1–20 (2016)
30. Huang, A.: Similarity measures for text document clustering. In: Proceedings of the Sixth New Zealand Computer Science Research Student Conference (NZCSRSC2008), Christchurch, New Zealand, pp. 49–56 (2008)
31. Salloum, S.A., Al-Emran, M., Shaalan, K.: Mining social media text: extracting knowledge from facebook. Int. J. Comput. Digit. Syst. **6**(2), 73–81 (2017)
32. Mhamdi, C.: Framing "the Other" in Times of Conflicts: CNN's Coverage of the 2003 Iraq War. Mediterr. J. Soc. Sci. **8**(2), 147 (2017)

A Survey of Arabic Text Mining

Said A. Salloum, Ahmad Qasim AlHamad, Mostafa Al-Emran and Khaled Shaalan

Abstract Recently, text mining has become an interesting research field due to the huge amount of existing text on the web. Text mining is an essential field in the context of data mining for discovering interesting patterns in textual data. Examining and extracting of such information patterns from huge datasets is considered as a crucial process. A lot of survey studies were conducted for the purpose of using various text mining methods for unstructured datasets. It has been noticed that comprehensive survey studies in the Arabic context were neglected. This study aims to give a broad review of various studies related to the Arabic text mining with more focus on the Holy Quran, sentiment analysis, and web documents. Furthermore, the synthesis of the research problems and methodologies of the surveyed studies will help the text mining scholars in pursuing their future studies.

Keywords Text mining · Arabic text · Opinion mining
Sentiment analysis · Social media

S.A. Salloum (✉) · K. Shaalan
Faculty of Engineering & IT, The British University in Dubai, Dubai, UAE
e-mail: ssalloum@uof.ac.ae

K. Shaalan
e-mail: Khaled.shaalan@buid.ac.ae

S.A. Salloum · A.Q. AlHamad
University of Fujairah, Fujairah, UAE
e-mail: aqd14@yahoo.com

M. Al-Emran
Faculty of Computer Systems and Software Engineering, Universiti Malaysia Pahang, Pahang, Malaysia
e-mail: malemran@buc.edu.om

M. Al-Emran
Al Buraimi University College, Al Buraimi, Oman

K. Shaalan et al. (eds.), *Intelligent Natural Language Processing: Trends and Applications*, Studies in Computational Intelligence 740,
https://doi.org/10.1007/978-3-319-67056-0_20

1 Introduction

Since there is a lot of information presented in different versions of text documents, text mining is a critically important procedure. The process of making this text readable for machines is very challenging. The task covers all aspects like extracting linguistic features from the text to make it readable like at human level so that it can be mined. This is especially true for the Arabic language. Text mining is related to a certain degree to data mining [1] in which interesting patterns in huge databases are found [2]. Text mining, also called Intelligent Text Analysis, Text Data Mining or Knowledge-Discovery in Text (KDT), is mainly used to define the procedure of extracting interesting and non-trivial data and knowledge from unstructured text [3]. The field of text mining is relatively new and linked to various research fields like information retrieval, data mining, machine learning, statistics, and computational linguistics. Since as much as 80% data or even more is saved in the form of text, text mining can potentially be said to hold great monetary value [4]. In addition, knowledge could be discovered from various sources; especially, unstructured texts, which are still considered the greatest easily accessible source of knowledge. In Knowledge Discovery from Text (KDT), the main problem arises when trying to extract explicit and implicit ideas and semantic links among different ideas using Natural language processing (NLP) methods [5]. The objective is to obtain a full understanding of vast amounts of text data. KDT is greatly connected to NLP, but it also takes from processes in statistics, machine learning, information extraction, handling of information etc. During its procedure of finding out hidden secrets, KDT has a greatly important part in upcoming applications, like Text Understanding.

Arabic is counted among world's most popular languages. As many as 280 million people use it as a first language while at least 250 million consider it a second language. Although Arabic is a widely used language, there is still very little research done on the retrieval or mining of written documents in Arabic literature. The reason is the one-of-a-kind morphological rules of the language. It is a difficult language due to a lot of factors. Orthographic alongside diacritics tends to be more phonetic and very clear in Arabic and some combinations of the letters can be written in unique ways [6–8]. Furthermore, text mining is an interrelated field with Natural Language Processing (NLP). NLP is one of the hot topics that is concerned with the interrelation among the huge amount of unstructured available text [9], besides the analysis and interpretation of human-being languages [10, 11]. This study aims to carry out a comprehensive review of studies regarding text mining in the Arabic context. We categorized the collected studies into three parts namely, the Holy Quran, sentiment analysis, and web documents. The method of categorizing those studies was in line with other review studies such as [12–14].

2 Background

As discussed earlier, a sub-field of data mining is known as the text mining, which refers to the extraction of knowledge from text documents. As far as web documents are concerned and most of the documents are in a textual format, text mining is considered as a sub-field of web mining or more precisely web content mining [15]. The patterns of text-mining applications are referred to as the information extraction, text clustering, and text classification, which have been well designed and purposeful to web documents. Plain text documents are most likely to witness and receive the techniques of information extraction. However, a different problem domain is presented when information is extracted from HTML web pages. Because many markup tags are included in the HTML documents, through which useful information from the web could be identified. The Arabic language is ranked as the fifth among other languages in terms of usage. In recent times, a noticeable growth is observed in the amount of data shared associated with everyday activities. It can have data that interprets opinions of the users. It is not possible to examine such vast amount of information without the help of further digital means. As a result, this refers to another new area of research that is known as Sentiment analysis and opinion mining [16]. Both sentiment analysis as well as opinion mining evaluates the standing of the writer related to the subject. The feature of opinion mining includes the authentication of the writer's position that is used to categorize a specified comment in the subjective class. All subjective comments that express the viewpoint of the author are supposed to be classified and due to the difficulty of Arabic language, the process will be difficult.

2.1 Text Mining Versus Data Mining

Text mining is somewhat like data mining. The difference is actually that in data mining [17] methods are created to tackle structured data derived from databases. However, text mining can be used for unstructured or semi-structured data sets like emails, full-text documents, HTML files and others. Therefore, text mining is a more efficient method for organizations. However, to this day, research studies have been more focused on data mining techniques using structured data. The major difficulty related to text mining is simply that natural language was created so that people could interact and to keep track of information, but Machines are still far behind in understanding natural language. Text mining is different from data mining [18] concerning the target data. Data mining looks into "structured" data from databases. Text mining on the other hand, aims at mining knowledge hidden in "unstructured" data inside free texts. Therefore, in text mining, there is an additional level of difficulty where the unstructured text must be prepared for data mining algorithms and techniques such as clustering, classification, and visualization as per (Fig. 1). A great effort in accomplishing this is in the difficulty and

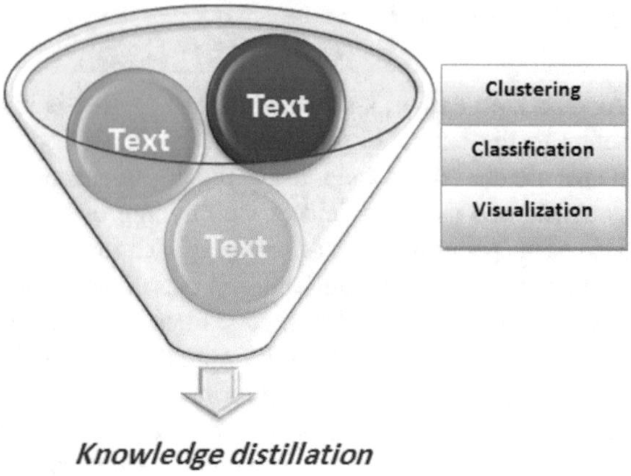

Fig. 1 Knowledge extraction from unstructured text

lengthy task of adding footnotes to the text that is being scanned using different linguistic information before implementing machine learning and text mining algorithms [19].

2.2 Natural Language Processing and Text Mining

Text mining has been recently developed as a field of study and is characterized by the streamlining of irregular data sets, which are typed in human language [20, 21]. The main mode of interaction and communication between people is text and not all of the people are used to share structured datasets. Therefore, text mining is the best possible option under such circumstances. NLP is said to be the most outstanding subject for study in comparison to others. The basic objective of NLP [9] is concerned with collecting information about the way computers evaluate and derive information from human languages in order to develop sophisticated software. The idea of sharing important data by using unique and apparently pointless information is quite inspiring. Text mining techniques analyse the data to extract the useful information that can be utilized for particular issues. It is predicted that text mining will undertake the entire NLP structure [22] to successfully analyse the human language as a structure which is appropriately given to unstructured data types [23]. With the advancement in technology, text-mining systems will be improved with the time and everyone has their eyes on it.

3 Text Mining of the Holy Quran

Around 1.6 billion Muslims over the whole world refer to the Holy Quran for guidance. Extracting information and knowledge out of the Holy Quran will reap great advantages for both, the students of Islamic studies and the public alike.

The Holy Quran is considered as the direct words of God and therefore it must be tackled carefully when automated methods of machine learning and NLP are used. The language of the Holy Quran is Arabic. Arabic is said to be one of the most difficult languages to handle using NLP and machine language. This can be attributed to a few of its special characteristics like the diacritic, several derivations of some words, complex Diglossia etc. [24, 25]. Therefore, handling the Arabic language using machine language and artificial language methods is not an easy job. A number of text mining studies have been conducted on the Arabic text of Holy Quran.

A research study by [26] was based on applying the most frequent words, word cloud, document-term matrix, and clustering techniques to evaluate the Arabic text in order to provide statistical and factual information about the Holy Quran. Another research study by [27] has used different methods that help in enabling the analysis of unstructured information by searching the knowledge embedded within the Quran text. These methods include vector space model, the pronoun tagging, part-of-speech tags (POS), the verse relatedness dataset, and concept clouds. Finally, a thesis research that was conducted by [28] to discover and extract the interesting, non-trivial knowledge out of the Holy Qur'an uses methods like information extraction, text categorization, concept linkage and discovery of associations. Table 1 summarized the work described earlier including the results they obtained.

Table 1 Research studies conducted on Arabic text mining in terms of the Holy Quran

Study	Problem	Methodology	Results
[26]	Evaluating Arabic text in order to present and provide statistical information and factual details that will assist the researchers in this field of study	Most Frequent Words, Word Cloud, Document-Term Matrix, and Clustering	The obtained results from this study carry different features of the Holy Quran like some significant words, its word cloud and different chapters with high term frequencies
[27]	Creating a platform that enables the analysis of unstructured information in searching of interesting knowledge embedded within the Qur'an text	Vector space model, The pronoun tagging, the part-of-speech tags (POS), the verse relatedness dataset, and concept clouds	Results revealed that world knowledge is required to reach the human upper bound in certain computational tasks such as detecting text relatedness, question answering, and textual entailment
[28]	Discovering and extracting of interesting, non-trivial knowledge out of the Holy Qur'an by applying text-mining methods	Information extraction, text categorization, concept linkage, and discovery of associations	POS tagged corpus improved the results over the raw text and the intention of this project to improve the results further through added layers of annotation

4 Sentiment Analysis of Arabic Text (Opinion Mining)

A study by [29] stated that sentiment analysis is a research field that is linked to different areas of study like NLP, text mining, and computational linguistics. It is related to the extraction of particular data from individuals from textual data. Sentiment analysis is sometimes called as opinion mining, subjective opinion analysis or emotion extraction [30]. Every term has a specific objective. Opinion mining is the process of classifying the negative and positive opinions that are presented in the form of text. Emotion extraction, on the other hand, is related to differentiation among different emotions (ex. happy, angry, and sad). The task concentrates on sentiment analysis depending upon opinion mining. The main objective of sentiment analysis is to investigate the emotions of people, opinions, their behaviors, feelings, and judgments about entities like facilities, items, problems, occasions, people, firms, and topics [31]. In this section, studies that related to sentiment analysis are presented. These studies differ in the methods of preprocessing, analysis techniques, and the design of reviews. A few has utilized the supervised way, and some has utilized the unsupervised way of knowledge. Multi-stages approach that was based on semantic orientation (Lexical-based classifier to handle unlabeled tweets) and a machine learning (SVM classifier) was proposed by [29] in order to recognize the polarity of Arabic tweets. However, the main challenge of this hybrid approach was to handle the practice of tweeting in dialectical Arabic. [32] proposed another approach that was based on sentiment classification and Support Vector Machines to support the Arabic language syntactic and grammatical complexity and to analyze the Arabic reviews and comments. The study relied on a dataset that is composed of 625 Arabic reviews and opinions of the public obtained from the official website of Trip Advisor. On a similar study, [33] suggested an Arabic sentiment lexicon that allocates sentiment marks to the words from the Arabic WordNet. Semi-supervised learning, Machine-based learning classifier were used in order to conduct the study. An approach proposed by [34] suggested that the method of preprocessing on the reviews can enhance the performance of the classifiers namely, Support Vector Machines, Naive Bayes, and K-nearest neighbor classifiers. Another important issue that was tackled by [35] was to mine the Arabic slang language. This study requires efficient techniques to be used like a Slang Sentimental Words and Idioms Lexicon (SSWIL) and the Gaussian kernel SVM classifier. The SVM classifier was tested by using Facebook news comments. A study by [36] proposed a new sentiment analysis tool called the Colloquial Non-Standard Arabic—Modern Standard Arabic—Sentiment Analysis Tool (CNSA-MSA-SAT) which was developed to analyze the colloquial Arabic and Modern Standard. The main part of this new tool was the polarity lexicons which were built by tokenizing a vast quantity of comments and reviews from social media, and then the CNSA-MSA-SAT was allowed to learn on these values and comments and to judge on opinions. The lack of a pre-conceived and open-source tested and judged dataset on sentiment analysis was the biggest hurdle, especially in Arabic. The judgment was also not standardized.

Hence, this approach relied on manual opinion evaluation to find accuracy and increasing the level of confidence in the results. A study by [37] presented a new Arabic corpus for the Opinion Mining (OM) task that has been made available to the scientific community for research purposes. The study that uses the Machine-learning algorithms like support vector machines and Naïve Bayes was based on 500 movie reviews which were extracted from various websites and blogs in Arabic. The following Table 2 summarizes the studies described above.

Table 2 Research studies conducted on Arabic text mining in terms of sentiment analysis

Study	Problem	Methodology	Results
[29]	Recognizing the polarity of Arabic tweets and the practice of tweeting in dialectical Arabic	Hybrid classifier, Lexical-based classifier, Feature extraction, and Support Vector Machines classifier	The general quality of the obtained results in this study from hybrid classifier quantified by F-measure is 84% and accuracy is 84.01%
[32]	Evaluating the influence of Arabic grammatical richness on opinion mining accuracy, building a new accurate statistical approach that supports the Arabic language syntactic and grammatical complexity, and analyzing Arabic reviews and comments more accurately	Sentiment classification, Support Vector Machines, a dataset composed of 625 Arabic reviews and opinions of the public obtained from the official website of Trip Advisor	Results obtained were rooted in Support Vector Machines depicted that this method greatly affects the identification of opinions
[34]	The effects of stemming feature correlation and n-gram models for Arabic text on sentiment analysis	Support Vector Machines, Naive Bayes, and K-nearest neighbor classifiers	The results of the experiments suggested that choosing the method of preprocessing on the reviews will enhance the performance of the classifiers
[33]	Development of an Arabic sentiment lexicon	Semi-supervised learning, Machine learning-based classifier	The tests were performed on every Arabic sentiment corpora, and the results were obtained to 96% classification accuracy
[38]	Problems in analyzing Arabic slang; novel representative words and idioms (opinions) and the unstructured format. Extracting opinions of the young generation on diverse issues from	Support Vector Machines, a Slang Sentimental Words and Idioms Lexicon (SSWIL) of opinion words	Using SSWIL makes the classification on the youth's comments better from 75.35% classified to 86.86%. The precision and recall by using only classical opinions and SSWIL were

(continued)

Table 2 (continued)

Study	Problem	Methodology	Results
	different websites (i.e. news websites)		82.4, 59.3, 88.63, and 78 respectively
[35]	Arabic slang language suffers from the new expressive (opinion) words and idioms as well as the unstructured format. Mining Arabic slang language requires efficient techniques to extract youth opinions on various issues, such as news websites	A Slang Sentimental Words and Idioms Lexicon (SSWIL) of opinion words, Gaussian kernel SVM classifier	The results indicated that the accuracy is 86.86%; precision and recall are 88.63% and 78% respectively
[36]	Analyzing the colloquial Arabic and Modern Standard Arabic. The lack of a pre-conceived and open-source tested and judged dataset on sentiment analysis was the biggest hurdle, especially in Arabic	CNSA-MSA-SAT algorithm	The tool's judgment on opinions was compared to the polarity values from human judgments and it was found that it is 90% accurate for reviews and comments.
[37]	The resources available for opinion mining (OM) in other languages are still limited. In this article, a new Arabic corpus for the OM task was made available to the scientific community for research purposes	Machine learning algorithms like support vector machines and Naïve Bayes	The best result (0.90) using SVM over the OCA improves on the best result obtained with the Pang corpus (0.8535) using trigrams to generate the word vectors. This improvement is 5.45%. Moreover, it should be noted that for both corpora, the use of the trigram and bigram models overcome the use of unigram model

5 Text Mining of Arabic Web Documents

The available amount of information on the web has been considerably increased in current times. Whereas much of textual information exists on the web, it is easy for the user to retrieve the information appropriately only if the user figures out the language that is used in the web document. It is significant in the situation where users are trying to find out certain information that is made up of Arabic scripts and the words mentioned are the same in different languages. Due to this, several researchers have carried out studies in order to figure out the piece of information present on the web document. A study by [39] suggested an innovative process for constructing an extensive Arabic dictionary by making use of linguistic methods to obtain relevant compound and single Frequently Association (FA) terms from field-related corpora utilizing Arabic POS. However, the study relied on corpora that were extracted from the Arabic Wikipedia data along with Alhayah news. This contributed to text mining for web document in the Arabic context. However, further enhancement could be made on the used technique by including a document classification module that can categorize the documents automatically. Furthermore, text summarization by making use of information from Arabic FA dictionary can also be studied.

A study by [40] proposed an Arabic script web page language identification scheme using a decision tree-ARTMAP approach (DTA) in order to discover the general identities within a web document which in turn will lead to optimize the performance of the language. Similarly, authors of [41] focused on various text stemming tactics for the Arabic topic modeling. The application of the lemma-based stemming for topic modeling enables the users to find out interesting topics published in the press articles between 2007 and 2009. In addition, Dormant Dirichlet Allocation model was used for the extraction of the latent topics from the three Arabic real-world corpora. A study by [42] highlighted some crucial research issues that include effectually management of large volumes of the content present in the Arabic web pages as well as enabling the classification of relevant information. However, the researchers had noted that researches related to sentiment text mining are extremely limited in the context of Arabic language for the reason that they require the involvement of in-depth semantic processing. The aim was to improve the document representation models that are required for the text clustering that is dependent on semantic similarities. In terms of Arabic text classification, [43] emphasized on the need to do more work on the Arabic language compared to English languages. The study was based on support vector machine, decision tree algorithm, and Naïve Bayes. The study introduces the first and advanced assessment of work performed in the area of Arabic text classification and a large dataset which can be utilized for evaluating the Arabic text classification algorithms. A similar research was then conducted by [34] that studied the sentiment analysis for public Arabic tweets and comments in social media using classification models. The study also relied on using the Naïve Bayes, SVM, and K-Nearest Neighbor classifiers and applied to an in-house developed dataset through the use of three datasets that were made from the tweets/comments. Some enhancements could add value to this work by decreasing the amount of the dataset in order to reach concrete conclusions in case no large datasets will be required.

A study by [44] used Cosine similarity for examining the performance of Arabic language text classification. Support Vector Machine, k-Nearest Neighbors, Naive Bayes, Random Forest, classification tree and Neural Network are the methods used in this study. Another study by [45] managed to speed up the information retrieval process and to gain high-speed access in Arabic document base by using a root-based hierarchical indexing model. A study by [46] suggested a new approach for Arabic Text classification. Rational kernels were used to evaluate the effect of using sequences of terms for an ATC system. Methods used in this study include Decision Trees, K-Nearest Neighbors, Naive Bayes, and Support Vector Machines (SV M). Text detection in videos is a challenging problem due to the variety of text specificities, the presence of complex background, and anti-aliasing/compression artifacts. In order to tackle this problem, authors of [47] used the Stroke Width Transform (SWT) algorithm and a convolutional auto-encoder (CAE) in order to detect and extract passages or sequences of words containing relevant information from the prophetic narrations texts. Table 3 summarizes all the above-described work.

Table 3 Research studies conducted on Arabic text mining in terms of web documents

Study	Problem	Methodology	Results
[39]	Lacking of an effective method to automatically extract relevant FA terms to build a comprehensive dictionary	Text classification methodology using Arabic FA terms, Naïve Bayes (NB) and KNN classifiers	The new approach achieved a precision of (80.65%) followed by NB (72.79%) and KNN (36.15%)
[40]	Optimizing the performance of language identification based on Arabic script web documents by proposing a DT-ARTMAP (DTA) approach	Decision tree and ARTMAP approaches.	The results revealed that the proposed approach has outperformed both the decision tree and the default ARTMAP approaches
[41]	In the tasks related to semantic analysis, it is preferable to directly deal with texts in their original language Studies on topic models which provide a good way to automatically deal with semantic embedded in texts are not complete enough to assess the effectiveness of the approach on Arabic texts	Support vector machine (SVM), The Latent Dirichlet Allocation model	The results indicated a high effectiveness for the approach. The BBW lemma-based stemmer reduces significantly vocabulary dimension and under- and over-stemming errors. In addition, classification performance is improved slightly compared to classification of raw and light stemmed texts
[42]	Enabling the classification of relevant information is a crucial problem. Studies on sentiment text mining have been very limited in the Arabic language because of the need to involve deep semantic processing	K-means document clustering with semantic feature extraction and document vectorization	The results showed that the proposed solutions are reasonably accurate and fast. Through the proposed document, vectorization solutions with k-means, the study have succeeded in increasing the purity and decreasing the MICD and BDI compared to the standard k-means algorithm

(continued)

Table 3 (continued)

Study	Problem	Methodology	Results
[43]	How to build large Arabic corpora for text classification? Investigating a variety of text classification techniques using the same datasets	Support vector machine, the decision tree algorithm, and Naive Bayes	The best classification accuracy was 97% for the Islamic Topics dataset and the least accurate was 61% for the Arabic Poems dataset
[34]	Studying the sentiment analysis for public Arabic tweets and comments in social media using classification models	The Naïve Bayes, SVM, and K-Nearest Neighbor classifiers.	Results indicated that SVM gives the highest precision while KNN (K = 10) gives the highest Recall
[45]	Speeding up the information retrieval process in Arabic document-based by using a root-based hierarchical indexing model	Root-based hierarchical indexing model	The hierarchical approach is proven to be useful in reducing the space requirement
[48]	Analyzing the behaviors of the spammers in the content-based Arabic Web pages through analyzing the weights of the most ten popular Arabic words used by Arab users in their queries	Decision Tree classifier	Results showed that the behavior of the spammers in the Arabic Web pages can be unique and distinguished in comparison to other languages. Decision Tree was used to evaluate this behavior and it obtained the degree of accuracy which is equal to 90%
[46]	Introducing a new approach for Arabic Text classification. Rational kernels are used to evaluate the effect of using sequences of terms for an ATC system	Decision Trees, K-Nearest Neighbours, Rocchio, Naive Bayes, Support Vector Machines (SVM)	Results showed an improvement of the classification in terms of precision
[47]	Text detection in videos is a challenging problem due to variety of text specificities, the presence of complex background, and anti-aliasing/compression artifacts	The Stroke Width Transform (SWT) algorithm and convolutional auto-encoder (CAE)	Experiments indicated that the use of learned features significantly improves the text detection results

(continued)

Table 3 (continued)

Study	Problem	Methodology	Results
[49]	Detecting and extracting passages or sequences of words containing relevant information from the prophetic narrations texts	Named-entity extraction techniques, finite state techniques	Experimental evaluation results demonstrated that our approach is feasible. The system achieved an encouraging precision and recall rates, the overall precision and recall are 71% and 39% respectively
[44]	Investigating the performance of Arabic language text classification, the comparison between eight text classification methods	Vector space model (VSM), Latent Semantic Indexing (LSI), the singular value decomposition (SVD), Naive Bayes, k-Nearest Neighbors, Neural Network, Random Forest, Support Vector Machine, classification tree, Term Frequency, and Inverse Document Frequency (TF. IDF)	This study revealed that the classification methods that use LSI features significantly outperform the TF. IDF-based methods. Results also revealed that k-Nearest Neighbors (based on cosine measure) and support vector machine are the best performing classifiers
[50]	Proposing a preliminary classification of Arabic information extraction methods	Stemmer, N-gram, and "Stemmer" and "Ngram"	These studies and tests will permit to designate more appropriate Arabic information extraction method
[51]	Proposing an Arabic text classification system called Arabic Text Classifier (ATC). The main goal of the ATC is to compare the results between both classifiers used (CKNN, CNB) and selects the best average accuracy result rates to start a retrieving process	K-Nearest Neighbor, Naïve Bayes, and Arabic Text Classifier (ATC)	The results demonstrated that two algorithms were applied to the Arabic text A satisfactory number of patterns for each category were indicated. The selection of the feature space and the training data set used was also included. The accuracy was measured by the use of *k*-fold cross-validation method to test the accuracy of the system. The value of *k* can extremely influence the accuracy of any text classification algorithm

6 Conclusion and Future Work

The rapid emergence of the Web has significantly resulted in the Electronic textual documents. For the extraction of information from huge collections of textual data, researchers have conducted a number of approaches through different text mining techniques. When the textual information is not structured in line with the grammatical convention, information extraction becomes a difficult task. As far as this unstructured form is concerned, it is really hard to extract logical patterns with accurate information. This survey gains an insight into different text mining techniques besides their implementation in the Holy Quran, sentiment analysis (i.e. opinion mining), and web documents. The recent advancement in the field of intelligent computing is explored by the survey and it provides a complete synopsis of the existing text mining techniques, which can be used for the extraction of logical patterns from the grammatically incorrect and unstructured textual data. New ways for researchers would be drawn from this survey so that they could proceed further and develop clustering techniques or novel classification methods that will be helpful for the analysis of text in large-scale systems in the Arabic context. As a future work, some enhancement for analyzing the unstructured text presented on widespread systems (like the social web) is highly required. From the surveyed literature, we have noticed that researchers have paid less attention to sentiment analysis in the Arabic text. As a future work, we are currently working on investigating the text mining techniques on Arabic textual dataset from different gulf newspapers pages on Facebook. In addition, sentiment analysis (i.e. opinion mining) of the Arabic embedded text in these pages will be taken into consideration.

References

1. Hung, J.L., Zhang, K.: Examining mobile learning trends 2003–2008: a categorical meta-trend analysis using text mining techniques. J. Comput. High. Educ. **24**(1), 1–17 (2012)
2. Zaza, S., Al-Emran, M.:. Mining and exploration of credit cards data in UAE. In: 2015 Fifth International Conference on e-Learning (econf), pp. 275–279. IEEE (2015, October)
3. Gök, A., Waterworth, A., Shapira, P.: Use of web mining in studying innovation. Scientometrics **102**(1), 653–671 (2015)
4. Fan, W., Wallace, L., Rich, S., Zhang, Z.: Tapping into the power of text mining (2005)
5. Zhang, J.Q., Craciun, G., Shin, D.: When does electronic word-of-mouth matter? A study of consumer product reviews. J. Bus. Res. **63**(12), 1336–1341 (2010)
6. Shaalan, K.: A survey of arabic named entity recognition and classification. Comput. Linguist. **40**(2), 469–510 (2014)
7. Ray, S.K., Shaalan, K.: A review and future perspectives of arabic question answering systems. IEEE Trans. Knowl. Data Eng. **28**(12), 3169–3190 (2016)
8. Oudah, M., Shaalan, K.: NERA 2.0: improving coverage and performance of rule-based named entity recognition for Arabic. Nat. Lang. Eng. 1–32 (2016)
9. Salloum, S.A., Al-Emran, M., Shaalan, K.: A survey of lexical functional grammar in the Arabic context. Int. J. Com. Net. Tech, **4**(3)

10. Al Emran, M., Shaalan, K.: A survey of intelligent language tutoring systems. In: 2014 International Conference on Advances in Computing, Communications and Informatics (ICACCI), pp. 393–399. IEEE (2014)
11. Al-Emran, M., Zaza, S., Shaalan, K.: Parsing modern standard Arabic using Treebank resources. In: 2015 International Conference on Information and Communication Technology Research (ICTRC), pp. 80–83. IEEE (2015)
12. Al-Emran, M.: Hierarchical reinforcement learning: a survey. Int. J. Comput. Dig. Syst. **4**(2), (2015)
13. Al-Emran, M., Malik, S.I.: The impact of google apps at work: higher educational perspective. Int. J. Interact. Mob. Technol. (iJIM) **10**(4), 85–88 (2016)
14. Al-Emran, M., Shaalan, K.: Learners and educators attitudes towards mobile learning in higher education: state of the art. In: 2015 International Conference on Advances in Computing, Communications and Informatics (ICACCI), pp. 907–913. IEEE (2015, August)
15. Chen, X., Vorvoreanu, M., Madhavan, K.: Mining social media data for understanding students' learning experiences. IEEE Trans. Learn. Technol. **7**(3), 246–259 (2014)
16. Al-Radaideh, Q.A., Twaiq, L.M.: Rough set theory for Arabic sentiment classification. In: 2014 International Conference on Future Internet of Things and Cloud (FiCloud), pp. 559–564. IEEE (2014, August)
17. Gupta, V., Lehal, G.S.: A survey of text mining techniques and applications. J. Emerg. Technol. Web Intell. **1**(1), 60–76 (2009)
18. Navathe, S.B., Ramez, E.: Data warehousing and data mining. Fundam. Database Syst. 841–872 (2000)
19. Sukanya, M., Biruntha, S.: Techniques on text mining. In: 2012 IEEE International Conference on Advanced Communication Control and Computing Technologies (ICACCCT), pp. 269–271. IEEE (2012, August)
20. Witten, I.H.: Text mining. Practical handbook of internet computing, 14-1 (2005)
21. Salloum, S.A., Al-Emran, M., Monem, A.A., Shaalan, K.: A survey of text mining in social media: facebook and twitter perspectives. Adv. Sci. Technol. Eng. Syst. J. (2017)
22. Schoder, D., Gloor, P.A., Metaxas, P.T.: Spec. Issue Soc. Med. KI **27**(1), 5–8 (2013)
23. Steinberger, R.: A survey of methods to ease the development of highly multilingual text mining applications. Lang. Resour. Eval. **46**(2), 155–176 (2012)
24. Abdul-Baquee, S., Atwell, E.S.: Knowledge representation of the Quran through frame semantics: a corpus-based approach. In: Proceedings of the Fifth Corpus Linguistics Conference. University of Liverpool (2009)
25. Farghaly, A., Shaalan, K.: Arabic natural language processing: challenges and solutions. ACM Trans. Asian Lang. Inf. Process. (TALIP) **8**(4), 14 (2009)
26. Alhawarat, M., Hegazi, M., Hilal, A.: Processing the text of the Holy Quran: a text mining study. Int. J. Adv. Comput. Sci. Appl. (IJACSA) **6**(2), 262–267 (2015)
27. Muhammad, A.B.: Annotation of conceptual co-reference and text mining the Qur'an. University of Leeds (2012)
28. Sharaf, A.M.: The Qur'an annotation for text mining. First year transfer report. School of Computing, Leeds University, December 2009
29. Aldayel, H.K., Azmi, A.M.: Arabic tweets sentiment analysis–a hybrid scheme. J. Inf. Sci. **42**(6), 782–797 (2016)
30. Pang, B., Lee, L.: Opinion mining and sentiment analysis. Found. Trends® Inf. Retr. **2**(1–2), 1–135 (2008)
31. Liu, B.: Sentiment analysis and opinion mining. Synth. Lect. Hum. Lang. Technol. **5**(1), 1–167 (2012)
32. Cherif, W., Madani, A., Kissi, M.: A new modeling approach for Arabic opinion mining recognition. In: Intelligent Systems and Computer Vision (ISCV), pp. 1–6. IEEE (2015, March)
33. Mahyoub, F.H., Siddiqui, M.A., Dahab, M.Y.: Building an Arabic sentiment lexicon using semi-supervised learning. J. King Saud Univ. Comput. Inf. Sci. **26**(4), 417–424 (2014)

34. Duwairi, R.M., Qarqaz, I.: Arabic sentiment analysis using supervised classification. In: 2014 International Conference on Future Internet of Things and Cloud (FiCloud), pp. 579–583. IEEE (2014, August)
35. Soliman, T.H., Elmasry, M.A., Hedar, A., Doss, M.M.: Sentiment analysis of Arabic slang comments on facebook. Int. J. Comput. Technol. **12**(5), 3470–3478 (2014)
36. Al-Kabi, M., Gigieh, A., Alsmadi, I., Wahsheh, H., Haidar, M.: An opinion analysis tool for colloquial and standard Arabic. In: The Fourth International Conference on Information and Communication Systems (ICICS 2013), pp. 23–25 (2013)
37. Rushdi-Saleh, M., Martín-Valdivia, M.T., Ureña-López, L.A., Perea-Ortega, J.M.: OCA: opinion corpus for Arabic. J. Am. Soc. Inform. Sci. Technol. **62**(10), 2045–2054 (2011)
38. Hedar, A.R., Doss, M.: Mining social networks Arabic slang comments. In: IEEE Symposium on Computational Intelligence and Data Mining (CIDM) (2013)
39. Atlam, E.S., Morita, K., Fuketa, M., Aoe, J.I.: A new approach for Arabic text classification using Arabic field-association terms. J. Am. Soc. Inform. Sci. Technol. **62**(11), 2266–2276 (2011)
40. Selamat, A., Ng, C.C.: Arabic script web page language identifications using decision tree neural networks. Pattern Recogn. **44**(1), 133–144 (2011)
41. Brahmi, A., Ech-Cherif, A., Benyettou, A.: Arabic texts analysis for topic modeling evaluation. Inf. Retr. **15**(1), 33–53 (2012)
42. Alghamdi, H.M., Selamat, A., Karim, N.S.A.: Arabic web pages clustering and annotation using semantic class features. J. King Saud Univ. Comput. Inf. Sci. **26**(4), 388–397 (2014)
43. Khorsheed, M.S., Al-Thubaity, A.O.: Comparative evaluation of text classification techniques using a large diverse Arabic dataset. Lang. Resour. Eval. **47**(2), 513–538 (2013)
44. Al-Anzi, F.S., AbuZeina, D.: Toward an enhanced Arabic text classification using cosine similarity and latent semantic indexing. J. King Saud Univ. Comput. Inf. Sci.
45. Eldos, T.M.: Arabic text data mining: a root-based hierarchical indexing model. Int. J. Model. Simul. **23**(3), 158–166 (2003)
46. Nehar, A., Benmessaoud, A., Cherroun, H., Ziadi, D.: Subsequence kernels-based Arabic text classification. In: 2014 IEEE/ACS 11th International Conference on Computer Systems and Applications (AICCSA), pp. 206–213. IEEE (2014, November)
47. Zayene, O., Seuret, M., Touj, S.M., Hennebert, J., Ingold, R., Amara, N.E.B.: Text detection in Arabic news video based on swt operator and convolutional auto-encoders. In: 12th IAPR Workshop on Document Analysis Systems (DAS), IEEE, pp. 13–18, 2016
48. Wahsheh, H., Alsmadi, I., Al-Kabi, M.: Analyzing the popular words to evaluate spam in Arabic web pages. IJJ Res. Bull. JORDAN ACM–ISWSA, **2**(2), 22–26
49. Harrag, F.: Text mining approach for knowledge extraction in Sahîh Al-Bukhari. Comput. Hum. Behav. **30**, 558–566 (2014)
50. A-Brahimi, B., Touahria, M., Tari, A.: Data and text mining techniques for classifying Arabic tweet polarity. J. Dig. Inf. Manage. **14**(1), 15
51. Zubi, Z.S.: Using some web content mining techniques for Arabic text classification. Recent Advances on Data Networks, Communications, Computers (2009)

Part VII
Text Summarization

TALAA-ATSF: A Global Operation-Based Arabic Text Summarization Framework

Riadh Belkebir and Ahmed Guessoum

Abstract Text summarization is one of the most challenging and difficult tasks in natural language processing, and artificial intelligence more generally. Various approaches have been proposed in the literature. Text summarization is classified into two categories: extractive text summarization and abstractive text summarization. The vast majority of work in the literature followed the extractive approach, probably due to the complexity of the abstractive one. To the best of our knowledge, the work presented here is the first work on Arabic that handles both the extractive and abstractive aspects. Indeed, while the literature lacks summarization frameworks that allow the integration of various operations within the same system, this work proposes a novel approach where we design a general framework which integrates several operations within the same system. It also provides a mechanism that allows the assignment of the suitable operation to each portion of the source text which is to be summarized, and this is achieved in an iterative process.

Keywords Artificial intelligence · Natural language processing
Text summarization

R. Belkebir (✉) · A. Guessoum (✉)
Natural Language Processing and Machine Learning Research Group,
Laboratory for Research in Artificial Intelligence, Computer Science
Department, University of Science and Technology Houari Boumediene
(USTHB), BP 32 El-Alia, 16111 Bab Ezzouar, Algiers, Algeria
e-mail: belkebir.riadh@gmail.com

A. Guessoum
e-mail: aguessoum@usthb.dz

K. Shaalan et al. (eds.), *Intelligent Natural Language Processing: Trends and Applications*, Studies in Computational Intelligence 740,
https://doi.org/10.1007/978-3-319-67056-0_21

1 Introduction

Text summarization is a cognitive effort that requires a deep understanding of the source document. According to [32, 35], humans tend to use a number of operations to generate summaries. From our investigation of the aforementioned and other text summarization studies, we have come to the conclusion that these operations fall into one of two categories: *single-sentence operations* or *multi-sentence operations*. A *single-sentence operation* is applied to a single sentence. These are for instance *sentence compression*, *concept generalization and fusion*, *sentence selection*, etc. *Sentence compression* is a *single-sentence operation* because it is applied to a single sentence, while *sentence fusion* is a *multi-sentence operation* because it needs more than one sentence to be applied. On the other hand, a *multi-sentence operation* is applied in a setting where two sentences or more are merged.

Most of the work that we have found in the literature followed the extractive approach because of the difficulty of the abstractive one. We also found a lack of summarization frameworks that can integrate the various operations proposed in the literature within the same system.

In this work, we propose a framework that tries to cover all the possible basic operations of text summarization that have been developed in the literature. Our approach allows the integration of these operations within the same framework. It sets a strategy to select the most suitable operation for a given portion of text. It is also adaptable to the different operations. For instance, we find in the literature operations that have mainly relied on templates, others that select the best sentences, etc.

We have used different mathematical and computer science theories to develop our framework. First, we use *number theory* and specifically *Bell numbers theory* [13] to generate what we will define later in this paper as *partitions of the document*. We also use *graph theory* for the *multigraph* that we define later. A machine learning approach has also been used in this work to assign the suitable operations to the different portions of the source document.

The remainder of this paper is organized as follows. Section 2 presents the related work. Section 3 discusses the challenges and difficulties in Arabic text summarization. Section 4 presents the problem statement and introduces various definitions. Section 5 explains the system design. The experimentation work is presented in Sect. 6. Section 7 discusses the results and Sect. 8 gives a conclusion and some possible directions for the development of Arabic text summarization based on this work.

2 Related Work

In this section, we present various studies that have tackled the text summarization problem. We give an overview of the main operations that have been used as well as the text summarization frameworks that have been proposed in the literature. We conclude the section by reviewing the main works that have been devoted to Arabic text summarization.

2.1 Text Summarization Operations

In order to study how human abstractors generate summaries, [35] has analyzed the process of generating a set of articles. The set consisted of fifteen articles on telecommunications, ten articles about the legal domain and ten articles dealing with medical issues. The styles and the structures of the documents vary even within the same domain. Based on the corpus that was studied, it was reported that human abstractors most of the time reuse text from the original document to generate a summary. After analysis of the operations that the human experts used, [35] defined six operations that can be used to transform a source document into a summary document. These are sentence reduction, sentence combination, syntactic transformation, lexical paraphrasing, generalization or specification, and reordering.

It was stated in [32] that to generate English summaries humans tend to copy and paste snippets from the original document after some slight modifications. In the same study [32], based on the analysis of a summarization corpus, the summarization operations were classified into five classes. They were defined as either atomic or complex operations. The atomic operations are insertion and deletion of words, while the complex operations include replacement and reordering of words and merging of sentences. It was stated that atomic operations cannot be further divided into other operations, whereas complex operations can.

In the literature, text summarization has been addressed using different operations. Among these, we mention *sentence compression* [15, 16, 27, 33, 70, 72, 73] where the goal is to transform a given source sentence by means of reduction and/or paraphrasing operations while preserving the important information found in the source sentence [15]. *Text simplification* has also been proposed [18, 26, 37, 63, 67–69], the focus being on rewriting operations applied to source sentences so as to decrease the syntactic or lexical level of complexity and at the same time to preserve their meaning [63]. Some studies have considered *sentence revision* [54, 60, 65] where the aim is to insert and substitute phrases [65]. We also mention that *sentence fusion* has been tackled in some studies [9, 50]. It is a text-to-text generation technique that aims to synthesize common information between documents.

2.2 Text Summarization Frameworks

A text summarization framework tries to combine different operations within the same system. If we look at the summarization frameworks that have been developed, we find just a few works that have tried to combine different summarization operations within a single framework, and even fewer ones have considered abstractive text summarization probably due to its difficulty. Among these, one can mention the work of [39] where a semantic role labeling approach has been proposed to represent documents. The authors presented an abstractive summarization framework that deals with multi-document summarization and where a ranking of the predicate argument structures is used to select the content of the summary document. [41] have used a semantic representation called abstractive meaning representation (AMR) to construct an abstractive framework where original documents are represented as a set of AMR graphs. Then, the summaries graphs are generated from the AMR graphs using a graph-to-graph transformation operation. Finally, the text summary is generated from the graph. [28] have proposed an approach that uses the notion of Information Item (INIT), which is defined as the smallest element of coherent information in a sentence or a text. This system selects the summary content from the proposed representation. The output of a syntactic parser for a sentence is used and part of a sentence is regenerated using a natural language generation engine. The system selects sentences among the regenerated sentences based on the document frequency of the contained words. In [42] an approach that uses concept-level semantic representation has been proposed to generate summaries. The approach generates ultra-concise opinion summaries by combining several operations. These operations are sentence regeneration, internal concept representation, and the integration of syntactic sentence simplification. The authors have tested two versions of the approach. The first version used sentence regeneration while the second version is without sentence regeneration. The authors reported that the two versions are reliable in generating summaries. Moreover, the version that uses sentence regeneration was more robust in the case of the presence of noisy data, while the one without sentence regeneration produced better results and outperformed many state-of-the-art systems.

From our survey of the existing works we have concluded that it is needed to develop a system that allows the integration of all of the operations that humans perform when generating text summaries. In fact, many operations have been developed in the literature, but we have realised the need for a global theoretical framework within which the integration of heterogeneous operations is possible. Indeed, we have realised that the existing frameworks suffer from the limitation that their systems are not flexible in the sense that it is difficult to integrate new operations into their systems. Our proposed framework handles this problem, among others, and it should be able to scale up and include any new operation with minimal changes to the summarization process.

2.3 *Arabic Text Summarization*

Many automatic summarization approaches have been developed for English, and other European languages like French and Spanish [43, 55, 61]. A lot of effort is also underway for languages like Indian, Chinese and Japanese. Some attempts have been made to promote research on Arabic text summarization though research in this area is still underdeveloped compared to that on the abovementioned languages. We review in this section the main approaches that have tackled Arabic text summarization. The aim here is not to give an exhaustive review of the work that has been done on Arabic text summarization, since there exists an interesting study [2] that presented a survey of several research works on Arabic text summarization and which concluded that there exists a great opportunity for further research on Arabic text summarization.

It is worth mentioning that all of the work that has been done on Arabic text summarization follows an extractive approach; this is probably due to the difficulty of the abstractive one, especially for a language like Arabic. [3] have presented AKEA, an unsupervised algorithm that was used to extract keyphrases from a document using linguistic patterns like statistical knowledge, Part-Of-Speech (POS) tags, and the internal structural patterns of terms. In [4] the authors proposed an automatic extractive Arabic text summarization system that uses Rhetorical Structure Theory (RST) combined with a sentence scoring scheme to generate summaries. [38] presented a comparative study between three automatic summarization approaches: a symbolic approach, a numerical approach and a hybrid between the two. The conclusion was that the numerical approach outperformed the symbolic one, and the hybrid approach outperformed the other two. In [1] probabilistic neural networks (PNN) have been used with several features. The effect of each feature was studied and, then, all the features were combined to train the probabilistic neural network (PNN) so as to generate a summary. Lakhas [19] is a summarization system that produces 10-word summaries from news articles. First, the system summarizes the source articles; then it translates them into English. This approach has given good results for very short summaries (headlines). [23] proposed a query-driven summarization system which extracts the most relevant paragraphs from a source document by using a vector space model and cosine similarity. Based on these, top ranked sentences are selected to generate the summary. [21] have proposed a clustering-based approach to handle Arabic text summarization; TF-IDF was used to create the different instances that represent the document sentences. [62] have proposed CLASSY (Clustering, Linguistics, And Statistics for Summarization Yield), a generic and query-driven approach. CLASSY is a multilingual (English and Arabic) extractive system which allows the generation of both single- and multi-document summaries. [22] have proposed a language-independent, single- and multi-document extractive summarization system which handles both English and Arabic. [56] designed a single- and multi-document summarization system. The system is based on the minimal-redundancy maximal-relevance (mRMR) [58] approach as well as a discriminant analysis method. In [14], Support Vector Machines (SVM) have been proposed to summarize documents. In [64] two different methods to generate an extractive summary were proposed. The first method uses a Naive Bayes classifier while the second approach uses Genetic Programming (GP). The final summary is then generated by either the union or the intersection of sentences of the two summaries generated by the two methods. [20] have designed a machine learning-based system that summarizes Egyptian dialect Twitter posts. [34] have proposed OSSAD, an Ontology-based Summarizer for the Arabic language. This summarizer is query-driven and relies on a machine learning approach to generate summaries.

To the best of our knowledge, there is no Arabic text summarization framework in the literature. In fact, the approach proposed in this work is the first that proposes a solution to integrate various operations within the same system. Besides, this approach is the first one that deals with both the extractive and the abstractive aspects in Arabic text summarization.

3 Challenges and Difficulties in Arabic Text Summarization

As mentioned above, the work that has been done on Arabic text summarization is very much underdeveloped compared to other languages. This is due to various factors among which we can mention the following:

- The Arabic language lacks validated gold standard corpora. In fact, most studies use their own corpora. This makes the comparison between Arabic summarization systems difficult.
- The community working on Arabic text summarization is quite small.
- Arabic can be considered a difficult language because of its complex spelling, vocabulary, morphology and syntax.
- The problem of diacritics (tashkyl) in the Arabic language makes it even more complex to handle, and most modern writings do not use diacritics; these are inferred from the context.

However, it is worth pointing out that recently some efforts have started. For instance, [21] have designed the EASC (parallel) corpus. Nevertheless, there is still a lot of effort to be done to produce Arabic summarization systems with a quality similar to the ones produced for other languages.

To improve the quality of the current Arabic summarization systems, the community needs to develop and improve NLP tools for Arabic since summarization systems rely mainly on these NLP tools to generate summaries. They include Part-of-Speech tagging, named entity recognition, syntactic and semantic parsing, etc.

Let us emphasize once more that research work done on Arabic text summarization so far has followed the extractive approach; to the best of our knowledge, our work is the first to adopt the abstractive approach.

In the sequel, we define the problem and the main concepts that will be used in this work.

4 Problem Statement

Having analyzed the available literature, we have come to the conclusion that there is a need for a full summarization framework. In fact, many summarization operations[1] have been developed, but we do not have a system that can integrate all of them in a compositional manner. In this work, we try to respond to the following research questions:

- **Q1:** What representation model should we consider to represent the document to be summarized?
- **Q2:** How should we assign summarization operations to the different portions of the source document?
- **Q3:** After applying a sequence of such operations, what is the stopping condition in the summarization process?

When humans summarize documents, they tend to do it in an iterative process. After applying the operations to the source document, one ends up with the first version of a summary. At this moment, one (or the system) should decide whether it should summarize the summary again or stop the process. The most common way that has widely been used in text summarization was to set a retention rate upon which a decision is made whether to stop summarizing or not. In our case, we should come up with a more convincing approach to stop summarizing because the retention rate is not fixed for all the texts; it rather varies according to the type and the content of the text.

In the sequel, we give definitions and examples of the different concepts that are used in the proposed framework.

4.1 Definitions

We start from the intuitive idea here that before summarizing a document, we should find the different combinations (partitions) of sentences that we could generate from the source document. This will later allow the association of operations to the different partitions.

Definition 1 (*Partition of a document*) Given a document D, a Partition P of D is a set of non-empty subsets of sentences $SP_i \subseteq D$, $1 \leqslant i \leqslant k$ (the (subsets) of the partition), such that $\mathrm{U}_{\mathrm{i}=1}^{\mathrm{k}} SP_i = D$ and for every $i <> j$, $SP_i \cap SP_j = \emptyset$. In other words, a partition is a set of non-empty subsets of sentences of D whose union is equal to D.

[1]In the remainder of this paper, whenever we use the term "operation(s)" it will obviously stand for "summarization operation(s)".

The notion of *subset of sentences of a partition of a document* is introduced because the summarization operations will be associated with these subsets of sentences.

Definition 2 (*Subset of sentences of a partition of a document*) Given a partition $P = \{sp_1, sp_2, ..., sp_n\}$ of a document D. An element sp_i of P is a subset of sentences of the partition P of the document D.

A document could have many possible partitions. Below we define the notion of *set of partitions of a document*. Understanding the possible partitions is an important step as it allows us to understand the various possible summaries and hence the operations that need to be applied to the *subsets of sentences of the partition* to produce the summaries. As we will explain later, we select the best partition from the set of partitions of a document and also the best operations to be applied to this partition.

Definition 3 (*Set of partitions of a document* (*SPD*)) Given a document D. The set of partitions of a document D is the set SPD of all possible partitions that could be derived from D. $SPD = \{P$, where P is a partition of $D\}$.

It is known in combinatorics that the number of all partitions of a set of n elements is given by the *Bell number* B_n [13]. As such, the number of partitions of a document D with n sentences is *Bell number* B_n, where $B_0 = 1$, $B_1 = 1$, $B_2 = 2$ and $B_3 = 5$. The Bell numbers satisfy the following relation:

$$B_{n+1} = \sum_{k=0}^{n} \binom{n}{k} B_k \tag{1}$$

Example 1 Let us consider a document D which has three sentences: a, b, c. There are five possible partitions of D:

1. $\{\{a\}, \{b\}, \{c\}\}$
2. $\{\{a, b\}, \{c\}\}$
3. $\{\{a, c\}, \{b\}\}$
4. $\{\{b, c\}, \{a\}\}$
5. $\{\{a, b, c\}\}$

For example, the first partition contains three subsets of sentences each one having one single sentence. The second, third, and fourth partitions have two subsets of sentences, where the first subset has two sentences and the second has one sentence. The fifth partition has one subset of sentences which contains all three sentences.

Several taxonomies have been proposed in the literature [36, 44]. These works classified text summarization systems according to a number of factors: input, output, language, etc. But none of these taxonomies mentioned the nature of the operations that a summarization system performs. In this work, we propose a classification of text summarization operations with respect to the nature of the operation that is performed in the summarization process. We distinguish two classes of operations: *single-sentence operations* and *multi-sentence operations*.

We define below *single-sentence operation, multi-sentence operation, set of single-sentence operations* and the *set of multi-sentence operations*. These concepts are related to the nature of the operations that are used in the summarization process. For example, *sentence compression* is considered as a *single-sentence operation* since it acts on only one sentence at a time, while *sentence fusion* is a *multi-sentence operation* since it merges sentences and needs at least two sentences for its application.

Definition 4 (*Single-sentence operation* ($\odot$)) Given a subset of sentences *SP* of a partition *P* of a document *D*, a *single-sentence operation*, denoted by $\odot(SP)$, is an operation that is applied to a singleton subset of sentences of a partition of D, i.e. one that has **only one sentence**.

Definition 5 (*Set of single-sentence operations* (*SG*)) The set of single-sentence operations of a partition *P* of a document *D* is denoted by $SG = \{\odot_1, \odot_2, ..., \odot_n\}$.

Definition 6 (*Multi-sentence operation* ($\oplus$)) Given a subset of sentences *SP* of a partition *P* of a document *D*. A multi-sentence operation denoted by $\oplus(SP)$ is an operation that is applied to a subset of sentences of a partition that has **two or more sentences.**

Definition 7 (*Set of multi-sentence operations* (*SM*)) The set of multi-sentence operations of a partition *P* of a document *D* is denoted by $SM = \{\oplus_1, \oplus_2, ..., \oplus_n\}$.

The process of mapping operations to the different subsets of sentences which make the partition of the document is used to get a summary document at a given iteration.

Definition 8 (*Mapping* (Θ) *of operations to a partition of a document*) Let $P = \{sp_1, sp_2, ..., sp_n\}$ be a partition of a document *D* and let $\odot_i \in SG$ and $\oplus_j \in SM$. The process of mapping operations to *P* is the association of operations to all the subsets of sentences of the partition *P*. This is denoted by $\Theta(P) = \{\Phi(sp_1), \Phi(sp_2), ..., \Phi(sp_n)\}$. Where $\Phi(sp)$ is defined as follows:

$$\Phi(sp) = \begin{cases} \odot_i(sp), & \text{if } |sp| = 1 \\ \oplus_j(sp), & \text{otherwise} \end{cases} \quad (2)$$

If the number of sentences of the chosen subset of sentences of the partition is equal to one, then we should select an operation that belongs to the set of *single-sentence operations*. Otherwise, we select an operation that belongs to the set of *multi-sentence operations*.

We note that $\Theta(P)$ represents at the end a summary document since it performs two things: (1) it associates the operations to the subsets of sentences of the partition P and (2) generates a summary document which is the result of the application of theses operations to the subsets of sentences.

Below we define the *space of mappings of operations to a partition of a document*. For a given partition, we could generate many possible mappings. If we consider a partition $P = \{sp_1, sp_2, ..., sp_n\}$ of a document D, a *set of single-sentence operations* $SG = \{\odot_1, \odot_2, ..., \odot_{n_1}\}$, a *set of multi-sentence operations* $SM = \{\oplus_1, \oplus_2, ..., \oplus_{n_2}\}$, $m = n_1 + n_2$ the total number of operations, and n the total number of *subsets of sentences of the partition* P, the worst-case complexity of generating this space is $O(m^n)$.

Definition 9 (*Space of mappings of operations to a partition of a document*) Let P be a partition of a document D. The space of mappings of operations to the partition P of D, denoted by $SMOP(P) = \{\Theta_1(P), \Theta_2(P), ..., \Theta_{nb}(P)\}$, is the set of all the possible combinations of the operations that could be assigned to P.

Definition 10 (*Best mapping* ($\bar{\Theta}$) *of operations to a partition of a document*) Let $SMOP(P) = \{\Theta_1(P), \Theta_2(P), ..., \Theta_n(P)\}$ be the *space of mappings of operations to a partition* P of a document D. The best mapping of operations to a partition of a document D, denoted by ($\bar{\Theta}$), is defined as the mapping of operations that yields a summary document which maximizes a fitness function that represents the quality of the document.

$$\bar{\Theta}(P) = \arg\max_{\Theta} f(SMOP(P)) \tag{3}$$

We have chosen to represent a document using a *multigraph*, i.e. consisting of multiple layers each of which represents one relation between the sentences. The representation of a document as a *multigraph* will set the ground for the application of the different operations of text summarization. A more detailed discussion about this *multigraph* is provided in the System Design Sect. 5.

Definition 11 (*Multigraph of a document*) The *multigraph* $G = (V, E)$ over a document D represents the different relations between the sentences of this document, where:

- V is a set of vertices or nodes (each node corresponding to a sentence in the document D)
- E is a multiset of edges (an edge representing a relation between two sentences)

This *multigraph* has several layers, each one representing a relation between sentences.

The summarization task is an iterative process such that, after mapping the operations to the partition, we generate a summary for the current iteration which represents a new summary.

Definition 12 (*Summarization function*) Given a document D, the summarization function $S(D)$ represents the sequence of transformations of the document D. A function Γ_i of the document D is the result of applying the best mapping of operations of all the partitions of document D at iteration i. $P_1, P_2, ..., P_n$ are the partitions of the document D.

$$\Gamma_i(D) = \arg\max_{\Theta} (f(U_{\mathrm{k}=1}^{\mathrm{n}} SMOP(P_k))) \tag{4}$$

$\Gamma_i(D)$ takes the best mapping of operations of all the partitions of the document D and generates the current summary of iteration i. The summarization function $S(D)$ is the composition of the results of all the iterations, which is denoted by:

$$S(D) = \Gamma_n \circ \Gamma_{n-1} \circ \cdots \Gamma_2 \circ \Gamma_1(D) \tag{5}$$

i.e.:

$$S(D) = \Gamma_n(\Gamma_{n-1}(...(\Gamma_2(\Gamma_1(D))))) \tag{6}$$

We note that initially the best summary is the first generated summary. After the generation of a summary at every iteration $i > 1$, its quality is assessed and, if it outperforms the quality of the best generated summary of the previous iterations, then the current summary will be the best summary. At each iteration $i > 1$ the document that is given as an argument to the function Γ_i is the best summary.

5 System Design

5.1 General Idea

In our approach, we consider text summarization in a graph-theoretic setting. A document is decomposed into a set of subsets of sentences where the number of subsets of sentences is given by a *Bell number* [13] as explained above. We also use a *multigraph* representation to represent the source document.

One novel and important aspect of this work is the use of a *multigraph*. The graph nodes represent the different sentences, and the edges between sentences represent the different relations that hold between them. Each layer of the graph represents a semantic relation between the sentences of the document. These relations could include, for instance, the semantic relations between sentences as well as the rhetorical relations that might exist between them. A more detailed discussion about this issue will be given in the sequel. This representation will also prepare the ground for the operations that will be applied to generate the summary.

The mapping of operations to a partition uses a machine learning approach. To select the operation that will be applied to each subset of sentences of a partition, a probability is associated with each operation and the operation that has the highest probability is selected. This probability is estimated by a machine learning algorithm that has already been trained on an annotated corpus. The annotated corpus would contain the subset of sentences and the operation that should be applied on this subset of sentences. The machine learning system is trained to learn the operations that the system should associate with a partition.

5.2 *Why A Multi-layer Graph?*

In the text summarization literature, several representation models have been adopted to represent the text document (graphs, trees, term vectors, lexical representation, etc.). To justify the choice that we will consider in this approach, we start by providing the representation models that have been used in the literature. Then, we present the representation model that has been chosen in this work.

5.2.1 Rhetorical Structure Theory

The work presented in [45] has gained a lot of interest for text summarization. The approach therein analyzes the input text and provides a tree representation that represents the text rhetorical relations. This representation has been used in several studies [30, 46–49].

5.2.2 Lexical Chains

Another approach is to use lexical or coreference chains. This approach has first been used in [6]. The idea of this approach is to select the longest coreference chain. The assumption here is that the longest coreference chain contains the main topic of the document while the shorter chains indicate subtopics. To generate coherent summaries, this approach selects the longest coreference chain. Lexical-chains-based approaches aim to determine sequences of words which are semantically related to each other (for example, synonyms). It is assumed that the main topic could be detected just by selecting the longest chain. This technique has been used in a number of studies [7, 8, 24, 51].

5.2.3 Graph-based Representation

Graph-based summarizers have been proposed in the literature. In this approach, a node represents a sentence and an edge represents the relation between two sentences. These relations include semantic relations, semantic distances between sentences, etc. The assumption here is that the topology of the graph could provide interesting insights into the most important content of the text. Among the most relevant approaches in this category, we mention [25, 29, 52, 53, 59].

The use of a *multigraph* aims to build a semantic representation of the document. This *multigraph* will prepare the ground for the operations that the mapping process will select. After an iterative process, we generate a summary at each iteration. If the quality of the newly generated summary at the current iteration is better than the best summary thus far, then the newly generated summary will be the best summary.

The choice of a multi-layer graph instead of a simple graph is motivated by the fact that many operations in the literature have relied on several relations (a simple graph cannot represent all the relations). For instance, we find in the literature operations that have mainly relied on the centrality of a sentence [25, 57] which is expressed by the number of connections between a sentence and the other sentences of the document. To represent this aspect, we need to define a relation which expresses the distance between sentences (aggregate similarity is an example of these relations). Another relation that might be considered as a representation layer in the *multigraph* is the RST relation [45] which shows the rhetorical structure of a document. This kind of relations has been useful in various summarization systems. Moreover, RST relations could be used by a number of operations. It could help both the *single-sentence* and *multi-sentence operations*. Other relations that could be represented by the *multigraph* are lexical chains [6], latent semantic analysis [71], and coreference information [5]. We also mention that several studies have reported that using a graph representation is very useful in text summarization [25, 29, 52, 66] since it allows to generate coherent documents.

5.3 The Process of Summary Generation

Algorithm 1 gives the steps of the general algorithm for the generation of a summary. It takes as input a source document and the maximum number of iterations and outputs the summary of this document. The algorithm follows an iterative process. First, it generates the partitions of the document and the *multigraph*. For each partition of the document, it builds the space of mappings of operations of the partition denoted by *SMOP*(*P*) and selects the best mapping of operations of the partition $\bar{\Theta}(P)$. If the summary of the mapping $\bar{\Theta}(P)$ is better than the best summary that far according to the fitness function being used, then the best summary will be the summary of $\bar{\Theta}(P)$. At the end, Algorithm 1 outputs the best summary. The whole process implements the summarization function $S(D)$ and each iteration i of the process is the function $\Gamma_i(D)$.

Algorithm 1 General algorithm for summary generation

```
Input Source document D, maxIter
Output bestSummary
i =1;
bestSummary = D;
while i <= maxIter do
    SPD = Generate the partitions of the bestSummary;
    multiGraph = Construct the multigraph;
    for all P in SPD do
        SMOP(P) = Generate the space of mappings of operations of P
        Θ̄(P) = Select the best mapping of operations from SMOP(P)
        summary = Generate the summary from the best mapping Θ̄(P) of operations of P;
        if fitness(summary) > fitness(bestSummary) then
            bestSummary = summary;
        end if
    end for
    i = i + 1
end while
Return bestSummary;
```

5.3.1 Selection of the Best Operation by Machine Learning

To select the best operation to be applied to generate the summary, Algorithm 2 takes as input a subset of sentences. The algorithm uses features like the position of the subset of sentences, the number of keywords of the subset of sentences, the number of words shared with the title, whether the subset of sentences is the first or the last subset of sentences and the number of words in the subset of sentences. The algorithm returns the operation that has the highest probability.

Algorithm 2 Algorithm for the selection of best operation using machine learning

```
Input A subset of sentences sp
Output best_operation
F= extract_features (sp);
best_operation = Machine_learning_prediction (F);
Return best_operation;
```

Once we generate all the partitions of a document, we calculate the quality of each partition which is expressed by a fitness value. The fitness measures the quality of the summary document which would be the result of applying the mapping of operations to obtain the considered partition.

To generate the *multigraph*, Algorithm 3 captures the relations that exist between sentences. Let us consider an instance of the general problem and let us suppose that we are interested in tackling the problem taking into account only two relations, the similarity between sentences and the semantic relations that could exist between sentences represented by the RST relations. In this case, we have only two matrices

A and B. Matrix A stores the similarity distances between sentences, while Matrix B stores the RST relations that might exist between the different sentences. These two matrices will be needed for the operations that will be applied to generate the summary.

Algorithm 3 Algorithm for the construction of the *multigraph*

Input *listNodes* is a list of n nodes which represent the sentences of the source document
Output The *multigraph* represented by *nb* matrices
for k := 1 to *nb* **do**
  for i:= 1 to n **do**
    for j := i+1 to n **do**
      $M_k[i,j] = Relation_k(listNodes[i], listNodes[j])$;
    end for
  end for
end for
Return $M_1, M_2, \ldots, M_{nb}$ matrices ;

5.3.2 Defining the Fitness Function

In order to evaluate the quality of the summary (S), we define a fitness function (f) which combines the grammaticality and the concentration of information to evaluate the importance of a summary. The concentration of information is a semantic measure of the quality of the summary as will be explained below.

$$fitness(S) = information\_concentration(S) + grammaticality(S) \quad (7)$$

Information concentration The information concentration is measured by the amount of important content present in the summary. We define two parameters: the aggregate similarity and the *tf*. The aggregate similarity gives high value to summaries with central sentences in the text. While the *tf* allows to give high ranks to documents that have important words.

$$information\_concentration(S) = \alpha * (Aggregate\_similarity(S)) + \beta * (TF(S)) \quad (8)$$

Aggregate similarity Let $S = \{s1, s2, s3, s4, s5, s6, s7\}$ be a summary document. The aggregate similarity technique allows to assign weights to the different sentences of the document to assess its quality. The weight of any given sentence is calculated as the sum of the weights of its connections. For example, the weight of Sentence s_6 is equal to 0.65 + 0.84 + 0.12, while the average weight of the sentence s_6 is $(0.65 + 0.84 + 0.12)/3$

In our case, the aggregate similarity of a summary of a document is calculated by summing up the average weights of all the sentences of the summary and dividing this sum by the number of sentences of the document (See Fig. 1).

Fig. 1 Similarity graph of a document

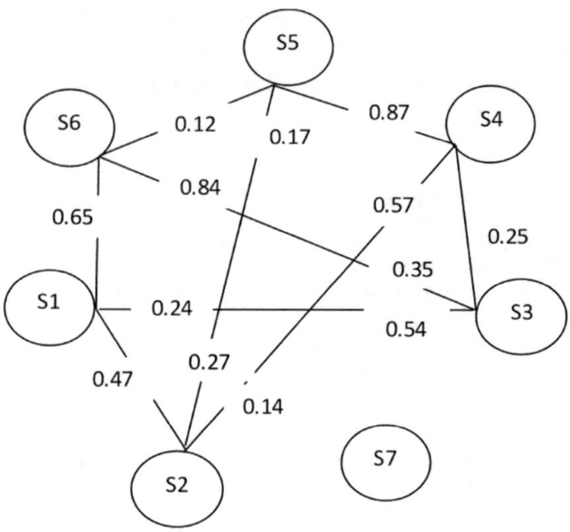

We note that the values of the TF, aggregate similarity and the grammaticality are between 0 and 1.

where α, β and γ are set empirically.

TF

$$TF(term, D) = \frac{freq(term, D)}{freq(term, D) + 0.5 + 1.5 * \frac{lengthdoc}{avrlengthdoc}} \tag{9}$$

where:
D: A summary document
term: A term in the summary document D
TF: term frequency
IDF: Inverse Document Frequency
freq: the frequency of a term in the document
lengthdoc: the document length
avrlengthdoc: the average length of training documents.
The *tf* value is calculated for each term of the document; the global value of the *tf* of the document is the sum of all the *tf* values of the terms divided by the total number of the terms of the document.

Grammaticality The grammaticality of the summary should also be maximized. For this reason, we propose to use a trigram language model. [17] stated that the trigram assumption is arguably quite strong, and linguistically naive. However, it leads to models that are very useful in practice. The grammaticality of the document is represented as the sum of all trigrams of the document over the total number of trigrams of this document. The value of the grammaticality ranges from 0 to 1.

$$Grammaticality(S) = \gamma * Trigrams(S) \tag{10}$$

6 Evaluation of the Global Summarization System

6.1 Evaluation of the Algorithm for Affectation of Operations to Document Partitions

6.1.1 Operations Used in the Evaluation

We have used three operations in the experimentation. All the operations used belong to the *single-sentence operation* category.

- **Sentence extraction using AdaBoost** [10]

  This operation is based on the AdaBoost Machine Learning-based technique to generate Arabic summaries. This technique is used to predict whether a new sentence is likely to be included in the summary or not.

- **Concept generalization and fusion for abstractive sentence generation** [12]

  This operation addresses the problem of abstractive text summarization by fusing and generalizing concepts. It tackles the problem of generalization of sentences, i.e. such that the system will be able to generate from a sentence like *"Sue ate bananas, apples and potatoes"* an output like *"Sue ate fruits and vegetables"* or *"Sue ate some food"*. This task requiring the use of world knowledge, the operation uses WordNet to generalize the concepts. This operation has been basically proposed for the English language in [12] and we have ported it to Arabic for the purposes of this work.

- **Sentence compression**

  This operation tries to generate a single sentence that preserves the most salient information from the original sentence while ensuring that it is grammatically well-formed. We have modeled sentence compression as an optimization problem using differential evolution. This method allows to search the space of compressions in a polynomial time.

6.1.2 Discussion of the Learning Process and the Corpus Annotation

The approach adopted in this work requires the use of a mapping process which is responsible for assigning operations to the different portions of the text (subsets of sentences of a partition). After segmenting the source document into sentences and generating all the possible partitions, we need to select the best partition, hence the best mapping of operations for this partition of the source document. At this stage, a prediction function that assigns a certain probability that represents the suitability of the operation to each subset of sentences of the partition is needed.

To assign a probability to any operation, a machine learning system is trained to learn the mapping between the subsets of sentences of the partition and the operations. This mapping function takes as input the different features of the problem and outputs an estimated value that gives hints about the suitability of this operation to this portion of text.

To train the system we have developed a parallel corpus of pairs <subsets of sentences, operation>. Figure 2 shows an example from the TALAA-ASC corpus [11]. The document is represented using the XML format. It contains packets each of which represents a subset of sentences of the document. Inside each packet tag, we have defined the "id" and the "operation" attribute. For the operation attribute, we have used the following codes:

- S: Select a sentence.
- R: Remove a sentence.
- C: Compress a sentence.
- G: Generalize a sentence.
- F: fuse sentences.


```xml
<?xml version="1.0" encoding="UTF-8"?>
<file>
<packet  id="1" operation="S">
<source>حث وزير الخزانة الامريكي جاك لو دول منطقة اليورو على "تعزيز الطلب" على منتوجاتها من
أجل تخفيف نسبة البطالة وتجنب الدخول في مرحلة الانكماش الاقتصادي. </source>
</packet>
<packet  id="2" operation="S">
<source>وقال لو خلال اجتماع الدول العشرين التي تحتضن العديد من كبار الدول الاقتصادية
إن " أوروبا بحاجة الى حل مشاكلها والقضاء على اختلافاتها داخل منطقة اليورو"، مضيفاً
ما يبدو واضحاً من التجربة الامريكية هو ضرورة الجمع بين إتخاذ خطوات لدفع عجلة الاقتصاد على المدى القصير "
وإجراء تغييرات هيكلية على المدى الطويل". </source>
</packet>
<packet  id="3" operation="S">
<source>وأشار "على الاوروبيين العمل على تطبيق هاتين الخطوتين". </source>
</packet>
<packet  id="4" operation="R">
<source>وعبر وزير الخزانة الامريكي عن "قلقه من التوتر القائم بين الدول الاوروبية،
 منبهاً إلى مضار الدفع بسياسات عاجلة". </source>
</packet>
<packet  id="5" operation="R">
<source>وأشار إلى أن ما يقلقه هو "تأجيل الجهود لدفع عجلة الاقتصاد مما سينعكس سلباً
 على حاجات بلدان منطقة اليورو". </source>
</packet>
<packet  id="6" operation="F6_7">
<source>وكان البنك المركزي الاوروبي قد فرض تدابير جديدة لإنعاش الاقتصاد في منطقة اليورو.</source>
<fusion>و بسبب الضغوط لدفع عجلة الاقتصاد في منطقة اليورو فرض البنك المركزي الاوروبي تدابير
 جديدة لإنعاش الاقتصاد في منطقة اليورو</fusion>
</packet>
<packet  id="7" operation="F6_7">
<source>ويواجه البنك الكثر من الضغوط لدفع عجلة الاقتصاد في منطقة اليورو جراء التباطؤ الذي
 يشهده قطاع الانتاج والتضخم الذي هبط فقط 3 في المئة. </source>
<fusion>و بسبب الضغوط لدفع عجلة الاقتصاد في منطقة اليورو فرض البنك المركزي الاوروبي تدابير
 جديدة لإنعاش الاقتصاد في منطقة اليورو</fusion>
</packet>
</file>
```


Fig. 2 Example of a document from the annotated TALAA-ASC corpus

Table 1 Evaluation of the algorithm for affectation of operations

Evaluation	j48	SVM	Logistic	MLP	Nb	Random forest
Weighted precision	0.639	0.623	0.668	0.638	0.715	0.597
Weighted recall	0.645	0.618	0.654	0.633	0.630	0.610
Weighted f-measure	0.623	0.587	0.623	0.607	0.585	0.598

6.1.3 Comparison

In order to evaluate the performance of the system in associating the operations, we have used the annotated TALAA-ASC corpus. We have used 10-fold cross validation. Table 1 presents the results of the recall, precision and F1-score.

We have used WEKA [31] to asses the performance of the different machine learning algorithms. We have tested the system with the following machine learning techniques: J48, SVM, Logistic regression and Naive Bayes; and have used the default values for their parameters. These techniques have been trained on the TALAA-ASC [11] corpus which we have later enriched and annotated with the following operations: *sentence selection*, *sentence compression*, *concept generalization and fusion* and *sentence fusion*. We note that the sentence fusion operation which merges sentences is not used when training the machine learning system that selects the operations to be applied, this is one of our future work directions.

Table 1 shows that the machine learning techniques that have given the best results are the J48 and Logistic regression with F-Measure equal to 0.623. SVM comes next with an F-Measure of 0.587 followed by Naive Bayes with an F-Measure of 0.585.

6.2 Evaluation of the Summarization Framework

ROUGE-N is an n-gram recall measure that compares between a summary and a set of reference summaries. It is computed as follows:

$$ROUGE-N = \frac{\sum_{S\in\{ReferenceSummaries\}} \sum_{gram_n \in S} Count_{match}(gram_n)}{\sum_{S\in\{ReferenceSummaries\}} \sum_{gram_n \in S} Count(gram_n)} \tag{11}$$

where n is the length of the n-gram. $Count_{match}(gram_n)$ is the maximum number of n-grams co-occurring in a set of reference summaries and a summary.

Table 2 Evaluation of the summarization framework

Evaluation		AdaBoost	j48	SVM	Logistic	ML Op	Nb
Rouge 1	R	0.43308	0.13690	0.29116	0.14934	**0.74566**	0.15340
	P	**0.53312**	0.32110	0.41174	0.36549	0.48694	0.32575
	F	0.44071	0.17442	0.31470	0.19166	**0.55995**	0.18813
Rouge 2	R	0.15690	0.00935	0.10429	0.01641	**0.37068**	0.02339
	P	0.18658	0.04216	0.12997	0.05356	**0.28130**	0.07482
	F	0.15771	0.01503	0.10458	0.02471	**0.31244**	0.03400
Rouge SU4	R	0.17887	0.00631	0.12812	0.01172	**0.44232**	0.02927
	P	0.22046	0.04911	0.17082	0.05981	**0.28683**	0.08999
	F	0.16966	0.01092	0.12331	0.01912	**0.33117**	0.03511

In order to evaluate the framework, we have used a number of machine learning algorithms and the Rouge package [40] to evaluate the system (See Table 2).

We have compared the performance of the proposed system against the performance of a number of machine learning systems. All of these existing systems are extractive and try to select the most relevant sentences from a document. These systems have been trained on an external corpus of [10] which contains 30 documents and then tested on the enriched TALAA-ASC corpus [11].

We have compared the performance of our Summarization framework against the AdaBoost system proposed in [10], J48, SVM logistic and Naive Bayes. The proposed system has a better performance compared to the other existing systems since it has the ability to select the suitable operation for a given portion of text. In our case, for the purposes of testing the proposed framework we have included only three operations which are *sentence selection/deletion*, *sentence compression* and *concept generalization and fusion*. We note that the performance of the system could be improved if we include other operations from the *multi-sentence operation* category. We are currently working on the implementation of some *multi-sentence operations* so we can include them into the framework and test it.

7 Discussion

We now discuss some of the most relevant points about the proposed approach.

The approach could automatically select the portions of the document (set of subsets of sentences) and the most suitable operations that we should apply to them. In our context, a document portion represents a set of subsets of sentences. We select an operations that belongs to the set of *single-sentence operation* if the set of subsets contains one sentence or, in case we are dealing with multiple sentences (the set of subsets contains more than one sentence), we should use an operation from the *set of multi-sentence operations*. In this case, we have to summarize several sentences using one operation.

The *multigraph* has been constructed to facilitate the task for the other operations that will be applied to generate the summary. For instance, the *single-sentence operation* of deleting sentences could benefit from the use of the *multigraph* and some relations that the *multigraph* conveys, like the semantic similarity between sentences. This information could be used as a feature for this operation (delete sentence). Among the other operations that could benefit from the *multigraph* are the fusion and generalization of concepts, especially, if concepts that we intend to generalize do not belong to the same sentence. This is being investigated in a separate research work. In this situation, the *multigraph* provides information about the RST relation, and here we use some rules that will give us the ability to decide whether two adjacent sentences could be fused or not.

An important issue that has to be emphasized is that the efficiency of the approach will mainly depend on the quality of the operations that make up the entire framework. To have a high-quality summarizer, all of the operations should be of good quality. The framework is only responsible for selecting the best operation(s) to apply to the best portion of text, and it is not responsible for the quality of the operations which are external to the framework.

We note that this framework is the first that handles the Arabic language using both abstractive and extractive operations. Combining the two operations has improved the results reached by other existing machine learning systems which have been used in the literature to tackle Arabic text summarization.

8 Conclusion

In this work, we have presented an approach that can integrate several operations into one single framework which has benefitted from several aspects. It uses concepts inspired from number theory. In this sense, we have generated the partitions of a document where the number of partitions is the *Bell numbers* [13]. We also use probabilities to select the most likely partition and the set of operations that will be applied to the source text to produce the summary. Another interesting issue in this approach is the use of a *multigraph* which embodies several relations represented by different matrices. The *multigraph* could also be useful for the operations selected by the machine learning system.

This work could be extended in several directions. One possible improvement is to propose an active learning approach that will have the ability to adapt new operations in real time. Currently, if a new operation has to be integrated into our summarization framework, we must retrain the machine learning system to learn which operation to use among the available ones.

It will also be interesting to integrate in an enrichment of the implementation an operations that belongs to the set of *multi-sentence operation*. For the moment, though the theoretical framework includes two categories of operations, we have included in the implementation only operations that belong to the *set of single sentence operations*; we are currently working on the suggested expansion of the system.

We state that the approach presented in this paper is very novel and gives different dimensions to research in text summarization. The main strength of this framework is its ability to seemlessly include any text summarization operation.

References

1. Fattah, M.A., Ren, F.: Probabilistic neural network based text summarization. In: International Conference of Natural Language Processing and Knowledge Engineering, 2008. NLP-KE'08, pp. 1–6. IEEE (2008)
2. Al-Saleh, A.B., Menai, M.E.B.: Automatic Arabic text summarization: a survey. Artif. Intell. Rev. **45**(2), 203–234 (2016)
3. Amer, E., Foad, K.: Akea: an Arabic keyphrase extraction algorithm. In: International Conference on Advanced Intelligent Systems and Informatics, pp. 137–146. Springer (2016)
4. Azmi, A., Al-thanyyan, S.: Ikhtasir—a user selected compression ratio Arabic text summarization system. In: Proceeding of International Conference of Natural Language Processing and Knowledge Engineering (NLP-KE 2009), pp. 1–7
5. Azzam, S., Humphreys, K., Gaizauskas, R.: Using coreference chains for text summarization. In: Proceedings of the Workshop on Coreference and its Applications, pp. 77–84. Association for Computational Linguistics (1999)
6. Baldwin, B., Morton, T.S.: Dynamic coreference-based summarization. In: EMNLP, pp. 1–6 (1998)
7. Barzilay, R., Elhadad, M.: Using lexical chains for text summarization. Adv. Autom. Text Summ. 111–121 (1999)
8. Barzilay, R., Elhadad, M., McKeown, K.R.: Text summarizations with lexical chains. Adv. Autom. Text Summ. 111–121 (1999)
9. Barzilay, R., McKeown, K.R.: Sentence fusion for multidocument news summarization. Comput. Linguist. **31**(3), 297–328 (2005)
10. Belkebir, R., Guessoum. A.: A supervised approach to Arabic text summarization using adaboost. In: New Contributions in Information Systems and Technologies, pp. 227–236. Springer (2015)
11. Belkebir, R., Guessoum, A.: Talaa-asc: a sentence compression corpus for Arabic. In: 2015 IEEE/ACS 12th International Conference of Computer Systems and Applications (AICCSA), pp. 1–8. IEEE (2015)
12. Belkebir, R., Guessoum, A.: Concept generalization and fusion for abstractive sentence generation. Expert Syst. Appl. **53**, 43–56 (2016)
13. Bell, E.T.: Exponential numbers. Am. Math. Mon. **41**(7), 411–419 (1934)
14. Boudabous, M.M., Maaloul, M.H., Belguith, L.H.: Digital learning for summarizing Arabic documents. In: International Conference on Natural Language Processing, pp. 79–84. Springer (2010)
15. Clarke, J., Lapata, M.: Global inference for sentence compression: an integer linear programming approach. J. Artif. Intell. Res. **31**, 399–429 (2008)
16. Cohn, T., Lapata, M.: An abstractive approach to sentence compression. ACM Trans. Intell. Syst. Technol. **4**(3) 41:1–41:35 (July 2013)
17. Collins, M.: Course notes for nlp by Michael Collins. Columbia University (2013)
18. Coster, W., Kauchak, D.: Simple english wikipedia: a new text simplification task. In: ACL (Short Papers), pp. 665–669 (2011)
19. Douzidia, F.S., Lapalme, G.: Lakhas, an Arabic summarization system. In: Proceedings of DUC2004 (2004)

20. El-Fishawy, N., Hamouda, A., Attiya, G.M., Atef, M.: Arabic summarization in twitter social network. Ain Shams Eng. J. **5**(2), 411–420 (2014)
21. El-Haj, M., Kruschwitz, U., Fox, C.: Exploring clustering for multi-document Arabic summarisation. In: Asia Information Retrieval Symposium, pp. 550–561. Springer (2011)
22. El-Haj, M., Rayson, P.: Using a keyness metric for single and multi document summarisation. Association for Computational Linguistics (2013)
23. El-Haj, M.O., Hammo, B.H.: Evaluation of query-based Arabic text summarization system. In: International Conference on Natural Language Processing and Knowledge Engineering, 2008. NLP-KE'08, pp. 1–7. IEEE (2008)
24. Ercan, G., Cicekli, I.: Lexical cohesion based topic modeling for summarization. In: International Conference on Intelligent Text Processing and Computational Linguistics, pp. 582–592. Springer (2008)
25. Erkan, G., Radev, D.R.: Lexrank: graph-based lexical centrality as salience in text summarization. J. Artif. Intell. Res. **22**, 457–479 (2004)
26. Feblowitz, D., Kauchak, D.: Sentence simplification as tree transduction. In: Proceedings of the Second Workshop on Predicting and Improving Text Readability for Target Reader Populations, pp. 1–10, Sofia, Bulgaria (August 2013). Association for Computational Linguistics
27. Filippova, K., Alfonseca, E., Colmenares, A.C., Kaiser, L., Vinyals, O.: Sentence compression by deletion with lstms. In: Proceedings of the 2015 Conference on Empirical Methods in Natural Language Processing, pp. 360–368. Association for Computational Linguistics (2015)
28. Genest, P.-E., Lapalme, G.: Framework for abstractive summarization using text-to-text generation. In: Proceedings of the Workshop on Monolingual Text-To-Text Generation, pp. 64–73. Association for Computational Linguistics (2011)
29. Giannakopoulos, G., Karkaletsis, V., Vouros, G.: Testing the use of n-gram graphs in summarization sub-tasks. In: Proceedings of the Text Analysis Conference (TAC) (2008)
30. Gonçalves, P.N., Rino, L., Vieira, R.: Summarizing and referring: towards cohesive extracts. In: Proceedings of the Eighth ACM Symposium on Document Engineering, pp. 253–256. ACM (2008)
31. Hall, M., Frank, E., Holmes, G., Pfahringer, B., Reutemann, P., Witten, I.H.: The weka data mining software: an update. ACM SIGKDD explor. newsl. **11**(1), 10–18 (2009)
32. Hasler, L.: From extracts to abstracts: human summary production operations for computer-aided summarisation. Ph.D. thesis, University of Wolverhampton (2007)
33. Huang, M., Shi, X., Jin, F., Zhu, X.: Using first-order logic to compress sentences. In: AAAI (2012)
34. Imam, I., Nounou, N., Hamouda, A., Khalek, H.A.A.: An ontology-based summarization system for Arabic documents (ossad). Int. J. Comput. Appl. **74**(17) (2013)
35. Jing, H.: Using hidden markov modeling to decompose human-written summaries. Comput. Linguist. **28**(4), 527–543 (2002)
36. Jones, K.S. et al. Automatic summarizing: factors and directions. Adv. Autom. Text Summ. pp. 1–12 (1999)
37. Kauchak, D.: Improving text simplification language modeling using unsimplified text data. In: Proceedings of the 51st Annual Meeting of the Association for Computational Linguistics (Vol. 1: Long Papers), pp. 1537–1546, Sofia, Bulgaria (August 2013). Association for Computational Linguistics
38. Keskes, I., Boudabous, M.M., Maaloul, M.H., Belguith, L.H.: Étude comparative entre trois approches de résumé automatique de documents Arabes. In: Proceedings of the Joint Conference JEP-TALN-RECITAL 2012, vol. 2: TALN, pp. 225–238, Grenoble, France (June 2012). ATALA/AFCP
39. Khan, A., Salim, N., Kumar, Y.J.: A framework for multi-document abstractive summarization based on semantic role labelling. Appl. Soft Comput. **30**, 737–747 (2015)
40. Lin, C.-Y.: Rouge: a package for automatic evaluation of summaries. In: Text Summarization Branches Out: Proceedings of the ACL-04 workshop
41. Liu, F., Flanigan, J., Thomson, S., Sadeh, N., Smith, N.A.: Toward abstractive summarization using semantic representations. In: Proceedings of the 2015 Conference of the North American

Chapter of the Association for Computational Linguistics: Human Language Technologies, pp. 1077–1086, Denver, CO (May–June 2015). Association for Computational Linguistics
42. Lloret, E., Boldrini, E., Vodolazova, T., Martínez-Barco, P., Muñoz, R., Palomar, M.: A novel concept-level approach for ultra-concise opinion summarization. Expert Syst. Appl. **42**(20), 7148–7156 (2015)
43. Lloret, E., Palomar, M.: Text summarisation in progress: a literature review. Artif. Intell. Rev. **37**(1), 1–41 (2012)
44. Mani, I., Maybury, M.T.: Advances in Automatic Text Summarization. the MIT Press (1999)
45. Mann, W.C., Thompson S.A.: Rhetorical structure theory: toward a functional theory of text organization. Text-Interdiscip. J. Study Discourse **8**(3), 243–281 (1988)
46. Marcu, D.: From discourse structures to text summaries. In: Proceedings of the ACL, Vol. 97, pp. 82–88. Citeseer (1997)
47. Marcu, D.: To build text summaries of high quality, nuclearity is not sufficient. In: Working Notes of the AAAI-98 Spring Symposium on Intelligent Text Summarization, pp. 1–8 (1998)
48. Marcu, D.: Discourse trees are good indicators of importance in text. Adv. Autom. Text Summ. pp. 123–136 (1999)
49. Marcu, D.: The Theory and Practice of Discourse Parsing and Summarization. MIT Press, Cambridge, MA, USA (2000)
50. McKeown, K., Rosenthal, S., Thadani, K., Moore, C.: Time-efficient creation of an accurate sentence fusion corpus. In: Human Language Technologies: The 2010 Annual Conference of the North American Chapter of the Association for Computational Linguistics, pp. 317–320. Association for Computational Linguistics (2010)
51. Medelyan, O.: Computing lexical chains with graph clustering. In: Proceedings of the 45th Annual Meeting of the ACL: Student Research Workshop, pp. 85–90. Association for Computational Linguistics (2007)
52. Mihalcea, R.: Graph-based ranking algorithms for sentence extraction, applied to text summarization. In: Proceedings of the ACL 2004 on Interactive Poster and Demonstration Sessions, p. 20. Association for Computational Linguistics (2004)
53. Mihalcea, R., Tarau, P.: Textrank: bringing order into texts. Association for Computational Linguistics (2004)
54. Nenkova, A.: Entity-driven rewrite for multidocument summarization. In: Proceedings of IJCNLP08 (2008)
55. Nenkova, A., McKeown, K.: Automatic summarization. Found. Trends Inf. Retr. **5**(23), 103–233 (2011)
56. Oufaida, H., Nouali, O., Blache, P.: Minimum redundancy and maximum relevance for single and multi-document arabic text summarization. J. King Saud Univ. Comput. Inf. Sci. **26**(4), 450–461 (2014)
57. Patil, K., Brazdil, P.: Text summarization: using centrality in the pathfinder network. Int. J. Comput. Sci. Inf. Syst. **2**, 18–32 (2007)
58. Peng, H., Long, F., Ding, C.: Feature selection based on mutual information criteria of max-dependency, max-relevance, and min-redundancy. IEEE Trans. Pattern Anal. Mach. Intell. **27**(8), 1226–1238 (2005)
59. Qazvinian, V., Radev, D.R.: Scientific paper summarization using citation summary networks. In: Proceedings of the 22nd International Conference on Computational Linguistics-Vol. 1, pp. 689–696. Association for Computational Linguistics (2008)
60. Saggion, H.: A classification algorithm for predicting the structure of summaries. In: Proceedings of the 2009 Workshop on Language Generation and Summarisation, pp. 31–38. Association for Computational Linguistics (2009)
61. Saggion, H., Poibeau, T.: Automatic text summarization: past, present and future. In: Multi-source, Multilingual Information Extraction and Summarization, pp. 3–21. Springer (2013)
62. Schlesinger, J.D., Oleary, D.P., Conroy, J.M.: Arabic/english multi-document summarization with classythe past and the future. In: International Conference on Intelligent Text Processing and Computational Linguistics, pp. 568–581. Springer (2008)

63. Siddharthan, A.: An architecture for a text simplification system. In: Language Engineering Conference, 2002. Proceedings, pp. 64–71. IEEE (2002)
64. Mohammed, I., Sobh, A.H.: An optimized dual classification system for Arabic extractive generic text summarization. Ph.D. thesis, Citeseer (2009)
65. Tanaka, H., Kinoshita, A., Kobayakawa, T., Kumano, T., Kato, N.: Syntax-driven sentence revision for broadcast news summarization. In: Proceedings of the 2009 Workshop on Language Generation and Summarisation, pp. 39–47. Association for Computational Linguistics (2009)
66. Wan, X., Xiao, J.: Towards a unified approach based on affinity graph to various multi-document summarizations. In: International Conference on Theory and Practice of Digital Libraries, pp. 297–308. Springer (2007)
67. Wang, T., Chen, P., Amaral, K., Qiang, J.: An experimental study of lstm encoder-decoder model for text simplification. arXiv:1609.03663 (2016)
68. Wang, T., Chen, P., Rochford, J., Qiang, J.: Text simplification using neural machine translation. In: Thirtieth AAAI Conference on Artificial Intelligence (2016)
69. Woodsend, K., Lapata, M.: Wikisimple: automatic simplification of wikipedia articles. In: AAAI (2011)
70. Yamangil, E., Shieber, S.M.: Bayesian synchronous tree-substitution grammar induction and its application to sentence compression. In: Proceedings of the 48th Annual Meeting of the Association for Computational Linguistics, pp. 937–947. Association for Computational Linguistics (2010)
71. Yeh, J.-Y., Ke, H.-R., Yang, W.-P., Meng, I.-H.: Text summarization using a trainable summarizer and latent semantic analysis. Inf. Process. Manage. **41**(1), 75–95 (2005)
72. Yoshikawa, K., Hirao, T., Iida, R., Okumura, M.: Sentence compression with semantic role constraints. In: Proceedings of the 50th Annual Meeting of the Association for Computational Linguistics: Short Papers-Vol. 2, pp. 349–353. Association for Computational Linguistics (2012)
73. Zajic, D., Dorr, B.J., Lin, J., Schwartz, R.: Multi-candidate reduction: sentence compression as a tool for document summarization tasks. Inf. Process. Manage. **43**(6), 1549–1570 (2007)

Multi-document Summarizer

Hazem Bakkar, Asma Al-Hamad and Mohammed Bakar

Abstract In this study, we address the multi-document summarization challenge. We proposed a summarizer application that implements three well-known multi-document summarization techniques; Topic-word summarizer, LexPageRank summarizer and Centroid summarizer. The contribution in this study is demonstrated by proposing a fourth summarization technique that is built on the previous acquired knowledge and experiments performed on the previously mentioned summarization techniques. Evaluating the system-generated summaries is performed using ROUGE [1], results showed that the new summarizer outperforms the other summarization techniques, and it takes a relatively short time to generate summaries comparing to other summarizers. However, LexPageRank summarizer evaluation performed better than the new summarizer evaluation, the cost of achieving a better evaluation using this technique was the time needed to generate the summaries, LexPageRank summarizer needs a long time to generate summaries comparing to other summarizers. In this study, DUC04 is used as a corpus in testing and implementing the proposed application.

1 Introduction

The massive size of textual data that exits online, and the huge number of documents that are available offline within various bodies of organizations raised the need to find effective techniques to extract the important information out of this enormous number of online and offline documents.

H. Bakkar (✉)
British University in Dubai, Dubai, UAE
e-mail: hazem.bakkar@hotmail.com

A. Al-Hamad · M. Bakar
Dammam University, Dammam, Saudi Arabia
e-mail: asma_toubassy@yahoo.com

M. Bakar
e-mail: mwb_sw@hotmail.com

K. Shaalan et al. (eds.), *Intelligent Natural Language Processing: Trends and Applications*, Studies in Computational Intelligence 740,
https://doi.org/10.1007/978-3-319-67056-0_22

IR systems usually retrieve several number of documents that are related to the user query input, some other systems contain a document relevancy-assessment sub- system that retrieves several documents that are related to a user query. It is usual that the result of a query is hundreds of related documents. If we want to extract the informative text out of these documents, we will need an automatic summarizer. If we use a single document summarizer then it will most likely generate very similar summaries from the retrieved documents since most of them contain similar textual information. Here comes the benefit of using a multi-document summarizer, which will use the shared information that exist in the similar documents only once and then it will focus on the unique information that is spread in the cluster of the summarized documents.

2 Background

What is text summarization?

Text Summarization is generating a short text from a source longer text, while maintaining the main informative sentences of the source text or document. Text summarization methods can be classified into two types; abstractive summary and extractive summary.

2.1 Abstractive Summary

Abstractive summary is generating a summary depending on understanding the main concepts and information that exists in the source text. Abstract summarization uses Linguistics tools to examine and understand the text, these tools are used also to search for new expressions and concepts related to the information content in the source text, these expressions and concepts will be used to generate a short summary [2, 4]. In other words, Abstractive summary is about understanding the original text and generate a shorter version of it using other words.

The main problem in Abstractive summaries involves in the representation of the information that lies beneath the source text. The main concept of the abstractive summary is to understand the source text and then re-phrase it using new concepts and expressions. This task depends on the representation of the source text, which reflects the system understanding of the text content, the capability of generating a descent summary depends on the system's ability to capture the representation of the source text. It is known that until now, there are no systems that have an acceptable ability to understand natural language. Thus, abstractive summaries development is directly affected by improving the technology of intelligent under-standing of natural language [5].

2.2 Extractive Summary

This type of summaries depends on extracting key words or sentences from the original text using statistical methods to determine key text segments [6]. Extractive systems usually perform analysis for different surface level features such as term frequency, inverse document frequency, word location in the text, and indicators that point to the text segments that should be extracted [5, 6].

Extractive summary systems identify the most important text segments (word, sentence or paragraph) using two approaches: (1) by identifying the most frequent segment in the source text or (2) by being located in the most preferable position in the source text [6]. These approaches guarantee the simplicity of the extractive summary systems compared to the abstractive summary systems in both implementation and conceptual prospective [5].

Generating extractive summary includes two major stages [7]: (1) the preprocessing stage, and (2) the processing stage.

In the preprocessing stage, text is examined and analyzed by applying the following steps: firstly, the sentences boundary is identified. Different languages have different sentences boundaries, in English language, for example, the sentence boundary is identified by the dot at the end of the sentence. Secondly, removing stop-words that exist in the source text and removing any word with no semantics or words that do not represent relative information to the summary targeted content. Finally, stemming the words in the text; generating stems will insures the validation of the words semantics [7].

In the processing stage, features that determine the relevancy of sentences are chosen and computed, and then they are weighted using specific weight methods and equations. The sentences are given scores; the highest scored sentences will be added to the final summary [5].

Extractive text summaries have some drawbacks [8, 9]: (1) Extractive text summaries generally extract long sentences, which includes unnecessary or extra parts, this usually shrinks the available space allocated to more informative text segments which includes words or sentences. (2) Informative contents generally spread between sentences across the original text, this sparse of sentences makes it hard to capture the targeted content if the summary length is not long enough to capture and process these sentences. (3) Extractive summaries are weak in combining conflict information when generating the final summary. (4) Overall integrity in the final summary has some frequent problems, for example, sentences that contain pronouns; these pronouns usually lose their referents whenever they are extracted from their context. Another problem that faces the extractive summaries is in joining the sentences that are from different context, this will generate a faulty and inaccurate interpretation of the original text. Temporal expressions also lose their meaning when they are extracted from their context.

3 Multi-document Extractive Summarization

This approach uses a cluster of text documents that shares the same topic to generate a summary using extractive summarization techniques. The main goal of the multi-document summarization is to allow professionals and individuals to have an over-view about the topics and important information that exist in clusters of large number of documents within relatively a short time [5].

Single document summarization gives the reader a short compact summary about the contents of one summarized document. This type of summarization is limited by one source of information that is limited by one document. However, multi-document summarization generates a comprehensive summary from multiple sources; this type of summarization is considered more challengeable because of its variation of information sources. In 2003, Single document summarization track was removed from the Document Understanding Conference (DUC03), since that time the quantity of researches about single document summarization is remarkably shrinking [10]. Researches in the last decade proposed many multi-document summarization techniques, these techniques were heavily experimented and studied; in the following section, an implementation of a multi-document techniques will be discussed.

4 Implemented Summarization Techniques

In the proposed application three extractive summarization techniques were implemented and studied, in the next section, a theoretical review is presented to show the major parts of the summarization techniques that are crucial for the success of the proposed application.

4.1 Topic-Word Summary

Topic-Words Summarization is one approach of the Topic representation several approaches [11]. This type of summarization depends on identifying the most descriptive words in the input text, [12] used in his work a frequency threshold to determine the most descriptive words in the document in order to be selected for summarization. [12] ignored in his approach the most frequent words because it is most likely to be prepositions or determines. Also, he did not consider the words which appeared a few times in the document as these words represent the least important words in the text [11]. [13] proposed an enhanced statistical approach built on Luhun's approach, [13] used log-likelihood ratio test to determine the most descriptive words in the text, these words were called topic signatures as proposed in [14].

Topic signatures played a major rule in identifying the most important news content in multi document summarization as detailed in [15, 16]. The most important contribution in [13] approach that it determines a threshold that classifies the words in the input into two classes: descriptive and none descriptive. The classification process is accomplished by determining the statistical significance for the words in the input; the statistical significance will eliminate the need to use arbitrary thresholds in the proposed approach.

In order to determine the topic signature words, it is necessary to have a sufficient static information about the background corpus; this is mainly achieved by computing the frequency of mentioning the words in the background corpus. The likelihood of an input A and background corpus is calculated by assuming two cases: (1) The first case (H1) is when the probability of a word in the input text is equal to the probability of this word in the background corpus (B). (2) The second case (H2) is when a word in the input text has a different probability than the probability of the same word in the background corpus, usually this means that the probability of a word in the input text is higher than the probability of the same word in the background corpus [11]. The following is an arithmetic representation for the two cases [11]:

$$\mathrm{H1: P(w|A) = P(w|B) = p \ (w \text{ is not descriptive})}$$
$$\mathrm{H2: P(w|A) = pI \text{ and } P(w|B) = pB \text{ and } pI > pB \ (w \text{ is descriptive})}.$$

To compute the likelihood of a text with reference to a descriptive word, the automatic summarizer uses binomial distribution. The input text and the background corpus are represented as a series of words w_i: w_1, w_2, w_n. The frequency of occurrences for each word is identified by Bernoulli trial with a success probability = p, the success is occurring only when w_i = w, the following distribution represents the over-all probability of observing the word w occurring k times in N trials [11].

$$b(k, N, p) = \binom{N}{k} p^k (1-p)^{N-k} \tag{1}$$

In H1, p is computed from the input text and the background corpus together, while in H2, p1 is computed from the input text only, and p2 is computed from the back-ground corpus. Now all the elements are ready to compute the likelihood ratio using the following equation:

$$\lambda = \frac{b(k, N, p)}{b(k_I, N_I, p_I) \cdot b(k_B, N_B, p_B)} \tag{2}$$

where the terms with subscript I is calculated from the input text, and the terms with index B are calculated from the background corpus [11].

Using Eq. 2, we can represent $-2\log \lambda$ by the statistical distribution X^2, this will make it possible to identify the words that are considered as topic signatures [11]. Regarding the term topic signatures, we can determine the importance of a sentence in the summarizer input by calculating the number of topic signatures that exists in the sentence or by identifying the part of the sentence that is occupied by topic signatures [11]. These two ways of identifying the importance of a sentence are forming the scoring functions that are used to weight the sentences in the summarization process. It is noted that [12] approach gives a higher score for long sentences, while [13] approach gives higher score to sentences that have a higher density of topic signatures [11].

4.2 Centroid Summary

A centroid is a collection of words that are considered descriptive and important for a cluster of documents. Centroids main benefit is that they can determine whether a document in a cluster is relevant or irrelevant, also Centroids can identify the salient sentences in a cluster of documents [17].

Normally, documents main subjects are describing different topics in a sequential manner, a topic after another. This arrangement of topics is also implemented when generating a summary for any collection of documents. In order to build a meaningful summary; documents are grouped in clusters depending on their topic, so that every cluster of documents is addressing a relevant topic [5]. Documents in these clusters are represented as Term-Frequency-Inverse Document frequency (TF-IDF) vectors [18]. Term Frequency (TF) is representing the average of times a term occurs in each document in the cluster while Inverse Document Frequency (IDF) is calculated using the documents in all the clusters that make the corpus. Each cluster of documents in the corpus is considered a separate theme, and these clusters are representing the required input for the automatic summarizer that is under study. The theme that is derived from the cluster consists the words that have the highest TF-IDF scores in that specified cluster.

In Centroid summary, there are three main factors that determine whether sentences will be selected to be added to the final summary or not [5]; the first factor is computing the similarity of the sentences with the theme of the cluster (C_i). The second factor is the location of the sentence in the original input text (L_i). Different contexts of documents have different weights for sentence location importance in the body of the text, for example, in the news context, the sentences that are closer to the beginning of the text gain higher scores as they have higher importance. These sentences will likely have more chances to be added to the final summary. The third factor that determine the score that is assigned to a sentence is the similarity between this sentence and the first sentence of the document that contains both sentences (F_i) [5]. The following equation calculates the overall score (S_i) of the sentence (i) [5]:

$$S_i = W_1 * C_i + W_2 * F_i + W_3 * L_i \tag{3}$$

Ci, Li and *Fi* are the scores that are described in the previous paragraph, while W_1, W_2 and W_3 are the weights for the linear combination of the *Ci, Li, Fi* scores.

In centroid summary each document in the cluster is identified as a weighted TF-IDF vector, then a centroid is created using the first document in the cluster, after that the TF-IDF for each other documents that will be processed by the summarizer will be compared with the centroid generated from the first document of the cluster. If the similarity measure between a document and the centroid is within a specified threshold then the document under process will be added to the cluster for further processing, otherwise the document will be ignored. The following equation is used to calculate the similarity between the centroid and the processed document [19]:

$$sim(D, C) = \frac{\sum_k (d_k * c_k * idf(k))}{\sqrt{\sum_k (d_k)^2}\sqrt{\sum_k (c_k)^2}} \tag{4}$$

4.3 *LexPageRank Summary*

In this type of summaries, the selection of sentences that build up the summary is based on selecting the most central sentences in the cluster of documents that contains the most informative words. This approach of selecting the sentences is the same in principle of the selection procedure in centroid summary. The difference between LexPageRank summary and centroid summary is in the way of measuring the sentence centrality. LexPageRank summary is using Prestige principle to determine the most central sentences in a cluster of documents.

In LexPageRank summary, the sentences of a document are represented as a graph of TF-IDF vectors [2, 3] and a cluster of documents is represented as a network graph of sentences. Some of these sentences are similar to each other, and some other sentences have a small degree of similarity between them, which means that these sentences share a little amount of information between each other [2, 3]. Prestigious sentences are the sentences that are similar to many other sentences in the cluster; these sentences are considered as the most central sentences with reference to the topic [2, 3]. In order to identify the most prestigious sentences, similarity metric is used to define centrality degree for a sentence; formally, cosine similarity is used to achieve this goal. A cluster of documents can be represented by a cosine similarity matrix, where each element in the matrix represents a cosine similarity value between a sentence and its corresponding sentence. There are two methods to compute sentence prestige using the cosine similarity matrix: 1—Degree centrality. 2—Eigenvector centrality and LexPageRank [2, 3].

Degree Centrality.

It is most likely to observe relevancy between sentences in the related documents in a certain cluster, this means that cosine similarity values will be mostly more than zero in the cosine similarity matrix. A threshold value is determined in order to ignore the low values of cosine similarity in the matrix; this threshold will guarantee considering the significant cosine similarity values for sentences, these sentences contain the information that will most likely be considered for adding in the final summary. After deleting the low values of cosine similarity, the cluster can be implemented as an undirected graph where the sentences are represented as nodes, and these nodes have connections between each other, these connections represent the significant similarity between the connected sentences. Now degree centrality of a sentence can be identified as the degree of each node in the similarity graph [2, 3]. The value of cosine similarity threshold has a direct effect on the generated summary. If this threshold is too small then weak and less informative sentences can appear in the produced summary, while choosing a high value of cosine similarity will cause losing informative and important sentences that should be added to the summary.

Eigenvector Centrality and LexPageRank.

In computing the degree centrality in a cluster of documents, each edge is considered as a vote, a node is considered a prestigious one when it has high number of votes. The voting approach has one important drawback, this drawback appears when there is a cluster of related documents; and this cluster contains one document which is not related to the main topic of the cluster. When using voting approach within the document, some of the sentences in this document will get high voting degree and will be considered prestigious, this voting degree will qualify these sentences to be added to the final summary. This means that the produced summary will contain some unrelated sentences, which should not be considered in this summary. This drawback can be avoided by considering the prestigious degree of the node that did the voting, and considering the node prestige in weighting other nodes [2, 3].

PageRank is an important well-known implementation of the prestige principle [20]. PageRank is the tool for assigning a prestige score for each webpage in the web, the score in the PageRank is calculated by counting the number of pages that link to that page and the individual scores of the linking pages [2, 3]. The PageRank of page A is calculated as follows:

$$PR(A) = (1-d) + d(\frac{PR(T_1)}{C(T_1)} + \cdots + \frac{PR(T_n)}{C(T_n)}) \tag{5}$$

whereT_1 … T_n are pages that link to page A, $C(T_i)$ is the number of the outbound links from page T_i, d is the damping factor which ranges from 0 to 1 [2, 3]. To calculate the value of PageRank, a binary adjunct matrix M is constructed, where M (u, v) = 1 when there is a link from page u to page v, then the matrix is normalized so that the summation of each row is equal 1, after that the principal eigenvector of the matrix is calculated. Thus, PageRank of i-th page equal to the i-th element in the eigenvector [2, 3].

The previous procedure can be implemented on the cosine similarity matrix to identify the most prestigious sentences in the document. PageRank is used to estimate the weight for each vote by considering the prestige degree of a sentence, the more a sentence is prestigious the higher weight its vote will gain. Since cosine similarity is symmetric then the graph that represent the sentences is undirected graph, this will have no influence on the way of calculating the principal eigenvector [2, 3]. Principal Eigenvector is computed using a simple iterative power method [2, 3].

5 The Proposed Summarizer Application

In Sect. 4, three well-known automatic summarization techniques—Centroid summary, Topic-word summary and LexPageRank summary—were reviewed. In our project, these three techniques will be implemented in addition to a fourth summarization technique that is designed by the authors depending on the knowledge acquired from studying several summarization techniques, the following section will describe the implementation of these techniques.

5.1 Centroid Summarizer

Centroid summarizer is based on identifying the most similar sentences to the original document, for this purpose, the following equation is used in our project:

$$Centrality = \frac{1}{N} \sum\nolimits_{y \neq x} sim(x, y) \tag{6}$$

where x and y are vectors that represent each sentence in the input. Words in the input represents features, the value of the feature is equal to the weight of this word in the vector. Weights are calculated using Term Frequency (TF) or (TF-IDF) or by using binary representation in which the value will be 1 if the term appears in the sentences or 0 if the term does not exist in the sentences. To correctly implement any summary technique in the proposed project, it is necessary to determine the following parameters and methods: (1) Vector feature representation. (2) Similarity approach. (3) Sentence length limitation. (4) Redundancy mitigating method.

In implementing the centroid summarizer, Binary representation is used to represent vector feature weights, because it is simple to compute and it showed strong results in the testing phase. In this summarizer, cosine similarity metric is used to identify the similarity between the vectors. The sentence length in this summarizer is ranged between 15 and 50 words with a total summary length no more than 200 words maximum; these words were tokenized using NLTK.

To decrease the redundancy of sentences, any sentence has a cosine similarity more than 0.75 with a sentence in the summary is rejected. The values of these parameters were chosen carefully after performing several trials on the proposed application.

After calculating the centrality score, the program will select the sentences that will form the centroid summary, for this purpose the program will use a greedy algorithm that will choose the sentences that have the highest score then the second highest, and so on until the formed summary reaches the words count number limitation. The following pseudo code describes the greedy algorithm that is used to generate the final summary. This algorithm is also used in the other summarization techniques that are implemented in this application [21].

```
Greedy (Sentences, threshold), Begin
Centrality = [Sim (Sen, Doc) for Sen in Sentences]
Build the Sentence + Centrality Dictionary Diction
(we call it Diction)
Sort Diction according to Centrality in decreasing order.
Current_Summary = []; length = 0;
while((len < threshold)and(i <= Diction.size())

{if (Valid(Diction[i], Current_Summary))
{Current_Summary.append(Diction[i].Sentence)
length += len(Diction[i].Sentence)
}

i += 1
}
print(CurrentSummary) End
```

5.2 *Topic-Word Summarizer*

In this summarizer, the importance of sentences is computed by counting the topic words in each sentence, this will determine the weights for words. To score the sentences with respect to topic signatures, several equations can be used:

$$TWeights\ (S) = \#\ of\ topic\ words\ in\ sentence\ x \tag{7}$$

$$TWeights\ (S) = \frac{\#\ of\ topic\ words\ in\ sentence}{\#\ of\ words\ in\ sentence\ x} \tag{8}$$

In this project, the following equation is used to represent the sentence vector Weights:

$$TWeights\ (S) = \frac{\#\ of\ topic\ words\ in\ sentence\ x}{\#\ of\ nonstopwords\ in\ sentence\ x} \tag{9}$$

The reason for selecting Eq. 9 as a representation method in this summarizer is that it avoids the negative effect of counting stop words in calculating the weights. Also, this equation will maximize the productive use of the allowed number of words in the generated summary.

A topic tool algorithm is used to determine the topic words in the input text, this tool is developed by [21], and it is implemented in the *topics.py* file. This tool uses a cut off parameter for Topic words, the default value for this parameter is 0.1, this value was tested and showed better results than other values like 0.2 and 0.3.

The generated summary will include no more than 200 words with sentences' lengths between 15 and 50 words maximum, any sentence with cosine similarity more than 0.75 will be ignored.

5.3 LexPageRank Summarizer

Every sentence in this type of summarizers is represented as a node. The similarity between two nodes is represented by an edge if this similarity is exceeding a predetermined threshold, otherwise, there will be no edge between these two nodes.

In the proposed project the LexPageRank summarizer uses TF-IDF to represent the vectors, this generates more accurate vectors and relatively better results in the evaluation phase using ROUGE [1]. The similarity threshold used in this summarizer is equal to 0.2, this value showed good results when used in the experiment performed in [2, 3], another reason for using this threshold is that building LexPageRank summary is a time-consuming process, it is too slow, so that it was difficult to try several similarity thresholds.

The ending condition in the LexPageRank summarizer is that the iteration will end only if all values become less than 0.001 between iterations. This value showed that it is a reasonable value that creates a balance between performance and good results. It is noted that the quality of results did not noticeably changed when decreasing the threshold.

The sentence length in this summarizer is ranged between 15 and 50 words; the words were tokenized using NLTK. To decrease the redundancy of sentences, any sentence has a cosine similarity more than 0.75 with a sentence in the summary was rejected, and the generated summary is limited to maximum 200 words.

5.4 The Proposed Summarizer

After studying several techniques about automatic multi-documents summarizers, an idea about an additional summarization technique is formed, the proposed summarization technique uses the first and the last sentences of each input document as a reference to build the final summary. The proposed summarizer weights and scores sentences using the same technique in Centroid Summarizer, also the proposed summarizer is using topic-words approach to determine which words are considered topic signatures, it uses the same technique in Topic-words summarizer. The binary representation is the method that is used for representing sentences in this summarizer, and Eq. (9) is used to weight the sentences exactly the same way performed in the topic-words summary.

In this summarizer, the similarity measurement and the scoring processes are performed with regard to the entire cluster not only the selected subset of sentences that was collected at the beginning of the summarization operation.

After representing the sentences and scoring them, the sentences validity is verified by counting the number of words that exists in each sentence, which should be between 15 and 50 words, any sentence has a number of words that is not within this limitation will be ignored, also, any sentence with Cosine similarity value more than 0.75 will be ignored. Since the main idea behind this project is to implement extractive summarizers, it was a good idea to add some aspects of the other type of summarization methods, which is the abstractive summarization technique. This idea was achieved by replacing the least frequent nouns and verbs with their synonyms, by doing this; the final summary will have an extra benefit especially in readability and quality. It is meant to keep the frequent verbs and nouns unchanged because these words have a strong effect on the final summary, changing these words maybe changes the meaning or context of the sentences which can mislead the summarizer to generate an inaccurate summary with wrong meanings and context.

The following steps describe the proposed summarizer's operation:

1. Remove the stop words from the input text.
2. Tokenize sentences, and then extract the first and last sentence of the input document.
3. Load topic-words from *.ts, these files are generated using topic identifying tool called TopicS which is implemented in the file Topics.py [21].
4. Represent the sentences as vectors.
5. Get the Cosine similarity between the extracted vectors and the rest of the sentences in the entire cluster.
6. Scoring the sentences using the Cosine similarity metric.
7. Weighting and scoring the sentences using Eq. 9.
8. Combine both metrics results in one value that represents the overall score.
9. Sort the scored sentences descending.

10. Greedily select the sentences with the highest score then the second highest and so on until the generated summary reaches the words allowed limit which is 200 words. Also, all sentences with Cosine similarity more than 0.75 will be ignored.
11. Replace the least frequent nouns and verbs with their synonyms.
12. Generate the final summary.

6 Running the Application

To run the application, the user has to choose which type of summarizer he wants to use; this can be done by activating the desired instruction by deleting the hash sign (#) from the code line in the file project.py.

The following code lines exist at the end of the file project.py:

```
# summary = centrality_sum(path)
#summary = topic_word_sum(path, topicFile)
# summary = lex_rank_sum(path)
# summary = custom_summarizer(path, topicFile)
```

The system-generated summaries will be saved automatically in a folder called summaries; this folder exists in the parent folder that contains the file project.py.

7 Software, Dataset, and Tools Used in the Project

In the following sections, the main software and tools used in the project are generally reviewed, in addition to a description of the corpus used in implementing the summarizer application.

7.1 Python 2.7.6

Python is one of the most flexible programming languages that is known with the following aspects and specifications:

1. Python is an open source language that is free for everyone; also, it is an object-oriented language.
2. Python is an interpreted language that can be easily understood, it is also a dynamic language that can represent and manipulate massive types of variables and elements.

Python is usually competing with other languages like PERL, Java or TCL.

Python is very flexible language, because it has dynamic data types that are accompanied by dynamic typing scheme. In addition, Python can use Modules that are created by other languages like C++ [22].

Python most valuable aspects are: (1) it is readable. (2) It is clear and a powerful language. (3) It is used in a lot of different domains like web applications, system administration, desktop applications, windows applications and scientific research [23].

7.2 Natural Language Toolkit (NLTK)

NLTK is a premier platform for developing applications and systems that process human natural language data. NLTK contains more than 50 corpora and lexical resources. NLTK contains a large number of text-processing libraries for semantic reasoning, tokenization, and stemming. It also has many other libraries and functions that manipulate natural language data [24].

NLTK is a free text-processing platform that can be used by different operating systems like Windows, Mac, and Linux [24].

7.3 Dataset Used in the Project

DUC04 stands for Document Understanding Conference. This is one of the most famous conferences for researches about summarization related topics, it is held annually by the National Institute of Standards and Technology (NIST) [25]. In the proposed project DUC04 corpus is used to test the application functionality; it is also used to evaluate the quality of the generated summary by the four summarization techniques that are implemented in the project [26].

DUC04 corpus contains 50 clusters of news text documents; the documents in each cluster mainly discuss the same topic. Every cluster contains an average of 10-text documents. DUC04 also contains Model summaries that are made by human summarizers; these models are used for evaluating the quality of the system-generated summaries.

7.4 Rouge

In this research, ROUGE is used to perform the evaluation for the summaries generated using the proposed summarizer [1].

ROUGE is the abbreviation for (Recall Oriented Understudy for Gist Evaluation), it is the official scoring technique for Document Understanding Conference (DUC) 2004 [1].

ROUGE is used for English Multi-document Summarization. In ROUGE the accuracy of the system-generated summaries are measured by comparing the overlap between the Model summaries that are written by professional human summarizers and the system-generated summaries [1].

ROUGE uses different measures, ROUGE-N uses N-Grams to measure the overlap, ROUGE-L uses Longest Common Subsequence, and ROUGE-W uses Weighted Longest Common Subsequence [1].

8 Summarizer Performance

The quality of the automatic generated summary is identified by comparing this summary with the human-made summaries that share the same input documents.

ROUGE is used to evaluate the performance of the application. ROUGE uses several parameters to evaluate the summarizer performance. The same parameters will be used for all of the summarizing techniques implemented in this project. Previous researches used ROUGE for evaluating summarizers showed that the best settings for ROUGE in evaluating summaries are ROUGE-2 Average-R score (Recall), with words stemming and without removing the stop words [27]. This setting is implemented using the following instruction in ROUGE:

/ROUGE-1.5.5.pl –c 95 –r 1000 –n 2 –m –a -1 100 –x config.xml

The evaluation of the application performance was executed using two stages, the first stage was testing the summarizers using one folder of DUC04 corpus, the benefit of this stage is to get an overview about how well the system summarizers are performing. Evaluating one folder gives us quick results, while performing the evaluation on the entire corpus will take at least 6 hours, which is a non-practical process in our case; because an evaluation using ROUGE should be done everytime after changing the parameters in the summarizers in the sake of improving the quality of the system-generated summaries.

The results of the tests that were performed using ROUGE on a single folder from DUC04 corpus are shown in Table 1, Higher scores means a higher similarity to model summaries which are prepared by human judges, therefore, higher scores mean a better summary quality.

The results of the tests that were performed using ROUGE on the entire corpus (DUC04) are shown in Table 2:

Table 1 ROUGE results based on testing one folder from the corpus

Technique	ROUGE-2 recall	ROUGE-1 recall
Centroid tech	0.11414	0.44226
Topic-word tech	0.10670	0.43735
LexPageRank	0.11911	0.43243
New technique	0.04963	0.32896
Baseline summaries	0.04963	0.41278

Table 2 ROUGE results based on testing the entire corpus

Technique	ROUGE-2 recall	ROUGE-1 recall
Centroid tech	0.04409	0.30884
Topic-word tech	0.03547	0.2916
LexPageRank	0.6987	0.34126
New technique	0.05882	0.32896

The results of the trials showed that testing the entire corpus gave better results than the trials performed on a single directory. The proposed summarizer outperforms all other techniques presented in this project except for LexPageRank summarizer which showed better results than the new summarization technique. However, testing using LexPageRank summarizer took more than 4 hours to test only 14 folders out of 50 folders, the same applies for generating summaries using Centroid summarization technique. Therefore, it is clear that there is a penalty for producing high quality summaries; this penalty is consuming a long time to generate the summary.

9 Conclusion

In this project, we implemented three famous summarization techniques Topic-word summarizer, LexPageRank Summarizer and Centroid summarizer. Then a new fourth summarization technique is proposed and implemented. The evaluation of the system-generated summaries using the four summarization techniques-shown in Tables 1 and 2—clearly indicated an excellent ROUGE score when compared to the baseline summaries which are made by human summarizers. Also, it is noted that evaluating the summarizers' performance using a single folder instead of the entire corpus showed a less quality than performing the same evaluation using the entire corpus. This means that the overlap between the system-generated summaries and the model summaries is increased after we extend the size of the summarized documents to include the entire corpus; this is because of the use of the unique information that is spread between the documents in the entire corpus.

ROUGE evaluation results also showed that LexPageRank summarizer achieved the best score between the other three summarizers, however LexPageRank summarizer also showed that it needs a long time to generate the summaries from the corpus that is used in the project, it took more than 6 hours to generate the summaries. In the other hand, the new proposed summarizer and the topic-word summarizer took less than 10 min to generate the summaries out of the entire (DUC04) corpus. Centroid summarizer also needs a long time-relatively-than Topic-word summarizer and the new proposed summarizer to build its summaries.

10 Future Work

The summarizer project has many aspects that can be modified and extended. The proposed project summarizes only English language text documents, as a future work the proposed application can be modified to summarize Arabic text documents as a first step, then it can be extended to summarize other languages texts. Another addition that is considered crucial for the summarizer project is to develop a graphical user interface; this will make the use of the application easier for end users. If the proposed application is equipped with a graphical user interface, then it will be easy to put it online and make it available for public to use and test.

Evaluating the system-generated summaries is very important to determine the quality of these summaries, so it will be helpful if the application can make an automatic summary evaluation, so that the user can estimate which technique can give better results for summarizing his cluster of documents.

References

1. Lin, C.: Rouge: a package for automatic evaluation of summaries, pp. 74–81 (2004)
2. Erkan, G., Radev, D.: LexPageRank: prestige in multi-document. Text Summ. **4**, 365–371 (2004)
3. Erkan, G., Radev, D.: LexRank: graph-based lexical centrality as salience in text summarization. J. Artif. Intell. Res. (JAIR) **22**(1), 457–479 (2004)
4. Hahn, U., Romacker, M.: The SYNDIKATE text knowledge base generator, pp. 1–6 (2001)
5. Gupta, V., Lehal, G.: A survey of text summarization extractive techniques. J. Em. Technol. Web Intel. **2**(3), 258–268 (2001)
6. Kyoomarsi, F., Khosravi, H., Eslami, E., Dehkordy, P., Tajoddin, A.: Optimizing text summarization based on fuzzy logic, pp. 347–352 (2008)
7. Gupta, V., Lehal, G.: A survey of text mining techniques and applications. J. Em. Technol. Web Intel. **1**(1), 60–76 (2009)
8. Lin, J.: Summarization. In: Encyclopedia of Database Systems, 1st edn. Springer, Heidelberg, Germany (2009)
9. Cheung, J.: Comparing abstractive and extractive summarization of evaluative text: controversiality and content selection. B. Sc. (Hons.). University of British Columbia (2008)
10. Alliheedi, M.: Multi-document Summarization System Using Rhetorical Information. Master of Mathematics. The University of Waterloo
11. Nenkova, A., McKeown, K.: A survey of text summarization techniques. Springer, pp. 43–76 (2012)
12. Luhun, H.: The automatic creation of literature abstracts. IBM J. Res. Dev. **2**(2), 159–165 (1958)
13. Dunning, T.: Accurate methods for the statistics of surprise and coincidence. Comput. Linguist. **19**(1), 61–74 (1993)
14. Lin, C., Hovy, E.: The automated acquisition of topic signatures for text summarization, pp. 495–501 (2000)
15. Conroy, J., Schlesinger, J., O'Leary, D.: Topic-focused multi-document summarization using an approximate oracle score. In: the COLING/ACL on Main conference poster sessions. Association for Computational Linguistics, pp. 152–159 (2006)

16. Harabagiu, S., Lacatusu, F.: Topic themes for multi-document summarization, pp. 202–209 (2005)
17. Radev, D., Jing, H., Styś, M., Tam, D.: Centroid-based summarization of multiple documents. Inf. Process. Manag. **40**(6), 919—938 (2004)
18. Zhang, Y., Zincir-Heywood, N., Milios, E.: Narrative text classification for automatic key phrase extraction in web document corpora, pp. 51–58 (2005)
19. Radev, D., Hatzivassiloglou, V., McKeown, K.: A description of the CIDR system as used for TDT-2, p. 205 (1999)
20. Page, L., Brin, S., Motwani, R., Winograd, T.: The PageRank citation ranking: bringing order to the web. Stanford InfoLab (1999)
21. Nenkova, A.: Computational Linguistics (2012)
22. Haas, J.: Python. http://linux.about.com/cs/linux101/g/python.htm
23. Voidspace.org.uk: A Very Brief Introduction to Python. http://www.voidspace.org.uk/python/articles/python_datatypes.shtml#introducti
24. Nltk.org: Natural Language Toolkit—NLTK 3.0 documentation. http://www.nltk.org/
25. www-nlpir.nist.gov.: Document Understanding Conferences—Introduction. http://www-nlpir.nist.gov/projects/duc/intro.html
26. Hong, K., Conroy, J., Favre, B., Kulesza, A., Lin, H., Nenkova, A.: A repository of state of the art and competitive baseline summaries for generic news summarization. In: Proceedings of LREC, May (2014)
27. Owczarzak, K., Conroy, J., Dang, H., Nenkova, A.: An assessment of the accuracy of automatic evaluation in summarization, pp. 1–9 (2012)

Part VIII
Character and Speech Recognition

Features Extraction and On-line Recognition of Isolated Arabic Characters

Benbakreti Samir and Boukelif Aoued

Abstract In this paper, we present an introduction for the conception of a recognizer for on-line Arabic handwriting. On-line handwriting recognition of Arabic script is a difficult problem, since it is naturally both cursive and unconstrained. This recognizer permits to interpret a script represented by the pen trajectory. This technique is used notably in the PDA electronic diaries. First of all, we will construct a data base with several scripters. Afterwards, and before attacking the recognition phase, there is a constructional phase of samples of the Arabic characters acquired from an electronic tablet to digitize (NOUN DATABASE). This work presents the survey, the implementation and the test for a particular neural network: TDNN (Time Delay Neural Network), applied to the recognition of on-line handwritten writing.

Keywords Isolated handwritten characters recognition • On-line recognition
Convolution neural network • TDNN

1 Introduction

Cursive texts recognition remains always an open problem in both printed or handwritten styles. This is due to the difficulties in which researchers and developers have confronted, such as the variability of the shape, the style, and the slant of the script. The Arabic handwritten script is naturally cursive, difficult to process, and presents wide variability.

The emergence of new devices of seizure, like the coupled pens for numeric papers, permits to generate on-line documents in a very efficient way. Genuine

B. Samir
Department of Transmission, National Institute of Telecommunications, Oran, Algeria
e-mail: sbenbakreti@ito.dz

B. Aoued (✉)
Electronics Department, University Djillali Liabes, Sidi Bel Abbes, Algeria
e-mail: aboukelif@yahoo.fr

K. Shaalan et al. (eds.), *Intelligent Natural Language Processing: Trends and Applications*, Studies in Computational Intelligence 740,
https://doi.org/10.1007/978-3-319-67056-0_23

documents can be generated thanks to these devices; they can consist on note documents, course documents, exams copies, drafting, etc. It widens the application fields of the on-line writing seizure which is confined frequently to small size terminals (PDA, Smartphone) where only the recognition of the characters is justified.

On-line documents represent a new source of information in natural language in which few recognition-based applications exist. In the case of on-line handwritten document, it is about the sampled trajectory of the available writing instrument under the shape of (x(t); y(t)) points sequence in the space, neat in time. Therefore, it is possible to retrace a character stroke by stroke as illustrated in Fig. 1.

The use of statistical approaches permitted a huge progress in the handwriting recognition domain. Among the approaches that have been put to contribution, the connection-based approaches (neural networks) that possess a strong discriminating power and a capacity to construct borders of decisions in big dimension spaces, and on the other hand the modeling based on the hidden Markovs models (HMMs) [1] uses a parametric approach to model the observations sequences generated by stochastic processes (handwritten scripts for example), that are more obvious when it is about word recognition. The HMMs have a huge capacity to model the observations distribution for every shape class to recognize.

For the isolated characters, it is the global shape that is taken in consideration the neural networks are much more adapted, they present the advantage to be compatible with the real time approaches as well as the 2D pictorial nature of the script.

Considering the nature of the handwritten signal and our wish to process it (on line) where time is an inseparable data of the input signal, we bent on the most adapted neural networks for this kind of applications which use the connections with delays ("Time-Delay Neural Networks" TDNN) introduced by Alexander Waibel and Geoffrey Hinton. They already revealed their very high performances in the recognition of isolated letters [2, 3], they must be able to contribute to the recognition of words also [4], or even sentences [5].

It is the object of the works developed in the setting of this article. Several works have been conducted for hand written recognition, they present a lot of difficulties due to the complexity of the date and the unavailability of on-line data base.

Fig. 1 On-line writing, (numeric ink)

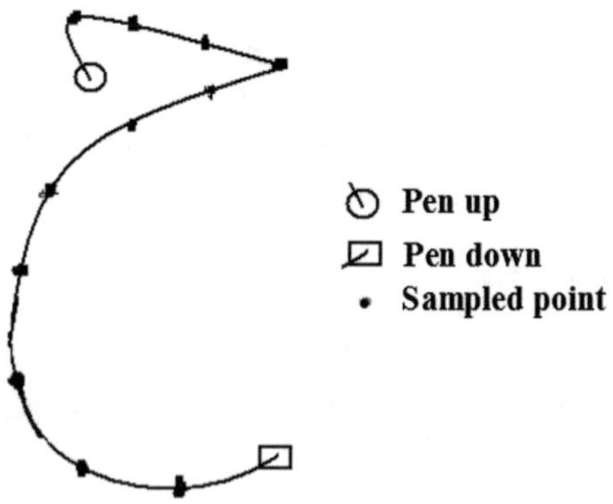

2 On-line Writing

The user writes naturally with a stiletto on a slate or a screen. The recognition software interprets the written characters or words to transform them into numerical characters.

Three properties characterize the on-line recognition: the writing order notion (strokes temporal sequence), the tracing dynamics (speed, acceleration, pen raise) and tracing skeleton (no stroke thickness).

The first commercial handwritten recognition softwares have been integrated in electronic organizers provided with a stiletto permitting characters or words and sentences seizure and sometimes provided of keyboard. Outside of this important market of small personal assistants, other applications have been developed from graphic tablet and lately of digital pens.

In the education world, to assist the teacher in his training task of the writing that can only supervise a single child at a time during his gesture of writing production. As much in the school as the medical environment, to detect quickly the different reasons related to the psychic and motor unrests (Parkinson, sclerosis) and to school failures (dyslexia).

Finally, in the meetings universe, with the possibility of holding notes, annotations, and conservations of all written or oral traces of the different intervening parties of the meeting.

The main research axes on on-line recognition can be summarized like follows:

Words, sentences or texts recognition while using contextual knowledge (specific to the document, linguistics, etc.);

Automatic adaptation to a writer's writing from a generic recognition system;

The presentation of the recognition results, the pen interface ergonomics, document edition;

The education tools that help in the writing training, the detection of unrests related to the writing;

The writer's authentication, the signatures recognition.

3 General Principles on the Recognition of the On-line Writing

Contrary to the documents paper that is digitalized as pictures, the on-line documents (specifically the characters and gestures) are stocked as electronic ink. The on-line documents can be seized while using several types of peripherals as illustrated in Fig. 2.

These different peripherals makes that the recorded electronic ink can be of different natures and qualities. The applicative domains are vast, from writing texts to diagrams seizure and even to forms replenishment or documents edition gestures.

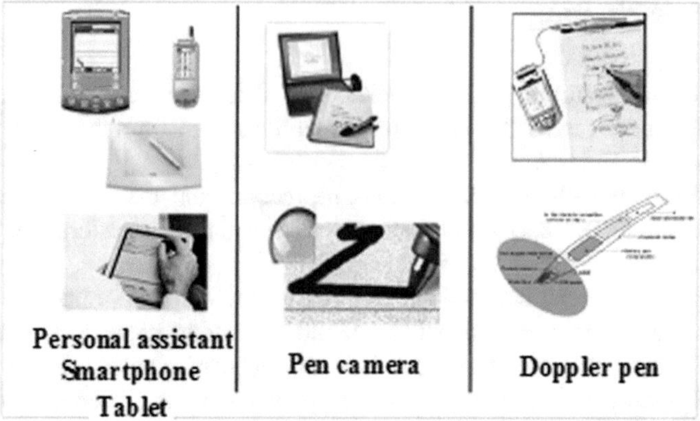

Fig. 2 Systems of on-line writing acquirement

But the main difficulty met in the recognition of the handwritten writing is the variability of the writing styles. Indeed, the shape of the handwritten characters varies a lot from a writer to another and even for a given writer according to the context of the character (the position in the word, the neighboring letters …). This variability is a source of ambiguousness between characters since one tracing can have different significances according to the context or to the writer. These properties make that the recognition of the handwritten writing is an applicative domain for very rich shape recognition in difficulties and in challenges.

In this work, we concentrate only on isolated character recognition (letters). Indeed, this problematic is the basis of many complex systems permitting words, sentences and texts recognition. A small improvement of the characters recognition efficiency can permit to decrease the complexity of the following steps.

The main difference between the two areas handwriting online and offline, often treated separately, lies in the nature of data to be processed (temporal or spatial) and relevant information that can be extracted in a object recognition.

We are interested here in the recognition in the on-line domain, this one knows an interest with the advent of the information society in which we enter with in particular the need of mobility and access to information without discontinuing. The recognition diagrams are globally common and decline themselves in two notions: the preprocesses and the features extraction.

The preprocesses: they concern the acquirement and the normalization of data and serve to suppress the noises led by the context at the time of the acquirement (sampling frequency of the graphic tablet, quality of digitalization of the scanner, extraction of the text in a document) and those generated by the human.

Features extraction: In this section, we present a preview of the methods of feature extraction used in character recognition. Actually, in addition to the taxonomy presented in Fig. 3, the character recognition domain can be described by the data collection method, feature extraction methods, classification or data representation format methods.

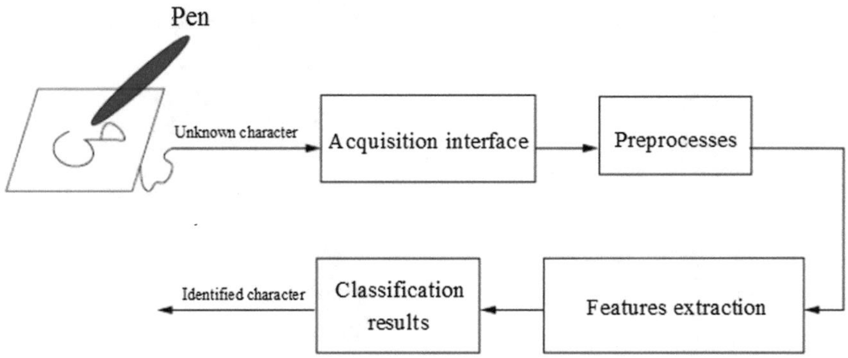

Fig. 3 General diagram of a handwritten writing recognition system

The proposed recognition system can be represented schematically according to Fig. 3. It includes a preprocess step permitting to normalize the size of character and to sample the tracing in a stationary number of equidistant points. The second step consists in the feature extraction from the previously gotten tracing, this description constitutes the input of the neural network (TDNN in our case, we will give more details in the following section). This one is going to provide on the output layer, after the attainment of training, the class of the character presented to the entry.

4 Time Delay Neural Networks TDNN

TDNN neural networks are convolutional networks out of their topology, they include a sliding window corresponding to a restricted vision field of the aggregate signal. They have been used initially in speech recognition [6], but they have been used successfully as well in the isolated character recognition like the numbers or words. The choice of this kind of neural network can be explained by the following points: it corresponds to our constraints; sturdiness toward the translation and also a big generalization capacity.

TDNN differ of the classic Multi-Layers Perceptron (MLP) by the fact that it takes a certain notion of time. Instead of taking all the input layer neurons at the same time, it accomplishes a temporal sweep. The TDNN input layer takes a specter window and sweeps the imprint. The TDNN permits also to recognize the signal less strictly than with the classic MLP [7] (in other words, it can hold small shifting) (Fig. 4).

Structure:

TDNNs are constituted like MLPs from an input layer, hidden layers and an output layer, but they differ by the inter-layer links organization. TDNNs introduce some

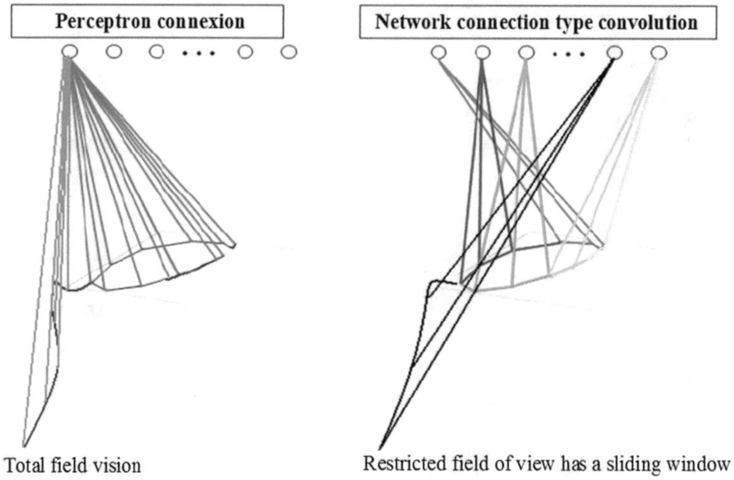

Fig. 4 Connections' illustrations in MLP and TDNN

constraints that permit them to have a certain reentrancy degree by temporal shift and distortion. These use three ideas: shared weights, temporal window and delay.

The shared weights permit to reduce the number of parameters of the neural network and induce an important generalization capacity. The weights are shared according to the temporal direction, for a characteristic data, the window associated to this data will have the same weights according to the temporal direction. This constraint entails a capacity to extract the differences with the sweep progression of the signal. This concept of shared weights is the behavior presumed of the human brain where several neurons calculate the same function on different inputs.

The concept of temporal window implies that every neuron of the layer L + 1 is only connected to one subset of the L layer (we don't have a total connectivity). The size of this window is the same between each two given layers. This temporal window permits that every neuron has only a local vision of the signal; it can be seen like a unit of detection of a local characteristic of the signal.

In addition to the two previous constraints, we introduce delays between two successive windows for a given layer.

On top of that, each layer has two directions: a temporal direction and a characteristic direction.

Functioning:

The goal of TDNN is not to learn the signal basically but to extract its features. The first layer acquires the signal, and then one or several hidden layers transform the signal into arrays of features. A neuron detects a local characteristic of the curve variation. The field of vision of the neuron is restricted to a limited temporal window. With the constraint of the shared weights, the same neuron is duplicated in the direction time (the same duplicated matrix of weight) to detect the presence or the absence of the same characteristic in different places along the signal. While using several neurons to every temporal position, the neuron network does the detection of different features: the output of the different neurons produce a new characteristic vector for the superior layer.

The temporal component of the signal is progressively eliminated with the progression of its transformation in characteristic by the superior layers, to compensate this loss, the number of neurons in the characteristic direction has been increased.

Training

To train the TDNN, we use the classic algorithm of back-propagation of the gradient but in its stochastic version (the weights are updated to every example). The phase of calculation are similar to those described earlier.

Implementation

Having been required in order to make a certain number of tests to determine the most adapted topology to our application that we desired, as for the multi-layers perceptron, to develop not a particular neural network but a neuronal simulator allowing us to model it. We followed the same concept for the MLP, but we encountered some difficulties for the back-propagation of the error, subtler than MLP.

For the implementation, we used a step of 0.01, initialized the slants and the weights between −0.1 and 0.1.

Implementing TDNN on scripting:

The majority of the architectures for the character recognition include two principle parts (as described in Fig. 5). The first, corresponding to the low layers, implements the successive convolutions permitting to transform the features progressively in a much more meaning sizes toward the problem (TDNN). The second corresponds in a classic MLP, it receives in the input the outputs set of the TDNN part.

Some remarks are essential, such as:

A neuron of a layer of a perceptron is connected to all neurons of the previous layer but for a convolutional neural network, a neuron is connected to a subset of neurons of the previous layer.

Every neuron can be seen as a detection unit of a local characteristic.

The two stated blocks are completely configurable; they are described by the variables presented below.

The Extraction part is characterized by:

The number of layers, nb_layer,

The number of neurons of every layer according to the temporal direction, window_T,

The number of neurons of every layer according to the characteristic direction, nb_feat,

The size of the temporal window seen by every layer (except the input layer), the number of neurons of the L layer seen by a neuron of the layer L + 1, field_T,

The temporal delay (number of neurons) between every window, delay.

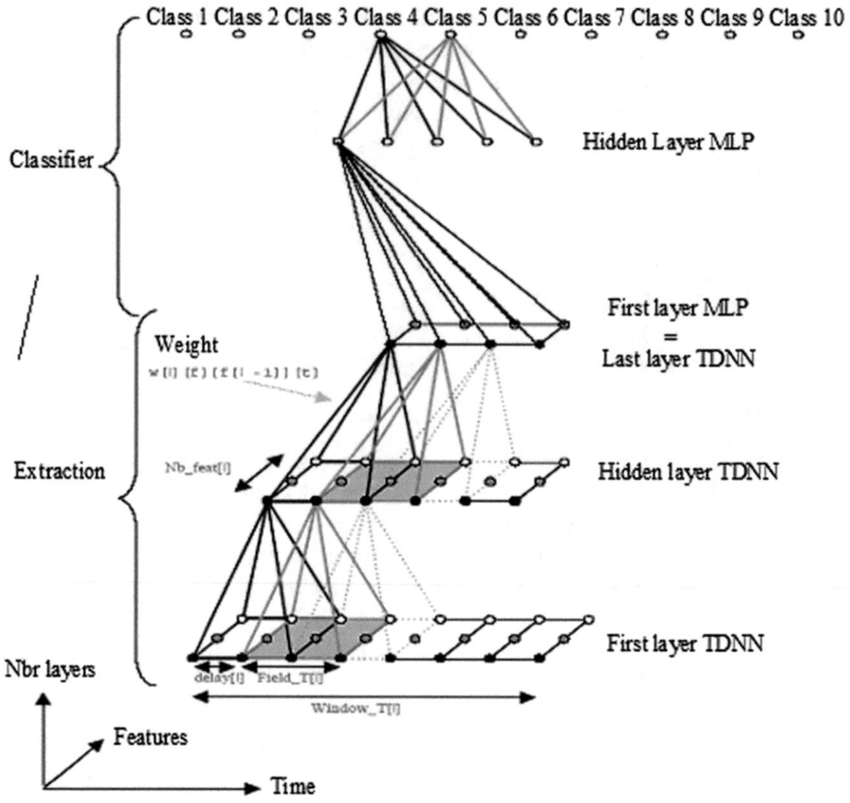

Fig. 5 TDNN architecture

A neuron is identified by its L layer, its characteristic f, and the emplacement temporal t. For every neuron are defined:
An output, or activation of the neuron, x[L][f][t]
A weights matrix of the outputs,
w[L][f][f[L − 1]] [t]
A weight vector of the slants, w_biais[L][f]
The weighted sum of the inputs, v[L][f][t]
The term of error for the back-propagation of the gradient, y[L][f][t]
The gradient, delta[L][f][f[L − 1]] [t].

And the Classifier part is characterized by:
The number of layers, NN_nb_layer,
The number of neurons of every layer NN_nb_neuron layer.
A neuron of the classifier is identified by its layer L, and its location t.
For every neuron of the classifier are defined:
An output, or activation of the neuron, x[L][t]
a matrix of the weights of the inputs,

w[L][t][t[L − 1]]
the weighted sum of the inputs, v[L][t]
the term of error for the back-propagation of the gradient, y[L][t]
the gradient, delta[L][t][t[L − 1]].

The first layer of the network acquires the features of the signal. One or several hidden layers of the neural network (extraction phase) transform a sequence of characteristic vectors in another sequence of characteristic vectors of superior order. A neuron detects a local topological characteristic of the trajectory of the stiletto. The field of vision of the neuron is restricted to a limited temporal window. With the constraint of the shared weights, the same neuron is duplicated in the time direction (the same duplicated matrix of the weights) to detect the presence or the absence of the same characteristic in different places along the trajectory of the signal. While using several neurons (nb_feat) on every temporal position, the neurons network does the detection of different features: the outputs of the different neurons produce a new characteristic vector for the superior layer.

The operations accomplished by a layer of the TDNN are of type convolution. Every k neuron of the layer L + 1 has a core of w size (number of neurons of the temporal window of the L layer) * f (number of feature of the L layer).

The temporal component of the representation of the signal is eliminated progressively in sampling the convolution to every layer. To compensate this loss of information, the number of features is multiplied. We have an architecture of type bipyramidal. This bipyramidal network converts temporal information progressively into information of type feature.

Finally, the first layer of the classifier part (entirely connected MLP) corresponds to the last layer of the extraction part.

5 Preprocess

The described previously architecture has for goal to classify and to recognize the Arabic isolated characters coming from our data base NOUN-DATABSE in its first version 1.0 containing the 28 letters of the Arabic alphabet, constructed with on line acquirement and the help of a WACOM BAMBOO version 5.08-6.

In this trial version, we limited a number of 20 writers, every writer seized the alphabet 5 times, for a total of 2800 characters.

In addition, of the information bound to the character himself (extractions of the different features), we will record some information on the writers selected to contribute to the construction of the basis. The information sum up on the first name, the last name and the date of the seizure, the attached character and its occurrence (Fig. 6).

The system describes previously will be implemented to classify the isolated characters coming from our basis NOUN DATABASE.

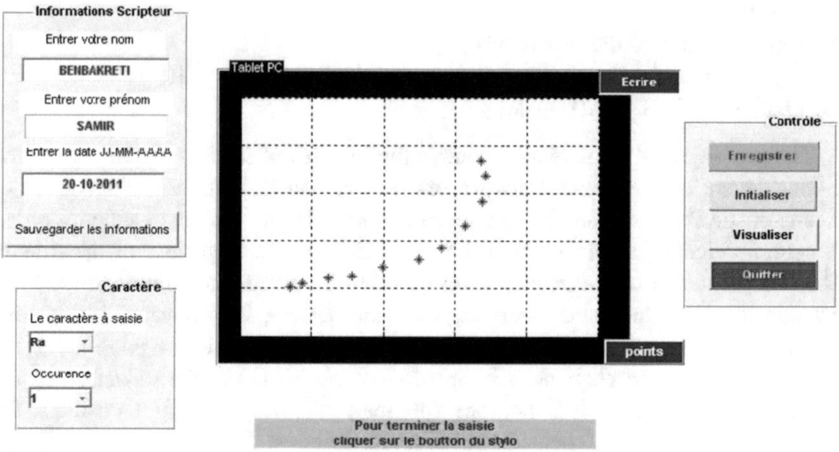

Fig. 6 On-line acquirement of the character “Ra”

From these data, we sample these points spatially to get equidistant points and so to become liberated from the variability of the speed of the tracing (Fig. 7). Thus, we will have to realize a normalization of the points of every character to what all characters possess the same number of points. Then to summarize, the features extraction for every point in the character permits to construct a matrix of 7 * 17 cells, the features are: the coordinates in x and y, the cosines of the direction (cos. and sin.), the cosines of the curvature (cos. and sin.) and the position pen up\pen down of the stiletto.

Once the process of extraction of the features is finished, the matrix of characteristic whose size results from the product between the number of feature and the number of points representing every character (stationary number gotten in an experimental manner) will be propagated in the TDNN to do the classification and to recognize the character (Fig. 8).

The training, i.e. the determination of the weights is the important points in the conception of a neuronal system; these calculations use the data set of the temporal set to search the best weights so that the neural network reproduces the behavior of the system. Ideally, these must permit to converge quickly to the global minimum of the cost function.

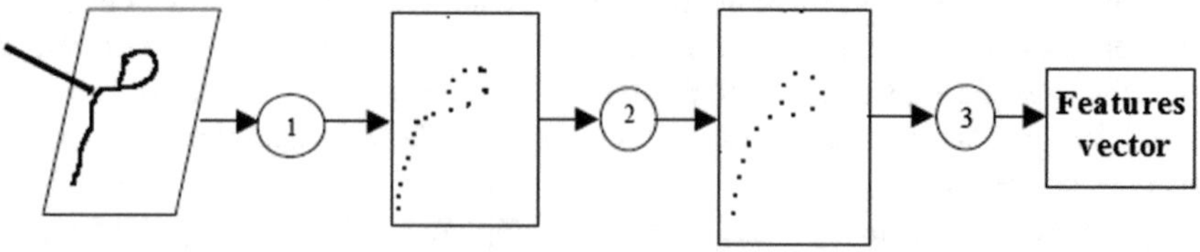

Fig. 7 Representation of the on-line extraction process (**1**: Acquirement of the signal. **2**: Sampling. **3**: Features extraction)

Features number =7

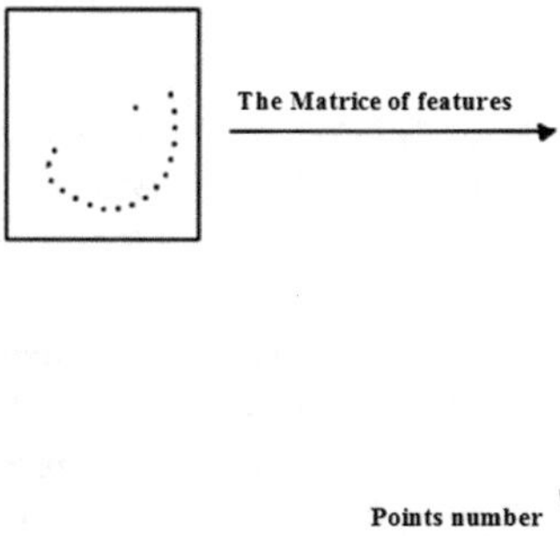

MATP =

-0.0610	0.5000	0	0	0	0	1.0000
-0.0604	0.4775	0.0309	-0.9995	0	0	0
-0.0596	0.4545	0.0379	-0.9993	0.9999	-0.0755	0
-0.0586	0.4310	0.0446	-0.9990	0.9999	-0.0890	0
-0.0575	0.4070	0.0512	-0.9987	0.9999	-0.1012	0
-0.0562	0.3827	0.0567	-0.9984	0.9999	-0.1135	0
-0.0547	0.3580	0.0625	-0.9980	0.9999	-0.1247	0
-0.0531	0.3329	0.0682	-0.9977	0.9999	-0.1355	0
-0.0513	0.3077	0.0733	-0.9973	0.9999	-0.1460	0
-0.0493	0.2822	0.0782	-0.9969	1.0000	-0.1560	0
-0.0472	0.2565	0.0831	-0.9965	1.0000	-0.1655	0
-0.0450	0.2306	0.0878	-0.9961	1.0000	-0.1751	0
-0.0427	0.2047	0.0927	-0.9957	1.0000	-0.1844	0
-0.0402	0.1787	0.0974	-0.9952	1.0000	-0.1937	0
-0.0376	0.1527	0.1019	-0.9948	1.0000	-0.2029	0
-0.0349	0.1268	0.1066	-0.9943	1.0000	-0.2116	0
-0.0320	0.1009	0.1109	-0.9938	1.0000	-0.2207	0
-0.0291	0.0751	0.1155	-0.9933	1.0000	-0.2294	0
-0.0261	0.0495	0.1201	-0.9928	1.0000	-0.2386	0
-0.0229	0.0241	0.1249	-0.9922	1.0000	-0.2477	0
-0.0197	-0.0010	0.1296	-0.9916	1.0000	-0.2568	0
-0.0164	-0.0259	0.1341	-0.9910	1.0000	-0.2660	0
-0.0130	-0.0505	0.1389	-0.9903	1.0000	-0.2745	0
-0.0095	-0.0749	0.1430	-0.9897	1.0000	-0.2827	0
-0.0060	-0.0992	0.1467	-0.9892	1.0000	-0.2903	0
-0.0023	-0.1234	0.1504	-0.9886	1.0000	-0.2971	0

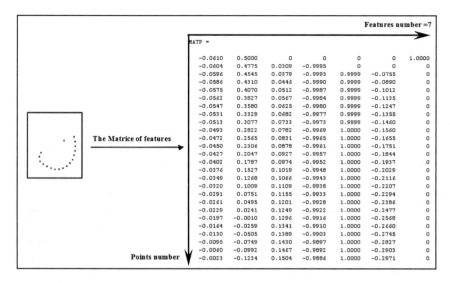

Fig. 8 The features extraction of the letter "Noun"

The TDNN is a sequence of two networks, the first is a dynamic network and the second is a static network (as shown previously). For the training of these two sub-networks, we can apply the method of training as the method of back-propagation of the gradient with the constraint of shared weights. However, the used version is the version said of the "stochastic gradient" that permits to converge more quickly than the true gradient, or the methods of second order as the algorithm of Lavenberg-Marquardt, the algorithm Newton, Quasi-Newton, Gauss-Newton. We tested first order methods which are the back-propagation of the gradient (for reasons of calculation and memory space) as well as the "resistant" back-propagation of the gradient: Rprops and two second order methods, the method of Lavenberg-Marquardt (LM) and that of the conjugated gradient (SCG).

Topology and parameters

The main property of the TDNN is the capacity to identify the local features regardless of their position in time. The TDNN is a dynamic feed-forward network, where information propagates itself from the input to the output without backward return and the dynamics is located in the layer of input, as delay. The TDNN stands out of a network of classic neurons, like the MLP by the fact that it takes in account a certain sense of time. That is to say that to the place to take at the same time in account all neurons of the input layer, it takes a window of the specter then does a temporal sweep.

What allows the network to take in consideration the local features of the temporal set of precipitations. The use of shared weights permits to reduce the number of parameters of the neural network while leading a more important generalization capacity.

The TDNN's tested architecture of basis is a network of 3 layers, the first layer is the input layer of the network, the second is the hidden layer of the extraction part and the first layer of the classification part (the function of activation being a sigmoid), the last is the output layer (the function of activation being a linear function). We give next an architecture example.

6 Experiments and Results

In our work, we stopped the training when the error (EQM) becomes minimal (10e-3), or the number of iteration reaches 100. Then the network is valued from the data that are different of those used during the training. By definition this last segment of data is called test set.

The step of gradient and training are the important factors in the convergence of the neural network, the learning time increases with the network complexity, it is necessary to find an optimal step. More the step is small, more the number of iterations of the training basis will be important. But more the step is big, more the necessary number of iterations will be less but the network risks to diverge. In our experience, we fixed the step of training to 0, 01 and the maximum number of iterations to 100. The results are mentioned in the following experiences:

Experiment 1:

In a first time and in order to test our model, we are going to launch the training only for 3 characters (alef, ba, ta) with the mentioned previously requisite architecture.

The performance matrix (Per) that regroups the characters badly classified and those well classified of every character, give us a preview on the general rate of the characters badly classified (C) (Table 1).

The general rate of characters badly classified, C = 0.0118 is equivalent to 1.18 and 98.82% of rate of training of the three characters.

The confusion matrix (CM) regrouping the number of well classified samples in the diagonal is the following: (Fig. 9).

Table 1 Performance matrices of the first three characters

Performance matrix (Per)	Negative false	Positive false	Positive true
Alef	0.0216	0.0078	0.9784
Ba	0.0078	0.0059	0.9922
Ta	0.0059	0.0216	0.9941

CM =	[499]	[0]	[11]
	[4]	[506]	[0]
	[0]	[3]	[507]

Fig. 9 Confusion matrices of (Alef, ba, ta)

Table 2 Results of the variations of the training function

(3 characters)	LM	RP	GD	SCG
EQM	6.33.10e-4	2.510e-2	0.29	1.10e-3
C (%)	0.0006	1.18	42.48	0.003
Time	12 m 03 s	6 m 47 s	6 m 43 s	14 m 12 s

Experiment 2:
The same previous architecture is renewed, but while varying the functions of training.

In this section, we have shown the influence of the method of the back-propagation of error gradient under several versions on the model and with the consideration of the time convergence.

It is very difficult to know what algorithm of training of a network "feedforward" will be the fastest for a problem. It depends on a lot of factors, including:

The complexity of the problem
The number of vectors (or points) of data on the set training
The number of weights and slants in the network

The goal of the network, used for the recognition of shapes (discriminative analysis) or the approximation of functions.

It is necessary to perform a training to determine the weights allowing in the output of the neural network to be as near as possible to the fixed objective.

This training takes place thanks to the minimization of a function, named cost function, calculated from the examples of the basis of training and the output of the neural network; this function determines the objective to reach.

We tested two algorithms of the first order: the back- propagation of the Descendant Gradient (GD) and the "resistante" back-propagation of the gradient (PR) and two second order algorithms: the Levenberg-Marquart algorithm (LM) and the algorithm of the conjugated gradient (SCG) (Table 2).

The following table sums up the different results:

7 Discussion

The LM algorithm got the lowest EQM as well as a very low rate of characters badly classified (C) neighboring 0%, it is more efficient than the other algorithms, with the detriment of its execution time which is more elevated. However, we cannot carry a definitive judgment on this algorithm unless its capacity of generalization is confirmed.

Therefore, we are going to increase the number of classes (of characters) to 7, here are the results: (Fig. 10).

After 100 iterations, the graphs of the EQM show us clearly that the algorithm of Lavenberg-Marquart converges more quickly and reaches the minimum after only some iterations (about 15 iterations) but is not the case of the other algorithms that

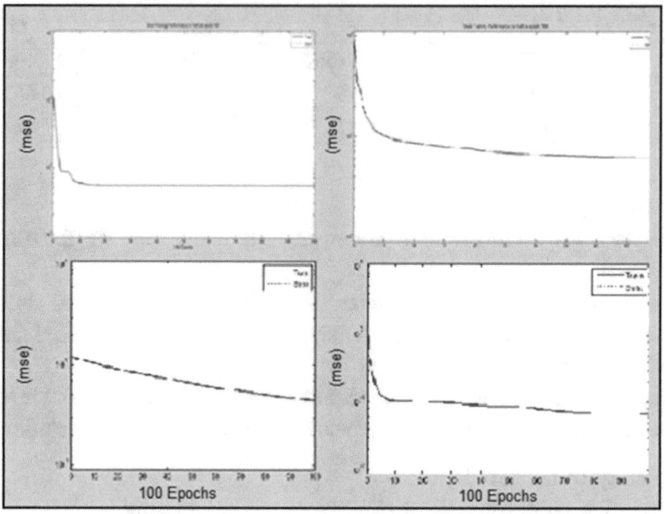

Fig. 10 Comparison between the EQM of different functions of training

converges slowly, especially the algorithm of descendant gradient GD whose error remains very big (the approximation is coarse of data) (Table 3).

When the size of the network increases, the performances of the LM algorithm weakened relatively, outside of the these enormous requirements of storage and time execution has more than 3 h, the algorithm doesn't possess an interesting generalization capacity when it is about solving a problem of shape recognition, even though the EQM is always as bass as the one of the other algorithms, the rate of samples badly classified (C = 35%) increased appreciably by previous experience.

The "PR" algorithm presents the best results concerning the rate (C) and the requirements of the memory for this algorithm are relatively small in comparison to the other algorithms, translate by the result in a small time execution (Time = 48 min).

Even though the second order algorithms are relatively more effective, we notices that a TDNN network is capable to do a good precision recognition by a simple training of the first order (PR).

The use of these training algorithms reappear of the virtual storage problems in the environment of Matlab programming, what didn't allow us to increase again the

Table 3 Variation results of in the training function after increasing the number of characters to 7

(7 characters)	LM(2)	RP	GD	SCG(2)
EQM	53.10e-3	61.10e-3	0.43	67.10e-3
C (%)	35	33	81	38
Time	3 h 22 m 11 s	48 m 13 s	53 m	1 h 25 m 04 s

numbers classes (of characters) to 28. On the other hand, we estimate that Matlab represents an excellent programming tool in the laboratory in the case of the matrix calculates (as in our case).

Experience 3: (influence of the size of the T window)
This experience is about the influence of the size of the temporal window on the results of the model. We recall that all previous experiences have been made with a temporal window of 4 neurons. This size will be increased until the obtaining of the best possible results: (Table 4)

Here is a summary table of the results obtained:

The results show that the EQM decreases while the width of the applied temporal window in the input of the TDNN is increasing. The lowest error corresponds to the window of 5 neurons. It means that we head toward better rates when we increase the width; however, when we pass the width of size 6, we find that the EQM increases. The same report is made with regard to the rate (C), except that it is better for T = 6.

Therefore the most suitable size resides between 5 and 6 neurons concerning c (cm and per) and of error rate (EQM) as the shows the Fig. 11:

To decide between 5 and 6, we increased the number of classes (7 characters), the results are mentioned in the Table 5:

This table shows that the best results have been obtained with a size of 5 neurons, and fix the optimal size of the temporal window applied to the input of the TDNN to five.

Therefore, the dimensioning of the size of the window influences the storage capacity of the TDNN. More one increases the size of the window, the number of free parameters increases and the more one increases the storage capacity. However, too much storage capacity can harm the network generalization power and plunges results in mediocrity.

Table 4 Results of the variation of the size of the window

(3 characters)	T = 4	T = 5	T = 6	T = 7
EQM	25.10e-3	8.10e-3	20.10e-3	22.10e-3
C (%)	1.18	1.05	0.85	1.11
Time	6 m 47 s	7 m 06 s	7 m 37 s	8 m 39 s

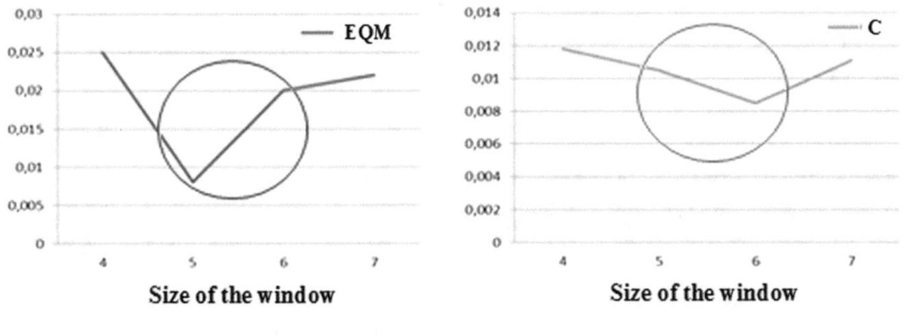

Fig. 11 Variation of (C) and (EQM) according to the size of the temporal window

Table 5 Variation of the size after increasing the number of classes

(7 characters)	T = 5	T = 6
EQM	61.10e-3	66.10e-3
C (%)	33	37.51
Time	52 m 49 s	1 h 01 m 43 s

Experiment4: (Training of the 28 characters)
The resistant back-propagation algorithm (Rprop) requires like all others algorithms seen previously the enormous needs of memory, the size of the input matrix that propagates in the architecture of the TDNN, which, as well, does not facilitate the task. These reasons pushed us to divide the general training in four parts. The division takes in consideration the ability to compare one character to another that resembles to it like (jim, ha, kha) (Fig. 12).

The results are summarized in the following table: (Table 6)

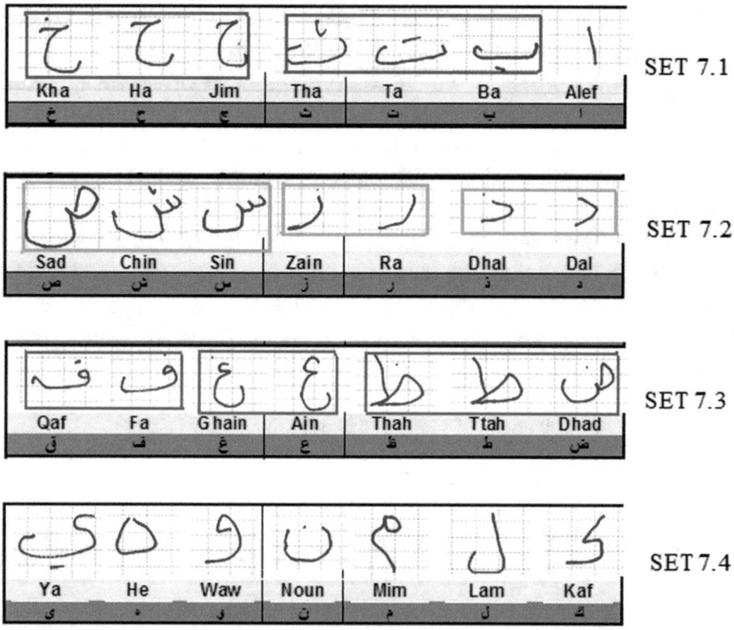

Fig. 12 Carving of the four sets of training and illustration of the resemblance between some characters

Table 6 The rates of C, EQM and the execution time for every group

Ideal architecture (7 characters)	The set 7.1	The set 7.2	The set 7.3	The set 7.4
EQM	61.10e-3	63.10e-3	72.10e-3	64.10e-3
C (%)	33	30	37.2	28
Time	49 m 07 s	49 m 21 s	48 m 21 s	52 m 48 s

7.1 Generalization Phase

The experiences made previously, allowed us to define the ideal architecture of our TDNN network. The constraints of memory, due to the used material in one side, and on the other side to the divergence of the algorithms of training after increasing the number of classes (of characters), didn't allow us to make the training of the 28 characters to the same time, from where the division imposed of the four sets regrouping each seven characters. Nevertheless, the division has been achieved in a way to be able to compare the character test to the other neighboring classes that resemble it, (example: Ra and Zin). These constraints reverberated on the phase of generalization and test.

For a better understanding, we elaborated a graphic interface, allowing the system to regroup the steps: of acquirement, of feature extraction matrix and especially the classification, as shown in the following figure (Fig. 13).

Let's recall that the goal of this work, is to make an acquirement, an extraction and recognition of Arabic characters in a dynamic way or rather to say, an on-line way and this is beyond the quality of the recognition rate of the questionable characters especially when it is about neighboring characters.

Figure 14 shows the generalization rates of all the characters:

The rates of recognition obtained in this figure vary, starting by 14,31% for the character Ta and can reach the maximum recognition rate of 99,61% for the Lam character.

To enhance the results, we compared our work to the one realized by Tlemsani and Al [8] that used the dynamic bayésiens networks (DBN) to conceive the system of characters recognizer. Let's note that the same data base NOUN-DATABASE has been used by the author.

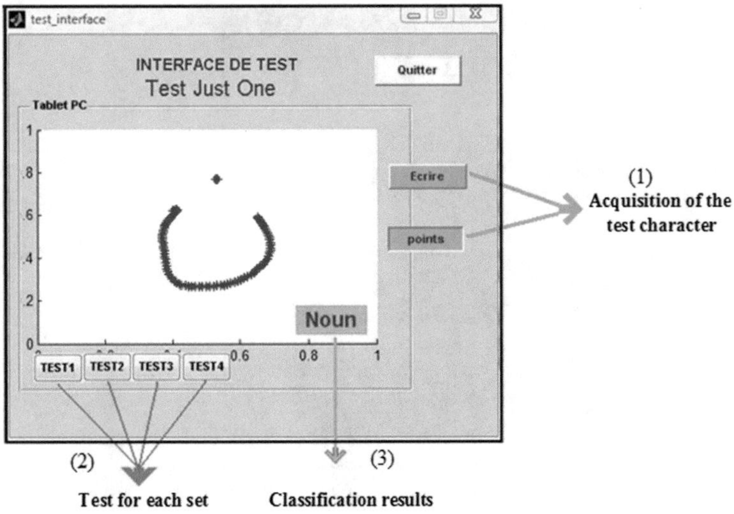

Fig. 13 Test of character on line "Noun"

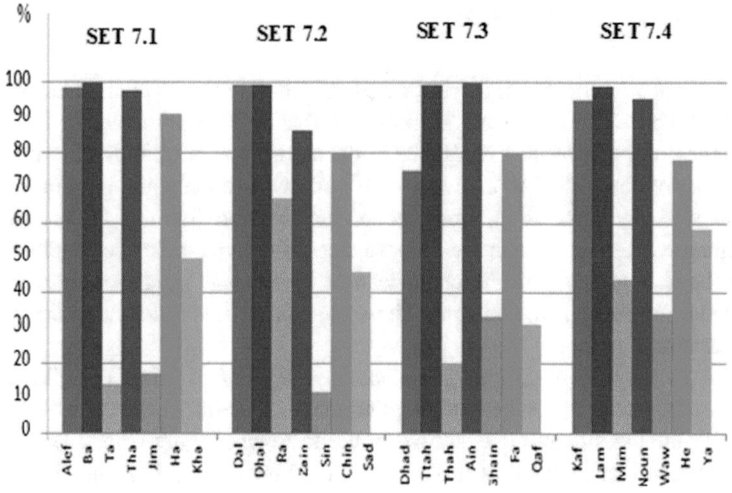

Fig. 14 The rates of generalization of the characters for the four groups

Presented otherwise, the following figure shows the difference of performance noted between our TDNN model and the DBN model: (Fig. 15).

We notice that the recognition rates for the groups 7.1 and 7.3 are relatively similar, while those of the other groups 7.2 and 7.4 are differed, with a special mention for the group 2 containing the letters (Dal, Dhal, Ra, Zain, Sin, Chin, Sad) that reached the 70%.

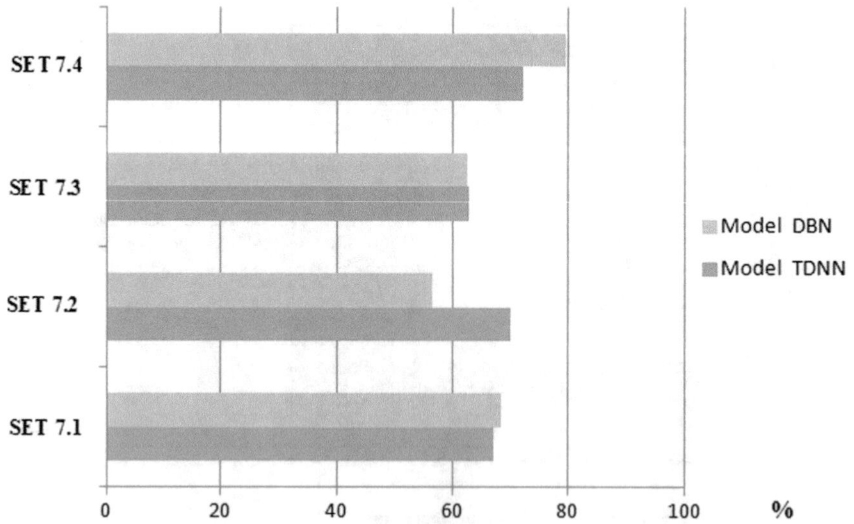

Fig. 15 Comparison of the recognition rates between TDNN models and DBN for the four groups

The advantage of our application founded on the TDNN model, is that it permits a faster on-line execution—goal of the works—than that of the model proposed in the DBN, nevertheless, this last possesses have the capacity to load all of the 28 characters to the same time.

8 Conclusion

The approach being proposed in this paper develops a solution based on TDNN neural networks for on-line recognition of dynamically acquired isolated handwritten characters, and to analyze all constraints that weigh on this network.

The neurons networks represent a new technique of data processing. Concretely, they result in algorithms putting in play the concepts associated to the nature of the human brain for the training notion. Let's note that the TDNN is a dynamic feed-forward network, this characteristic is located in the input layer as delay. The TDNN distinguished itself of a classic neural network as the MLP, by the fact that it takes in account a certain sense of time; in other words, instead of taking into account all the neurons in the input layers at the same time, it takes a spectrum window then does a temporal sweep.

In the goal to accomplish an on-line recognition of the isolated Arabic characters, we constructed our data base, which represents a very important support for all possible future works in the domain. After the accomplishment of the delicate step that is the features extraction, we did a series of experiment on our TDNN model. This network was trained with first order algorithms (descendant Gradient and resistant back-propagation) and others of second order (Lavenberg-Marquardt) in the goal to define the ideal architecture of our network.

The confusion matrices show that the TDNN being applied to the isolated Arabic character recognition deals reasonable results, we estimate that we can perfect the model by the means of hybrid methods, the present tendencies head toward neuro-markoviens systems.

As future work, the passage toward the second step while processing the words and the sentences especially represent an interesting challenge with an assured enrichment of our new data base of on-line Arabic writing. The goal of this paper is to contribute in a terrible development (minimal on the Arabic manuscript) that the domain of recognition of the non-constrained writing handwritten have ever knew. We intend to integrate this part also in more widened themes and more complex like the recognition of words while using the contextual and lexical constraints.

References

1. Adeyanju, A., Ojo, O.S., Omidiora, E.O.: Recognition of typewritten characters using hidden Markov models. Br. J. Math. Comput. Sci. **12**(4), 1–9 (2016)
2. Guyon, P. Albrecht, I., Le Cun, et al.:Design of a neural network character recognizer for a touch terminal. Pattern Recog. **24**(2) (1991)
3. Sakshica, D., Gupta, K.: Handwritten digit recognition using various neural network approaches. Int. J. Adv. Res. Comput. Commun. Eng. **4**(2) (2015)
4. Bengio, Y., Le Cun, Y.: World-level training of a handwritten word recognizer based on convolutional neural network. In: International Conference on Pattern Recognition, pp. 88–92 (1994)
5. Marukatat, S., Artieres, T., Dorizzi, B., Gallinari, P.: Sentence recognition through hybrid neuromarkovian modelling. In: Actes de ICDAR '01, 6th International Conference on Document Analysis and Recognition, vol. I (Seattle, 2001), pp. 731–735
6. Shaharin, R., Prodhan, U.K., Rahman, M.: Performance Study of TDNN training algorithm for speech recognition. Int. J. Adv. Res. Comput. Sci. Technol. (IJARCST, 2014) **2**(4) (2014)
7. Acharyya, A., Rakshit, S., Sarkar, R., Basu, S., Nasipuri, M.: Handwritten word recognition using MLP based classifier: a holistic approach. IJCSI Int. J. Comput. Sci. Iss. **10**(2) (2013)
8. Tlemsani, R.: Reconnaissance En-Ligne du manuscript Arabe. Theses of University of Sciences and Technologies-MOHAMED BOUDIAF-Oran (2012)

A Call Center Agent Productivity Modeling Using Discriminative Approaches

Abdelrahman Ahmed, Yasser Hifny, Sergio Toral and Khaled Shaalan

Abstract In this article, we present a novel framework for measuring productivity of customer service representative (CSR) in real estate call centers. The framework proposes a binary classification task for measuring CSR productivity. Generative and discriminative classifiers are compared in this study. The generative classifier is Naive Bayes (NB) versus the discriminative classifiers which are: logistic regression (LR) and linear support vector machine (LSVM). To train the classifiers, a speech corpus (7 h) is collected and annotated from three different call centers located in Egypt. The accuracy results on this corpus show that LSVM can lead to the best results (82%) and machine learning methods may successfully replace subjective evaluation methods commonly used in that domain.

Keywords Productivity measurement · Naïve Bayes · Logistic regression
Support vector machine

A. Ahmed (✉) · S. Toral
Electronic Engineering Department, University of Seville, Seville, Spain
e-mail: abdahm@alum.us.es

S. Toral
e-mail: storal@us.es

Y. Hifny
Department of Information Technology, University of Helwan, Cairo, Egypt
e-mail: yhifny@fci.helwan.edu.eg

K. Shaalan
School of Informatics, Edinburgh, UK
e-mail: Khaled.shaalan@buid.ac.ae

K. Shaalan
The British University in Dubai, Dubai, UAE

K. Shaalan et al. (eds.), *Intelligent Natural Language Processing: Trends and Applications*, Studies in Computational Intelligence 740,
https://doi.org/10.1007/978-3-319-67056-0_24

1 Introduction

Call centers are the front door of any organization where crucial interactions with the customers are handled [Reynolds (2010)]. The effective and efficient operations are the key ingredient to the overall success of in-source and outsource call center profitability and reputation. It is very difficult to measure productivity objectively because the agent output, as a firm worker, is the spoken words delivered to the customer over the phone. The evaluation mostly is handled in a subjective way. Subjective evaluation in call center is the essence of qualitative method which is performed through the monitoring and evaluating the interactions between the agent and the customer according to the evaluator perception [Sharp (2003)]. It is performed by listening to the agent recorded call, tapping a live call or making a test call by one of the quality team or anonymous caller [Rubingh (2013)]. The quality team listens to the agents recorded calls and uses predefined evaluation forms (evaluation check list) [Cleveland (2012)]. The evaluation process has many drawbacks. One of the reasons is that the quality teams evaluate the agents according to their perception and precedent experiences [Cleveland (2012)]. The subjective evaluation opens the door for favoritism due to what is called social ties [Breuer et al. (2013)]. A quantitative study of social ties and subjective performance evaluation [Breuer et al. (2013)] highlights a closer social attachment between supervisors and subordinates that leads to better performance rating when there are no differences in true performance. Another drawback of subjective evaluation is the resources limitation to evaluate all the agents consistently per time. For instance, some agents are evaluated in different day shifts (day shift/night shift) which leads to inconsistent or unfair evaluation from one agent to another. A typical challenge in performance evaluation studies is that the true performance is not observable to the researcher and hence it is hard to assess the gap and detect the evaluation distortions [Breuer et al. (2013)]. This means that the subjective evaluation may underestimate the agent when his performance could be higher. Conversely, the agent may be overestimated in evaluation because of other factors that may not be relevant to the true performance or the quality of service.

This paper proposes three classification methods for objectively measuring the agent's productivity through machine learning approach. The next section (Sect. 2) discusses the conceptual framework and gives an overview about the main building blocks. Section 3 discuss in some details the binary classification methods and parameters optimization methods. Section 4 explains the experiment carried out and Sect. 5 discusses the study results. Finally, Sect. 6 concludes the study and recommends research opportunities for future work.

2 Related Work and Proposed Framework

This section gives an overview about productivity measurement and highlights general concepts and methods of agent evaluation in call centers environment.

2.1 Productivity Measurement Definition

A productivity measure is commonly understood as a ratio of outputs produced divided by resources consumed (Eq. (1)) [Steemann Nielsen (1963)]. The observer has many different choices with respect to the scope and nature of both the outputs and resources considered [Card (2006)]. For example, the output can be measured in terms of delivered product or service, while the resource can be measured in terms of effort or monetary cost [Card (2006)]. An effective productivity measurement enables the establishment of a baseline against which performance improvements can be measured [Thomas and Zavrki (1999)]. This is the crucial part in productivity measurement because each call center has its own objectives and productivity criteria, which differs from one domain to another. Therefore, it requires a dynamic approach for grasping the eminent productivity characteristics for each call center, which helps the organizations to make better decisions about investments in processes, methods, tools, and outsourcing [Card (2006)]. The productivity measurement can be formulated using Eq. (1).

$$Productivity = \frac{\text{Agent Output}}{\text{Input Effort}} \tag{1}$$

The quality of service measurement in call centers is performed using quantitative and qualitative methods [Rubingh (2013)]. The quantitative method regards the first call resolution, average handling time of the call, the wrap up time and adherence time [Cleveland (2012)]. Every call center draws its own baseline that measures the overall performance objectively according to the ultimate call center objectives and strategies [Abbott (2004)]. The qualitative method is the process of monitoring and evaluating the call handling of the customer interactions subjectively [Cleveland (2012)]. The quantitative and qualitative methods aim at fulfilling the call center key performance indicators (KPIs), which are widely applied to measure agent performance as well as overall call center productivity [Reynolds (2010)]. [Taylor et al. (2002)] argues that qualitative and quantitative measures are not located simply in polar opposites but represent a continuum, along which operations can be located according to combination of the identified quality/quantity dimensions. As mentioned, the qualitative method in call center is subjected to evaluator perception and/or favoritism which impacts the agent evaluation. This study tries to replace the legacy qualitative methods by classifying the calls objectively using machine learning approaches.

The conceptual framework of the study is described in Fig. 1, where the evaluation process is automated from the beginning to the end. The block diagram includes a speaker diarization process, speech recognition, modeling and classification process. The next section discusses each building block in the framework.

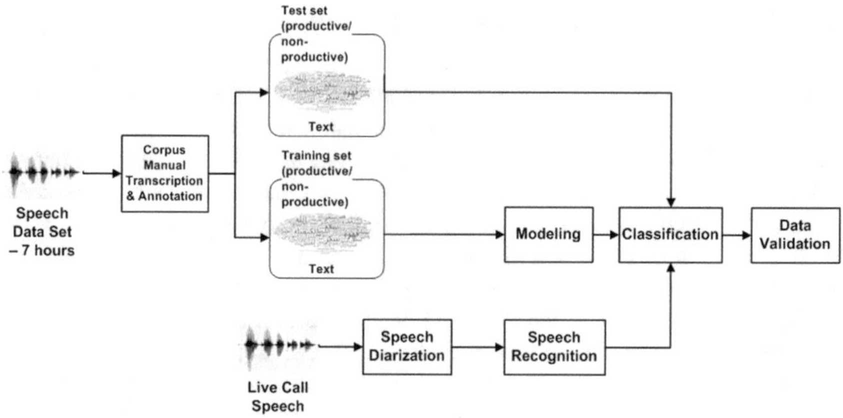

Fig. 1 The study frame work

2.2 *Speech Recognition and Speaker Diarization*

Speech recognition systems started in the 80s and achieved a significant improvement by the new era of machine learning using neural networks [Yu and Deng (2012)]. By transcribing the calls into text, the content analysis has become a powerful tool for features prediction and interpretation [Othman et al. (2004)]. The speech in Arabic language achieved a high accuracy in terms of word error rate (WER) [Ahmed et al. (2016)]. The word error rate is the main indicator of the speech recognition accuracy and performance [Young et al. (2015)]. The lower word error rate (WER), the higher performance of speech recognition [Woodland et al. (1994)]. The inbound or outbound call is divided into agent talk part, customer part, silences, music on hold and noise. As the agent part is the target of the analysis, it requires a diarization process. The diarization process is the process of using an acoustic model for sophisticated signal and speech processing to split the one channel mono recorded voices into different speakers [Tranter and Reynolds (2006)]. It removes silences, music as well as giving clear one speaker speech [Tranter and Reynolds (2006)].

The diarization process was intended to be performed using LIUM diarization toolkit [Meignier and Merlin (2010)]. It is a java based open source toolkit specialized in diarization using speech recognition models. It required Gaussian Mixture Model (GMM) for training voice and corresponding labels using two clustering states or more according to the number of speakers (number of states equal to the number of speakers). It uses GMM mono-phone to present the local probability for each speaker [Meignier and Merlin (2010)]. For speech recognition, we uses both of GMM and Hidden Markov Model (HMM) methods but in different configurations. In Arabic language, we use 3 HMM states for each phone (Arabic proposed phones are 40 phones), each state is presented by 16 Gaussian models. The Arabic speech recognition was presented in paper [Ahmed et al. (2016)]. We leave the diarization process for future work.

Table 1 Sample of Arabic letters, corresponding characters transliteration and its English equivalent

Arabic letter	أ	د	ش	ل	ك
Transliteration	ga	d	sh	l	k
Equivalent English	A	D	SH	L	K

2.3 *Converting the Arabic Text into Latin (Transliteration Process)*

This step is essential to convert the Arabic transcription into Latin for machine processing. The character set are 36 character as shown in Table 1.

The transliteration process maps each letter from Arabic to the corresponding Latin character. The next example shows a transliteration of an Arabic statement:

عليكم السلام ورحمة الله وبركاته

Îykm aslam w r@mt all* w brkat*

Example 1. Sample of Arabic statement transliteration in Buckwalter.

The transliteration shown above transforms the statement from right to left (Arabic writing direction) to left to right (Latin). Buckwalter[1] is a powerful open source tool to transliterate the Arabic to Latin, and it is used in various Arabic language processing applications.

2.4 *Sentiment Analysis and Binary Classification*

After the speech has been transcribed into text, the next step is to process the text using sentiment analysis. Sentiment analysis refers to the natural language processing by detecting and classifying the sentiment expressed by an opinion holder [Murphy (2012)]. Sentiment analysis, also called opinion mining, is the way to classify the text based on opinions and emotions regarding a particular topic [Richert et al. (2013); Chen and Goodman (1996)]. This technique classifies the text in polarity way (on/off/yes/no/good/bad), and it is used for assessing people opinion in books, movies etc. It deals with billions of words over the web and classifies the positive and negative opinions according to the most informative features (word) extracted from the text.

The sentiment analysis uses different binary classification methods in order to classify and predict the most informative features. The binary classification means that the classifier results will be only two results i.e. productive/non-productive

[1]http://www.qamus.org/transliteration.htm.

(in our study). This article has selected three classification methods to compare the classification performance applied on text (Sect. 3). Taking in consideration that agent productivity is different than opinion mining (emotions classification) because productivity is assessment of the output of the agent as mentioned in Eq. (1) regardless emotional words. However, this work is performed under assumption that the semantics of the call contents should be shaped with a positive meaning that tends to classify the call to a productive call [Ezpeleta et al. (2016)].

2.5 *Data Validity*

The classification accuracy is one of the most important factors in this study. The human accuracy shows that the level of agreement regarding sentiment is about 80%.[2] However, this percent cannot be considered as a baseline of the study because of two reasons. The first reason is that the accuracy is dependent on the domain of the collected text which varies from one domain to another. For example, the productivity features are perceived in different way than other domains like spam emails or movies review. The second reason is that the machine learning approach and human perception are incommensurable. Hence, The study presents unprecedented baseline of performance in real-estate call centers located in Egypt. For the data validation of the study, the classifier should be able to classify the test set accurately as intended. The accuracy calculation is given by Eq. (5).

$$Accuracy = \frac{F^{c}_{cor}}{F_{tot}}, \tag{2}$$

where F^{c}_{cor} is the correctly classified features per class and F_{tot} is the total features extracted.

3 Binary Classification Methods

This section describes the classification using Naïve Bayes denoted by (NB), logistic regression denoted by (LR) and linear support vector machine (LSVC).

3.1 *Naïve Bayes Classifier (NB)*

The Naïve Bayes classifier is built on Bayes theorem by considering that features are independent from each other. Naïve Bayes satisfies the following equation:

[2]http://www.webmetricsguru.com/archives/2010/04/sentiment-analysis-best-done-by-humans.

$$p(c|x) = \frac{p(x|c)p(c)}{p(x)}, \tag{3}$$

where c is the class type (productive/non-productive), and $x_1, x_2, \ldots, x_n$ are the text features. The $p(x|c)$ is the likelihood of the features given the class in the training process. We ignore $p(x)$ in the equation denominator as it is a constant value that never change. Accordingly, we are looking for the maximum value of both of the likelihood value $p(x|c)$ and prior probability $p(c)$ to predict the class probability given input features—Eq. (4):

$$p(c|x) = argmax[p(x|c)p(c)] \tag{4}$$

We calculate the joint probability by multiplying the probability of words given class $p(x|c)$ with class probability $p(c)$ to get the highest probability as follows [Murphy (2006)]:

$$p(C_k|x_1, x_2, \ldots, x_n) = p(C_k) \prod_{i=1}^{n} p(x_i|C_k) \tag{5}$$

The data set is manually transcribed and classified into productive/non-productive. The learning process is to find the probability of maximum likelihood of $p(x|c)$ and $p(c)$. The $p(c)$ is simply calculated as following:

$$p(c) = \frac{N_c}{N_{tot}}, \tag{6}$$

where N_c is the number of the words (features) annotated per class divided by total number of features in both classes N_{tot}. To calculate the maximum likelihood, we count the frequency of word per class and divide it to overall words counted per the same class [Jurafsky and Martin (2014)] as following:

$$p(x_i|c) = \frac{count(x_i|c)}{\sum count(x|c)} \tag{7}$$

As some words may not exist in one of the classes (productive/non-productive), the result of $count(x_i, c)$ will be zero. The total multiplied probabilities will be zero as well. A Laplace smoothing [Jurafsky and Martin (2014)] is used to avoid this problem by adding one:

$$p(x_i|c) = \frac{count(x_i|c) + 1}{\sum count(x|c) + 1} \tag{8}$$

To avoid underflow and to increase the speed of the code processing, we use $log(p)$ in Naïve Bayes calculations [Yu and Deng (2012)].

3.2 Logistic Regression (LR)

Logistic regression (also called logit regression) is a binary classification method when the classes are linearly separable classes [Jurafsky and Martin (2014)]. It is required to give value 1 when the feature is classified as productive and 0 when it is non-productive. Referring to linear regression and model training, the probability of the class c given feature x is presented in Eq. (9):

$$p(c|x) = w_0x_0 + w_1x_1 + \cdots + w_nx_n = \sum_{i=1}^{N} w_ix_i = w^Tx \tag{9}$$

where w is the weight of each feature per class.

The model mentioned in Eq. (9) gives values ranges from $-\infty$ to ∞ which does not present the required probability distribution from 0 to 1. The **odd-ratio** is a mathematical assumption to limit the results from 0 to ∞ very close to linear manner in Eq. (10) by dividing the probability of success (productive) by probability of failure (non-productivity).

$$Odd - Ratio = \frac{p(c|x)}{1 - p(c|x)} \tag{10}$$

However, the original Eq. (9) lies from $-\infty$ to ∞, hence, this can be achieved by taking the natural log:

$$log(\frac{p(c|x)}{1 - p(c|x)}) = z = w^Tx \tag{11}$$

By using a mathematical derivation, the logit equation $\Phi(z)$ will be as following:

$$p(c|x) = \Phi(z) = \frac{1}{1 + e^{-z}} \tag{12}$$

where z is the net input of the features and their weights:

$$z = w_0x_0 + w_1x_1 + \cdots + w_nx_n = \sum_{i=1}^{N} w_ix_i = w^Tx$$

The logit function $\Phi(z)$ is sigmoid function that limits the classification output from 0 to 1 (probability function). If we assume the classification threshold is 0.5 and logit function output is above the 0.5 (i.e. $\Phi(z) = 0.8$), then it means that the probability of a particular sample features of x and w being productive is 80% as shown in Fig. 2. The threshold value is also called hyperplane where $z = 0$ or $\sum_{i=0}^{N} w_ix_i = 0$ [Jurafsky and Martin (2014)].

Equation (12) predicts the class given the features. Then we have to train the model for best value of w which is required for optimum classification. The following figure illustrates the network diagram by defining an initial value of the weights w,

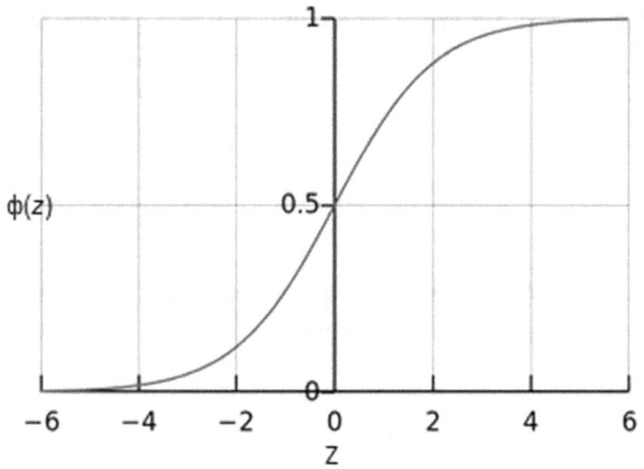

Fig. 2 Logit regression function—Wikipedia

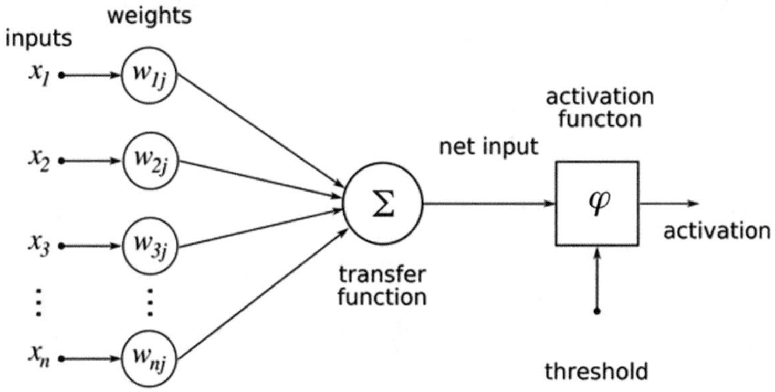

Fig. 3 Logit regression training model

summing the net input, classifying then adjust the weights according to error detected [Raschka (2015)] (Fig. 3).

In linear regression, the error is determined by sum-squared cost function (SSE). Equation (13) is the cost function (SSE).

$$J(w) = \sum_i (\Phi(z)^{(i)} - y^{(i)})^2 \tag{13}$$

The optimum value of w can be estimated by cost function minimization. The gradient based methods are used in language processing and machine learning to predict the optimum value that reduces the cost function to the minimum. In logistic regression, the cost function minimization is determined by conditional maximum likelihood estimation [Jurafsky and Martin (2014)]. This means that we choose the parameters w that make the probability of the observed class y in the training data to be the highest given feature x.

Assuming that the features are independent each other so that the maximum joint probability of the class c given the feature x is the probability product of all the class observations given the input features. The likelihood definition of logistic function in Eq. (14).

$$L(w) = \prod_{i=1}^{n} p(y^{(i)}|x^{(i)}, w) \tag{14}$$

The logit function $\Phi(z)$ is a probability function that can be governed by probability rules. This function is similar to Bernoulli function as stated in Eq. (15):

$$L(w) = \prod_{i=1}^{n} (\Phi(z^{(i)}))^{y^{(i)}} (1 - \Phi(z^{(i)}))^{1-y^{(i)}} \tag{15}$$

Taking log helps in calculating the small values by summation rather than multiplication. Furthermore, it eliminates the exponents for easier manipulation, the likelihood becomes:

$$l(w) = logL(w) = \sum_{i=1}^{n} y^{(i)} log(\Phi(z^{(i)})) + (1 - y^{(i)}) log(1 - \Phi(z^{(i)})) \tag{16}$$

Therefore, the objective function (cost function) is minimized by taking the negative log of Eq. (16).

$$J(w) = -logL(w) = -\left[\sum_{i=1}^{n} y^{(i)} log(\Phi(z^{(i)})) + (1 - y^{(i)}) log(1 - \Phi(z^{(i)}))\right] \tag{17}$$

It is required to estimate the w to maximize the likelihood (minimize the cost function) by differentiating Eq. (16):

$$\frac{\partial l(w)}{\partial w} = \sum_{i=1}^{n} \left(y^{(i)} \frac{1}{\Phi(z^{(i)})} - (1 - y^{(i)}) \frac{1}{1 - \Phi(z^{(i)})} \right) \frac{\partial \Phi(z^{(i)})}{\partial w} \tag{18}$$

The partial derivative of the sigmoid function is:

$$\begin{aligned} \frac{\partial \Phi(z)}{\partial z} = \frac{\partial}{\partial z}\left(\frac{1}{1 + e^{-z}}\right) &= \frac{e^{-z}}{(1 + e^{-z})^2} = \frac{1}{1 + e^{-z}}\left(1 - \frac{1}{1 + e^{-z}}\right) \\ &= \Phi(z)(1 - \Phi(z)) \end{aligned} \tag{19}$$

Now, by substituting Eq. (18) in Eq. (19):

$$\frac{\partial l(w)}{\partial w} = \sum_{i=1}^{n} \left(y^{(i)} \frac{1}{\Phi(z^{(i)})} - (1 - y^{(i)}) \frac{1}{1 - \Phi(z^{(i)})} \right) \frac{\partial \Phi(z^{(i)})}{\partial w}$$

$$\frac{\partial l(w)}{\partial w} = \sum_{i=1}^{n} \left(y^{(i)} \frac{1}{\Phi(z^{(i)})} - (1 - y^{(i)}) \frac{1}{1 - \Phi(z^{(i)})} \right) \Phi(z^{(i)})(1 - \Phi(z^{(i)})) \frac{\partial}{\partial w} z^{(i)}$$

$$= \sum_{i=1}^{n} \left(y^{(i)}(1 - \Phi(z^{(i)})) - (1 - y^{(i)})\Phi(z^{(i)}) \right) x^{(i)}$$

Hence, the predicted weight of the learning process:

$$\Delta w = \sum_{i=1}^{n} \left(y^{(i)} - \Phi(z^{(i)}) \right) x^{(i)} \tag{20}$$

The final value of w is the previous value added to the predicted one in Eq. (19).

$$w_{new} := w_{old} + \Delta w = w_{old} + \sum_{i=1}^{n} (y^{(i)} - \Phi(z^{(i)}))x^{(i)} = w_{old} - \eta \nabla J(w) \tag{21}$$

One way to update the parameters is to use gradient descent as shown in Eq. (21) but usually Newton's method is used to improve the speed of the training [Fan et al. (2008)]. η is the learning rate (a constant between 0 to 1) that controls the step size or the amount of the effect of Δw. It is adjusted to decide the speed (number of steps) for reaching the optimum value or may overfitting the model. Overfitting is one of the famous problems in machine learning when the model fits the training set but cannot be generalized well for unseen data [Murphy (2012)]. The overfitted model has a high variance and it is too complex. Similarly, the model that suffers from underfitting when it fails to capture the pattern in the training data and suffers from low performance [Raschka (2015)].

Regularization (called L2 regularization) is the method to control the collinearity (high correlation among features) and prevent overfitting [Raschka (2015)]. It penalizes the extreme parameter weights and the model complexity [Murphy (2012)]. *L*2 regularization is the summation of weights energy over the parameters described in Eq. (22).

$$L2 = \lambda \, \|w\|^2 = \lambda \sum_{i=1}^{n} w_i^2 \tag{22}$$

The cost function after adding *L*2 regularization is illustrated in Eq. (23).

$$J(w) = - \left[\sum_{i=1}^{n} y^{(i)} log(\Phi(z^{(i)})) + (1 - y^{(i)}) log(1 - \Phi(z^{(i)})) \right] + \lambda \, \|w\|^2 \tag{23}$$

where λ is the regularization parameter.

3.3 Linear Support Vector Machine (LSVM)

Support Vector Machine (SVM) is the most powerful and widely used in machine learning and binary classification [Raschka (2015)]. The margin is defined as the distance between the separating hyperplane (decision boundary) and the training samples that are closest to this hyperplane [Murphy (2012)]. This method is hypothesized on the maximization of the margin that it gives the best hyperplane position and linearly classifies the samples. The support vectors are the nearest samples (almost touching) to the margin that bounds the hyperplane. Figure 4 illustrates the hyperplane and the margin for binary classes $x1, x2$.

The hyperplane equation in terms of weight vector w^T apart from b and vector x.

$$w^T x + b = 0 \tag{24}$$

This means that any vector resides on the hyperplane, the equation results zero. In Fig. 4, it is assumed that the first support vector point of the positive class resides on the margin with distance +1 from the hyperplane and the negative point of negative class on −1. So we have two equations:

$$w^T x_{pos} + b = +1 \tag{25}$$

$$w^T x_{neg} + b = -1 \tag{26}$$

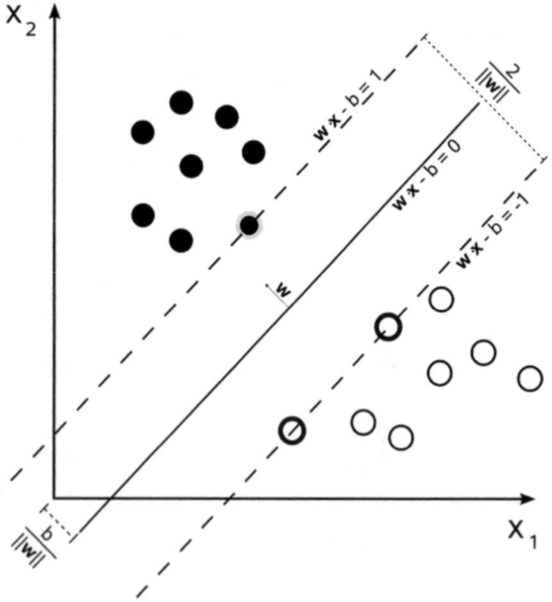

Fig. 4 Support vector machine—margin maximization—Wikipedia

When we subtract both equations and normalize them by $\|w\|$ where $\|w\| = \sqrt{\sum_{i=1}^{n} w_i^2}$ [Raschka (2015)]:

$$\frac{w^T(x_{pos} - x_{neg})}{\|w\|} = \frac{2}{\|w\|} \tag{27}$$

The left side of the Eq. (27) is the distance between the positive and negative boundary (the margin). The objective function of SVM becomes maximum by minimizing $\|w\|$ (or $\frac{1}{2}\|w\|^2$ for mathematical convenience) [Murphy (2012)]. This is under the constraint that the samples are classified according to the following:

$$w^T x_{pos} + b \geq +1, \quad \textit{if} \quad y = 1 \quad (\textit{productive-class}) \tag{28}$$

$$w^T x_{neg} + b < -1, \quad \textit{if} \quad y = -1 \quad (\textit{non-productive-class}) \tag{29}$$

where y is the raw output data located in the training set.

The general form of the last two equations:

$$y(w^T x + b) \geq +1 \tag{30}$$

Now, we have to use Lagrange equation to combine both of w and the constraint $y(w^T x + b) \geq +1$ in one equation (a slack variable) [Gunn et al. (1998)]. By adding Lagrange multiplier α, the Lagrangian equation will be as following:

$$\ell(w, b, \alpha) = \frac{1}{2} w^T w - \sum_{n=1}^{N} \alpha_n (y_n (w^T x_n + b) - 1) \tag{31}$$

By differentiation and substituting the equations that are equal to zero, the final equation of α minimization:

$$\nabla \ell(\alpha) = \sum_{n=1}^{N} \alpha_n - \frac{1}{2} \sum_{n=1}^{N} \sum_{m=1}^{N} y_n y_m \alpha_n \alpha_m x_n^T x_m \tag{32}$$

where $\alpha_n \geq 0$ for $n = 1, \ldots, N$ and $\sum_{n=1}^{N} \alpha_n y_n = 0$

It is obvious that w has been disappeared from the first derivative of Lagrange equation, however, it is concluded from the partial derivative with respect to w that α is proportionally related to w.

$$w = \sum_{n=1}^{N} \alpha_n y_n x_n \tag{33}$$

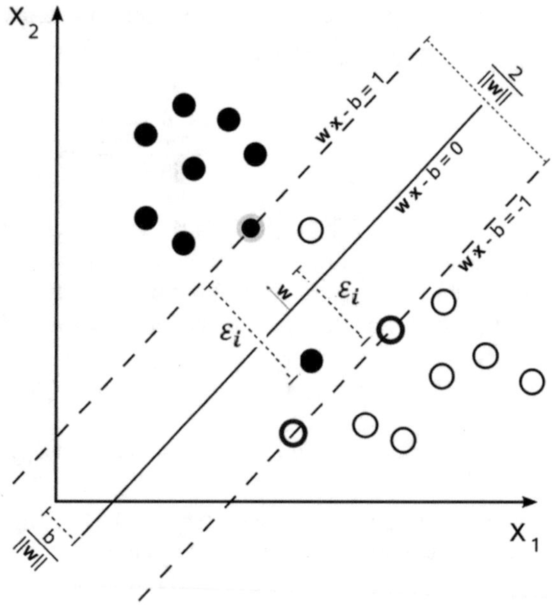

Fig. 5 Support vector machine—non-linearly separable data—Wikipedia

Equation (31) is the dual form that is used in case of hard margin. Hard margin is the maximum distance between the hyperplane and the margin when the data is linearly separable [Friedman et al. (2001)]. There are other cases that the data is almost linearly separable but there are some data points that makes few mistakes and violates the margin. That data points reside either within the margin (correctly classified) or with the margin in the opposite side (misclassified) as they are shown in Fig. 5.

Referring to Eq. (29), an error ξ should be added as shown in Eqs. (30) and (34).

$$y(w^T x + b) \geq +1 - \xi \tag{34}$$

ξ is the error distance of the vector data (slack variable) that violates the margin and takes the value range $\xi \geq 0$ [Fan et al. (2008)]. Hence, $\xi = 0$ means there is no error and vector data are correctly classified and reside out of the margin (the decision boundary). When $0 \leq \xi \leq 1$, it means that the data vector reside in the margin but still correctly classified (The vector resides between the margin and hyperplane). When $\xi \geq 1$ means that the data vector is misclassified (in the wrong side of the hyperplane). The soft margin is the case when it is accepted to have some errors or misclassified data points. The primal form of the loss function in this case is in Eq. (35).

$$\min_{w} \frac{1}{2} w^T w + C \sum_{n=1}^{N} \xi_n \tag{35}$$

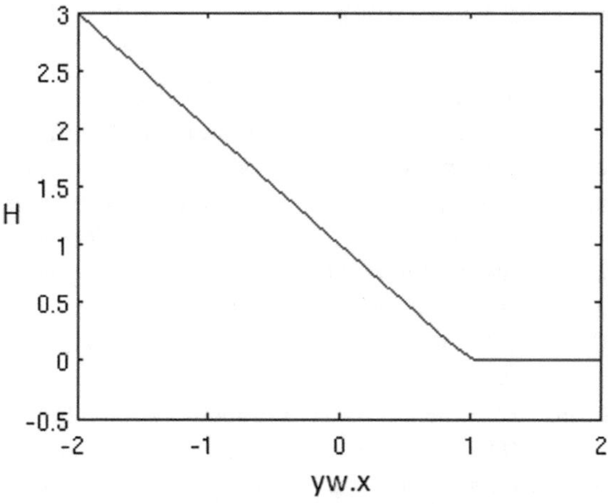

Fig. 6 Hinge loss function—CVX Research Inc

Constant C is a penalty parameter ($C = \frac{1}{\lambda}$) which controls the accepted amount of errors. The errors are added to the constraints of Lagrangian equation (dual form) and determined by finding the derivative of $\ell(w, b, \alpha, \xi)$ with respect to ξ. Referring to the primal form in Eq. (35), $\xi = 0$ whenever $y(w^T x + b) > 1$. On the other hand, $\xi = 1 - y(w^T x + b)$ whenever $y(w^T x + b) < 1$ [Murphy (2012)]. In short, $\xi = max(0, 1 - y(w^T x + b)$ which is called hinge function denoted by H—Eq. (35).

$$H = \begin{cases} 0 & if\ yw^T x \geq 1 \\ 1 - yw^T x & if\ yw^T x < 1 \end{cases} \tag{36}$$

The hinge function is plotted in Fig. 6.

Then the primal form after adding hinge function of Eq. (35).

$$\min_w \frac{1}{2} w^T w + C \sum_{n=1}^{N} max(0, 1 - y_i w^T x_i) \tag{37}$$

The form in Eq. (37) is called L1-SVM. The quadratic smoothed form is L2-regularization form of SVM (L2-SVM), which is used in our experiment, in Eq. (38) [Fan et al. (2008); Rennie and Srebro (2005)].

$$\min_w \frac{1}{2} w^T w + C \sum_{n=1}^{N} max(0, 1 - y_i w^T x_i)^2 \tag{38}$$

It is worth mentioning that both of dual and primal form in LSVM can be used independently to get the optimum classification margin [Wang (2005)] which is applied in python libraries as will be mentioned in next section.

4 The Experiment

The corpus consists of 7 h real estate call centers hosted in Egypt. A call center quality team has listened to the calls carefully (30 calls) and categorized subjectively the data set into productive/non-productive. The criterion used for file annotation is the ability of the agent or CSR to respond to the customer inquiries with right answers and fulfill the answer items. The evaluator gives score for each item fulfilled out of the total items required. For example, the customer asks about an apartment for sale, assuming the answer of the agent is set of 5 main items (Ask the customer name, the contact number, the budget available, the city, number of bed rooms). When the agent misses one of the answer items, the score is deducted by one point. Each call center draws the baseline of productivity according to the call center ultimate objectives. Referring to previous example, when the agent misses one item out of 5, this may be considered non-productive in call center (X) and still productive in call center (Y). Here it comes the power of machine learning for drawing the baseline for every different call center environment. The 30 calls are split into smaller audio chunks each audio file duration is around 20–50 s. This process is performed to simulate the output files that are produced by both of the diarization and speech recognition decoding process [Meignier and Merlin (2010)]. Therefore, the total files are 500 files divided into training set of 400 files and test set of 10% of the files (100 files). The quality team subjectively has annotated the files in to productive (400 files) and non-productive (100 files). This unbalanced annotation biases the results of probability of each class $p(c)$ in Eq. (3) which deviates from one class to the other. It should be for example 400 productive versus 400 non-productive. In this case, we may have to use a scaling factor to adjust the probability per class. The text has been converted from Arabic into Latin using Buckwalter script for machine processing then converted it back into Arabic. The code is developed in python version 2.7 using natural language toolkit (NLTK) Naïve classifier, Scikit learning library [Raschka (2015)]. The code uses bag-of-words which is an unordered set of words frequency regardless to its position in the sentence [Jurafsky and Martin (2014)]. Both Logistic regression and Linear support vector machine are classified using liblinear/scikit library instead of libsvm. This is because liblinear library has more flexibility for penalties and loss function parameters adjustment and better scaling for large numbers of samples.[3]

The scope of the study is to perform the classification methods: Naïve Bayes (NB), Logistic regression (LR) and linear support vector machine (LSVM). The comparison is to show the best classification performance among the classification methods.

[3]http://scikit-learn.org/stable/modules/generated/sklearn.svm.LinearSVC.html.

5 Results

We generated the data model and applied different binary classification approaches on the whole data set to compare their performance. The classification results accuracy is shown the Table 2.

The data set and the test set has been classified in order to extract the most informative features (MIF). Most informative features (MIF) is presented by ratio of occurrence in which the feature is much appearance in one of the classes than other. The features have been converted from Buckwalter to Arabic, then into English as following.

The NB classifier provides the ratios for comparing different features with respect to training set that are more often repeated in one of the classes which is called likelihood ratios [Murphy (2012)]. As shown in Table 3, NB presented the 10 most

Table 2 Binary classification methods and accuracy

Classification method	Accuracy (%)
Naïve Bayes	67.3
Logistic regression	80.769
Linear support vector machine	82.692

Table 3 The most informative features

Word index	Feature in BuckWalter	Feature in Arabic	Translation into English	Classification tendency	Likelihood ratios
1	trw@	تروح	To go	Non Productive	19.3
2	ambar@	امبارح	Yesterday	Non Productive	16.3
3	M∧rf$	معرفش	I have no idea	Non Productive	13.4
4	Fwq	فوق	Upper	Productive	13.4
5	Alsaylz	السايلز	Sales	Productive	13.4
6	Nmrty	نمرتي	My mobile number	Non Productive	10.4
7	Aktwbr	أكتوبر	October (City in Egypt)	Productive	10.4
8	m%*r	مظهر	The view	Productive	10.4
10	Ktyr	كتير	Expensive	Non Productive	9.8
9	alsT@	السطح	The roof	Productive	8.3

informative features out of 100 features extracted and have high tendency to specific class. We tried subjectivity to explore the meaning behind the classification for better understanding the definition of productivity in real estate call centers. For feature number 3 - معرفش or in English saying **(I have no idea)** is non-productive according to lack of awareness of the product or the service. In feature number 6, the agent dictates his/her mobile number over the phone, and this is considered non-productive as it consumes much time. The prices for the feature number 10—**(expensive)** is classified non-productive feature because it drags the CSR in useless debate and consumes much time. Furthermore, it might be an unjustified answer by agent which may indicate less awareness to the market changes and prices. This feature may be categorized under the same feature number 3. For productive agents, they mentioned the key features of the apartments or villas such as **(the view)**, **(the roof)** and the city **(October)** can be considered as product awareness. Nevertheless, the feature in itself works perfectly in evaluating the agent for mentioning some selling points through the call. There are other features meaningless, for example, **(to go)**, **(yesterday)**, **(Upper)**, and **(Sales)**. These features may be related to the error percentage which is expected from the beginning. The accuracy is expected to be improved by training larger corpus and getting balanced training set for productive and non-productive.

Referring to Table 2, the experiment results can be summarized in a two main themes. The first theme is the generative versus discriminative models. In the previous experiment, the Naïve Bayes gives the lowest classification performance compared to other methods, which is expected, because it is generative method. It is well-known in machine learning that generative models gives less accuracy in classification compared to discriminative one [Jurafsky and Martin (2014)]. The second theme is about the linear support vector classification. It shows that the margin maximization approach is the most appropriate method in this experiment.

6 Conclusion

The article proposed the call center agents performance evaluation using different machine learning approaches. Three methods are developed in this work: Naïve Bayes classifier, Logistic regression and support vector machine for binary classification (productive/non-productive). The annotation was performed by a call center quality team that depends on the scoring the itemized answers. The resulted accuracy of classification shows that discriminative models (logistic regression/support vector machine) could give high accuracy (80, 82%) compared to generative model (Naïve Bayes accuracy 67%). There are still research gap in productivity measurement for better extracting the productivity features (the most informative features—MIF). Furthermore, extending the productivity measurement in a range of scales (rather than binary classification) and considering the conversation context may help better understanding objectively the evaluation gap.

References

Abbott, J.C.: The Executive Guide to Call Center Metrics. Robert Houston Smith Publishers (2004)

Ahmed, A., Hifny, Y., Shaalan, K., Toral, S.: Lexicon free Arabic speech recognition recipe. In: International Conference on Advanced Intelligent Systems and Informatics, pp. 147–159. Springer (2016)

Breuer, K., Nieken, P., Sliwka, D.: Socialities and subjective performance evaluations: an empirical investigation. Rev. Manag. Sci. **7**(2), 141–157 (2013)

Card, D.N.: The challenge of productivity measurement. In: Proceedings of the Pacific Northwest Software Quality Conference (2006)

Chen, S.F., Goodman, J.: An empirical study of smoothing techniques for language modeling. In: Proceedings of the 34th Annual Meeting on Association for Computational Linguistics, Association for Computational Linguistics, pp. 310–318 (1996)

Cleveland, B.: Call Center Management on Fast Forward: Succeeding in the New Era of Customer Relationships. ICMI Press (2012)

Ezpeleta, E., Zurutuza, U., Hidalgo, J.M.G.: Does sentiment analysis help in Bayesian spam filtering? In: International Conference on Hybrid Artificial Intelligence Systems, pp. 79–90. Springer (2016)

Fan, R.E., Chang, K.W., Hsieh, C.J., Wang, X.R., Lin, C.J.: Liblinear: a library for large linear classification. J. Mach. Learn. Res. **9**, 1871–1874 (2008)

Friedman, J., Hastie, T., Tibshirani, R.: The Elements of Statistical Learning, vol. 1. Springer Series in Statistics. Springer, Berlin (2001)

Gunn, S.R., et al.: Support vector machines for classification and regression. ISIS Tech. Rep. **14**, 85–86 (1998)

Jurafsky, D., Martin, J.H.: Speech and Language Processing. Pearson (2014)

Meignier, S., Merlin, T.: LIUM SpkDiarization: an open source toolkit for diarization. In: CMU SPUD Workshop, vol. 2010 (2010)

Murphy, K.P.: Naive Bayes Classifiers. University of British Columbia (2006)

Murphy, K.P.: Machine Learning: A Probabilistic Perspective. MIT Press (2012)

Othman, E., Shaalan, K., Rafea, A.: Towards resolving ambiguity in understanding Arabic sentence. In: International Conference on Arabic Language Resources and Tools, pp. 118–122. NEMLAR, Citeseer (2004)

Raschka, S.: Python Machine Learning. Packt Publishing Ltd (2015)

Rennie, J.D., Srebro, N.: Loss functions for preference levels: regression with discrete ordered labels. In: Proceedings of the IJCAI Multidisciplinary Workshop on Advances in Preference Handling, pp. 180–186. Kluwer, Norwell, MA (2005)

Reynolds, P.: Call center metrics: best practices in performance measurement and management to maximize quitline efficiency and quality. North American Quitline Consortium (2010)

Richert, W., Chaffer, J., Swedberg, K., Coelho, L.: Building Machine Learning Systems with Python, vol. 1. Packt Publishing, GB (2013)

Rubingh, R.: Call Center Rocket Science: 110 Tips to Creating a World Class Customer Service Organization. CreateSpace Independent Publishing Platform. https://books.google.ae/books?id=IknGmgEACAAJ (2013)

Sharp, D.: Call Center Operation: Design, Operation, and Maintenance. Digital Press (2003)

Steemann Nielsen, E.: Productivity, definition and measurement. The Sea **2**, 129–164 (1963)

Taylor, P., Mulvey, G., Hyman, J., Bain, P.: Work organization, control and the experience of work in call centres. Work Employ. Soc. **16**(1), 133–150 (2002)

Thomas, H.R., Zavrki, I.: Construction baseline productivity: theory and practice. J. Construct. Eng. Manag. **125**(5), 295–303 (1999)

Tranter, S.E., Reynolds, D.A.: An overview of automatic speaker diarization systems. IEEE Trans. Audio Speech Lang. Process. **14**(5), 1557–1565 (2006)

Wang, L.: Support Vector Machines: Theory and Applications, vol. 177. Springer Science & Business Media (2005)

Woodland, P.C., Odell, J.J., Valtchev, V., Young, S.J.: Large vocabulary continuous speech recognition using HTK. In: 1994 IEEE International Conference on Acoustics, Speech, and Signal Processing, 1994. ICASSP-94. IEEE, vol. 2, pp. II/125–II/128 (1994)

Young, S., Evermann, G., Gales, M., Hain, T., Kershaw, D., Liu, X., Moore, G., Odell, J., Ollason, D., Povey, D.: The HTK Book (for HTK version 3.5). Cambridge University Engineering Department, Cambridge, UK (2015)

Yu, D., Deng, L.: Automatic Speech Recognition. Springer (2012)

Part IX
Morphological, Syntactic, and Semantic Processing

Alserag: An Automatic Diacritization System for Arabic

Sameh Alansary

Abstract Diacritization of written text has a significant impact on Arabic NLP applications. We present an approach to Arabic automatic diacritization that integrates morphological analysis with shallow syntactic analysis. The developed system (Alserag) is a rule based system. The system depends on three modules in order to provide fully diacritized Arabic words namely, morphological analysis module, syntactic analysis module and morph-phonological processing module. To evaluate the performance of the system, we used the benchmark LDC Arabic Treebank datasets used by the state-of-the-art systems (Metwally et al. 2016; Zitouni 2006) and (Shahrour et al. 2015). The proposed system achieved a morphological WER of 5.6%, and a syntactic WER of 10.1%.

1 Introduction

Diacritizing Arabic written text is crucial for many NLP tasks, translation can be enumerated among a longer list of applications that vitally benefit from automatic diacritization [SMR05], [RAR09] and [AMM09]. Arabic diacritics are superscript and subscript diacritical marks (vocalization or voweling), defined as the full or partial representation of short vowels, shadda (gemination), nunation, and hamza [MBK06]. Diacritization helps the reader in disambiguating the text or simply in articulating it correctly. As Arabic is a language where the intended pronunciation of a written word cannot be completely determined by its standard orthographic representation; it rather depends on a set of special diacritics. The absence of these diacritics in Arabic text increases lexical and morphological ambiguity, because one written form can have several vocalizations, each vocalization may have different meaning(s)

S. Alansary (✉)
Bibliotheca Alexandrina, Alexandria, Egypt
e-mail: sameh.alansary@bibalex.org

S. Alansary
Faculty of Arts, Phonetics and Linguistics Department, Alexandria University, Alexandria, Egypt

K. Shaalan et al. (eds.), *Intelligent Natural Language Processing: Trends and Applications*, Studies in Computational Intelligence 740,
https://doi.org/10.1007/978-3-319-67056-0_25

[BZDOGH15] and [EFS12]. The word form "ذكر" '*kr' can have many possible pronunciations like '*akar' (male) and '*akara' (to mention). However, these diacritics are generally left out in most genres of written Arabic which results in widespread ambiguities in vocalizations and meaning.

Although native speakers are able to disambiguate the intended meaning and pronunciation from the surrounding context with minimal difficulty, it is not the case with automatic processing of Arabic which is often hampered by the lack of diacritics. Several applications can radically benefit from automatic diacritization, such as Text-to-speech (TTS), Part-Of-Speech (POS) tagging, Word Sense Disambiguation (WSD), and Machine Translation [EFS12]. Text diacritization in Arabic can be divided into two types the morphological diacritization and the syntactic diacritization [MRA16]. For complete diacritization, each word must be diacritized both morphologically and syntactically. Actually, there is a big difference between the two types of diacritizations. On the one hand, the morphological diacritization pertains to differentiating between the different words or word forms that share the same orthographic form. On the other hand, the syntactic diacritics depend on the syntactic role of the word in the sentence. However, determining the correct syntactic marks is not always an easy task since it depends on syntactic rules which may even be difficult to native speakers.

2 Related Work

Much work has been done on Arabic diacritization. The actually implemented systems can be divided into two categories [AGA15]: Systems implemented by individuals as part of their academic activities and systems implemented by commercial organizations for realizing market applications. One of the advantages of the first type is that they present some good ideas as well as some formalization. The weak point about these systems is that they are mostly partial demo systems [Alb09]. The following are examples of these systems: [RAR09], [AMM09], [RARR14], [RARR15], [MRA16], [AGA15], [VK04], [ANB05], [ZSS06], [HR05], [HR07], [DGH07] and [RRHD08]. For the second category, the most representative commercial Arabic morphological processors are Sakhr, Xerox, and RDI [Alb09].

There are also other available systems as Mishkal Arabic diacritizer,[1] Harakat Arabic diacritizer[2] and Farasa[3]; they are free Arabic diacritizers which are available online. Finally, on March Google has launched an innovative new Google Labs Arabic tool called Tashkeel, a tool that adds the missing diacritics to Arabic text. Unfortunately, the tool is not available now. Researchers have proposed several approaches for handling Arabic text diacritization:

---

[1] http://tahadz.com/mishkal.

[2] http://harakat.ae/.

[3] http://qatsdemo.cloudapp.net/farasa/.

- **Statistical approach**: it requires a large corpus of fully diacritized text for extracting the case ending. One of theses systems that used statistical approach while developing their system is [SAZ08]. They made the training for detecting the case ending diacritics for each token based on its Part Of Speech (POS) and BP-chunk position as well as the position of token in the statement. The case ending diacritics is then efficiently obtained using the SVM technique.
- **Sequence labelling approach** as [ZSS06] who used an approach based on maximum entropy to restore the missing diacritics. The most important advantage of this approach is that by using maximum entropy it is easier to combine the information from diverse sources without considering how these pieces of information interact, which enabled them to use a mix of lexical, segment-based, and POS features. Then they used a maximum entropy classifier to label the characters with the missing diacritics. Instead of labelling each character independently, the labelling is chosen such that the conditional probability of the sequence of the labelled characters is maximized.
- **Morphological analysis approach** as [HR05] whose used an external morphological analyzer to produce all possible analyses for every input word. Each analysis consists of a possible diacritization and the corresponding POS Tag information which are organized as ten features for each input word. They also made use of ten trained SVM classifiers to predict the expected POS Tag features of the word. To select one analysis, for each word, from the output of the morphological analyzer, they chose the analysis that best conforms to the POS Tag features as predicted by the SVM classifiers [MRA16].
- **Hybrid approach**: it combines more than one approach to achieve both high accuracy and high coverage. Among theses systems that used hybrid approach while developing their systems are [Alb09] and [RAR09], they introduced a system of two layers in which the first layer deals with words that have occurred in the training data using the machine translation approach and the second layer handles the out-of-vocabulary (OOV) words using a morphological analysis based approach. In the first layer, all possible diacritized words corresponding to each input word, that occurred during training, are retrieved and A* search is used to select the most probable path of diacritized words that corresponds to the input sequence. Then, in the second layer, OOV words are passed to a morphological analyzer to get their possible analyses and diacritics. [SAZ09]; this system has integrated three different proposed techniques, each one has its own advantages and disadvantages. They are lexicon retrieval, diacritized bigram, and SVM statistical-based diacritizer.

Most of the previous approaches cited above utilize different sequence modeling techniques that use varying degrees of knowledge from shallow letter and word forms to deeper morphological information. None of the previous systems make use of syntax with the exception of [SKH15] which have integrated syntactic analysis.

Alserag is essentially a rule based Arabic diacritizer. Alserag is based on different steps: retrieval of unambiguous lexicon entries, disambiguating between the different stored possible solutions of the words to realize their internal diacritization through the morphological analysis step (the system tokenizes a text and provides a solution for each token and restores the appropriate internal diacritics from the dictionary), the syntactic processing step that is responsible for the case ending detection is based on shallow parsing and finally the morpho-phonological step that is developed to fulfill the requirements of vowel harmony and assimilation. Section 3 demonstrates the system architecture. Section 4 explains the different applied modules to fully diacritize texts. Section 5 evaluates the output and discusses the results and benchmarking process. Finally, Sect. 6 is the conclusion.

3 System Architecture

In this system, a rule-based approach was adopted. In this section, the different processes that took place in order to convert a plain text into a fully diacritized text will be described. Figure 1 presents the system overall architecture, where the diacritization is achieved through 7 main phases: (i) Preprocessing which is responsible for auto-correcting the raw text and segmenting the Arabic text into sentences. (ii) Tokenization which is the process of splitting the natural language input into lexical items. (iii) Disambiguation which is a process of choosing the right internal diacritization for the word from the dictionary. (iv) Name entity recognition (stored in the dictionary and have been obtained from the UNLarium[4] [Ala14]). (v) Syntactic shallow parsing which is an analysis of a sentence by identifying its constituents (NPs, JPs—etc.). (vi) Case ending module which is responsible for predicting the arguments of the predicate and assigning the diacritical marks that are attached to the ends of words to indicate their grammatical function. (vii) Morph-phonological module which is a series of rules that focus on the sound changes that take place in morphemes (minimal meaningful units) when they are combined to form words.

There are two engines that are used in Alserag, the first is Interactive ANalyzer (IAN)[5] which is used in the analysis process, it includes a grammar for natural language analysis. The syntactic processing is done automatically through the natural language analysis grammar, the second is dEep-to-sUrface natural language GENErator engine (EUGENE)[6] which is used in the generation process, It receives the analyzed input and provides a diacritized output without any human intervention, for more details see [Ala13].

4 Development of the System Resources

Alserag depends on two resources; the Arabic diacritized dictionary and a set of linguistics rules. Each one will be described in details in the following subsections.

[4]http://www.unlweb.net/unlarium/index.php?unlarium=dictionary.

[5]It is a web application developed in Java and available at http://dev.undlfoundation.org/index.jsp.

[6]It is a web application developed in Java and available at http://dev.undlfoundation.org/index.

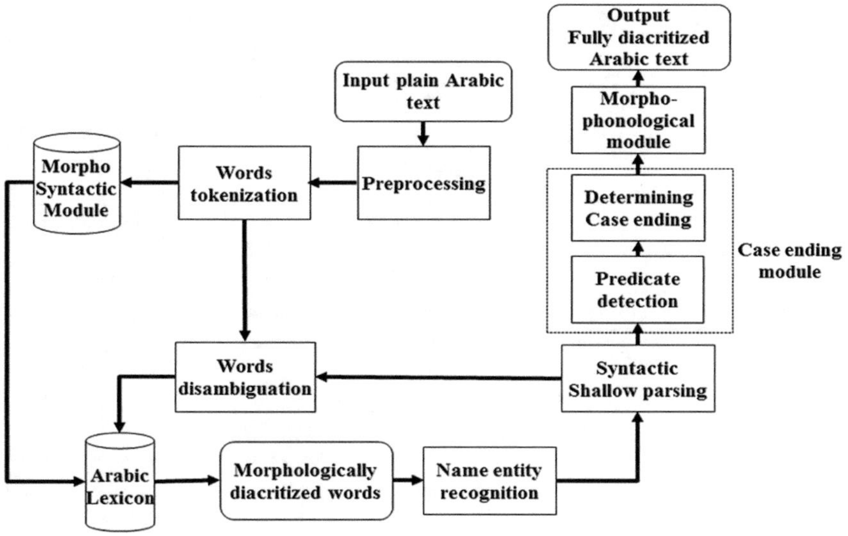

Fig. 1 Architecture of Alserag system

4.1 Dictionary

The Arabic diacritized dictionary is a dictionary where Arabic natural language words exist with their diacritics, along with the corresponding linguistic features which describe the Arabic word morphologically, syntactically and semantically. For example, the Arabic word "كتب" 'ktb' write its diacritics "كَتَبَ" 'kataba' and a list of linguistic features such as part of speech, tense, transitivity, person, gender, number, etc. are included in the dictionary. It is a word-form dictionary, for example the dictionary lists all the word forms of the verb "حضر" "HaDara" come' as "يحضر" "yaHoDuru" has comes', "يحضرون" "yaHoDuruwna" they are coming', "حضرت" "HaDarat" she came' and so forth as shown in Fig. 2.

The words in the Arabic diacritized dictionary are extracted from the Arabic dictionary in UNLarium. The process of diacritizing the entries mainly depends on two resources: BAMA and Alkhalil Arabic Morphological Analyzer. The diacritizing process begins with Buckwalter's analysis. Some words have only one solution, other words have more than one solution and some words couldn't be analyzed in Buckwalter. These are analyzed by Alkhalil which also suggests different solutions to some words. Then, these words are verified manually to select their correct diacritization. Not all of the Arabic diacritized dictionary entries are fully diacritized. Nouns, adjectives, subjunctive and indicative verbs are partially diacritized, since their case endings depend on the context. By default, a present tense verb is marked by a short /o/ (الضمة), in this case it is called indicative (المضارع المرفوع). However,

الأمر	المضارع المنصوب	المضارع المجزوم	المضارع المعلوم	الماضي المعلوم	
	أَحْضُرَ	أَحْضُرْ	أَحْضُرُ	حَضَرْتُ	أنا
	نَحْضُرَ	نَحْضُرْ	نَحْضُرُ	حَضَرْنَا	نحن
أُحْضُرْ	تَحْضُرَ	تَحْضُرْ	تَحْضُرُ	حَضَرْتَ	أنت
أُحْضُرِي	تَحْضُرِي	تَحْضُرِي	تَحْضُرِينَ	حَضَرْتِ	أنتِ
أُحْضُرَا	تَحْضُرَا	تَحْضُرَا	تَحْضُرَانِ	حَضَرْتُمَا	أنتما
أُحْضُرَا	تَحْضُرَا	تَحْضُرَا	تَحْضُرَانِ	حَضَرْتُمَا	أنتما مؤ
أُحْضُرُوا	تَحْضُرُوا	تَحْضُرُوا	تَحْضُرُونَ	حَضَرْتُم	أنتم
أُحْضُرْنَ	تَحْضُرْنَ	تَحْضُرْنَ	تَحْضُرْنَ	حَضَرْتُنَّ	أنتن
	يَحْضُرَ	يَحْضُرْ	يَحْضُرُ	حَضَرَ	هو
	تَحْضُرَ	تَحْضُرْ	تَحْضُرُ	حَضَرَتْ	هي
	يَحْضُرَا	يَحْضُرَا	يَحْضُرَانِ	حَضَرَا	هما
	تَحْضُرَا	تَحْضُرَا	تَحْضُرَانِ	حَضَرَتَا	هما مؤ
	يَحْضُرُوا	يَحْضُرُوا	يَحْضُرُونَ	حَضَرُوا	هم
	يَحْضُرْنَ	يَحْضُرْنَ	يَحْضُرْنَ	حَضَرْنَ	هن

Fig. 2 The different forms of the root "حضر" "HDr" 'come'

if a present verb is preceded by certain particles, the verb will be marked by a short /a/ (الفتحة), and if the verb ends by one of the three suffixes (ون-ين-ان), the final (ن) will be deleted, in this case it is called subjunctive (المضارع المنصوب). Nevertheless, imperative verb forms and past verb forms are fully diacritized, because their case endings are not affected by the context. Some enhancements have to be made in the Buckwalter solutions. For example, some solutions have a missing vocalization "َ" before "ا" as in "عَالِم" 'EAlim' (scientist), "مَكتَبات' 'makotabAt' (libraries). So, these missing vocalizations have been added manually.

4.2 The Linguistic Rules

Alserag depends on three modules in order to provide fully diacritized Arabic words namely, morphological analysis module, syntactic analysis module and morph-phonological processing module. Morphological analysis: is responsible for the morphological analysis of Arabic words and assigning the correct POS and the internal diacritization of words which is achieved through two processes; tokenization process and disambiguation process. However, before the tokenization process began, a preprocessing phase should take place over the string stream to fix the most common spelling mistakes, if needed. First, the tokenization algorithm is based on the entries of the dictionary. It starts from left to right trying to match the longest possible string with dictionary entries. The process starts with preventing joined lexical items. Then, it identifies the different suffixes and prefixes that could be attached to each lexical category. Disambiguation rules apply over the natural language list structure to constrain word selection and to correctly disambiguate the POS. They have the following format: (node 1) (node 2) (…) (node n) = P; Where (node 1), (node 2) and (node n) are nodes, and P is an integer expressing the possibility of occurrence. The engine is able to tokenize automatically some words correctly based on the dictionary and assign the correct POS to words. On the other hand, the larger the number of entries in the dictionary, the more the ambiguity during tokenization increases. For example, the sequence "بالفيضانات" "biAlfayaDAnAti" by the floods would be automatically segmented as "بال" bAl(worn) + "فيضانات" fayaDAnAt (floods), according to the longest match algorithm, given the fact that the dictionary includes [ART] Al (the), [ADJ بال] bAl (worn), [ب] bi (by) and [N فيضانات] 'fayaDAnAt' (floods). Rule in (1) states that adjectives can only be followed by a blank space (BLK), suffix (SFX) or to occur at the end of the sentence (STAIL), where (^) means not. So, [ب] + [ال] + [فيضانات] will be chosen as the appropriate combination.

1. (ADJ)(^SFX,^BLK,^STAIL)=0;

If words have spelling mistakes or undergo morpho-syntactic changes, rules will investigate the morphological form of those words. For example, the most common mistake in Arabic writing is /Hamza/ in the initial position as in "أنتزع" >anotaziE. Rules will investigate the morphological pattern of the wrongly spelled word by the regular expression techniques. For example, if a five-letters word begins with the sequence "/(أ|إ|آ|ا). ت../", the wrong written /Hamza/ ("إ"or"أ" or "آ") will be modified to the correct, according to the Arabic grammar, by the rule in (2), as in the pattern "افتعل" "{ifotaEala". Then the correct diacritized form will be retrieved from the dictionary "اِنْتَزَعَ" "{inotazaE".

2. ([/(أ|إ|آ|ا). ت../], ^Hamza_modified) (%y,PUT) := ("ا" >1,Hamza_modified) (%y);

Second, disambiguation is concerned with preventing the wrong automatic lexical choices and obtaining the right internally diacritized words. Some linguistic indicators can help in solving the lexical ambiguity which are morphological and adjacency indicators. Morphological indicators: affixation has an important role as the first level of part of speech disambiguation, as prefixes and suffixes are the smallest processing units rules can begin with. Prefixes can help as indicators in determining correct lexical choices. For example, in the word "لدفع" liDafoE, the noun "دَفْع" DafoE(push) is chosen instead of the verb (V) "دَفَع" 'DafaEa' (to push), since it is preceded by the preposition "ل" 'li' (to) by the rule in (3).

3. (P)(V)=0;

Adjacency indicators: After disambiguating the POS on the word level, the role of the adjacent word will take its effect as the second level of disambiguation. In this level, disambiguating the part of speech could be controlled by many qualifiers. Number and Gender qualifiers: as in "وهم يسمون" "wahum yusam~uwna" (and they call). According to the longest match algorithm, the engine will automatically choose the noun "وَهْم" wahom (illusion). But, because it is followed by a plural verb "يُسَمُّونَ" 'yusam~uwna' (they call) and subject and verb should agree in number and gender in Arabic, this tokenization will be rejected and will be retokenized as "وَ" 'wa' (and) + "هُمْ" 'hum' (they) by the rule in (4).

4. (SHEAD) (وهم, %x) (BLK) (V, ^NUM=%x)=0;

Functional word qualifier: Particles could be used as indicators for disambiguating the part of speech, as there are particles for verbs and others for nouns. For example, the particle "أي" '>ay~' (any) is a noun particle. Therefore, in the combination "أي شرط" (any condition), rule in (5) will reject the word "شرط" if it is chosen as a verb "شَرَّطَ" '$ar aTa' (slit), since it is preceded by the particle (PTC) "أي" '>ay~' (any). Then, it will backtrack it to the noun "شَرْط" '$aroT' (condition).

5. (PTC, "أَيّ") (BLK) (V)=0;

The co-occurrence of specific words with words with specific semantic features is used as an indicator. The word "تقلع" "tqlE" has different internal diacritizations that depend on the different meanings, such as "تُقْلِع" 'tuqoliE' (take off) with the semantic feature motion (MOT) and "تَقْلَعَ" 'taqolaE' (strip) which has the semantic feature contact (CTC). If the verb "تقلع" 'taqolaE' (strip) is followed by a noun such as "طائرة" 'TA}irap' (airplane) which has the semantic feature artifact (ARF) (Nouns denoting man-made objects). Rule in (6) will reject "تَقْلَعَ" (strip) "تُقْلِع" 'tuqoliE' (take off).

6. (V, SEM=CTC) (BLK)(ART)(N, SEM=ARF)=0;

Syntactic analysis: Shallow parsing is considered necessary for case ending assignment. Transformation rules have been developed to group words under the different phrasal categories. The rules follow the very general formalism := where the left side is a condition statement, and the right side is an action to be performed over. Phrasal grouping is necessary for identifying the sentence components and linking them by predicate. Then, the different functions of the sentence components can be identified and assigned the suitable case ending. This process will be illustrated in the following. Rules were developed to syntactically mark the phrasal units of the partially diacritized sentence in (1).

Sentence (1):

وَلِذَلِكَ لَمْ تَبْعَث اَلدِّرَاسَة اَلْمَدْرَسِيَّة لِتَارِيخ اَلْفَرَاعِنَة أَيّ شَوْق بَيْن اَلطَّلَبَة أَوْ اَلْخِرِّيجِين لِلْإِسْتِزَادَة

wali*'lika lamo taboEavo Ald~irAsap Almadorasiy~ap litAriyx AlfarAEinap >ay~ $awoq bayon AlT~alabap >awo Alxir~iyjiyna lil{isotizAdap
'Therefore, the school study for the Pharaohs history did not provoke any urge between the students or the graduates to increase.'

In sentence (1), different NPs structures are established. The first is established by rule in (7a); it combines the definite article "ال" "Al" (the) and the following noun to project a noun phrase (NP) "اَلدِّرَاسَة" 'Ald~irAsap' (the school study), "اَلْفَرَاعِنَة" 'AlfarAEinap' (the Pharaohs), "اَلطَّلَبَة" AlT~albap the students and "اَلْخِرِّيجِين" 'Alxir~iyjiyna' (the graduates). The second NP structure is formed by rule in (7b); it combines the indefinite noun "تَارِيخ" 'tAriyx' (history) and the NP "اَلْفَرَاعِنَة" "AlfarAEinap" (the Pharaohs), the composed NP "تَارِيخ اَلْفَرَاعِنَة" is automatically assigned with the features of its head "تَارِيخ" such as gender, number, animacy and semantic class that are necessary to describe the NP. The third NP structure consists of two coordinated elements and a conjunction; "اَلطَّلَبَة أَو اَلْخِرِّيجِين" 'AlT~alabap >aw Alxir~iyjiyna' (the students or the graduates) by rule in (7c). Moreover, adverbial phrase (AP) consists of the adverb "بين" 'bayona' (between) and the coordination NP; "اَلطَّلَبَة أَو اَلْخِرِّيجِين" 'AlT~alabap >aw Alxir~iyjiyna' is established by rule in (7d). However, the AP "بين الطلبة أو الخريجين" 'between the students or the graduates' is considered as an optional argument in the sentence in (1). Next, prepositional phrases (PPs) will be established; the two previously composed NPs "اَلْإِسْتِزَادَة" 'Al{isotizAdap' and "تَارِيخ اَلْفَرَاعِنَة" 'tAriyx AlfarAEinap' will be combined with the preceding preposition "لِ" 'li' (to) to form the prepositional phrases (PPs) "للاستزادة" (to increase) and "لتاريخ الفراعنة" (for the Pharaohs history) by rule in (7e).

7.
(a) (ART,%a)(%y,N) := ((%a)(%y),NP,ANI=%y,GEN=%y,NUM= %y, SEM = %y);
(b) (^ART,%a)(%y,N)(NP,%x) := (%a)((%y,np)(%x),NP,ANI=%y,GEN=%y, NUM = %y, SEM = %y);

(c) (NP,%a)(%y,COO)(NP,%x) := ((%a,np)(%y)(%x),NP,ANI=%a,GEN = %a, NUM = %a);
(d) (ADV,%a)(NP,%x)(%j) := ((%a)(%x),AP)(%j);
(e) (%x,P,^pp)(%n,NP) := ((%x,pp)(%n),PP);

Different syntactic functions of the predicate arguments should be identified in order to assign the case ending after the shallow parsing stage. In (1), the arguments of the verb should be identified which will be illustrated in the following. Verbs and their Arguments Diacritization: The sentence in (1) contains a verb, it is considered as the core of the sentence, since it is the verb that answers the three most important elements of any message—the what, who and when. In terms of the importance of the verb in the diacritization process, verb decides the case ending of the sentence elements. In sentence in (1), the verb "تَبْعَث" taboEavo provokeis a transitive verb that requires two arguments, one to function as a subject "اَلدِّرَاسَة" 'Ald~irAsap' (study) and another as an object "أَيّ" '>ay~ any'. After identifying the phrasal constructions, grammar rules have been developed to assign the function and the case ending of the composed verb arguments by rule in (8a). The rule states that, if a verb is followed by two noun phrases and there is gender agreement between the verb and the following noun phrase (NP, GEN=%v), this noun phrase will be considered as the subject of the verb (SBJ). The second will be considered as the object (OBJ). Once the functions of the arguments have been determined, the case ending will be assigned to each noun phrase; the nominative case (NOM) will be assigned to the subject and the accusative case (ACC) will be assigned to the object.

8 .
(a) (V,TSTD,%v)(NP,GEN=%v,^CAS,%n)(PP,%a)(NP,^CAS,%n2) : = (%v) (SBJ,CAS=NOM,%n)(%a) (OBJ,CAS=ACC,%n2);
(b) (لَمْ,%a)(%x, PRS, ^MOO) := (%a)(MOO=JUS,%x);

In the rule in (8a), the nominative and the accusative cases have been assigned to the heads of the two composed NPs; the words "اَلدِّرَاسَة" (CAS=NOM) and "أَيّ" (CAS=ACC). Rule in (8b) assigns the mode of the verb "تَبْعَث" as jussive (JUS) "تَبْعَثْ", because it is preceded by a jussive particle, as illustrated in sentence in (2).

Sentence (2):

وَلِذَلِكَ لَمْ تَبْعَثْ اَلدِّرَاسَةُ اَلْمَدْرَسِيَّة لِتَارِيخ اَلْفَرَاعِنَة أَيَّ شَوْق بَيْن اَلطَّلَبَة أَوْ اَلْخِرِّيجِين لِلْاِسْتِزَادَة

wali*'lika lamo taboEavo Ald~irAsapu Alomadorasiy~ap litAriyx AlofarAEinap >ay~a $awoq bayon AlT~alabap >awo Aloxir~iyjiyna lilo{isotizAdap

The modifiers are diacritized accordingly. The genitives such as ‘muDAf <iliyh’ “مضاف إليه” and the constituents after prepositions do not depend on the case ending of the preceding elements. In sentence in (1), genitive case (GNT) is assigned to genitives, as “اَلْفَرَاعِنَة”, “شَوْق” and “اَلطَّلَبَة”, “اِسْتِزَادَة” and “تَارِيخ”. Adjectives, coordinated elements and nouns in apposition are assigned the same case ending of the preceding element. The adjective “اَلْمَدْرَسِيَّة” Alomadorasiy~ap is assigned with the same case ending of the preceding noun “اَلدِّرَاسَةُ” ‘Ald~irAsap’, so it is assigned nominative case “اَلْمَدْرَسِيَّةُ”. As for the coordinated elements, as in “اَلطَّلَبَة أَو اَلْخِرِّيجِين”, the case of the NP “اَلطَّلَبَةِ” which is genitive is assigned to the NP. However, in Arabic, masculine plural noun ending with “ين” suffix, does not permit the genitive case marker (kasra), its genitive case is marked by fatha “ـَ” a. The final diacritization for the sentence in (1) is as in sentence in (3).

Sentence (3):

وَلِذَلِكَ لَمْ تَبْعَثْ اَلدِّرَاسَةُ اَلْمَدْرَسِيَّةُ لِتَارِيخِ اَلْفَرَاعِنَةِ أَيَّ شَوْقٍ بَيْنَ اَلطَّلَبَةِ أَوْ اَلْخِرِّيجِينَ لِلْاِسْتِزَادَةِ

wali*‘lika lamo taboEavo Ald~irAsapu Alomadorasiy~apu litAriyxi AlofarAEinapi >ay~a $awoqK bayona AlT~alabapi >awo Aloxir~iyjiyna lilo{isotizAdapi

Nominal sentences Diacritization: Nominative case is directly assigned to the topic of the sentence (noun or noun phrase in the beginning of sentences), because it is considered as “مبتدأ” ‘mubtadaa’. Since Arabic is a free word-order language, comment may precede topic in nominal sentences such as the sentence in (4).

Sentence (4):

فِي اَلْحَدِيقَة بَيْت

fiy AloHadiyqap bayot
A house in the garden

The case of the topic “بَيْت” ‘A house’ can be detected in the system in the case of the prepositional phrase “خبر شبه الجملة” “فِي اَلْحَدِيقَة” “fiy AlHadiyqap” ‘in the garden’ precedes it by rule (9).

9. (SHEAD,%x)(%c,PP)(NP,^CAS,%y) := (%x)(%c)(%y,mobtadaa,CAS=NOM);

Rule in (9) states that if a prepositional phrase “فِي اَلْحَدِيقَة” comes in the beginning of the sentence is followed by a noun phrase “بَيْت”, where (SHEAD) means beginning of the sentence. This noun phrase “بَيْت” is assigned with nominal case (NOM). However, these nominal phrases cases change if Anna and her sisters precede them. In the example, “إِنَّ فِي اَلْحَدِيقَة بَيْتًا” ‘< in~a fiy AloHadiyqap bayotAF’ (A house in the garden), the NP “بَيْت” ‘bayot’ (house) became accusative.

Building constituents boundaries is a very hard task during the parsing process, syntactic constituents are groups of words linked on the basis of their relationship with other words in the sentence; two or more words in a sentence can serve and function as one single unit. A set of rules has been developed in order to group the related words together to form different phrasal categories. In order to have a comprehensive idea about building the different constituents boundaries, let‘s consider the following example in sentence (5). Sentence (6) represents the output after applying the morphological analysis module to the sentence in (5); each word is assigned with the right internal diacritization. Rule (10a) starts by projecting all the adjectives in the sentence to the intermediate constituent J-bar (JB). So, by rule (10a) both the adjective “دولية” “dawoliy~ap” ‘international’ and “أهم” “>aham~” ‘most important’ are projected to be the intermediate constituent J-bar (JB). Then, the adjective “دولية” “dawoliy~ap” ‘international’ will be projected to the maximal projection JP as it is not followed by a complement by rule (10b).

Sentence (5):

بطولة كأس العالم لكرة القدم هي أهم مسابقة كرة قدم دولية تقيمها الفيفا

btwlp k>s AlEAlm lkrp Alqdm hy >hm msAbqp krp qdm dwlyp tqymhA AlfyfA
Football World Cup Championship is the most important international football contest held by the FIFA

Sentence (6):

بُطُولَة كَأْس اَلْعَالَم لِكُرَة اَلْقَدَم هِيَ أَهَمّ مُسَابَقَة كُرَة قَدَم دَوْلِيَّة تُقِيمهَا اَلْفِيفَا

butuwalap ka>os AloEaAlam likurap Aloqadam hiya >aham~ musaAbaqap kurap qadam dawoliy~ap tuqiymhA AlofiyfaA

10. **(a)** (J,^JB,%j) := (%j,JB);
(b) (JB,%j,^JP)(ÂRT|STAIL,%z) := (JP, –JB, %j)(%z);
(c) (NOU,^ NB, %x) := (%x,NB);
(d) (ART,%z)(NB,^np,%n)(^ART,^NB|STAIL,%y):=((%z,)(%n,np), POS = %n, SEM=%n,NP,DEF=def,GEN=%n,NUM=%n,LEMMA=%n,ARS = %n) (%y);
(e) (^ART|SHEAD,%z)(NB,^np,^def,^indef,%n)(NP,def,%y)(^COO|COO, SKIP |STAIL,%c):=(%z) ((%n, np)(%y,casatt=GEN),NB,DEF=def,GEN = %n, NUM=%n,LEMMA=%n,SEM=%n,ARS=%n,synf=edafa)(%c);
(f) (SHEAD|^ART,%s)(NB,^np,%n)(JP,GEN=%n,^def,%adjc): = (%s) ((%n, np) (np,%adjc), NB,SEM=%n, GEN=%n, NUM=%n,DEF=def, LEMMA = %n, ARS=%n);

(g) (^ART,^DEM|SHEAD,%z)(NB,def,^NP,%n)(^NB,^SPR,^ART,^DEM|STAIL,%y):=(%z)(NP, %n, NB) (%y);
(h) (JB,SUP|CMP,^jp,%j)(NP,%y)(^COO|COO,SKIP|STAIL,%c):= ((%j, jp) (%y,casatt=GEN),JB,GEN=%j,DEF=%j,ACAS=%j,FRA = %j, DEG =%j, LEMMA=%j)(%c);
(i) (%x, P,^pp)(%n,NP)(STAIL|PUT,^BLK,%s):=((%x,pp) (%n,casatt = GEN), PP)(%s);

Then, the heads of the noun phrases "بطولة" "buTuwlap" 'championship', "كأس" "ka>os" 'cup', "كرة" "kurap" 'ball', "قدم" "qadam" 'foot', "مسابقة" "musAbaqap" 'contest' and "فيفا" "fiyfA" 'FIFA' will be projected to the intermediate constituent N-bar (NB) by rule (10c). Then, rule (10d) starts to build the different NPs in the sentence by combining the definite article "ال" "Al" 'the' with the following noun "قدم" "qadam" 'foot' to project a noun phrase (NP), the same rule will be applied recursively to link both "ال" 'the' with "فيفا" "fiyfA" 'FIFA' and "ال" "Al" 'the' with "عالم" "EAlam" 'world'. The composed NPs are automatically assigned with the features of their heads such as gender, number, animacy and semantic class that are necessary to describe those NPs. Rules are able to build a bigger NBs. For example, rule (10e) combines the indefinite noun "كرة" "kurap" 'ball' with the definite NP "القدم" "Alqadam" 'foot' and assigns the NP "القدم" 'foot' with the genitive case. Rule (10e) will be applied recursively to link "كأس" 'cup' with "عالم" "EAlam" 'world' and "buTuwlap" championship with the definite NP "كأس العالم" "ka>os AlEAlam" 'the world cup' and finally "مسابقة" "musAbaqap" 'contest' with "كرة قدم" 'football'. Since that the adjectival phrase "دولية" "dawoliy/ ap" 'international' agrees with the preceding NB in gender and definiteness, they will be linked to form a bigger NB by rule (10f). Then, if there is any other complements or modifiers that are still not linked with their heads, rule (10g) will project all the NBs to NPs. Some other phrases require NPs as their complement, so after building all the NPs, it is their turn to be built. The superlative adjective "أهم" ">aham " 'most important' requires a complement to complete its meaning. So, rule (10h) will combine the JB "أهم" 'most important' with the NP "مسابقة كرة قدم دولية" "musAbaqap kurap qadam dawoliy/ ap" 'international football contest' to build a bigger JB and assign the NP with the genitive case, then it will be projected to the maximal projection JP by rule (10b). Next, prepositional phrases (PPs) will be built. The composed NP "كرة القدم" "kurap Alqadam" 'football' will be combined with the preceding preposition "لِ" "li" 'for' to form the prepositional phrase (PP) by rule in (10i) and

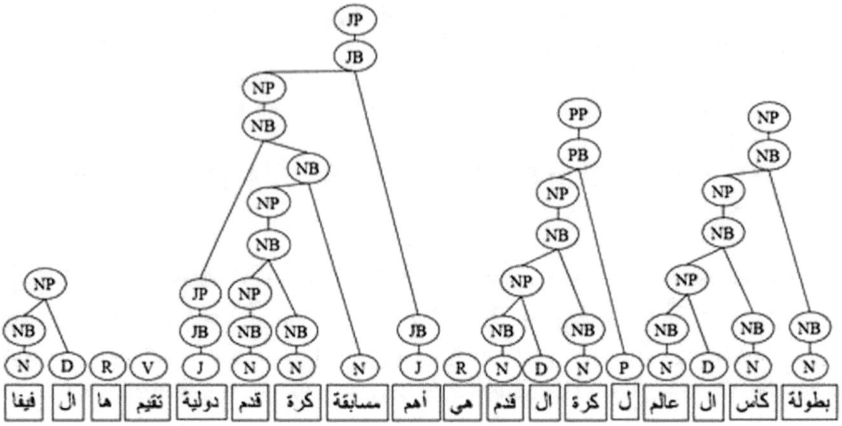

Fig. 3 shows the different composed phrases in the sentence (5)

also this rule will assign the genitive case to the NP "كرة القدم" "kurap Alqadam" 'football'. Figure 3 shows the composed phrases of the sentence (5).

Nominal sentences diacritization: Nominative case is automatically assigned to noun phrases in the beginning of sentences, because it is considered as the topic "مبتدأ" 'mobtadaa'. So, in sentence (5) the phrase "بطولة كأس العالم" "buTuwlap ka>os AlEAlam" 'world cup championship' will be assigned with the nominative case (NOM) by rule (11a). In Arabic, there are different types of comments "خبر" 'khabar'; single, the noun phrase, or the prepositional phrase comment. In sentence (5), the nominal phrase "هي أهم مسابقة كرة قدم دولية" "hiya >aham~ musAbaqap kurap qadam dawoliy ap" 'it is the most important international football contest' is the comment of the sentence. It consists of a noun phrase and an adjectival phrase. The noun phrase "هي" "hiya" 'it' is also considered as the topic so, it is assigned with (NOM) case, and the adjectival phrase "أهم مسابقة كرة قدم دولية" ">aham~ musAbaqap kurap qadam dawoliy ap" 'the most important international football contest' is considered as the comment so, it is assigned with (NOM) case, by rule (11b).

11 .

(a) (SHEAD,%s)(NP,^casatt,%n)(^COO|STAIL,%c):=(%s)(%n,topic,casatt = NOM) (%c);

(b) (SHEAD|^V|PSV, %s)(PPR,NP,%np)(NP|JP,^casatt,%w):=(%s)(%np)(%w,comment,casatt=NOM);

In sentence (5), the verb "تقيم" "tuqiym" 'hold' is a transitive verb that requires two arguments, one functions as a subject and the other as an object. After identifying the phrasal constructions in the sentence as in Fig. 3, grammar rules have been developed to assign the function and the case ending of the composed verb arguments by rule in (12). The rule states that, if a verb is preceded by a noun phrase and is followed by a pronoun and a noun phrase, the preceding noun phrase will be considered as the object of the verb if it agrees with pronoun "ها" and there is gender agreement between the verb and the following noun phrase. The second NP will be considered as the subject. Once the functions of the arguments have been determined, the case ending will be assigned to each noun phrase; the nominative case (NOM) will be assigned to the subject and the accusative case (ACC) will be assigned to the object if there is no other diacritic sign assigned earlier. The final diacritized sentences is shown in sentence (7).

12. (NP,^casatt,^subj,%n1)(V,TSTD,^TST2,%v)(SPR,GEN=%n1,NUM = %n1, %n) (NP,GEN=%v,^obj,%n2):=(obj,%n1)(%v)(%n,casatt=ACC)(subj,casatt= NOM, %n2);

Sentence (7):

بُطُولَةُ كَأْسِ اَلْعَالَمِ لِكُرَةِ اَلْقَدَمِ هِيَ أَهَمُّ مُسَابَقَةِ كُرَةِ قَدَمٍ دَوْلِيَّةٍ تُقِيمُهَا اَلْفِيفَا

butuwalapu ka>osi AloEaAlami likurapi Aloqadami hiya >aham~u musaAbaqapi kuraip qadamK dawoliy~apK tuqiymuhA AlofiyfaA

In fact, determining constituents boundaries isn't an easy task, many challenges have been faced. If there is a mistake in building the constituents, it will cause problems in assigning case endings and nunation. In what follows the focus will be on how constituents boundaries affect case endings assignment and nunation, as well as the challenges that were faced in determining the correct boundaries.

Challenges in nunation: Before describing the challenges, the criteria that determine the assigning of nunation should be described. Since Arabic does not have indefinite article, indefinite nouns and adjectives are marked for indefiniteness with nunation. So, in order to correctly assign the nunation the noun phrase should be correctly built first and then its definiteness should be determined. In sentence in (5), rules succeeded in determining the right boundaries and the definiteness of each noun phrase. So, only "قدم" "qadam" 'foot' is assigned with nunation, because it is the only indefinite noun phrase, while the other noun phrases are definite by edafa or by the definite article "Al" the. In sentences in (8) there are three examples that have the same syntactic pattern which is (N COO+N ART+N); however, they function differently which poses a challenge.

Sentence (8):

(a) بتفوق وسيادة الجنس الأبيض

btfwq wsyAdp Aljns Al>byD

by the white race superiority and predominance

(b) ما بين ساعة ونصف الساعة
mA byn sAEp wnSf AlsAEp
between an hour and half an hour
(c) على مختلف فئات وطبقات المجتمع
ElY mxtlf f}At wTbqAt AlmjtmE
on the different categories and classes of the society

Both sentences in (8a) and (8c) should be linked in the same way. In sentence (8a) the NB "تفوق" 'superiority' and the NB "سيادة" "siyAdap" 'predominance' should be linked with the coordinator "و" "wa" 'and', in (8c) the NB "فئات" "fi}At" 'categories' and the NB "طبقات" "TabaqAt" 'classes' should also be linked with the coordinator "و" "wa" 'and' to form bigger NBs. Then the generated NBs should be linked with the NPs "الجنس الأبيض" "Aljinos Al>aboyaD" 'the white race' in (8a) and "المجتمع" "AlmujotamaE" 'the society' in sentence (8c) to form yet bigger NBs. Finally, rules will project them to the maximal projection NPs and assign them with the feature definite. While in sentence (8b), first the NB "" "sAEap" 'hour' should be projected to the maximal projection NP and be assigned with the feature indefinite 'indef', then it should be linked with the other NP "نصف الساعة" "niSof AlsAEap" 'half an hour' by the coordinator "و" "wa" 'and' to form a bigger NP. The challenge is how rules could determine which constituents should be linked together and which NB should be projected directly to the maximal projection as in (8b).

The boundaries will be correctly determined for sentence (8b) and the nunation will be assigned correctly as in sentence (9b). The boundaries are also correctly determined for sentence (8a), since the two coordinated; NBs "تفوق" "tafaw~uq" 'superiority' and "سيادة" "siyAdap" 'predominance', have the same semantic feature which is state. So, the boundaries will be correctly determined as in sentence (9a). But, in sentence (8c) the two coordinated NBs; "فئات" "fi}At" 'categories' and "طبقات" "TabaqAt" 'classes', have different semantic features. Therefore, the boundaries will be wrongly determined and the nunation will be assigned wrongly as in (9c); it still a limitation in the current version of Alserag.

Sentence (9):

(a) بِتَفَوُّقِ وَسِيَادَةِ اَلْجِنْسِ اَلْأَبْيَضِ
bitafaw~uqi wasiyAdapi Alojinosi Alo>aboyaDi
(b) مَا بَيْنَ سَاعَةٍ وَنِصْفِ اَلسَّاعَةِ
maA bayona saAEapK waniSofi Als aAEapi
(c) عَلَى مُخْتَلَفِ فِئَاتٍ وَطَبَقَاتِ اَلْمُجْتَمَعِ
EalaY muxotalafi fi}Ati waTabaqaAti AlomujotamaEi

Challenges in verb argument diacritization: Determining the verb's arguments is based on some criteria such as the number of constituents that follow the verb, verb's transitivity, and the semantic features of the following noun phrases. So, any mistake in one of these factors will cause wrong case ending assignment. In sentences (10) there are two sentences that have the same syntactic pattern V ART+N. However, the NP in (10a) should be assigned with NOM case because it is the subject of the verb, while in sentence (10b) the NP should be assigned with ACC case because it is the object of the verb. The developed rules were able to differentiate between them by using the semantic feature of the NPs as a clue. In (10a) the NP "Alwalad" the boy is assigned with the semantic feature human (HUM) which implies that this NP is the doer of the verb. In sentence (10b) the NP "التفاحة" "Alt~uf~AHap" 'the apple' is assigned with the semantic feature food (FOO) which implies that NP is the object that have been eaten. So, it will be correctly diacritized as shown in sentences (11).

Sentence (10):

(a) أكل الولد
>kl Alwld
The boy ate

(b) أكل التفاحة
>kl AltfAHp
He ate the apple

Sentence (11):

(a) أَكَلَ اَلْوَلَدُ
>akala Alowaladu

(b) أَكَلَ اَلتُّفَّاحَةَ
>akala Alt~uf~AHapa

Although the system was able to overcome many challenges, it still has some limitations. The system has been improved in this phase and still there is more potential for further improvements. One of the most difficult problems that faces a parser is structural ambiguity, since it leads to problems in determining the boundaries between constituents. Despite these limitations, the system is proved to be promising when tested and demonstrated.

Morpho-phonological process: Many morpho-phonological alternations occur in Arabic due to the concatenative nature of Arabic morphology, the interaction between morphological and phonological processes is usual. There are two cases where morpho-phonological change is necessary; vowel harmony and assimilation necessity. Vowel harmony takes place in the diacritization process (i.e. phonological). For example, a morpho-phonological rule is necessary for "له" lahu for him that consists of two morphemes "لِ" 'li' + "هُ" 'hu', to change the vowel "ـِ" in "لِ" to "ـَ" to be more harmonious with the vowel "ـُ" on the suffix "ةُ". Moreover, the phonological Arabic system doesn't permit the moon letters "(أ-ب-غ-ح-ج-ك-و-خ-ف-ع-ق-ي-م-ه)", to be assimilated with the /l/ of the definite article "ال" 'Al' 'the', as they are not near in the place of articulation, but they can assimilate with the other Arabic alphabets which are called sun letters. When the definite article is followed by a sun letter, the /l/ of the Arabic definite article Al—assimilates to the initial consonant of the following noun, resulting in a doubled consonant (phonologically) which is orthographically expressed by putting a shaddah "ـّ" '~' on the consonant after /ل/. For example, the word "الصباح" "AlS~abAH" the morning before applying the morphonological rule, is diacritized as "اَلْصَبَاح" 'AlSabAH'. Another rule adds a shaddah before the vowel ("ـَ", "ـُ" or "ـِ"), if the diacritical mark is on a sun letter, it would be diacritized "اَلصَّبَاح" 'AlS~abAH'.

5 Evaluation and Benchmarking

The corpus has been selected from the International Corpus of Arabic (ICA). The selected corpus size is 400,000 Modern Standard Arabic words; they are divided into 300,000 words as tuning data and 100,000 words as testing data. The selected texts are from different sources; Newspapers, Net Articles and Books representing the following genres; politics: 148,211, miscellaneous: 100,253, child stories: 57,174, economy: 34,930, society: 32,955 and sports: 26,477.

The results were evaluated automatically for accuracy against the reference which is a fully diacritized texts by Arabic linguist using the following two metrics; diacritization error rate (DER) which is the proportion of characters with incorrectly restored diacritics. Word error rate (WER) which is the percentage of incorrectly diacritized white-space delimited words: in order to be counted as incorrect, at least one letter in the word must have a diacritization error.

These two metrics were calculated as: (1) all words are counted excluding numbers and punctuators, (2) each letter in a word is a potential host for a set of diacritics, and (3) all diacritics on a single letter are counted as a single binary (True or False) choice. Moreover, the target letter that is not diacritized is taken into consideration, as the output is compared to the reference. In addition to calculating DER and WER, the evaluation system calculates internal diacritics and case ending separately. The DER measurement was 3.6% while total WER measurement was 13.3%.

Moreover, our proposed system results were compared with the results of other three known state-of-the-art approaches; (Metwally et al. 2016; Zitouni 2006), and (Shahrour et al. 2015). The outputs of these three systems were evaluated using the

Table 1 Comparison between the results of Alserag approach and some state-of-the-art approaches

Approach	Morphological WER	Syntactic WER	Total WER (%)
Alserag	5.6%	10.1 %	15.7
Metwally et al. (2016)	4.3%	9.4 %	13.7
Zaitouni	7.9%	10.1 %	18
Shahrour et al. (2015)	NA	NA	9.4

same testing data (LDC Arabic Treebank). It is a part of Arabic Tree Bank part 3 (ATB3) from "An-Nahar" Lebanese News Agency. It consists of 91 articles (about 52.000 word). Table 1 summarizes the results of the current proposed system in comparison with other systems.

The results were evaluated automatically for accuracy against the reference which is a fully diacritized (morphologically and syntactically) using the following three metrics; morphological (WER) which is the percentage of words that have at least one morphological diacritic error. Syntactic (WER) which is the percentage of words that have one syntactic error and don't have any morphological errors. Total Word Error Rate: the percentage of words that have at least one error (morphological or syntactic). The comparison indicates that Alserag results outperforms (Zitouni 2006) in terms of the morphological WER while they scored the same syntactic WER. On the other hand both (Metwally et al. 2016) and (Shahrour et al. 2015) outperform our system in terms of morphological WER and syntactic WER.

We manually investigated the types of errors. Some of the errors are related to the morphological analysis and they are classified as the following: 2.5% verb person disambiguation, 2.1% active and passive participle, and 1.1% plural and dual noun disambiguation. The other types of errors are related to the syntactic tree such as: 15% verb arguments detection, 15% the adjectival modifiers of the verb arguments, 7% is related to the nunation, 1.5% is related to apposition, and 2.2% is related to name entity recognition.

6 Conclusion

Alserag system is developed based on the rule-based approach which is considered as our contribution to the subject of automatic diacritization. All of the other available systems that were mentioned are statistical based. The results of the system were evaluated against the reference. The proposed system achieved a morphological WER of 5.6%, and a syntactic WER of 10.1%.

References

[SMR05] Smr, O.: Yet another intro to Arabic. In: NLP. http://ufal.mff.cuni.cz/~smrz/ANLP/anlp-lecture-notes.pdf (2005)

[RAR09] Rashwan, M., Abdou, S., Rafea, A.: Stochastic Arabic hybrid diacritizer. IEEE Trans. Nat. Lang. Process. Knowl. Eng. (2009)

[AMM09] Attia, M., Rashwan, M.A., Al-Badrashiny, M.A.: Fassieh, a semi-automatic visual interactive tool for morphological, PoS-Tags, phonetic, and semantic annotation of Arabic text corpora. IEEE Trans. Audio Speech Lang. Process. **17**(5), 916–925 (2009)

[RARR14] Rashwan, M.A., Al Sallab, A.A., Raafat, H.M., Rafea, A.: Automatic Arabic diacritics restoration based on deep nets. In: Proceedings of the EMNLP 2014 Workshop on Arabic Natural Langauge Processing (ANLP), pp. 65–72 (2014)

[RARR15] Rashwan, M.A., Al Sallab, A.A., Raafat, H.M., Rafea, A.: Deep learning framework with confused sub-set resolution architecture for automatic Arabic diacritization. IEEE/ACM Trans. Audio Speech Lang. Process. **23**(3), 505–516 (2015)

[MRA16] Metwally, A.S., Rashwan, M.A., Atiya F.A.: A multi-layered approach for Arabic text diacritization. In: Proceeding of 2016 IEEE International Conference on Cloud Computing and Big Data Analysis (ICCCBDA), pp. 389–393 (2016)

[AGA15] Abandah, G.A., Graves, A., Al-Shagoor, B., Arabiyat, A., Jamour, F., Al-Taee, M.: Automatic diacritization of Arabic text using recurrent neural networks. Int. J. Doc. Anal. Recogn. (IJDAR) **18**(2), 183–197 (2015)

[MBK06] Maamouri, M., Bies, A., Kulick, S.: A challenge to Arabic Treebank annotation and parsing. In: Linguistic Data Consortium. University of Pennsylvania, USA (2006)

[BZDOGH15] Bouamor, H., Zaghouani, W., Diab, M., Obeid, O., Oflazer, K., Ghoneim, M., Hawwari, A.: A pilot study on Arabic multi-genre corpus diacritization annotation. In: Proceedings of the Second Workshop on Arabic Natural Language Processing, 2014, pp. 80–88. Association for Computational Linguistics, Beijing, China (2015)

[EFS12] EL-Desoky, A., Fayz, M., Samir, D.: A smart dictionary for the Arabic full-form words. (IJSCE) **2**(5) (2012). ISSN: 2231-2307

[Alb09] Al Badrashiny, M.: Automatic diacritizer for Arabic text. MA thesis, Faculty of Engineering, Cairo University, Egypt (2009)

[VK04] Vergyri, D., Kirchhoff, K.: Automatic diacritization of Arabic for acoustic modeling in speech recognition. In: COLING Workshop, Geneva, Switzerland (2004)

[ANB05] Ananthakrishnan, S., Narayanan, S., Bangalore, S.: Automatic diacritization of Arabic transcripts for ASR. In: Proceedings of ICON-05, Kanpur, India (2005)

[ZSS06] Zitouni, I., Sorensen, J.S., Sarikaya, R.: Maximum entropy based restoration of Arabic diacritics. In: Proceedings of the 21st International Conference on Computational Linguistics and 44th Annual Meeting of Association for Computational Linguistics, pp. 577–584. Association for Computational Linguistics (2006)

[HR05] Habash, N., Rambow, O.: Arabic tokenization, part-of-speech tagging and morphological disambiguation in one fell swoop. In: Proceedings of the 43rd Annual Meeting on Association for Computational Linguistics, pp. 573–580. Association for Computational Linguistics, Ann Arbor (2005)

[HR07] Habash, N., Rambow, O.: Arabic diacritization through full morphological tagging. Human Language Technologies 2007: The Conference of the North American Chapter of the Association for Computational Linguistics Companion Volume. Short Papers, pp. 53–56. Association for Computational Linguistics, Rochester, NY (2007)

[DGH07] Diab, M., Ghoneim, M., Habash, N.: Arabic diacritization in the context of statistical machine translation. In: Proceeding of MT-Summit, Copenhagen, Denmark (2007)

[RRHD08] Roth, R., Rambow, O., Habash, N., Diab, M., Rudin, C.: Arabic morphological tagging, diacritization, and lemmatization using Lexeme models and feature ranking. In: Proceedings of the 46th Annual Meeting of the Association for Computational Linguistics on Human Language Technologies: Short Papers, pp. 117–120. Association for Computational Linguistics, Columbus, Ohio, USA (2008)

[SAZ08] Shaalan, K., Abo Bakr, H.M., Ziedan, I.A.: A statistical method for adding case ending diacritics for Arabic text. In: Proceedings of Language Engineering Conference, pp. 225–234. Cairo, Egypt (2008)

[SAZ09] Shaalan, K., Abo Bakr, H.M., Ziedan, I.: A hybrid approach for building Arabic diacritizer. In: Proceedings of the 9th EACL Workshop on Computational Approaches to Semitic Languages, pp. 27–35. Association for Computational Linguistics (2009)

[SKH15] Shahrour, A., Khalifa, S., Habash, N.: Improving Arabic diacritization through syntactic analysis. In: Proceedings of Empirical Methods in Natural Language Processing Conference (EMNLP), pp. 1309–1315. Association for Computational Linguistics (2015)

[Ala14] Alansary, S.: MUHIT: a multilingual harmonized dictionary. In: The 9th Edition of the Language Resources and Evaluation Conference, Reykjavik, Iceland (2014)

[Ala13] Alansary, S.: A suite of tools for Arabic natural language processing: a UNL approach. In: The Special Session on Arabic Natural Language Processing: Algorithms, Resources, Tools, Techniques and Applications, (ICCSPA'13), Sharjah, UAE (2013)

Learning Context-Integration in a Dependency Parser for Natural Language

Amr Rekaby Salama and Wolfgang Menzel

Abstract Dependency parsing aims at finding an optimal tree representation of the input sentence in terms of word-to-word relationships, which are important for many Natural Language Processing tasks. This requires to diambiguate the linguistic input, a task that might profit considerably from external knowledge about the context. To study the role of the visual context on language comprehension, we developed a multi-modal dependency parser which is able to learn the correspondence between different input channels. Our system is base on the state-of-the-art RBG parser, but it builds a model that is able to interpret the linguistic input using the clues provided through different modalities. We build two small-scale datasets to verify the effectiveness of our model and were able to show a 20% relative improvement of sentence accuracy compared to the original, single-modality RBG parser.

Keywords Natural language parsing · Multi-modal integration
Dependency parsing · RBG parser · MALT parser

1 Introduction

Parsing is an essential task for many applications of Natural Language Processing (NLP) like information retrieval, information extraction, machine translation, and others. Recently, dependency parsing attracted increasing attention. It represents the structure of the linguistic input as a labelled tree consisting of binary relationships between the lexical items in the sentence.

In general, three main approaches to parsing can be distinguished:

- Grammar-based approaches where the parser uses a formal grammar to analyze the input. The grammar typically consists of some kind of rewriting rules. A large

A.R. Salama (✉) · W. Menzel
Department of Informatics, University of Hamburg, 22527 Hamburg, Germany
e-mail: salama@informatik.uni-hamburg.de

W. Menzel
e-mail: menzel@informatik.uni-hamburg.de

K. Shaalan et al. (eds.), *Intelligent Natural Language Processing: Trends and Applications*, Studies in Computational Intelligence 740,
https://doi.org/10.1007/978-3-319-67056-0_26

set of rules is needed to reach a significant coverage of all the possible linguistic variations. Manual compilation of the grammar is labor intensive and error prone, hence, a large coverage is difficult to achieve.
- Constraint-based approaches which make use of a set of conditions that each relationship in the structural description has to meet. Constraints can be soft, which allows the parser to deal with inconsistencies and incomplete knowledge. This significantly contributes to a high degree of accuracy and improves the overall robustness against a limited coverage and ungrammatical input. Still, manual compilation of the constraint set is extremely cumbersome.
- Data-driven approaches where the parser is trained on a set of annotated data, containing pairs of sentences and their structural descriptions (namely syntax trees). Training creates a model that is able to establish the correspondence of the input sentence to its structural description. This highly simplifies to port the parser to another language as long as the necessary training data are available. Modern data-driven approaches to parsing clearly outperform any grammar or constraint-based ones both in terms of accuracy and parsing speed.

Dependency structures lend themselves particularly well to the application of machine learning techniques. Local configurations of parsing hypotheses, i.e. single word-to-word relationships or locally restricted combinations of them, can easily be evaluated by the model and compared to each other. That led to the development of a great number of different dependency parsers over the last few couple of years.

A notorious difficulty in parsing is the high degree of lexical and structural ambiguity that potentially leads to large numbers of alternative analyses which might correspond to different interpretations of the input sentence. Ideally, the parser would be able to decide which interpretation is the best one. Taking such a decision, however requires a sophisticated evaluation scheme that goes beyond a simple correct/incorrect distinction and thus allows the parser to rank and select partial parsing hypotheses. Having available such a mechanism, the parser becomes able to fully disambiguate a sentence, instead of only enumerating the possible alternative interpretations, which is the typical behavior of a rule-based system. Different measures to score partial hypotheses have been devised. They are based e.g. on probabilities, the distance from a decision boundary, or the severity of a constraint violation. The most frequent cases in the training set, for instance, drive the parser towards a particular decision. The quality of the parsing results depends crucially on the amount and the quality of the available training data, but also on the ability of the parser to extract the relevant model information from the training examples.

What a classical parser usually lacks, is the possibility to condition its decisions on features of the extra-linguistic context into which the sentence is embedded. The entities in the surrounding world and their relationships, can substantially affect the plausibility of different parsing results. Moreover, changes of the state-of-affairs in the world, might exert a dynamically changing influence on the interpretation of the sentence, while it is being uttered. It has been shown that soft constraints at least provide the possibility to capture such an influence (see Sect. 3). It remains to be

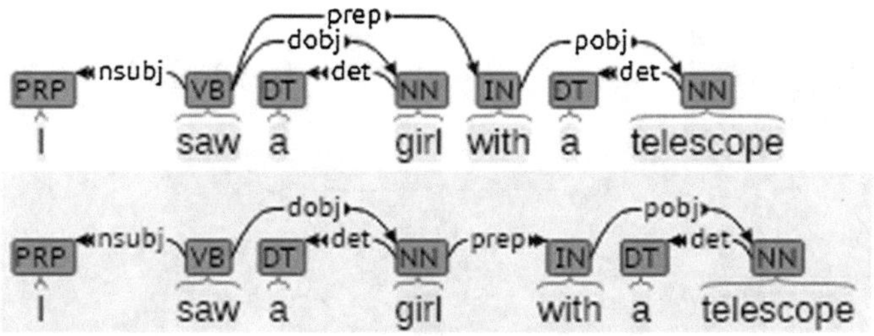

Fig. 1 Two syntactic dependency trees with **a** high attachment and **b** low attachment of a prepositional phrase

investigated whether state-of-the-art machine learning approaches can be equipped with similar capabilities.

Prepositional phrase (PP) attachment is not only a major source of ambiguity in natural language processing, but also a phenomenon where the consequences of the different attachment options crucially affect the understanding of an utterance. A well-known example of such an ambiguity is presented in Fig. 1. The sentence "I saw a girl with a telescope" is globally ambiguous although all its word have a unique interpretation. The possible two interpretations are the result of simply attaching the prepositional phase "with a telescope" differently: either to the verb "saw" (where it becomes the instrument of the seeing event) or to the noun "girl" (where it indicates the actual possession of the telescope). Thus, the different attachment possibilities give rise to two different meanings: I used a telescope to watch a girl, or I noticed a girl who had a telescope with her.

If we push the ambiguity one step further e.g. with the sentence "I saw the girl on the hill with a telescope.", the set of alternatives grows even bigger (cf. Fig. 2).

Which of these interpretations happens to be more plausible cannot be determined from the sentence alone, but requires to include information from the situation the sentence refers to. Note also, that this kind of preference cannot be learned from a purely linguistic annotation, as it is usually found in a treebank.

Human sentence comprehension solves this problem since it is exposed to a variety of additional information pieces that help in the disambiguation. Visual information, in particular, plays a crucial role to supply the kind of extra-linguistic guidance and support that is badly needed. Also information from the preceding discourse would be of great value here. Integrating information from such a variety of different channels eventually makes the brain able to disambiguate the linguistic input seemingly without any noticeable effort.
Inspired by the model of human cognition we aim at developing a data-driven multimodal parser. Unlike the standard linguistics-only approach the system shall be able to process multi-modal information, to integrate it into a coherent picture of the world and to take its decisions according to the information available. Specifically, our parser has two input channels:

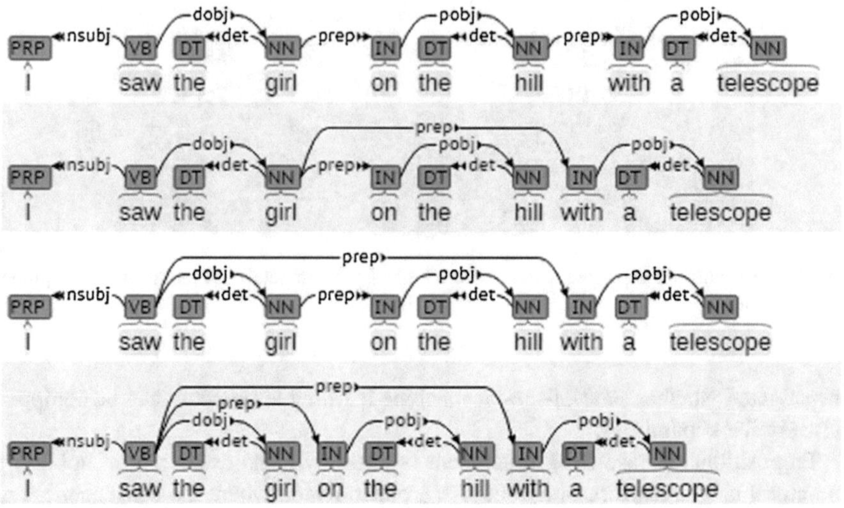

Fig. 2 Four different interpretations of a structurally ambiguous sentence

- Text input (linguistic modality) which is the standard input for dependency parsing, and an additional
- Context input (non-linguistic modality) which provides information about the relationships between the elements in the reality as they could could be provided by the visual channel. This information is then used to disambiguate the textual input.

Since we are primarily interested to study the mechanisms that would allow us to modulate the parsing behaviour by means of extra-linguistic cues, we assume that the visually conveyed context information is available as a symbolic decription on a conceptual level. We currently do not work on components for image or video understanding, but we inject the corresponding information directly into the parser in the form of semantic relationships between entities and events. However, collecting a large dataset that systematically combines sentences with a description of the corresponding situation is very time-consuming. Thus, we built up a relatively small dataset to use it in our research. More details are discussed in the following sections.

The following chapter starts with a literature review of the different approaches to dependency parsing. Previous work of integrating context information into the parsing of written input is discussed in Sect. 3. In Sect. 4, we illustrate our approach to build a multi-modal learning parser. Its system architecture is presented in Sect. 5. Finally, we evaluate our approach and compare it to the state-of-the-art in dependency parsing. We demonstrate the effectiveness of integrating the context information, and how it helps to disambiguate the linguistic input.

2 Dependency Parsing

Dependency parsing aims at identifying the binary syntactic relationships that hold between pairs of words in a given textual input sentence $input = w_1, w_2, ..., w_n$. The relationships are arranged into a dependency tree, which is a directed acyclic graph, where the nodes represent the input tokens. The edges between the nodes describe the syntactic relationships between the words by attaching a node to its head which then is said to dominate its modifiers. The edges of such a tree can be decorated with labels indicating the type of the syntactic relationship. Thus, a dependency tree $D = (W, E)$ combines the set of lexical items W taken from the input sentence with E, the set of edges $e(w_i, w_j), w_i, w_j \in W$.

A dependency tree is a valid one when it meets the following criteria [13]:

- Connected: The graph is connected.
- Acyclic: If $\exists\, e(w_i, w_j)$ then $\nexists\, path(w_j \rightarrow w_i)$.
- Single head: Each node has only a single head.
- Single label: Each edge carries only a single label (in case of a labeled D).

In our research, we focus on data-driven approaches where the parser builds its model from the training data. In the test phase, the model is applied to the input data and determines the best fitting dependency tree for the input sentence according to the mapping it learns.

Approaches to data-driven dependency parsing can be classified into two groups: transition-based procedures, and graph-based ones. In the next subsections, we investigate both methods and outline the state-of-the-art. Some of these parsers will later be used in our experiments (c.f. Sect. 6).

2.1 Transition-Based Dependency Parsing

In the transition-based approach, the sentence is incrementally processed word by word. That simulates the cognitive ability to process the input words immediately after they become available, like in listening or reading an utterance. The parser does not wait for the end of the sentence before parsing commences. Such an incremental processing mode is particularly useful in real-time applications like speech recognition. Ideally, the parser can guarantee to provide a single connected structure of the input sentence at any time during processing, even though the utterance is not yet complete. But this is difficult to achieve. Therefore, this structure usually does not contain the tokens still waiting for a proper head to be attached to.

Transition-based parsing adopts a set of actions to link the tokens with each other. Assuming we have a set of words $(w_1, w_2, \ldots, w_n)$ as an input, the parser uses a stack and a buffer to conduct the actions on them that eventuall will produce the desired dependency tree. The generic algorithm for the transition-based approach is:

```
Initialize the stack S←null
Initialize the queue Q←(w_1, w_2, .., w_n)
Initialize the dependency tree D←null
while D is not final (not yet complete)
    Transition t=get_next_transition(S,Q,D)
    apply(t,S,Q,D)
return the tree
```

The trained classifier learns how to select the proper transition in each input situation. The main actions in the arc-standard algorithm are:

- Shift: Move the next word from the input buffer to the top of the stack S.
  $S \rightarrow S|w_i$
  $Q = w_i, w_{i+1}, ..w_n \rightarrow Q = w_{i+1}, ..w_n$
  $D \rightarrow D$
- Left reduce: Link the two top tokens in the stack and assign the head of the stack to the last token.
  $S = w_1, w_2, ..w_i, w_j \rightarrow S = w_1, w_2, ..w_j$
  $Q \rightarrow Q$
  $D \rightarrow D \cup e(w_j, w_i) w_j(parent), w_i(child)$
- Right reduce: Link the two top tokens in the stack and assign the head of the stack to the second last token.
  $S = w_1, w_2, ..w_i, w_j \rightarrow S = w_1, w_2, ..w_i$
  $Q \rightarrow Q$
  $D \rightarrow D \cup e(w_i, w_j)$.

Because of the deterministic nature of this greedy approach it is not guaranteed to fetch the best tree with the highest score. In particular, a transition can not be corrected if the parser decided to connect two nodes but later finds evidence that would require to reconsider this step.

2.1.1 MALT Parser

MALT parser is one of the most popular transition-based parsers. It uses an arc-eager algorithm (Nivre's Algorithm) [14] based on a different set of transitions under the same framework of transition-based parsing. The transition system distinguishes two different actions for shift and reduce. That means, the parser can learn just shifting a word to the stack (without linking it), or to pop an existing word from the stack without a need to link it to other elements. Instead of left-reduce and right-reduce, Nivre also proposes two other actions: left-arc, and right-arc. Comparing to the arc-standard algorithm, MALT parser is not a strictly bottom-up approach which makes it able to create some dependency connections earlier than the arc-standard algorithm. We represent the actions as follows:

- Shift: Move the next word from the input buffer to the top of the stack S.
  $S \rightarrow S|w_i$

$Q = w_i, w_{i+1}, ..w_n \rightarrow Q = w_{i+1}, ..w_n$
$D \rightarrow D$

- Reduce: Pop operation from the stack S.
  $S = w_1, w_2, ..w_n \rightarrow S = w_1, w_2, ..w_{n-1}$
  $Q \rightarrow Q$
  $D \rightarrow D$
- Left-arc: Link the top token of the stack with the top of the buffer, and pop up the top of the stack.
  $S = w_1, w_2, ..w_{i-1}, w_i \rightarrow S = w_1, w_2, ..w_{i-1}$
  $Q = w_j, w_{j+1} \cdots \rightarrow Q = w_j, w_{j+1} \cdots$
  $D \rightarrow D \cup e(w_j, w_i) w_j(parent), w_i(child)$
- Right-arc: Link the top token of the stack with the top of the queue, and push the top of the buffer into the stack.
  $S = w_1, w_2, ..w_i \rightarrow S = w_1, w_2, ..w_i, w_j$
  $Q = w_j, w_{j+1} \cdots \rightarrow Q = w_{j+1} \cdots$
  $D \rightarrow D \cup e(w_i, w_j)$

The training procedure for MALT parser uses a range of different features in the training process. These features include the part-of-speech tags of all of the words in the sentence, the words themselves, the top element on stack S, the top element in the queue Q, and the current state of the dependency tree. On the other hand, the four mentioned transitions are the target classes of the learning process. Different methods like large-margin SVM learning algorithms, or Memory-based learning are used for training.

2.2 Graph-Based Dependency Parsing

Graph-based parsing represents the input sentence as a graph of words [24]. Each node in this graph corresponds to a word and the graph is fully connected (c.f. Fig. 3). Each link carries two scores, one for each direction. The task consists of extracting the best dependency tree from all the combinations of dependencies in the fully connected graph (solid arcs in Fig. 3). The extracted tree should have the highest total score [4].

The score of each dependency edge is computed as a function of the linguistic features of the tokens and their governors according to the learning algorithm and the features considered as shown in Eq. 1 (Fig. 4)

$$score(D, dep) = \sum score(e(w_i, w_j)) \quad where \quad e(w_i, w_j) \in dep, \quad dep \in E \tag{1}$$

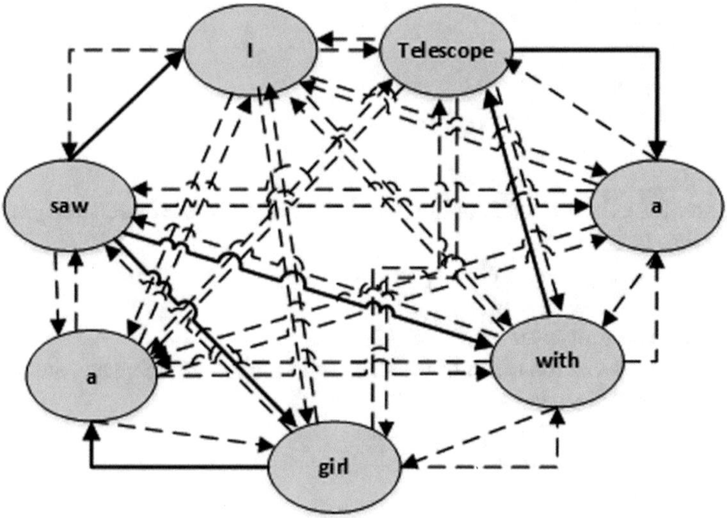

Fig. 3 A fully bidirectional connected graph of the input sentence, solid arcs represent the high-score dependencies

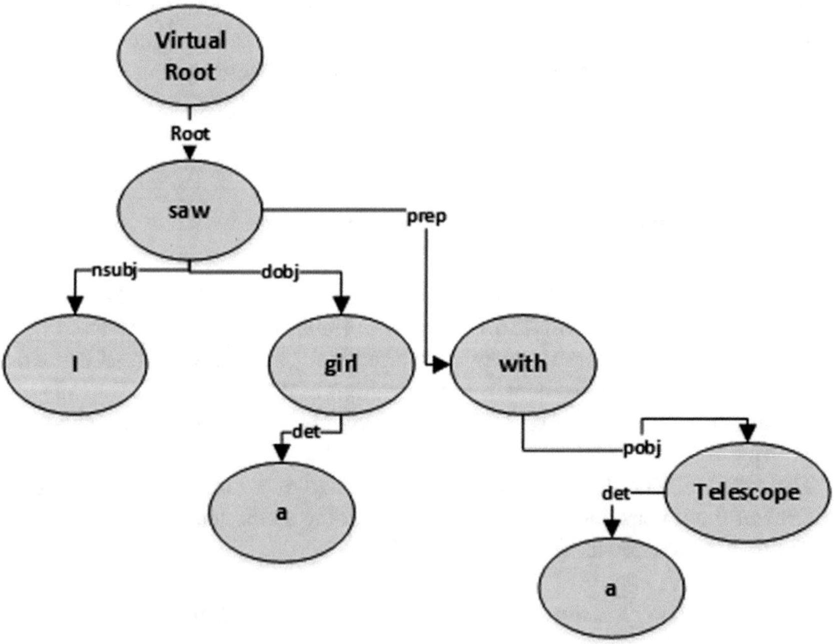

Fig. 4 The final dependency tree

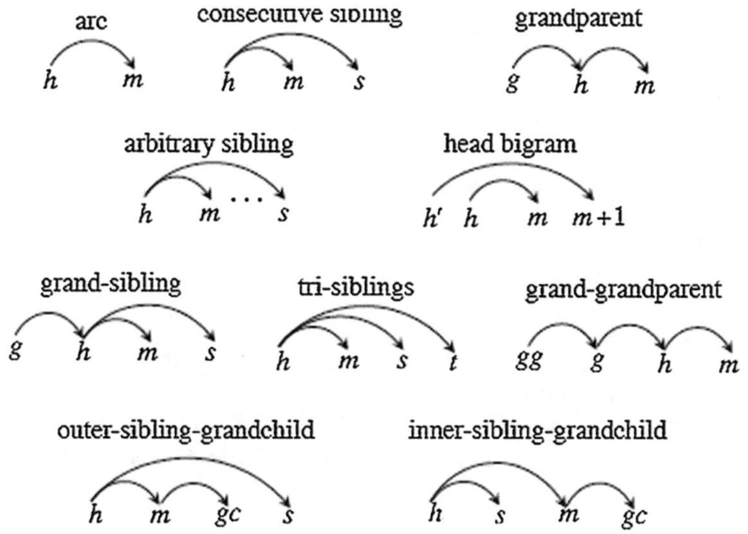

Fig. 5 First to third order features [24]

2.2.1 RBG Parser

The state-of-the-art of graph-based dependency parsing (and dependency parsing in general) is defined by RBG parser, a system that was introduced in 2014 [11]. RBG attaches a feature vector to each dependency edge in the original fully connected graph. It can use higher-order linguistic features that decribe more complex partial dependency structures (c.f. Fig. 5). The weights of these features are dynamically learned using the online passive aggressive algorithm (MIRA). This learning algorithm deals with the training data in a sequential manner and updates the weights after each prediction according to the distance between the prediction and the gold standard. RBG uses an averaging technique to manipulate the weights of a feature vectors [7].

Unlike other parsers who consider the tree extraction as a classical optimization problem, RBG uses a randomized greedy algorithm to find the best combination of dependency edges. It starts with a random dependency tree and tries to do arc factorization to achieve a higher fitness score. To avoid getting trapped in a local minimum, RBG runs multiple random restarts. Different decoding algorithms, such as hill climbing [23] or the Chu-Liu/Edmond-algorithm [6], are used to fetch a minimum spanning tree using the overall edges' scores. The different decoding algorithms are either restricted to projective dependency trees, or are able to also deal with non-projective dependencies.

```
Random initialize tree y^0
t=0
repeat
    list=bottom up node list of y^t
    For each word in list
        y^{t+1}= argmax score(y),∀y ∈ T(y^t, m)
        t=t+1
    end for
till no more changes in this iteration
return y^t
```

where $T(y^t, m)$ is a dependency tree list created by changing a node attachment to an alternative one (arc factorization).
In our research, we implemented our parser based on RBG as described in Sects. 4 and 5. We show how we modified the approach into "RBG-2" to enable multi-modal integration.

3 Previous Work on Context Integration

McCrae [12] introduced a multi-modal parser. He proposed using shallow semantics as a representation of the visual (context) modality. Context information was introduced by means of thematic relations into the process of sentence parsing. The system was only applied to German using a constraint-based parser (Weighted Constraint Dependency Grammar WCDG) [19]. Linguistic concepts from the sentence are mapped to the concepts in the visual context using a taxonomy defined by means of Web Ontology Language. McCrae adopted four thematic roles (Agent, Theme, Instrument, and Owner) for the representation of the context information. From his dataset we created an extended corpus to be used in our research. Therefore, we provide a more detailed discussion of these thematic roles in the next section. The main difference between our approach and the one of McCrae consists in the use of a learning (data-driven) parser. That improves the independence from the grammar of a particular language.

Baumgärtner [2] extended the work of McCrae to establish a bi-directional influence between the linguistic input and the visual reference resolution. Inspired by psycholinguistic research, he used an online integration of linguistic and visual information to model the shift of the visual attention as the sentence evolves and was able to show essential similarities to human behaviour [3].

In the next subsection, we review the common approaches for the description and annotation of semantic relations which could be utilized as a suitable context representation.

3.1 Previous Work on Semantic Representation

3.1.1 FrameNet

Frame semantics is one of the most influential semantic theories. It aims at representing the meaning of an utterance based on a cognitive model of sentence comprehension, assuming that human listeners understand the meaning of a complex linguistic construction by linking an action to a particular frame that has a set of influencing actors [1]. Mapping these actors would produce a complete understanding of the described situation [8].

Verbs like to buy, to sell, to pay, and to cost describe similar situations centering around a commercial operation. Despite the differences in the meaning of these verbs, they share the same semantic elements (such as a buyer, a seller, and a product). Therefore, they belong to the same semantic frame [18]. The FrameNet project aims at defining a set of semantic frames; each consisting of a defined set of elements, illustrating the meaning of the verbs that belong to this frame. The current published version of the FrameNet project comprises 1222 semantic frames, 89% are lexical frames, and 11% are non-lexical frames with an average of 9.7 frame elements per lexical frame.

3.1.2 VerbNet

In contrast to FrameNet, VerbNet defines a set of general thematic roles [9]. The verbs are grouped into classes based on their senses. For each class, VerbNet identifies the relevant thematic roles. For example, the thematic role "Agent" is filled by the active doer of an action. Instead of having specialized roles like in FrameNet (buyer, seller), VerbNet uses "Agent" to capture the meaning contribution in a more general way: If the role occurs with the verb "to buy", it is equivalent to the "Buyer"-role in FrameNet. It is equivalent to "Seller" with the verb "To sell" [16]. The current version of VerbNet (Extended VN) makes use of 23 thematic roles for 274 classes of verbs, comprising 5257 verb senses.

In our research, we use thematic roles as a representation of context information. Currently we focus on the four thematic roles as described in Sect. 4, three of them are taken from VerbNet, while the fourth is not listed there.

4 RGB-2: A Context Integrating Dependency Parser

Our multi-modal parser is based on RBG which has been extended to make it able to deal with multi-modal input. The original RBG parser as described above can deal only with linguistic input. We modified its learning and parsing capabilities to also consider the newly introduced context information.

Later in the experiment section, we compare our context integrating parser against the original version of RBG that uses only the linguistic information, as well as against MALT parser arguably one of the most popular transition-based parsers.

4.1 Context Representation

In this section, we illustrate the adopted methodology to describe the context information. We use binary predicates for that purpose, which consist of three sections:

$$relation\_type(predicate, argument)$$

Here, the predicate is the lexical item in the input sentence that denotes the main event in the triple, relation type is one of the predefined set of the possible relations, and the argument is an utterance that completes the meaning of the triple (relation). Following McCrae [12], we use thematic roles to describe the type of context relations.

In contrast to VerbNet which restricts itself to verbs as predicates, in our context representation, the predicate could also be a noun. We make this assumption, because otherwise we would miss some context information like the ownership (part-to-all relation) between different nominal constituents (phrases).

As one could see in the examples presented in Figs. 6 and 7, the input for the parser has two components: the linguistic utterance, and a formal description of the state-of-affairs in the world consisting of the relations between the entities. The thematic roles considered at this stage of research are shown in Table 1.

To introduce the multi-level data (consisting of syntactic dependencies as well as thematic relationships describing the visual contex) into RBG parser, the standard (language-only) CONLL-2005 format is not well suited. Although the extended version that was used in the CONLL-2008 shared task [20] already contained semantic structures in addition to the syntactic ones, both kinds of information have been combined into a single representation using the Propbank annotation style [15] for the

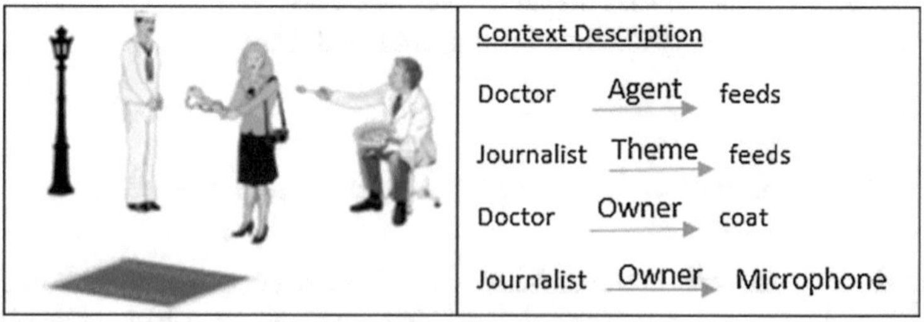

Fig. 6 The context information (image taken from [10]) Linguistic input: "The doctor with a coat feeds at this moment the journalist with a microphone."

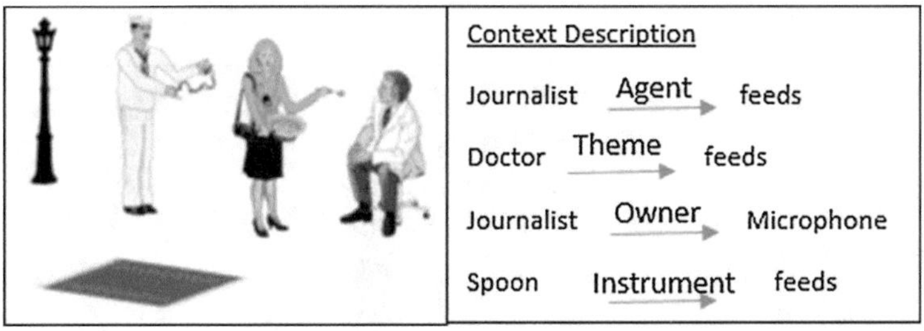

Fig. 7 The context information (image taken from [10]) Linguistic input: "The journalist with a microphone feeds at this moment the doctor with a spoon."

Table 1 Thematic roles used as relationship's description in the context input

Thematic role	Description	Properties
Agent	The human or entity that performs the action intentionally. Ex: The man kicks the ball. Agent (kicks, man)	From VerbNet, attached to a verb predicate
Theme	The entity, affected by the action. Ex: The racer examines the motor. Theme (examines, motor)	From VerbNet, attached to a verb predicate
Instrument	The instrument used during the performance of the action. Ex: He eats the desert with a spoon. Instrument (eating, spoon)	From VerbNet, attached to a verb predicate
Owner	It states a property/ownership of other object(s). Ex: A girl with red hair came along. Owner (girl, red hair)	non-VerbNet relation, attached to a noun predicate

semantic relationships. We modify this annotation in a way that it consists of the syntactic structure as in the original CONLL format, together with additional columns to specify the predicates and their thematic relations. Each predicate then has a separate column to accomodate its argument as shown in Fig. 8. We call this modified version CONLL-Context format.

Linguistic Layer								Context Layer				
ID	FORM	LEMMA	CPOS	POS	FEAT	HEAD	DEPREL	Predicate	Arg(P1)	Arg(P2)	...	Arg(PX)
1	A	a	DT	DT	-	2	det	-				
2	man	man	NN	NN	-	8	nsubj	man.01		(Agent)		
3	in	in	IN	IN	-	6	case	-	(Property			
4	street	street	NN	NN	-	6	compound	-	*			
5	racer	racer	NN	NN	-	6	compound	-	*			
6	armor	armor	NN	NN	-	2	nmod	-	*)			
7	is	be	VB	VBZ	-	8	aux	-				
8	examining	examine	VB	VBG	-	0	ROOT	examine.01				
9	the	the	DT	DT	-	10	det	-				
10	tire	tire	NN	NN	-	8	dobj	-		(Patient)		
11	of	of	IN	IN	-	15	case	-				
12	another	another	DT	DT	-	15	det	-				(Owner
13	racers	racer	NN	NNS	-	15	compound	-				*
14	motor	motor	NN	NN	-	15	compound	-				*)
15	Bike	Bike	NN	NN	-	10	nmod	bike.01				

Fig. 8 CONLL-Context format declares multi-level input

4.2 Feature Representation

As RBG is a graph-based parser, it starts with building edges between each pair of words in the input sentence. For every edge, there is a vector that consists of weighted linguistic features. This feature vector is used in the calculation of the scoring function. The weights of the features are learned during the training phase. To introduce the context information into the scoring function, we add an additional vector of context features to all edges that connect a pair of words which have an context relation between them.

As illustrated in Fig. 9, the resulting graph consists of the input words and both kind of feature vectors. If the input has t words, m linguistic feature vectors $fv_{a:1\to m}$, and p context feature vectors $cfv_{a:1\to p}$, for each pair of words (i, j), we could state the feature vectors as follows:

Fig. 9 Context-integrating dependency graph

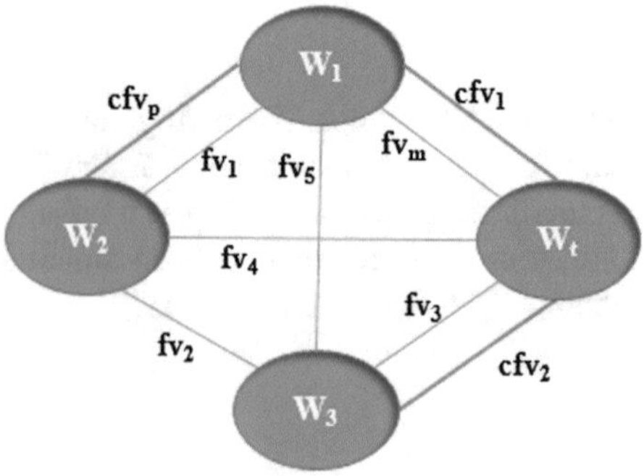

$$fv_{a:1\rightarrow m} = \{f_{i,j,1} \cdots f_{i,j,n}\} \tag{2}$$

where n is the number of linguistic features between each pair of words in the textual input. Similarly we represent the context feature vectors as

$$cfv_{a:1\rightarrow p} = \{cf_{i,j,1} \cdots cf_{i,j,q}\} \tag{3}$$

where q is the number of context features between each pair of words (that have context relations).

During training, for each input sentence x_i, the corresponding context description c_i is available. For each input, there is a set of candidate dependency trees $T(x_i)$, and one gold standard tree $\hat{y}$. The linguistic feature vector is $f(x_i, y)$ with weights θ. The context feature vector is $\hat{f}(c_i, y)$ with weights ω. During the training phase, the optimal feature weights θ and ω are determined according to:

$$\bar{y} = \max_{y \in T(x_i)} \{\theta.f(x_i, y) + \omega.\hat{f}(c_i, y) + \|y - \hat{y}_i\|\} \tag{4}$$

During the testing phase, a tree can be extracted according to the following equation

$$\bar{y} = \max_{y \in T(x)} \{\theta.f(x_i, y) + \omega.\hat{f}(c_i, y)\} \tag{5}$$

There are three different types of context features used in RBG-2 (c.f. Table 2). To construct a visual feature, we use combinations of properties of the visual predicate and argument: POS tag, word, and lemma for all the different feature types shown in Table 2.

Thus, bigram context features, for example, are complex combinations constructed out of four different components:

- HPp: The part of speech (POS) of the word preceding the head (left context).
- HP: The POS of the head word (predicate).

Table 2 Types of the visual features in RBG-2

Type	Type description
Core features (Predicate/Argument)	The entities described by a single word (node), i.e. the predicate or the argument of a visual relation
Unigram features (Predicate-Argument)	The relationship described by a single edge, i.e. between the predicate and the argument
Bigram features (Predicate-Argument)	The relationship described by a single edge including immediate left or right (single-word) context

Table 3 Sample feature coding

Feature	Meaning	Feature code
ID of the thematic role	Agent	1100
POS ID of coat	NN	0010
POS ID of feeds	VB	0100
POS ID of Doctor	NN	0010
POS ID of with	IN	0011

- MXP: The POS of the modifier (argument).
- MXPn: The POS of the word following the modifier (right context).

where X has to be substituted by one of the four thematic roles described in Table 1, e.g. AG for the agent role. Applied to the example presented in Fig. 6, where Doctor is the Agent of the feeds action (predicate), the resulting feature encodes agent as the relation type, feeds as the head of the context relation, and Doctor as its modifier (argument). Accordingly, HPp refers to the POS of the word preceding feeds. (namely the noun coat), HP is the POS of feeds. (which is a verb), MAGP is the POS of Doctor. (which is a noun) and MAGPn is the POS of the word next to Doctor (namely the preposition with). Unigram features are simplified versions of bigram features abstracting away the context information, while the core features only consider either the head, or the modifier.

RBG-2 uses a binary encoding of the visual context features as presented in Table 3, which results a complete feature embedding 1100-0010-0100-0011.

5 System Description

In this section, we discuss the architecture of the proposed multi-modal context-integrating parser which has been developed as an extra layer over the original RBG implementation [22]. Fig. 10 shows the modules RBG-2 consists of. We divide them into three groups:

- Original RBG modules: These are the modules we use as they are from the original RBG parser implementation.
- Additional RBG-2 modules: To introduce the visual input into the parser, we implemented new modules to allow the system to deal with the additional input channel.
- Modified RBG modules: We had to adapt the learning modules including the management of feature vectors, and the modification of the weights to enable the learning process to consider both, the linguistic input, and the input from the visual context.

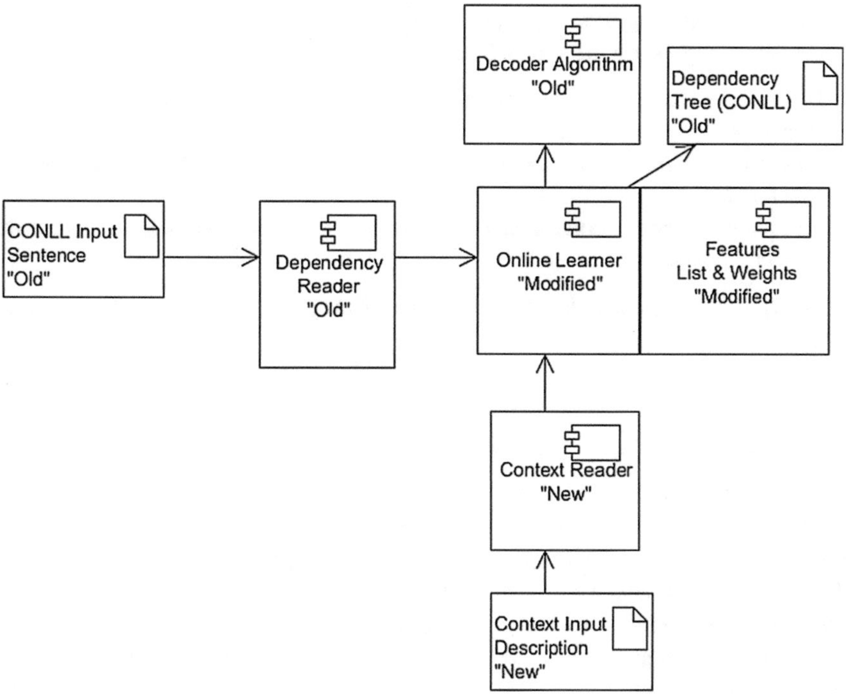

Fig. 10 RBG-2 architecture

Since we only modified the learning process and the calculation of the scoring function without deviating from the main steps of the parsing process, we do not have to change the decoder module of the RBG parser which is responsible for extracting the dependency tree from the fully connected graph.

6 Evaluation

In this section, we present our evaluation of the proposed multi-modal parser RBG-2 against MALT and RBG parsers. Due to the acceptance of multi-input channels in RBG-2, we implemented two versions of it:

- RBG-2 where the parser gives equal confidence to the different input channels (linguistic and visual context). The final score of the dependency arc is calculated based on equal weights for the linguistic and the context feature vectors.
- RBG-2(Con-Biased) where the parser trusts the context input more than the linguistic features. This is implemented by increasing the weights of the context features in the overall dependency arc calculation by a factor of three.

We use the following metrics in our experiment to determine the effectiveness of dependency parsing on the syntactic level:

- Unlabeled attached score (UAS): The percentage of correct token-parent attachments.
- Labeled Attached Score (LAS): The percentage of correct labeled token-parent attachments.
- Complete Attached Score (CAS): The percentage of complete sentences that are correctly analyzed.

For each POS, we present detailed results for the accuracy of the dependency attachment.

6.1 Datasets

Our research requires special datasets for the experiments. We need a multi-level dataset that comprises a syntactic dependency level, and a context level encoded in the thematic roles format as shown in Table 1. Moreover, we are particularly interested to consider utterances that are linguistically ambiguous, but where the ambiguity can be resolved using the additional visual context information. Therefore, we compiled two datasets:

- HAM-DS: This dataset in an extension of Baumgärtner's [2] data. The original corpus contained 96 sentences describing 24 images complemented by the context information used for disambiguation. The German sentences were constructed in a systematic form (subject, verb, object and an optional adverbial modifier). Originally the corpus was developed for German. Therefore, we translated it into English and increased the complexity of the utterances by adding prepositional modifiers in different places. This produced 480 English belonging to one of the following equally sized groups:
  - Group-1: The translated sentences of the original German corpus.
    Ex: "The princess washes obviously the pirate."
  - Group-2: Sentences with subject and object, both modified with a prepositional phrase.
    Ex: "The princess with long hair washes obviously the pirate with a woody leg."
  - Group-3: Sentences with a complex subject, an object, and the instrument of the action.
    Ex: "The princess with long hair washes obviously the pirate with a brush."
  - Group-4: Sentences with a complex subject, a complex object and the instrument of the action.
    Ex: "The princess with long hair washes obviously the pirate with a woody leg with a brush."

- Group-5: Sentences in passive voice with an unmodified subject and an unmodified object.
  Ex: "The pirate is washed by the princess."

- Illions-100DS: The main weak point of HAM-DS is that it has been artificially created. Because we also want to test our model on more natural sentences we adopted the ILLIONS image corpus [21] for that purpose. The corpus consists of images each described with five different sentences which, however, are unannotated. Thus, we selected a small sample of 100 sentences describing 33 images and attached the linguistic and visual dependencies to them. This dataset has the variety of sentences structure that we miss in HAM-DS.

To syntactically annotate the sentences, we used the online version of Turbo parser [17] and verified its CONLL output format against the "Stanford dependency manual" [5]. The context representation of each sentence was created manually using the thematic roles mentioned in Table 1.

6.2 *Experiments and Results*

6.2.1 Experiment 1

In Experiment 1, we work with HAM-DS described in Sect. 6.1. We performed a 5-fold cross validation where each group of the sentences mentioned in Sect. 6.1 is used as a separate fold. Fig. 11 shows the average results of all the folds. Clearly, RBG-2 outperforms both, RBG and MALT parser in all the different accuracy measures.
Although the RBG parser achieves competitive UAS, and LAS scores, it can not solve more ambiguous sentences because it only relies on the statistical information provided by the static linguistic training data, which are unable to reflect the case-specific contribution of the visual context. In contrast, the RBG-2 models were able to disambiguate such cases and therefore reach a remarkably higher CAS score due to the use of the multi-modal input.
Comparing RBG-2 with RBG-2(Con-Biased), we find that a higher trust in the context information leads to over-fitting. Over-fitting was most severe when the test fold contained all the passive sentences, while the training data consisted of all the other four sentence groups. Despite the higher UAS and LAS scores for RBG-2 comparied to RBG-2(Con-Biased), the CAS score for RBG-2 is inferior. Obviously, RGB-2(Con-Biased) was able to produce a higher share of completely correct results, in particular for the passive sentences. This is clearly shown in Table 4 where we see a noticeable gap between the UAS and LAS scores for the two different RBG-2 models.

Table 4 presents a breakdown of the parsing accuracy scores for the four most important POS categories nouns (NN), verbs (VB), adjectives (JJ) and prepositions (JJ). It shows that RGB-2 achieves a higher performance for nouns compared to the

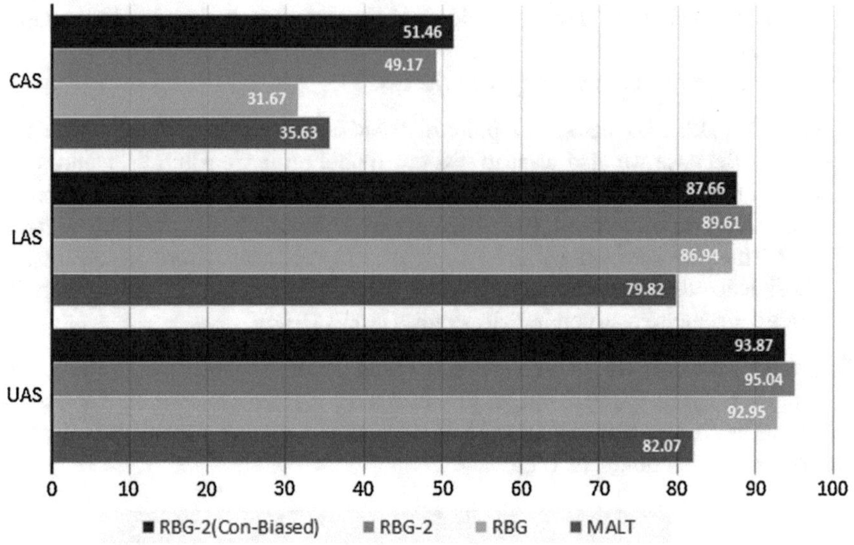

Fig. 11 HAM-DS 5-folds Exp

Table 4 HAM-DS 5-folds POS results

POS	UAS				LAS			
	MALT	RBG	RBG-2	RBG-2(Biased)	MALT	RBG	RBG-2	RBG-2(Biased)
NN	0.81	0.94	0.96	0.94	0.78	0.84	0.86	0.82
JJ	0.88	0.88	0.93	0.90	0.82	0.79	0.79	0.79
VB	0.63	0.95	0.94	0.95	0.61	0.88	0.88	0.88
IN	0.79	0.86	0.95	0.89	0.79	0.86	0.95	0.89

other parsers. Over-fitting can be observed best for the prepositions where RBG-2 achieves higher scores than RBG-2(Con-Biased).

6.2.2 Experiment 2

To overcome the negative impact of having all the passive sentences in a separate fold we repeat Experiment 1 using a 10-folds cross-validation. This way we make sure that the training data always contains at least a part of all the five different sentence groups.

In Table 5 we can see that the gap between UAS and LAS scores for the different RBG-2 models vanishes. The impact of over-fitting that appeared in the 5-folds experiment especially for the prepositions disappeared because the model now is trained on data that has a more appropriate coverage of all the sample sentences. As expected the CAS, UAS, and LAS scores improved due to the larger training

Table 5 HAM-DS 10-folds POS results

POS	UAS				LAS			
	MALT	RBG	RBG-2	RBG-2(Biased)	MALT	RBG	RBG-2	RBG-2(Biased)
NN	0.97	0.98	0.99	0.99	0.96	0.974	0.976	0.977
JJ	0.968	0.881	0.895	0.87	0.86	0.81	0.825	0.80
VB	0.981	0.985	0.985	0.985	0.981	0.982	0.982	0.982
IN	0.91	0.867	0.951	0.954	0.9098	0.866	0.95	0.953

set (compared to the size of the test data). Here, the most noticeable result is the big improvement of MALT parser which ranked last in the 5-folds experiments, but now is able to compete with the RBG-2 models. Although MALT parser profited most from the exposure to all the different groups of input sentences during training, adjectives are the only ones that achieved better scores for MALT parser than for the RBG-2 models. This is particularly interesting, because adjectives have been introduced into the input data just to increase the complexity of the sentences, but without any context information which could directly help to correctly attach them (Fig. 12).

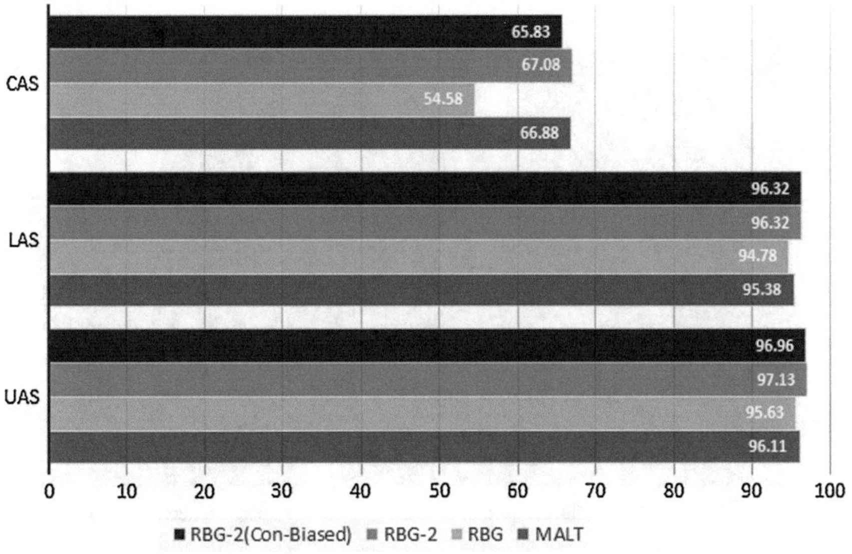

Fig. 12 HAM-DS 10-folds Exp

6.2.3 Experiment 3

As mentioned in Sect. 6.1, we created the ILLIONS-100DS data to also have available natural input sentences for testing the proposed multi-modal parser. The small size of the dataset does not give us the option to apply cross-validation, but fortunately we are able to use the whole dataset for training as well as for testing because of the nature of the RBG learning algorithm. Applying online learning, RBG parser deals with the incoming sentences one by one and adapts the feature weights taking into account only the current sentence without considering all the other already processed training samples. That allows us to use the complete data set both for training and testing purposes, thus at least partly compensating its small size.

The main challenge in this experiment was the diversity of grammatical structures in the naturally created sentences which have been collected by means of crowd-sourcing the annotation task. Additionally, crowd-sourcing also contributed a certain amount of ungrammatical utterances among the collected sentences.

As we can see in Fig. 13 and Table 6 RBG-2 outperforms the other parsers, indicating that the context information once again helps to guide the dependency parsing towards the most plausible interpretation. We see a particular improvement for the prepositions which means that the attachment of prepositional phrases gained most from the visual contribution.

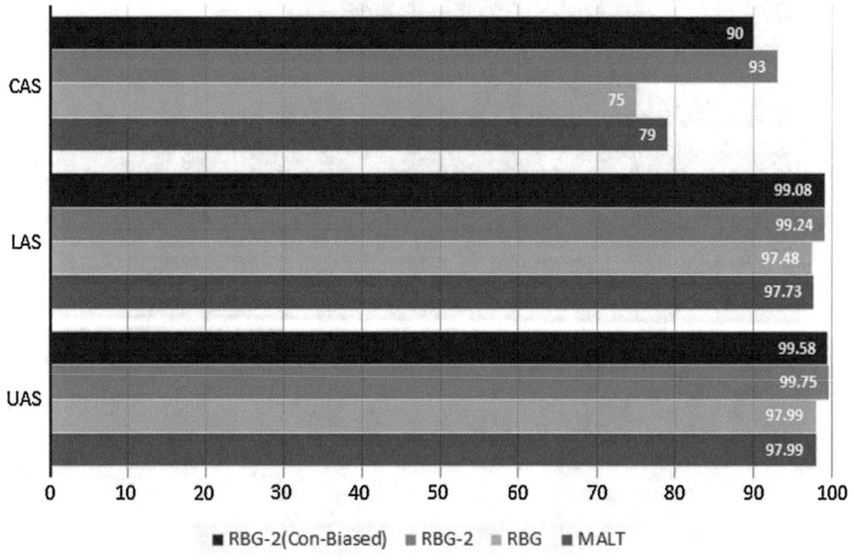

Fig. 13 Illions-100DS Exp

Table 6 Illions-100DS POS results

POS	UAS				LAS			
	MALT	RBG	RBG-2	RBG-2(Biased)	MALT	RBG	RBG-2	RBG-2(Biased)
NN	0.99	0.99	1.0	1.0	0.98	0.994	0.997	0.997
VB	0.958	0.976	0.994	1.0	0.959	0.959	0.976	0.982
IN	0.928	0.908	0.9897	0.9795	0.928	0.903	0.985	0.974

7 Conclusion and Future Work

In our research, we developed an approach for parsing natural language utterances into dependency trees which is able to utilize multi-modal input information. In addition to the input sentences the multi-modal architecture is also supplied with an abstract description of crucial visual relationships about the state-of-affairs in the world which is related to the different lexical items in the sentence. Our approach is based on a state-of-the-art graph-based dependency parser (RBG) which we developed into a multi-modal version (named RBG-2) that accepts input from the additional channel and features an extended learning algorithm capable to deal with this new kind of information.

We successfully verified the effectiveness of having available such an additional channel for the analysis of the linguistic input. Experiments show that RBG-2 learns the correlation between the input sentence and the context information and therefore is able to parse the unseen test sentences more reliably. Although the parser has been developed for the case when information from both channels is available, it is also able to accept language-only input (i.e. sentences without any additional clues through the context channel) and then performs exactly like the original RBG parser does.

We built two small datasets to validate the effectiveness of multi-modal parsing. Through different experiments, we observed a significant improvement of the parsing quality due to the multi-modal support. In some conditions, the proposed approach achieves around 20% improvement in the number of complete sentences that have been parsed correctly. On the level of tokens in some cases the attachment score improves by 10%.

As a line of future research we will apply the approach to a larger dataset. However, the creation of a dataset annotated with the necessary multi-level representations is extremely time-consuming. Therefore, we currently try to find possibilities to benefit as much as possible from existing datasets applying only minor-effort modifications to them in order to be able to verify our methodology on more mature-size datasets. In the long run we aim at extending the approach to a truly bi-directional information flow. This approach does not restrict the visual information to only appear as input features, but to reconstruct a possible world's description if no visual cues are available, in order to be able to effectively interact with a component for active vision.

References

1. Baker, C.F., Charles, J.F., John, B.L.: The berkeley framenet project. In: Proceedings of the 36th Annual Meeting of the Association for Computational Linguistics and 17th International Conference on Computational Linguistics, vol. 1, ACL '98, Association for Computational Linguistics, pp. 86–90, Stroudsburg, PA, USA, (1998). https://doi.org/10.3115/980845.980860
2. Baumgartner, C.: On-Line Cross-Modal Context Integration for Natural Language Parsing. PhD thesis, Universitat Hamburg, (2013)
3. Baumgartner, C., Wolfgang, M.: Integrating a model for visual attention into a system for natural language parsing. In: Proceedings of the 9th International Workshop on Natural Language Processing and Cognitive Science, vol. 1: NLPCS, (ICEIS 2012), pp. 24–33, (2012). ISBN 978-989-8565-16-7. https://doi.org/10.5220/0004088100240033
4. Bernd, B.: Very high accuracy and fast dependency parsing is not a contradiction. In: Proceedings of the 23rd International Conference on Computational Linguistics, COLING '10, pp. 89–97, Stroudsburg, PA, USA, (2010). Association for Computational Linguistics. http://dl.acm.org/citation.cfmid=1873781.1873792
5. Catherine, De M., Christopher, D.M.: Stanford typed dependencies manual, (2008)
6. Chu, Y.J., Liu, T.H.: On the shortest arborescence of a directed graph. Science Sinica, 14, (1965)
7. Crammer, K., Ofer, D., Joseph, K., Shai S-S., Yoram, S.: Online passive-aggressive algorithms. J. Mach. Learn. Res. 7:551–585, (December 2006). ISSN 1532-4435. http://dl.acm.org/citation.cfm?id=1248547.1248566
8. Cliff, G.: Semantic Analysis. Oxford University Press, 2nd edition, (2011). ISBN 978-0-19-956028-8
9. Karen, K., Anna, K., Neville, R., Martha, P.: Extending verbnet with novel verb classes. In: Proceedings of the Fifth International Conference on Language Resources and Evaluation—LREC'06. (http://verbs.colorado.edu/mpalmer/projects/verbnet.html) (2006). http://www.cl.cam.ac.uk/alk23/lrec06.pdf
10. Knoeferle, K.S.: *The role of visual scenes in spoken language comprehension : evidence from eye-tracking*. PhD thesis, Universitat des Saarlandes, Postfach 151141, 66041 Saarbrucken, (2005). http://scidok.sulb.uni-saarland.de/volltexte/2005/438
11. Lei, T., Yuan, Z., Regina, B., Tommi, J.: Low-rank tensors for scoring dependency structures. Association for Computational Linguistics, (2014)
12. McCrae, P.: A model for the cross-modal influence of visual context upon language processing. In: RANLP, pp. 230–235, (2009)
13. Nivre, J.: Incrementality in deterministic dependency parsing. In: Proceedings of the Workshop on Incremental Parsing: Bringing Engineering and Cognition Together, IncrementParsing '04, pp. 50–57, Stroudsburg, PA, USA, (2004). Association for Computational Linguistics. http://dl.acm.org/citation.cfm?id=1613148.1613156
14. Nivre, J., Jens, N.: Memory-based dependency parsing. In Proceedings of CoNLL, (2004)
15. Palmer, M., Daniel, G., Paul, K.: The proposition bank: an annotated corpus of semantic roles. Comput. Linguist. 31(1):71–106, (March 2005). ISSN 0891-2017. https://doi.org/10.1162/0891201053630264
16. Palmer, M., Daniel, G., Nianwen, X.: Semantic Role Labeling. Morgan and Claypool Publishers, 1st edition, (2010). ISBN 1598298313, 9781598298314
17. Noahs ARK research group. Ark Syntactic Semantic Parsing Demo. URL http://demo.ark.cs.cmu.edu/parse. Last accessed 01 Dec 2014
18. Josef, R., Michael, E., Miriam, R.L.P., Christopher R.J., Jan, S.: FrameNet II: extended theory and practice. In: International Computer Science Institute, Berkeley, California, (2006). Distributed with the FrameNet data
19. Schrder, I.: Natural Language Parsing with Graded Constraints. PhD thesis, Universitat Hamburg, (2002)

20. Surdeanu, M., Richard, J., Adam M., Lluís Màrquez, Joakim, N.: The conll-2008 shared task on joint parsing of syntactic and semantic dependencies. In: Proceedings of the Twelfth Conference on Computational Natural Language Learning, CoNLL '08, Association for Computational Linguistics, pp. 159–177, Stroudsburg, PA, USA, (2008). ISBN 978-1-905593-48-4. http://dl.acm.org/citation.cfm?id=1596324.1596352
21. Young, P., Lai, A., Hodosh, M., Hockenmaier, J.: From image descriptions to visual denotations: new similarity metrics for semantic inference over event descriptions. Trans. Assoc. Comput. Linguist. **2**, 67–78 (2014)
22. Yuan, Z., Tao, L., Regina B., Tommi, J.: RBG Parser. URL https://github.com/taolei87/RBGParser. Last accessed 15 Oct 2015
23. Yuan, Z., Tao, L., Regina B., Tommi, J.: New Inference for Dependency Parsing, Greed is Good if Randomized (2014a)
24. Yuan, Z., Tao, L., Regina, B., Tommi, J., Amir, G.: Steps to excellence: simple inference with refined scoring of dependency trees. Assoc. Comput. Linguist. (2014b)

Fast, Accurate, Multilingual Semantic Relatedness Measurement Using Wikipedia Links

Dante Degl'Innocenti, Dario De Nart, M. Helmy and C. Tasso

Abstract In this chapter we present a fast, accurate, and elegant metric to assess semantic relatedness among entities included in an hypertextual corpus building an novel language independent Vector Space Model. Such a technique is based upon the Jaccard similarity coefficient, approximated with the MinHash technique to generate a constant-size vector fingerprint for each entity in the considered corpus. This strategy allows evaluation of pairwise semantic relatedness in constant time, no matter how many entities are included in the data and how dense the internal link structure is. Being semantic relatedness a subtle and somewhat subjective matter, we evaluated our approach by running user tests on a crowdsourcing platform. To achieve a better evaluation we considered two collaboratively built corpora: the English Wikipedia and the Italian Wikipedia, which differ significantly in size, topology, and user base. The evaluation suggests that the proposed technique is able to generate satisfactory results, outperforming commercial baseline systems regardless of the employed data and the cultural differences of the considered test users.

Keywords Semantic networks · Vector space · Text processing theory Multilinguality

Mathematics Subject Classification (2010) Primary 68T30 · Secondary 68T50

D. Degl'Innocenti (✉) · D. De Nart · M. Helmy · C. Tasso
Department of Mathematics and Computer Science, University of Udine, Via delle Scienze 206, Udine, Italy
e-mail: deglinnocenti.dante@spes.uniud.it

D. De Nart
e-mail: dario.denart@uniud.it

M. Helmy
e-mail: alameldien.muhammad@spes.uniud.it

C. Tasso
e-mail: carlo.tasso@uniud.it

K. Shaalan et al. (eds.), *Intelligent Natural Language Processing: Trends and Applications*, Studies in Computational Intelligence 740,
https://doi.org/10.1007/978-3-319-67056-0_27

1 Introduction

Measuring semantic likeness between items such as words, texts, or DBpedia entities is a vital component of several Artificial Intelligence applications, supporting tasks such as question answering, ontology alignment, Word Sense Disambiguation, and exploratory search. The concept of semantic likeness over the years has attracted the interest of the Natural Language Processing (NLP), Semantic Web, and Information Retrieval (IR) communities [8]. Two variants have been thoroughly discussed: *Semantic Similarity* which can be defined as the likeness of the meaning of two items, for instance "king" and "president" though not being synonyms have an high semantic similarity because they share the same function, and *Semantic Relatedness* which can be considered as a looser version of semantic similarity since it takes into account any kind of relationship, for instance "king" is semantically related to "Nation" because a king rules over a nation. Due to the high ambiguity of the very definition of these semantic relationships it is not uncommon to evaluate similarity and relatedness metrics upon their performance in a specific, well-defined and reproducible task [3].

Many metrics have been introduced in the literature, surveyed in Sect. 2, relying mostly upon word distribution or graph traversing over linked data. Such approaches, however present several shortcomings, most notably their evaluation tends to be demanding from a computational point of view, thus preventing their usage in scenarios where a very large number of comparisons must be made.

In this work we tackle the problem of assessing the degree of semantic likeness between entities from an exploratory search point of view, i.e. with the goal of retrieving for a given entity a neighbourhood of other entities which might be relevant from the point of view of an user who wants to learn more about the searched entity. With this task in mind, we focus on assessing semantic relatedness rather than semantic similarity.

We introduce therefore a new strategy to assess semantic relatedness between entities leveraging the link structure of a corpus of hypertextual documents. The employed similarity metric relies on the Jaccard similarity coefficient and exploits the Minhash optimisation to perform dimensionality reduction and therefore allow an efficient relatedness assessment. The presented model is then trained on both English and Italian Wikipedia and benchmarked against Google's and Bing's exploratory search tools which rely primarily on DBpedia and Freebase, allowing the comparison of search results. Our contribution is twofold: we introduce an efficient strategy to evaluate semantic similarity and we assess its performance upon the task of related entity retrieval over two distinct data sets written in different languages.

2 Related Work

A broad range of measures for assessing similarity and relatedness between entities has been proposed in the literature; such measures are grounded into set theory [19], statistics [22], and graph theory [18]. One of the best known semantic relatedness measures is the *Google Distance* [5] which exploits a search engine to estimate pairwise similarity between words or phrases. Such a metric has proven to be effective for a number of knowledge intensive tasks such as evaluating approximate ontology matching [7]. However the implied intensive usage of the underlying search engine makes this metric impractical or too expensive for most applications. Other strategies rely on structured knowledge bases such as taxonomies and ontologies. Wordnet[1] is among the first and still most used resources to estimate semantic similarity with a variety of techniques including graph search algorithms and machine learning. An extensive survey of semantic similarity metrics built upon Wordnet is presented in [3, 4]. The LOD cloud has also been widely exploited and several authors proposed strategies to evaluate similarity and relatedness among entities included in such a cloud. Most LOD-based techniques rely on the selection of a limited number of features among the multitude of properties present in the cloud, to perform this task techniques such as Personalized Page Rank are commonly used in the literature [18]. These techniques, despite being particularly demanding from a computational point of view are often used in the field of semantic-based personalisation [16]. Wikipedia has been used as well to compute semantic relatedness metrics: the authors of [6] introduce Explicit Semantic Analysis (ESA), a technique using machine learning to build vectorial representations of Wikipedia items based upon the textual contents of their corresponding articles. The authors of [23] propose an alternative to ESA which leverages the links included in Wikipedia articles to achieve similar performance but at a sensibly lower cost both in terms of computational complexity and of required data. The similarity metric therein presented is the combination of two metrics, one for incoming links and one for outgoing ones, the former one being closely related to the aforementioned Google Distance.

Vector Space Model (herein VSM) approaches are an alternative to explicit and formal knowledge representations such as the one provided by LOD. In a VSM entities, instead of being described by a set of predicates, are represented as a vector in a space with a finite number of dimensions. VSM leverage the *distributional hypothesis* of linguistics, which claims that words that occur in similar contexts tend to have similar meanings [9]. Some authors [17] in fact define the meaning of a concept as the set of all propositions including that concept. VSMs are commonly used to support several NLP and IR tasks, such as document retrieval, document clustering, document classification, word similarity, word clustering, word sense disambiguation, and many others. The most notable advantage of these techniques over formal representations is that vector spaces can be built in a totally automated and unsupervised way. For a deeper and more exhaustive survey of vector spaces and their usage in state of the art systems, we address the interested reader to [20], [15], and [14].

[1] http://wordnet.princeton.edu.

3 Similarity Assessment and Neighborhood Retrieval

As shown by the authors of [23] hypertextual connections between Web pages alone can carry a great deal of semantics at a reasonable computational cost. However their proposed method involving the combination of two distinct metrics for incoming and outgoing links can be still too demanding when a very large content base must be scouted to find related items. Wikipedia, which includes over 8 million items is a perfect example of such a situation. To overcome this limitations and to set up a minimal theoretical framework, we introduce a new hypothesis: the *Reference Hypothesis*. We assume that entities that are referenced in a similar set of documents might yield strong semantic affinity. For instance, in Fig. 1 two entities (*A* and *B*) are referenced by three different documents: this implies a semantic affinity between A and B.

This assumption is motivated by the fact that intuitively referencing something in a document implies the referenced item to be relevant in the context of the document, therefore entities that get constantly referenced together are relevant in the same contexts, hence they might be semantically related. This hypothesis can be seen as a generalised version of the aforementioned distributional hypothesis, however we would like to stress how even though words can be seen as entities, entities can be intended as way more abstract items, for instance other documents or ontology entries. For instance, the reference hypothesis applies to the scientific literature since articles citing similar sources are very likely to deal with similar topics. Other works in literature embrace this assumption though not formalising it, such as [10] wherein a scientific paper recommender system exploiting co-citation networks is presented. Building a vector space exploiting the Reference Hypothesis is straightforward once a large enough corpus of documents annotated with hyperlinks is provided. Within the corpus, two sets must be identified: the *entity set E* and the *document set D*; the first includes all the referenced entities, while the latter is the considered annotated documents. The vector space is represented with an $E \times D$ matrix that initially is a zero matrix. Iteratively, for each $d \in D$ all the references to elements in E are considered, and for each $e \in E$ referenced in d, the (e, d) cell of the matrix is set to 1. Since referencing a given entity only once in a document is a typical best practice

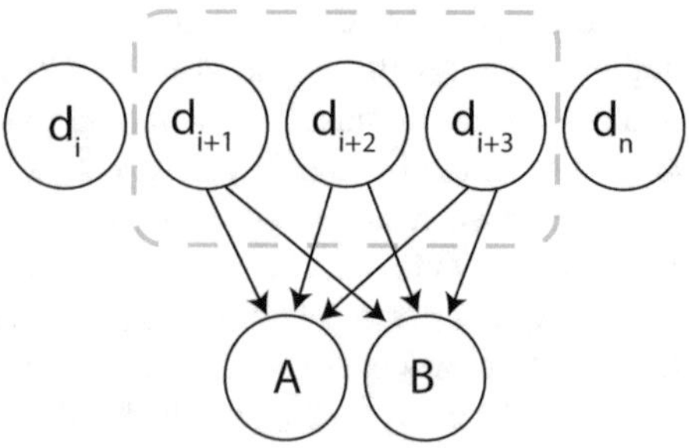

Fig. 1 Two entities referenced by the same set of documents

in several domains[2] we are not considering how many times e is referenced in d. Once all documents are processed we obtain a matrix where each row represents all the references to a given entity: we call such matrix *Reference Matrix* and the vector space it generates *referential space*.

Evaluating the similarity of two entities in such a vector space reduces to computing the distance between their vectors. Countless distance metrics exist in the literature such as norms, cosine similarity, hamming distance, and many others surveyed in [21]. All these metrics can be used in the Reference Matrix, however we prefer the Jaccard similarity coefficient (also known as Jaccard index [11]), defined as:

$$J(A,B) = \frac{|A \cap B|}{|A \cup B|} = \frac{|A \cap B|}{|A| + |B| - |A \cap B|} \quad (3.1)$$

where A and B are sets of items. Since each entity $e_i \in E$ can be considered a binary vector, it can also be expressed as the set that contains all the document $d_j \in D$ such that $(e_i, d_j) = 1$ in the Reference Matrix. The similarity of two equal sets is one, whereas the similarity between two sets that have no elements in common is zero. This choice is motivated by the intimate simplicity of such a metric and by the evidence presented in the literature that the Jaccard index performs better than other methods for finding word similarities in VSM approaches [12, 15].

However, evaluating the Jaccard index is linear in the size of the considered vectors, which can be extremely large when considering large corporas such as Wikipedia. The computation of the Jaccard index can be reduced to constant time using the MinHash optimisation [2]. Such a technique allows to efficiently compute the similarity between sets without explicitly computing their intersection and union. Its most common form consists in using an hash function to map each element of the set to an integer number and then selecting the minimum as a representative of the whole set. The probability that two different sets share the same minimum with respect to the hash function tends to the Jaccard similarity coefficient between the two sets [13]. The more hash functions are used, the closer the estimate gets to the real Jaccard similarity coefficient value. In this work we used 256 distinct hash functions to achieve a fine enough approximation of the Jaccard similarity coefficient. This translates to representing each entity as a 256 positions vector. Such a vector can be considered as an entity's fingerprint in the considered text corpus and implies a significant dimensionality reduction with respect to the initial vectorial space which may count millions of dimensions. This optimisations allows our method to scale up as the number of considered entities grows: being the number of positions of the fingerprint vector constant, checking semantic similarity between two entities will take constant time. With respect to other solutions presented in the literature such as [23] wherein the evaluation of semantic similarity is polynomial with respect to the size of the considered knowledge base, the MinHash optimisation significantly reduces the complexity of such an operation. As a matter of fact, checking which items are

[2]For instance in Wikipedia only the first time an entity is referenced it is annotated with an hyperlink, and in literature bibliographies have no duplicate entries.

the closest ones to a given entity implies checking the target entity against all items present in the knowledge base. With our solution this operation is linear with respect to the knowledge base's size, with other solutions it is quadratic in the best case.

4 Task Based Evaluation

Similarly to [3], we evaluated our system upon a specific application, in this case the retrieval of a set of neighbour entities for exploratory search purposes. Our evaluation activity, due to the intrinsic subjectivity of the very concept of semantic relatedness, was user-based. Two experiments are presented: in the first one we asked users to give an overall ranking to a list of related items, while in the second one we asked users to assess the relatedness of each item included in a given list to a target entity. Such an evaluation was performed over two datasets with different characteristic features and with two substantially different user groups to test the effectiveness of our methodology in different situations, thus preventing data overfitting and cultural biases in the presented conclusions.

4.1 Experimental Setting

Two hyperlinked text corpora were considered: the English Wikipedia and the Italian Wikipedia. The English Wikipedia is a well known and massive collaborative encyclopedia, counting over 8 million articles contributed by users from all around the world. On the other hand, the Italian one is a substantially smaller corpus, counting around 1.3 million articles and curated by users that mostly reside in Italy. We considered these two dataset because they differ significantly in size, in language, and in the user base that generated them.

Using the technique described in Sect. 3, a testbed system, herein named Referential Space Model (RSM), was developed and trained on Wikipedia, associating to each of its items a representative vector. Building on the results of [23] that provides evidence of the importance of both incoming and outgoing links, we also developed an alternative model relaxing the distributional hypothesis and considering outgoing links, i.e. the items mentioned in the article corresponding to a give item. We refer to this second testbed system as *RSM.outnode*. We chose as baseline two of the most popular search engines on the market[3]: *Google* and *Bing*. One of the most prominent features of said search engines is in fact the ability to leverage the LOD cloud to improve search results, more specifically they can retrieve a neighborhood of items closely related to the search query given by the user. To obtain fair and generic search results i.e. not influenced by the recorded browsing history, prefer-

[3]http://www.alexa.com/.

ences, and location, Google and Bing search process was depersonalized to prevent the search engines from customizing the final result.

To assess the quality of our two alternative approaches we constructed a dataset of the top visited Wikipedia pages. As a reliable source of data we used the list of Wikipedia Popular Pages[4] that maintains a set of the most accessed 5000 articles on the English Wikipedia and it is updated weekly. For our data set we focused, for both English and Italian, on the most *stable articles* during the year 2015. We define the stable articles as the Wikipedia pages that constantly appear in every weekly version of that list throughout the year, and so receiving constant interest from the visitors of Wikipedia. A set of 1583 stable items were identified for the English language, and a set of 4361 for the Italian. Four evaluation datasets, two for English, and two for Italian, were built by randomly selecting from each language's stable articles list 100 items (used in experiment 1) and 25 items (used in experiment 2) upon which all of the four systems are able to retrieve related items.

4.2 Overall Relevance Assessment

The goal of our first experiment was to assess which one of the four systems produces the overall best set of related items given one search key. To this extent, we considered datasets of 100 items. The crowdsourcing experiment was designed as follows: for each of the considered items a page was generated including the name of the item, a brief description, a picture, and a box including the results produced by the four systems i.e. four lists of five semantically related items. We decided to show only five results for two reasons: firstly both Bing and Google show at least five related items, which means that for some search queries no more than five items will be shown, secondly it is a known fact that users typically pay attention only to the top spots of search results lists, with the top five items attracting most of the attention.[5] To avoid cognitive bias, the names of the systems were not shown and the presentation order was randomized, so that the worker had no means of identifying the source of the presented item lists and couldn't be biased by personal preference or previous evaluations. The workers were then asked to rate the four item lists according to their perceived quality in terms of relatedness on a discrete scale from 1 to 5 where 1 meant total randomness and 5 that all presented items where perceived as strongly related. Each one of the 100 items in the data set was shown with the same related items lists to 5 distinct users and their judgements were averaged per system to mitigate subjectivity of judgement. The experiment was performed using the popular crowd sourcing platform *Crowdflower*[6] and iterated twice: once for the English

[4]https://en.wikipedia.org/wiki/User:West.andrew.g/Popular_pages.

[5]https://chitika.com/google-positioning-value.

[6]http://www.crowdflower.com/.

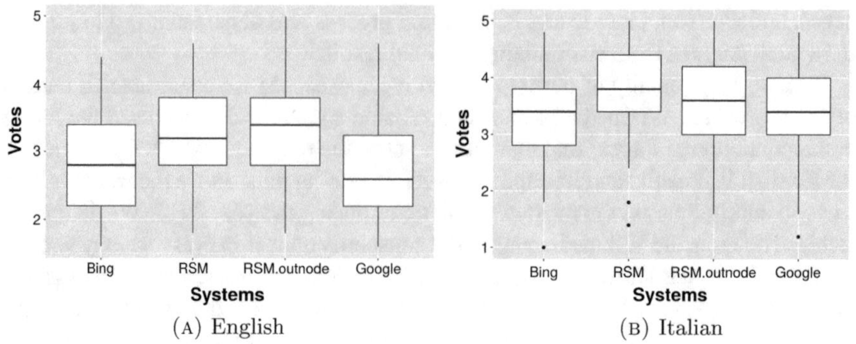

Fig. 2 Experiment 1 distribution of worker's judgement

dataset and one for the Italian one. In the English iteration 32 users from 18 different countries were involved, with an average of 15.62 judgements per user. In the Italian iteration, instead, were involved 59 users from 8 countries, with an average of 8.47 judgements per user). The distribution of the worker's judgement is shown in Fig. 2.

4.3 Item by Item Relevance Assessment

The goal of our second experiment was to assess the perceived quality of each item included in the related items list. To this extent we considered datasets of 25 items. The experimental setup was similar to the previous experiment, using the same platform and displaying the same information about the target entity (i.e. title, description, and picture). Instead of four lists, this time the workers were shown a single list generated by one system only and were asked to rate each item in the list on a scale from 1 to 5 where 1 implied complete unrelatedness and 5 a very high perceived relatedness. The name of the system that generated the list was not shown to avoid bias. A hundred related items lists where therefore generated and human-rated item by item. Again, each item was judged by five distinct users to mitigate subjectivity of judgement. This second experiment was again iterated twice and involved by design substantially more workers to further abstract over subjective experience and thus obtain a more impartial judgement. In the end 146 workers from 38 countries were involved with an average of 3.42 judgements per user in the English experiment and 109 workers from 14 countries with an average of 4.59 judgements in the Italian one. In Fig. 3 the distribution of workers' judgements is shown.

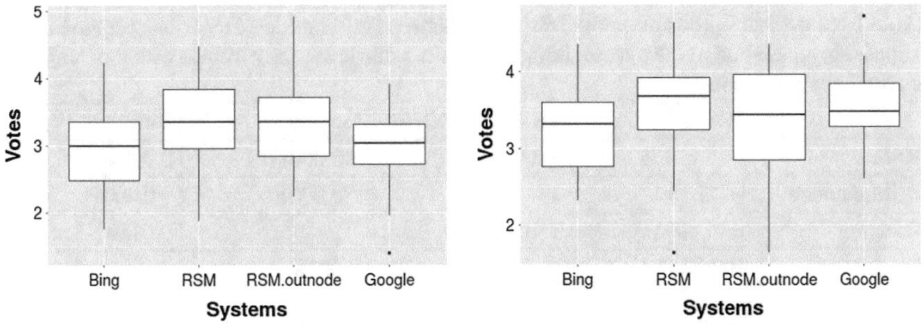

(A) English (B) Italian

Fig. 3 Experiment 2 distribution of worker's judgement

5 Discussion

The data gathered with the experiments described in Sect. 4.1 provide some interesting insights on the effectiveness of the proposed technique.

5.1 Overall List Quality

The results of experiment one showed how our testbed systems RSM and RSM.outnode can achieve satisfactory performance in both the considered scenarios. In the English part of the experiment RSM and RSM.outnode achieved, on a scale from 1 to 5, respectively a 3.20 and 3.33 average perceived quality, while Google and Bing respectively 2.79 and 2.82. The statistical significance of the judgement distributions shown in Fig. 2 was evaluated as well showing how while there is a substantial difference between the perceived quality of our systems and the baseline ones (Bing and Google), between RSM and RSM.outnode there is no statistically significant difference. More specifically the Welch Two Sample t-test was used and produced the results shown in Table 1, where in the upper right half of the matrix are shown the p-values produced by the test, and in the lower left half the same values recalculated with the Benjamini & Hochberg correction for multiple hypotesis testing [1]. According to these results, Google's and Bing's related items lists are perceived almost as identical in terms of quality, while our testbed systems' outputs receive a significantly higher likely by the crowdsourced workers. Moreover, while RSM.outnode appears to achieve an higher perceived quality than RSM on average, the statistical significance analysis shows that such a difference is unlikely to be significant in the current experimental setting. In terms of overall perceived quality the neighbourhoods of related items to a given search key produced by RSM and RSM.outnode do not differ significantly in terms of perceived quality, but there is evidence that consistently outperform the benchmark systems offered by Google and Bing.

Table 1 Statistical significance of the difference between the considered systems over the English corpus. The upper half of the matrix shows the p-values, the lower the p-values with the Benjamini & Hochberg correction

	RSM	RSM.outnode	Google	Bing
RSM	–	0.1896	<0.0001	0.0001
RSM.outnode	0.2275	–	<0.0001	<0.0001
Google	<0.0001	<0.0001	–	0.6838
Bing	0.0003	<0.0001	0.6838	–

Table 2 Statistical significance of the difference between the considered systems over the Italian corpus. The upper half of the matrix shows the p-values, the lower the p-values with the Benjamini & Hochberg correction

	RSM	RSM.outnode	Google	Bing
RSM	–	0.0079	0.0013	<0.0001
RSM.outnode	0.0158	–	0.6835	0.0141
Google	0.0039	0.6835	–	0.0308
Bing	<0.0001	0.0125	0.0369	–

The Italian part of the experiment a similar outcome was observed, with two notable differences: expressed scores were substantially higher for all systems and in particular results produced by Google received a generally more favourable reception with respect to the English part of the experiment. While the former outcome may be ascribed to cultural factors, since the whole judgement distribution is skewed towards higher scores, the latter suggests that the localised versions of Google and Bing may differ in the used data or retrieval technique. As a matter of fact, the English Bing and Google received very similar judgements, see Table 1, and the provenance of the related items lists was unknown to workers to avoid confirmation bias, thus the significant difference sported in the Italian experiment, shown in Table 2, implies substantial differences between the English and the Italian versions of the two search engines. On the other hand, the RSM model appears the one producing the best received related items lists, while RSM.outnode and Google present no statistically significant difference. The statistically significant difference between the perceived quality of the lists generated by RSM and RSM.outnode in this setting can be ascribed to substantial reduction in the size of the training data. Overall, RSM is perceived as the best system, RSM.outnode and Google are on par, and Bing is perceived as the worst one.

5.2 *Information Gain Analysis*

The results of experiment two support the evidence provided by the previous one. In the English part of the experiment, items retrieved by RSM and RSM.outnode on average score a 3.41 out of 5 on perceived quality while Bing and Google stop at 2.93 out of 5. In the Italian part of the experiment, instead, items retrieved by RSM score an average of 3.6 out of 5, RSM.outnode and Google are tied around 3.5, and Bing scores around 3.4 on average. These numbers, however, provide little information being average values of perceived quality of item ranked in different positions. Looking at the whole distribution of judgements shown in Fig. 3, the high variance of the four distributions can be easily noticed. Such a variance can be justified by the fact that all items included in the generated lists are considered and rated. However, not all positions of a result list are equal to the extents of exploratory search. To address this issue we evaluated the Normalized Discounted Cumulative Gain (NDCG) of the four considered systems. NDCG is a metric commonly used in IR to assess a search engine's performance basing on the comparison between an ideal list of the most relevant retrievable items and the actual list produced by the evaluated system. Its core idea is that the higher the position of an item in the result list the more important the quality of that item should be in the quality evaluation of the system, therefore the presence of scarcely relevant items in the top spots tends to "punish" the evaluated system. The ideal list was computed by considering, for both parts of the experiment, for each of the 25 search keys, all the items retrieved by the four systems, picking the five ones that on average received the highest user ratings and ordering them in descending average rating order. The distribution of the NDCG values scored by the four considered systems over the search queries included in the data sets is shown in Fig. 4 and its detailed statistics are presented in Tables 3 and 4. These results support the evidence brought by the first experiment as well, with RSM and RSM.outnode providing consistently results perceived as more relevant than the ones brought by Google's and Bing's tools in the English part of the experiment. Again, there is no statistically significant difference in the average perceived quality between RSM and RSM.outnode (p-value = 0.68) and between Google and Bing as well (p-value = 0.88). On the other hand, the statistical significance between RSM and Google, RSM and Bing, RSM.outnode and Google, and RSM.outnode and Bing is high with p-values below 0.0001. Finally, the NDCG analysis shows how, despite scoring being on average on par with its RSM.outnode counterpart, the RSM system has the smallest variance in the perceived relevance of its results, implying that it is less likely to produce results perceived as poor on a single-try basis. In the Italian part of the experiment, instead, RSM achieves substantially higher nDCG scores than its RSM.outnode counterpart, which, again, presents a vary large nDCG score distribution and, on average, performs slightly worse than Google's related items search, though its median nDCG value is higher than Google's. Like in the previous experiment, the RSM model appears to be able to cope better with changes in training data.

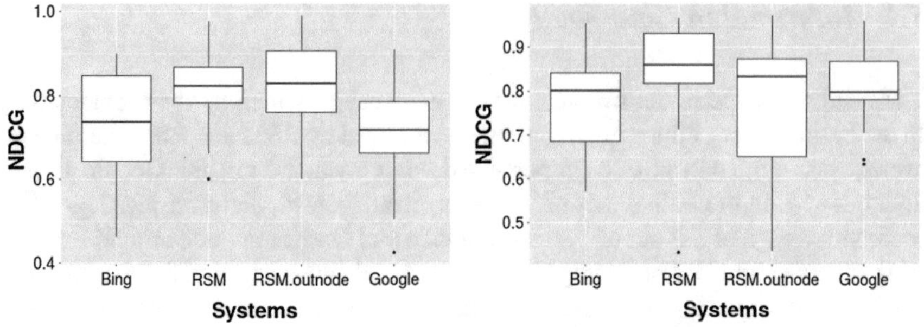

(A) English (B) Italian

Fig. 4 NDCG values distribution evaluated on the results of experiment 2

Table 3 Distribution statistics on NDCG evaluation—English

	Bing	RSM	RSM.outnode	Google
Minimum	0.4629	0.6009	0.6006	0.4250
1st Quartile	0.6423	0.7829	0.7601	0.6631
Median	0.7376	0.8232	0.8293	0.7186
Mean	0.7247	0.8113	0.8226	0.7196
3rd Quartile	0.8475	0.8678	0.9066	0.7855
Maximum	0.9010	0.9771	0.9910	0.9102

Table 4 Distribution statistics on NDCG evaluation—Italian

	Bing	RSM	RSM.outnode	Google
Minimum	0.5714	0.4319	0.4352	0.6329
1st Quartile	0.6859	0.8177	0.6511	0.7804
Median	0.8015	0.8602	0.8338	0.7980
Mean	0.7793	0.8493	0.7664	0.8121
3rd Quartile	0.8418	0.9313	0.8733	0.8677
Maximum	0.9546	0.9664	0.9726	0.9598

Finally, it is important to stress how the MinHash optimisation allowed us to move the complexity of a pairwise similarity measurement from linear to constant. This means that without the said optimisation it would be computationally demanding to retrieve items semantically related to one with a lot of connections. Consider for instance the Wikipedia article about Barack Obama which, at the time this article being written, contained over 250 links was referenced over 9900 times by other Wikipedia articles: without MinHash it takes over 300 s on our test machine[7] to generate a list of semantically related items, while with that optimisation it takes less

[7]An Intel I7 with eight cores and 32 GB RAM.

than a second on the same machine. Moreover, the constant complexity of MinHash allows it to seamlessly scale up to larger knowledge bases. While our approach allows this optimisation to be made retaining quality results, other metrics, such as the ones presented in [6, 23], do not.

6 Conclusions

Our evaluation provided concrete evidence that our approach is able to achieve results consistently perceived as satisfactory by the crowdsourced workers. In particular, the referential model built upon considering references rather than outgoing links appears to be more robust as the training data and the users change. The referential hypothesis thus allowed us to build a sound VSM that captures semantic relatedness among the considered items, and the usage of the Jaccard index and the MinHash optimisation allowed us to handle the over 8 million items included in Wikipedia with ease. Providing semantic tools that can efficiently scale up to the large volumes of data involved in nowadays information access applications such as personalised information retrieval and personalised recommendation, is in our opinion a critical step towards fully accomplishing the potential of the Web of data. It is well known that the graph nature of the LOD cloud implies high computational costs when exploration and reasoning tasks must be performed and many current state of the art algorithms involve extensive graph traversing. For instance, it took over 2,400 min to the authors of [16] to train a state of the art semantics-based recommender system on a data set such as *DBbook*[8] which is relatively small when compared to real-world scenarios that may include millions of items and users. Though training is typically a batch-time operation, an excessive complexity may discourage its field usage since in Adaptive Personalisation applications it typically needs to be frequently repeated because it is likely to users to regularly update their preferences, new items to be included and new users to register.

References

1. Benjamini, Y., Hochberg, Y.: Controlling the false discovery rate: a practical and powerful approach to multiple testing. J. Royal Statist. Soc. Ser. B Methodol. **57**(1), 289–300 (1995)
2. Broder, A.Z.: On the resemblance and containment of documents. In: Proceedings of Compression and Complexity of Sequences (SEQUENCES'97), pp. 21–29. IEEE, June 1997
3. Alexander, B., Graeme, H.: Semantic distance in wordnet: an experimental, application-oriented evaluation of five measures. In: Workshop on WordNet and Other Lexical Resources, Second Meeting of the North American Chapter of the Association for Computational Linguistics (NAACL), pp. 29–34 (2001)
4. Budanitsky, A., Hirst, G.: Evaluating wordnet-based measures of lexical semantic relatedness. Comput. Linguist. **32**(1), 13–47 (2006)

[8]http://challenges.2014.eswc-conferences.org/index.php/RecSys#DATASET.

5. Rudi, L.C., Paul, M.B.V.: The google similarity distance. IEEE Trans. Knowled. Data Eng. **19**(3), 370–383 (2007)
6. Gabrilovich, E., Markovitch, S.: Computing semantic relatedness using wikipedia-based explicit semantic analysis. IJcAI **7**, 1606–1611 (2007)
7. Risto, G., Warner ten K., Zharko, A., Frank Van H.: Using google distance to weight approximate ontology matches. In: The 16th International Conference on World Wide Web, pp. 767–776. ACM, (2007)
8. Sebastien, H., Sylvie, R., Stefan, J., Jacky, M.: Semantic similarity from natural language and ontology analysis. Synth. Lect. Human Lang. Technol. **8**(1), 1–254 (2015)
9. Harris, Z.: Distributional structure. Word **10**(23), 146–162 (1954)
10. Tin, H., Kiem, H., Loc, Do, Huong, T., Hiep, L., Susan, G.: Scientific publication recommendations based on collaborative citation networks. In: Collaboration Technologies and Systems (CTS), 2012 International Conference on, pp. 316–321. IEEE, (2012)
11. Jaccard, P.: Lois de distribution florale. Bulletin de la Socíeté Vaudoise des Sciences Naturelles **38**, 67–130 (1902)
12. Lillian, L.: Measures of distributional similarity. In: Proceedings of the 37th Annual Meeting of the Association for Computational Linguistics on Computational Linguistics (ACL), (199)
13. Leskovec, J., Rajaraman, A., Ullman, J.D.: Mining of Massive Datasets. Cambridge University Press, (2014)
14. Levy, O., Goldberg, Y., Dagan, I.: Improving distributional similarity with lessons learned from word embeddings. Trans. Assoc. Comput. Linguist. **3**, 211–225 (2015)
15. Christopher, D.M., Prabhakar, R., Hinrich, S.: Introduction to Information Retrieval. Cambridge University Press, New York, NY, USA (2008)
16. Cataldo, M., Pasquale, L., Pierpaolo, B., Marco de G., Giovanni, S.: Semantics-aware graph-based recommender systems exploiting linked open data. In: Proceedings of the 2016 Conference on User Modeling Adaptation and Personalization, pp. 229–237. ACM, (2016)
17. Novak, J.D.: Learning, Creating, and Using Knowledge: Concept Maps as Facilitative Tools in Schools and Corporations. Taylor & Francis, London, United Kingdom (2010)
18. Mohammad, T.P. Roberto, N.: From senses to texts: an all-in-one graph-based approach for measuring semantic similarity. Artific. Intell. **228**, 95–128 (2015)
19. Rodríguez, M.A., Egenhofer, M.J.: Determining semantic similarity among entity classes from different ontologies. IEEE Trans. Knowled. Data Eng. **15**(2), 442–456 (2003)
20. Turney, Peter D.: Pantel, Patrick: from frequency to meaning: vector space models of semantics. J. Artif. Int. Res. **37**(1), 141–188 (2010)
21. Jingdong, W., Heng, T.S., Jingkuan, S., Jianqiu, J.: Hashing for similarity search: a survey. arXiv:1408.2927, (2014)
22. Weeds, Julie: Weir, D.: Co-occurrence retrieval: a flexible framework for lexical distributional similarity. Comput. Linguist. **31**(4), 439–475 (2005)
23. Ian, W., David, M.: An effective, low-cost measure of semantic relatedness obtained from wikipedia links. In: Proceeding of AAAI Workshop on Wikipedia and Artificial Intelligence: An Evolving Synergy, AAAI Press, Chicago, USA, pp. 25–30(2008)

A Survey and Comparative Study of Arabic NLP Architectures

Younes Jaafar and Karim Bouzoubaa

Abstract Arabic Natural Language Processing (ANLP) has known a significant progress during the last years. As a result, several ANLP tools and applications have been developed such as tokenizers, Part Of Speech taggers, morphological analyzers, syntactic parsers, etc. However, most of these tools are heterogeneous and can hardly be reused in the context of other projects without modifying their source code. This problem is known to be common to all languages, that is why some advanced NLP language independent architectures have emerged such as GATE (Cunningham et al. ACL, 2002) [1] and UIMA (Apache UIMA Manuals and Guides, 2015) [2]. These architectures have significantly changed the way NLP applications are designed and developed. They provide homogenous structures for applications, better reusability and faster deployment. In this article, we present a comparative study of NLP architectures in order to specify which ones can suitably deal with Arabic language and its specificities.

Keywords Arabic NLP · Architectures · Interoperability

1 Introduction

Arabic is the largest living Semitic language in terms of number of speakers. It is the native language of over 300 million people according to the British council.[1] According to a study conducted by the Internet World Stats in June 2015 [3], Arabic is fourth among the top ten languages of Internet users. The recent rise in the number of Arab internet users has generated a rapid increase in Arabic digital

[1]https://www.britishcouncil.org/voices-magazine/surprising-facts-about-arabic-language.

Y. Jaafar (✉) · K. Bouzoubaa
Mohammadia School of Engineers, Mohammed Vth University in Rabat, Rabat, Morocco
e-mail: younes.jfr@gmail.com

K. Bouzoubaa
e-mail: karim.bouzoubaa@emi.ac.ma

K. Shaalan et al. (eds.), *Intelligent Natural Language Processing: Trends and Applications*, Studies in Computational Intelligence 740,
https://doi.org/10.1007/978-3-319-67056-0_28

content, which encourages researchers in Natural Language Processing (NLP) to intensify research on Arabic NLP (ANLP) and develop appropriate tools.

Indeed, ANLP researchers have produced various basic tools to process the digital Arabic content such as tokenizers, sentence splitters, light stemmers, etc. These are small tools that process data in order to prepare it for high level processing stages. They are frequently used and represent an essential step for almost any NLP application. However, researchers find themselves continuously developing these same basic tools from scratch for their projects since, in most of the time, they cannot be reused easily in other contexts due to their heterogeneity. The ANLP community has also produced several high level ANLP tools and applications such as search engines, question/answering systems, etc. Similarly to basic tools, these applications are not always encapsulated in homogeneous and interoperable entities that can be used in one single project. Researchers are usually brought to adapt outputs of tools in order to exploit them in other processes. This complexity is due to the diversity of architectures and development languages used by each tool and also the lack of models and standards that govern their implementations. Indeed, most of researchers do not take the basic software engineering (SE) principles into account while developing their tools. They focus mainly on results of tools and neglect the internal/external modelization and interoperability issues. Hence, launching complex processes such as pipelines is very difficult without changing the source code each time to suit a specific need of a researcher.

It should be noted that the above problems are common not only to Arabic but to all languages. To deal with such challenges, some advanced language independent architectures for NLP have emerged such as GATE [1] and UIMA [2]. They have generic structures that can handle basic tools and high level applications of various languages. The philosophy behind these architectures is to gather and develop several NLP tools within a single and homogeneous structure that is flexible, extensible and modular. These architectures save time for researchers since they contain several tools to use and reuse without having to develop them from scratch every time. Moreover, they facilitate the development of high level applications such as question/answering, opinion mining, sentiment analysis, etc. since all tools and applications are homogenous regarding their inputs/outputs.

Therefore, using such architectures by the ANLP community surely helps in the standardization of the various tools and applications and overcome the previous mentioned problems for Arabic. Indeed, some Arabic NLP researchers have exploited language independent architectures such as GATE [1], while others have devoted their efforts in developing software architectures specifically for the Arabic language such as AraNLP [4] or MADAMIRA [5]. However, given the large number of available architectures (as it is detailed in Sect. 3), ANLP researchers can be confused when selecting the appropriate one according to their needs. To help them making the choice, comparative studies should be conducted in order to present specificities, advantages and drawbacks of each architecture. To our knowledge, no such comparative studies have been addressed so far for the context of Arabic. Thus, our objective in this article is to present a survey and comparative study of the most known NLP architectures in order to specify which ones can suitably deal with the Arabic language and its specificities. To make such comparison, it is necessary however to set a reference consisting of a set of characteristics we consider to be ideal of an Arabic NLP architecture:

- Diversity of Arabic tools, to provide as much as possible pre-built and ready to use tools;
- Diversity of Arabic resources, to provide large range of useful resources such as lexicon and corpora.
- Flexibility of exploitation, to better use results and call the desired modules inside and outside the architecture, such as variety of input/outputs, to be able to program pipelines, to exploit using web services, etc.;
- Availability of maintenance and support, to know if the architecture is well supported, continually up to date, and proposing new modules and extensions.

The rest of this paper is organized as follows. The next section presents related works concerning the comparison and the benchmark of Arabic NLP architectures. In Sect. 3 we present a set of existing NLP architectures able to deal with Arabic. In Sect. 4 we present a benchmark of these architectures according to the previous four mentioned characteristics. In Sect. 5 we present a discussion of the obtained results of our benchmark. Finally, we present the conclusion.

2 Related Works

Many surveys have been addressed for tools in the context of Arabic language such as [6]. However, we have not found surveys or comparative studies concerning NLP architectures for the specific context of Arabic. That is why we present below some related works dealing with NLP architectures and Software Engineering from a general perspective.

Prasanth and Nakul [7] presented how Software Engineering and NLP can be combined and integrated to increase the chances of universal programmability. Authors presented also how Software Engineering and NLP can be seen in context of each other.

Leidner [8] presented in his article some issues in Software Engineering for Natural Language Processing. He describes some factors that add complexity to the task of engineering reusable NLP systems such as: accuracy, efficiency, productivity, flexibility, scalability, etc. He focuses then on the reuse of components and properties that lead to reuse. He concluded that NLP frameworks are a valuable asset and that researchers should strive toward development of component APIs rather than prototypes.

The international conference on Language Resources and Evaluation (LREC) organized in its 2010 edition[2] a workshop dedicated specifically to NLP

[2]http://www.lrec-conf.org/lrec2010/.

frameworks, in order to present works and exchange successful patterns of NLP engineering. However, there were no works for benchmarking and comparing existing frameworks. Researchers presented mainly the integration of their projects within existing frameworks or they presented new domain-dependent environments of development.

To the best of our knowledge, the book chapter entitled "Combining Natural Language Processing Engines" [9] is the most comprehensive study addressing the benchmark of NLP architectures. Authors address the issue of combining language processing engines such as speech-to-text or translation. These engines are attaining sufficient accuracy to combine and include them in more advanced processing tasks. According to authors, it is better to implement existing engines as a sequence of components, which will facilitate their development and avoid the complexity of debugging larger applications. The heterogeneity of input/output formats from one engine to another makes the combination more complex. However, this can be overcome by creating an aggregate processor which automatically reformats inputs/outputs as needed by each component. Authors address then the issues of the aggregation which can be summarized in four elements: heterogeneous computing environments, remote operations, data formats and exception handling. They listed also some desired attributes for architectures for aggregating speech and natural langue processing engines. They categorize them in four areas: modular components, computational efficiency, data management, and robustness. Then, they give examples of existing architectures that support the above attributes: UIMA, GATE and InphoSphere.

3 NLP Architectures

In this section we present some known and commonly used NLP architectures that support partially/entirely Arabic, namely: GATE [1], UIMA [2], LIMA [10], LingPipe [11], OpenNLP [12], NLTK [13], NooJ [14], ATKS [15], AraNLP [4], MADAMIRA [16] and SAFAR [17]. Since we want to focus mainly on Arabic language, we break these architectures into two main categories: (1) Independent language NLP architectures that support many NLP tasks and handle many languages including Arabic, (2) ANLP architectures that are specifically designed for processing Arabic.

3.1 Independent Language NLP Architectures

In this subsection we give general presentations of some independent language NLP architectures without focusing on Arabic language. These general presentations aim to introduce each architecture and get a clear idea about its structure and functionalities. This way, we will focus on benchmarking these architectures regarding to their support of Arabic in Sect. 4.

UIMA,[3] Unstructured Information Management Architecture, is a software architecture for the analysis of unstructured data using a variety of analysis technologies, such as rule-based language processing, machine learning, information retrieval, and ontology [18]. UIMA originated at IBM Research,[4] but is now an open source project. There are two implementations of the UIMA framework written in Java and C++ and hosted by the Apache Software Foundation [19]. UIMA accepts input in different forms, including documents, audio files, and video streams.

The UIMA framework provides a run-time environment that allows developers to plug in their UIMA component implementations and to build analytic applications. It also provides tools to create new text analysis algorithms or to enable existing algorithms from multiple projects to operate within the framework. Tools within UIMA are treated as pluggable, composable objects that house the core analysis algorithms for analyzing documents and recording analysis results. A simple or primitive tool contains a single annotator (i.e., analysis algorithm that analyzes documents and outputs annotations) while complex tools may contain a large number of annotators and/or other tools organized in a pipeline. For example, a tool that conducts named entity detection may include a pipeline of annotators starting with language detection feeding tokenization, then part-of-speech tagging, then deep grammatical parsing and finally, named-entity recognition [2].

A UIMA pipeline is made through a three stage cascade consisting of (1) a Collection Reader that imports the data to process and turns it into a Common Analysis Structure (CAS), (2) a sequence of Tools that are composed to automatically analyze a document and add meta-data, often in the form of text annotations, to the CAS, and (3) a CAS Consumer that formats the final CAS data and exports it to output.

GATE,[5] General Architecture for Text Engineering, is a free infrastructure for developing applications that process human language. It is nearly 20 years old, having been launched in 1996 at the University of Sheffield. GATE architecture relies on components, reusable chunks of software with well-defined interfaces that may be used in a variety of contexts. These components can be one of three types: (1) Language Resources (LRs) that represent entities, such as gazetteers, corpora and ontologies; (2) Processing Resources (PRs) that represent algorithmic entities, such as parsers, POS tagger and ngram modelers; and (3) Visual Resources (VRs) that represent visualization and editing components in GUIs. In the GATE framework, components may be implemented by different programming languages and databases, but in each case they are represented to the system as a Java class. So, GATE utilizes Java class conventions to construct and configure resources at runtime, and defines interfaces that different component types must implement [1, 20].

---

[3]https://uima.apache.org/.

[4]http://www-01.ibm.com/software/ecm/content-analytics/uima.html.

[5]https://gate.ac.uk/.

The GATE family has grown over the last 20 years to include an integrated development environment for language processing components (GATE Developer), a pipeline-based web application (GATE Teamware), an object library for inclusion in diverse applications (GATE Embedded), a cloud computing solution for hosted large-scale text processing (GATE Cloud), and a multi-paradigm search repository (GATE Mimir). The GATE Developer implements a very widely used information extraction system called ANNIE (A Nearly-New Information Extraction System), which is a set of modules comprised of a tokenizer, a gazetteer, a sentence splitter, a part of speech tagger, a named entities transducer, and a coreference tagger. GATE also incorporates a large number of plugins in order to integrate and leverage other projects and tools. GATE currently deals with the following languages: Arabic, Bulgarian, Cebuano, Chinese, English, French, German, Hindi, Italian, Romanian, Russian and Spanish [20, 21].

GATE, by default, accepts input documents in different formats, including plain text, HTML, SGML, XML, RTF, RDF, UIMA CAS XML, and CoNLL format. Annotations resulting from processing text can be saved in the XML format in one of two ways: (1) The original format can be preserved and selected annotations added, or (2) All of the data in a GATE document can be encoded in GATE's own XML serialization format.

OpenNLP,[6] is a Java machine learning toolkit for the processing of natural language text. The main goal of the OpenNLP project is to create a mature toolkit for the most common NLP tasks: tokenization, sentence detection, part-of-speech (POS) tagging, named entity recognition, parsing, chunking, and coreference resolution. An additional goal is to provide a large number of pre-built model files for the aforementioned tasks. Danish, German, English, Spanish, Dutch, and Portuguese are the languages that currently have pre-trained models for OpenNLP [22]. OpenNLP includes maximum entropy and perceptron based machine learning algorithms. It also contains several components in order to build a full natural language processing pipeline. Each component (e.g., sentence detector, tokenizer, POS tagger) contains parts that enable users to execute the relevant natural language processing task, to train a model, and also to evaluate a model. These components are easily integrated into a Java application using OpenNLP's API. A command line interface is also available for experimenting and training [23]. OpenNLP tools can be used by themselves or as plugins integrated into other NLP projects, such as Apache UIMA, GATE, Apache Stanbol, Apache cTAKES, and KNIME. The output of OpenNLP tools, when they are used by themselves, is in a simple text format. When the OpenNLP tools are used as plugins in other projects, the output corresponds with the format of these projects [19].

NooJ,[7] developed by Max Silberztein, is a linguistic development environment that processes texts and corpora at the orthographical, lexical, morphological, syntactic and semantic levels. It can process texts written in over 20 languages,

[6]https://opennlp.apache.org/.

[7]http://www.nooj-association.org/.

including some Roman, Germanic, Slavic, Semitic (such as Arabic) and Asian languages, as well as Hungarian. NooJ has been in development for many years and has a solid community of computational linguists. It has recently been turned open source and cross-platform by producing a Java and a MONO version.

At various stages of the analyses, NooJ constructs a Text Annotation Structure (TAS) in which linguistic units are represented by annotations. In fact, an annotation corresponds to several linguistic information for each text unit. NooJ's annotations can represent prefixes, suffixes, simple words, multi-words units, frozen expressions as well as named entities [24].

NooJ's architecture consists of three modules:

(1) Corpus handling module, which consists in processing, indexing, annotating and querying of corpora. This module can also handle pre-annotated corpora.
(2) Lexical module, a robust dictionary module that is designed to handle simple words, affixes and multi-word units. Dictionary entries consist of Part-Of-Speech (POS) and any number of typed features representing the morphological, syntactic or semantic characteristics of the lemma.
(3) Syntactic module, an interesting feature that helps users build grammars in graph and apply them to corpora as a query [25]. All these three modules are integrated into one single intuitive graphical user interface, with the possibility of using also command lines.

NooJ allows users to create, edit, debug and maintain a large number of grammars that belong to the four classes of generative grammars in the Chomsky-Schützenberger hierarchy: finite-state grammars, context-free grammars, context-sensitive grammars and unrestricted grammars. Moreover, NooJ's dictionaries and grammars are simple objects to build; there is no complex formalism to learn. They are represented by finite-state transducers and can represent simple words, compound words, phrasal verbs, as well as support verb/predicative noun associations. NooJ users can also develop extractors to identify semantic units such as names of persons, locations, dates, etc.

It should be noted that NooJ recognizes over 100+ text file formats, including MS-WORD, HTML, XML, etc. Hence, input texts units can be either simple paragraphs or XML nodes. In addition, all NooJ objects are XML/SOAP compliant.

NLTK,[8] Natural Language ToolKit, is a platform for natural language processing and text analytics. The NLTK has originally been designed to support teaching in NLP and closely related areas, including cognitive science, information retrieval, and machine learning [26]. However, the NLTK has not only been used successfully for teaching, but also for prototyping and building Python programs to work with natural language data. It provides researchers and developers access to easy-to-use interfaces of over 50 corpora and lexical resources, such as WordNet. In addition, the NLTK provides wrappers for industrial-strength NLP libraries and a suite of text processing libraries for tokenization, POS tagging, parsing,

[8]http://www.nltk.org.

classification, stemming, etc. With the help of some Python libraries, the NLTK can access and extract texts from electronic books (e.g., books from Project Gutenberg[9]), html web pages, blog contents, PDF, MSWord and other binary formats, as well as local text files. It also supports different languages and different character sets. The NLTK toolkit is available for Windows, Mac OS X, and Linux [27, 28].

LingPipe,[10] is a Natural Language Processing tool kit developed in Java by Alias-I[11] company. It is designed to be effective, extensible, reusable and robust. LingPipe performs tasks such as tokenization, Part-Of-Speech tagging, named entity recognition, clustering, database text mining etc.

It includes a Java API with source code and unit tests, multilingual models, N-best outputs with statistical estimations, etc. However, compared to GATE and UIMA, LingPipe provides very few features and options.

LingPipe supports Unicode and can process texts in any language with a range of character sets and input formats, including HTML, XML and plain text. Its output is in XML format.

LIMA,[12] a multilingual framework for linguistic analysis and linguistic resources development and evaluation [10], is an NLP framework developed by the LVIC laboratory of CEA LIST.[13] LIMA can handle different languages such as French, English, German and Arabic. Authors claim that LIMA has a flexible architecture that is designed to develop NLP applications and integrate various processes and resources. Moreover, it can be used also to test and evaluate linguistic modules and to produce new linguistic resources. LIMA has been developed in order to be extensible and efficient covering a large diversity of applications.

3.2 Arabic NLP Architectures

The previous subsection presented some language independent NLP architectures. In this subsection, we introduce Arabic NLP architectures that are intended to process only Arabic. We present their internal structures, components and functionalities. It should be noted that we have found only few number of such architectures for Arabic.

ATKS,[14] Arabic Toolkit Service, is a set of NLP components targeting Arabic language that can be exploited as web services. It was developed by Microsoft within the Advanced Technology Lab in Cairo and consists of eight ANLP components: a morphological analyzer (SARF), a spell-checker, an auto corrector, a

---

[9]https://www.gutenberg.org/.

[10]http://www.alias-i.com/lingpipe/.

[11]http://alias-i.com/.

[12]https://github.com/aymara/lima.

[13]http://www.kalisteo.eu/en/index.htm.

[14]https://www.microsoft.com/en-us/research/project/arabic-toolkit-service-atks/.

diacritizer, a named entity recognizer (NER), a colloquial to Arabic converter, a parser and a part-of-speech (POS) tagger. It should be noted also that these components are integrated into several Microsoft products and services such as Windows, Office and Bing.

MADAMIRA,[15], is a toolkit for morphological analysis and disambiguation of Arabic and its dialects [5], written in Java. MADAMIRA combines two previously used systems for Arabic processing, MADA [29] and AMIRA [30]. This toolkit provides seven ANLP processing tasks: Tokenization, morphological disambiguation for full range of morphological features, Part-of-Speech tagging, lemmatization, diacritization, named entity recognition and base phrase chunking. MADAMIRA uses machine-learning algorithms to select the linguistic features of a word according to its context. It should be noted that MADAMIRA is considered in our point of view as a fixed pipeline rather than a flexible toolkit that separate components from each other and give at the same time the possibility to create other types of pipelines. Figure 1 describes the pipeline of MADAMIRA to generate an analysis.

AraNLP,[16] is a Java-based architecture that contains several Arabic text basic tools [4]. AraNLP attempts to bring together most of the vital Arabic text pre-processing tools into one single structure that can be accessed easily by end users. So far, AraNLP includes a sentence detector, tokenizer, light stemmer, root-based stemmer, part-of-speech tagger (POS-tagger), word segmenter, normalizer, and a punctuation and diacritic remover. These tools are developed within AraNLP as Java classes which may integrate the existing tools or develop the whole tool from scratch. The library tools are compiled by incorporating existing tools and by developing new ones when required. All AraNLP tools outputs consist of simple texts that can be saved to files using the UTF-8 encoding. Users can build customized pipelines by piping output from one tool into the next. For example, a customized pipeline can forward output from the sentence detector to the word segmenter into the POS tagger. Figure 2 illustrates an example of processing pipeline using the tools provided by AraNLP.

SAFAR,[17] is a Java-based framework dedicated to Arabic Natural Language Processing (ANLP). It brings together all layers of ANLP: resources, pre-processing, morphology, syntax and semantics. The general idea behind SAFAR is to gather, within a single homogeneous architecture, the available set of Arabic tools that are already developed, and develop new ones if necessary. Application builders can then realize many benefits by reusing components and avoid problems of interoperability as long as all components within the framework share the same architecture. It is an effective way for standardization, optimization efforts and accelerating developments of new tools [31]. All tools within SAFAR

[15] http://innovation.columbia.edu/technologies/cu14012_arabic-language-disambiguation-for-natural-language-processing-applications.

[16] https://sites.google.com/site/mahajalthobaiti/resources.

[17] http://arabic.emi.ac.ma/safar/.

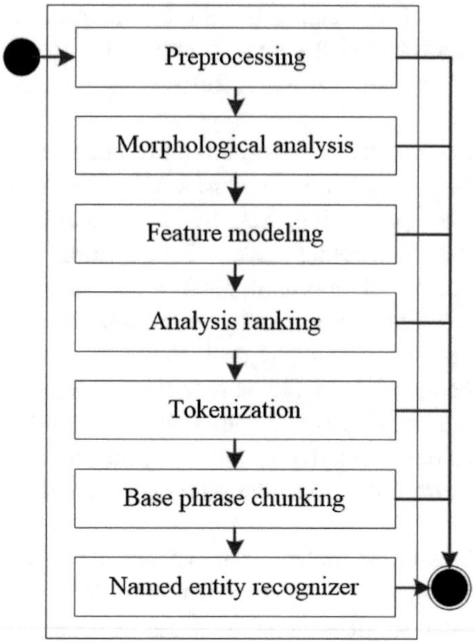

Fig. 1 MADAMIRA pipeline

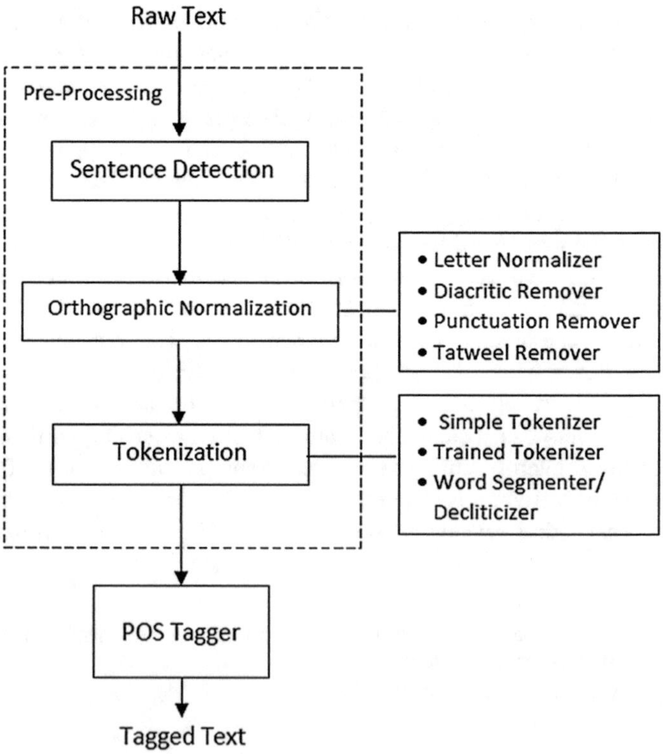

Fig. 2 Typical processing pipeline of AraNLP [4]

are standardized according to several Java interfaces. Therefore, users can add new implementations of any family tools just by implementing the appropriate interface.

SAFAR interfaces provide a variety of inputs and outputs. Researcher would have a large choice when calling methods. They could also easily create customizable pipelines where the output of one tool is the input of another. SAFAR outputs can be either memory objects or output files in XML format.

So far, SAFAR contains several tools and resources such as morphological analyzers, stemmers, parsers, utilities, etc.

As shown in Fig. 3, SAFAR has several layers. Each one is developed as a set of reusable Java APIs that provide services directly usable by other layers in accordance with the use relationship modeled with arrows in the figure. SAFAR layers consist of models and interfaces [32] that help standardizing the various aspects shared by tools belonging to the same family. Several implementations of various tools are already integrated within SAFAR.

Morphology: deals with three type of tools namely, morphological analyzers, stemmers and generators. So far, four morphological analyzers are integrated within SAFAR: Alkhalil [33], BAMA [34], MADAMIRA [5] and ATKS analyzer [15]. Along with five stemmers: Khoja's stemmer [35], Light10 [36], ISRI [37], Motaz's stemmer [38], Tashaphye [39] and SAFAR stemmer [40].

Syntax: contains implementations of Arabic syntactic parsers. Stanford parser [41] and ATKS parser [15] are integrated in this layer. Let us mention that the ANLP community lacks of available syntactic parsers.

Semantic: designed to implement tools dealing with semantic. A module for transforming Arabic texts into Conceptual Graphs is in the phase of being integrated within this layer.

Utilities: includes a set of technical services and pre-processing tools: (1) The *NormalizerTool* allows normalization of an input text by deleting some elements such as non-Arabic letters, special characters, etc. (2) The *tokenizer tool* returns all tokens in a specific text. (3) The *sentence splitter* tool splits a text into several sentences based on punctuations or custom delimiters specified by users. (4) The *transliterator* tool transliterates a text from Arabic to Latin and vice versa using some transliteration encodings such as Buckwalter [42]. (5) The *benchmark* tool performs the evaluation and comparison of Arabic morphological analyzers by

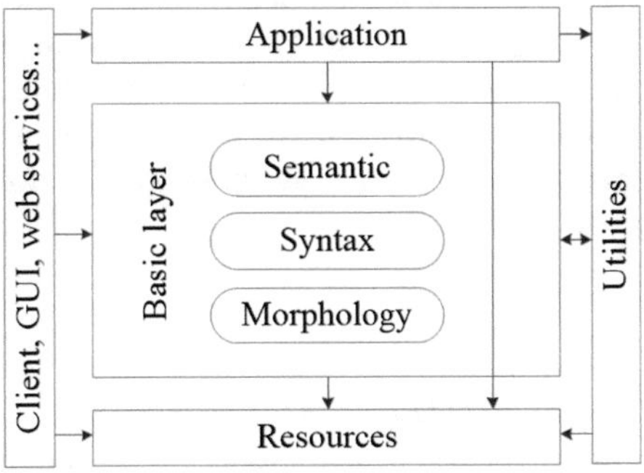

Fig. 3 SAFAR framework architecture

returning some common metrics such as accuracy, precision, execution time, etc. [43].

Resources: provides services for maintaining, consulting and managing Arabic language resources such as corpora, alphabets, clitics, dictionaries, ontologies and particles [44].

Application: contains high-level applications such as Question/Answering.

Client layer: interacts with all other layers and provides web applications,[18] web services, etc.

More information about tools available in SAFAR framework can be found in the official website.

4 Benchmarking NLP Architectures

4.1 Comparison of Frameworks, Platforms and Toolkits

Throughout this article we have used the word "architecture" as a generic word meaning at the same time: "toolkit", "platform" and "Framework". In the literature, these three words are used randomly and sometimes do not reflect their true meaning while categorizing NLP architectures. This is why we believe we should first define and compare these three types of architectures before going through the benchmark. We should also present advantages and disadvantages of each one of them regarding the manipulation and processing of Arabic. All the definitions below are taken from the context of Software Engineering but adapted to the context of NLP.

Toolkit: A toolkit is a set of tools (generally small tools) within a single box that is used for a particular purpose. This architecture is simple compared to frameworks and platforms as it is explained below. Tools within such architecture are not necessarily interoperable. It may consist of heterogeneous tools within the same structure. The main advantage of a toolkit is being simple and easy to use. However it is hard to be extended and difficulties of interoperability and reuse may occur.

Platform: Consists of several tools and programs within a single and homogenous structure. We can consider platforms as black boxes where we do not need to be aware of their internal architectures. Programs and tools within a platform are generally designed to be flexible and interoperable. Techtarget,[19] one of the most known websites of information technology, considers platforms to be advanced than toolkits, less complicated than frameworks and at the same time they improve quality, productivity, and flexibility. However, they do not provide APIs to extend their components.

[18]http://arabic.emi.ac.ma:8080/SafarWeb_V2/.

[19]http://searchservervirtualization.techtarget.com/definition/platform.

Table 1 Classification of NLP architectures according to their types

Type	Architectures	Number
Toolkit	MADAMIRA, ATKS, NLTK, LingPipe, OpenNLP	5
Platform	AraNLP	1
Framework	UIMA, GATE, SAFAR, LIMA, NooJ	5

Framework: According to Techtarget,[20] a framework is a real or conceptual structure that is developed in order to be used as a support and guide to build other useful programs and tools. It is generally a layered structure defining what types of programs can be implemented and how these programs can interact with each other. The real purpose behind frameworks is to improve the efficiency while creating new components. It can also improve the quality and robustness of new programs. Some of frameworks advantages is that they provide APIs to create new advanced programs; they also improve quality, reliability, robustness, productivity, and flexibility. However, they are complicated to handle and require some effort before being able to add new components.

According to the above definitions, architectures presented in Sect. 2 can then be classified as illustrated in Table 1. First of all, we observe that the number of toolkits and frameworks exceed largely the number of platforms. Indeed, toolkits are simple to construct, this is because it is easy to bring several components together as a toolkit rather than dealing with advanced architectures such as frameworks. This encourages many researchers to use and develop toolkits for simple and quick usage. On the other hand, frameworks are robust and provide the possibility to be extended and used by large number of other researchers in several contexts. That is to say, toolkits and frameworks are widely used because of their simplicity and robustness respectively, which justifies the number of available toolkits and frameworks. Concerning platforms, they are more complex than toolkits, which involves more effort in their constructions and development. However, they are still less robust than frameworks and cannot be extended. In fact, researchers prefer to provide effort of development in robust architectures that can be widely used and extended rather than dealing with fixed ones.

4.2 Benchmarking NLP Architectures

Now that we have presented the differences between NLP frameworks, platforms and toolkits, we address hereafter the benchmark of our selected NLP architectures. The main objective of this benchmark is to specify which NLP architectures can suitably deal with Arabic language and its specificities.

For this benchmark, we have selected several criteria that we have grouped into four categories (explained in the Sect. 1 as characteristics we consider to be ideal of

[20]http://whatis.techtarget.com/definition/framework.

an Arabic NLP architecture): (1) Arabic integrated tools, (2) Arabic integrated resources, (3) flexibility of exploitation and (4) maintenance and support. These categories are respectively presented in Tables 2, 3, 4 and 5. It should be noted that this benchmark concerns the Arabic side within each architecture and not all its

Table 2 Comparison of NLP architectures according to their Arabic integrated tools

	Architectures	Dedicated to Arabic?	Arabic tools		Score
			Morphology, syntax and semantic	Utilities and applications	
Frameworks	UIMA	No	–	Language detection	1
	GATE	No	–	Gazetter collector, Arabic IE, OrthoMatcher, Infered gazetteer, Tokeniser, Transducer	6
	LIMA	No	Clitic stemmer	–	1
	NooJ	No	–	–	0
	SAFAR	Yes	**Analyzers** Alkhalil, MADAMIRA, BAMA, ATKS analyzer **Stemmers** Light10, Tashaphyne, Khoja, SAFAR Stemmer, ISRI, Motaz stemmer **Parsers** Stanford and ATKS parsers,	**Utilities** SAFAR normalizer, SAFAR sentence splitter, SAFAR tokenizer, SAFAR transliterator, ATKS transliterator SAFAR analyzers Benchmark, SAFAR stemmers Benchmark	23
			Semantic Arabic sentences to CG	**Applications** Light summarizer, Stem counter, Morphosyntactic processor	
Platforms	AraNLP	Yes	Light stemmer, root stemmer, part-of-speech tagger	Sentence detector, tokenizer, normalizer, punctuation and diacritic remover, word segmenter	8
Toolkits	OpenNLP	No	–	–	0
	NLTK	No	ISRI	–	1
	LingPipe	No	–	Named entities	1
	ATKS	Yes			8

(continued)

Table 2 (continued)

	Architectures	Dedicated to Arabic?	Arabic tools		Score
			Morphology, syntax and semantic	Utilities and applications	
			Part of speech tagger, SARF (morphological analyzer), ATKS parser	Diacritizer, NER, speller, transliterator, colloquial to arabic converter	
	MADAMIRA	Yes	Morphological disambiguation, POS tagging, lemmatization, base phrase chunking	Tokenization, diacritization, named entity recognition	7

Table 3 Comparison of NLP architectures according to their Arabic integrated resources

	Architectures	Dedicated to Arabic?	Arabic resources			Score
			Lexicon	Corpora	Ontology	
Frameworks	UIMA	No	–	–	–	0
	GATE	No	–	–	–	0
	LIMA	No	Clitic dictionary	–	–	1
	NooJ	No	–	ELectronic DICtionary for ARabic (EL-DICAR)	–	1
	SAFAR	Yes	Characters lexicon, Clitics lexicon, Stop-words lexicon, "Al wassit" dictionary, Contemporary Arabic dictionary	NAFIS Stemming Gold Standard Corpus, Morphological analyzers Gold standard corpus	Arabic Wordnet (second release)	8
Platforms	AraNLP	Yes	–	–	–	0
Toolkits	OpenNLP	No	–	–	–	0
	NLTK	No	–	–	–	0
	LingPipe	No	–	–	–	0
	ATKS	Yes	–	–	–	0
	MADAMIRA	Yes	–	–	–	0

Table 4 Comparison of NLP architectures according to their flexibility of exploitation

		Data format		Flexibility		Exploitation			Portability		Score
	Architectures	Input	Output	Extensibility	Pipeline	API	Web services	Online demo	Dev. language	Cross-platform	
Frameworks	UIMA	Text audio video…	Text, XML	Yes	Yes	Yes	Yes	–	Java, C ++	Yes	10
	GATE	Text, HTM XML…	Text, XML	Yes	Yes	Yes	Yes	–	Java	Yes	10
	LIMA	Text	Text, XML	Yes	Fixed	Yes	No	–	C ++	No	7
	NooJ	Text XML	XML	Yes	No	No	No	–	.NET, Java	Yes	6
	SAFAR	Text	Text, XML	Yes	Yes	Yes	Yes	Yes[a]	Java	Yes	10
Platforms	AraNLP	Text	Text	No	Yes	No	No	–	Java	Yes	4
Toolkits	OpenNLP	Text	Text	No	Yes	Yes	No	–	Java	Yes	5
	NLTK	Text	Text	No	No	Yes	No	–	Python	No	3
	LingPipe	Text XML HTML	XML	No	No	Yes	No	–	Java	Yes	5
	ATKS	Text	Text	No	No	No	Yes	–	–	Yes	4
	MADAMIRA	Text	Text, XML	No	Fixed	No	No	Yes[b]	Java	Yes	6

[a]http://arabic.emi.ac.ma:8080/SafarWeb_V2/
[b]http://camel.abudhabi.nyu.edu/madamira/

Table 5 Comparison of NLP architectures according to their maintenance and support

	Architectures	First release	Last release	Documentation	Published articles	Last article	Score
Frameworks	UIMA	2006	2016	Yes	–	–	3
	GATE	1995	2017	Yes[a]	–	–	2
	LIMA	2010	2017	–	1	2010	2
	NooJ	2005	2015	–	1	2011	2
	SAFAR	2008	2016	Yes[b]	14	2016	16
Platforms	AraNLP	2014	2015	–	1	2014	2
Toolkits	OpenNLP	2010	2017	–	–	–	1
	NLTK	2005	2016	Yes[c]	–	–	2
	LingPipe	2003	2011	–	–	–	1
	ATKS	2013	2013	–	–	–	1
	MADAMIRA	2014	2014	–	1	2014	2

[a]https://gate.ac.uk/gate/doc/plugins.html#Lang_Arabic
[b]http://arabic.emi.ac.ma/safar/
[c]http://www.nltk.org/api/nltk.stem.html#module-nltk.stem.isri

aspects. For example, UIMA has too many published articles, but we are interested only on those that concern Arabic, the same goes for integrated tools and the other criteria. That is to say, tables below aim to present how these architectures are concerned by Arabic and how they handle it.

As it is shown in Table 2, NLP architectures dedicated to Arabic exceed largely independent language architectures in terms of Arabic integrated tools. This is obvious because ANLP architectures are intended to contain only Arabic tools unlike others. Indeed, the ANLP community is not encouraged to integrate its works within such language-independent architectures. This justifies the lack of Arabic language processing components within these architectures. SAFAR framework comes in the first place since it implements various tools within all its layers. OpenNLP and NooJ come in the last position with no Arabic pre-built tools even if they are very known for other languages. The "score" is a metric that we have proposed in order to rate architectures. For the case of tools, each architecture gets one point of an integrated Arabic tool. We give equal weights for all tools since we believe that all of them are useful for the ANLP community, a tool that may interest one researcher may not interest another and vice versa. Hence, the score is calculated by counting the number of Arabic tools that are ready to use within each architecture. SAFAR gets the highest score with 23 integrated tools.

Table 3 presents Arabic integrated resources within each architecture. We notice that unlike tools, resources come with fewer numbers. This is because they are time consuming when developing them comparing to some tools such as toknizers and light stemmers. Resources also require the cooperation of computer scientists as well as linguists, which complicates the task especially for huge resources. Table 3 shows that each of LIMA and NooJ provide only one resource which are respectively a lexicon and corpora. SAFAR is the only architecture that provides many

Arabic resources. Some architectures provide no Arabic resource, while others provide resources (such as clitics, roots, etc.) but used in the context of their programs and it is up to the programmer to understand the workflow of the program and extract the corresponding resource. The score of architectures is calculated the same way as tools. That is to say, by counting the number of Arabic resources that are ready to use within the architecture.

Table 4 concentrates on how much an architecture is flexible in order, for instance, to be extended and how much easy/hard it is to exploit it from different computing contexts to assess for instance its interoperability. According to Table 4, frameworks are more likely to be extended by end users than platforms and toolkits. Indeed, frameworks provide development environments and complex architectures that can facilitate their extensibility as well as the creation of complex processes such as pipelines. That is to say, even with few integrated Arabic tools, a framework has the possibility to be extended to implement new ones. Frameworks provide also several formats of inputs and outputs. Moreover, UIMA, GATE and SAFAR frameworks provide web services, which can improve their exploitation from heterogeneous contexts. Most of the presented NLP architectures are developed using Java, this makes them cross-platform and can be executed within several operating systems. Table 4 shows also that SAFAR and MADAMIRA are the only architectures that provide online executions of their components. Indeed, online executions can be very useful for linguists and users that do not have enough knowledge on how to execute processes using programming languages; they are also useful in case of quick processing tests and quick results. Scores are calculated as follow: each architecture gets two points if it provides more than one type of input data formats (the same goes as well for output data formats) because we believe that researchers may need to exploit different types of data such as xml and not only text files. Two points if the architecture is extensible, this characteristic is very important to let users add new components as needed. One point for each criterion that is taken into account by the architecture namely: pipeline, API, web services, online demo, and the cross-platform feature.

Table 5 shows also that most of the presented architectures are up to date regarding their releases. However, SAFAR is the only architecture that provides many published articles concerning Arabic. Each one of these articles addresses one or many aspects of processing Arabic within SAFAR. Other architectures do not provide any articles or provide few ones focusing on Arabic language as for LIMA, NooJ, AraNLP and MADAMIRA. Concerning the documentation, UIMA, GATE, SAFAR and NLTK have extensive ones. This can be very helpful to get started and be familiar with these architectures with a minimum effort from end users. Indeed, less documentation leads to more effort to discover how to manipulate it, and vice versa. Scores are calculated by attributing one point for each architecture that has a release within the last five years, one point if it has documentation and one point for every Arabic published article. We give equal weights to these criteria because we consider all of them as important to maintain an architecture up to date.

5 Discussion

According to the benchmark of NLP architectures conducted in this article and especially Tables 2, 3, 4 and 5, we present an overview to sum up results of this benchmark in Figs. 4, 5 and 6 and discuss them in order to specify which architectures can be used according to researchers needs. Scales in the following figures go from 0 to 8 for Figs. 4 and 5, and from 0 to 23 for Fig. 6. These numbers correspond to minimum and maximum scores obtained by architectures for a given category.

Figure 4 gives an overview of toolkits scores previously obtained. MADAMIRA and ATKS toolkits get the highest scores concerning integrated Arabic tools, maintenance/support and flexibility of exploitation, with nearly the same values. This is because they are dedicated specifically to Arabic. All other toolkits have reduced scores. One interesting remark is that none of the presented toolkits incorporates Arabic resources to be used by end users. This can be explained by two factors: (1) some toolkits are created for specific purposes such as the case of MADAMIRA, which is a morphological disambiguation system, these toolkits do use resources but they are used for the sake of the tools they were developed for and not as independent resources to be used in other similar contexts. The only remaining solution to use such resources is to get their sources from the project and adapt them to suit a new researcher's need. (2) Toolkits are not intended to be extended by end users; therefore resources created by other researchers are difficult

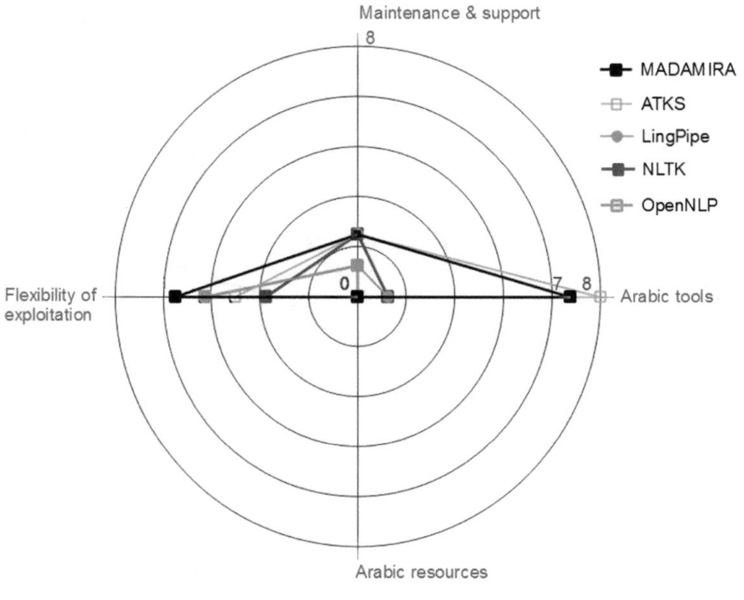

Fig. 4 Overview of the benchmark of toolkits

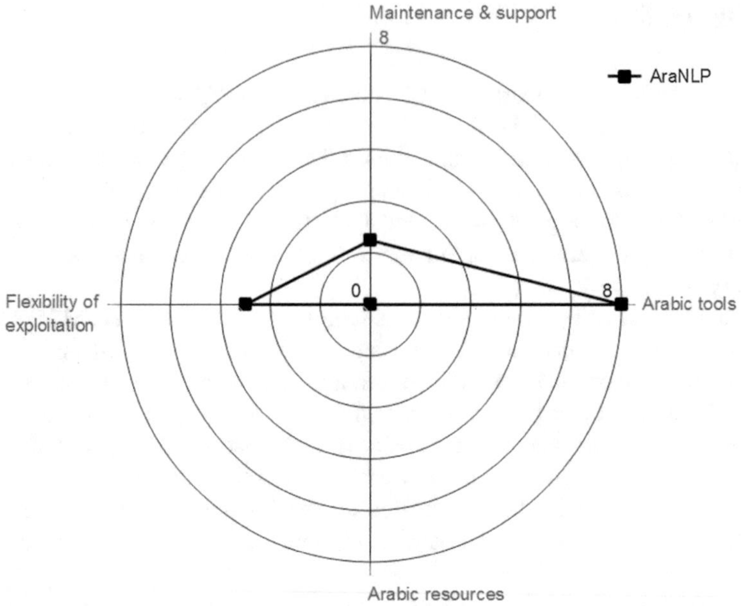

Fig. 5 Overview of the benchmark of platforms

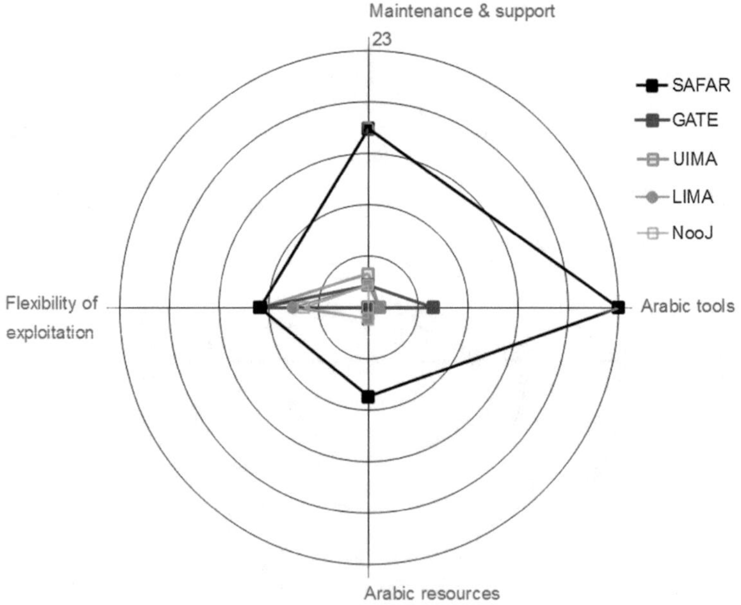

Fig. 6 Overview of the benchmark of frameworks

to be integrated within these toolkits, which limits their usage. These factors do not promote the integration of ANLP tools and resources within these architectures.

Figure 5 gives an overview of the obtained scores for the unique platform presented in this article. Unfortunately we have found only one architecture for Arabic that can be considered as a platform, namely AraNLP. This limits a little bit the comparison of platforms against each other as we have done for toolkits above. AraNLP has nearly the same scores as MADAMIRA and ATKS toolkits since it is dedicated to Arabic. Moreover, it does not include any Arabic resource.

Figure 6 summarizes frameworks scores. SAFAR exceeds other frameworks in handling Arabic tools and resources and also in the number of support provided. Indeed, SAFAR is the only framework that is dedicated specifically to Arabic among the five presented frameworks. This justifies its highest scores concerning Arabic tools and resources. For the maintenance and support, SAFAR gets the highest score mainly due to its number of Arabic publications that exceed others. The obtained scores for the flexibility of exploitation are nearly the same for the five frameworks. In fact, flexibility is one of the main objectives behind the creation of frameworks. That is why they give a special attention to this criterion.

All figures above show that architectures dedicated to Arabic get highest scores concerning handling Arabic. However, are these scores enough to claim that those architectures are suitable for processing Arabic? Indeed, this question cannot be answered out of the context of researchers needs. According to the benchmark presented in this article, we can categorize researchers into three categories:

1. Researchers who need specific Arabic processing without the need of extensibility or advanced processing. For example, a researcher who needs to call a morphological analyzer or a syntactic parser to process an Arabic text without any need of further treatment. For such situations, it is more suitable to call a toolkit such as MADAMIRA, ATKS or even a platform such as AraNLP which includes the needed tool. These kinds of architectures are simple to use in general. However, a researcher cannot use a toolkit such as OpenNLP to process Arabic since OpenNLP does not incorporate tools for this purpose.
2. Researchers who need advanced processing including but not limited to Arabic, such as developing a translation system. In such cases, researchers should deal with Arabic along with other languages. They also should deal with all language levels starting from basic tools to semantic, which involves calling several tools within the same project. For this type of researchers, using a language independent framework such as GATE and UIMA will be helpful. Indeed, these frameworks can be used to handle Arabic processing despite they don't concern it directly. Moreover, they provide APIs and data models that facilitate the integration and developments of new components. They are also flexible and help building/manipulating complex processes such as pipelines, which makes them suitable for advanced processing.
3. Researchers who need advanced architectures which are dedicated to Arabic and containing as much as possible tools and resources. For such cases, researchers could use frameworks such as SAFAR. Unlike GATE and UIMA, SAFAR

framework focuses only on Arabic and provides a wide range of Arabic processing tools and resources. SAFAR provides an API and clear models to help researchers adding new components and reuse existing ones by simple calls. Moreover, all its layers are developed in compliance to the Arabic language properties.

From the above researchers categorization, we can classify all architectures into four main categories namely zone "a", "b", "c" and "d" which are presented in Fig. 7.

As shown in Fig. 7, zone "1", "2" and "3" correspond respectively to the previous researcher's needs categorization: zone "1" represents all architectures which are dedicated to specific processing including but not limited to Arabic. Zone "2" represents all advanced architectures used to handle complex processing either for Arabic or other languages. Zone "3" concerns all architectures which are dedicated specifically to Arabic. These three zones are general and can be intersected to produce new zones for more detailed and specific needs. For example, zone "a" is equal to zone "1" minus zone "b", which represents all architectures for specific processing that are not dedicated to Arabic namely: NooJ, OpenNLP, NLTK and LingPipe. The same goes as well for zone "b", which is the intersection of zone "1" and zone "3" and which represents all architectures for specific processing that are dedicated specifically to Arabic, namely: AraNLP, ATKS and MADAMIRA. In the other side, zone "c" is equal to zone "2" minus zone "d", which represents all

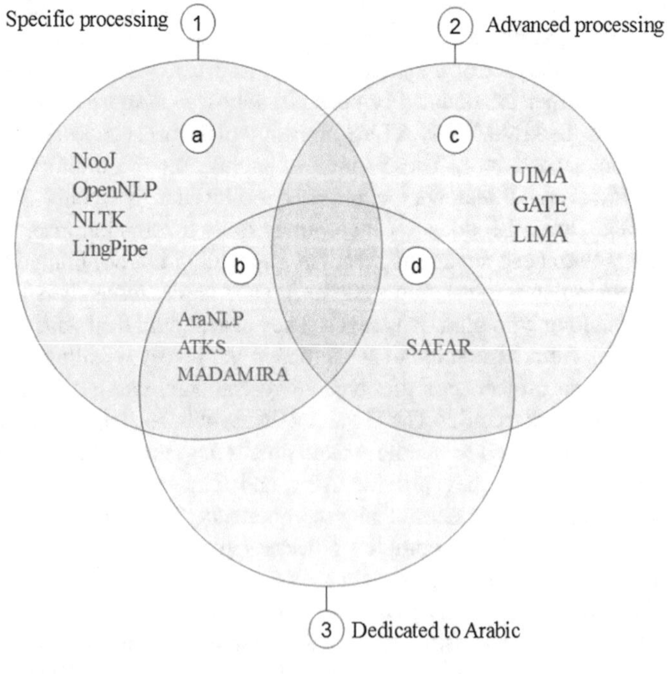

Fig. 7 Categorization of architectures according to researchers needs

advanced architectures that are not dedicated to Arabic, namely UIMA, GATE and LIMA. Finally, zone "d" is the intersection of zone "2" and zone "3" and which represents architectures for advanced processing that are dedicated specifically to Arabic, namely SAFAR.

Hence, we conclude that all the presented architectures can be used according to researches needs and specific contexts. However we believe that frameworks are an important step towards standardization, the resolution of interoperability issues, reusability and integration of all development efforts in the field of Arabic language processing.

6 Conclusion

In this paper, we presented a comparative study of Natural Language Processing architectures that can handle Arabic language. Indeed these architectures represent for us a way to standardize the various aspects shared by Arabic processing tools in order to promote interoperability. For this, we have highlighted several architectures among which we cite UIMA, GATE, AraNLP and SAFAR.

We divided these architectures into three types: (1) toolkits such as LingPipe, (2) platforms such as AraNLP and (3) frameworks such as UIMA and SAFAR. Except SAFAR, AraNLP, MADAMIRA and ATKS, all other presented architectures are language independent. That is to say, they have a general architecture to support any language. We have then specified four general characteristics that we consider important for any architecture to better handle Arabic, namely: (1) diversity of integrated Arabic tools, (2) diversity of integrated Arabic resources, (3) flexibility of exploitation and (4) availability of maintenance and support. It is worth mentioning that there are many other software engineering criteria that are used for benchmarking tools such as robustness and efficiency. We have not used such criteria in this article because we wanted to adjust the discussion toward Arabic NLP rather than a deep software engineering problem.

According to the conducted comparative study, SAFAR, MADAMIRA, ATKS and AraNLP get the highest scores regarding handling Arabic. This is because all of them are dedicated specifically to Arabic and incorporate many pre-built and ready to use tools that help researchers focusing on their objectives rather than dealing with the development of these tools from scratch or integrate them in frameworks such as GATE. This gives researchers a good base to start from to develop new applications. However, we consider that there is no perfect architecture for all situations and cases. Researchers can use these architectures according to their needs and level of complexity of their tasks. For example, if the main objective of a researcher is to get only morphological analyses of Arabic words without any further processing, it would be better to call a toolkit such as MADAMIRA or ATKS. In the other hand, for advanced processing and complex manipulation of data, it would be better to use a framework such as UIMA, GATE or SAFAR rather than a platform or a toolkit. Indeed, frameworks provide advanced architectures to handle complex NLP processing tasks and provide also APIs to add new implementations and call existing one with ease.

Acknowledgements We would like to thank Professor Mohamed Issam Kabbaj (Mohammadia School of Engineers, Mohammed Vth University in Rabat, Morocco) for his feedback on the work presented in Sect. 4.2.

References

1. Cunningham, H., Maynard, D., Bontcheva, K., Tablan,V.: A framework and graphical development environment for robust NLP tools and applications, In: ACL (2002)
2. Apache UIMA Manuals and Guides. https://uima.apache.org/d/uimaj-current/index.html. Last Accessed 11 Nov 2015
3. Internet World Users By Language: Top 10 Languages. http://www.internetworldstats.com/stats7.htm. Last Accessed 11 Nov 2015
4. Althobaiti, M., Kruschwitz, U., Poesio, M.: AraNLP: a Java-based library for the processing of Arabic text, In: Proceedings of the 9th Language Resources and Evaluation Conference (LREC), Reykjavik (2014)
5. Pasha, A., Al-Badrashiny, M., Diab, M., El Kholy, A., Eskander, R., Habash, N., Pooleery, M., Rambow, O., Roth, R.M.: MADAMIRA: a fast, comprehensive tool for morphological analysis and disambiguation of Arabic, In LREC'14, Reykjavik (2014)
6. Shaalan, K.: A survey of arabic named entity recognition and classification. Comput. Linguist. **40**(2), 469–510 (2014)
7. Prasanth, Y., Nakul, S.: Integrating natural language processing and software engineering. Int. J. Softw. Eng. Appl. **9**(11), 127–136 (2015)
8. Leidner, J.L.: Current issues in software engineering for natural language processing. In: Proceedings of the HLT-NAACL 2003 Workshop on Software Engineering and Architecture of Language Technology Systems, Stroudsburg (2003)
9. Bikel, D.M., Zitouni, I.: Combining natural language processing engines. In: Multilingual Natural Language Processing Applications: From Theory to Practice, pp. 523–542. IBM Press (2012)
10. Besançon, R., De Chalendar, G., Ferret, O., Gara, F., Mesnard, O., Laïb, M., Semmar, N.: LIMA: a multilingual framework for linguistic analysis and linguistic resources development and evaluation. In: Proceedings of the Seventh International Conference on Language Resources and Evaluation (LREC'10), Valletta (2010)
11. Alias-i, LingPipe. http://alias-i.com/lingpipe. Last Accessed 01 Mar 2017
12. Apache. OpenNLP. https://opennlp.apache.org/. Last Accessed 01 Mar 2017
13. NLTK. Natural language toolkit. http://www.nltk.org/. Last Accessed 01 Mar 2017
14. Silberztein, M.: NooJ: a linguistic development environment. http://www.nooj-association.org/. Last Accessed 01 Mar 2017
15. Microsoft. Arabic toolkit service (ATKS). https://www.microsoft.com/en-us/research/project/arabic-toolkit-service-atks/. Last Accessed 01 Mar 2017
16. Diab,M., Habash, N., Rambow, O.: Arabic language disambiguation for natural language processing applications. http://innovation.columbia.edu/technologies/cu14012_arabic-language-disambiguation-for-natural-language-processing-applications. Last Accessed 01 Mar 2017
17. Jaafar, Y., Bouzoubaa, K.: SAFAR: software architecture for Arabic language processing. http://arabic.emi.ac.ma/safar/. Last Accessed 01 Mar 2017

18. Ferrucci, D., Lally, A.: Building an example application with the unstructured information management architecture. IBM Syst. J. **43**(3), 455–475 (2004)
19. Wilcock, G.: Introduction to linguistic annotation and text analytics. Synth. Lect. Hum. Lang. Technol. **2**(1), 1–159 (2009)
20. Cunningham,H., Maynard, D., Bontcheva, K., et al.: Text Processing with Gate. Gateway Press, CA (2011)
21. Cunningham, H., Maynard, D., Bontcheva, K., et al.: Developing language processing components with GATE version 8 (a User Guide). https://gate.ac.uk/sale/tao/split.html. Last Accessed 11 Nov 2015
22. Ingersoll, G.S., Morton, T.S., Farris, A.L.: Taming text: how to find, organize, and manipulate it. Manning Publications Co. (2013)
23. Buyko, E., Wermter, J., Poprat, M., Hahn, U.: Automatically adapting an NLP core engine to the biology domain. In: Proceedings of the Joint BioLINK-Bio-Ontologies Meeting. A Joint Meeting of the ISMB Special Interest Group on Bio-Ontologies and the BioLINK Special Interest Group on Text Data Mining in Association with ISMB (2006)
24. Silberztein, M.: Complex annotations with NooJ. In: International NooJ Conference, Barcelone (2007)
25. Silberztein, M., Váradi, T., Tadic, M.: Open source multi-platform NooJ for NLP. In: COLING (Demos), Mumbai (2012)
26. Bird, S., Klein, E., Loper, E., Baldridge, J.: Multidisciplinary instruction with the natural language toolkit. In: Proceedings of the Third Workshop on Issues in Teaching Computational Linguistics (2008)
27. Perkins, J.: Python 3 Text Processing with NLTK 3 Cookbook, Packt Publishing Ltd (2014)
28. Bird, S., Klein, E., Loper, E.: Natural language processing with Python. O'Reilly Media, Inc. (2009)
29. Habash, N., Rambow, O., Roth, R.: Mada+ tokan: a toolkit for Arabic tokenization, diacritization, morphological disambiguation, pos tagging, stemming and lemmatization. In: Proceedings of the 2nd International Conference on Arabic Language Resources and Tools (MEDAR), Cairo (2009)
30. Mona, D.: Second generation AMIRA tools for Arabic processing: Fast and robust tokenization, POS tagging, and base phrase chunking. In: chez 2nd International Conference on Arabic Language Resources and Tools (2009)
31. Souteh, Y., Bouzoubaa, K.: SAFAR platform and its morphological layer. In: Proceeding of the Eleventh Conference on Language Engineering (ESOLEC'2011), Cairo, Egypt (2011)
32. Jaafar, Y., Bouzoubaa, K.: Arabic natural language processing from software engineering to complex pipelines. In: Conference on Intelligent Text Processing and Computational Linguistics (CICLing'2015), Cairo, Egypt (2015)
33. Alkhalil Morpho Sys (2013). http://sourceforge.net/projects/alkhalil/. Last Accessed 23 Apr 2015
34. Buckwalter, T.: Buckwalter Arabic Morphological Analyzer Version 1.0 (2002)
35. Khoja, S., Garside, R.: Stemming Arabic Text. Lancaster, UK, Computing Department, Lancaster University (1999)
36. Larkey, L.S., Ballesteros, L., Connell, M.E.: Light stemming for Arabic information retrieval. In: Arabic Computational Morphology: Knowledge-Based and Empirical Methods, pp. 221–243. Springer, Netherlands (2007)
37. Algasaier, H.: The ISRI Arabic stemmer. http://www.nltk.org/_modules/nltk/stem/isri.html. Last Accessed 11 Nov 2015
38. Motaz, S.: Arabic computational linguistics. http://sourceforge.net/projects/ar-text-mining/. Last Accès le 11 Nov 2015
39. Zerrouki, T.: Tashaphyne 0.2. https://pypi.python.org/pypi/Tashaphyne. Last Accessed 11 Nov 2015
40. Jaafar, Y., Namly, D., Bouzoubaa, K., Yousfi, A.: Enhancing Arabic stemming process using resources and benchmarking tools. J King Saud Univ.—Comput. Inf. Sci. (2016)

41. Spence Green, C.D.M.: Better Arabic parsing: baselines, evaluations, and analysis. In: Chez the 23rd International Conference on Computational Linguistics (COLING 2010), Beijing (2010)
42. Buckwalter, T.: Arabic transliteration/encoding chart. http://languagelog.ldc.upenn.edu/myl/ldc/morph/buckwalter.html. Last Accessed 12 Nov 2015
43. Jaafar, Y., Bouzoubaa, K.: Benchmark of Arabic morphological analyzers: challenges and solutions. In: 9th International Conference on Intelligent Systems: Theories and Applications (SITA'14), Rabat, Morocco (2014)
44. Namly, D., Bouzoubaa, K., Tahir, Y., Khamar, H.: Development of Arabic particles lexicon using the LMF framework. In: Colloque pour les Etudiants Chercheurs en Traitement Automatique du Langage Naturel et ses applications (CEC-TAL 2015), Sousse, Tunisia (2015)

Part X
Building and Evaluating Linguistic Resources

Arabic Corpus Linguistics: Major Progress, but Still a Long Way to Go

Imad Zeroual and Abdelhak Lakhouaja

Abstract Arabic is an old Semitic language, the standardization of its lexicon and grammar are deeply rooted and well established a long time ago in history. Arabic is a morphologically rich language characterized by the phenomenon of derivation and inflection. It is an international language with over 500 million native speakers around 29 countries. In the last 15 years, Arabic has achieved the highest growth of the ten top online languages. Consequently, the volume of stored electronic information increases rapidly. Despite this proud heritage, lexical richness, and online user growth, Arabic is relatively an under-resourced language compared to other languages with less or similar population size (e.g., French and German). The boundaries of this chapter cover the major progress that has been made in Arabic linguistic resources, primarily corpora compilation and the challenges that researchers face in the development of such process. It is hoped that this overall view of the Arabic corpus linguistics would guide current and future research directions.

Keywords Corpus linguistics · Arabic language · Linguistic resources
Corpus compilation · Natural language processing

1 Introduction

According to the Expert Advisory Group on Language Engineering Standards (EAGLES) (Sinclair 2004), a corpus is a collection of naturally occurring samples of a language. These samples are selected and stored in electronic format; they are ordered according to external criteria to represent, as far as possible, a language as a

I. Zeroual (✉) · A. Lakhouaja
Computer Sciences Laboratory, Faculty of Sciences, Mohammed First University, Oujda, Morocco
e-mail: mr.imadine@gmail.com

A. Lakhouaja
e-mail: abdel.lakh@gmail.com

K. Shaalan et al. (eds.), *Intelligent Natural Language Processing: Trends and Applications*, Studies in Computational Intelligence 740,
https://doi.org/10.1007/978-3-319-67056-0_29

source of linguistic research. To compile a corpus or combine texts (of both written and spoken language) into a corpus is called corpus compilation. The design, compilation, and analysis of corpora have led to the creation of a new scholarly field known as corpus linguistics. The roots of developing a corpus can be traced back to the German linguist Kaeding, who, in 1897, assembled a large corpus of German that contains 11 million words (Khorsheed et al. 2009). Based on this corpus, Kaeding was able to publish the first known word frequency list based on word counting. In doing so, he needed the help of over five thousand assistants over a period of years to process the corpus to undertake his analysis (Bongers 1947; Kennedy 2014).

As stated by (O'Keeffe and McCarthy 2010; Kennedy 2014), the real appearance of corpus linguistics started with the revolution of the computer since the 1980s which came with the access to machine-readable texts. In fact, the first corpus to be given that name was the Brown Corpus (Francis and Kucera 1982; Hunston 2013), developed and released in 1964. It consisted of texts amounting to over one million tokens. The earlier corpora that occurred before the 1960s were not computerized and they are generally called pre-electronic and "consisted of work in five main areas: biblical and literary studies, lexicography, dialect studies, language education studies, and grammatical studies".

The corpus linguistics has an impact on other fields, primarily lexicography (Teubert 2015) so that corpora can be used to describe reliably the lexicon and the grammar of languages. Other researchers have been concerned with the use of corpora in a variety of natural language processing applications (Armstrong et al. 2013) such as machine translation (Hu et al. 2016) to develop statistical algorithms and models. In a similar concept, (Boulton and Landure 2016; Bertels 2017) recently emphasize the potential relevance of corpus linguistics for language learning and teaching in all its forms and uses. In term of pedagogy, they believe that corpus linguistics should be considered for use in education to reduce the time that would be necessary to learn a language. Further, corpora have been successfully used as a reference resource by both advanced learners majoring in the language as well as learners with lower levels of proficiency needing language for specific purposes.

Typically, several researchers have developed corpora that comply with their suitable objectives; however, some corpora have been designed for general purposes. Thus, to be able to compile or use corpora successfully, some factors are necessary to be taken into consideration. Otherwise, the results may be different from the expected. Knowing the motivations and aims of the corpus development, and understanding the corpus nature, are among a few factors to name. The main tasks of the corpus linguistics are corpus design and compilation in addition to the study and analysis of data extracted from the corpus. Further, several issues are related to each task. For instance, corpus designers focus on design criteria in order to create a well-defined corpus that meets the standards. To systematically develop a corpus, researchers tend to balance the corpus considering the kind of texts included, the topics and domains covered, and the size of the corpus, among others.

Other researchers work on methods for text analysis and processing, and on the corpus-based linguistic description.

Since the appearance of corpus linguistics, many satisfying studies on corpora and the amount of available data grew significantly. Unfortunately, not all languages have benefited equally from this growth. An example of such a language is Arabic. The Arabic language is expanding in the world, with more than 500 million native speakers around 29 countries. The presence of the Arabic language on the internet grew around 6.091% in the last 15 years (2000–2015), it is the highest growth of the ten top online languages, and it is the fourth most spoken language on the internet (Jurida et al. 2016). Further, teaching Arabic as a foreign language has become a global educational enterprise (Sakho 2012). Although all these achievements, Arabic is still a resource-poor language relative to other languages with a similar spreading rate. For these reasons, many researchers and academia have become more interested in the corpus linguistics field in order to bridge the gap between Arabic and other resource-rich languages (e.g., English). Some considerable efforts have been (and are) performed; yet, the obtained results successfully reach the state-of-the-art of other languages such as English and Spanish.

The remainder of this chapter is arranged in five main sections focusing on the basic aspects of corpus linguistics related to the Arabic language. In Sect. 2, we provide background information regarding the scope of corpus linguistics, mentioning some factors that affect the compilation and the manipulation of corpora. We have addressed issues related to corpus design, compilation, and analysis in Sect. 3. In Sect. 4, we attempt to present a general view about Arabic corpus linguistics focusing on relevant corpus-based researches and studies, followed by an overview of the major progress achieved in this field. Finally, we conclude this chapter in Sect. 5.

2 The Scope and Aims of Corpora

Corpus linguistics, in essence, is a source of evidence for linguistic description and argumentation. Grammarians have always needed sources of evidence as a basis to illustrate grammatical features such as the nature, the structure and the functions of language. For instance, Watson (2002) has noted that the Arabic language was codified primarily in the Quran; yet, it was based on the language of the western Hijazi tribe of Quraysh, with some interference from pre-Islamic poetic koiné and eastern dialects. Thus, these latter have been the sources for the compilation of descriptive grammars of Arabic. Arabic grammarians use these sources as examples to illustrate grammatical features or construction. Further, Leech (1992a) claims that the focus of corpus linguistics is on performance rather than competence. i.e., the focus is on observation of language in use leading to theory rather than vice versa. Halliday et al. (2014) stresses that corpus-based analysis is an important source of insight into the nature of language and is specifically geared to investigate frequencies in corpora to establish probabilities in the grammatical system in order to understand the language variation and the grammatical change across registers.

Alongside the linguistic description, corpora have been used in lexicography. In fact, corpora exceedingly support many aspects of dictionary creation (Kilgarriff 2013) such as, at first stage, development of headword list, writing individual entries and identifying their syntactic behaviour, discovering words senses, providing examples and translations. Moreover, the major revision of relevant dictionaries is systematically based on corpora (Milfull 2009). For example, the Dictionary of the Older Scottish Tongue, the Middle English Dictionary, the Dictionary of Old English, and the Oxford English Dictionary which partly or completely draw on corpora.

In addition to linguistic description and lexicography, corpora have been widely used and significantly affect a wide range of research activities in several fields. One of the first influential corpus-based researches had a pedagogical purpose. Word frequency lists, which are generated from corpora, are intended to gather statistical information on the use of words and letters of a language. These lists are a quick guide and better curricula materials for teaching and learning vocabulary. Nation (2013) believes that the high-frequency vocabulary is important for the learners and need to focus on its learning burden and ensure that the learners will come back to it again. Whereas some low frequent vocabulary may not need to become a part of the learners' output or the teacher may give some brief attention to it.

In recent years, corpora have been increasingly used for language learning and teaching. For instance, a case study (Sahragard et al. 2013) has been conducted for Iranian EFL (English as a Foreign Language) learners to prove the advantage of using corpora for teaching grammar, primarily English relative clauses. The authors suggest that applying corpora is an effective way to aware students of their errors which ultimately leads to self-correction. Hyland (2015) concerned the increase of features number in student essays such as relative clauses, modality, and passives. He emphasized the vital role of corpora in measuring the students' writing improvement.

The progress in most Natural Language Processing (NLP) applications is driven by available data. Thus, large and high-quality corpora are valuable resources for various NLP research disciplines (Tiedemann 2007). Over the last years, various corpora are built to support NLP research that aimed to extract meaningful information to enable their use for developing applications such as machine translation. The overlapping between corpus linguistics and descriptive translation studies have contributed to the birth and rise of the corpus-based translation studies (CTS). CTS have become a major paradigm and research methodology. It applies statistical analysis of words or phrases in parallel or comparable corpora in different languages to obtain probabilities of translations. Hu (2016) has discussed in more details the implication of CTS as a new research methodology in translation studies. Moreover, He has explained how corpora can be used in translation teaching, primarily on the establishment of the corpus-based mode of translation teaching and the use of corpora in compiling translation textbooks. On the other hand, query translation and multilingual corpora can be combined to enhance the performance of Cross-Lingual Information Retrieval (CLIR) (Bhattacharya et al. 2016). CLIR is a task used to search and retrieve the relevant information between source documents and user queries. CLIR models can be trained with document-aligned comparable and parallel corpora, or they can include a translation mechanism followed by monolingual Information Retrieval (IR). Magdy and Jones (2014) claim that training a machine translation system using corpora pre-processed for IR can lead to an effective retrieval. In the same way, corpora can be useful for other NLP tasks such as word sense disambiguation (Lefever and Hoste 2013), summarization (Li et al. 2013), syntactic annotation (Xing et al. 2016), and named entity recognition (Nothman et al. 2013).

3 Stages in Corpus Building

The aim of this section is to outline certain stages in corpus building, from corpus design and compilation to corpus processing and analysis. In general, the purpose to which a corpus is compiled influences its design, size, and nature. Basically, corpora differ from other electronic representations such as archives and databases. Archives are normally unstructured repositories of texts, whereas databases are collections of an entire population of data. This latter are designed to facilitate data entry and retrieval. In fact, corpora are a subset of databases. They are designed and compiled according to some explicit criteria defined by their developers in order to be representative samples of languages (Leech 1991). Further, the term corpus comes from Latin that means "body"; i.e., it is systematic, planned and structured compilation of text.

Corpora, in one hand, may consist of a single book like the ones developed and used by Baneyx et al. (2007) to build an ontology of pulmonary diseases, or for common-sense knowledge enhanced embeddings to solve pronoun disambiguation problems (Liua et al. 2016). At the other hand, they are usually developed based on a number of several books (e.g., Shamela (Belinkov et al. 2016)), or editions of a particular newspaper (e.g., Maamouri et al. 2013). Recently, Internet technology revolution paves the way for building corpora based on website content (Nakov 2014).

In the following subsections, we describe essential stages of the corpus building task. These stages are:

- Corpus design criteria;
- Selection of sources and corpus compilation;
- Corpus processing and text handling;
- Corpus analysis.

3.1 Corpus Design and Compilation

Since the 60s, the design and compilation of corpora have been a moot point for corpus linguists. However, in the 90s, primarily with the completion of the British National Corpus (BNC) (Leech 1992b) and the Penn Treebank Corpus (Marcus et al. 1993), basic guidelines in corpus design and compilation have been set by relevant authors namely (Leech 1992b; McEnery and Wilson 1996; Sinclair 1996; Biber et al. 1998). Later, these guidelines were expanded by Sinclair (2005) in ten fundamental criteria. It worth mentioning that corpus design and compilation succeeded to identify at least what type of corpus is being constructed (see (Atkins et al. 1992), Corpus Typology).

Leech (1992a) made clear that corpora are not haphazard collections of textual material. Thus, a great care must be taken during the compilation process; otherwise, the developed corpora will lead to different results from the expected. The corpus design usually starts from identifying the appropriate criteria which mean that the corpus likely seeks to be representative with respect to the phenomena under investigation (Ball 1994). If there are no specific criteria, the corpus should be designed for a general use to suit most corpus-based studies.

As mentioned, Sinclair (2005) formulates the overall instructions proposed by the previous authors in ten fundamental criteria to follow in the design and the compilation of a general corpus:

1. *The contents of a corpus should be selected without regard for the language they contain, but according to their communicative function in the community in which they arise.*
2. *Corpus builders should strive to make their corpus as representative as possible of the language from which it is chosen.*
3. *Only those components of corpora which have been designed to be independently contrastive should be contrasted.*
4. *Criteria for determining the structure of a corpus should be small in number, clearly separate from each other, and efficient as a group in delineating a corpus that is representative of the language or variety under examination.*
5. *Any information about a text other than the alphanumeric string of its words and punctuation should be stored separately from the plain text and merged when required in applications.*
6. *Samples of language for a corpus should wherever possible consist of entire documents or transcriptions of complete speech events, or should get as close to this target as possible. This means that samples will differ substantially in size.*
7. *The design and composition of a corpus should be documented fully with information about the contents and arguments in justification of the decisions taken.*
8. *The corpus builder should retain, as target notions, representativeness and balance. While these are not precisely definable and attainable goals, they must be used to guide the design of a corpus and the selection of its components.*

9. *Any control of subject matter in a corpus should be imposed by the use of external, and not internal, criteria.*
10. *A corpus should aim for homogeneity in its components while maintaining adequate coverage, and rogue texts should be avoided.*

Although the mentioned ten guidelines are core principles to design and compile a general corpus, they may not always suit every potential corpus builder. Sinclair (2008) himself notes that some of these guidelines can be difficult to uphold because of the nature of language itself. However, the representativeness, balance, and homogeneity in the design process are necessarily idealistic. Alternatively, specialized corpora are designed relatively to individual research aims such as creating a dictionary, studying and providing analysis of the language used in a specific subject domain. However, it is cautioned to design a specialized corpus with regard to Sinclair's guidelines as ultimately a reliable or generalizable result can be derived from the analysis.

In addition to corpus design and sources selection, text encoding or markup is one of the main tasks in corpus compilation. Typically, corpora consist of electronic versions of texts taken from various sources. Therefore, a confusion may arise due to different codes used for markup. Since the 1980s, the Standard Generalized Markup Language (SGML) has become increasingly accepted as a standard way of encoding electronic texts. Using SGML is considered as a basis for corpus preparation; it facilitates the portability of corpora, enabling them to be reused in different contexts on different equipment, thus saving the cost of repeated typesetting. Since SGML can be complex for some corpus builders and users, an Extensible Markup Language (XML) that was derived from SGML contains a limited feature set to make it simpler to use.

3.2 *Corpus Processing and Analysis*

Working just on corpus compilation will not satisfy the demand for corpus-based researches. A set of general analysis and text processing are essential. Further, some degree of automatic processing greatly reduces the human intervention and vastly expands the empirical basis. Over the last years, great efforts have been put into developing procedures for automatic text processing. Among the most corpus analytical procedures are:

- Concordance: It aims to search the text and finds every occurrence of each word and displays all the occurrences with its immediate context in a corpus. The occurrences are usually sorted according to alphabetical order. Further, a concordance can be made for every target item in the corpus, or alternatively for selected ones. Concordances can be produced in several formats. The most usual form is the Key Word in Context (KWIC) concordance (Kennedy 2014).
- Word frequency procedure: It aims to produce lists of words and their frequency in the corpus. Based on such lists, it is possible to indicate the distribution of

words across the text categories and provide graphical displays to summarise the lists in easily assimilated form. Moreover, it is feasible to produce word frequency lists using a part-of-speech tagged corpus not merely their orthographic status. For example, Leech et al. (2014) use the tagged BNC corpus to build the word frequencies list in written and spoken English.

- Collocation statistics: It is a procedure to calculate statistical information about the association, the strength of collocation, and the comparative frequencies of word forms in a corpus or multiple corpora. Unlike the previous analytical procedures, collocation requires a big-sized corpus. Otherwise, the analyst cannot cope with the available data.

The following text processing tasks are considered more sophisticated to implement and their design is more contentious. These tasks are normally performed automatically to allow an extensive and difficult analysis to be carried out on corpora. To name a few:

- Lemmatization: There is an overall agreement on the concept of lemmatization process in various languages. In fact, lemmatization is detecting semantically equivalent surface words written in different syntactic forms and relates them to their canonical base representation (i.e., lemma). This latter is a dictionary lookup form which can relate different word forms that have the same meaning (Attia and Van Genabith 2013).
- Part of speech tagging: It is a basic task in many fields primarily corpus linguistics and natural language processing. It aims to automatically assign a morpho-syntactical label to every word in the corpus. Further, this task allows simple syntactic searches to be performed.
- Parsing: A natural successor to part of speech tagging is parsing. Basically, it provides a dependency tree as an output. Here, the goal is predicting for each sentence or clause an abstract representation of the grammatical entities and the relations between them. Consequently, the parser assigns a fully labelled syntactic tree or bracketing of constituents to sentences of the corpus (Tsarfaty et al. 2013).

The cited tasks are some of the key features of corpus annotation. One of the main aims being addressed in corpus linguistics is to develop new forms of annotation and improve the accuracy of automatic annotation.

4 Overview of Progress in Arabic Corpus Linguistics

Arabic is an interesting language and fruitful area of research to corpus linguistics as much as it is a challenging language to existing NLP applications because of its characteristics. The Arabic lexicon and grammar are deeply rooted and well established a long time ago and its morphology differs from Indo-European languages (Gharaibeh and Gharaibeh 2012). Further, Arabic is a high derivational and inflectional language (Alsaedi et al. 2016); often, a word represents a whole sentence through sequential concatenation. For example, the Arabic word "أَنُلْزِمُكُمُوهَا" from the 28th verse of Chap. 11 (sūrat Hud) <AanulozimukumuwhaA> means in English "Should we compel you to accept it". Basically, the Arabic language consists of three main categories (Al-Dahdah 1989; Ghalayini 2013): Noun "اسم" <Asm>, Verb "فعل" <fEl> and Particle "حرف" <Hrf>. In addition, each one of these categories has dozens of subcategories (Zeroual et al. 2017). The omission of diacritics (short vowels) primarily in written modern standard Arabic has posed some difficulties to several automatic processing systems (Chennoufi and Mazroui 2016); when these vowels are omitted it is left for the reader to infer, knowing that the vowels can encode grammatical category or feature information. In addition, most Arabic roots consist of three consonants and the vowels add grammatical information when attached to these consonants. What's more, it is estimated that the average number of possible part of speech tags for a word in most languages is 2.3, whereas in modern standard Arabic is 19.2 (Farghaly and Shaalan 2009). For example, the three consonants "كتب" ktb can stand for the verb "كَتَبَ" <kataba> "he wrote", or for the plural noun "كُتُبٌ" <kutub> "books", among other meanings. The free word order nature in Arabic sentences is another feature that makes parsing one of the most difficult tasks. i.e., we could easily change the order between the subject and the verb without the need for an agreement between one another in number (singular or plural). For instance, the following sentences are both correct in Arabic: "الأَوْلاَدُ يَلْعَبُونَ" and "يَلْعَبُ الأَوْلاَدُ" which literally mean "The boys play" and "plays the boys", respectively. These Arabic language features are generally the most common challenges faced by researchers in the Arabic corpus linguistics and NLP fields.

In this section, we provide an overview of the state-of-the-art of Arabic corpus linguistics. There have been over the past few years a tremendous growth in interest and activity in Arabic corpus building and analysis area. Yet, encouraging works are undertaken recently. Next, we list a number of relevant and recent corpora regarding their objectives and scope. The listed corpora are mainly classified based on their target language and mode, mentioning some of their characteristics such as their designated purpose, availability, size, text domain, and the presence of annotations.

4.1 Quranic Corpora

The Arabic language was codified primarily in the Quran (Watson 2002) and based on the language of the western Hijazi tribe of Quraysh, with some interference from pre-Islamic poetic *koiné* and eastern dialects. The Quranic scripture is used to guide the lives of 1.6 billion Muslims worldwide and they use it to perform their daily prayers (Yassein and Wahsheh 2016). The Quran is the finest piece of literature in the Arabic Language, and the number of non-Arabic speaker that learn Arabic language with the objective to understand the Quran is significantly increasing.

The Quran contains over 77,000 words, it is divided into 114 chapters where each chapter is divided into verses, adding up to a total of 6,243 verses. Some relevant corpora are created from the original text of the holy Quran, namely:

- Quran Corpus of Haifa (Dror et al. 2004): This corpus has been built using an automatic morphological analysis on the Quranic text. However, the work is not complete, it remains manually unverified and has multiple possible analyses for each word in the final published data set. Considering a random sample, the authors of the Haifa corpus estimate the final accuracy of annotation using an F-measure of 86%. Further, approximately 40% of the roots in Haifa's corpus are missing and the word's lemmas are not given.
- The Quranic Arabic Corpus (Dukes and Habash 2010): It is an online-annotated corpus with multiple layers of annotation including morphological segmentation, part of speech tagging, syntactic analysis using dependency grammar and a semantic ontology. Despite that this corpus is manually verified, it has some problems on the level of lemmas and roots, and has not sufficient grammatical information; yet, the patterns are not given.
- QurAna corpus (Sharaf and Atwell 2012a): In this corpus, only the personal pronouns are tagged with antecedent information (over 24,500 pronouns). These antecedents are maintained as an ontological list of concepts. The Quranic Arabic Corpus was used to identify the targeted segments that contain pronouns, and for each pronoun, the starting and ending IDs of the text span that represents antecedents were recorded manually through forms developed using PHP scripting language.
- QurSim corpus (Sharaf and Atwell 2012b): It is an annotated corpus where semantically similar or related verses are linked together. With the help of domain experts, the authors adopt the same methodology of *Ibn Kathir*, a Muslim scholar who is known for his classic book of Quran commentary (or Tafsir in Arabic). In fact, the principle of this method is to link two verses if one of them was cited while commenting on the other. The size of the dataset is over 7,600 pairs of related verses and the authors claimed that this dataset could be extended to over 13,500 pairs of related verses observing the commutative property of strongly related pairs.
- The Boundary-Annotated Quran Corpus (Sawalha et al. 2014): Unlike the other Quranic corpora, the words in this corpus are tagged with prosodic and boundary annotation rather than morphological or syntactical annotation. It was built by gathering and tracking boundary stops from the "Tanzil Quran project",[1] the part of speech tags from the Quranic Arabic Corpus, and the prosodic annotation scheme from Tajwid (recitation) mark-up in the Quran.
- Qurany[2]: the Quranic text is augmented with an ontology or index of key concepts that were imported from "Mushaf Al Tajweed", a recognized expert source which is compiled by Dr. Mohamed Habash, Director of the Islamic

[1]http://tanzil.net/.

[2]http://quranytopics.appspot.com/.

Studies Centre in Damascus, published by Dar Al-Maarifah in Syria and authenticated by the Al-Azhar Islamic Research Academy in Egypt. The Qurany allows users to search in the Holy Quran for abstract concepts via an ontology browser. Users can use this browser to identify a precise concept among nearly 1200 concepts and find the related verses to this concept; yet, the corpus includes 8 variant English translations.
- Al-Mus'haf corpus (Zeroual and Lakhouaja 2016b): It is an enriched corpus with morphosyntactical information. The process of building this corpus consists of a semi-automatic technique by using "AL-Khalil Morpho Sys" (Boudchiche et al. 2016), then manual processes. The corpus has 1770 roots, vowelled patterns for each stem and lemma, over 100 part of speech tags used, and true lemma (1554 patterns).

4.2 Classical Arabic Corpora

The Classical Arabic (CA) is the form of Arabic language particularly used in literary texts and applied on the academic and religious levels. The Quran is considered to be the highest form of CA texts. The amount of published CA texts is higher than the texts published in Modern Standard Arabic. Consequently, the free and large linguistic resources published by the Arabic corpus linguistic community are available in CA. For instance:

- the King Saud University Corpus of Classical Arabic (KSUCCA) (Alrabiah et al. 2013): the corpus contains 50M+ words. The data of the corpus includes only pure CA texts, the resources dated back to the period of the pre-Islamic era until the end of the 4th Hijri century (equivalent to the period from the 7th to early 11th century CE). The corpus covers six broad genres which are most of the topics that were popular in that period. These are: Religion, Linguistics, Literature, Science, Sociology, and Biography. The major resources of the corpus were extracted from the Shamela[3] library. Recently, the corpus has been tagged using MADA + Tokan (Habash et al. 2009) with a tagset that consist of 41 basic tags where the estimated accuracies are: 87.80% for lemmas, 84.90% for roots, 83.40% for part of speech tagging, 89.90% for masculinity and femininity, and finally, 90.10% for singularity and plurality. KSUCCA[4] is available for download.
- Shamela (Belinkov et al. 2016): It is a large-scale historical Arabic corpus from diverse periods of time (from the 7th century to the modern era). The corpus is drawn from the Shamela library, it is a voluntary project accomplished by the cooperation between the site's owners and Alrawdhah[5] cooperative Office for

[3]http://shamela.ws/.

[4]http://ksucorpus.ksu.edu.sa.

[5]http://www.arrawdah.com.

Call and Guidance. The corpus is cleaned and organized with a metadata information in a semi-automatic process. Yet the entire corpus is processed with Madamira (Pasha et al. 2014), a state-of-the-art morphosyntactical analyser and disambiguator. The result is a full analysis per word, including tokenization, lemmatization, part-of-speech-tagging, and various morphological features. The corpus contains over 6,000 texts, totalling around 1 billion words, of which 800 million words are from dated texts and the rest parts are automatically dated by building a 5-gram language model with Kneser-Ney smoothing, using the SRILM toolkit (Stolcke et al. 2011).

- Tashkeela (Zerrouki and Balla 2017): It is a recent corpus of Arabic raw and diacritized texts that contains over 75 million of fully vocalized words obtained from 97 Islamic books filtered from 7079 books from Shamela Library. These classical books present 98.85% of corpus data, while 1.25% are data collected from 20 modern books and texts crawled from Internet websites such as Aljazeera Learning Arabic.[6] Tashkeela corpus is available for download from its project website.[7]

4.3 Modern Standard Arabic Corpora

The Modern Standard Arabic (MSA) is the form used in contemporary scholarly published works as well as in the media. MSA does not differ from CA in morphology or syntax, but richness of stylistic and lexis usage is apparent in Classical works. Most researchers on Arabic corpus linguistics has concentrated on MSA, and they have an aspect of making their works publicly available which can be used for different purposes. In 2010, The Mediterranean Arabic Language and Speech Technology (MEDAR) conducted a survey[8] to list the projects carried out for developing Arabic language resources. Unfortunately, the produced list of this survey is no longer updated. Another and probably the very recent survey (Zaghouani 2017) is conducted to identify the list of the freely available Arabic corpora and language resources. The survey published an initial list of 66 sources. A few major examples are mentioned here to give an idea of the variety of what is available.

- Open Source Arabic Corpora (OSAC) (Saad and Ashour 2010): It is a collection of largest free accessible raw corpora and a large number of web documents extracted from over 25 Arabic websites using the open source offline explorer, *HTTrack*. The corpus compiling includes converting html/xml files into UTF-8 encoding using "Text Encoding Converter" by *WebKeySoft* and removing the

[6]http://learning.aljazeera.net.

[7]https://sourceforge.net/projects/tashkeela/.

[8]http://www.medar.info/MEDAR_Survey_III.pdf.

html/xml tags. The corpus contains about 113 million words and covers several topics: Economy, History, Education, Religion, Sport, Health, Astronomy, Law, Stories, and Cooking Recipes.

- Alwatan-2004 (Abbas et al. 2011): The main purpose of compiling this corpus is to evaluate topic identification methods, but it is suitable for other Arabic NLP tasks. The corpus contains 20,291 documents that correspond to 10 million words and organized in six topics (religion, economy, local news, international news and sport). These documents were downloaded from the Saudi newspaper Alwatan.[9] The texts were prepared by removing punctuation marks, digits, and Stop-List words. For the next stage of treatment, the corpus was tagged via the KALIMAT Corpus (El-Haj and Koulali 2013) using the Stanford Arabic part of speech tagger (Toutanova et al. 2003) (33 basic tags).
- El-Haj et al. (2015) creates several resources (e.g., KALIMAT, EASC, and ABMC) for those working on computational methods to analyse and study languages primary Arabic. These resources are articles collected from the Arabic language version of Wikipedia and two Arabic newspapers (Alrai[10] and Alwatan (Footnote 9)), and human-generated extractive summaries of those articles. A group of students was asked to search for Wikipedia and select articles within ten given subject areas (art and music, the environment, politics, sports, health, finance and insurance, science and technology, tourism, religion, and education). El-Haj et al. use Amazon's Mechanical Turk to recruit appropriately skilled human participants to create the gold-standard summaries of the collected articles. The final corpus includes a total of 153 documents, containing 18,264 words and each document consists an average of 380 words, with a minimum of 116 words and a maximum of 971 words.

Several MSA corpora were designed and built for more specific purposes but they can be used for other purposes if the relevant research question(s) can be answered. The Corpus of Contemporary Arabic (CCA[11]) (Al-Sulaiti and Atwell 2006) and the Arabic Learner Corpus v2 (ALC) (Alfaifi et al. 2014) are generally compiled for language teaching and learning research. The ALC is a written and spoken MSA corpus developed at Leeds University between 2012 and 2013. The data were produced by 942 learners of Arabic, from 67 different nationalities studying at pre-university and university levels in Saudi Arabia. The available corpus comprises 282,732 words stored in TXT and XML formats, hand-written sheets which are in PDF format as well as the audio recordings which are available in MP3 format. The CCA corpus is designed to resemble the American National Corpus, it is also a written and spoken MSA corpus collected from 1990s up to 2005 and covers several topics; For the written part, the six major topics were Fiction, Arts, Science, Business, Miscellaneous; and for the spoken part the data were derived from TV, Radio, Conversation. Concerning the resources selection

[9]http://www.alwatan.com.sa.

[10]http://alrai.com/.

[11]http://www.comp.leeds.ac.uk/eric/latifa/research.htm.

phase, the authors carried out a survey of language teachers and language engineers to get their opinions on the texts that might be of use to them. As a result, they managed to compile a corpus of 1 million words. Finally, It may deserve to mention here, the Qatar Arabic Language Bank (QALB), an ongoing project to build a large error-annotated corpus of Arabic text with manual corrections (Zaghouani et al. 2014; Rozovskaya et al. 2015). The QALB corpus will be beneficial for corpus-based studies of errors and for design and develop Arabic automatic spelling and text correction tools. The corpus includes texts gathered from online user comments written to Aljazeera articles, it also includes texts produced by learners of Arabic as a foreign language. Next, these texts were processed with the morphosyntactical analyser Madamira. The texts are then manually annotated for errors. The errors include spelling, punctuation, word choice, morphology, syntax, and dialectal usage.

4.4 General Corpora

The general corpora can be used for general corpus-based studies or for inferring the general patterns of the Arabic language for NLP applications. Typically, they include a large number of data targeting more than one language and mode (e.g., MSA and CA). However, this category of corpora is rare and usually not available for free. As examples, we state:

- arTenTen (Arts et al. 2014): It is a member of the TenTen Corpus Family (Jakubíček et al. 2013). The arTenTen is a web-crawled corpus of Arabic, it was crawled using Spiderling (Suchomel et al. 2012) gathered in 2012. The arTenTen corpus is partially tagged, one sample of the corpus comprises roughly 30 million that were tagged using the Stanford Arabic PoS tagger; and another sample that contains over 115 million words were tokenised, lemmatised, and part of speech tagged using MADA system. The arTenTen comprises 5.8 billion words but it can only be explored by paying a fee via the Sketch Engine website.[12]
- King Abdulaziz City for Science and Technology (KACST) Arabic corpus (Al-Thubaity 2015): It comprises over 731 million words from pre-Islam until 2013 (more than 1,500 years). The corpus aims to represent the two main forms of Arabic language (CA and MSA) and the transience between them. The resources were collected mainly from all Arab countries, but also Arabic publications from other regions. The KACST includes a wide diversity of texts content covering 10 mediums namely Old Manuscripts, Books, Newspapers, Magazines, Curricula, University Theses, Websites, Refereed Periodicals, Official Prints, and News Agencies. Each text has been further classified more specifically into 80 domains (e.g., Islamic and Arabic Poems) and 481 topics

[12]https://www.sketchengine.co.uk/.

(e.g., Hadeeth, love, and wisdom). In order to allow the corpus users to study the Arabic language and its many varieties in both general and specific ways, and across many different levels, thereby allowing for more accurate language models to be constructed, the following metadata were assigned to each text: title, year of publication, time period, author name and gender, region, medium, domain, and topic.

4.5 Multilingual Corpora Including Arabic Language

The Arabic language has been included in relevant pioneer multilingual corpora such as the open source parallel corpus (OPUS) (Tiedemann 2012), the largest collection of freely available parallel corpora in more than 90 languages and includes data from several domains. This corpus comprises over 40 billion tokens in 2.7 billion parallel units (aligned sentences and sentence fragments). The largest sources of OPUS are legislative and administrative texts (mostly from the European Union and associated institutions), translated movie subtitles, newspapers, and localization data from open-source software projects. Arabic is one of the top languages that OPUS sub-corpora are well above 100 million tokens. For instance, the MultiUN corpus (Multilingual Corpus from United Nation) (Chen and Eisele 2012), which is available in all six official languages of the United Nation and the German language, comprises 271.5 million Arabic tokens. Table 1 provides more statistics about MultiUN corpus.

More details are provided by Steinberger et al. in an overview (Steinberger et al. 2014) of these and other European multilingual parallel corpora. Except these European resources, the Arabic language is covered by a small number of bilingual and multilingual corpora such as the tiny Arabic-English parallel corpus (10 K sentences) used to build an Arabic stemmer based on statistical machine translation using an English stemmer (Rogati et al. 2003). A similar Arabic-English parallel corpus has been adopted to handle the word translation disambiguation (Ahmed and Nürnberger 2008). In addition, a multilingual named entity corpus for Arabic, English, and French has been developed based on comparable newswires from the "Agence France Presse" covering the period 2004–2006 (Mostefa et al. 2009). Finally, a free Arabic-English parallel corpus has been built within the project MEDAR (Maegaard et al. 2009) which has been running from 2008 to 2010.

On the other hand, the Arabic language is more presented in multilingual corpora which their data are based on video's subtitles. For example, the AMARA corpus (Abdelali et al. 2014), a parallel corpus of educational video subtitles, multilingually aligned for 20 languages including Arabic. The data of this corpus are collected in cooperation with Amara platform.[13] 3000 videos have available subtitles in at least six languages and 1000 videos have available subtitles in 25

[13]http://amara.org.

Table 1 Statistics about the MultiUN corpus

Languages	Files	Tokens	Sentences
ar	68,870	271.5M	11.1M
de	4,034	6.7M	0.2M
en	100,373	443.5M	17.2M
es	5,683	30.1M	1.0M
fr	90,826	474.3M	14.9M
ru	81,258	328.5M	13.9M
zh	69,360	83.1M	10.9M

languages. A similar project[14] called WIT^3 (Cettolo et al. 2012), an acronym for Web Inventory of Transcribed and Translated Talks. It is a collection of lecture translations that have been automatically crawled from the TED talks[15] in 109 languages. The purpose of this project is to support the machine translation evaluations campaigns of the International Workshop on Spoken Language Translation (IWSLT) (Paul et al. 2010). As of October 2011, 17 thousand transcripts corresponding to translations of around 1000 talks have been collected. The WIT^3 comprises 2.4 million Arabic tokens. Recently, an ongoing project (Zeroual and Lakhouaja 2016a) seeks to build a new multilingual aligned and part of speech tagged corpus based on TED talks. This corpus includes 1100 talks where 100 videos were selected from different 11 topics which are "Architecture and Design", "Art and Creativity", "Culture and Stories", "Economy and Innovation", "Education and Learning", "Global Issues", "Health and Medicine", "Nature and Environment", "Science and Tech", "Social Issues" and "Sports and Adventure". According to the authors, Arabic is the second most used language in the 1100 translated talks. Figure 1 presents the overall distribution of the top 30 languages by the number of talks.

The corpus will be released in an XML file for each talk which will include all sentences paired, tagged, and aligned with different languages. Moreover, the XML files contain a metadata of the talk, for instance, talk id, title, category, translator, time slot, speaker, and language id.

4.6 Dialectal Corpora

The colloquial Arabic dialects (or Al-'ammiyya) is the form of Arabic used in everyday oral communication. It differs significantly in each Arab country; nevertheless, they are all mutually intelligible. Dialectal Arabic varieties, notably Egyptian (Maamouri et al. 2014) and Gulf Arabic (Khalifa et al. 2016), have lately received some attention and have a growing collection of resources that include

[14]https://wit3.fbk.eu/.

[15]http://www.ted.com/.

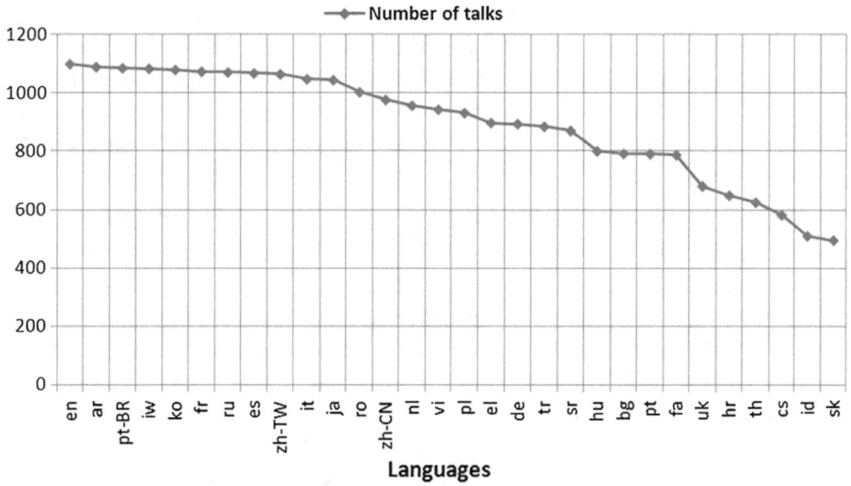

Fig. 1 Distribution of the top 30 languages by the number of talks

annotated corpora (Cotterell and Callison-Burch 2014), a neural architecture for Dialectal Arabic Segmentation (Samih et al. 2017), sentiment analysis (Mdhaffar et al. 2017), and morphological analysers (Salloum and Habash 2014; Khalifa et al. 2017). Additionally, the first project to build a multidialectal Arabic parallel corpus (Bouamor et al. 2014) has been launched three years ago; the corpus contains a collection of 2,000 sentences in Standard Arabic, Egyptian, Tunisian, Jordanian, Palestinian, Syrian Arabic, and English.

4.7 Tools and Analysers

After building corpora, it is just a matter of time until preliminary tools appear to analyse these data such as the aConCorde (Roberts et al. 2006): a proper concordance of Arabic; and the Arabic module for NooJ (Silberztein 2016): a corpus processing system. On the other hand, some improvements have been made in the analysis of corpora primarily in the accuracy of the automatic grammatical tagging and parsing of texts. These achievements come from two robust morpho-syntactical analysers:

1. Madamira (Pasha et al. 2014): It is an analyser that assigns the suitable morpho-syntactical tags for each input word according to the context. It is the result of combining two morphological analysis systems, MADA (Habash et al. 2009) and AMIRA (Diab 2009). The system analyses the input word out of the context which usually results more than one output. Then, to estimate only one solution from among the multiple outputs, a disambiguation task based on support vector machines and N-gram language models is performed;

2. AlKhalil Morpho Sys (Boudchiche et al. 2016, 2). It comes with very high coverage since the percentage of analysed words exceeds 99%. In addition to the new morphological features included like lemma and its pattern, the number of analysed words per second is 631 words. Moreover, the analyser provides supplementary functionalities such as indexing and the ability to search by the root.

MORPH2 (Kammoun et al. 2010), Qutuf (Altabba et al. 2010), SALMA (Sawalha et al. 2013), and ASMA (Abdul-Mageed et al. 2013) are also relevant morphological analysers that achieved satisfactory results as their developers claimed. However, they are not available for reuse or evaluation.

The Part of Speech (PoS) tagging is a key feature in the development of very valuable corpora which provides useful information for many advanced NLP applications, information retrieving, and text-to-speech synthesis systems (Abumalloh et al. 2016). The Arabic PoS taggers have been successfully yielded the results of the state-of-the-art. Regarding the availability, the novelty, and the efficiency, we can state (Ababou and Mazroui 2016) and (Darwish et al. 2014) which target the fine-grained PoS tagging; whereas the tagger (Imad and Abdelhak 2016) is for a standard use.

Dukes (2015), in his thesis work, aims to study the Arabic traditional grammar to develop a representation for parsing as it describes syntax using a hybrid of phrase structure and dependency relations. The main contributions were the first treebank for Classical Arabic and the first statistical dependency-based parser in any language for ellipsis, dropped pronouns, and hybrid representations. A similar study has been conducted but for modern standard Arabic. The study addressed the problem of exploiting the Pen Arabic Treebank (Maamouri et al. 2013) for statistical parsing (Al-Emran et al. 2015). For more details about the state-of-the-art Arabic parsers, a survey (Zaki et al. 2016) is just conducted that covers 22 developed syntactic parsers within the last 10 years.

4.8 Situation, Challenge and Recommendation

Typically, A resource-poor language refers to the language that lacks "the basic resources that are fundamental to computational linguistics" and has a few and small corpora (Zamin et al. 2012). Despite the huge effort made in Arabic linguistics field, Arabic is relatively a resource-poor language, at least until 2011, when Rabiee (2011) reported that not a single modern standard Arabic tagged corpus was freely or publicly available. This grabs the attention of researchers and academia to this field in order to bridge the gap between Arabic and other resource-rich languages such as English, German, and French, which they have a vast number of resources and natural language processing tools.

Several factors may explain the limits of Arabic corpora building projects such as the morphology of Arabic, which differs in the structure of affixes from

Indo-European languages (Gharaibeh and Gharaibeh 2012) (e.g., free word-order syntax, Diacritics, and elliptical personal pronoun). Arabic is a rich Semitic language and highly productive both derivationally and inflectionally (Alsaedi et al. 2016). These two complex paradigms are based on the interaction between roots and patterns which have intrigued lexicographers and morphologists for centuries. Further, most researchers have often proposed corpora or provided analysis tools that comply with their suitable objectives without considering the standardization and the international aspects. It is well known that the creation of valuable corpora is expensive, time-consuming, and requires specialized personnel. Therefore, another factor is the absence of funding and investment for the development of large and annotated Arabic corpora. Mansour (2013) points out that launching such huge projects needs the collaboration of different national institutions primarily Universities and research centers as well as the governments' fund and support.

5 Conclusion

In the main part of this chapter, we provide the reader with the necessary background information for understanding the corpus linguistics. Further, we describe how corpus linguistics is related to other fields which contribute significantly to their development. As seen, corpora are considered as a source of evidence for linguistic description and argumentation and play a vital role in many aspects of lexicography such as in dictionary creation. Besides, they are one of the influence factors in pedagogical research, language learning and teaching, natural language processing, and information retrieval, among others.

Through discussion of corpus linguistics and reviewing the history of corpora development, we present essential stages of the corpus building procedure, and what make it different from another kind of data storage like archives and database. Corpus linguistics is not limited to the process of corpus design and compilation but also analysing the data collected and present it in a suitable format. Further, some of the key features of corpus annotation, lemmatization, PoS tagging, and parsing are highlighted as advanced automatic text processing tasks and relevant forms of annotation.

The chief purpose of this study is to shed lights on the major progress achievement in Arabic corpus linguistics, the challenges, and the opportunities. Despite the proud heritage of the Arabic language, its lexical richness, the significant online user growth, and the huge effort made recently in this field, Arabic is relatively a resource-poor language. This calls the corpus linguistics society to put more emphasis on Arabic. It worth mentioning, there are several challenges imposed by Arabic language nature, the absence of standardization and the international aspects in most available corpora. Further, the need is great for a collaboration of different national institutions primarily Universities and research centers as well as the governments' fund and support.

References

Ababou, N., Mazroui, A.: A hybrid Arabic POS tagging for simple and compound morphosyntactic tags. Int. J. Speech Technol. **19**, 289–302 (2016)

Abbas, M., Smaïli, K., Berkani, D.: Evaluation of topic identification methods on Arabic corpora. JDIM **9**, 185–192 (2011)

Abdelali, Ahmed, Guzman, Francisco, Sajjad, Hassan, Vogel, Stephan: The AMARA corpus: building parallel language resources for the educational domain. In LREC **14**, 1044–1054 (2014)

Abdul-Mageed, M., Diab, M.T., Kübler, S.: ASMA: a system for automatic segmentation and morpho-syntactic disambiguation of modern standard Arabic. In: RANLP, pp. 1–8 (2013)

Abumalloh, R.A., Al-Sarhan, H.M., Ibrahim, O., Abu-Ulbeh, W.: Arabic part-of-speech tagging. J: Soft Comput. Decis. Support Syst. **3**, 45–52 (2016)

Ahmed, F., Nürnberger, A.: Arabic/english word translation disambiguation using parallel corpora and matching schemes. In: Proceedings of EAMT, vol. 8, p. 28 (2008)

Al-Dahdah, A.: The Grammar of the Arabic Language in Tables And Lists. Maktabat Lebnan, Beirut (1989). [in Arabic]

Al-Emran, M., Zaza, S., Shaalan, K.: Parsing modern standard Arabic using Treebank resources. In: 2015 International Conference on Information and Communication Technology Research (ICTRC), pp. 80–83. IEEE (2015)

Alfaifi, A.Y.G., Atwell, Eric, Hedaya, I.: Arabic learner corpus (ALC) v2: a new written and spoken corpus of Arabic learners. Proc. Learn. Corpus Stud. Asia World **2014**(2), 77–89 (2014)

Alrabiah, M., Al-Salman, A., Atwell, E.S.: The design and construction of the 50 million words KSUCCA. In: Proceedings of WACL'2 Second Workshop on Arabic Corpus Linguistics, pp. 5–8. The University of Leeds (2013)

Alsaedi, N., Peter B., Rana, O.F.: Sensing real-world events using Arabic Twitter posts (2016)

Al-Sulaiti, L., Atwell, E.S.: The design of a corpus of contemporary Arabic. Int. J. Corpus Linguist. **11**, 135–171 (2006)

Altabba, M., Al-Zaraee, A., Shukairy, M.A.: An Arabic morphological analyzer and part-of-speech tagger. A Thesis Presented to the Faculty of Informatics Engineering, Arab International University, Damascus, Syria (2010)

Al-Thubaity, A.O.: A 700M+ Arabic corpus: KACST Arabic corpus design and construction. Lang. Resour. Eval. **49**, 721–751 (2015)

Armstrong, S., Church, K., Isabelle, P., Manzi, S., Tzoukermann, E., Yarowsky, D.: Natural Language Processing Using Very Large Corpora, vol. 11. Springer Science & Business Media (2013)

Arts, T., Belinkov, Y., Habash, N., Kilgarriff, A., Suchomel, V.: arTenTen: Arabic corpus and word sketches. J. King Saud Univ.—Comput. Inf. Sci. **26**, Special Issue on Arabic NLP, 357–371 (2014). https://doi.org/10.1016/j.jksuci.2014.06.009

Atkins, S., Clear, J., Ostler, N.: Corpus design criteria. Lit. Linguist. Comput. **7**, 1–16 (1992)

Attia, M., Van Genabith, J.: A jellyfish dictionary for Arabic. In: Electronic Lexicography in the 21st Century: Thinking Outside the Paper: Proceedings of the eLex 2013 Conference, 17–19 October 2013, Tallinn, Estonia, pp. 195–212 (2013)

Ball, C.N.: Automated text analysis: Cautionary tales. Lit. Linguist. Comput. **9**, 295–302 (1994)

Baneyx, A., Charlet, J., Jaulent, M.-C.: Building an ontology of pulmonary diseases with natural language processing tools using textual corpora. Int. J. Med. Inform. **76**, 208–215 (2007)

Belinkov, Y., Magidow, A., Romanov, M., Shmidman, A., Koppel, M.: Shamela: A Large-Scale Historical Arabic Corpus (2016). arXiv:1612.08989:45

Bertels, A.: Corpus Linguistics for Language Teaching and LSP (2017)

Bhattacharya, P., Goyal, P., Sarkar, Sudeshna: Query translation for cross-language information retrieval using multilingual word clusters. WSSANLP **2016**, 152 (2016)

Biber, D., Conrad, S., Reppen, R.: Corpus Linguistics: Investigating Language Structure and Use. Cambridge University Press (1998)

Bongers, H.: The History and Principles of Vocabulary Control: As It Affects in General and of English in Particular. 3. The KLM-List. Wocopi (1947)

Bouamor, H., Habash, N., Oflazer, K.: A multidialectal parallel corpus of Arabic. In LREC, pp. 1240–1245 (2014)

Boudchiche, M., Mazroui, A., Bebah, M.O.A.O., Lakhouaja, A., Boudlal, A.: AlKhalil morpho sys 2: a robust Arabic morpho-syntactic analyzer. J. King Saud Univ.—Comput. Inf. Sci. (2016). https://doi.org/10.1016/j.jksuci.2016.05.002

Boulton, A., Landure, C.: Using corpora in language teaching, learning and use. Recherche et pratiques pédagogiques en langues de spécialité. Cahiers de l'Apliut **35** (2016)

Cettolo, M., Girardi, C., Federico, M.: Wit3: web inventory of transcribed and translated talks. In: Proceedings of the 16th Conference of the European Association for Machine Translation (EAMT), pp. 261–268 (2012)

Chen, Y., Eisele, A.: MultiUN v2: un documents with multilingual alignments. In: LREC, pp. 2500–2504 (2012)

Chennoufi, A., Mazroui, A.: Impact of morphological analysis and a large training corpus on the performances of Arabic diacritization. Int. J. Speech Technol. **19**, 269–280 (2016)

Cotterell, R., Callison-Burch, C.: A multi-dialect, multi-genre corpus of informal written Arabic. In: LREC, pp. 241–245 (2014)

Darwish, K., Abdelali, A., Mubarak, H.: Using stem-templates to improve Arabic POS and gender/number tagging. In: LREC, pp. 2926–2931. Citeseer (2014)

Diab, M.: Second generation AMIRA tools for Arabic processing: fast and robust tokenization, POS tagging, and base phrase chunking. In: 2nd International Conference on Arabic Language Resources and Tools (2009)

Dror, Judith, Shaharabani, Dudu, Talmon, Rafi, Wintner, Shuly: Morphological analysis of the Qur'an. Lit. Linguist. Comput. **19**, 431–452 (2004)

Dukes, K.: Statistical parsing by machine learning from a classical Arabic treebank (2015). arXiv:1510.07193

Dukes, K., Habash, N.: Morphological annotation of Quranic Arabic. In: *LREC* (2010)

El-Haj, M., Koulali, R.: KALIMAT a multipurpose Arabic Corpus. In: Second Workshop on Arabic Corpus Linguistics (WACL-2), pp. 22–25 (2013)

El-Haj, M., Kruschwitz, U., Fox, C.: Creating language resources for under-resourced languages: methodologies, and experiments with Arabic. Lang. Resour. Eval. **49**, 549–580 (2015)

Farghaly, Ali, Shaalan, Khaled: Arabic natural language processing: challenges and solutions. ACM Trans. Asian Lang. Inf. Process. (TALIP) **8**, 14 (2009)

Francis, W., Kucera, H.: Frequency analysis of English usage (1982)

Ghalayini, M.I.M.S.: Jami'al-durus al-'arabiyah. Turath For Solutions (2013)

Gharaibeh, I.K., Gharaibeh, N.K.: Towards Arabic noun phrase extractor (ANPE) using information retrieval techniques. Softw. Eng. **2**, 36–42 (2012)

Habash, N., Rambow, O., Roth, R.: MADA + TOKAN: a toolkit for Arabic tokenization, diacritization, morphological disambiguation, POS tagging, stemming and lemmatization. In: Proceedings of the 2nd International Conference on Arabic Language Resources and Tools (MEDAR), Cairo, Egypt, pp. 102–109 (2009)

Halliday, M., Matthiessen, C.M.I.M., Matthiessen, C.: An Introduction to Functional Grammar. Routledge (2014)

Hu, K.: Corpus-based translation studies: problems and prospects. In: Introducing Corpus-based Translation Studies, pp. 223–233. Springer

Hu, K., et al.: Introducing Corpus-Based Translation Studies. Springer (2016)

Hunston, S.: Corpus linguistics: historical development. In: The Encyclopedia of Applied Linguistics (2013)

Hyland, K.: Teaching and Researching Writing. Routledge (2015)

Imad, Z., Abdelhak, L.: Adapting a decision tree based tagger for Arabic, pp. 1–6. IEEE (2016). https://doi.org/10.1109/IT4OD.2016.7479306

Jakubíček, M., Kilgarriff, A., Kovář, V., Rychlý, P., Suchomel, V.: The tenten corpus family. In: 7th International Corpus Linguistics Conference CL, pp. 125–127 (2013)
Jurida, H.S., Džanić, M., Pavlović, T., Jahić, A., Hanić, J.: Netspeak: linguistic properties and aspects of online communication in postponed time. J. Foreign Lang. Teach. Appl. Linguist. **3**, 1–19 (2016)
Kammoun, N.C., Belguith, L.H., Hamadou, A.B.: The MORPH2 new version: a robust morphological analyzer for Arabic texts. In: JADT 2010: 10th International Conference on Statistical Analysis of Textual Data (2010)
Kennedy, G.: An Introduction to Corpus Linguistics. Routledge (2014)
Khalifa, S., Habash, N., Abdulrahim, D., Hassan, S.: A large scale corpus of Gulf Arabic (2016). arXiv:1609.02960
Khalifa, S., Hassan, S., Habash, N.: A morphological analyzer for Gulf Arabic verbs. WANLP 2017 (co-located with EACL 2017), 35 (2017)
Khorsheed, M.S., Alhazmi, K.M., Asiri, A.M.: Developing typewritten Arabic corpus with multi-fonts (TRACOM). In: Proceedings of the International Workshop on Multilingual OCR, p. 16. ACM (2009)
Kilgarriff, A.: Using corpora as data source for dictionaries. In: The Bloomsbury Companion to Lexicography, pp. 77–96. Bloomsbury, London (2013)
Leech, G.N.: The state of the art in corpus linguistics. In: Aijmer, K., Altenberg, B. (eds.) English Corpus Linguistics: Studies in Honor of Jan Svartuk. Longman, London (1991)
Leech, G.: Corpora and theories of linguistic performance. In: Directions in Corpus Linguistics, pp. 105–122 (1992a)
Leech, G.: 100 million words of English: the British National Corpus (BNC). Lang. Res. **28**, 1–13 (1992b)
Leech, G., Rayson, P., et al.: Word Frequencies in Written and Spoken English: Based on the British National Corpus. Routledge (2014)
Lefever, E., Hoste, V.: Semeval-2013 task 10: Cross-lingual word sense disambiguation. In: Proceedings of SemEval, pp. 158–166 (2013)
Li, L., Forascu, C., El-Haj, M., Giannakopoulos, G.: Multi-document multilingual summarization corpus preparation, part 1: Arabic, English, Greek, Chinese, Romanian. In: Association for Computational Linguistics (2013)
Liua, Q., Jiangb, H., Linga, Z.-H., Zhuc, X., Weid, S., Hua, Y.: Commonsense Knowledge Enhanced Embeddings for Solving Pronoun Disambiguation Problems in Winograd Schema Challenge (2016). arXiv:1611.04146
Maamouri, M., Bies, A., Kulick, S., Ciul, M., Habash, N., Eskander, R.: Developing an Egyptian Arabic treebank: impact of dialectal morphology on annotation and tool development. In: LREC, pp. 2348–2354 (2014)
Maamouri, M., Bies, A., Kulick, S., Gaddeche, F., Mekki, W., Krouna, S., Bouziri, B., Zaghouani, W.: Arabic Treebank: Part 1 v 4.1 (2013)
Maegaard, B., Attia, M., Choukri, K., Krauwer, S., Mokbel, C., Yaseen, M.: MEDAR: Arabic language technology, state-of-the-art and a cooperation roadmap. In: Proceedings of the Second International Conference on Arabic Language Resources and Tools. Citeseer (2009)
Magdy, W., Jones, G.J.F.: Studying machine translation technologies for large-data CLIR tasks: a patent prior-art search case study. Inf. Retr. **17**, 492–519 (2014)
Mansour, M.: The absence of Arabic corpus linguistics: a call for creating an Arabic national corpus. Int. J. Human. Soc. Sci. **3**, 81–90 (2013)
Marcus, M.P., Marcinkiewicz, M.A., Santorini, B.: Building a large annotated corpus of English: the Penn Treebank. Comput. Linguist. **19**, 313–330 (1993)
McEnery, T., Wilson, A.: Corpus linguistics. Edinburgh University Press, Edinburgh (1996)
Mdhaffar, S., Bougares, F., Esteve, Y., Hadrich-Belguith, L.: Sentiment analysis of tunisian dialect: linguistic resources and experiments. In: WANLP 2017 (co-located with EACL 2017), pp. 55 (2017)
Milfull, Inge: Mutual Illumination: the dictionary of old English and the ongoing revision of the oxford english dictionary (OED3). Florilegium **26**, 235–264 (2009)

Mostefa, D., Laïb, M., Chaudiron, S., Choukri, K., Chalendar, G.: A multilingual named entity corpus for Arabic, English and French. In: MEDAR 2009, 2nd (2009)
Nakov, P.: Web as a corpus: going beyond the n-gram. In: Russian Summer School in Information Retrieval, pp. 185–228. Springer (2014)
Nation, I.S.P.: Teaching & learning vocabulary. Heinle Cengage Learning, Boston (2013)
Nothman, J., Ringland, N., Radford, W., Murphy, T., Curran, J.R.: Learning multilingual named entity recognition from Wikipedia. Artif. Intell. **194**, 151–175 (2013)
O'Keeffe, A., McCarthy, M.: The Routledge Handbook of Corpus Linguistics. Routledge (2010)
Pasha, A., Al-Badrashiny, M., Diab, M., El Kholy, A., Eskander, R., Habash, N., Pooleery, M., Rambow, O., Roth, R.M.: Madamira: a fast, comprehensive tool for morphological analysis and disambiguation of arabic. In: Proceedings of the Language Resources and Evaluation Conference (LREC), Reykjavik, Iceland (2014)
Paul, M., Federico, M., Stüker, S.: Overview of the IWSLT 2010 evaluation campaign. In: IWSLT, vol. 10, pp. 3–27 (2010)
Rabiee, H.S.: Adapting standard open-source resources to tagging a morphologically rich language: a case study with Arabic. In: RANLP Student Research Workshop, pp. 127–132 (2011)
Roberts, A., Al-Sulaiti, L., Atwell, E.: aConCorde: towards an open-source, extendable concordancer for Arabic. Corpora **1**, 39–60 (2006)
Rogati, M., McCarley, S., Yang, Y.: Unsupervised learning of arabic stemming using a parallel corpus. In: Proceedings of the 41st Annual Meeting on Association for Computational Linguistics—Volume 1, pp. 391–398. Association for Computational Linguistics, ACL '03. Stroudsburg, PA, USA (2003). https://doi.org/10.3115/1075096.1075146
Rozovskaya, A., Bouamor, H., Habash, N., Zaghouani, W., Obeid, O., Mohit, B.: The second QALB shared task on automatic text correction for Arabic. In: ANLP Workshop 2015, pp. 26 (2015)
Saad, M.K., Ashour, W.: Osac: open source arabic corpora. In: 6th ArchEng Int. Symposiums, EEECS, vol. 10 (2010)
Sahragard, R., Kushki, A., Ansaripour, E.: The application of corpora in teaching grammar: the case of English relative clause. J. Pan-Pac. Assoc. Appl. Linguist. **17**, 79–93 (2013)
Sakho, M.L.: Teaching Arabic as a Second Language in International School in Dubai A Case Study Exploring New Perspectives in Learning Materials Design and Development. British University in Dubai (2012)
Salloum, W., Habash, N.: Adam: analyzer for dialectal arabic morphology. J. King Saud Univ.-Comput. Inf. Sci. **26**, 372–378 (2014)
Samih, Y., Attia, M., Eldesouki, M., Mubarak, H., Abdelali, A., Kallmeyer, L., Darwish, K.: A neural architecture for dialectal Arabic segmentation. In: WANLP 2017 (co-located with EACL 2017), pp. 46 (2017)
Sawalha, M., Atwell, E., Abushariah, M.A.M.: SALMA: standard Arabic language morphological analysis. In: 2013 1st International Conference on Communications, Signal Processing, and their Applications (ICCSPA), pp. 1–6. IEEE (2013)
Sawalha, M., Brierley, C., Atwell, E.: Automatically generated, phonemic Arabic-IPA pronunciation tiers for the boundary annotated Qur'an dataset for machine learning (version 2.0). In: Proceedings of LRE-Rel 2: 2nd Workshop on Language Resource and Evaluation for Religious Texts, LREC 2014 Post-Conference Workshop 31st May 2014, Reykjavik, Iceland, pp. 42–47. The University of Leeds (2014)
Sharaf, A.-B.M., Atwell, E.: QurAna: corpus of the Quran annotated with pronominal anaphora. In: LREC, pp. 130–137. Citeseer (2012a)
Sharaf, A.-B.M., Atwell, E.: QurSim: a corpus for evaluation of relatedness in short texts. In: LREC, 2295–2302 (2012b)
Silberztein, M.: Formalizing Natural Languages: The NooJ Approach. Wiley (2016)
Sinclair, J.: Preliminary recommendations on corpus typology. In: EAGLES Document TCWG-CTYP/P. http://www.ilc.pi.cnr.it/EAGLES/corpustyp/corpustyp.html
Sinclair, J.: Intuition and annotation—the discussion continues. Lang. Comput. **49**, 39–59 (2004)

Sinclair, J.: Corpus and text-basic principles. In: Developing Linguistic Corpora: A Guide to Good Practice, pp. 1–16 (2005)

Sinclair, J.: Borrowed ideas. Lang. Comput. Stud. Pract. Linguist. **64**, 21 (2008)

Steinberger, R., Ebrahim, M., Poulis, A., Carrasco-Benitez, M., Schlüter, P., Przybyszewski, M., Gilbro, S.: An overview of the European Union's highly multilingual parallel corpora. Lang. Res. Eval. **48**, 679–707 (2014)

Stolcke, A., Zheng, J., Wang, W., Abrash, V.: SRILM at sixteen: update and outlook. In: Proceedings of IEEE Automatic Speech Recognition and Understanding Workshop, vol. 5 (2011)

Suchomel, V., Pomikálek, J., et al.: Efficient web crawling for large text corpora. In: Proceedings of the seventh Web as Corpus Workshop (WAC7), pp. 39–43 (2012)

Teubert, W.: Corpus Linguistics and Lexicography: The Beginning of a Beautiful Friendship, Issues 31 (2015)

Tiedemann, J.: Building a multilingual parallel subtitle corpus. Proc. CLIN, 14 (2007)

Tiedemann, J.: Parallel data, tools and interfaces in OPUS. In: LREC, pp. 2214–2218 (2012)

Toutanova, K., Klein, D., Manning, C.D., Singer, Y.: Feature-rich part-of-speech tagging with a cyclic dependency network. In: Proceedings of the 2003 Conference of the North American Chapter of the Association for Computational Linguistics on Human Language Technology—Volume 1, pp. 173–180. Association for Computational Linguistics (2003)

Tsarfaty, R., Seddah, D., Kübler, S., Nivre, J.: Parsing morphologically rich languages: Introduction to the special issue. Comput. Linguist. **39**, 15–22 (2013)

Watson, J.C.E.: The Phonology and Morphology of Arabic. Oxford University Press on Demand (2002)

Xing, J., Wong, D.F., Chao, L.S., Leal, A.L.V., Schmaltz, M., Lu, C.: Syntaxtree aligner: a web-based parallel tree alignment toolkit. In: 2016 International Conference on Wavelet Analysis and Pattern Recognition (ICWAPR), pp. 37–42. IEEE (2016)

Yassein, M.B., Wahsheh, Y.A.: HQTP v. 2: holy Quran transfer protocol version 2. In: 2016 7th International Conference on Computer Science and Information Technology (CSIT), pp. 1–5. IEEE (2016)

Zaghouani, W.: Critical Survey of the Freely Available Arabic Corpora (2017). arXiv:1702.07835

Zaghouani, W., Habash, N., Mohit, B.: The qatar arabic language bank guidelines. Technical Report CMU-CS-QTR-124, School of Computer Science, Carnegie Mellon University, Pittsburgh, PA, September, 2014

Zaki, Y., Hajjar, H., Hajjar, M., Bernard, G.: A survey of syntactic parsers of arabic language. In: Proceedings of the International Conference on Big Data and Advanced Wireless Technologies, p. 31. ACM (2016)

Zamin, N., Oxley, A., Bakar, Z.A., Farhan, S.A.: A statistical dictionary-based word alignment algorithm: an unsupervised approach. In: 2012 International Conference on Computer & Information Science (ICCIS), vol. 1, pp. 396–402. IEEE (2012)

Zeroual, I., Lakhouaja, A.: Towards a multilingual aligned parallel corpus. In: Proceedings of the International Conference of High Innovation in Computer Science, Kenitra, Morocco (2016a)

Zeroual, I., Lakhouaja, A.: A new Quranic corpus rich in morphosyntactical information. Int. J. Speech Technol., 1–8 (2016b). https://doi.org/10.1007/s10772-016-9335-7

Zeroual, I., Lakhouaja, A., Belahbib, R.: Towards a standard part of Speech tagset for the Arabic language. J. King Saud Univ.—Comput. Inf. Sci. **29**, 174–181 (2017). https://doi.org/10.1016/j.jksuci.2017.01.006

Zerrouki, T., Balla, A.: Tashkeela: novel corpus of Arabic vocalized texts, data for auto-diacritization systems. Data Brief **11**, 147–151 (2017)

An Evaluation of the Morphological Analysis of Egyptian Arabic TreeBank

Reham Marzouk and Seham El Kareh

Abstract This research is a corpus-based study which aims to reveal the ability of the recent morphological analyzers to handle the ambiguities that apear in the Egyptian Arabic electronic texts written in Social Media. The research evaluates the automatic annotation of the Egyptian Arabic Penn-Treebank ARZ ATB using CALIMA, the Columbian Arabic diaLectal Morphological Analyzer. The corpus is collected by Linguistic Consortium Data as a part of BOLT project, which aims to develop a technology that enables English speakers to retrieve and understand information from informal foreign language sources including chat, text messaging, and spoken conversations. In order to reach better results, the research concentrated on the nouns category. For achieving the research task, a gold standard was built by using the most frequent 1723 nominal word types from 6543 word types of 16226 words selected randomly from the ARZ ATB corpus. The total number of the collected morphemes was 2798. Recall, Precision, F-score, and accuracy of the tool performance were calculated, the recall was 89%, the precision was 94.5%, F-score was 93.7% and the accuracy reached to 93%. The errors were classified to reveal the main morphological ambiguities that the tool couldn't handle due to the development of the written form of the Egyptian dialect in social media. According to the results, the Orthographic variations that appeared in the Egyptian Arabic dialects reflected the lack of an authorized writing system governs the using of the dialect in its written form. Thus, gathering and describing the main orthographic variations is imperative to handle the ambiguities that are revealed in the study.

Keywords Arabic TreeBank · Egyptian Arabic dialect · Morphological analysis Social media

R. Marzouk (✉) · S.E. Kareh
Faculty of Arts, Phonetics and Linguistics Department, Alexandria University,
P.O. BOX 21526, Alexandria, Egypt
e-mail: marzoukreham@gmail.com

S.E. Kareh
e-mail: sehamelkareh@gmail.com

K. Shaalan et al. (eds.), *Intelligent Natural Language Processing: Trends and Applications*, Studies in Computational Intelligence 740,
https://doi.org/10.1007/978-3-319-67056-0_30

1 Introduction

Social media has become an increasingly important resource for natural language processing (NLP). The users of the social media usually use their own dialects and accents in communication, which means that, the language used in social media contains linguistic challenges including non-standard orthography, and syntax. Besides, social media contains certain characteristics that can increase the difficulty of processing it. Some of these characteristics which are selected by linguists to be an evidence of the uniqueness of the social media language are: phrasal abbreviation, expressive lengthening, emoticons, merging and splitting words, and speech effects [1]. Due to this complexity, series of papers showed that natural language processing (NLP) performs less accuracy with social media in other languages such as English. In part-of-speech tagging, the accuracy of Stanford tagger falls from 97% accuracy on Wall Street Journal text to 85% accuracy on Twitter. In named entity recognition, the CoNLL-trained Stanford recognizer achieves 44% F-measure down from 86% on the CoNLLtestset' [2].

Therefore, due to these constant occurring diversities in the Arabic language and its dialects, different Treebanks were constructed to cover these diversities and to represent the development of the language. Egyptian Arabic Treebank ARZ-ATB is the latest Treebank created by Linguistic Data Consortium (LDC) to represent the spoken and written Egyptian dialect. Different genres are included in ARZ-ATB, some of these genres represent the social media with its different mediums, such as discussion forums, chatting, SMSs [3]. These genres include a real sample of the recent Egyptian written form, and the morphological ambiguity that is caused by this particular written form.

This research is a corpus based study, whose objective is to reveal the main morphological ambiguities in the Egyptian Arabic Dialect used in the social media and the ability of morphological analyzers to handle such ambiguities. We concentrated on the nouns category. The nouns, in this research, are defined as all those elements that have nominal inflections and functions. This category includes also adjectives, adverbials and prepositional, which are formally nouns in a particular function and demonstratives, relatives and pronouns in all types which are nouns in status but not in form' [4]. The researchs task has been achieved by evaluating the morphological analysis of about 2 k nominal words selected under certain criteria from a blind test data. The test data sized about 20 k words and extracted from SMSs which were collected by LDC during the Penn ARZ ATB project. This research involves an original scientific study which is conducted to reach to an answer of the question: how far can the recent development in the written Arabic dialect used in social media affect the accuracy of the morphological analysis and increase the domain of the morphological ambiguity by adding new ambiguous forms to the language? Hence, it aims to reach the following goals:

Shedding light on the phenomena of the social media language and its effect on the Arabic Language.
Evaluating the performance of the dialectal morphological analyzer in analyzing social media language.
Classifying the ambiguities that decrease the morphological analysis accuracy. The remaining of the work is structured as following: Sect. 2 is an overview of the main previous attempts to develop dialectal morphological analyzers and thier performances' accuracy, Sect. 3 is a description of the role of the social media in the natural language processing, Sects. 4 and 5 introduce the procedures of processing and annotating ARZ ATB corpus, Sect. 6 is an evaluation of the Egyptian Arabic Morphological Analyser (CALIMA) that are used to annotate the corpus, Sect. 7 displays the results of the evaluation with an analysis of the errors, and finally Sect. 8 is a conclusion of the work.

2 Related Studies

In the last decades, a number of morphological analyzers have been developed as practical application for different approaches. Most applications are applied on MSA. However, in the recent years some morphological analyzers and generators have been developed in order to handle the different Arabic dialects such as Levantine Arabic and Egyptian Arabic. Some of these morphological analyzers were evaluated by their developers and others evaluated by other associations, such as the morpho-challenge annual evaluation campaign for the unsupervised morphological analysis.

Attia [5] investigated different methodologies to manage the problem of morphological ambiguities in MSA. He showed how to control morphological ambiguity by building ruled-based system that takes the stem as the base form, and it uses finite state technology. He conducted a small scale evaluation to test the ambiguity rate and the precision of his analyzer in comparison with Buckwalter and Xerox analyzers. Attia's Morphological analyzer achieved precision reached to 79% while Buckwalter's achieved only 64%. The ambiguity rates of the three analyzers were as follows, Attia was 1.74, Xerox was 4.32, and Buckwalter was 2.6 [5].

Thereafter, Habash et al. [6] built MADA+TOKEN, a versatile and freely available system that can derive extensive morphological and contextual information from raw Arabic texts, and then use this information for NLP tasks. Applications include high-accuracy part-of-speech tagging, diacritization, lemmatization, disambiguation, stemming, and glossing. MADA+TOKEN operates by examining a list of all possible analyses for each word, and then selecting the analysis that matches the current context best. The toolkit consists of two components: MADA is a utility that adds lexical and morphological information through disambiguating part of speech tags and lexeme, diacritization and full morphological analysis.

TOKAN is a utility that generates a tokenization to tokenize the words and identify its stem. MADA+TOKEN depends on three resources: Buckwalter morphological analyzer, SRILM a disambiguating toolkit to disambiguate through constructing lexeme n-gram, and SVMTool package to operate the SVM used in MADA to predict the orthographic features. According to Habash et al., the tool has over 96% accuracy in morphological analysis and lemmatization, and over 86% accuracy in predicting full diacritization [6].

Pasha et al. [7] presented MADAMIRA, a system for morphological analysis and disambiguation of Arabic that combines the best aspects of previous two systems: MADA+AMIRA, it has the same general design of MADA with additional components inspired by AMIRA. MADAMIRA is designed to analyze MSA and EGY. The input text passes through different stages of analysis and disambiguation. First, the input text is converted into a Buckwalter representation used in MADAMIRA, and then a morphological analysis is performed to develop a list of all the possible analyses. The text and the analyses are passed to a Features Modeling component which uses language models and SVM to derive the analyses for words' morphological features such as lemma, and diacritic features. An analysis ranking model is used to score; the scores were ranked according to how accurate they matched the predictions. To evaluate MADAMIRA a blind test data set was created and compared to a gold annotated version and several accuracy metrics were calculated to cover the diacritics, Lemma, POS, morphological features and the tokenization, for both MSA and EGY. The accuracy average of the previous values was 80% for MSA and 76.4% for EGY [7].

3 Social Media Usage in Arab World

'Social media refers to the means of interaction among people in which they create, share, and exchange information and ideas in virtual communities and networks (like Twitter and Facebook)' [8]. Electronic texts that are produced through these channels, were used as resources for various language processing applications as information extractions, linking, classification, POS tagging, .. etc.

The Annual Arab social media report, produced by the Dubai school of Government's governance and innovation program, is a recurring series that highlights and analyzes usage trends of online social networking across the Arab region. It provides information about the usage of the social media among the Arab countries. The report is based on studies and surveys among the Arab youth whose ages are above 15 years old, and it reveals the essentiality of studying the language used in social media as representative form of the Arabic dialects. Additionally, it assists in determining the most used devices to communicate among the users. The results of report (2015) concluded that:

- More than half of the users in the Arab world use primarily the social media to communicate, while gaining information, reading books, listening to music, etc. came in the lower positions.
- Facebook and Whatsapp are the most social media channels used across the Arab regions. And Whatsapp is the most preferred channel in most Arab regions.
- More than 4 out 5 social media users in the Arab world use Whatsapp. Besides, Syrian are considered as the highest users of Whatsapp with (96%), followed by Egyptians with (94%) (Figs. 1 and 2).

Egyptian Arabic used in social media is much different from other written genres, since 'its vocabulary is informal with intentional deviations from standard orthography such as repeated letters for emphasis; typos and non-standard abbreviations are common; and non-linguistic content is written out, such as laughter, sound

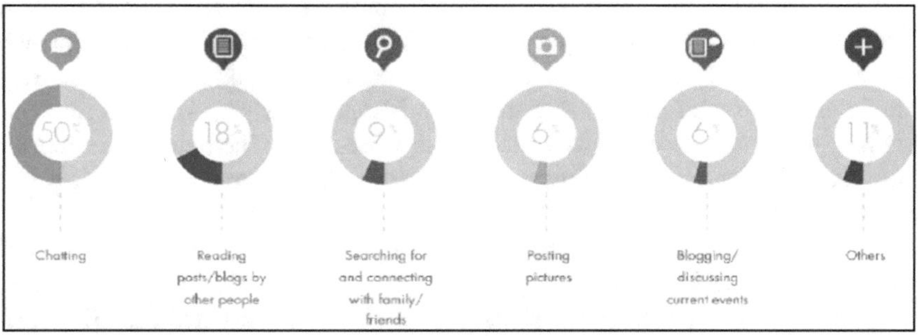

Fig. 1 The preferable activities when using social media. (Arab social Media 2015)

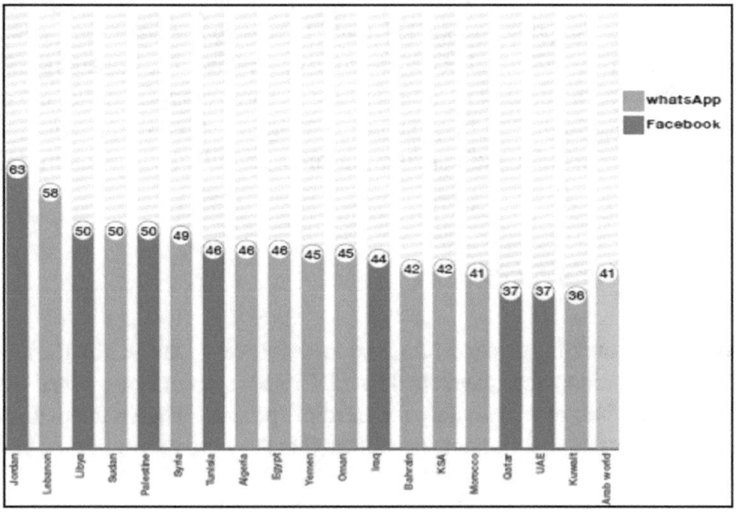

Fig. 2 The most preferred social media channels by country. (Arab social Media 2015)

representations, and emoticons' [1]. Thus, processing Egyptian Arabic used in social required pre-processing procedures to reach to semi standardized forms that allows language applications such as machine translation to accept these forms.

4 Processing Social Media Text

Broad Operational Language Translation (BOLT) program is one of the recent projects that constructed to develop technology to enable non-native speakers to understand information from informal foreign language sources such as chat, text messaging and spoken conversations. For that purpose, LDC collected and annotated informal linguistic data of the English, Chinese, Arabic, and Egyptian Arabic as representative of the Arabic language family.

A stage in achieving this project was developing an annotated Egyptian Arabic TreeBank (ARZ ATB). The corpus is collected from different social media channels such as discussion forums, Short Messaging Systems (SMS), and online chats. EGY Arabic texts in social media are a mixture of English scripts, Arabic scripts and Arabizi (using roman letters to write Arabic words) [1]. Therefore, the collected date had to follow pre-processing stages to convert Arabizi to Arabic script before starting the annotation.

For that purpose, LDC collected a large volume of naturally occurring informal texts of the SMS, and chat massaging that are mix between Arabic script and Arabizi. 46 Egyptian native speaker participants have offered 26 contributed data. Each participant contributed with 48 k words. 'The Egyptian Arabic SMS and Chat collection consisted of 2,140 conversations in a total of 475 K words after manual reviewing by native speakers of Egyptian Arabic to exclude inappropriate messages and messages that were not Egyptian Arabic' [1]. Only 15% of conversations are completely written in standard Arabic, while 66% are entirely Arabizi. The remaining 19% is a mixture of the two, in addition to the existence of some cross-linguistic effects, for example emoticon, emoji, and frequent use of written out representation of speech effects such as representation of laughter (hahaha), and filled pause (umm) [1].

4.1 Annotating Arabizi

The process of converting Arabizi into Arabic script were achieved by the help of a web-based automatic system developed by LDC. This system is shortened the time of the converting into approximately the third of the time. Then a manual revision was undertaken by native annotators to confirm that the word was translated correctly [1], (Fig. 3).

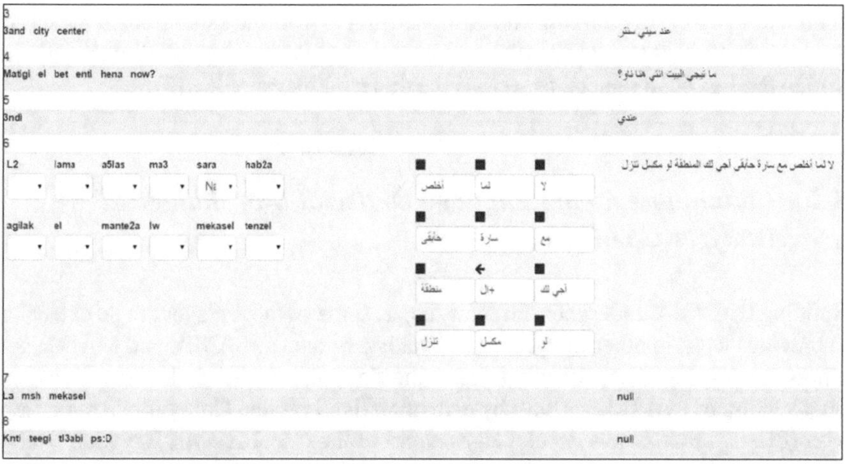

Fig. 3 Arabizi translation tool. Reprinted from Transliteration of Arabizi into Arabic Orthography: Developing a Parallel Annotated Arabizi-Arabic Script SMS/Chat Corpus, by Bies et al. [1], EMNLP 2014 Proceedings

5 Stages of Annotation Egyptian Arabic Penn TreeBank (ARZ ATB) CORPUS

By transliterating the non Arabic script. ARZ PennTreeBank became ready for annotation which goes through two stages: POS/morphological annotation, and the syntactic annotation. In the first stage, a morphological analyzer is needed. CALIMA is the Egyptian Arabic morphological analyzer which is used for the automatic morphological annotation of the Egyptian Arabic Treebank EGY ARZ ATB. CALIMA is built by a team from Columbia University in 2010 and it has four versions [3, 9].

5.1 Egyptian Arabic Morphological Analysis

Annotating Egyptian Arabic Treebank, using CALIMA, follows the Part Of Speech (POS) guidelines used by the Linguistic Data Consortium (LDC) for EGY. CALIMA accepts multiple orthographic variants and normalizes them into CODA, the Conventional Orthography for Dialect Arabic. CODA is a standard writing language created by Columbia University for computational purposes. It aims to serve as a unified framework for all the Arabic dialects [9].

CODA uses MSA-like orthographic decisions (rules, exceptions and ad hoc choices), e.g., cliticizing single letter particles, using Shadda for phonological gemination, using Ta-Marbuta and Alif-Maqsura [9]. At the same time, it preserves the phonological forms of the dialectal words, e.g. /q/ is substituted with a glottal stop in the Egyptian dialect, as well as the dialectal morphology [9].

5.2 Columbia Arabic Language Dialectal Morphological Analyzer CALIMA

Building CALIMA undergoes different stages, some of these stages are an extension to previous work in other morphological analyzers such as SAMA, and MADA, and others are developed specifically to meet the Egyptian Arabic dialect requirements. First version of CALIMA is built by extending the Egyptian Colloquial Arabic Lexicon (ECAL). ECAL was developed as a part of the CALLHOME Egyptian Arabic corpus, which contains 140 telephone conversations and their transcriptions [10]. ECAL has about 66 k words: 27 k verbs, 36 k nouns and adjectives, 1.5 k proper nouns, and 1 k closed classes. For each word the lexicon provides the phonological form, the lemma, the undiacritized script, and the morphological features [9]. Finite state transducers were used to map the phonological undiacritized forms of EGY script in ECAL to a diacritized form. Then, rewrite rules have been written to convert the diacritized entries into CODA and LDC EGY POS compliant tags. These rules fall into three categories [10]:

- Ignore rules: to specify which of the ECAL entries should be excluded due to errors.
- Correction rules: to correct some of the ECAL incorrect entries.
- Prefix/suffix/stem rules: to identify specific pairs of prefix/suffix/stem substrings and morphological features to map them to appropriate prefix/suffix/stem morphemes.

The following examples clarify how the prefix rules look like:
$lil\ + article + prep\_prefix => !!\&\&li + +lli/PREP + Al/DET + l$
$lil + article + prep\_prefix => !!\&\&li + l + li/PREP + Al/DET +$
$lil + prep\_prefix + article => !!\&\&li + l + li/PREP + Al/DET +$
$li + prep\_prefix => !!\&\&li + li/PREP +$
In the example the input of the rule processor represents the surface form (in the left side) and the morphological features (in the right side). Each rule generates substring of the target token which is the last token in the previous string.

The system includes 4,632 rules covering all POS: 1,248 ignore rules, 1,451 correction rules, 83 prefix rules, 441 suffixes rules, and 1,409 stem rules to map core POS tags. Some rules were semi-automatically created, and all the rules were manually checked [9].

Following to the Standard Arabic Morphological Analyzer SAMA, CALIMA contains six tables, three tables specify the complex prefix/suffix and stems, and complex prefix/suffix tables which are sets of prefix/suffix morphemes treated as

wl	wali	NPref-Li	and + for/to <pos>wa/CONJ+li/PREP+</pos>
ll	lil	NPref-Li	to/for + the <pos>li/PREP+Al/DET+</pos>
wll	walil	NPref-Lil	and + to/for + the <pos>wa/CONJ+li/PREP+Al/DET+</pos>
wbAl	wabiAl	NPref-BiAl	and + with/by the <pos>wa/CONJ+bi/PREP+Al/DET+</pos>

Fig. 4 A sample of the complex prefix table

single database entries [9], e.g. wi+Ha+yi is a complex prefix made of three prefix morphemes. These last three tables specify compatibility across the class categories (prefix-stem, prefix-suffix and stem-suffix) as seen in (Fig. 4).

Some extensions were occurred in the further versions of CALIMA:

1. Adding new clitics such as prepositional clitics /Ea/, 'on', as in «التربيزة , and multiple POS tags for the proclitic /fa/ as in فالكتاب.
(CONJ, SUB_CONJ, CONNEC_PART and PRE)
2. Adding common non CODA variants such as adding /u/ for the suffixes /h/ to be /uh/ and /ha/ was also added as a variant for the future particle /Ha/.
3. An extension of some frequent forms, such as the adverb برده, which have different variation e.g. بردو, برضو and, برضه were added.
4. The orthographic variations were supported by adding 16 prefix cases, 41 stem cases, and eight suffix cases. Thus, the total number of recognizable word forms increased in CALIMA from 4M to 48M [9].

5.3 Annotating ARZ ATB Corpus Using CALIMA

First version of CALIMA had many "holes", for which it either did not provide a solution for a given input word, or did not provide the desired solution [3]. Therefore, the human intervention was imperative. The annotator was permitted to enter the "proposed solution" which appears in a drop list, such proposed solutions were saved to be used later in the next versions. That process continued till the last version of the tool 2.0.5 (the version used in the analysis). WILDCARD is a modified solution in case the analyzed word doesn't match any of the proposed solutions. In this card, the analyzers output also includes solutions, in which the stem of the word would be unvocalized, but the prefixes and suffixes exactly matched the possibilities elsewhere in CALIMA [3] (Fig. 5).

Finally, during the revision, if the annotated solution didn't match the correct one, the annotator adds potential entries; these entries go through a process of arbitration and normalization before being added to the tables [3].

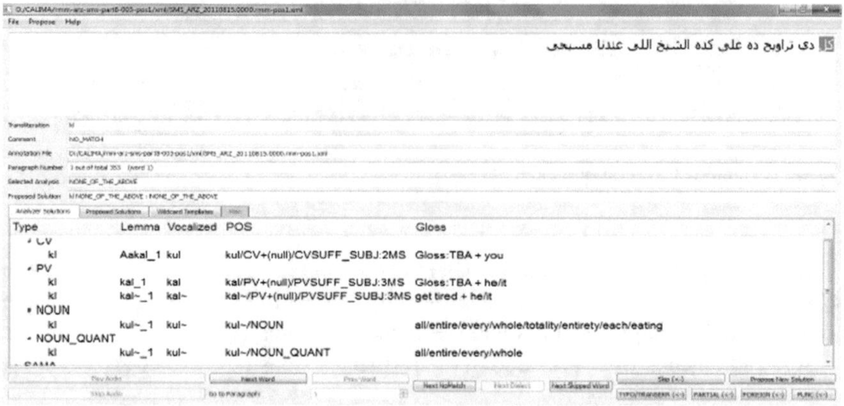

Fig. 5 CALIMA interface

6 Evaluation of Egyptian Morphological Analysis

This section is a description of the evaluation of CALIMA using a corpus which is collected from Whatsapp SMSs. The evaluation is a Gold standard based evaluation, in which the system output is compared with a previously prepared Gold Standard. The evaluation reported the accuracy, the recall, the precision, and the F-score for each morphological entry.

6.1 The Gold Standard

Gold standard is a corpus whose annotation is checked and corrected manually for the purpose of evaluating, and measuring the accuracy of automatic annotation systems. Moreover, a gold standard can be used to determine the specifications of the morphological analyzers by specifying which morphological features it can or cannot handle [11]. Sawalha, 2012, put some criteria that are essential to construct a gold standard for evaluation, the corpus to be used as gold standard, the format of the gold standard, its size, the script used, transliteration scheme, and the phases of constructing the gold standard [12].

This gold standard is created using ARZ ATB Corpus (Egyptian Arabic Penn TreeBank) in specific, the Whatsapp messages. These messages are mainly informal messages written to friends and relatives. Thus, the users used the daily common spoken language to write his messages. In contrast toWhatsapp messages, the forum discussion messages are directed to different people and they are composed of the

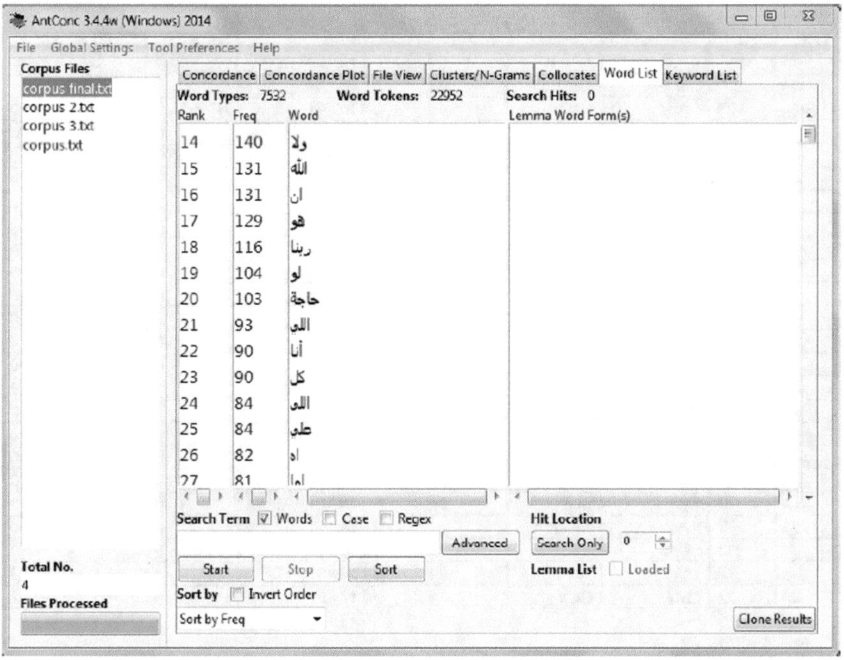

Fig. 6 Word frequency of the Gold Standard

informal Egyptian dialect and MSA. It was built by using the most frequent 1723 nominal word types from 6543 word types of 16226 words selected randomly from the ARZ ATB corpus. The total number of the morphemes together is 2798. AntConc 3.4.4 is the tool used to find out the most frequent words in the corpus, and the least frequency of the included words was four times (Fig. 6).

6.1.1 Gold Standard Format

Gold standards should include detailed morphosyntactic information for each word. Thus, words were divided into their morphemes: proclitics, prefixes, stem, suffixes and enclitics.

The gold standard is prepared in the same format of the results of CALIMA, and the words are included in separated lines. Besides, the tag set of CALIMA was used to encode the morphological features in the gold standard, as well as the basic morphological information such as: the lemma, the vocalization of the word. The visual

Word Type	Freq	Vocalization	Lemma	Proclitic	Prefix	POS	Suffix	Enclitics
الله	131	*Al~ah*	*Al~ah*	0	DET	NOUN-PROP	0	0
ربنا	116	*rab~inA*	*rab~*	0	0	NOUN	0	PRON_1P
حاجة	103	*Hajap*	*Hajap*	0	0	NOUN	NSUFF_SG	0
كل	90	*kul~*	*kul~*	0	0	NOUN_QUANT	0	0
مع	79	*maEa*	*MaEa*	0	0	NOUN	0	0
كده	70	*Kidah*	*Kidah*	0	0	NOUN	0	0
حد	55	*Had~*	*Had~*	0	0	NOUN	0	0
كدا	55	*Kidah*	*Kidah*	0	0	NOUN	0	0
ماما	54	*mAmA*	*MAmA*	0	0	NOUN	0	0
ممكن	55	*Mumkin*	*mumkin*	0	0	ADJ	0	0
جدا	47	*jid~AF*	*jid~*	0	0	NOUN	0	CASE_ACC_INDEF
غير	46	*Giyr*	*Giyr*	0	0	NOUN	0	0
كتير	45	*Kitiyr*	*Kitiyr*	0	0	ADJ	0	0
طيب	44	*Tayib*	*Tayib*	0	0	ADJ	0	0
والله	44	*wal~ahi*	*Al~ah*	CONJ	DET	NOUN_PROP	0	0
بعد	43	*baEd*	*baEd*	0	0	NOUN	0	0
محمد	43	*maHamad*	*maHamad*	0	0	NOUN_PROP	0	0
الناس	41	*Ain~As*	*Nas*	0	DET	NOUN	0	0
كمان	39	*Kaman*	*kaman*	0	0	NOUN	0	0
واحد	36	*waHid*	*waHid*	0	0	NOUN_QUANT	0	0
كويس	35	*Kuwayis*	*kuwayis*	0	0	ADJ	0	0
تاني	32	*tAniy*	*vAniy*	0	0	ADJ	0	0
عمرو	32	*>amr*	*>amr*	0	0	NOUN_PROP	0	0
قبل	32	*Qabl*	*Qabl*	0	0	NOUN	0	0

Fig. 7 Gold Standard

representation of the gold standard allows the end user to see all the morphological information. It is stored in separated columns, (following the Morpho-Challenge competition instructions for creating gold standards), each word and its analysis are stored in one line, where the word occupied the first column followed by its lemma, vocalization, and the analyzed morphemes (Fig. 7).

6.2 *The Test Set*

The 2798 morphemes have been analyzed by CALIMA and the results were sorted according to the morphological feature in different tables. The data set is divided into different lists, each one presents a certain morphological features e.g. lemma, stem, gender and number affixes... etc. Each morphological feature was evaluated separately to provide more detailed results (Figs. 8, 9, 10 and 11).

Word	LEMMA1	LEMMA2	LEMMA3	LEMMA4
متالية	*mivAliy*	0	0	0
محمود	*maHmuwd*	maHmuwd	0	0
محضر	*miHaD~ar*	maHDar	HaD~ar	HaD~ar
محبس	*maHbas*	Habas	0	0
مرات	*Marap*	0	0	0
محظوظ	*maHZuwZ*	maHZuwZ	0	0
مستخبي	*mistaxab~iy*	0	0	0
مريض	*mariD*	ray~aD	0	0
مسكن	*musak~in*	0	0	0
مشغولة	*ma$guwl*	0	0	0
مصورين	*muSaw~ir*	miSaw~ar	0	0
مصحصحة	*miSaHSaH*	0	0	0
معروف	*maEruwf*	maEruwf	0	0
معلش	*maEli$*	0	0	0

Fig. 8 A sample of the LEMMA tables

Word	POS 1	POS 2	POS 3	POS 4	POS 5	POS 6
ناوي	ADJ	NOUN	0	0	0	0
نافع	ADJ	NOUN	0	0	0	0
نادي	NOUN	CV	PV	NOUN_PROP	IV	ADJ
نفسيا	ADJ	0	0	0	0	0
هادي	ADJ	PV	IV	NOUN	0	0
كلية	NOUN	PV	0	0	0	0
هدايا	NOUN	0	0	0	0	0
هاديه	ADJ	NOUN	IV	0	0	0

Fig. 9 A sample of the in POS data set

Word	NSUFF 1	NSUFF 2	NSUFF 3	NSUFF 4
متَعودين	NSUFF_PL	IVSUFF_SUB	PVSUFF_SUB	NSUFF_PL
متَعصبين	NSUFF_PL	PVSUFF_SUB		
محجبَة	NSUFF_FEM_SING	0	0	0
معيِنين	NSUFF_PL	0	0	0
معيِشتك	NSUFF_FEM_SING	PVSUFF_SUB	0	0
معيِنه	NSUFF_FEM_SING	CVSUFF_OBJ	PVSUFF_OBJ	0
مفيِده	NSUFF_FEM_SING	NSUFF_SING_FEM	0	0
مفضلَة	NSUFF_FEM_SING	0	0	0
مقَاسات	NSUFF_FEM_PL	0	0	0

Fig. 10 A sample of the number and gender suffix in the data set

WORD	PROCLITIC 1	PROCLITIC 2	PROCLITIC3
صدري	POSS_PRON	CVSUFF_SUB	PVSUFF_SUB
اكلي	POSS_PRON	CVSUFF_SUB	PVSUFF_SUB
اهلي	POSS_PRON	0	0
اهله	POSS_PRON	CVSUFF_SUB	PVSUFF_SUB
بطنه	POSS_PRON	CVSUFF_SUB	PVSUFF_SUB
شغلك	POSS_PRON	0	PVSUFF_SUB
شغلهه	POSS_PRON	CVSUFF_SUB	PVSUFF_SUB

Fig. 11 A sample of the proclitics in the data set

6.3 *The Confusion Matrix*

'Accuracy, precision, recall and F-measure are applicable to measure the accuracy of the individual morphological categories of the morpheme tags' [12]. The computed metrix were used to measure systems proficiency by determining the ability of the morphological analyzer to predict the morphological features of the analyzed word. Defining the confusion matrix depends on two conditions applicability and correctness. And the confusion elements are: TP(true Positive), TN(True Negative), FP(False Positive), FN(False Negative) [12] (Fig. 12).

According to these elements we could specify the following measurements: Recall: it is defined as the percentage of applicable cases that are correctly predicted from the total number of actual positive cases in the gold standard.

$$TPR = \frac{numberofapplicablecasescorrectlypredicted}{numberofactualpositivecasesinthegoldstandard} = \frac{TP}{TP+FN}$$

Precision: it is defined as the percentage of applicable cases which are correctly predicted from the total number of positive predictions.

$$PPV = \frac{numberofapplicablecasecorrectlypredicted}{totalnumberofpositiveprediction} = \frac{TP}{TP+FP}$$

The F-measure: it is the harmonic mean of precision and recall, was selected as the final evaluation measure. It is interpreted as a weighted average of the precision and recall. F1score reaches its best value at 1 (100%) and worst score at 0 (0%) [12].

$$F1Score = 2 \times \frac{precision.recall}{precision+recall}$$

Confusion Matrix	Predicted positive	Predicted negative
Positive cases	TP	FN
Negative cases	FP	TN

Fig. 12 The four elements of recall and presicion

Accuracy: it is the percentage of the correct predictions made for a certain morphological feature.

$$ACC = \frac{TP+TN}{totalnumberofmorphemes}$$

7 Results

7.1 The Quantitative Analysis

Different measurements were applied to evaluate categories extracted from the data sets in comparison with the gold standard, these categories are: lemma, stem, POS, vocalizations,inflection (number, gender, & person), proclitic particles, enclitic genitive pronouns, and definiteness.

7.1.1 Measuring the Lemma, Stem & Vocalization Coverage

The first measurement evaluates the tool's ability to cover lemma, stem and vocalization by measuring the correctness of its output. A comparison of the results between the data set and the gold standard was performed, and the accuracy was measured. The resulting values were sorted as in Table 1:
The second column "errors" indicates to the percentage of the words whose analysis returned without matching the correct answer. The third column indicates to the percentage of the words whose analysis matched the correct answer. The fourth column indicates to the words returned with no analysis (Fig. 13).

Table 1 Lemma and stem correctness

Feature	Error (%)	Correct (%)	No analysis (%)
Lemma	0.98	93.6	5.4
Vocalization	2.5	92.2	5.3
Stem & Lemma	2.5	92.2	5.3

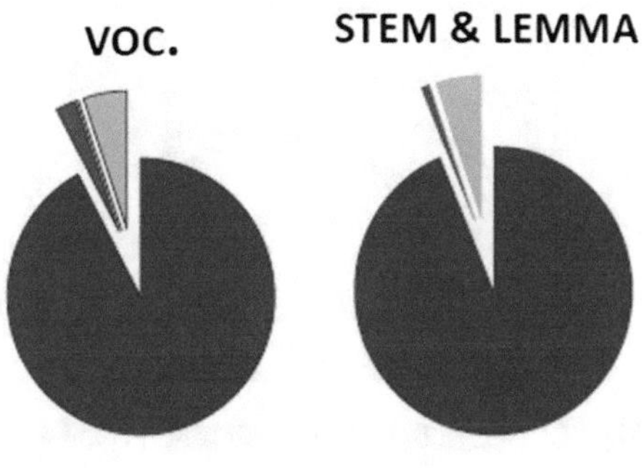

Fig. 13 Lemma & POS coverage

7.1.2 Evaluation of the Inflections and Definiteness

The most common device of defining a standard noun is the morphological marker /l/. It is prefixed to the noun and assimilated to any initial solar, i.e. coronal consonant by the l-Assimilation rule. Nouns also can be defined by being suffixed with a possessive pronoun.

The second measurement was designed to compare the affixation output of the analyzer, with the manually annotated gold standard. The data set was used to measure the tools ability in predicting affixes that represent the inflections, and the definiteness. 1087 affixes that are extracted from the data set to present the inflections, and the definiteness, are divided as the following: 605 suffixes to present gender, number, and persons, and 482 prefixes to present definiteness.

7.1.3 Evaluation of the Cliticization

In the third measurement, the recall, precision, accuracy and the F-measure of the proclitics, as well as of the enclitics were evaluated. The total number of the measured clitics is 526. The aim of the experiment is to measure the ability of the analyzer in tokenizing and predicting the proclitic particles such as conjunctions, vocative particles, ...etc, and the enclitic pronouns such as possessive pronouns. The recall, precision, accuracy and the F-score of the previous features were calculated and the results were displayed.

A model to calculate the four values of the confusion matrix TP/ FP/ TN/ FN was built, the model accepts manually designed Gold Standard and data sets, calculate the values of the desired morphological features then it measures the following matrixes: recall, precision, F-measure, accuracy and specificity (Figs. 14 and 15). The mean average of the final results was calculated for all the values to provide detailed information about the accuracy and the ability of the analyzer to predict the correct analysis (Fig. 16 and Table 2).

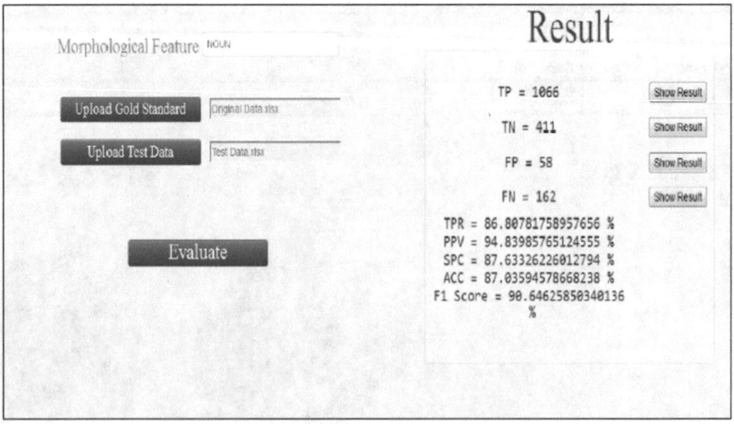

Fig. 14 Description for the results of the nouns evaluation

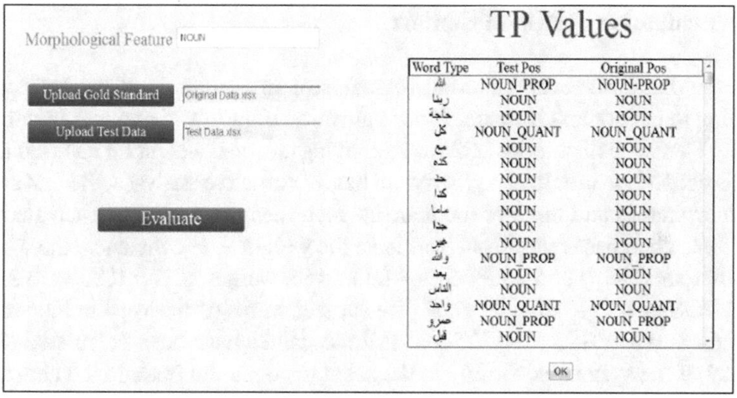

Fig. 15 True Positive TP nouns as shownd in the gold standard and data set

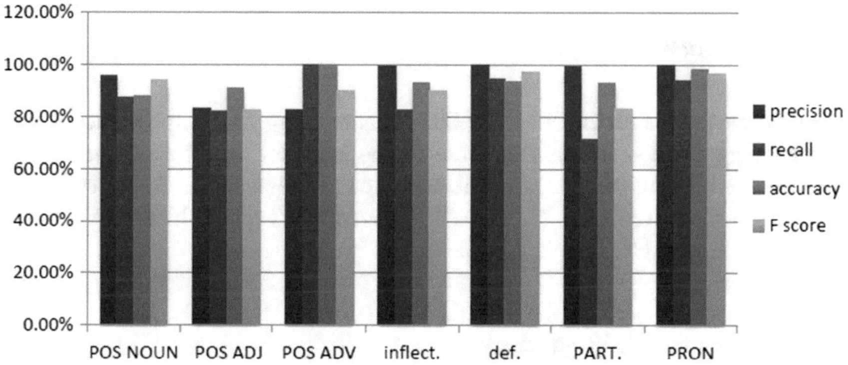

Fig. 16 Recall and precision percentages of the morphological features

Table 2 Recall and Precision percentage

Feature	Precision (%)	Recall (%)	Accuracy (%)	F-score (%)
Part of speech of noun	95.7	86.9	87.7	91
Part of speech of adjective	83	82.2	90.9	82.5
Part of speech of adverbs	82.3	100	99.8	90
Inflections	99.3	82.5	93.3	90.1
Definiteness	100	94.5	93.8	97.2
Proclitic particles	99.2	71.6	93.3	
Enclitic pronouns	100	93.7	99.3	96.7
Average	94.5	89	93	93.7

7.1.4 Evaluation of POS in Context

For words that have several alternative analyses, an evaluation of the correct word according to the context is made. A normalization of points to one was important to be done. The normalization is achieved by having an equal weight for each alternative analysis which is from 0 to 1, e.g. if a word has 3 alternative analyses, the first analysis has 4 morphemes and three of them are correct, then the weight of each analysis is $1/3 * 3/4$. The unacceptable analysis have the weight 0. For instance, the word has four different POS. The first POS is ADJ and its weight is $1 * 0.25 = 0.25$. The second POS is *CV + VSUFF_SUB* (the subject suffix of the verb is hidden), and its weight is $0.5 * 0.24 = 0.125$ .etc. All the values have been calculated and the accuracy of the system, according to the accuracy equation (mentioned above), was determined as 84%.

7.2 *Qualitative Analysis*

After measuring the accuracy of CALIMA in annotating and morphologically analyzing the Egyptian Arabic dialect written in the social media, the errors which have been detected during the analysis were classified as following: Orthographic variations 40.4%, Gold tags errors 23.3%, Lack of coverage of broken plural 20%, Homography 11%, Typo 3.6%, Others 1.70%.

7.2.1 Orthographic Variations

In EGY 40.3% of the orthographic variations are still uncovered in the morphological analysis phase; this percentage is a result of several reasons:

- Hamza: most of the Egyptian writers do not differentiate between the Alif with Hamza and Alif without Hamza. Consequently, Hamza in EGY is considered as a variation and not a mis-spelling, but some words are confused with others if the Hamza is deleted e.g. /bArid/, بارد which may be interpreted as an adjective with the meaning (cold), or as a verb with the meaning (I reply).
  Taa Marbouta: the miswriting of /h/, "ه" instead of /ap/, "ة" has increased the error rate, and caused wrong stem tagging, e.g. the noun /kalimap/ كلمة (a word) is tagged as verb when it is written /kallimuh/ كلمه (talked to him).
- Shortening of long vowels: the inclination of the Egyptian speakers to shorten long vowels, due to the phonological alternation rules that govern the dialect, has been transfered to the written form of the dialect. Hence, words with shortened vowel are found during the analysis. These changes may cause confusion with other words. For instance, the confusion between the adjective سمعة /samEap/, (be hearing), and the noun سمعة /sumEap/ (reputation).

- Speech effects: It became one of the main characteristic of the social media texts, such as writing laughing sound, e.g. /hahaha/ هاهاها , repeating a certain phoneme e.g./jAAAAmdap/جاااامدة, or repeating a whole morpheme e.g. /lAlAlAlA/ لالالالا .
- Replacing emphatic letters with non-emphatic letters: Some Egyptian words that do not have equivalents in MSA can be written either emphasized or not e.g. /Tarabiyzap/, ترابيزة , طرابيزة and /AuwDap/, أودة , أوضة .
- Compound words: such as the numerical numbers from eleven to nineteen which are turned to be compound words in EGY e.g. ستاشر , أربعتاشر .
- Merging words: e.g. /AinshA'allah/, إنشالله instead of /Ain+ShA'+Allah/, شاء الله إن.
- Splitting words: e.g. /yA/+/riyt/, يا ريت instead of /yAriyt/, ياريت. .
- Abbreviations e.g. /AlH/, الح .

7.2.2 Tags Errors

Tags errors came in the second position among the morphological errors with 23.3%, due to the following reasons: The genre that has been chosen to be analyzed from the ARZ ATB Corpus revealed differences in the usage of the Egyptian written Arabic dialect. Some of these differences appeared to suit the media in which they are used, such as SMS, forums, and chatting where short abbreviated written texts are preferred. These differences were summarized as following:
Adjectives with different patterns: some adjectives forms have been added recently side by side with the original form. One of these forms is the form CaCCu:C whose usage is limited to certain gender and social state. For example:
غلوس , طيوب , حبوب , للذوذ

Nouns with different patterns: some nouns also underwent changes from their original form and created new forms for the same words, such as:
نصاية , ربعاية , دبدوب , بنوتة

Borrowings: the spreading of new borrowings has accompanied the usage of the technology with its media, some of the borrowings underwent the Arabic morphological inflections. Hence they are joined with Arabic morphological features such as definite articles, plural suffixes, etc., as well as, they took different forms to express certain Arabic morphological categories such as adjectives. These borrowings are like:
... , مشير , الكمنت , مهنج , الفيس , الفيبر , الليكات

New words: languages continuously undergo extension of their vocabulary as a part of the development process, the extension in EGY's vocabulary included new nominal forms. Some of them are primitives and other are derived from either new verbs or existent verbs. For example:

...، محمد , روش , مروشن , مأنتخ , فاكس , تيت

Adverbs: Adverbs in EGY are non-inflected forms of nouns in specific function [13]. CALIMA follows the LDC guidelines. LDC guidelines are based on "A Dictionary of Egyptian" for Badawi & Hinds, 1986. Although, adverbs were limited in a small list without mentioning other adverbs such as: بلوشي , بشويش , امبارح , كده which are existed in the dictionary.

7.2.3 Lack of Coverage of Broken Plural

By evaluating the analyer's performance during annotating the word's number (singular, dual, plural), it became clear that the performance is low. That is a result of several reasons: broken plurals were one of the reasons for decreasing CALIMA accuracy, since 73% of the broken plurals have been annotated with their singular lemmas, but with no signification of their numbers. While 27% have neither singular lemmas nor significations of their numbers. Other reason is the suffix /ap/, /ة /, which is always tagged as a feminine singular suffix. Therefore the analyzer is not able to differentiate between the /ة / as a singular feminine suffix, and /ة / as a part of the broken plural word such as نشاط , whose plural is أنشطة .

7.2.4 Homographic Words

Homographic words are the words that have the same graphemes, but with different function and meaning, and they cause a problem because of the lack of diacritics in the EGY written texts. CALIMA failed to cover 11% of these homographic words, such as /Eam Al/, عمال as an adjective which means (continuous), and /Eum Al/, عمال as a noun which means (workers).

7.2.5 Typography

Typography forms 3.6% of the errors, and its occurrence is related to the spelling mistakes. The spelling mistakes are mainly a result of the users' ignorance of the written system or due to the careless usage of the language. The preprocessing stage is responsible for decreasing the error rate caused by typography during the correction of the misspelled words.

7.2.6 Morphotactic Ambiguity

Some of the errors which don't form a high percentage, although they affected the accuracy, were the unpermitted concatenations of some morphemes. For instance, annotating the possessive pronoun /iy/ as an object suffix of the perfect verb instead of the suffix /niy/, e.g. the suffix /iy/ in the word is annotated as VSUFF_OBJ.

8 Conclusion

A range of four fair and precise evaluation experiments was conducted using a gold standard consisting of 1723 nominal words extracted from Egyptian Arabic corpus. The results show that the Egyptian Morphological Analyzer CALAMA has the ability to cover the majority of the morphological ambiguities, but the Egyptian Arabic Dialect still includes morphological ambiguities that are not handled by morphological analyzers. Thus the study attempts to provide a valuable resource for improving Egyptian Morphological analyzers through Classifying the reasons beyond the ambiguities that may rise during the morphological analysis process as a step toward rendering specific solutions to handle these ambiguities. Moreover, the study explored that the main cause of the ambiguity in such types of texts is the orthographic variations which returns to the inexistence of a consistent written form for the Egyptian dialect. Therefore, a preprocessing stage to normalize such variations before the analysis became essential to avoid such ambiguities.

References

1. Bies, A., Song, Z., Maamouri, M., Grimes, S., Lee, H., Wright, J., Strassel, S., Habash, N., Eskander, R., Rambow, O.: Transliteration of Arabizi into Arabic orthography: developing a parallel annotated Arabizi-Arabic script sms/chat corpus. In: Proceedings of the EMNLP 2014 Workshop on Arabic Natural Langauge Processing (ANLP), pp. 93–103 (2014)
2. Eisenstein, J.: What to do about bad language on the internet. In: HLT-NAACL, pp. 359–369 (2013)
3. Maamouri, M., Bies, A., Kulick, S., Ciul, M., Habash, N., Eskander, R.: Developing an Egyptian Arabic treebank: impact of dialectal morphology on annotation and tool development. In: LREC, pp. 2348–2354 (2014)
4. Badawi, E.S., Carter, M., Gully, A.: Modern Written Arabic: a Comprehensive Grammar. Routledge (2013)
5. Attia, M.A.: Handling Arabic morphological and syntactic ambiguity within the LFG framework with a view to machine translation. Ph.D. thesis, University of Manchester (2008)
6. Habash, N., Rambow, O., Roth, R.: Mada+ tokan: A toolkit for Arabic tokenization, diacritization, morphological disambiguation, pos tagging, stemming and lemmatization. In: Proceedings of the 2nd International Conference on Arabic Language Resources and Tools (MEDAR), Cairo, Egypt, pp. 102–109 (2009)

7. Pasha, A., Al-Badrashiny, M., Diab, M.T., El Kholy, A., Eskander, R., Habash, N., Pooleery, M., Rambow, O., Roth, R.: Madamira: a fast, comprehensive tool for morphological analysis and disambiguation of Arabic. LREC **14**, 1094–1101 (2014)
8. Habib, M.B., Van Keulen, M.: Information extraction for social media. Association for Computational Linguistics (2014)
9. Habash, N., Eskander, R., Hawwari, A.: A morphological analyzer for Egyptian Arabic. In: Proceedings of the Twelfth Meeting of the Special Interest Group on Computational Morphology and Phonology, pp. 1–9. Association for Computational Linguistics (2012)
10. Habash, N., Diab, M.T., Rambow, O.: Conventional orthography for dialectal Arabic. In: LREC, pp. 711–718 (2012)
11. Faaß, G., Heid, U., Schmid, H.: Design and application of a gold standard for morphological analysis: smor as an example of morphological evaluation. In: LREC (2010)
12. Sawalha, M.S.S.: Open-source Resources and Standards for Arabic Word Structure Analysis: Fine Grained Morphological Analysis of Arabic Text Corpora. University of Leeds (2011)
13. Gadalla, H.A.: Comparative Morphology of Standard and Egyptian Arabic, vol. 5. Lincom Europa Munich (2000)

Building and Exploiting Domain-Specific Comparable Corpora for Statistical Machine Translation

Rahma Sellami, Fatiha Sadat and Lamia Hadrich Beluith

Abstract In this paper we address the problem of mining domain-specific comparable and parallel data to improve the accuracy of a Statistical Machine Translation system. First, we present a novel strategy for building domain-specific comparable corpora from Wikipedia. Our strategy exploits the categorization and the multilingualism of Wikipedia documents in order to extract domain-specific comparable corpora. Second, we describe a combined anchor-point-based method for comparable sentences alignment. Third, we present a compositional-based approach for parallel phrase mining. We conducted multiple evaluations to qualify the extracted comparable and parallel data. Applied to Arabic and French languages pair, we extract 81 domain-specific comparable and parallel corpora. The extracted parallel data are used to adapt an Arabic to French domain-generic SMT system to a specific domain one. This additional training data provided significant improvements of the translation quality in terms of BLEU and OOV scores over the baseline system.

Keywords Comparable corpora · Compositional-based approach
Domain-specific · SMT · Anchor point

1 Introduction

Statistical Machine Translation (SMT) use comparable or parallel corpora as essential resources to train translation models. These corpora are widely available for general-domain but not for specific domains such as art, society and media.

R. Sellami (✉) · L.H. Beluith
ANLP Research Group, MIRACL Laboratory, Sfax University, Sfax, Tunisia
e-mail: rahma.sellami@fsegs.rnu.tn

L.H. Beluith
e-mail: l.belguith@fsegs.rnu.tn

F. Sadat
University of Quebec in Montreal, Montreal, Canada
e-mail: sadat.fatiha@uqam.ca

K. Shaalan et al. (eds.), *Intelligent Natural Language Processing: Trends and Applications*, Studies in Computational Intelligence 740,
https://doi.org/10.1007/978-3-319-67056-0_31

Systems specialized in specific domains require in-domain training data to give the best performance.

Very productive methods for creating domain-generic comparable and parallel corpora have been proposed [1, 17, 24]. Nevertheless, very few researches have been done for domain-specific data [2]. In this paper, we first present a novel strategy, based on category tags and inter-language links, for mining many domain-specific comparable corpora from Wikipedia. Then, a combined anchor point-based-method is proposed for comparable sentences mining. Anchor points are elements aligned with trust and which methods can be based to reduce the search space in order to align their neighbor [4]. Various types of anchor points are proposed and combined for comparable sentences alignment and thus complete a compositional-based approach for parallel phrase mining. The compositional translation based approach consists of the fact that the translation of an expression is a function of the translation of the parts [11]. This approach has proved its effectiveness for bilingual lexicon mining from comparable corpora [7]. Nevertheless, no works have been done in the field of parallel phrase mining. In this paper, we propose to exploit the compositional translation based approach for parallel phrase mining from comparable corpora.

This paper is organized as follows. Section 2 presents previous works on mining domain-specific comparable and parallel corpora. Section 3 describes the proposed strategy for domain-specific comparable corpora extraction from Wikipedia. Section 4 illustrates an anchor-point-based method for comparable sentences alignment. Section 5 presents a compositional-based approach for parallel phrase mining. Section 6 evaluates the resulting comparable and parallel domain-specific corpora applied to Arabic and French languages pair. The last section concludes the present paper with future extensions.

2 Related Works

Very few works have studied the mining of domain-specific parallel corpora from domain-generic multilingual resources. Most of these works are based on information retrieval approaches. Plamada and Volk [22] proposed an approach for mining Alpine domain parallel corpora from Wikipedia. They exploited inter-language links to extract comparable domain-generic articles. The extracted corpus is subsequently used for information retrieval queries aiming to identify the articles belonging to the Alpine domain. Parallel sentences are then selected by means of similarity metric [34] developed a configurable Focused Monolingual Crawler for collecting domain-specific corpora from the Web.Then, they presented a method for extracting bilingual named entities, phrases and sentences from the collected corpora. Pal et al. [20] designed a crawler to collect comparable corpora from Wikipedia, based on an initial seed keyword list and inter-language links. Textual entailment techniques are then used to extract parallel phrases from these comparable corpora. The parallel text fragments extracted thus were able to bring some improvements in the performance of an existing MT system on the tourism domain. Gamallo and Loopez [10] proposed a strategy to extract CorpusPedia, bilingual comparable corpora, from Wikipedia. They specified some categories to make the collected corpus comparable according to some specific topics. Also, a measure of comparability is used to verify whether the corpora are lowly or highly comparable. The difference with respect to our strategy is that they only consider the articles associated to one specific category and not to an entire domain.

Barrón-Cedeño et al. [2] proposed a simple model for extracting comparable corpora from Wikipedia based on the category graph. Our strategy for domain-specific comparable corpora mining is close to [2]. The difference between our proposal and the Barrón-Cedeño et al.'s proposal relies in the fact that we explore the whole category graph. However, [2] used a stopping criterion based on a domain vocabulary list. We assume that the vocabulary list could not be complete and this hypothesis can reduce the coverage of the resulting comparable corpus. Recently, [6] proposed an integrated system to extract both parallel sentences and fragments from comparable corpora. They first applied parallel sentence extraction to identify parallel sentences from comparable sentences. Then they extracted parallel fragments from the comparable sentences. Parallel sentence extraction is based on a filter and a classifier. Chu et al. [6] improved this method by proposing a novel filtering strategy and three novel feature sets for classification. They demonstrated that the extracted parallel data significantly improves SMT performance. Wolk et al. [38] proposed a method of automatic web crawling in order to build topic-aligned comparable corpora. They developed methods of obtaining parallel sentences from comparable data and proposed methods of filtration of corpora capable of selecting inconsistent or only partially equivalent translations. Evaluation of the quality of the created corpora was performed by analyzing the impact of their use on statistical machine translation systems.

3 Domain-Specific Comparable Corpora Building

The domain-specific comparable corpora strategy we propose in this paper is designed to exploit Wikipedia categorization and inter-language links.

Wikipedia articles form a network of semantically related terms called Wikipedia Articles Graph (WAG), while the categories are organized in a taxonomy-like structure called Wikipedia Category Graph (WCG) [39]. Articles are usually not placed in the most general category they logically belong to and are tagged as a sub-category thereof which forms a Category-Article Graph (CAG). This is the concatenation of WAC and WCG (Fig. 1). Cycles and shortcuts occur among the different categories. We first extracted the main categories of Wikipedia, the sub-categories and all articles associated to each category. We extracted 6 main categories (art, society, science, technology, space and time, people) and a total

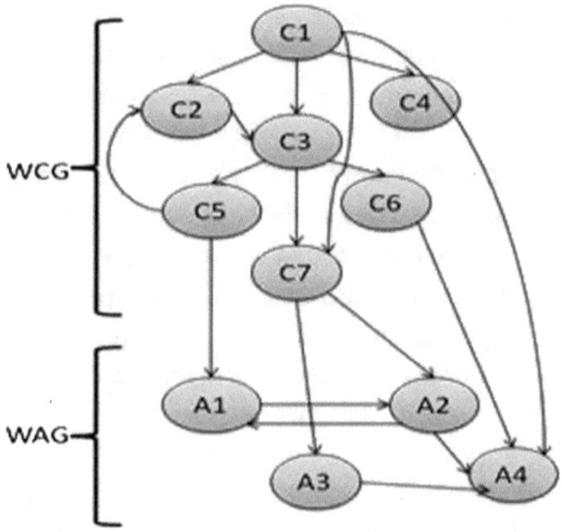

Fig. 1 Category-Article Graph (CAG)

of 81 sub-categories (architecture, film, language, media, tourism, biology, agriculture, robotics, etc.). Thus, we constructed a CAG for each sub-category. The root is the domain name and the endpoints are the titles of the articles associated to one domain. Once the CAG is constructed for each domain, we parsed the graph and the Arabic and French Wikipedia dumps and we extract bilingual articles (related by inter-language links). The output is a set of comparable corpora classified in Wikipedia domains and aligned at article level.

4 Combined Anchor-Point-Based Method for Comparable Sentences Alignment

The main idea of the comparable sentences alignment process is to find correlative elements, also called anchor points, in comparable sentences.

Anchor points can be structural information associated with the document title, subtitle, caption, etc. [25]. They can also be lexical [14] and be extracted based on a bilingual dictionary [12] or transliterations properties of languages. Prochasson et al. [23] have defined some properties of the anchor points: they should be easily identified, relevant regarding corpora topics and not polysemous.

We start with some pre-processings. It consists of tokenization, normalization, lemmatization of source and target sides, truecasing the French letters and stop-words removing.

We propose to combine four types of anchor points for comparable sentence alignment.

- **Word frequency**

Word frequencies have been used in many previous works in information retrieval. Lardilleux and Lepage [15] investigated the use of hapaxs (words that occur only once in a single document) for word alignment and concluded that they can safely be aligned in most cases. This notion is also used in [22] for parallel document alignment. In contrast, [9] exploit high frequency words and their translations for aligning noisy parallel corpora.

In this paper, we propose to exploit words occurring less than four times and the most frequent words as anchor points for comparable sentences alignment.

- **Bilingual fragments**

We extract article titles related by inter-language links. Also, files, images and videos in Wikipedia are often stored in a central source across different languages. This allows the identification of captions, which are most of the time parallel [29, 35].

We exploit the fact that these titles and captions will appear in the text body of articles. If a pair of candidate sentences contains such bilingual fragments (title or caption), it is most likely comparable.

- **Named entities**

Named Entities (NEs) are expressions commonly used and are frequent in all kinds of texts. Bilingual NEs were previously used in many works. Samy et al. [27] used bilingual NE as anchor points for parallel corpus alignment. Semmar and Saadane [32] exploited NE transliteration to improve the results of a linguistic word alignment approach from parallel text corpora.

We make the following assumption: if a NE co-occurs in two sentences, they are very likely to talk about the same event. In this work, two sources of NE translations are exploited. Wikipedia and United Nation corpora are used for person, location and organization named entity translation mining [28, 31].

- **Cognates**

Cognates, words that have similar spelling between two languages, are easy to discover in similar spelling languages. Otherwise, authors use transliteration to close language spelling. Many authors used cognate-based features for alignment of parallel or comparable corpus [1, 32].

We propose two methods for cognate's detection from bilingual sentences. The first one selects cognate's type foreign language words, digits, alphanumerical symbols or punctuation marks. These strings appear reliably in comparable sentences. The second method is based on the transliteration of Arabic words; it can select only words of similar length with a large number of common characters regardless of the order. For this purpose, we define two scores Distance_Score and Length_Score:

$$\text{Distance\_Score} = \frac{\text{editDistance}(\text{ar}, \text{fr})}{(\max(|\text{ar}|, |\text{fr}|))} \tag{1}$$

$$\text{Length\_Score} = \frac{\max(|\text{ar}|, |\text{fr}|)}{\min(|\text{ar}|, |\text{fr}|)} \tag{2}$$

where max(|ar|, |fr|) is the number of characters of the longest string and Min(|ar|, |fr|) is the number of characters of the smallest string. EditDistance is the Editex technique [40], based on a variant of Levenshtein edit distance algorithm [16]. Editex combines edit distance with the use of a group of similar letters (aeiouy, bp, ckq, dt, lr, mn, gj, fpv, sxz, csz); such letters in a similar group frequently correspond to a similar pronunciation. As in Levenshtein distance, the minimal number of insertions, deletions, and replacements necessary to transform one string to another is computed. However, edits that replace a letter with another letter from a different group are weighted more heavily, and deletions of letters that are frequently silent (h and w) are weighted less heavily than other deletions. According to these scores, two words are cognates if Distance_Score is lower than 0.6 and Length_Score is lower than 1.5. These two values are fixed empirically.

All cited anchor points are combined for comparable sentences alignment. At the end of this step, a similarity score Sim_Anch is attributed for each pair of sentences.

$$\text{Sim\_Anch} = \sum_{i=1}^{n=4} \text{Count}(\text{Anchor}(i)) \tag{3}$$

Only pairs of sentences with a similarity score equal or greater than a threshold Anch are included in the sentence aligned comparable corpus.

5 Compositional-Based Approach for Parallel Fragment Generation

In this section, a new approach for mining parallel phrases from comparable sentences pairs based on the compositional translation [11] is proposed.

The input of this step is pairs of pre-processed comparable sentences (in form of bag of lemma of lexical words). Phrase generation, phrase translation, re-composition and filter steps are executed in order to generate fragment translations.

- **Phrase generation**

In order to generate source and target phrases from comparable sentences we extract all n-grams up to length 5 from each lemmatized sentence.

- **Phrase translation**

Compositional translation consists of translating each lemma in the source phrase. The translation step considers all alternative translations based on lexicons and anthologies. For each lemma in the source phrase, the following steps are considered.

1. First, semantic relations (such as synonyms and hyponyms) are defined using an Arabic ontology [8].
2. Second, lemma and its semantic relations are translated based on bilingual lexicons. For this purpose, we use GLOB-LEX, a bilingual lexicon based on many resources: a bilingual lexicon extracted from Wikipedia titles and based on a hybrid approach [30], anchor points which contribute to the comparable sentences mining and some dictionaries (the universal dictionary, wiktionary and Omegawiki). All these data are in dictionary format (lemma). GLOB-LEX is composed of 2 219 509 pairs of terms. We consider all hypothesis translations.
3. Third, semantic relations for each translation are added based on a French ontology [3].
4. Finally, all generated translations are revised to delete any duplicated translations.

- **Re-composition**

We re-compose the translation candidates of a fragment, taking into account all possible combinations. In order to overcome distortion phenomena, fragments are treated as bags of translated lemmas. In the end of this step, N translation candidates are produced.

$$\mathrm{N} = \prod_{i=1}^{i=m} T_i \tag{4}$$

where Ti is the number of generated translations of a source lemma i. m is the number of lemmas in the source fragment. Ti = 1 if no translation was generated for a lemma i.

- **Translation filter**

This step consists of matching the sequence of translated lemmas with tokens lemmas in the comparable sentence (in target language). Ideally, we should select only pairs of sequences that co-occur exactly in the comparable sentences. But, due to translation phenomenon (insertion and deletion) that appear in comparable sentences and the fact that lexicons cannot cover all lemma in our comparable corpus, we propose to accept pairs of sequences containing some insertions and deletions.

The process of translation filter consists of the following steps. Given a sequence of lemmas in the source language and various translation candidates produced by the previous step, we select the best translation sequence based on the target side of the comparable sentence and using a lemma-overlap score. This score is calculated based on the translated lemmas and all sequences of lemmas generated from the target comparable sentence.

6 Evaluation

6.1 Domain-Specific Comparable Corpora Evaluation

In this section, we conduct an evaluation of the degree of comparability of comparable corpora based on a quantitative comparability measure C_{LG}.

Note that C_{LG} is a comparability measure, proposed by [18] based on vocabulary similarity. Table 1 shows the degree of comparability of Arabic–French domain-specific documents extracted from Wikipedia. First, it presents the number of Arabic documents, French documents and bilingual comparable documents. Second, it presents the percentage of comparable documents in many cases: C_{LG} is equal to 0, between 0 and 0.2, between 0.2 and 0.3 and greater than 0.3. For example, considering the Eating topic, we notice that 73.68% of documents have a comparability measure, C_{LG}, equal to 0 (these documents can be characterized as semi-comparables). Whereas, C_{LG} is between 0 and 0.2 in 13.53% of documents and it is between 0.2 and 0.3 in 9.77% of documents. While, only 3% of documents have a comparability measure which is greater than 0.3. These documents can

Table 1 Comparability evaluation of domain-specific comparable corpora

Domain	# ar doc	# fr doc	# comp doc	$C_{LG} = 0$ (%)	$C_{LG} > 0$ $C_{LG} < 0.2$ (%)	C_{LG} >=0.2 $C_{LG} < 0.3$ (%)	C_{LG} >=0.3 (%)
Eating	235	145	133	73.68	13.53	9.77	3.01
Media	893	854	843	82.1	2.62	4.46	10.81
Belief	1121	1149	1067	75.52	10.52	3.35	10.61
Astronomy	1259	1725	1220	84.74	2.15	4.52	8.59

be characterized as strongly comparables. This demonstrates the difficulty of identifying parallel sentences in the extracted comparable corpora. Nevertheless, we can locate comparable sentences and then select subsequently parallel segments. In the following sub-sections, we will focus only on the evaluation of the media domain corpus due to lack of space.

6.2 Comparable Sentences Evaluation

In order to evaluate the effectiveness of the combined anchor-point-based method for comparable sentences alignment, we conducted a manual evaluation of the aligned sentence pairs. This evaluation is based on 1000 pairs of automatically selected comparable sentences randomly chosen. It describes the data distribution for different values of the threshold "Anch". Figure 2 characterizes the aligned sentences (parallel, semi-parallel, comparable and semi-comparable) with different threshold values for the similarity measure.

We define the following terms:

- **Parallel sentences**: pairs of translated sentences.
- **Semi-parallel sentences**: pairs of translated sentences with some insertion or deletion.
- **Comparable sentences**: pairs of non translated sentences but sharing a same topic.
- **Semi-comparable sentences**: pairs of non comparable sentences containing some translated terms.

It is clear that when we increase the threshold the percentage of semi-comparable sentences decreases and the percentage of comparable sentences increases. With Anch = 2, the percentage of parallel and semi-parallel sentences decreases. This is due to some short parallel sentences (e.g. Titles of documents) containing only one anchor point.

In order to maintain a maximum coverage value, we chose a threshold value equal to one for the rest of evaluations.

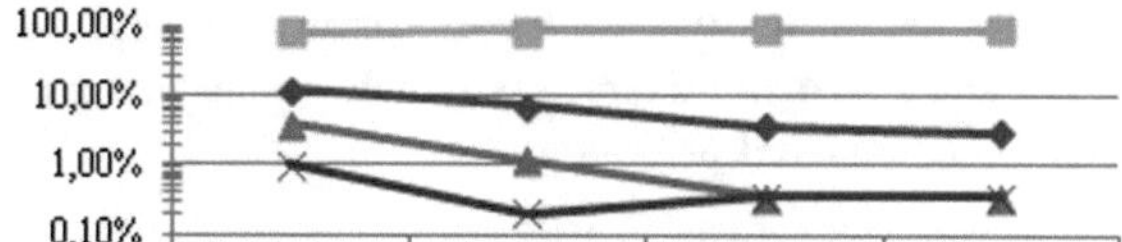

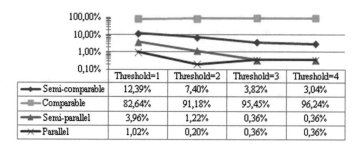

	Threshold=1	Threshold=2	Threshold=3	Threshold=4
Semi-comparable	12,39%	7,40%	3,82%	3,04%
Comparable	82,64%	91,18%	95,45%	96,24%
Semi-parallel	3,96%	1,22%	0,36%	0,36%
Parallel	1,02%	0,20%	0,36%	0,36%

Fig. 2 Evaluation of comparable sentences alignment process

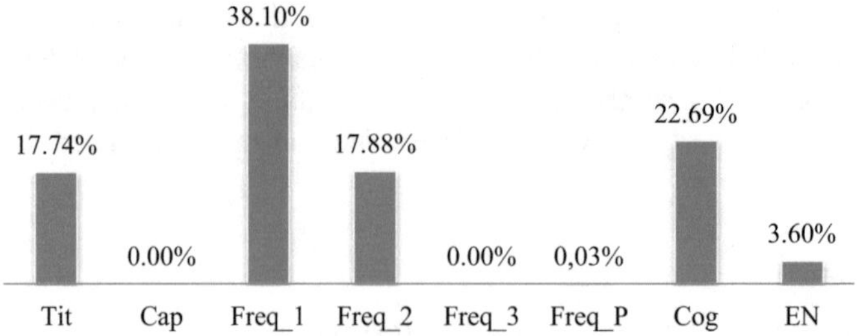

Fig. 3 Anchor point distribution in comparable sentences

Figure 3 shows the distribution of different anchor point types (title, caption, words that occur only once, words that occur twice, words that occur three times, most frequent words, named entities and cognates) in aligned comparable sentences. We notice that Freq_3 and Caption do not match any pair of sentences, whereas Freq_1 (words that occur only once) is the most used feature.

An error analyses of the detected sentences pairs shows that errors are mainly caused by:

- **Wrong lemmatization of rich morphological words**. An example is the anchor point "نرد/nrd[1] - dé (dice)". In this example, the Arabic word "نرد" in the non-vocalized form have two potential tokenizations: the first one is "ن + رد" (we respond). In this case, the lemma is "رد". The second one is "نرد" (dice). In this case, the lemma is "نرد". MADA toolkit which we use for lemmatization returns wrongly the lemma "رد". Then two non-comparable sentences are aligned; the first one contains the lemma رد"" and the second sentence contains the lemma "dé/dice".

- **Ambiguity of non-vocalized arabic words**. For example, the Arabic word "أم/ Om" has two morphological analyses: which refers to the disjunction "or" and which refers to the noun "mother" in English. Our system wrongly matches the anchor point "أم"-"mère/mother" to a pair of non-comparable sentences; the French one contains the word "mère/mother" and the Arabic one contains the preposition "أم/or".

- **Date and number anchor points can wrongly accept non-comparable sentences**. The same numeral characters which appear in a pair of non-comparable sentences are considered as cognates.

- **Erroneous cognates**. For example, the Arabic-French pair of "معبر/mEbr"-"membre". In this example, the Arabic word means "a passageway" whereas the

[1]All Arabic transliterations are provided using the Buckwalter transliteration scheme.

French word means "a member". These words look as cognates but they are semantically different.

- **Translation errors of hapax and words that occur twice in a document**. An example is the word "1er" (1st) which is considered as hapax in a document. This word is translated into "1" whereas "1" may be used in other contexts.

6.3 Parallel Phrases Evaluation

We have extracted 31 842 translations (in media domain). In order to get an idea about the extracted data quality, we took randomly 1000 pairs of fragments and classified them into three classes:

- **Parallel**: perfect translation.
- **Comparable**: there are some insertion, suppression or relations of hyponymy or hyperonymy.
- **Non comparable**: phrases are independents.

Table 2 shows the distribution of the extracted translations by degree of parallelism, in both cases of translation filter:

- **−Tol**: select only pairs of sequences that co-occur exactly in a comparable sentences.
- **+Tol**: accept pairs of translated sequences containing insertions or deletions.

We note that taking into account a certain tolerance in translation filter improves the value of precision. This makes it possible to improve the percentage of parallel segments by 5%, decrease the percentages of comparable segments by 1% and the rate of non-comparable segments by 4%.

Our principle purpose is to use the extracted data for the adaptation of a generic machine translation to a specific domain. We used phrase-based SMT systems trained with the Moses toolkit [14]. Word alignment is done with GIZA++ [19]. We implemented a 5-gram language model using the SRILM toolkit [36]. We tokenized the Arabic side of the training, development and test data using the MADA + TOKAN morphological disambiguation system [26]. French preprocessing of the training, tuning and test data simply included down-casing and separating punctuation from words.

A summary of the size of the used data sets is given in Table 3. Our domain-generic parallel corpora are composed of many parallel data: sentence aligned multiUN contains 2 769 361 pairs of sentences, news-commentary contains

Table 2 Distribution of translation hypotheses

	Parallel (%)	Comparable (%)	Non comparable (%)
+Tol	49.2	12.4	38.4
−Tol	54.31	11.42	34.26

Table 3 Sizes of Arabic–French data sets

Corpus	Nb of sentences	Nb of tokens (ar–fr)
Specific tuning corpus	0.6 K	8.2 K–8.4 K
Specific test corpus	0.4 K	3.5 K–3.4 K
Specific language modeling corpus	22.5 K	61.3 K
Specific parallel corpus based on compositional approach +Tol	31.8 K (fragments)	48.3 K–43.5 K
Specific parallel corpus based on compositional approach −Tol	26.8 K (fragments)	39.4 K–36.2 K
Specific parallel corpus based on [37] approach	15.89 K	348.5 K–90.3 K

90 753 pairs of sentences, nist08 contains 813 pairs of sentences and 15 500 pairs of NEs and a bilingual lexicon composed of 235 938 pairs of terms. The French side of these general-domain parallel corpora with the French Euronews corpus are used for general-domain language modeling. The generic-domain tuning data is the test data of the first edition of TRAD 2012. It is composed of different issues of the Arabic newspaper "Le Monde Diplomatique". It contains 423 pairs of sentences. Because of the lack of a media domain parallel corpus, we constructed manually our domain-specific parallel corpora for tuning and testing as follows: after mining comparable sentences from the media domain comparable corpus, the rest of sentences are used for tuning and testing. We select randomly 600 sentences for tuning and 400 sentences for testing from the Arabic side. We used human translators to translate these Arabic sentences into French. In this way we guarantee that tuning and test data are totally different from training data as they are manually translated. Furthermore, the cosine similarity is calculated to verify the distance between tuning and test data, we obtained 0.025 of similarity. The domain-specific monolingual corpus is extracted from the French Wikipedia articles in media domain. This corpus is constructed automatically using the method of domain-specific comparable corpora building, except the constraint of inter-language link to an Arabic article.

In addition, we implement a hybrid length and lexical based approach [35] to detect parallel sentences from comparable one. Our baseline system is a domain-generic SMT, trained with the domain-generic data described in Table 3. Several experiments are done in order to adapt the domain-generic SMT to the media domain.

- **Baseline**: trained with domain-generic training and tuning data described in Table 3.
- **Dev-sp**: trained with domain-generic training data and domain-specific tuning data.
- **Dev + LM-sp**: trained with the same translation model and tuning data of Dev-sp and use a domain-specific corpus in the target language added to the domain-generic data for language modeling.

- **Dev + LM + TMC + Tol**: trained with the same tuning data and language model as Dev + LM-sp system and use parallel phrases based on compositional +Tol based approach and the domain-generic parallel corpus to train the translation model.
- **Dev + LM + TMC − Tol**: trained with the same tuning data and language model as Dev + LM-sp system and use parallel phrases based on compositional −Tol based approach and the domain-generic parallel corpus to train the translation model.
- **Dev + LM + TMS**: trained with the same tuning data and language model as Dev + LM-sp system and use parallel sentences based on the [37] approach and the domain-generic parallel corpus to train the translation model.
- **Dev + LM + TMC + S**: trained with the same tuning data and language model of Dev + LM-sp system and combine parallel phrases based on compositional-based approach, parallel sentences based on the [37] approach and the domain-generic parallel corpus for translation modeling.

We should note that language model adaptation consists of adding the specialized data to the initial general data. A linear or log-linear interpolation of the two language models are impossible due to the reduced size of the domain-specific data.

Considering the domain-generic and domain-specific corpora, different adaptation strategies of the translation models are explored.

(a) Concatenation of new data to the initial generic data to construct a single translation model.
(b) Linear interpolation with adopting the same weights for the two models [33].
(c) Linear interpolation by favoring the specific translation model against the generic model [33].

Translation results obtained on the Specific test set are reported in terms of BLEU and OOV scores in Table 4.

Table 4 shows that the best results are obtained with a domain-specific tuning data and domain-specific translation and language models adapted with the strategy (a).

When integrating a domain-specific tuning data and a domain-specific language model, we observed a relative improvement of 5.56% of the BLUE score and 1.6 points of the OOV score compared to the basic system (Baseline). Furthermore, the Dev + LM + TM + Tol system reaches a BLUE score of 33.87%, when using the specialized parallel corpus based on the compositional approach (+Tol) concatenated to the domain-generic data; which introduces a relative improvement of 10.72% of the BLUE score compared to the baseline system and an improvement of 4.9% in the BLUE score compared to the DEV + LM-sp system. Thus, the Dev + LM + TM + Tol system is considered as the best system in terms of BLUE score.

Although the extracted data is not very large, the percentage of out of vocabulary words decreases by 5.39 points when integrating these specialized data into the translation model. This percentage decreases further when using of the data based on the hybrid approach [37]. This is due to the large coverage of this corpus which

Table 4 SMT results for Arabic to French

Combinaision	Adaptation strategy	% BLEU	% OOV
Baseline	–	30.59	9.89
Dev-sp	–	31.12	9.43
Dev + LM-sp	–	32.29	8.23
Dev + LM + TM-Tol	(a)	33.01	3.72
Dev + LM + TM + Tol	(a)	33.87	2.84
	(b)	30.17	4.11
	(c)	25.33	3.63
Dev + LM + TMS	(a)	32.18	1.3
	(b)	29.36	4.68
	(c)	21.21	2.67
Dev + LM + TMS + Tol	(a)	33.30	1.84
	(b)	30.28	4.68
	(c)	27.5	4.68

Table 5 Significance of SMT improvements in terms of P-Value

Systems	Blue	P-Value	Nist	P-Value
Dev + LM-sp	32.29		6.57	
Dev + LM + TM − Tol	33.01	0.15	6.57	0.41
Dev + LM + TM + Tol	33.87	0.02	6.71	0.04
Dev + LM + TMS + Tol	33.3	0.08	6.50	0.16

is larger than the corpus based on our compositional approach. Note that this improvement in OOV score was not accompanied by an improvement in BLUE score. This demonstrates the noisy of the data based on the hybrid approach. Fusion of the two corpus improves the results of the Dev + LM + TMS system in terms of BLUE score without reaching the Dev + LM + TM + Tol system BLEU score.

Table 5 evaluates the significance of the improvement obtained with Dev + LM + TM−Tol, Dev + LM + TM + Tol and Dev + LM + TMS + Tol systems against the Dev + LM-sp system. The most statistically significant improvement is obtained with the Dev + LM + TM + Tol system, in terms of BLUE and NIST score (P-Value < 0.05).

A first analysis of the Out Of Vocabulary words of the Dev + LM + TM + Tol system showed that 26.4% of these words were not translated due to tokenization errors (pre-processing of the corpus of text). Most specialized terms attached to a punctuation mark (e.g. (mbc), DRAMA., HD.) are not recognized by the tokenization process and subsequently are not translated. Manual tokenization of OOV words from the test corpus before the decoding process improves the blue score by 0.18 points and the OOV score by 0.28. In a second analysis of OOV words, we found that 32% of these words are written in foreign languages most of which are in English (e.g. Broadcasting, Sylvanas).

Table 6 shows an example of an Arabic sentence with the French reference, taken from the test corpus, in addition to different translations produced by various

Table 6 Example of translations of an Arabic sentence produced by various implemented systems

Arabic sentence	إس إم إنترتينمنت هي شركة تسجيلات كورية مستقلة ووكالة مواهب ومنتج وناشر لموسيقى البوب.
Buckwalter translitteration	As Am Antrtynmnt hy $rkp tsjylAt kwryp mstqlp wwkAlp mwAhb wmntj wnA$r lmwsyqy Albwb
Reference	SM Entertainment est une entreprise coréenne d'enregistrements indépendante, agence de talents, producteur et éditeur de la musique pop
Dev + LM + TM − Tol	Las ou enregistrements est une entreprise coréenne indépendant et l'Agence de talents et de producteur et éditeur de musique pop
Dev + LM + TM + Tol	SM Entertainment est une entreprise coréenne enregistrements indépendant et l'Agence de talents et de producteur et éditeur de musique pop
Dev + LM + TMS	Avex Trax coréenne indépendant et l'Agence de talents et productive et éditeur de musique pop
Dev + LM-sp	Wallace mre est une entreprise coréenne enregistrements indépendant et l'Agence de talents et productif et éditeur de musique pop
Dev-sp	Wallace ou des enregistrements est une entreprise coréenne indépendant et l'Agence de talents et de produit et éditeur de musique pop
Baseline	Wallace ou enregistrements est une entreprise coréenne indépendant et l'Agence de talents et productif et éditeur de musique pop

systems we have implemented. In this example, we observe the gradual improvement of translations when adding new specific data. Thus, the Dev + LM + TM + Tol produces the best translation, which is very close to the reference.

7 Conclusion

We have presented a novel model for mining domain-specific parallel data from Wikipedia. This model combine the use of (i) the taxonomy structure of Wikipedia articles to extract domain-specific comparable data, (ii) the concept of anchor points for comparable sentences alignment and (iii) the compositional based approach for parallel data mining. Experimental results, obtained using Arabic and French Wikipedia encyclopedia allow to jointly validate the extraction of domain-specific comparable and parallel corpora and the proposed adaptation methods. The best adapted system, trained on a combination of the baseline and the extracted data, improves the baseline by 3.3 BLEU points. Preliminary experiments with self-training also demonstrate the potential of this technique.

As a follow-up, we intend to investigate the evolution of the translation results as a function of the precision/recall quality of the extracted corpus, and of the quality of the automatically translated data. Furthermore, we plan to address the problem of

Out Of Vocabulary (OOV) words using word embedding. We have also only focused here on the adaptation of the translation model. We expect to achieve further gains when combining these techniques with LM adaptation techniques.

References

1. Aker, A., Feng, Y., Gaizauskas, R.: Automatic bilingual phrase extraction from comparable corpora. In: Proceedings of CICLING 2012, pp. 23–32, Mumbai (2012)
2. Barrón-Cedeño, A., España-Bonet, C., Boldoba, J., Marquez, L.A: Factory of comparable corpora from wikipedia. In: Proceeding of 8th Workshop on Building and Using Comparable Corpora, pp. 3–13, Beijing, China (2015)
3. Benoit, S., Darla, F.: Building a free french wordnet from multilingual resources. In: International Language Resources and Evaluation (LREC'08), Marrakech, Morocco (2008)
4. Brown, P., Pietra, S., Pietra, V., Mercer, R.: The mathematic of statistical machine translation: parameter estimation. Comput. Linguist. **19** (1993)
5. Buckwalter, T.: Buckwalter Arabic Morphological Analyzer Version 1.0. Linguistic Data Consortium, University of Pennsylvania. Catalog: LDC2002L49 (2002)
6. Chu, C., Nakazawa, T., Kurohashi, S.: Integrated parallel sentence and fragment extraction from comparable corpora: a case study on Chinese–Japanese wikipedia. ACM Trans. Asian Low-Resour. Lang. Inf. Process. **15**(2), 101–1022 (2016)
7. Delpech, E.: Traduction assistée par ordinateur et corpus comparables: contributions la traduction compositionnelle. PhD thesis, Univérsité de Nante, France (2013)
8. Elkateb, S., Black, W., Vossen, P., Farwell, D., Pease, A., Fellbaum, C.: Arabic wordnet and the challenges of arabic. In: The Challenge of Arabic for NLP/MT, pp. 15–24, London (2006)
9. Fung, P., Cheung, P.: Mining very-non-parallel corpora: parallel sentence and lexicon extraction via bootstrapping and em. In: Proceedings of Empirical Methods on Natural Language Processing (EMNLP), pp. 57–63, Barcelona, Spain (2004)
10. Gamallo, O.P., Loopez, I.G.: Wikipedia as multilingual source of comparable corpora. In: Proceedings of the 3rd Workshop on Building and Using Comparable Corpora, pp. 21–25 (2010)
11. Grefenstette, G.: The world wide web as a resource for example-based machine translation tasks. In: ASLIB99 Translating and the Computer, vol. 21 (1999)
12. Haruno, M., Yamazaki, T.: High-performance bilingual text alignment using statistical and dictionary information. In: Proceedings of the 34th Annual Meeting of the Association for Computational Linguistics (ACL96), pp. 131–138 (1996)
13. Kay, M., Roscheisen, M.: Text-translation alignment. Comput. Linguist. **19**, 121–142 (1988)
14. Koehn, P., Hoang, H., Birch, A., Callison-Burch, C., Federico, M., Bertoldi, N., Cowan, B., Shen, W., Moran, C., Zens, R., Dyer, C., Bojar, O., Constantin, A., Herbst, E.: Moses: open source toolkit for statistical machine translation. In: Proceedings of the 45th Annual Meeting of the ACL on Interactive Poster and Demonstration Sessions, ACL '07, pp. 177–180. Association for Computational Linguistics, Stroudsburg, PA, USA (2007)
15. Lardilleux, A., Lepage, Y.: Hapax legomena: their contribution in number and efficiency to word alignment. In: Proceedings of Human Language Technology. Challenges of the Information Society, Third Language and Technology Conference. Lecture Notes in Computer Science 5603, pp. 440–450, Poznan, Poland (2007)
16. Levenshtein, V.: Binary codes capable of correcting deletions, insertions, and reversals. Soviet Phys. Dokl. **10** (1966)
17. Li, B., Gaussier, E.: Improving corpus comparability for bilingual lexicon extraction from comparable corpora. In: Proceedings of the International Conference on Computational Linguistics (COLING10), pp. 644–652 (2010)

18. Munteanu, D.S., Marcu, D.: Extracting parallel sub-sentential fragments from non-parallel corpora. In: Proceedings of the 21st International Conference on Computational Linguistics and 44th Annual Meeting of the ACL, pp. 81–88 (2006)
19. Och, F.J., Ney, H.: A systematic comparison of various statistical alignment models. Comput. Linguist. **29**, pp. 19–51 (2003)
20. Pal, S., Pakray, P., Naskar, S.K.: Automatic building and using parallel resources for smt from comparable corpora. In: Proceedings of the 3rd Workshop on Hybrid Approaches to Translation (HyTra), pp. 47–56, Gothenburg, Sweden (2014)
21. Patry, A., Langlais, P.: Paradocs: l'entremetteur de documents paralèlles indpendant de la langue. Traitement Automatique des Langues **51**, 41–63 (2010)
22. Plamada, M., Volk, M.: Towards a wikipedia-extracted alpine corpus. In: The Fifth Workshop on Building and Using Comparable Corpora, Istanbul, Turkey (2012)
23. Prochasson, E., Morin, E., Kageura, K.: Anchor points for bilingual lexicon extraction from small comparable corpora. In: Proceedings of 12th Conference on Machine Translation Summit (MT Summit XII), pp. 284–291, Ottawa, Ontario, Canada (2009)
24. Rapp, R., Sharoff, S., Zweigenbaum, P.: Recent advances in machine translation using comparable corpora. Nat. Lang. Eng. **22**, 4 (2016)
25. Romary, L., Bonhomme, P.: Parallel alignment of structured documents. In: J. Vronis (ed.) Parallel Text Processing, pp. 201–218 (2000)
26. Sadat, F., Habash, N.: Combination of arabic preprocessing schemes for statistical machine. In: Proceedings of the 21st International Conference on Computational Linguistics and the 44th annual meeting of the Association for Computational Linguistics, pp. 1–8 (2006)
27. Samy, D., Moreno, A., Guirao, J.: A proposal for an arabic named entity tagger leveraging a parallel corpus (Spanish–Arabic). In: Proceedings of Recent Advances In Natural Language Processing RANLP, pp. 459–465 (2005)
28. Sellami, R., Deffaf, F., Sadat, F., Hadrich Belguith, L.: Improved statistical machine translation by cross-linguistic projection of named entities recognition and translation. Computacion y Sistemas **19**, 4 (2015)
29. Sellami, R., Sadat, F., Hadrich Belguith, L.: Exploiting multiple resources for japanese to english patent translation. In: Proceedings of MT Summit XI Workshop on Patent Translation, pp. 34–39, Nice, France (2013)
30. Sellami, R., Sadat, F., Hadrich Belguith, L.: Traduction automatique statistique partir de corpus comparable: application au couple de langues arabe-français. In: Proceedings of CORIA 2013, pp. 431–440, Neuchtel, Switzerland (2013)
31. Sellami, R., Sadat, F., Hadrich Belguith, L.: Improving named entity translation by exploiting noisy parallel corpora. In: Izwaini, S. (ed.) Paper in Translation Studies, Newcastle upon Tyn, pp. 179–198, Cambridge Scholars Publishing (2015)
32. Semmar, N., Saadane, H.: Etude de l'impact de la translittération de noms propres sur la qualité de l'alignement de mots partir de corpus parallèles franais-arabe. In: Actes de la 21e confrence sur le Traitement Automatique des Langues Naturelles, pp. 268–279, Marseille, France (2014)
33. Sennrich, R., Schwenk, H., Aransa, W.: A multi-domain: translation model framework for statistical machine translation. In: Proceedings of the 51st Annual Meeting of the Association for Computational Linguistics, Sofia, Bulgaria (2012)
34. Skadia, I.: Analysis and evaluation of comparable corpora for under-resourced areas of machine translation. In: The 5th Workshop on Building and Using Comparable Corpora, LREC 2012, pp. 17–19 (2012)
35. Smith, J.R., Quirk, C., Kristina, T.: Extracting parallel sentences from comparable corpora using document level alignment. In: Human Language Technologies: The 2010 Annual Conference of the North American Chapter conference of the Association for Computational Linguistics, pp. 403–411, Los Angeles, California (2010)
36. Stolcke, A.: Srilman extensible language modeling toolkit. In: Proceeding of ICSLP, pp. 901–904, Denver (2002)

37. Varga, D., Lszl, N., Peter, H., Andrs, K., Viktor, T., Viktor, N.: Parallel corpora for medium density languages. In: Proceedings of the RANLP, pp. 590–596 (2005)
38. Wolk, K., Rejmund, E., Marasek, K.: Multi-domaIn: machine translation enhancements by parallel data extraction from comparable corpora (2016). arXiv:1603.06785
39. Zesch, T., Gurevych, I., Mhlhuser, M.: Comparing wikipedia and german word-net by evaluating semantic relatedness on multiple datasets. In Proceedings of NAACL, pp. 205–208 (2007)
40. Zobel, J., Dart, P.: Phonetic string matching: lessons from information retrieval. In: Proceedings of the Eighteenth ACM SIGIR International Conference on Research and Development in Information Retrieval, pp. 166–173, Zurich, Switzerland (1996)

A Cross-Cultural Corpus Study of the Use of Hedging Markers and Dogmatism in Postgraduate Writing of Native and Non-native Speakers of English

Rawy A. Thabet

Abstract This study investigates the frequency of hedged propositions in academic writing, which are produced by both native (NSs) and non-native speakers (NNSs). To this end, two corpora, which represent native and non-native writings respectively, are compiled and investigated using contrastive interlanguage analysis (CIA). This computer-aided investigation, which involves comparing quantitative and qualitative data, is adopted to identify what the most frequent hedging markers, used by native and non-native writers, are, and whether there is any significant difference between the frequencies of these markers in both writings. This research is an attempt to fill a gap in literature, as there is a paucity of studies written on corpus analysis in the Middle East. The findings suggest that non-native speakers underuse hedges and the quality of these hedges is usually not so high as those of the native speakers.

Keywords Corpus analysis • Native speakers • Non-native speakers
Hedges • Modality • Overuse • Underuse

1 Introduction

One of the main problems that is faced by non-native speakers is the inability to express their stance or point of view without being dogmatic/hyperbolic. Scarcella and Brunak [1] admitted that the Arab (as an example of non-native speakers) learners lack the competence of using hedges. However, when the literature of modality was reviewed, it was found that this incompetence is not confined to Arabs but rather it is a common feature among L2 learners, such as French, German and Dutch learners [2, 3].

Many researchers have investigated this aspect of uncertainty (imprecision) and certainty (precision) [4, 5] by analysing texts produced by non-native speakers and

R.A. Thabet (✉)
Faculty of Education, The British University in Dubai, Dubai, UAE
e-mail: rawy.sabet@buid.ac.ae

K. Shaalan et al. (eds.), *Intelligent Natural Language Processing: Trends and Applications*, Studies in Computational Intelligence 740,
https://doi.org/10.1007/978-3-319-67056-0_32

contrasting them with native speakers' writing using certain software (viz., concordance software, such as WordSmith and Wmatrix3). The main approach used to hold this comparison between native and non-native features is called 'Contrastive Interlingual analysis' (CIA) [6]. This computer-aided method entails many functions, such as wordlist, which helps to find the words/phrases of high, medium frequency and even hapax legomena (i.e., words that are located only once in a corpus) [7].

In this current study, the researcher investigates the hedging markers used by the British University in Dubai (BUiD) students when they wrote their assignments for two modules (i.e., Research Methods in Education and TESOL Syllabus Design). 90 assignments, which formed the experimental corpus, were retrieved from Blackboard. All hedging markers and devices, which show the writer's uncertainty and certainty, were quantified and compared to another corpus written by native speakers who were at the same educational level. The two main questions that the study tries to answer are:

1. What are the most frequent hedging markers used by native and non-native writers?
2. Is there any significant difference between the frequencies of these markers in both writings?

Although there is an extensive literature on corpus analysis in other parts of the world, little research and investigation has been undertaken into postgraduate writing in the Arab Word, so this study seems to be one of the few sizeable corpora of tertiary English writing from the Middle East.

2 Literature Review

2.1 Introduction

First, the researcher discusses how the shift from accuracy to appropriacy has paved the way to the introduction of metadiscourse and corpus analysis. Then, the most relevant and seminal studies that discussed metadiscourse and how it is categorized are investigated. After that, hedges, as central exemplars of metadiscourse, are defined. Finally, the researcher explains how different researchers have approached hedges.

2.2 Metadiscourse

When there was a shift of focus from the mere study of language grammar to language function, metadiscourse found its way into this field of applied linguistics. The term metadiscourse was first introduced by Zelling Harris in 1959 (cited in 8) to show how the writer guides the recipient to understand the text or speech in a certain way. The term has been used with other linguistic terms, such as connectives and hedges.

2.3 *Metadiscourse Signals*

Hyland [8] critically analysed the work done on metadiscourse and tried to present a more robust model, but ended up with a model that is very similar to Vande Kopple [9]. Hyland based his taxonomy on two main dimensions:

The interactive plane: On this plane, the writer is aware of the reader's anticipations and seeks hard to satisfy his/her needs and expectations using some resources (devices), which could be used to constrain/control what can be unfolded (understood) from the text by the reader [10]. This plane entails five categories: transition markers (e.g., and), frame markers (e.g., finally), evidentials (e.g., Z states, according to X), code glosses (e.g., such.) and endophoric markers (e.g., noted above.).

The interactional plane: On the interactional plane, the writer's stance and judgment can be clearly identified by the reader. The writer also creates an imagined dialogue with the reader and responds to the questions that the reader would raise. This plane entails five categories: hedges (e.g., about.), boosters (e.g., definitely), attitude markers (e.g., surprisingly), self-mentions (e.g., my), and engagement markers. It seems that the distinction between these two dimensions is vague and carries many interpretations. Both Hyland and Vande Kopple's models are very similar, but Hyland's model included more subcategories than Vande Kopple's (10 and 7, respectively). Additionally, it is more detailed and pays more attention to certain features such as how writers explain their stances and how they can engage readers. Hyland's list of hedges is of great importance to the researcher as he uses the same list and applies it to the two corpora. To be more precise, this list will be searched for in the two corpora to find how frequent each hedge is in the two corpora (native and non-native). Hyland's list of hedges consists of 101 hedges, but these devices were randomly mentioned on a list, so the researcher decided to improve this list by categorizing hedges according their part of speech (see Table 1).

2.4 *Hedges*

According to the Cambridge Dictionary [11], a hedge is "a word or phrase that makes what you say less strong". Hedging is a feature of academic writing which distinguishes it from other genres. There are some epistemic devices, such as *perhaps* and *may*, that show the open mindedness of the writer and that he/she does not have full commitment to what he/she is proposing. In other words, the presence

Table 1 Hyland's list of 101 hedges

Adverb		Adjectvie	Verb		Adjective phrase	Adverbial phrase	Modal auxiliaries	Noun
About = approximately	Plausibly	Apparent	Appear	Indicate	Certain amount	From my perspective	Could	Doubt
Almost = nearly	Possibly	Doubtful	Appeared	Indicated	Certain extent	From our perspective	Couldn't	
Apparently	Presumably	Plausible	Appears	Indicates	Certain level	From this perspective	May	
Approximately	Probably	Possible	Argue	Postulate		In general	Might	
Around = approximately	Quite	Presumable	Argued	Postulated		In most cases	Ought	
Broadly	Rather	Probable	Argues	Postulates		In most instances	Should	
Essentially	Relatively	Typical	Assume	Seems		In my opinion	Would	
Fairly	Roughly	Uncertain	Assumed	Suggest		In my view	Wouldn't	
Frequently	Sometimes	Unclear	Claim	Suggested		In this view		
Generally	Somewhat		Claimed	Suggests		In our opinion		
Largely	Typically		Claims	Suppose		In our view		
Likely	Uncertainly		Estimate	Supposed		On the whole		
Mainly	Unclearly		Estimated	Supposes		To my knowledge		
Maybe	Unlikely		Feel	Suspect				
Mostly	Usually		Feels	Suspects				
Often			Felt	Tend to				
Perhaps			Guess	Tends to				
				Tended to				

of hedges in writing proves that the information is subjective because it is given as a personal view rather than a fact [8]. Poos and Simpson [12] asserted that hedges can serve many pragmatic functions, for example, the hedging markers, such as *kind of* and *sort of*, can be used to show inexactitude (lack of precision) or to reduce the force of an ascertain. In a similar vein, Lakoff [13] describes hedges as those devices which make the writer's proposition fuzzier or less fuzzy. Learners of language should be taught how to strike a balance in their writing in order not to sound either arrogant or excessively tentative [14].

2.5 Meyer's Taxonomy

In a similar vein, Salagar-Meyer [15] proposed a taxonomy of hedging markers, which consists of five categories:

(1) **Shields**: this group consists of all auxiliary verbs (communicating possibility); lexical verbs with modal meaning such as *appear* and *seem*; adverbials of probability such as *likely*; adjectives of probability such as *probable*; epistemic verbs which are identified with the probability of a proposition such *to propose* or *to suggest*.
(2) **Approximators**: this category includes adverbs of degree, time and frequency, such as *approximately, roughly* and *often*. They are used to make things obscure or when the precise figures are inaccessible.
(3) **Author's personal point of view** (personal doubt), such as *I believe*.
(4) **Intensifiers** (emotional), such as *extremely interesting*.
(5) **Compound hedges**, such as *it could be suggested*. This subcategory can include compound hedges up to quadruple hedges or more, for example, *it would seem somewhat unlikely that*. Murniato [16] did a better job than Salagar as she (i.e. Murniato) divided the compound hedges into two categories: (1) a modal auxiliary with a lexical verb, which has a sense of hedging, for example *would appear* (2) a lexical verb with an adjective carrying a meaning of a hedge, for example *seem acceptable*. She added that these compound hedges could consist of double, treble or quadruple words.

In summary, all of these categories [Vande Kopple, Meyer, Lakoff and Hyland's) cause confusion and many of them overlap. For example, in Salager-Meyer's taxonomy and with a deep look at approximators and shields, it can be easily discovered that most of the approximators can do the same job of shields. In addition to that, many of the compound hedges consist of at least one main modal auxiliary, which is part of the shields. Koutsantoni [17] confirms that the examples of intensifiers given by Salager-Meyer are no more than examples of attitude markers and not hedges. She adds that the third category, which is the 'author's personal doubt', can include any item from the other four categories.

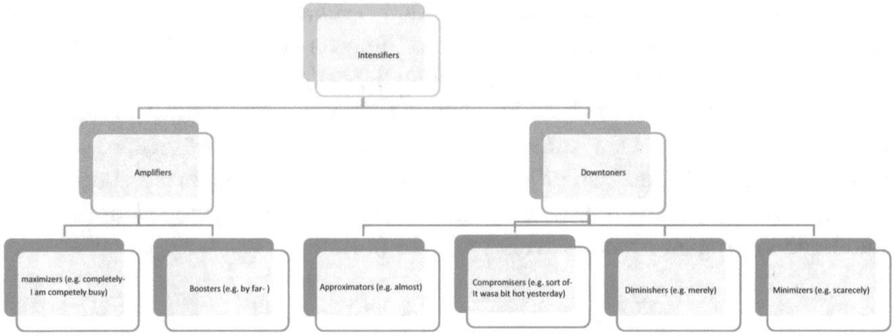

Fig. 1 Quirk et al.'s modal of intensifiers. Adopted from [18, p. 589]

Hedges can be expressed in different ways using different devices. Some of the devices that express the writer's engagement are boosters, diminishers and minimizers (adverbials of degree) (Fig. 1).

2.6 *Intensifiers*

Quirk et al. [18] distinguished between two main categories that show the writer's degree of commitment. These two categories are amplifiers (e.g., maximizers and boosters) and downtoners (e.g., approximators, compromisers, diminishers and minimizers-negative maximizers). Amplifiers are qualifiers or word intensifying expressions that reinforce the significance of adjacent expressions and show accentuation. Words that are usually used as intensifiers may include some adverbs, such as *completely* and *really*. If these intensifiers were ordered and distributed on an inverted triangle according the degree of emphasis, the maximisers sit at the top and the minimizers at the bottom (see Fig. 2). However, downtoners are words or expressions, which weaken the power of another word or expression. According to Quirk's categorization, the overstating is expressed by the amplifiers that show the

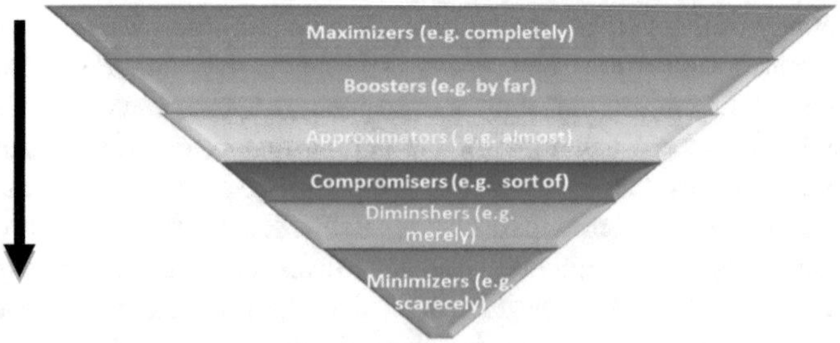

Fig. 2 Intensifiers pyramid

positive emphasis using emphatic devices while the downtoners are used to show the writer's caution. This caution is one the features that distinguishes the native speakers' writing.

2.7 Modality

Modality is usually connected with modal auxiliaries even though there are many other forms that would do the same function of modals, for example, modality could be expressed by some adverbs, such as *probably* and *possibly*; some verbs would also serve as modals, such as *I think* and *I feel* [19]. The previous studies, which were not based on corpus analysis, showed that non-native speakers tend to overuse or underuse certain modal auxiliaries/meanings [20]. Aijmer [19] used a computer-aided approach to compare between argumentative writings produced by Swedish L2 and English natives. What distinguishes her study is that she did not only compare between the Swedish L2's writing and native English speakers' writing, but she also held a comparison to the writing of other languages (i.e., French and German) regularly.

According to Papafragou [21] epistemic modality is defined as the "assessment of probability and predictability". Aijmer [19] used "degrees of likelihood" to explain the meaning of epistemic modality while root or deontic modality refers to the degree shown by the writer to express obligation, ability, power of deciding (volition), necessity, permission and necessity. Root modality has been referred to by many researchers, using different titles; for example, Halliday [22] refers to root modality as modulation.

2.8 Bundles

As mentioned earlier, these words, which usually occur together, have been given different titles such as clusters, bundles and multi-word expressions. This sequence of words helps us to identify the different registers, for example, *as can be discerned* refers to academic field and a bundle like *in pursuance of* refers to a legal document. The more proficient the writers become, the more bundles will be incorporated in their texts [23]. Wray [24] suggests that these formulaic patterns are overlooked in language acquisition. It is worth mentioning that these collocations can help to strengthen the relationship between the receiver of the text and sender because the presence of certain collocations helps the reader to know the register of the text.

2.9 Collocational Frame (It is … that)

Learners need to be equipped with the hedging devices that help them to strike "a balance between authority and concession" [12, p. 4]. As mentioned earlier in the literature review, in order for interlocutors to show their precision or imprecision, there are many approaches that they can use such as hedging markers or intensifiers (amplifiers and downtoners). What distinguishes an expert writer from an apprentice is the ability to vary the degree of precision to the extent that suites the context. Whether the interlocutor is hedging or boosting, his/her main objective is to comment on the proposition given by him/her. This comment could show how he/she feels towards what he/she is writing. This feeling could be related to the likelihood, the desirability or the seriousness of a proposition [25]. One of the evaluative forms that Lemke [25] studied was the sentences that include *It is…that*. Lemke [25] explained the use of *that* as a conjunction comes before a noun clause, whereas the extraposed *it is* precedes an adjective. This adjective could fall into one of seven semantic classes (probability, appropriateness, importance, seriousness, etc.). These adjectives are, in essence, evaluative epithets. He added that the noun clause that is introduced by *that* could represent a proposition or fact (if realis-) or a possibility (if irrealis). There is a variety of forms that this collocational frame could take, for example:

It + verb to be (functioning as a copula) + evaluative epithets (adjectives) + that…

Or *It + passive voice (to be + past participle) + that…*

These evaluative forms are very important for the study as the researcher examines them in both corpora and finally deducts some findings about their use, frequencies and varieties.

2.10 Lexical and Functional Words

There are two classes of words: lexical (also referred to as content or substantive words) and functional words. The former includes these words that carry meaning such as nouns, verbs, adjectives and prepositions [26]. They are also referred to as open-class category because it is possible for this category to be extended indefinitely by adding more items to it [27]. This idea can be supported by the fact that new words are coined and added to dictionaries almost every day. The second category includes the functional words, which is considered a closed-class list because there is a specific number of them and it is rare that new words are added to them. They serve as the mortar that sticks lexical words together [28]. When counting the elements that could be added to the functional word list, Hinojosa et al. [27] mentioned conjunctions, determiners, pronouns and prepositions. If the readers just go a few lines up, they will find that prepositions were counted among the lexical words by Corver and Van Riemsdijk [28]. This discrepancy in the categorization corroborates the fact that the distinction between these two categories is not easy because some lexical words would serve as functional words and vice versa.

2.11 Corpus Linguistics Definition and Potential

Granger [6] states that corpus linguistics and second language research were two different fields, but with the advent of the new branch of knowledge known as learner corpus research in the 1980s, these two branches have been linked together. This new methodology has enabled researchers to explore different areas of language and make recommendations for better ways of learning a second language.

Corpus linguistics is defined as the analysis of electronic collections of authentic texts (i.e., naturally occurred). This authenticity feature was also mentioned by Halliday [29] as he enumerated three advantages and one disadvantage of corpus analysis. One of these advantages is that corpus enabled scholars to study grammar quantitatively. This quantitativeness is based on the ability of researchers to count the frequency of language items in texts [6].

Corpus linguistics is not a new method because it has been there for a long time, but with the advent of computers, this branch of study has enabled scholars to explore some areas that were very difficult to investigate without this magnificent device. The same idea of the added advantage of computers has been raised by Stubbs [30, p. 232] as he said, "the heuristic power of corpus methods is no longer in doubt". Although the focus of the corpus-based studies, conducted over the past two decades, was only on the features of the native English speaker, such as describing the registers and different dialects of Americans, British and Australians, this trend did not last for long as the focus had also been directed to non-native English. This change of focus started in the 1980s and the material collected from non-native English has been called learner corpora [6].

Halliday [29, p. 29] also defined corpus as "a large collection of instances of spoken and written texts". He added that the two main inventions that radically changed the work of grammarians are tape recorders and computers as the former was used to record the spoken discourse and the latter for saving the written texts. He continues to say that in the 1950s, when the two American scholars Randolph Quirk and W. Freeman Twaddell, started analysing their first corpus manually, they realized that the whole process would be computerized soon. Similarly, Schmitt [31] asserts that corpus analysis has recently gained significant popularity for two reasons: first, it focuses on the real language (spoken or written) produced by people; secondly, its outcomes can help in designing curricula.

3 Methodology

3.1 Participants

The participants of this study are students joining Master of Education programme, TESOL concentration. Each student has to study six modules (three elective and three core). The core modules are 'Teaching and Learning', 'Research Methods in Education' and 'Educational Policy'. The elective modules are 'Discourse for Language Teachers', 'TESOL Syllabus and Design', and 'Second Language Teaching and Learning'. The final written assignments, which were submitted to one of the core modules (i.e., Research Methods in Education) and to one elective module (i.e., TESOL Syllabus and Design), were uploaded to the corpus analysis software to be analysed. These two modules were carefully selected for the following reasons. First, the main question of this study is to find the frequency and quality of hedging devices used by the BUiD's students and comparing this frequency and quality to that of the BAWE writers. In 'Research Methods' module, students are required to write a research proposal while in 'TESOL Syllabus and Design' students are required to critically evaluate some syllabi. Therefore, in both modules students are expected to criticize the existing teaching material and methodology or to convince their study supervisor or funding institutions of the validity of their proposals. The total number of students that participated in this study is 70, who combined submitted 90 assignments. The number of assignments exceeds the number of students because some of them (20 students) submitted one assignment to each of the two modules. The majority of the participants are Arabs (85%) and 15% are from other nationalities, such as Indian (6%), British (2%), Bangladeshi, French, Nigerian and Pakistani with 1% each. The British participants were not raised in Britain, but were naturalized when they were adults. Finally, both genders were almost equally represented, as the male participants was accounted for 55% and female participants 45% of the study group. Ninety assignments were submitted to two modules—Research Methods in Education & Syllabus Design—between 3,000 and 4,000 words in length and with about 300,000 words in total. This number decreased to less than 300,000 when the text was formatted and converted to a text-only version, which is the appropriate format that can be uploaded to corpus analysis software.

3.2 Contrasting and Analysing the Two Corpora

According to Granger [6] contrastive interlingual analysis includes comparing NS to NNS. In this type, the contrast is held between writing features in both native (control) and non-native (experimental) English. The main concerns related to this type are the different varieties of native languages, such as the different dialects, spellings and the level of professionalism [32] of these native people whose writing form the body of the control corpus. McKenny [14] stressed that for the two corpora to be successfully compared, the number of words and the purposes for which the texts of both corpora were written should match. This condition is met in this current study as both corpora have the same length and their texts are written to serve the same purpose. Contrasting native and non-native writing makes it possible to spot not only the misuse of some language features, but it also enables linguistics to determine the overuse and/or the underuse of some specific features when compared with native writing as a reference.

3.3 Motive Behind Writing and Corpus Compilation

McKenny [14] ascertains that most of the texts in the native language corpora were compiled for purposes other than corpus analysis. Similarly, BAWE corpus was made up of students' papers, submitted to their modules and not for corpus analysis. Generally, the subjects who contributed to BAWE and BUiD corpora were post-graduate students undertaking their master degrees. However, in BAWE case, students' papers were added to the corpus provided they gained a distinction. In addition, the authors of the selected papers were paid an amount of money and signed a disclaimer forms so that their universities could use their submitted paper for research purposes.

As for the compilation of the BUiD corpus, all word documents were converted to plain text because most tagging software works perfectly with texts that have no formatting (44). Since all section headings in the control corpus (BAWE) are encoded as < heading > … < /heading >, the researcher did the same thing in the experimental corpus. When the 90 assignments were joined using the WordSmith tool, each one of these assignments was given a specific number, for example, the first assignment was given the number 11, the second was given the number 22 and the last assignment was given the number 9090. Assigning numbers to each assignment would help the researcher to know in which assignment a specific language feature or concordance occurs.

3.4 Control Corpus Compilation

In order to obtain a full version of the British Academic Written English corpus, an online application form was completed and sent to the University of Oxford Text Archive. The request was soon approved and the researcher was given a full copy of the BAWE corpus. This corpus was compiled over a period of 3 years (2004–2007) and it consisted of 2,761 assignments written by students joining three universities; Oxford Brookes, Warwick and Reading [33]. All these writings were deemed as proficient writings (graded Merit or Distinction) and the authors were predominantly English native speakers (80%) and non-native English speakers (20%) [14].

The length of the texts ranged from 1,000 to 5,000 words. These written texts were classified into four disciplinary groups (DG), which are Arts and Humanities, Life sciences, Physical Sciences and Social Sciences. Then, the texts, submitted to each disciplinary group, were subcategorized into disciplines. Each disciplinary group consists of about 4–9 disciplines; for example, Arts and Humanities consists of 8 disciplines including Archaeology, Classics, etc. From all these contributions, the researcher selected texts submitted to Arts and Humanities and Social Sciences DGs. As the experimental corpus consists of assignments submitted to the Master of Education programme, the researcher tried to be very selective and had three main criteria when choosing the texts from BAWE. First, the topic had to be closely related to the educational field, such as English, History, Linguistics, and Sociology. Second, the more argumentative and text-oriented the piece of writing was, the more suitable it was deemed to be included for contrasting. Based on the previous criterion and based on the length of the experimental corpus (300,000), 101 texts were selected from BAWE with 300,000 words in total. All these key issues, such as the length and purpose of writing, should be considered when comparing the two corpora so that the only difference between them would be the level of proficiency and authorial expertise [34].

The focus of this study is the assignments written by 70 postgraduate students who undertook their master degrees at the British University in Dubai. The experimental corpus is referred to as the BUiD corpus. The methodology adopted in conducting this research is mainly empirical as it is based on direct observation of certain features in the two corpora (experimental and control). These features and language items have been quantified in the non-native corpus and then compared to the corpus written by native speakers. This method of contrast is called Contrastive Interlingual Analysis. As a starting point, all hedging markers, suggested by Hyland, were typed in a notepad to be searched for in both corpora. Homonyms, which do not serve as hedging markers, have been excluded. In other words, all language items that do not represent the writer's stance or degree of commitment are culled. In this regard, Aijmer [19] said that sometimes the manual analysis is necessary to avoid disambiguation. The manual filtering of both corpora, in this current study, resulted in deleting some markers that were mistakenly included within the list of hedges generated by WordSmith; for example, the epistemic meaning of the adverb *around* is approximately, but in concordance 1, it was used as a preposition which meant 'in this direction', so it was deleted. In concordance 3, the word 'May' served as the name of the fifth month of the year and not as a hedge, so it was deleted as well.

Concordance 1: *This gives more .... of learning and competing **around** the world.*
Concordance 2: *There is .... **about** the tendency in ....*
Concordance 3: *April 2014–**May** 2014 literature Review ....*
Concordance 4: *This reflected on ....; I felt helpless and defenseless.*

Annotating corpus is another solution to removing disambiguation, for example, tagging the word 'can' as a modal auxiliary when it serves as a modal and tagging it as a noun when it serves as a noun would help to distinguish between the auxiliary verb can and its homonym. To overcome the problem of unneeded language features, the researcher prepared a list of all search-words (hedging markers suggested by Hyland) and uploaded this list to WordSmith. A list of concordances of search-words was generated. The next step was filtering this list by deleting all irrelevant language markers or the markers that did not serve as hedging devices. Only the devices that showed tentativeness and degrees of un/certainty were included [35]. This step of weeding out devices that did not serve as hedging markers had been neglected by many studies as most of them followed "wanton frequency count" [12].

As mentioned in the literature review, Salagar-Meyer [15] did not develop a list of hedges for her proposed taxonomies, so the researcher referred to other studies to create a list for each taxonomy; for example, while reviewing the work of Hyland [8], it was found that the list of hedges entitled 'attitude markers', developed by Hyland, is very similar to the examples of intensifiers suggested by Salagar-Meyer. In the same vein, the researcher referred to the work of Holmes [36] to create a list of lexical verbs with epistemic meaning. Actually, this list was a merge of Holmes [36] and Hyland's (14) lists. Generally, most of these lists, used to search for concordances of Salagar-Meyer's taxonomies, were created in a similar way, i.e., merging the lists of hedges developed by other researchers to create one list for each taxonomy.

4 Findings and Discussion

4.1 The Most Frequent Single Words

As a starting point, the 40 most frequent single words were identified and compared in the two corpora (the experimental and control). The researcher started with single words and then moved on to compound forms. This sequence of steps is a representation of the bottom-up approach, which the researcher would like to follow in the beginning. According to Scott and Tribble [7] the most frequent words are found at the top while the tail of this list is full of hapax legomena. They also ascertain that once the text has been transformed into a wordlist, all the functional words, such as *the* and *of* are sent to the top of this list. As can be seen in Table 2, the first column contains the serial number of concordances; the second column shows the word itself; the third shows the number of tokens of each type of the words in the whole texts; and the extreme right-hand column shows the percentage of these tokens in texts as a whole. For instance, the word-type *the* has 22,979 tokens, which represents 7.55% of the whole running words in the BUiD Corpus. It can be easily discerned that there is a divergence in the use of the definite article *the* in both corpora: in the BUiD Corpus, the frequency of this article makes up 7.55% while in BAWE, it represents 6.88%. This finding contradicts McKenny's [14]

Table 2 The forty most frequent words in both corpora

N	BUiD			N	BAWE		
	Word	Freq.	%		Word	Freq.	%
1	THE	22,979	7.55	1	THE	20,781.00	6.88
2	AND	10,281	3.38	2	OF	12,543.00	4.15
3	OF	10,260	3.37	3	AND	9,104.00	3.01
4	TO	9,955	3.27	4	TO	8,271.00	2.74
5	IN	8,056	2.65	5	#	7,801.00	2.58
6	#	5,920	1.94	6	IN	7,042.00	2.33
7	A	5,486	1.80	7	A	6,192.00	2.05
8	IS	4,750	1.56	8	IS	5,010.00	1.66
9	THAT	4,055	1.33	9	THAT	3,511.00	1.16
10	STUDENTS	3,212	1.06	10	AS	3,226.00	1.07
11	BE	2,948	0.97	11	IT	2,321.00	0.77
12	FOR	2,823	0.93	12	FOR	2,182.00	0.72
13	THIS	2,685	0.88	13	BE	2,112.00	0.70
14	AS	2,616	0.86	14	THIS	2,062.00	0.68
15	ARE	2,430	0.80	15	WITH	1,881.00	0.62
16	IT	2,236	0.73	16	ARE	1,672.00	0.55
17	ON	2,113	0.69	17	BY	1,648.00	0.55
18	*WILL*	1,947	0.64	18	ON	1,643.00	0.54
19	WITH	1,929	0.63	19	NOT	1,553.00	0.51
20	TEACHERS	1,798	0.59	20	WHICH	1,521.00	0.50
21	THEIR	1,721	0.57	21	AN	1,384.00	0.46
22	THEY	1,429	0.47	22	FROM	1,276.00	0.42
23	BY	1,382	0.45	23	OR	1,185.00	0.39
24	LEARNING	1,382	0.45	24	WAS	1,087.00	0.36
25	LANGUAGE	1,358	0.45	25	*CAN*	993.00	0.33
26	STUDY	1,259	0.41	26	THEIR	968.00	0.32
27	RESEARCH	1,188	0.39	27	HAVE	907.00	0.30
28	WHICH	1,123	0.37	28	I	891.00	0.29
29	NOT	1,115	0.37	29	HIS	852.00	0.28
30	TEACHING	1,078	0.35	30	BUT	834.00	0.28
31	HAVE	1,071	0.35	31	HAS	827.00	0.27
32	AN	1,063	0.35	32	P	813.00	0.27
33	OR	1,054	0.35	33	AT	800.00	0.26
34	FROM	1,010	0.33	34	MORE	781.00	0.26
35	*CAN*	990	0.33	35	THEY	781.00	0.26
36	LEARNERS	933	0.31	36	ONE	768.00	0.25
37	TEXTBOOK	906	0.30	37	HE	698.00	0.23
38	SCHOOL	872	0.29	38	ITS	660.00	0.22
39	BOOK	866	0.28	39	*WILL*	655.00	0.22
40	TEACHER	846	0.28	40	ALSO	643.00	0.21

conclusions as he reports that the non-native speakers in his study significantly underused the definite article when compared to the native speakers. The definite article usually collocates with nouns. To prove that, when the definite article is searched for in the BUiD corpus, it is found that it collocates with the word *STUDENTS* 871 times. This finding suggests that there would be an overuse of nouns in NNSs' corpus. This will prove right when the two corpora are tagged with the USAS tagset. As an ESL teacher with many years of experience teaching Arabs, the researcher can assume that the overuse of the definite article is due to its wrong use, which could be attributed to the L1 transfer.

As is expected, the most frequent words on the top of both lists are functional words such as *the*, *and*, *of* and *to*. It is worth mentioning that the top 9 most frequent words are almost the same in the two corpora. It is also interesting to notice that on the experimental list (BUiD Corpus), the first content word comes in the tenth position while there is no one content word among the 40 most frequent words in the reference corpus as all of these 40 most frequent words are functional words. It is equally interesting to notice that the frequency of the modal verb *will* is 1,947 while it is only 655 in the BAWE corpus.

4.2 Lexical Density

Lexical density is usually used as a measure of the level of proficiency of text. Kenny [37] developed a technique that is referred to as the type-token ratio (TTR). As the name indicates, the total number of word types is divided by the total number of the running words. Then, the result of this division (i.e. quotient) is converted to a percentage. This technique had a lot of criticism because of its sensitiveness to the length of the text, for example, if a text consists of 10,000 running words, it is said that this text has 10,000 tokens. This dependence on the size of the text is considered one of the limitations of this measure, which could have been firmly accepted if it had excluded the repeated words (Table 3).

In his endeavour to overcome the deficiency of the TTR, Scott [38] developed the standardized type/token ratio by dividing the texts into smaller segments and taking the average of the TTR of each of the segments. This approach was also criticized for not reflecting the reality of the text lexical density.

In reaction to the limitation of both techniques (i.e., TTR & STTR), scholars started to adopt another tool developed by Ure [39]. In order to measure the density of lexis in a text, Ure tried to find the proportion of the lexical words to the grammatical ones. As recommended by Ure [39] and Stubbs [30], the lexical density is calculated by dividing the lexical/content words by the total number of

Table 3 TTR & STTR of the two corpora

Corpus	BUiD	BAWE
Tokens (running words) in text	304409	302121
Tokens used for word list	298489	294320
Types	10904	17607
Type/token ratio (TTR)	3.65	5.98
Standardized type/token ratio (STTR)	37.53	40.54

tokens in the corpus. To create a list of content words, the researcher used a stoplist of the 100 most frequent words.

In the BUiD Corpus (using the 100 most frequent words)

Content words = 304,409 (tokens) − 143.445 (functional words removed) = 160.964....

Lexical density = 160.964 (content words)/304.409 (tokens) = 52.877%.

In the BAWE Corpus (using the 100 most frequent words)...Content words = 302.121 − 141.683 = 160.438.

Lexical density = 160.438/302.121 = 53.103%.

The percentages in the Table 4 suggest that the lexical density of the native speakers' corpus is slightly higher than the non-native's. This is not a surprising finding for the researcher because he expected that the lexical density of BUiD would be less than BAWE, because he is an Arab and was educated in an Arabic country where teaching is mainly grammar-oriented. However, the high lexical density is not evidence of the full command of the language as there are native speakers whose writing is not highly lexically dense [40]. In addition, the categorization of a text into lexical and functional items is not easy because some lexical words work as grammatical words and vice versa [41]. In other words, the function of each category (lexical and grammatical) may overlap.

4.3 Hyland's Taxonomy

As mentioned earlier in the methodology section, all hedging markers (101), suggested by Hyland [8], were typed in a notepad and searched for in both corpora using the function of 'get search-word from a file' in the corpus analysis software called 'WordSmith'. When adding all totals of hedging adverbs, verbs, adjectives,

Table 4 Lexical density using a stoplist of the 100 most frequent words

Corpus	BUiD (%)	BAWE (%)
Lexical density using stoplist	52.877	53.103

Table 5 Hedges according to Hyland taxonomy

Part of speech	BUiD	BAWE
Hedging modal auxiliary	1661	1617
Hedging adverbs	723	1096
Hedging verbs	691	1026
Hedging adjectives	94	200
Adverbial phrase	66	50
Hedging noun	11	20
Noun phrase	5	13
Total	3240 (1.07%)	4002 (1.33%)

Table 6 Expected contingency

Part of speech	BUID	BAWE
Hedging modal auxiliary	1460.29	1803.73
Hedging adverbs	810.334	1000.91
Hedging verbs	764.895	944.787
Hedging adjectives	130.972	161.775
Adverbial phrase	51.6761	63.8295
Hedging noun	13.81	17.0579
Noun phrase	8.0187	9.90458

modal auxiliary and compound hedges, it was found that, generally, the NSs used more hedges than NNS; 4,022 (1.33%) hedges and 3,251 (1.07%) hedges, respectively (Tables 5, 6).

Chi-square = 3.14… Degree of freedom (df) = (C−1) (r−1) = (2−1) (7 −1) = (1) (6) = 6 Probability = 0.05.

Based on the Chi-square results, there is a likelihood that there would be a statistically significant difference between the frequencies of the hedging markers in the two corpora. Looking closely at the frequencies of hedging markers in both corpora, it can easily be discerned that NSs employed more adverbs of probability than NNSs, especially, the adverb 'perhaps' which was used 81 times by NSs while the NNSs used it only eight times. Similarly, adverbs like 'possibly', 'likely', 'roughly' were far underused by the NNS.

4.4 Salagar-Meyer's Taxonomy

When applying Salagar-Meyer's [15] proposed taxonomy of hedging markers, which consists of five categories, it was also found that native speakers, overall, used more hedging devices than non-native speakers; 9,945 (3.37% of the total number of words) and 7,324 (2.42%), respectively (see Table 7). The two most frequent types of hedges in both native and non-native speakers' corpora are shields and author's personal point of view. These two types accounted for 52.81 and 36.65% of the total number of hedges used by native speakers whereas they

Table 7 Salagar-Meyer's proposed taxonomy of hedging markers found in BAWE and BUiD

Category	BAWE				BUiD			
	Freq	Percentage of this category to the total # of token	Percentage of this category to the total # of hedges	Expected contigency	Freq	Percentage of this category to the total # of token	Percentage of this category to the total # of hedges	Expected contigency
Shields	5252	1.8	52.81	5373.0297	4078	1.35	55.68	3956.9703
Approximators:	381	0.13	3.83	393.907	303	0.1	4.14	290.093
Author's personal point of view	3645	1.22	36.65	3268.737	2031	0.67	27.73	2407.263
Intensifies (Similar to attitude markers developed by Hyland 2005)	632	0.21	6.35	884.56309	904	0.3	12.34	651.43691
Compound hedges such as it could be suggested.	35	0.01	0.35	24.763159	8	0	0.11	18.236841
Total	9945	3.37			7324	2.42		

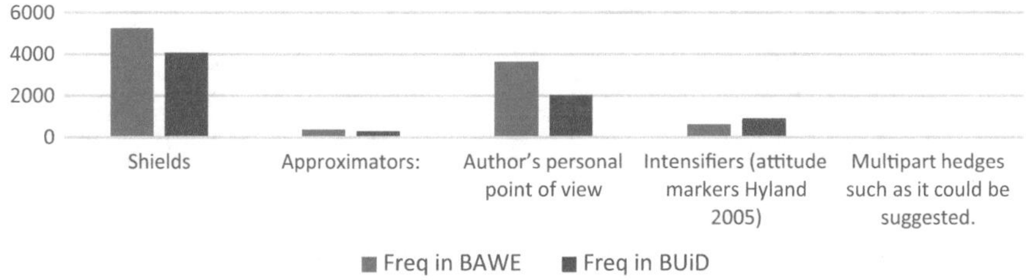

Fig. 3 Hedges in BAWE & BUiD (using Salagar-Meyer's taxonomy)

constituted 55.68 and 27.73% of the total number of hedges used by non-native speakers [42]. The native speakers exceeded the non-native speakers in the frequency of the shields, approximators, author's personal point of view and compound hedges. The order of the hedge types in both corpora is the same as shields come in the first place and Author's personal point of view come in the second place followed by intensifiers, Approximators and compound hedges (Fig. 3).

Chi-square = 1.9 > 0.05 Degree of freedom (df) = (C−1) (r−1) = (2−1) (5−1) = (1) (4) = 4 Probability = 0.05.

Based on Chi-square result, there is a significant difference between NS and NNS in their use of hedges. Generally, this taxonomy (i.e. Salagar) is problematic, especially, the category of intensifiers which was described by Koutsantoni [17] as vague and function as attitude markers more than as hedges. Based on this conclusion, the researcher used Hyland's list of attitude markers as intensifiers. This vagueness and lack of a list of lexical items led to a discrepancy in the counts of intensifiers calculated when Salager-Meyer [15] and Quirk et al.'s [18] models were applied.

4.5 *Syntactic and Semantic Tagging*

The researcher also used Wmatrix3 to identify the variety of parts of speech used in both corpora. The two corpora were uploaded to the Wmatrix3 tool and tagged with the UCREL CLAWS7 tagset. The main motive behind this step was to find whether the non-native speakers in the experimental corpus overused or underused some parts of speech. It is clear that NNSs used more verbs, nouns and fewer adverbs and adjectives than NSs (see Table 8). This finding is almost in line with Ringbom [43] who found that NNSs corpus included more verbs and fewer adjectives than NSs'.

Table 8 Parts of speech in both corpora

Total number	BUiD	%	BAWE	%
Verbs	53713	18.64	46476	16.14
Noun	84837	29.41	80320	27.89
Adverb	9998	3.42	13788	4.77
Adjective	23094	8	27555	9.56
Total	**171642**	**59.47**	**168139**	**58.36**

Table 9 Parts of speech-CLAWS tagset

Sr no	Item	O1 (BUiD)	%	O2 (BAWE)	%		LL	Logratio
1	NN2	27735	9.62	16877	5.86	+	2667.5	64.31
13	**VM**	**4748**	**1.65**	**3533**	**1.23**	+	**178.7**	**34.37**

The researcher was looking for the devices and parts of speech that were used to show tentativeness or degree of commitment. According to the CLAWS7 tagset, VM stands for modal auxiliary (e.g., can) and that was the first target for the researcher. Searching the tagged lists of the two corpora, where O1 stands for observed frequency in the BUiD corpus and O2 stands for observed frequency in the BAWE corpus, VM was the thirteenth item on the list (see Table 9).

Modal auxiliaries have significant importance, as their proper use by non-natives is considered to be a challenge. They are also important devices used by the writer to show tentativeness or hedging [44]. For these two main reasons, the researcher decided to investigate modal auxiliaries in both corpora. It is clear from Table 9 that BUiD students overused the modal auxiliaries as there are 4,748 occurrences of them, which represent 1.65% of the total words in BUiD while BAWE students used them 3,533 times, which represent 1.23% of the total words in BAWE. These numbers are different from Hyland's because Hyland's list of modal auxiliaries did not include can and will. This finding (i.e., the overuse of modal auxiliaries by non-native speakers) necessitates having a deeper look at the different modal auxiliaries and investigating them individually to find the reasons behind this tendency (Table 10).

Examining the frequency of the modals, it was found that *will, can, should* and *might* were overused by NNSs while would, *may, could, must, shall, can't* and the

Table 10 Frequency of modal auxiliaries

Word	Frequency in BUiD corpus	Relative frequency	Frequency in BAWE corpus	Relative frequency
will	1921**	0.67	620	0.22
can	1035	0.36	1097	0.38
should	530*	0.18	313	0.11
would	374	0.13	486	0.17
may	273	0.09	338	0.12
could	255	0.09	319	0.11
might	191*	0.07	125	0.04
must	137	0.05	192	0.07
shall	23	0.01	20	0.01
can't	4	0	11	0
need	2	0	10	0
can not stand	2	0	2	0
Shall/will	1	0	0	0
Total	**4748**	**1.65**	**3533**	**1.23**

semi-modal *need* were underused. In the BUiD corpus, the modal auxiliary *will* came in the first place with the highest number of frequencies (1,921 times) followed by *can* with 1,041 occurrences while in the BAWE corpus, the order is reversed where *can* occupied the first place with 1,097 occurrences and will the second place with 620 occurrences. Generally, within the global list of BUiD, it is the modal verb *will* that mostly stands out because BUiD students used this modal almost three times as often as BAWE students. This overuse could be attributed to either L1 transfer (interlingual), developmental factors, or speech-like writing (viz, students' writing is affected by the way they speak, i.e., register-interference). The last reason needs to be supported by referring to an Arabic corpus where this feature of modality can be checked. Another reason, I would suggest, could be that in one of the modules, Research Methods, students were requested to write a research proposal, and so they used the word *will* many times to talk about their plans even though the present simple could have been used to express future planned activities. For example, one of the students was discussing the approvals that he would get to be able to run his study said, "*an approval on the study will be obtained from the HCT research*"; *someone else who was explaining the stages of his research said* "*[t]he first stage will involve questionnaires to be collected*". A final potential reason for the overuse of *will* by NNSs is that it is teaching-induced. It has also been noticed that NSs used *may, could* and *would* (modals mainly express probability) more than NNSs. This could be attributed to the fact that NSs tend to use these modals when they wish to show their attenuation about their propositions (epistemic stance). Finally, the modal auxiliary *should* is one of the modals that was overused by BUiD students. When the researcher examined the occurrences of *should*, he found that students used this modal mainly to express the ethical code of conduct or norms that usually prevail in teaching and research contexts; for example, "*the teacher should aim to create a suitable psychological atmosphere in order to lower learners' anxiety arising from their increased autonomous roles*".

4.6 Modals with Deontic and Epistemic Meanings

The next step for the researcher is to find out how many of these modal verbs have deontic meaning and how many have epistemic meaning. As mentioned earlier, the researcher follows the bottom-up approach when analysing the devices used to express modality. In other words, he starts with analysing the single modal auxiliaries, then the modal adverbials and finally the harmonic modal combinations. Before exploring the different modal auxiliaries, used by both NSs and NNSs, it is important to discuss the root and epistemic meanings of modal verbs. Although there seems to be a unanimous agreement among researchers on the forms of modal verbs that are used to express modality, Coates [44] and Hermeren [45] compiled a list of modals other than the one agreed upon by most researchers. They adopted a different technique, which is based on ferreting out the frequency of the various modals. This approach required putting a lot of effort and time because they had to check every single form to find whether it serves the epistemic or root meaning [36]. Each one of the two corpora (used in this current study) consists of about 300,000 words and with this big size of the corpora, the researcher decided to investigate the function (epistemic or root) of only some of the modal. It was also very helpful to refer to Aijmer's [19] study in which she classified modal auxiliaries as follows: *Must, may, should* and *might* could have both root or epistemic meanings while *have to, must, ought to* and *should* usually serve as root modals; the remaining verbs like *will, would* and *could* usually serve as epistemic devices; the modal verb *will*, in particular, is used to express the future plans, but with some kind of certainty. Although the number of modal auxiliaries is few, it seems to be a challenging task to determine the function of these modals because they are polysemous, for example, *can* is used to express possibility, ability and permission [18]. In this study, the researcher intends to investigate only the epidemic and root meaning of the modal auxiliary *would*, which would somehow show the preference of native and non-native towards the use of epidemic and root meanings of modal auxiliaries.

Table 11 shows all the occurrences of *would.* The epistemic modal *would* followed by *be* was underused by BUiD students as they used it eight times less than BAWE. The combination of *would* and verb *to be* was usually followed by an adverb (*would be very useful*), or past participle (*would be given*) or present participle (*would be asking*) or an adjective (*would be ideal*) or prepositional phrase (*would be of great help*). *Would* was mainly deployed in both corpora as the epistemic modal except for some forms, such as *would like* which served as a polite way to request something. BUiD students significantly overused this form, which carries the root meaning of would. This finding corroborates the previously proven fact, which suggests that BUiD students tend to hedge less than their counterparts in BAWE. Additionally, the modal verb *would* can be used to express probability or the possibility per se, not to mention adding another lexical verb with epistemic meaning like *appear* or *seem*. This combination strengthens the meaning and shows that the writer is trying to be objective as much as possible. Examining this combination of *would* and some lexical verbs with epistemic meaning like *seem, appear*, and *need,* it can be easily discerned that BUiD students significantly underused this combination of double hedging.

Table 11 'Would' with root & epistemic meanings

Item Would	BUiD	BAWE	Root or epistemic
Would be	135	143	Epistemic
Would better	2	2	Epistemic
Would like	26	11	Root
Would + adverb	28	58	Epistemic
Would + seem/appear/need	3	25	Epistemic

4.7 Harmonic and Disharmonic Combinations of Modals and Adverbs

Sometimes the modal verbs interplay with other lexical verbs or other parts of speech, which perform the same function of modal auxiliaries [46, 49]. For example, the will certainly combination of the modal and adverb is considered harmonic because the modal auxiliary will is used to denote certainty in the future and the adverb certainly strengthens the certainty of the verb will. Examining Table 12, it can be easily discerned that both NNSs and NSs used the harmonic modals *would probably* and *would definitely* equally, but the NSs used *would surely* four times more than the NNSs. Similarly, the combination of *will likely* was used by NSs twice as much as NNS, but the combination of *will most likely* was not seen in the NSs' corpus. Generally, NSs used more combinations than NNSs. Contrary to this finding, Aijmer [19] concluded that NNSs used more combinations and with different types and she attributed that to either the influence of spoken language or the L1 transfer.

4.8 Intensifiers

In this current study, the researcher could identify these intensifiers (adverbials of degree) using the Semantic Tag function and USAS (UCLER Semantic Analysis System) on Wmatrix3. This tool helps to group word senses together and categorize them according to the generality they lie within [47]. According to USAS tagging, each word within the two corpora is assigned a semantic and syntactic tag. This approach makes it easy to identify the behaviour of words like adverbials of degree. When the semantic tags of the two corpora were juxtaposed, it was found that NNSs underused all of the adverbials of degree except for the approximators. It did not seem wise to conclude that the low count of adverbials of degree implies that non-native speakers' writing was less proficient. In other words, it was too early to judge that low/high frequency stood for low/high proficiency in writing, but it was worth having a deeper look at the different patterns of adverbials used by both NSs and NNSs and trying to justify their under- or over-use. As can be seen in Table 13, there is a statistically significant difference in the count of adverbials between NSs and NNSs. The former used some adverbials almost twice as often as the latter, but both NSs and NNSs used maximizers almost equally as there is no significant

Table 12 Harmonic and disharmonic of modal interplay

Modals interplay	BUiD	BAWE
Would		
Would probably	2	2
Would surely	1	4
Would definitely	2	2

Table 13 Adverbials of degree (USAS)

	Amplifiers		Downtoners			
Corpus	A13.2	A13.3	A13.4	A13.5	A13.6	A13.7
	Maximizers	Boosters	Approximators	Compromisers	Diminishers	Minimizers
BAWE	476	1496	189	129**	262**	122**
BUiD	461	1076	195	57	109	53

difference between them with the log-likelihood (LL) = 0.24 which is less than the LL cut-off at 6.63. However, the difference between the frequency of boosters is significant as the LL = 68.97 which is higher than the cut-off value. NNSs are often stigmatized for their overstatement and use of boosters, but in this case, it proves the opposite as the NNSs underused almost all the scalar intensifiers [18].

This finding (i.e., underuse of amplifiers) is congruent with Granger's [48]. In order to find the reason behind this underuse, she added up the total number of tokens of amplifiers (including both maximizers and boosters) and the total number of types of these amplifiers. To her astonishment, she found that NNSs underused both the types and tokens of amplifiers. The low number of types could be expected, as the NNSs, unlike the NSs, do not have a rich language variety at their disposal. However, the second finding, which is the low number of tokens, is surprising as this means that NNSs' language is less emphatic or hyperbolic than NSs. This last conclusion contradicts the well-known thought, which implies that NNSs tend to overstate issues more than NSs [32]. As mentioned earlier, the findings of this current research, pertaining the tokens and types of the amplifiers (see Table 14) found in both corpora, are consistent with Granger's. Therefore, the researcher decided to investigate the frequencies of boosters in the two corpora to find out which boosters the NNSs underused or which ones they did not use at all. Boosters, in particular, were focused on and investigated in detail because they were the main reason of the high frequency of amplifiers in both corpora. When the lists of boosters were compared, it was found that there are 22 types of boosters (with 38 frequencies in NSs' corpus) that were not deployed at all by the non-native speakers (e.g., remarkably, desperately and agonizingly). As mentioned before, this case of non existence of some boosters in the NNSs' corpus could be attributed to the "natural deficiency of non-native vocabulary" [32, p. 28]. Similarly, most of the boosters, underused by non-native speakers, were a combination of an intensifying adverb ending with the suffix -ly followed by an adjective (adv-adj-, e.g., increasingly difficult). This type of adverbial collocations requires high combinatory skill, which is not within the capabilities of the non-native speakers. To counteract this deficiency, NNSs resorted to use all-round/stereotyped boosters that

Table 14 Types and tokens of amplifiers

Amplifiers	Types		Tokens	
	NS	NNS	NS	NNS
Maximizers	32	24^{-}	476	461^{-}
Boosters	50	34^{-}	1496	1076^{-}
Total	82	58^{-}	1972	1537^{-}

can be used in many contexts, such as very. This booster was overused by NNSs as they used it 272 times while NSs used it 187 times only.

Looking closely at the occurrences of some other boosters, it is found that NSs used more complex forms of some boosters than NNSs; for example, more, the most frequent booster in both corpora with 531 occurrences in BAWE and 441 in BUiD, was used in compound forms with a sense of a downtoner, such as "which was no more than a form of collective identity" and "the world today is no more than a global triumph of free market". However, when the researcher examined all the occurrences of more in BUiD's corpus (NNSs), no one example of such a complex form was found. Most of, if not all, cases in which more was used, were comparisons, such as "the findings will be more reliable" and "to write more details". This means that NSs have the linguistic competence that enables them to use words in more varied and complex forms than that of the non-native.

In addition to that, in the NSs corpus, with close investigation of the occurrences and contexts in which more was used, it was found that most of the cases denoted understating more than overstating. In other words, Wmatrix3 misinterpreted these devices as boosters, but in reality, they were no more than expressions of understatement.

As for the frequencies of diminishers and minimizers, NSs far exceed the NNSs in the use of these downtoners. This means that NSs were more cautious than NNSs as the former used the downtoners devices to show some kind of vagueness, which is now considered one of the main characteristics of the native speakers' language [49]. However, the NNSs used more approximators (195) than NSs (189) (see Table 14). Although the difference was not great, it proved that NNSs sounded more tentative than NS.

It is also worth mentioning that NSs' use of compound downtoners far exceeded the NNSs, for example the diminisher to some extent was used by the NSs twice as much as the NNSs (11 times and 4 times, respectively). This corroborates the fact that NSs have the ability to form varied and complicated structures of language items, even the hedged ones.

4.9 State of Inexactitude

According to Quirk et al. [18], *sort of* and *kind of* are considered part of the compromisers, but they were not included in the list generated by Wmatrix3 (USAS function), so the researcher decided to search for them using Wordsmith and the results are shown below.

As can be seen in Tables 15 and 16, there are 17 concordances of sort of in the BAWE corpus and seven concordances in the BUiD corpus. Some concordances of *sort of* in the two corpora were culled in order to exclude all the examples, which did not serve as a hedging marker. For instance, in Table 16, line number 4 was deleted because the phrase *sort of* in this context was a synonym of type of and it did not have the sense of a hedging device [12]. It is worth mentioning that while

Table 15 Concordances of sort of in BUiD corpus

Concordances of sort of in BUiD corpus
1. with the receptive skills as a **sort of** warming up for the productive skills
2. assume that there should be a **sort of** reconsideration of the number of
3. that's implemented directly from **sort of** answers which will determine
4. learning L2 and establish some **sort of** a bridge between both language
5. vidual on the planet has some **sort of** a gadget that connects him/her to
6. n of the book therefore, such **sort of** question helped in establishing the
7. establishing ICTs within this **sort of** perform rather than other people

Table 16 Concordances of *sort of* in the BAWE corpus

Concordances of sort of in BAWE corpus
1. inspiration". The writer takes on a **sort** of god-like essence as Author
2. of literary production as "a **sort of** involuntary secretion" described by
3. stitutional change - causes a **sort of** national reappraisal of institutions
4. , such as nails, ironworks, a **sort of** mortar and some kind of candles.
5. to justifiably attribute any **sort of** idealism to Husserl, the evidence is
6. Scope ambiguityThis is the final **sort of** ambiguity which is caused by
7. the very heart by a pleasant **sort of** involuntary helplessness" and yet "
8. things." Correlatively, the same **sort of** optimism is just as comical
9. had been used to uphold some **sort of** roof of which just a few pieces
10. the way it is because of some **sort of** intending or pointing on behalf of
11. posed that "Children use some **sort of** nonsemantic procedure to
12. Nietzsche an intentional choice, the **sort of** absolute undecidability
13. offer prior justification for the **sort of** cognition that can come to know
14. guage barrier, is exactly the **sort of** reality people with hearing
15. with impairments (Oliver, 1990). The **sort of** approach which is evident
16. . The inference was that this **sort of** 'being inside something and looking
17. English (Roach 2000). Thus this **sort of** group is called tone unit which

the researcher was weeding out the examples of sort of in the BAWE corpus, which did not have the sense of a hedge, he did not find it a challenging task. However, when he carried out the same task in the BUiD corpus, it took him more time to distinguish between the examples of both types and meanings of sort of, which could induce a kind of unsuitability of the use of these hedges. After weeding out the non-hedging examples of sort of, it was found that NSs used this hedge twice as much as NNSs. This finding gives another evidence that NSs tend to show their tentativeness by using these expressions of inexactitude that would invite the reader to take part in the debate being initiated by the writer. In other words, the writer tries to play the role of the reader by judging his/her own stance and determining to what extent he/she (i.e., the writer) is true or false.

4.10 Collocational Frame (It Is … that)

The two forms below and any other form that represented the writer's stance were searched for in the two corpora on WordSmith, using the collocational frame *it ** that.*

It + verb to be (functioning as a copula) + evaluative epithets (adjectives) + that... Or.... *It + passive voice (to be + past participle) + that...*

Then all the concordances that did not represent the writer's stance, were culled using 'Delete' and 'zap' functions in the WordSmith tool. Here are some examples of the culled concordances below. The first example (Concordance 6) was mistakenly included because the tool did not distinguish between the extraposition it is…that and any other form that included the adjacent words 'it…that'; this was the reason for including the first example in the concordances on WordSmith. The second example (Concordance 7) suggests that this student was not aware of the different correct forms of the extraposition and this explains the reason for entering incorrectly the adverb 'clearly' in place of the adjective 'clear', which should have been used here. The other examples contain the pronoun it, which functioned as an object for a verb and not as a part of the extraposition collocational frame. Additionally, concordance number 7 represents a case of it-clefted (Table 17).

Concordance 6: *supported it with diagrams that*
Concordance 7: *It is clearly that through this method*
Concordance 8: *Define **it** as"...a process that*
Concordance 9: *Merely choosing a textbook without first evaluating **it** would mean that*
Concordance 7: ***It** is there that he writes*

NNSs used more extraposed collocational frames than NSs, with usages of 124 and 106, respectively (142 and 125 concordances before culling). However, the quality, variety and complexity of the structures that come after the expletive it in NSs' concordances, are more advanced than NNSs and show how competent the native speakers are. Some of the most advanced expressions used by the NSs are it is poignant that; it is ironic that and it is posited that. None of these adjectives (i.e., ironic and posited) were used by the NNSs. As can be seen in the tables above, the concordances were sorted by the percentage of frequency of each one of these extrapositions. In BUiD, the extraposition *It is found that* comes in the first place with 99% of the whole texts while the extraposition that occupied the first place in BAWE, is *it is likely that*. The modal adverb *likely* was defined by Salagar-Meyer

Table 17 The first 20 concordances of the extraposition 'it**that' in BUiD and BAWE

N	Concordance in BUiD	Word #	%	N	Concordance in BAWE	Word #	%
1	interaction in the classroom. **It is found that** there are actually several weak	4168	99.00	1	Generative Grammar framework, **it is likely that** the minor differences of perspective	1924	99.00
2	through the different tests. So **it is recommended that** this contradiction	4012	99.00	2	sensitise educators; however **it is doubtful that** students need to be aware of	3304	99.00
3	appropriate for them Secondly, **it is crucial that** the authors would use more	4090	98.00	3	s from other genres. However, **it is likely that** most texts will still aim to be	4029	99.00
4	to make this book more useful **it is recommended that:** 1) A needs analysis	4160	98.00	4	ace'. Thus, with this in mind **it is hoped that** with time some inroads may be made	5345	99.00
5	forts to reach it Generally, **it is thought that** adhering to the supplies of	2392	97.00	5	and interrogative sentences. **It is certain that** this area will present the logician	3225	99.00
6	impressive and meaningful. **It is said that** practice makes a man perfect	3945	97.00	6	tak, 1990: 351). In addition, **it is poignant that** Nisa herself chose the name	4981	98.00
7	in the textbook. Furthermore, **it was found that** the dominance of the listening	4227	96.00	7	very nature of its structure, **it seems unlikefy** that English will be ousted in favour of	1877	98.00
8	unspecified forms in instruction. **It was argued that** such way will cause	4010	96.00	8	biggest ever budget in 1944. **It is certain, that** before the war had ended	27 60	96.00
9	. 5.Conclusion To conclude, **it is clear that** whatever is called a paradigm	3878	93.00	9	the world. However, although **it is true that** Musil's descriptions of the Other	5363	96.00
10	reading, and writing. However, **it is hypnotized that** teachers employ the	3491	93.00	10	. From reading Shostak's text **it becomes apparent that** it was as much about her	4819	95.00
11	, rank it as totally lacking. **It is noticeable that** the	3472	93.00	11	ing styles in modern theatre. **It rings true**	3924	95.00

(continued)

Table 17 (continued)

N	Concordance in BUiD	Word #	%	N	Concordance in BAWE	Word #	%
	textbook does not allocate				**that** action is louder than words		
12	rom the result of this study, **it is concluded that** integrating such aids with	4155	92.00	12	s intellectual^ bankrupt and **it is claimed that** social identities are created by	2820	93.00
13	in a sentence. For all above, **it is concluded that** the UAE English skills textbook	3836	91.00	13	lly promoted to children, but **it was discovered that** it it appealed to both children and	3224	91.00
14	ve their progression. Likewise,**it was perceived that** using of blogs helps	3094	91.00	14	ted to insincere conclusions. **It is possible that** Bull weighted his analysis in favour of	5954	91.00
15	appendices C & D). Finally, **it was noticed that** the units'themes are of little	3936	90.00	15	is the "hypothesis testing". **It is assumed that** output provides learners with the opportunity	3670	90.00
16	to the cultural restrictions. **It is recommended that** this study can be carried	2899	90.00	16	ernet transactions." (URL). **It is ironic that** most of the content available on the Internet	1709	89.00
17	otions in effective teaching **It is argued that** assessment guidelines and	3356	90.00	17	qualsiasi are stressed), and **it would seem that** if these linguistic alternatives continue	2822	87.00
18	adictoiy to this approach. So **it is considered that** such an an experiment	3777	87.00	18	less, as Lyons (1977) argues, **it is clear that** there are strong semantic associations	1568	87.00
19	listening to writing Thus, **it is important that** teachers introduce lessons	3422	87.00	19	th Tyson's 'architect' model. **It was recognised that** the need for the roles of 'clerk of works'	2926	87.00
20	ned the problem faced in UAE. **It is evidenced that** most of the students	3624	86.00	20	oncrete groups as they stand. **It is clear that** whichever scenario is true, the Theban Magical	4208	86.00

Table 18 Examples of adjectives of importance and probability

Adjectives of	BUiD Corpus		BAWE Corpus	
	Examples	Frequency	Examples	Frequency
Importance	It is important that	4	It is important that	4
Total 1		4		4
Probability			It seems likely that	2
			it seems unlikely that	1
			It is possible that	4
			It seems possible that	1
			it is likely that	2
			it is unlikely that	1
			it is doubtful that	1
Total 2		0		12
Total of total		**4**		**16**

[15] as one of the shield markers that hedges the speaker or writer and gives the degree of commitment to the proposition so that this person is protected in case his proposition proves wrong. Lemke's distinction between strong adjectives (e.g., critical and crucial) and week adjectives (appropriate and convenient) is not duplicable as the researcher tried to apply his model to the concordances of the extrapositions, found in both corpora, but unfortunately, it somehow did not work, so the researcher started interpreting the meaning of the different adjectives in the extrapositions intuitively as follows.

Looking closely at the Table 18, it can be easily discerned that NSs used more probability adjectives than NNSs. However, both used the same number of the adjective of importance. The first finding provides further evidence of the fact that NSs tend to show some kind of tentativeness in their writings.

5 Conclusion

Hedging, as a rhetorical or persuasion strategy gained a lot of popularity over the past 25 years and numerous scholars conducted studies on how the hedging devices can be used in academic writing [15, 50]. Two hedging taxonomies or models proposed by Hyland [8] and Salagar-Meyer [15] are applied to the two corpora. This application yielded the same result which is that NSs use more hedges than NNSs – 4,022 (19%) and 3,251 (12%), respectively. This finding is in line with Rezanejad, Lari and Mosalli's [42] study.

Although, generally, there is an overuse of modal auxiliaries by NNSs, some of these modals were mistakenly used. Similarly, Holmes [36] suggests that the overuse of the modal auxiliary *will* could be attributed to one of three hypotheses: either L1 transfer (interlingual); or developmental factor; or teaching-induced; or speech-like writing (viz., students' writing is affected by the way they speak, i.e., register-interference).

It is interesting to notice that NSs use *may, could* and *would* (modals mainly expressing probability) more than NNSs. This could be attributed to the fact that NSs tends to use these modals when they want to show their attenuation about their propositions (epistemic stance). In other words, NSs prefer to use these probability modals when they give unproven truth in their proposition [50].

The second category of hedges (according to Hyland's model), employed by both native and non-native speakers, is the hedging adverbs. This time, the native speakers use more adverbs than non-native speakers; 1,096 and 723, respectively. Generally, within the global list of the hedging adverb, the difference between the two frequencies is statistically significant, particularly, the difference between the probability adverbs, such as '*perhaps*' which was used 81 times by NSs and eight times by NNSs. Similarly, adverbs like '*possibly*', '*likely*', '*roughly*' were greatly underused by the NNSs.

The finding of the underuse of hedging by non-native speakers was confirmed by Salagar-Meyer's [15] proposed taxonomy of hedging markers, which was applied to both corpora. The chi-square results suggest that there is a statistically significant difference between native and non-native speakers in their use of the hedging markers.

There seems to be unanimous agreement among researchers that NSs tend to 'downstate' while NNS tend to 'overstate' [51]. However, in this study and contrary to the expectations, NNS underused all scalar intensifiers (including both amplifiers and downtoners). This finding is in line with Granger [48]. The underpresentation of boosters in the non-native speakers' corpus is significant enough to be the cause of the underpresentation of amplifiers in general. However, the underuse of maximizers is ignored, as the difference is not significant.

The underuse of boosters is not confined to booster types but it includes the frequency of these boosters as well. When the lists of boosters compared, it is found that there are 22 types of boosters (with 38 frequencies in NSs' corpus) that are not deployed at all by the non-native speakers (e.g., *remarkably, desperately* and *agonizingly*). As mentioned before, this case of nonexistence of some boosters in the NNSs' corpus could be attributed to the "natural deficiency of non-native vocabulary" [32, p. 28]. Similarly, most of the boosters, underused by non-native speakers, were a combination of an intensifying adverb ending with the suffix -ly followed by an adjective (adv-adj-, e.g., increasingly difficult). This type of adverbial collocation requires high combinatory skill, which does not seem within the capabilities of the non-native speakers. To counteract this deficiency, NNSs resort to use all-round or stereotyped boosters that can be used in many contexts, such as *very*. This booster (i.e., very) is overused by NNSs, with 272 occurrences while NSs used it only 187 times. The non-native speakers' language deficiency is further corroborated by the lack of complex forms found in the native speakers' corpus such '*no more than*'. The core word of the previous phrase is the adverb '*more*'. This adverb in this context has been mistakenly classified by Wmatrix3 as a booster, but in reality and in this context, it is no more than a downtoner. If the frequency of this adverb is taken away from the total of boosters in the NSs' corpus, this would reduce the number of amplifiers greatly. As mentioned earlier, NNSs do not only underuse the amplifiers, but the downtoners as well. The corroborating evidence for this underuse is found in the significant difference between the frequencies of downtoners in both NSs and NNSs' corpora (702 times and 414 times, respectively). Generally, downstating, as a way of hedging, is used to express vagueness and attenuation, which are two rhetorical strategies that distinguish a native speakers' writing [14, 49]. One of the important hedges that lies within the compromisers (subcategory of downtoners) is *sort of.* This hedge, which shows the degree of commitment of the writer towards the truth in a proposition, is significantly underused by the NNSs who were not trained or taught to exploit the indirect meaning of this hedge. Although NNSs underused this hedge, which shows the degree of commitment to the truth in their propositions, they overused the extraposition '*it... that*' which they used to indirectly comment on their propositions. The collected data suggests that the overuse could be attributed to a combination of factors. The substantive one is that they found this formulaic structure easy to start the sentence with. Furthermore, this structure is usually used to show some kind of objective modality and since there is difference in the quantity and quality between native and non-native speakers, this suggests that both groups use different ways to express modality [52]. Pedagogically, hedging is one of the areas that needs to be focused on by both language instructors and curriculum designers [35]. Data-driven learning (DDL), which is defined as the use of corpus concordances in classrooms, is one of the important applications of learner corpora.

References

1. Scarcella, R., Brunak, J.: On speaking politely in a second language. Int. J. Sociol. Lang. **1981** (1981)
2. Kasper, G.: Communication strategies: Modality reduction. Interlang. Stud. Bull. **4**, 83–266 (1979)
3. Robberecht, P., Petegham, M.: A functional model for the description of modality. In: The Fifth International Conference on Contrastive Projects (1982)
4. Naess, A.: Communication and Argument; Elements Of Applied Semantics. Allen and Unwin Limited, London (1966)
5. Skelton, J.: The care and maintenance of hedges. ELT J. **42**, 37–43 (1988)
6. Granger, S.: Modality in advanced Swedish learners' written interlanguage. In: Granger, S., Hung, J., Petch-Tyson, S. (eds.) Computer Learner Corpora, Second Language Acquisition and Foreign Language Teaching. John Benjamins Publishing Co, Netherlands (2002)
7. Scott, M., Tribble, C.: Textual Patterns. J. Benjamins, Philadelphia (2006)
8. Hyland, K.: Metadiscourse: Exploring Interaction in Writing (Continuum discourse series). Continuum International Publishing Group Ltd (2005)
9. Kopple, W.: Some Exploratory Discourse on Metadiscourse. Coll. Compos. Commun. **36**, 82 (1985)
10. Tse, P., Hyland, K.: So what is the problem this book addresses? Interactions in academic book reviews. Text & Talk—An Interdiscip. J. Lang. Discourse Commun. Stud. **26**, 767–790 (2006)

11. Dictionary, h.: Hedge meaning in the Cambridge English Dictionary. http://dictionary.cambridge.org/dictionary/english/hedge
12. Poos, D., Simpson, R.: Cross-disciplinary comparisons of hedging some findings from the Michigan Corpus of Academic Spoken English. In: Reppen, R., Fitzmaurice, S., Biber, D. (eds.) Using Corpora to Explore Linguistic Variation. John Benjamins Publishing Co, Amsterdam (2002)
13. Lakoff, G.: Hedges: A study in meaning criteria and the logic of fuzzy concepts. J. Philos. Logic **2** (1973)
14. McKenny, J.: A corpus-based investigation of the phraseology in various genres of written English with applications to the teaching of English for academic purposes (2006)
15. Salager-Meyer, F.: Hedges and textual communicative function in medical English written discourse. Engl. Specif. Purp. **13**, 149–170 (1994)
16. Murniato, M.: Types and functions of hedges used in 'J. K. Rowling's' interview with Oprah Winfrey show. http://kim.ung.ac.id/index.php/KIMFSB/article/download/3287/3263
17. Koutsantoni, D.: Developing Academic Literacies. Peter Lang, Oxford[etc.] (2007)
18. Quirk, R., Leech, G., Greenbaum, S., Crystal, D.: A Comprehensive Grammar of the English Language. Longman, London (1985)
19. Aijmer, K.: Modality in advanced Swedish learners' written interlanguage. In: Granger, S., Hung, J., Petch-Tyson, S. (eds.) Computer Learner Corpora, Second Language Acquisition and Foreign Language Teaching. John Benjamins Publishing Co, Netherlands (2002)
20. Hinkel, E.: The use of model verbs as a reflection of cultural values. TESOL Q. **29**, 325 (1995)
21. Papafragou, A.: Epistemic modality and truth conditions. Lingua **116**, 1688–1702 (2006)
22. Halliday, M.: An Introduction to Functional Grammar. Routledge, London (2014)
23. Haswell, R.: Gaining Ground in College Writing. Southern Methodist University Press, Dallas, Tex (1991)
24. Wray, A.: Formulaic Language and the Lexicon. Cambridge University Press, Cambridge (2002)
25. Lemke, J.: Resources for attitudinal meaning: evaluative orientations in text semantics. Funct. Lang. **5**, 33–56 (1998)
26. Corver, N., Van Riemsdijk, H.: Semi-lexical Categories: the Function of Content Words and the Content of Function Words. De Gruyter Mouton, Germany (2001)
27. Hinojosa, J., Martín-Loeches, M., Casado, P., Muñoz, F., Carretié, L., Fernández-Frías, C., Pozo, M.: Semantic processing of open- and closed-class words: an event-related potentials study. Cogn. Brain. Res. **11**, 397–407 (2001)
28. Corver, N., Van Riemsdijk, H.: Semi-lexical Categories. Mouton de Gruyter, Berlin (2001)
29. Halliday, M.: An Introduction to Functional Grammar. Hodder Arnold, London (2004)
30. Stubbs, M.: Text and Corpus Analysis: Computer-assisted Studies of Language and Culture. Blackwell, Oxford (1996)
31. Schmitt, N.: An Introduction to Applied Linguistics. Arnold, London (2002)
32. Lorenz, G.: Adjective Intensification Learners Versus Native Speakers: A Corpus Study of Argumentative Writing (Language & Computers). Rodopi, Amsterdam (1999)
33. Coventry University: British Academic Written English Corpus (BAWE). http://www.coventry.ac.uk/research/research-directories/current-projects/2015/british-academic-written-english-corpus-bawe/
34. Ortmeier-Hooper, C.: The ELL Writer. United States, United States (2013)
35. Hyland, K.: Hedging in academic writing and EAF textbooks. Engl. Specif. Purp. **13**, 239–256 (1994)
36. Holmes, J.: Doubt and certainty in ESL textbooks. Appl. Linguist. **9**, 21–44 (1988)
37. Kenny, A.: The Computation of Style: an Introduction to Statistics for Students of Literature and Humanities. Pergamon Press, Oxford (1985)
38. Scott, M.: WordSmith Tools, Version 3. Oxford (1999)

39. Ure, J.: Lexical density and register differentiation. In: Perren, J., Trim (ed.) Applications of Linguistics: Selected Papers of the 2nd International Congress of Applied Linguistics. Cambridge University Press, Cambridge (2017)
40. Meunier, F.: Computer tools for the analysis of learner corpora. Presented at the (1998)
41. Hunston, S., Francis, G.: Pattern Grammar: A Corpus-driven Approach to the Lexical Grammar of English. Benjamins, Amsterdam [u.a.] (2000)
42. Rezanejad, A., Lari, Z., Mosalli, Z.: A cross-cultural analysis of the use of hedging devices in scientific research articles. J. Lang. Teach. Res. **6**, 1384 (2015)
43. Ringbom, H.: Vocabulary frequencies in advanced learner English: a cross-linguistic approach. In: Granger, S., Leech, G. (eds.) Learner English on Computer. Longman, New York (2017)
44. Coates, J.: Semantics of the Modal Auxiliaries. Croom Helm, London (1983)
45. Henneren, L.: On Modality in English: A Study of the Semantics of the Modality. CWK Gleerup, Lund (1978)
46. Halliday, M.: Functional diversity in language, as seen from a consideration of modality and mood in English. Found. Lang. 6 (1970)
47. Archer, D., Wilson, A., Rayson, P.: Introduction to the USAS category system. http://ucrel.lancs.ac.uk/usas/usas%20guide.pdf
48. Granger, S.: In: Cowie, A. (ed.) Phraseology: Theory, Analysis and Applications. Oxford University Press, Oxford (1998)
49. Channell.: Vague Language. Oxford University Press, Oxford (1994)
50. Hyland, K.: Nurturing hedges in the ESP curriculum. System **24**, 477–490 (1996)
51. Hyland, K., Milton, J.: Qualification and certainty in L1 and L2 students' writing. J. Second Lang. Writ. **6**, 183–205 (1997)
52. Johansson, M.: It-clefts and pseudo-clefts in Swedish advanced learner English. Moderna Sprak. **95**, 16–23 (2001)

Part XI
E-learning

Intelligent Text Processing to Help Readers with Autism

Constantin Orăsan, Richard Evans and Ruslan Mitkov

Abstract Autistic Spectrum Disorder (ASD) is a neurodevelopmental disorder which has a life-long impact on the lives of people diagnosed with the condition. In many cases, people with ASD are unable to derive the gist or meaning of written documents due to their inability to process complex sentences, understand non-literal text, and understand uncommon and technical terms. This paper presents FIRST, an innovative project which developed language technology (LT) to make documents more accessible to people with ASD. The project has produced a powerful editor which enables carers of people with ASD to prepare texts suitable for this population. Assessment of the texts generated using the editor showed that they are not less readable than those generated more slowly as a result of onerous unaided conversion and were significantly more readable than the originals. Evaluation of the tool shows that it can have a positive impact on the lives of people with ASD.

Keywords Language technology · Autism spectrum disorder
Text simplification · Text accessibility

Mathematics Subject Classification (2010) Primary 97R40 · Secondary 91F20

C. Orăsan (✉) · R. Evans · R. Mitkov
Research Institute in Information and Language Processing, University of Wolverhampton, Wolverhampton WV1 1LY, UK
e-mail: C.Orasan@wlv.ac.uk

R. Evans
e-mail: R.J.Evans@wlv.ac.uk

R. Mitkov
e-mail: R.Mitkov@wlv.ac.uk

K. Shaalan et al. (eds.), *Intelligent Natural Language Processing: Trends and Applications*, Studies in Computational Intelligence 740,
https://doi.org/10.1007/978-3-319-67056-0_33

1 Introduction

Autistic Spectrum Disorder (ASD) is a neurodevelopmental disorder characterised by qualitative impairment in communication and stereotyped repetitive behaviour. It is a serious disability that affects approximately 60 people out of every 10,000 in the EU. People with ASD usually have language deficits with a life-long impact on their psychosocial functioning. These deficits are in the comprehension of speech and writing, including misinterpretation of literal meanings and difficulty understanding complex instructions [39]. Complex sentences, referential expressions, uncommon or technical words and figures of speech also constitute obstacles to proper understanding of texts for people with ASD. In many cases, people with ASD are unable to derive the gist or meaning of written documents [23, 41, 42]. The difficulties in reading comprehension that ASD causes represent a significant barrier to inclusion. People require access to written material for many purposes and in many contexts, from searching for employment opportunities and obtaining information to support their education to communicating by email or learning about local entertainment or news. Several studies have indicated a link between reading comprehension (and more generally, literacy) and access to education, employment, culture, and communication [9, 43].

This paper presents FIRST, an innovative project which developed language technology (LT) to make documents more accessible to people with ASD. OpenBook, the conversion software developed in this project,[1] is able to automatically detect a range of language phenomena which are problematic for people with ASD and replace some of them to make the text more comprehensible. It also aims to simplify complex structure in the text and clarify ambiguity. Not relying purely on textual changes, the conversion software adds illustrative pictures to documents and concise document summaries. Evaluation of LT carried out in the project revealed that the language processing components developed make a relatively large number of errors when dealing with unrestricted text. This is a problem given that the end users of the tool have low tolerance for ungrammatical and erroneous text which may be generated by LT components. For this reason, the OpenBook tool offers powerful post-editing options to carers to enable them to prepare texts for end users. In this way, all the changes made to a document can be supervised by carers who will ensure that the simplification is correct and the appropriate type of simplification is applied for a particular user. All this is in addition to the personalisation features embedded in the software.

Given the size of the project and the variety of topics covered during the 3 years it ran for, we cannot provide a detailed account of all the research we carried out. Instead, in this paper we provide an overview of the main achievements of the project with references to papers that provide more details. This paper is structured as

[1]http://openbook.net.

follows: Sect. 2 reviews several similar projects and a survey of the most relevant literature. Section 3 presents a brief overview of FIRST and the language technology integrated in OpenBook, followed by an evaluation of the tool in Sect. 4. The paper ends with discussion and conclusions.

2 Related Work

The challenge of text simplification has been addressed in several lines of research since the 1990s. Text simplification systems have been developed to rewrite text using various lexical [6, 8, 16, 30, 57, 61], syntactic [10, 11, 13, 51], and other [53, 58] transformation operations and components for the generation of assistive content such as definitions [19], images [3, 7], and summaries [3]. In previous work, text simplification has been used to improve the accuracy of NLP applications in areas such as dependency parsing [26] biomedical information extraction [21, 49], semantic role labeling [56], and machine translation [12, 40].

Of more relevance to this paper, text simplification methods have also been developed with the goal of facilitating text processing by various populations of readers, including people with poor literacy [52] or numeracy [5], people with aphasia [35], dyslexia [48], autism [22], people who are non-native speakers [1, 45], and children and language learners [29].

Max [35] described the use of a syntactic parser for sentence rewriting to facilitate the reading comprehension of people with aphasia. In the PSET project, Canning [10] implemented a system which exploits a parser in order to rewrite compound sentences as sequences of simple sentences and to convert passive sentences into active ones. One weakness of this approach is that it depends on high levels of accuracy and granularity of automatic syntactic analysis. Research has shown that the accuracy of parsers is inversely proportional to the length and complexity of the sentences being analysed [36]. These are often the sentences for which simplification is most required.

More recently, the availability of resources such as Simple Wikipedia (SW) has enabled text simplification to be included in the paradigm of statistical machine translation [14, 54]. In this context, translation models are learned by aligning sentences in Wikipedia with their corresponding versions in SW [63]. Manifesting Basic English [44], the extent to which SW is accessible to people with reading difficulties has not yet been fully assessed. Effective SMT relies on the availability of representative pairs of texts in their original and converted forms. At present, these resources are scarce and are often designed with a particular readership in mind, such as children [4, 59], people with Down's syndrome [8], or people at particular reading grade levels [46]. As a result, there are currently only a limited number of contexts in which SMT approaches are likely to be effective. Xu et al. [59] are critical of the use of Simple English Wikipedia to support SMT-based text simplification.

The field of automatic text summarisation also includes approaches that exploit text simplification processes. For example, Cohn and Lapata [13] present a syntactic tree-to-tree transduction method to filter non-essential information from syntactically parsed sentences. This compression process often reduces the syntactic complexity of those sentences. One advantage of this approach is that it can identify elements for deletion in the absence of explicit lexical/punctuational markers. However, these methods are "destructive" in the sense that information is deleted rather than preserved as a result of compression. Although some information loss is inevitable in text simplification, the method that we exploit in the FIRST project is designed not to automatically delete parts of input sentences.

This survey of related work has demonstrated that the field of text simplification has received a significant amount of interest from researchers in computational linguistics. However, there are very few projects which brought together such researchers with end users who could benefit directly from the research at a scale that the FIRST project did.

3 Overview of the Project and the Language Technology

The FIRST project was funded by the EC through its FP7 ICT work programme and addressed the objective concerning *smart and personalised inclusion*. The purpose of the project was to implement an advanced ICT-enabled solution for the empowerment of people with disabilities who are at risk of social exclusion as a result of their low literacy, resulting from cognitive and mental impairments. In line with this, the main aim of the project was to implement, deploy, and evaluate intelligent technology to support the authoring of accessible content in Bulgarian, English, and Spanish for users with ASD with a view to widening inclusion and empowerment in Europe.

The project was coordinated by the University of Wolverhampton and consisted of nine partners representing all the relevant stakeholders: language technology experts (University of Wolverhampton, UK; University of Alicante, Spain and University of Jaen, Spain), clinical partners who work directly with people with ASD (Central and North West London NHS Foundation Trust, UK; Deletrea, Spain and Parallel World, Bulgaria), software developers (iWeb technologies, UK and Kodar, Bulgaria) and an organisation which promotes the rights of people with ASD (Autism Europe, Belgium).

One of the challenges that became quite obvious from the beginning of the project was the fact that there is no clear description of the needs that people with ASD have. Therefore one of the first tasks undertaken in the project was to derive a detailed and accurate description of the requirements of users with reading difficulties. A summary of the findings are presented in Sect. 3.1 and they informed the choice of the LT integrated in the tool. On the basis of the analysis carried out, we split the LT processing into three main components: a structural complexity processor (Sect. 3.3), a meaning disambiguator (Sect. 3.4) and a personalised document generator (Sect. 3.5). These components are integrated with the user interface and backend services using a three-tier architecture (Sect. 3.2).

3.1 User Requirements

One of the first tasks undertaken in the project was to understand the needs of people with ASD. In light of this, our first objective was to derive a detailed and accurate description of the requirements of users and to gain an understanding of the most significant obstacles to reading comprehension. We also tried to learn how best to convert texts containing such obstacles into a form suitable for end users. The resulting specification needed to be granular enough to support practical design of the core LT in the project.

The user requirements were derived on the basis of an extensive literature survey and as a result of consultations with end users and intermediaries. For the latter, a sample of 37 children (aged 12–16) and 57 adults (aged 16+), meeting strict DSM-IV-TR criteria for ASD,[2] with IQ > 70, whose mother tongue was Bulgarian, English or Spanish, with the ability to read and with basic computer skills, were consulted. A pilot study with 25 Spanish participants was carried out first in order to improve the design of questionnaires and face-to-face interviews. The questionnaires were used to analyse specific reading comprehension problems by presentation of a series of texts followed by closed questions related to the content of those texts. Interviews and open questions were used to explore subjective and qualitative information (perception of difficulties regarding reading comprehension and preferences). Responses to the questions consisted of judgements on a 5 point rating scale indicating difficulty of comprehension. Responses to open-ended questions raised several topics that had not been considered in the design of the study.

Intermediaries were also presented with a questionnaire in order to complement the information collected from end-users. The questionnaires focused mainly on their perception of the reading difficulties that they have and the strategies used to overcome them. All the interviews and questionnaires were handled by the clinical partners in the project who had expertise in recruitment of participants with ASD and organisation of meetings with our focus group.

The results of the analysis were categorised as (a) linguistic obstacles to be removed, (b) preferences regarding the format of the output document, (c) preferences regarding the look and feel of the interface and (d) users suggestions of additional features that may be incorporated into the tool. Overall, the results obtained were in line with the findings of similar previous studies. The study conducted in FIRST is innovative with regard to the wide range of areas about which end users were consulted, the types of participant, and the qualitative information obtained from those participants. A set of user requirements was derived from the data obtained and these requirements were then categorised according to the type of LT service addressing each of them (structural complexity processor, meaning disambiguator, personalised document generator). Each user requirement was associated with an editing operation ("assistive element") expected to remove the obstacle to reading comprehension to which the user requirement pertained ("obstacle"). The main user requirements for each type of processing attempted in the project are

[2]http://behavenet.com/apa-diagnostic-classification-dsm-iv-tr.

summarised in Sects. 3.3–3.5. More details of those requirements are presented by Martos et al. [34] and González-Navarro et al. [24].

In addition to providing the specifications of the LT components, issues that are specific to our end users such as data protection requirements, including privacy, safety, security, and identity management also had to be addressed. Ethical approval was obtained for all interactions with end users, regardless of whether their involvement was for the purpose of establishing the user requirements, or evaluating the system.

3.2 Architecture of the System

OpenBook, the system developed in the FIRST project, features a three-tier architecture which enabled us to develop a modular system that can be easily maintained, scaled up and extended. By using loose coupling we are able to easily replace modules with better ones over time. The three tiers are:

- **Presentation tier**: controls interaction between users and the system via a graphical user interface. While designing the user interface, best practice in terms of accessibility, the Windows User Experience Interaction Guidelines[3] and Designing UX for Apps were taken into consideration.[4]
- **Integration engine**: acts as a business tier and provides interaction between the presentation tier and the LT modules. It also controls user management, access control level, error and logging management.
- **Resources tier**: contains the language technology services which are responsible of identifying and processing obstacles to reading comprehension. The outputs of these LT services are consumed by an Aggregator Web Service which combines different types of annotation and deals with possible conflicts between them. The aggregator also communicates with the integration engine. The GATE document format [15] was used for communication between different web services. The GATE document format was selected because it allows re-use of existing tools, makes the system easily extensible, and provides flexibility.

Users of the tool are able to customise it depending on their needs. They can affect both the look and feel of the interface, and the way the LT components run. The next sections present in more detail the types of language processing attempted in the project.

Given the difficulties encountered in automatic processing of language, OpenBook features two interfaces. The first is a powerful editor that gives intermediaries the possibility to simplify texts using a host of LT components. Figure 1 shows this interface during the process of inserting an image to represent the term *windfarms*.

[3]http://msdn.microsoft.com/en-us/library/windows/desktop/aa511258.aspx.

[4]http://msdn.microsoft.com/en-us/library/windows/apps/hh779072.aspx.

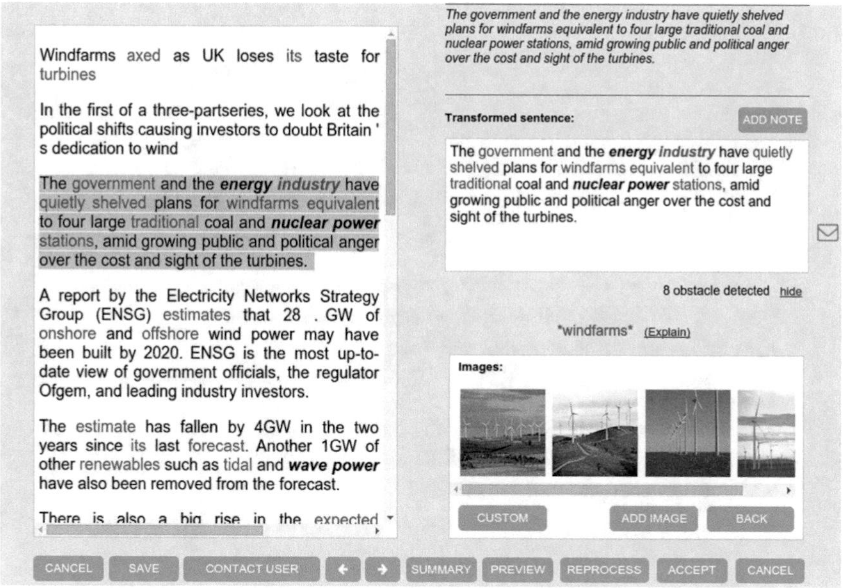

Fig. 1 The interface for intermediaries

The assumption is that intermediaries can easily identify mistakes made by the computer and correct them. Figure 2 presents the interface for end users which provides access to a more limited number of LT components and is meant mainly for reading documents prepared by intermediaries.

3.3 Processing Structural Complexity

In FIRST, we focus on structural complexity at the morphological and syntactic levels of language. The complexity of discourse structure is addressed to some extent by the meaning disambiguator (Sect. 3.4), which facilitates human processing of a limited set of discourse relations such as coherence and anaphora when reading. Our user requirements analysis led to the derivation of a set of user requirements of various levels of importance. Table 1 lists those with the highest level of importance. A few requirements linked to presentation (e.g. avoid cutting paragraphs at the end of page) and some that were dealt with by the meaning disambiguator (e.g. substitute rare conjunctions by more common ones) are omitted for brevity.

Examination of the user requirements revealed that many of them cannot be addressed automatically with sufficient accuracy using existing language technology. This is because either the existing research is not advanced enough to address

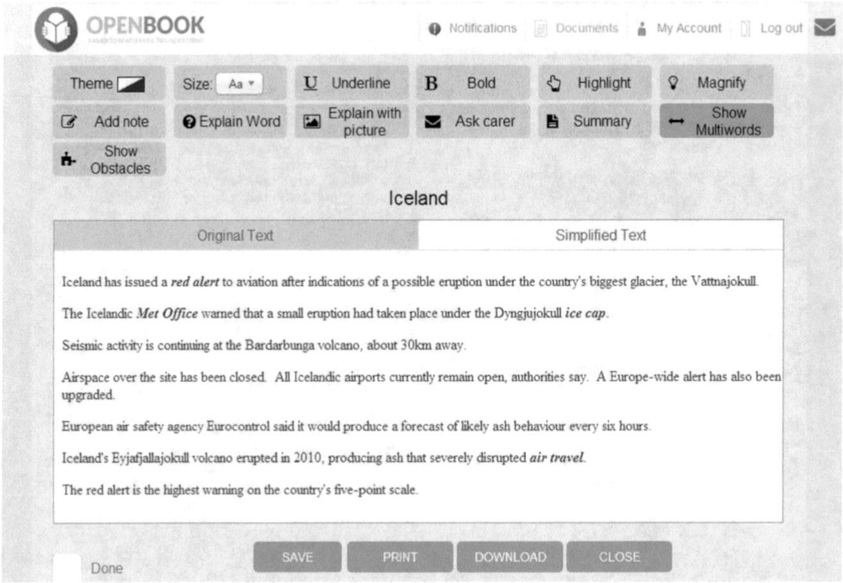

Fig. 2 The interface for end users

Table 1 User requirements relating to the processing of structural complexity

Obstacle	Text simplification operation	ID
Multiple copulative coordinated clauses	Substitute with sentences divided by periods	UR301
Subordinate adjectival/relative clause	Substitute by adjective or extract and rephrase as a sequence of sentences	UR302
Explicative clauses	Remove explicative clause	UR303
Adverbial clause after main clause	Place adverbial clauses before main clause	UR307
Conditional clause after main clause	Place conditional clauses before main clause	UR308
Long sentences	Rephrase into shorter sentences (less than 15 words)	UR309
Semicolon and suspension points	Avoid the use of semicolon and suspension points	UR310
Sentences in passive voice	Use active voice	UR313
Sentences with double negatives	Avoid sentences containing negatives and double negatives	UR314

the problems identified or, for languages such as Bulgarian and Spanish, there are no resources which can be used to adapt existing LT to these languages.

Most of the work carried out in this project on processing structural complexity focused on reducing the syntactic complexity of English sentences. In text, conjunctions, complementisers, *wh*-words, punctuation marks, and pairs consisting of a punctuation mark followed by one of these types of word are signs indicating the presence of syntactic complexity in a sentence. These signs link clauses and phrases together in coordination and also bound subordinate phrases and clauses embedded in complex constituents. To automatically reduce the syntactic complexity of English texts, we developed a method which can automatically identify and classify signs of syntactic complexity using a machine learning approach and rewrite complex sentences using a predefined set of rules. The method was designed in such a way that it can be adapted to other languages. Unfortunately, this was not possible during the FIRST project due to time constraints. Bulgarian and Spanish complex sentences were rewritten using a limited number of rules applied to the output of syntactic parsing. For this reason, coverage is limited.

As noted by Siddharthan [51], text simplification can be viewed as comprising three processes: analysis, transformation, and post-editing. In previous work, Evans [21] described a rule-based method for sentence rewriting. The main contributions of this method were a new approach to automatic sentence analysis and a method for rewriting sentences on the basis of that analysis. This analysis includes tokenization of input texts to enable identification of sentences, words, and potential coordinators, PoS tagging, and a machine learning method to categorize potential coordinators according to their specific coordinating functions. The method proved useful as a pre-processing step in biomedical information extraction [21]. We took the same approach in the FIRST project, but focused on a wider set of signs of syntactic complexity and addressed a wide range of types of both coordination and subordination. The method was also developed for a wider variety of texts.

The rest of this section briefly presents the processing applied to English sentences. These automatic processes address user requirements UR301-303 and UR309-310 (Table 1). Our method exploits Algorithm 1.

3.3.1 Identification of Signs of Syntactic Complexity

The identification of signs of syntactic complexity is achieved by using a supervised tagging approach which builds statistical models for predicting the functional class of signs of syntactic complexity. CRF tagging models [31, 55] were built and their intrinsic performance analysed. This approach was employed because, during development, we found that sequence based CRF tagging models provide better performance in the automatic tagging of signs than methods in which each sign is tagged independently of other signs in the same sentence. The mean accuracy of the best model, applied to texts of three different registers (health, literature, and news), was

Input: Sentence s_0, containing at least one sign of syntactic complexity of class c, where $c \in \{\text{CEV, SSEV}\}$.
Output: The set of sentences A derived from s_0, that have reduced propositional density.

$A \leftarrow \emptyset$;
$S \leftarrow \{s_0\}$;
while $S \neq \emptyset$ **do**
  $s_i \leftarrow pop(S)$;
  if s_i *contains a sign of syntactic complexity of class c (specified in Input)* **then**
    $\{s_{i_1}\} \leftarrow rewrite_c(s_i)$;
    $S \leftarrow S \cup \{s_{i_1}\}$;
  else
    $A \leftarrow A \cup \{s_i\}$
  end
end

Algorithm 1: Syntactic simplification algorithm

82.06%. More details on our approach and its evaluation are presented in the paper by Dornescu et al. [18].[5]

Information about the functions of different signs of syntactic complexity, in combination with patterns used to detect passive sentences, and gazetteers of complex and rare conjunctions were also used to provide carers with reports on the types of syntactic complexity detected within a sentence. These reports can be used to inform carers about how to rewrite sentences in order to make them more accessible. When processing the sentence. *The judge, Richard Walker, has ruled as a matter of law that the words are libellous, and the jury is being asked to decide the scale of damages, which Mr Burstein's solicitor advocate said should be between 20,000 and 50,000.* The system generates a report such as that shown in Table 2.

3.3.2 Sentence Rewriting

A rule-based approach was developed to convert compound sentences into sequences of sentences containing no coordinate clauses, and sentences containing complex noun phrases into sentences containing simple noun phrases. In our approach, sentences were tagged with information on the parts of speech of words and the syntactic functions of the signs of syntactic complexity that they contain. Sentence rewriting rules were then applied iteratively, each rule triggered by matching patterns in the tagged sentences. By counting how many times the application of a rule led to the generation of correct output we were able to determine the accuracy of the simplification process. Overall, the rules used to rewrite sentences containing compound clauses have an accuracy of 0.699. The rules used to rewrite sentences containing bound relative clauses have an accuracy of 0.583. The two primary sources of error were the specificity of the rules, which limits their coverage and the inability of the

[5] A demo of the English sign tagger is available at http://rgcl.wlv.ac.uk/demos/SignTaggerWebDemo/.

Table 2 Example of sentence analysis produced by the system

This sentence contains:

- 2 embedded clauses. These sentences may contain multiple facts. Texts are easier to read when each sentence contains one fact:
  - …as a matter of law [that] the words are…
  - …the scale of damages[, which] Mr Burstein's solicitor…
- 1 other embedded phrase. These sentences may contain multiple facts. Texts are easier to read when each sentence contains one fact.
  - …The judge[,] Richard Walker, has…
- 1 pair of linked clauses. These sentences may contain multiple facts. Texts are easier to read when each sentence contains one fact.
  - …the words are libellous[, and] [the jury is being asked…]
- 1 passive verb in a subordinate clause. These sentences contain multiple facts. Texts are easier to read when each sentence contains a single fact. When converting this sentence to a more readable form, try to ensure that the correct agent of the embedded verb is explicitly mentioned.
  - …libellous, and the jury [is being asked] to decide the scale of…

method to discriminate between conjunctions and commas linking coordinate bound relative clauses and those linking independent clauses [22].

Evaluation of the English structural complexity processor showed that the LT developed in the FIRST project performs at a level that compares favourably with the state of the art. However, in contexts where end users have a low tolerance for errors, direct access to some of these components should be limited. In FIRST, intermediaries have direct access to the full set of LT components via the carers' interface. They are the ones who process texts automatically and then post-edit the output to generate a more accessible form of the input text that can be consumed directly by end users, in this way addressing the errors introduced by LT.

The overall impact of the processing of syntactic complexity on users was not as great as initially hoped. There are two main reasons for this. In order to minimise the number of errors that could be introduced, the system integrated in OpenBook took a very conservative approach when applying rewriting rules. For this reason, a limited number of sentences were affected by processing. Secondly, when the rewriting rules generate correct sentences, users are not immediately aware that the text has been made more accessible. For this reason, we rarely found direct references to the benefits of reducing structural complexity in users' feedback. However, we believe this processing has contributed to the overall positive feedback received.

3.4 Processing Ambiguity in Meaning

A literature review together with research carried out in the project showed that the lexical component of language may be the one posing most difficulties for people with autism. Specific lexical items such as mental verbs, emotional language and figurative language constitute a difficult barrier to overcome for people with ASD. To address this issue, a set of LT modules were developed to detect semantic obstacles in input texts and, whenever possible, to resolve them. These modules, together with a collection of relevant resources were integrated into a framework for meaning disambiguation. While developing this framework, emphasis was, as much as possible, on the development of language independent modules. As can be seen in Table 3 the user requirements analysis revealed a large number of potential obstacles caused by semantic ambiguity. Given the difficulty of addressing these obstacles, the consortium agreed to focus on those with higher priority. For brevity, we present only these high priority obstacles here. They can be categorised into three broad groups: coreference, difficult words, and figurative language. The only aspect of emotional language that was tackled in the project is the one linked to mental verbs. The other recommendations, although marked as high priority, were not dealt with because they were considered too complicated to be reliably identified by an automatic system.

For all the obstacles, detection is first carried out. Then, depending on the type of obstacle, a different assistive element is provided, thus providing several strategies for the resolution of the obstacles. For example, for difficult words, the resolution comprises the extraction of definitions and synonyms of terms; for acronyms, their expanded form is provided; for infrequent slang, the expression is normalised, and for coreference, the antecedent of a pronoun, definite description, or omitted pronoun is provided.

In this section, we only provide a brief overview of the methods employed by the meaning disambiguator. A more detailed description is presented in the technical report by Lloret et al. [33].

3.4.1 Processing Figurative Language

Figurative language in general and idioms in particular present specific problems for our end users, as they are not able to grasp the meaning of these expressions. When reading a text they tend to construct the literal meaning of figurative expressions such as "calm before the storm" or "raining cats and dogs" and therefore misunderstand the meaning of the sentence that contains them. Even though some progress has been made in the field in recent years, the identification of conceptual metaphors for open domains is beyond the current state of the art in LT. For this reason the approach taken in the FIRST project to deal with figurative language was to compile dictionaries for each language. To allow more flexibility some of the entries in these dictionaries were encoded as JAPE expressions [15]. This allowed detection

Table 3 User requirements relating to the processing of ambiguity in meaning

Obstacle	Text simplification operation	ID
Polysemy	Provide easier synonyms	UR401
	Detect and highlight polysemy	UR425
Phraseological units (idioms, lexicalised metaphors)	Replace by a simple word	UR402
	Detect and highlight when replacement is not possible	UR425
	Provide simple definitions to explain phraseological units	UR410
Less common words	Replace infrequent words with simpler synonym	UR405
	Provide simple definitions to explain mental verbs	UR411
	Provide simple definitions to explain infrequent words	UR413
Emotional language	Replace complicated emotional adjectives with simpler synonyms	UR404
	Provide simple definitions to explain emotional adjectives	UR412
	Replace complex mental verbs with simpler synonyms	UR403
Slang	Normalize infrequent slang	UR407
	Provide simple definitions to explain infrequent slang	UR414
	Detect specialized slang belonging to a domain	UR423
	Provide simple definitions to explain specialized slang	UR424
Infrequent acronyms	Expand infrequent acronyms and abbreviations	UR415
Long numerical expressions	Express long numerical expressions with digits	UR417
Anaphors	Detect and leave anaphors with low resolution confidence unresolved	UR418
	Resolve pronominal anaphora	UR419
	Resolve definite descriptions	UR420
	Resolve ellipsis	UR421

of accepted morphological variations of these expressions and cases involving discontinuous expressions. Lloret et al. [33] and Barbu et al. [2] present more details of the methods developed in the project to detect figurative expressions (idioms).

3.4.2 Processing Difficult Words

To process difficult words, such as polysemous words, acronyms, abbreviations and slang, the recommended strategy was to provide definitions and/or synonyms. However, depending on the nature of the obstacle, in some cases the definition and/or synonym is considered from a broader perspective. In the case of acronyms or infrequent slang, the framework provides an expansion of the acronym or the normalised version of the slang expression, respectively, as assistive elements to facilitate comprehension of them. A set of resources was compiled to support this. After analysing and evaluating the existing available LT resources, it was noticed that the accuracy of disambiguation of polysemous words, specialised slang, less common words, and mental verbs is limited by the use of WordNet and WordNet-related resources.

3.4.3 Processing Coreference

The analysis and research carried out into coreference resolution for Bulgarian, English, and Spanish resulted in the development of three modules capable of detecting and resolving: pronominal anaphora for Bulgarian; pronominal anaphora for English; and pronominal anaphora, definite descriptions and ellipsis for Spanish. The types of coreference phenomena that could be addressed in each language depended on the availability of tools and resources for that language. For Bulgarian, an extension of the MSTParser [62] was employed by the coreference resolution system. For English, the Stanford Deterministic Coreference Resolution System [32] was used since our evaluation showed that this was the best performing system and the most appropriate to be used in the context of the project. For Spanish, the process was divided into three stages, in order to separate the functionalities of anaphor detection; potential candidate antecedent identification; and, finally, anaphora resolution. In this manner, depending on the type of coreferential phenomenon to be resolved, a different approach was used. The coreference module integrated in Open Book relied on Freeling for the detection of pronominal anaphora and definite descriptions, the Naïve Bayes algorithm for detecting ellipsis, a Voted Feature Interval algorithm for resolving pronominal anaphora and ellipsis, and the PART algorithm for resolving definite descriptions.

The pronoun resolver for English was evaluated on texts of the news, literature and health registers. This evaluation showed that if used as a tool to aid in post-editing of texts, then for more than 90% of the anaphoric pronouns, the system is able to present a list of options which contain the antecedent. A small scale evaluation of the Bulgarian pronoun resolver showed that its accuracy is around 50%. The evaluation results of the coreference resolver for Spanish, both for detection and resolution, shows that it achieves superior results to the best performing system in the Semeval-2010 task [47].

3.5 Generation of Personalised Documents

In addition to making the text more readable, an important aspect of making documents more accessible is the possibility to produce multimodal personalised documents. This is done by implementing modules which can supplement texts with images and summaries, and by providing a wide range of personalisable parameters derived from user requirements. The most important user requirements for this type of processing are presented in Table 4.

Table 4 User requirements relating to the generation of personalised documents

Obstacle	Text simplification operation	ID
Understanding of general meaning	Give relevant idea on top of text	UR501
	Show key words	UR502
	Post questions in or after the text to help monitor comprehension	UR503
	Give information on key concepts before reading text	UR504
	Support the overall meaning of the text with images	UR505
	Provide text with summaries	UR506
Phraseological units	Support the understanding of phraseological units with images	UR507
Non lexicalized metaphors	Support the understanding of metaphorical language with images	UR508
Less common words	Support the understanding less common words with images	UR509
Emotional language	Support the understanding emotional adjectives with images	UR510
Polysemy	Support the understanding of polysemy with images	UR511

3.5.1 Generation of Multimodal Documents

Images are very useful for explaining the meaning of a word, regardless of whether they are polysemous or rare. For this reason, two image retrieval systems were implemented to retrieve images from expressions automatically identified as difficult: an offline image retrieval module and an online image retrieval module. The offline image retrieval method uses information from the disambiguation module used to deal with ambiguity in meaning (see Sect. 3.4) to map single words and multiword expression to images extracted from the ImageNet database [17]. This database stores web images annotated with WordNet noun synsets. The ImageNet database links around 22,000 synsets with more than 14 million images, each checked using crowdsourcing to ensure the correctness of the association. In addition, Wikification [37] is used to link terms in the text with their corresponding Wikipedia page, which in turn is used to extract images that explain the terms. In cases where the offline image retrieval engine is unable to identify an appropriate image, the online image retrieval module queries Google and Bing for images related to the term. In line with expectation, the accuracy of this type of image retrieval is much lower than that obtained by the offline module. Despite this, end users of the OpenBook system very much appreciated the implemented image retrieval functions.

3.5.2 Generation of Summaries

Another way to help users of OpenBook is by preparing summaries of documents. OpenBook implements a sentence extraction algorithm inspired by TextRank [38] or LexRank [20] which identifies the most important sentences in the original text and builds a summary that includes only those important sentences, presented in the order in which they appeared in the original document. This algorithm was selected because it is language independent, it was deemed better than other general summarisation approaches, and it is fast. The size of the summary is controlled by a personalisable parameter and carers have the option to modify the summary to make it more comprehensive and to better fit users needs. Interviews with users (discussed in Sect. 4.3) revealed that, despite their relatively low accuracy, both image retrieval and automatic text summarisation were widely used and appreciated by carers and end users.

3.5.3 Personalisation

The personalisation of the produced documents can be controlled by users via a set of parameters decided on the basis of user requirements analysis. Users are able to control which LT components are run and to change their behaviour. For instance, users can indicate whether they want to obtain only definitions of words, rather than definitions and synonyms, or in the case of syntactic simplification indicate whether obstacles should only be detected or both detected and removed. End users and their carers can set these parameters for each individual document. Personalisation also includes parameters which control the look and feel of the interface, a very important aspect for users with ASD.

4 Evaluation of the FIRST Project

The validation of the impact of the FIRST technology on inclusion was made through quantitative evaluation via reading comprehension testing and qualitative research methodologies employing semi-structured interview techniques to compare the perceptions of people with ASD, their family members, and other relevant intermediaries [25]. Specifically, FIRST technologies are designed for people with high-functioning autism, defined as those with a formal diagnosis of autism and an IQ score greater than 70. Intermediaries (carers) also participated in the evaluation. The usefulness of the OpenBook software was also assessed through a benchmarking experiment where the times taken for carers to simplify texts when they had access to the tool and when they had no access to it were recorded.

4.1 Reading Comprehension Testing

4.1.1 Setting of the Experiment

Reading comprehension testing was used to test the effectiveness of OpenBook as a tool to convert texts into a more accessible form for end users. 243 participants (193 males and 50 females) with high-functioning autism were recruited in the UK, Spain and Bulgaria. In addition, a control group of 50 typical children were involved in the experiment in Bulgaria. Comprehension tests were conducted in a controlled environment under time-limited conditions. Each participant was presented with a battery of 6 texts followed by multiple-choice questions and a subjective text rating. Half of the texts were presented in their original form while the remainder were presented in a more accessible form generated by carers using OpenBook. The order of text presentation was random and both researchers and participants were blind to this order. The hypothesis investigated was that text simplification improves reading comprehension for participants with ASD, and therefore participants will be able to correctly answer more questions about converted texts than about texts in their original form.

Each clinical centre in the UK and Bulgaria identified 6 texts that were appropriate for respective age groups (adults and children) under examination. The research team in Spain identified 12 texts in total: 6 for children and 6 for adults. Texts for adults were selected from comprehension test batteries used to examine reading comprehension in proficient language learners. Texts for children were selected from childrens books and the internet. The texts for adults used in Spain and the UK were matched for word length and complexity. The same was done for the texts for children in Spain and Bulgaria. Text complexity was assessed using methods developed in the project and which focused on the obstacles identified in the user requirements. Additional evaluation of the readability of these texts, using both standard readability metrics and metrics predicted to be relevant for readers with ASD, was described in the paper by Yaneva and Evans [60].

The original texts were processed automatically using the OpenBook tool and then post-edited by the clinical teams, who acted as intermediaries for people with ASD. Post-editing operations were similar to those performed in the benchmarking experiment presented in the next section. Multiple-choice questions (MCQs) were generated by each clinical team for their respective texts, with the assistance of technical partners. MCQs were selected to test the general comprehension of the text, especially parts of the text containing identified obstacles to reading comprehension. Each adult text was followed by 6 MCQs, whilst children's texts had 4 MCQs. This selection was made to accommodate children's performance within the same timeframe as the adults. Both children and adults had 10 min per text to read and answer MCQs. Although we had planned to run reading comprehension sessions in large groups, we approached this activity with flexibility considering the social challenges and anxieties faced by people with ASD. Therefore, we ran the tests in smaller groups and sometimes in one-to-one sessions.

4.1.2 Evaluation Results

The reading comprehension tests indicated that participants performed better with MCQs based on versions of texts converted using OpenBook than on the original versions of the same texts ($t = 4.42$, $p < 0.001$, CI [0.63, 0.79]). However, the result was only of borderline statistical significance for Spanish children. Participants also blindly rated converted texts as easier to understand. ($t = 6.96$, $p < 0.001$, CI [0.71, 1.26]). This was consistent for UK, Spain and Bulgaria, for both adults and children.

Analysis of the data also suggests that there was no significant association between the comprehension scores and the subjective scores for either the scores for texts in their original form or the scores for texts in their simplified form. This means that although participants gave more correct answers in response to questions about texts converted using OpenBook, they did not identify them as being easier to understand. This suggests that the process of making text more accessible does not interfere with the way users regard the texts. A more detailed description of the testing procedure and discussion of the results is provided in the technical report of the project [28] and the paper by Jordanova and Cerga Pashoja [27].

4.2 *Evaluation of Text Conversion Using OpenBook: Readability Assessment*

In order to assess the extent to which the OpenBook software reduces the burden on carers converting texts into a more accessible form for end users, a benchmarking experiment was carried out. This experiment was conducted in two stages. In the first year of the project, professional carers were asked to convert 25 heterogeneous texts in Bulgarian, English, and Spanish into a more accessible form without the use of OpenBook. The second stage was semi-automatic in nature: carers used the interface to OpenBook when converting the texts, and were able to exploit LT functions to assist in the process. The same carers participated in both stages and the time it took them to perform the simplification was recorded. Participants in the conversion task were provided with the guidelines shown in Fig. 3. The guidelines also provided definitions of the linguistic terms used.

The mean time taken to convert texts with 250–350 words decreased significantly from 54 min for unaided conversion to 29 min for conversion using OpenBook. The time increased only for English, but this increase is not statistically significant. When using OpenBook carers inserted images in the simplified document, a process that they did not attempt during unaided conversion as it was considered too onerous.

Instructions to professionals: How to Simplify Texts

- You need to time yourself from the moment you start reading each text to the moment you finish the simplification. This should be done separately for each text.
- Should you need to interrupt the work, pause the timer and restart timing once you restart simplification task.
- Please **do not read the texts beforehand**.
- Read the instructions below carefully. Do not start simplifications if you are not clear or sure about any of the instructions (or parts of them). Please contact me if you do not understand or need clarification about the instructions below before starting the task.

After you have read the text carefully please:

1. **Detect infrequent words and substitute them with simpler synonyms or definitions (in case no simpler synonyms exist).**
2. **Identify figurative expressions such as idioms and metaphors and replace with simpler words or definitions.**
3. **Identify jargons or specialised terms and replace with specific definitions.**
4. **Identify phraseological units and polysemic words and replace with specific definitions.**
5. **Identify and divide long paragraphs.**
6. **Detect long sentences and divide them into shorter easier to understand chunks.**
7. **Rewrite complicated sentences to make them easier to understand.**
8. **Identify and resolve anaphora.**
9. **Replace abbreviations and acronyms with full definitions.**
10. **Use bullets points if necessary to break down the text in easier parts.**

- Email back both the original and simplified versions of each text.
- At the bottom of each text please make a note of the how long it took for the text to be simplified (in minutes).

Fig. 3 Text conversion guidelines

During unaided conversion, huge disparities in conversion rates between centres and carers were noted. These disparities diminished when OpenBook was used.

4.2.1 Readability

In addition to analysing the time taken to convert the text, we also carried out an analysis of the readability of the texts in order to check that the converted ones are indeed more accessible. The readability of each version of the texts was assessed using eight language independent readability metrics sensitive to variables such as the frequency of occurrence of commas, pronouns, metaphors, passive verbs, and polysemic words in the texts, the lengths of words and sentences in the texts, and the type token ratio of the text. One language specific readability metric combining and weighting information about the language independent metrics was also used. In terms of readability, we made the following observations about texts produced via semi-automatic text conversion in which carers exploited OpenBook:

- In Bulgarian, converted texts contain: fewer phraseological units and non -lexicalised metaphors than the originals (with statistical significance), and more polysemous words than the originals (with statistical significance), possibly due to increases in the diversity of vocabulary.
- In English, converted texts contain: fewer commas, shorter words, less diverse vocabulary, fewer phraseological units and non-lexicalised metaphors, and reduced syntactic complexity[6]
- In Spanish, converted texts contain: fewer metaphors, shorter sentences, less diverse vocabulary, more pronouns, and fewer polysemous words

We compared texts generated by carers exploiting OpenBook with texts generated by carers who were unaided. We noted that for Bulgarian texts, use of OpenBook led to smaller reductions in the numbers of phraseological units and non-lexicalised metaphors used, but smaller undesirable increases in the numbers of polysemous words used than was the case in unaided conversion. For English texts, use of OpenBook led to smaller reductions in the numbers of phraseological units and non-lexicalised metaphors and larger reductions in the diversity of vocabulary than unaided conversion. The explanation for use of OpenBook leading to smaller reductions in the occurrence of phraseological units and non-lexicalised metaphors in Bulgarian and English texts is that rather than re-phrasing and deleting these elements from sentences, as unaided editors do, OpenBook provides explanatory definitions of them. These figurative elements thus have a tendency to be preserved in the text. For Spanish texts, use of OpenBook led to larger reductions in sentence length and larger reductions in the number of metaphors used in the converted texts.

These findings are derived from information about readability metrics for which there were statistically significant differences between the scores for texts in their

[6]Measured using Scarborough's *index of productive syntax* [50].

original and converted versions. More detailed analysis and explanation of the findings of these experiments are reported in the technical report by Jordanova et al. [28].

4.3 User Feedback

Individual interviews were carried out in order to better understand the experiences of people using OpenBook and to explore its impact on better access to written information and improved social inclusion. Users with ASD and their carers were interviewed using questions related to topics deemed important for their social inclusion. The interviews lasted between 20 and 50 min and were recorded, transcribed and analysed using the Atlas.it data analysis package.[7] The interviews were used to obtain feedback from users about a period spent accessing the user interface to OpenBook, in their own time at home to access texts of their own choosing. In this context, end users exploited the LT to make fully automatic conversion of texts.

Thematic analysis was the framework used to analyse interview transcripts. Thematic analysis is a principal technique used by qualitative researchers to analyse data. It is a method for identifying, analysing, and reporting patterns (themes) within the data. In the case of the analysis carried out in the FIRST project, these were related to the effect OpenBook had on people with ASD and on their social inclusion).

Our thematic analysis led to the derivation of eight themes emerging from user feedback about use of OpenBook. These themes concerned positive changes in independence with regard to:

1. **Comprehension**: Improved reading comprehension as a result of using OpenBook was widely reported. The positive impact in comprehending written texts was spontaneously described as "obvious" and "encouraging" by one user. Other users reported improved comprehension when accessing complex information such as reading about formulae and mathematical curves and improvements in understanding subtext.
2. **Reading**: Improved comprehension seemed to have an impact on participants' reading skills. Improved reading abilities were reported by both adults with ASD and their carers. Adults with ASD reported that they were reading more as a result of using OpenBook and focusing better on reading. Individual carers reported improvements in the vocabulary of children with ASD.
3. **Writing**: One unintended impact of their use of Openbook was experienced by both adults and carers of children with ASD. They reported improvements in their writing skills, gaining confidence in writing emails and notes. Some participants began writing notes for the first time, while others felt more confident and consequently became more active writers.
4. **Communication**: Improved communication was another theme that emerged from our analysis, and was consistent for both adults and children. One

[7] http://atlasti.com/.

carer of a child with ASD reported improvements in their sociability, while adults with ASD reported improvements in sociability, vocabulary, and increased use of complex phrases and sentences. One teacher observed a student with ASD overcoming his shyness and becoming more sociable.

5. **Emotions**: Improvements in comprehension and communication were factors which seemed to have a positive impact on the emotions of both children and adults with ASD. One mother stated that her son's behaviour had improved and that her son no longer became angry when he failed to understand what his mother was saying.
6. **Self-efficacy and confidence**: Carers talked appreciatively about the fact that OpenBook appeared to make users more self-sufficient and consequently self-confident in their ability to look independently for information and communicate with others. Children involved in the study were reported to study longer and more effectively.
7. **Relationships**: In the interviews, children were reported to engage more frequently with their peers and carers and the quality of these interactions was consistently reported to have been enhanced. One teacher in Bulgaria stated that she expected her workload to be alleviated as a result of the student engaging independently with OpenBook. Adults also reported changes in relationships. To illustrate, one user explained that she started leaving notes for her work colleagues to facilitate her workload. Several users (adults and children) talked about having become less reliant on others since using OpenBook. However, some users felt that some professional carers (e.g. social workers) may be insufficiently computer literate to support their use of OpenBook.
8. **Independence**: The increased independence of users of OpenBook was a recurring theme throughout the interviews. Although some concern was expressed about the possibility of adults with ASD becoming more dependent on their carers, overall, statements described increased independence of both adults and children with ASD when accessing OpenBook and reading texts, when accessing technology, when reading physical books, and when writing notes and emails.

OpenBook was reported to have had a positive impact on reading comprehension of both adults and children with ASD. Improved reading comprehension seemed to improve the reading skills of people with ASD and their writing and communication abilities. Although some users found it disconcerting to use OpenBook, most users stated that it had a positive effect on their relationships, self-confidence and ultimately their independence. Full details of our thematic analysis and examples of user feedback addressing the eight themes are provided in the technical report by Jordanova et al. [28].

Generally, users said that the system is easy to use and not only improves the studies of children with ASD and relieves anxiety of adults with ASD regarding text comprehension but it also relieves the burden on teachers and carers. Users said that they would like to continue to use OpenBook in the future and will recommend it to friends and colleagues.

However, OpenBook did not work well for all users. One adult with autism found it difficult to process the obstacles highlighted by the software. His mother explained that he was made uncomfortable by the font colour used to highlight obstacles to reading comprehension and felt he had done something wrong. Some users criticised OpenBook due to the relatively low accuracy of some components. They commented on the tendency of the image retrieval tool to return inappropriate images for some ambiguous words and substitution of some ambiguous words with more complex definitions and synonyms. Most of these issues were resolved through the development phase of the project, and users reported improvement in processing times and synonym suggestions.

5 Discussion and Conclusions

This paper has presented a project which brought together a consortium of nine partners from academic, industrial, health, and charity sectors to develop ICT to convert electronic documents into a form facilitating reading comprehension for people with ASD. Research was conducted to gain insight into the specific user requirements of end users with ASD and the intermediaries working with end users to help them in accessing information in texts. The findings of this study were used to underpin the development of a tool integrating language technology components to convert texts into a more accessible form. LT was developed in the project to reduce the structural complexity caused by long and syntactically complex sentences and the ambiguity in meaning caused by difficult/rare terms, ambiguous words, anaphora, and figurative language. An LT component was also developed to generate additional content such as concise summaries and images to explain complex terms occurring in texts that end users seek to access.

Evaluation of OpenBook was complex and user-focused, exploiting qualitative and quantitative methods to assess the tool intrinsically and extrinsically. Intrinsic evaluation of the LT services integrated in OpenBook revealed that they were not accurate enough to support fully automatic conversion of text into a form facilitating reading comprehension for people with ASD. For this reason, those services are delivered by two different interfaces, each supporting a different text conversion service. In the first, end users (people with ASD) apply a restricted set of reliable LT components to automatically improve the accessibility of texts they are seeking to read. In the second, intermediaries (carers, educators, and health service providers) have access to the full set of LT components which can assist them in converting texts into a more accessible form for end users.

The first conversion service was evaluated by analysis of feedback from end users. The interface was found to be easy to use and there was enthusiasm for the concept underlying it. LT components providing users with explanations of complex words and idioms, retrieving images to explain those concepts, and generating summaries of input texts were all valued by end users (and carers). There was some criticism of the inaccuracy of some functions and the poor handling/lack of coverage of some domain specific terms.

The second service was evaluated by assessing the extent to which the use of OpenBook to convert texts improves the comprehension of converted texts by end users and by examining differences between the process of converting texts into a more accessible form using OpenBook and the process of making the conversion in an unaided fashion. Reading comprehension testing showed that texts converted by carers using OpenBook were understood better than texts in their original form. As an editing tool, it was found that OpenBook enabled more rapid conversion of text to a more accessible form. The texts converted using OpenBook were not significantly less readable than those generated more slowly as a result of onerous unaided conversion and were significantly more readable than the originals.

Overall, users recommended the use of OpenBook and were enthusiastic about using the system independently. They perceived that it made them more independent and keen to solve problems for themselves. Interviews with carers and end users revealed improvements in users' ability to comprehend texts, including improvements in understanding of subtext. Overall, use of OpenBook was associated with greater motivation to read, greater engagement in reading, and improved attention in reading as well as improvements in vocabulary. Improvements were also noted in the writing skills of both young and mature end users. Positive changes were also observed with regard to behaviour, communication and sociability of end users. There were anecdotal reports of collateral improvements in educational achievement.

One exciting aspect of the FIRST project is that the simplifying language technology developed can have wide-ranging applications beyond that of improving text accessibility for users with ASD. The software has potential to benefit other types of health service user, as well as groups in other sectors (e.g. language learners, migrants and readers of legal documentation). Certain LT components, such as the syntactic simplification module, were perceived to be of limited benefit for users in the FIRST project. However, the functionality of this module was restricted in accordance with the requirements of people with ASD and their low tolerance for errors in the system output. In its unrestricted mode, it is able to transform a wider range of syntactic constructions than was attempted in the project. We believe that these other types of transformation operation are applicable to other language processing applications, such as information extraction and text summarisation.

Acknowledgements We would like to acknowledge the contribution of all the partners to the project. This paper would not have been possible without their contribution to the various stages of the research carried out. We would also like to thank the carers and individuals with high-functioning autism who participated in the different evaluations reported in this paper. The research was partially funded by the EC under the 7th Framework Programme for Research and Technological Development (FP7- ICT-2011.5.5 FIRST 287607).

References

1. Angrosh, M., Nomoto, T., Siddharthan, A.: Lexico-syntactic text simplification and compression with typed dependencies. In: Proceedings of the 25th International Conference on Computational Linguistics: Technical Papers, COLING 2014, pp. 1996–2006 (2014)
2. Barbu, E., Martín-Valdivia, M., Alfonso, L., Lopez, U.: Open book: a tool for helping ASD users' semantic comprehension. In: Proceedings of the 2th Workshop of Natural Language Processing for Improving Textual Accessibility (NLP4ITA), pp. 11–19. Atlanta, US (2013)
3. Barbu, E., Martín-Valdivia, M.T., Martínez-Cámara, E., López, L.A.U.: Language technologies applied to document simplification for helping autistic people. Expert Syst. Appl. Int. J. **42**(12), 5076–5086 (2015)
4. Barzilay, R., Elhadad, N.: Sentence alignment for monolingual comparable corpora. In: Proceedings of the 2003 Conference on Empirical Methods in Natural Language Processing (EMNLP'03), pp. 25–32. Sapporo, Japan (2003)
5. Bautista, S., Saggion, H.: Can numerical expressions be simpler? Implementation and demonstration of a numerical simplification system for Spanish. In: Proceedings of the Ninth International Conference on Language Resources and Evaluation (LREC-2014) (2014)
6. Biran, O., Brody, S., Elhadad, N.: Putting it simply: a context-aware approach to lexical simplification. In: Proceedings of the 49th Annual Meeting of the Association for Computational Linguistics: short papers (ACL-2011), pp. 496–501. Portland, Oregon (2011)
7. Bosma, W.: Image Retrieval Supports Multimedia Authoring, pp. 89–94. ITC-irst, Trento, Italy (2005)
8. Bott, S., Rello, L., Drndarevic, B., Saggion, H.: Can spanish be simpler? LexSiS: Lexical simplification for Spanish. In: Proceedings of the 12th International Conference on Intelligent Text Processing and Computational Linguistics, Lecture Notes in Computer Science, pp. 8–15. Springer, Samos, Greece (2012)
9. Brugha, T., McManus, S., Meltzer, H., Smith, J., Scott, F.J., Purdon, S.: Autism spectrum disorders in adults living in households throughout England. Report from the Adult Psychiatric Morbidity Survey (2007)
10. Canning, Y.: Syntactic simplification of text. Ph.D. Thesis, University of Sunderland (2002)
11. Chandrasekar, R., Srinivas, B.: Automatic induction of rules for text simplification. Knowl.-Based Syst. **10**, 183–190 (1997)
12. Chen, H.B., Huang, H.H., Chen, H.H., Tan, C.T.: A simplification-translation-restoration framework for cross-domain SMT applications. Proc. COLING **2012**, 545–560 (2012)
13. Cohn, T., Lapata, M.: Sentence compression as tree transduction. J. Artif. Intell. Res. **20**(34), 637–74 (2009)
14. Coster, W., Kauchak, D.: Simple English Wikipedia: a new text simplification task. In: Proceedings of the 49th Annual Meeting of the Association for Computational Linguistics (ACL-2011), pp. 665–669. Portland, Oregon (2011)
15. Cunningham, H., Maynard, D., Bontcheva, K., Tablan, V., Aswani, N., Roberts, I., Gorrell, G., Funk, A., Roberts, A., Damljanovic, D., Heitz, T., Greenwood, M.A., Saggion, H., Petrak, J., Li, Y., Peters, W.: Text Processing with GATE (Version 6). http://tinyurl.com/gatebook (2011)
16. De Belder, J., Deschacht, K., Moens, M.F.: Lexical simplification. In: Proceedings of the 1st International Conference on Interdisciplinary Research on Technology, Education, and Communication (ITEC-2010). Kortrijk, Belgium (2010)
17. Deng, J., Dong, W., Socher, R., Li, L.J., Li, K., Fei-Fei, L.: ImageNet: a large-scale hierarchical image database. In: CVPR09 (2009)
18. Dornescu, I., Evans, R., Orasan, C.: A tagging approach to identify complex constituents for text simplification. In: Proceedings of Recent Advances in Natural Language Processing, pp. 221–229. Hissar, Bulgaria (2013)
19. Elhadad, N.: Comprehending technical texts: predicting and defining unfamiliar terms. In: AMIA Annual Symposium Proceedings, pp. 239–243 (2006)
20. Erkan, G., Radev, D.R.: Lexrank: Graph-based lexical centrality as salience in text summarization. J. Artif. Intell. Res. **22**(1), 457–479 (2004)

21. Evans, R.: Comparing methods for the syntactic simplification of sentences in information extraction. Lit. Linguist. Comput. **26**(4), 371–388 (2011)
22. Evans, R., Orasan, C., Dornescu, I.: An evaluation of syntactic simplification rules for people with autism. In: Proceedings of the 3rd Workshop on Predicting and Improving Text Readability for Target Reader Populations (PITR), pp. 131–140. Gothenburg, Sweden (2014)
23. Frith, U., Snowling, M.: Reading for meaning and reading for sound in autistic and dyslexic children. J. Dev. Psychol. **1**, 329–342 (1983)
24. González-Navarro, A., Freire-Prudencio, S., Gil, D., Martos-Pérez, J., Jordanova, V., Cerga-Pashoja, A., Shishkova, A., Evans, R.: FIRST: una herramienta para facilitar la comprensión lectora en el trastorno del espectro autista de alto funcionamiento. Rev. Neurol. **58**(Supl 1), 129–135 (2014)
25. Granizo, L., Naylor, P., del Barrio, C.: Analysis of the social relationships of pupils with Asperger syndrome in mainstream secondary schools: case studies. Rev. Psicodidactica 281–292 (2006)
26. Jelínek, T.: Improvements to dependency parsing using automatic simplification of data. In: Proceedings of the Ninth International Conference on Language Resources and Evaluation (LREC-2014) (2014)
27. Jordanova, V., Cerga Pashoja, A.: Effects of syntactic complexity, semantic reversibility, and explicitness on discourse comprehension in persons with aphasia and in healthy controls. BAOJ Psychol. **1**(003) (2016)
28. Jordanova, V., Evans, R., Cerga Pashoja, A.: D7.8: Final evaluation report. Technical Report, FIRST_D7.8_20141119, The FIRST Consortium (2014)
29. Kajiwara, T., Yamamoto, K.: Evaluation dataset and system for Japanese lexical simplification. In: Proceedings of the ACL-IJCNLP 2015 Student Research Workshop, pp. 35–40 (2015)
30. Kandula, S., Curtis, D., Zeng-Treitler, Q.: A semantic and syntactic text simplification tool for health content. In: AMIA Annual Symposium Proceedings, pp. 366–370 (2010)
31. Lafferty, J., McCallum, A., Pereira, F.C.: Conditional random fields: probabilistic models for segmenting and labeling sequence data. In: Proceedings of the 18th International Conference on Machine Learning, pp. 282–289. Morgan Kaufmann (2001)
32. Lee, H., Peirsman, Y., Chang, A., Chambers, N., Surdeanu, M., Jurafsky, D.: Stanford's multi-pass sieve coreference resolution system at the CoNLL-2011 shared task. In: Proceedings of the Fifteenth Conference on Computational Natural Language Learning: Shared Task, CoNLL Shared Task '11, pp. 28–34. Association for Computational Linguistics, Stroudsburg, PA, USA (2011)
33. Lloret, E., Moreda, P., Moreno, I., Canales, L.: D4.1: Meaning Disambiguator: v2.1. Technical Report, FIRST_D4.1_20140531, The FIRST Consortium (2014)
34. Martos, J., Freire, S., González, A., Gil, D., Evans, R., Jordanova, V., Cerga, A., Shishkova, A., Orasan, C.: User preferences: updated. Techbical Report D2.2, Deletrea, Madrid, Spain (2013)
35. Max, A.: Syntactic simplification—an application to text for aphasic readers. M.Phil in computer speech and language processing, University of Cambridge, Wolfson College (2000)
36. McDonald, R.T., Nivre, J.: Analyzing and integrating dependency parsers. Comput. Linguist. **37**(1), 197–230 (2011)
37. Mihalcea, R., Csomai, A.: Wikify!: linking documents to encyclopedic knowledge. CIKM **7**, 233–242 (2007)
38. Mihalcea, R., Tarau, P.: Textrank: bringing order into texts. In: Lin, D., Wu, D. (eds.) Proceedings of EMNLP 2004, pp. 404–411. Association for Computational Linguistics, Barcelona, Spain (2004)
39. Minshew, N., Goldstein, G.: Autism as a disorder of complex information processing. Ment. Retard. Dev. Disabil. Res. Rev. **4**, 129–136 (1998)
40. Mishra, K., Soni, A., Sharma, R., Sharma, D.: Exploring the effects of sentence simplification on Hindi to English machine translation system. In: Proceedings of the Workshop on Automatic Text Simplification—Methods and Applications in the Multilingual Society (ATS-MA 2014), pp. 21–29 (2014)

41. Nation, K., Clarke, P., Wright, B., Williams, C.: Patterns of reading ability in children with autism-spectrum disorder. J. Autism Dev. Disord. **36**, 911–919 (2006)
42. O'Connor, I.M., Klein, P.D.: Exploration of strategies for facilitating the reading comprehension of high-functioning students with autism spectrum disorders. J. Autism Dev. Disord. **34**(2), 115–127 (2004)
43. OECD: Literacy in the information age: final report of the international adult literacy survey. Technical Report, Organisation for Economic Co-operation and Development, Paris, France (2000)
44. Ogden, C.K.: Basic English: A General Introduction with Rules and Grammar. In: Paul, K. (ed.) Trench, Trubner & Co., Ltd., London (1932)
45. Paetzold, G.: Reliable lexical simplification for non-native speakers. In: Proceedings of the 2015 Conference of the North American Chapter of the Association for Computational Linguistics: Student Research Workshop, pp. 9–16 (2015)
46. Pellow, D., Eskenazi, M.: An open corpus of everyday documents for simplification tasks. In: Proceedings of the 3rd Workshop on Predicting and Improving Text Readability for Target Reader Populations (PITR), pp. 84–93. Gothenburg, Sweden (2014)
47. Recasens, M., Màrquez, L., Sapena, E., Martí, M.A., Taulé, M., Hoste, V., Poesio, M., Versley, Y.: Semeval-2010 task 1: Coreference resolution in multiple languages. In: Proceedings of the 5th International Workshop on Semantic Evaluation, SemEval@ACL 2010, pp. 1–8. Uppsala, Sweden (2010)
48. Rello, L., Baeza-Yates, R., Bott, S., Saggion, H.: Simplify or help?: Text simplification strategies for people with dyslexia. In: Proceedings of the 10th International Cross-Disciplinary Conference on Web Accessibility, pp. 15:1–15:10. New York, NY, USA (2013)
49. Rindflesch, T.C., Rajan, J.V., Hunter, L.: Extracting molecular binding relationships from biomedical text. In: Proceedings of the sixth conference on Applied natural language processing, pp. 188–195. Seattle, Washington (2000)
50. Scarborough, H.S.: Index of productive syntax. Appl. Psycholinguist. **11**, 1–22 (1990)
51. Siddharthan, A.: Syntactic simplification and text cohesion. Res. Lang. Comput. **4**(1), 77–109 (2006)
52. Siddharthan, A.: Text simplification using typed dependencies: a comparison of the robustness of different generation strategies. In: Proceedings of the 13th European Workshop on Natural Language Generation, pp. 2–11 (2011)
53. Specia, L.: Translating from complex to simplified sentences. In: Proceedings of the Conference on Computational Processing of the Portuguese Language, pp. 30–39. Springer, Porto Alegre, RS, Brazil (2010)
54. Štajner, S.: New data-driven approaches to text simplification. Ph.D. Thesis, University of Wolverhampton (2015)
55. Sutton, C., McCallum, A.: An introduction to conditional random fields (2010). arXiv:1011.4088
56. Vickrey, D., Koller, D.: Sentence simplification for semantic role labeling. In: Proceedings of ACL-08: HLT, pp. 344–352 (2008)
57. Walker, A., Siddharthan, A., Starkey, A.: Investigation into human preference between common and unambiguous lexical substitutions. In: Proceedings of the 13th European Workshop on Natural Language Generation (ENLG), pp. 176–180. Nancy, France (2011)
58. Wubben, S., van den Bosch, A., Krahmer, E.: Sentence simplification by monolingual machine translation. In: Proceedings of the 50th Annual Meeting of the Association for Computational Linguistics (ACL-12), pp. 1015–1024. Jeju, Republic of South Korea (2012)
59. Xu, W., Callison-Burch, C., Napoles, C.: Problems in current text simplification research: new data can help. Trans. Assoc. Comput. Linguist. **3**, 283–297 (2015)
60. Yaneva, V., Evans, R.: Six good predictors of autistic text comprehension. In: Proceedings of Recent Advances in Natural Language Processing (RANLP 2015), pp. 697–706. Hissar, Bulgaria (2015)
61. Zeng-Treitler, Q., Goryachev, S., Kim, H., Keselman, A., Rosendale, D.: Making texts in electronic health records comprehensible to consumers: a prototype translator. In: AMIA Annual Symposium Proceedings, pp. 846–850 (2007)

62. Zhikov, V., Georgiev, G., Simov, K., Osenova, P.: Combining POS tagging, dependency parsing and co-referential resolution for Bulgarian. In: Proceedings of RANLP'13. Bulgaria (2013)
63. Zhu, Z., Bernhard, D., Gurevych, I.: A monolingual tree-based translation model for sentence simplification. In: Proceedings of the 23rd International Conference on Computational Linguistics (COLING 2010), pp. 1353–1361. Beijing, China (2010)

Education and Knowledge Based Augmented Reality (AR)

Salwa Hamada

Abstract This paper gives a general idea about Augmented Reality (AR). It introduces AR and how it can be used to enhance the user and computer interface systems and turn it to an entertainment system. A sample AR system for teaching Arabic vocabulary Concepts to children in kindergarten will be represented in this paper too. Augmented Reality is a growing part of virtual reality area it is an integration of the real and virtual worlds. It considers as a part of a continuum of technologies. AR will change all computer process completely in future. Examples of using AR to accuse knowledge in general will be introduced too.

Keywords Augmented reality · Virtual reality · Language learning E-learning

1 Augmented Reality Technology

Augmented Reality AR technology is used to overlay real images with data or digital images to increase impact, to increase usability, or to enhance understanding [1, 2]. AR is the real-time delivery of digital info to enhance or enable a geographical/physical experience [3].

Augmented Reality enhances a user's perception of and interaction with the real world. The virtual objects display information that the user cannot directly detect with his own senses. The information conveyed by the virtual objects helps a user perform real-world tasks. AR is aspic example of what calls Intelligence Amplication (IA): using the computer as a tool to make a task easier for a human to perform. Over the last several years, AR applications have become portable and widely available on mobile devices. Moreover, AR is becoming visible in our audio-visual media (e.g., news, entertainment, sports) and is beginning to enter

S. Hamada (✉)
Electronics Research Institute ERI, P.O. Box 12611, Giza, Egypt
e-mail: hesalwa@hotmail.com

K. Shaalan et al. (eds.), *Intelligent Natural Language Processing: Trends and Applications*, Studies in Computational Intelligence 740,
https://doi.org/10.1007/978-3-319-67056-0_34

other aspects of our lives (e.g., e-commerce, travel, marketing) in tangible and exciting ways [3].

AR is a lead of computer research that combines actual scenes viewed by user, and virtual scenes generated by the computer. This augments the scene with additional information [4].

AR is used to overlay real images with data or digital images to increase impact and increase usability or to enhance understanding [1]. It is also the real-time delivery of digital information to enhance or enable a geographical or physical experience [1].

The assertion that AR could provide enhanced learning experiences is grounded in two interdependent theoretical frameworks:

1. Situated learning theory.
2. Constructivist and Interpretivist learning theory.

Situated learning theory posits that all learning takes place within a specific context and the quality of the learning is a result of interactions among the people, places, objects, processes, and culture within and relative to that given context. Constructivist Interpretivist theories of learning assume that meaning is imposed by the individual rather than existing in the world independently.

The ultimate goal of AR is to create a virtual environment that is inter-active with the user where a user cannot tell the difference between the real world and the virtual augmentation of it. AR superimposes graphics, audio, and other sense enhancements from computer screens into real time environments. It also combines virtual reality and user reality seamlessly together [4].

1.1 Goals of Augmented Reality

The goal of Augmented Reality is to create a system in which the user cannot tell the difference between the real world and the virtual augmentation of it. Another goal of augmented reality is to add information and meaning to a real object or place. Unlike virtual reality, augmented reality does not create a simulation of reality. Instead, it takes a real object or space as the foundation and incorporates technologies that add contextual data to deepen a person's understanding of the subject. Creating virtual environment for a more rich user experience is another goal of AR. AR can be used to achieve feats which are limited in real world. and to enhance imagination of youths and children.

1.2 Augmented Reality (AR) Versus Virtual Reality (VR)

One of the biggest confusions in the world of augmented reality is the difference between augmented reality and virtual reality. Both are earning a lot of media

Table 1 Augmented reality (AR) versus virtual reality (VR)

Augmented reality (AR)	Virtual reality (VR)
System augments the real world scene	Totally immersive environment
User maintains sense of presence in real world	Visual sense are under control of system
Needs a mechanism to combine virtual and real worlds	

attention and are promising tremendous growth. Augmented reality (AR) is closely tied to virtual reality (VR), since the concept of AR evolved as an extension, or variation, of VR [5]. Augmented Reality (AR) is a variation of Virtual Environments (VE), or Virtual Reality as it is more commonly called (Table 1).

(VR) technologies completely immerse a user inside a synthetic environment. While immersed, the user cannot see the real world around him. In contrast, (AR) allows the user to see the real world, with virtual objects superimposed upon or composited with the real world. Therefore, AR supplements reality, rather than completely replacing it. Ideally, it would appear to the user that the virtual and real objects coexisted in the same space.

2 Augmented Reality Applications

AR represents the cutting edge of modern society's social-technological development. AR applications are being created by independent groups and organizations all over the world for use within many disparate ends and it can be used in all programming applications that interact with users. In this section some applications of Augmented Reality will be described in next sections:

2.1 AR and Education

In education, AR has been used to complement a standard curriculum. Text, graphics, video and audio were superimposed into a student's real time environment. Textbooks, ash cards and other educational reading material contained embedded markers or triggers that, when scanned by an AR device, produced supplementary information to the student rendered in a multimedia format [2, 3, 6, 7].

Benefits of Using AR in Education

- It provides rich contextual learning.
- Appeals to constructivist notions of education where students take control of their own learning.

- Provides opportunities for more authentic learning and appeals to multiple learning styles.
- Provides each student with his/her own unique discovery path. Engages a learner in ways that have never been possible.
- No real consequences if mistakes are made during skills training.
- Engage, stimulate, and motivate students to explore class materials from different angles [8].
- Help teach subjects where students could not feasibly gain real-world experience [9].
- Enhance collaboration between students and instructors and among students.
- Foster student creativity and imagination.
- Help students take control of their learning at their own pace and on their own path.
- Create an authentic learning environment suitable to various learning styles.

How Can AR Support Educational Process

Presentation [10] lists some points that illustrate how can AR support the education process which are:

Engages kinaesthetic learning

- Object manipulation: such as Turning page, Moving objects, Moving users head.
- Further interaction: like Combining objects, Re-arranging objects, Ani-mating objects.

Application of AR in Education and Desktop Applications

AR will change the process of education to entertainment (education with enjoying). AR can support education with many applications such as AR gaming, skills training, AR text books, discovery based books, leaning books and colour and modelling objects books [3].

The idea of AR text books: "In simple terms, AR allows digital content to be seamlessly overlaid and mixed into our perceptions of the real world. In addition to the 2D and 3D objects which many may expect, digital assets such as audio and video les, textual information, and even olfactory or tac-tile information can be incorporated into users perceptions of the real world. Collectively, these augmentations can serve to aid and enhance individuals knowledge and understanding of what is going on around them. Rather than seeming out of place, the digital markups inherent in AR lets users perceive the real world, along with added data, as a single, seamless environment. In other words [11] denes it as: "A MagicBook": it is a real book where markers have been added to the pages. This allows the traditional content of the pages to be enhanced with new 3D virtual content [11]. See Figs. 1 and 2.

Fig. 1 A MagicBook an augmented reality text books

Fig. 2 Augmented reality education solar system. *Source* (https://www.youtube.com/watch?v=UkWuVVVUD4Q)

2.2 *AR Web Application and Discovery-Based Learning*

Augmentation is conventionally in real-time and in semantic context with environmental elements, such as sports scores on TV during a match. With the help of advanced AR technology (e.g. adding computer vision and object recognition) the information about the surrounding real world of the user becomes interactive and digitally manipulable. Artificial information about the environment and its objects can be overlaid on the real world [weki].

AR Applications that convey information about a real-world place open the door for discovery-based learning. Currently, many historic sites supply overlay maps and different points of historic information for their visitors. However, in the near future, AR will bring even more excitement into historic sites though various developing projects. For example, the EU-funded iTacitus AR project (www.itacitus.org) will allow visitors to pan across a location while hearing and seeing a historic event play out. Another AR tool, the TAT Augmented ID application, will use facial recognition technology to display certain, pre-approved information about a person when s/he is viewed through the camera of a mobile device. A third tool, SREngine, will use AR object recognition to display information about everyday items in the real world, allowing for easy price comparison while shopping, as well as identifying plants and animals [6].

2.3 AR Medical Applications

There have been really interesting advances in medical application of augmented reality. Medical students use the technology to practice surgery in a controlled environment. Visualizations aid in explaining complex medical conditions to patients. Augmented reality can reduce the risk of an operation by giving the surgeon improved sensory perception. This technology can be combined with MRI or X-ray systems and bring everything into a single view for the surgeon Fig. 3.

Doctors could use Augmented Reality as a visualization and training aid for surgery. It may be possible to collect 3-D datasets of a patient in real time, using noninvasive sensors like Magnetic Resonance Imaging (MRI), Computed Tomography scans (CT), or ultrasound imaging. These datasets could then be rendered and combined in real time with a view of the real patient. In e ect, this would give a doctor "X-ray vision" inside a patient. This would be very useful during minimally invasive surgery, which reduces the trauma of an operation by using small incisions or no incisions at all. A problem with minimally invasive techniques is that they reduce the doctor's ability to see inside the patient, making surgery more di cult. AR technology could provide an internal view without the

Fig. 3 Medical AR book. *Source* (http://press.mu-varna.bg/EN/augmenter-reality-medical-books/)

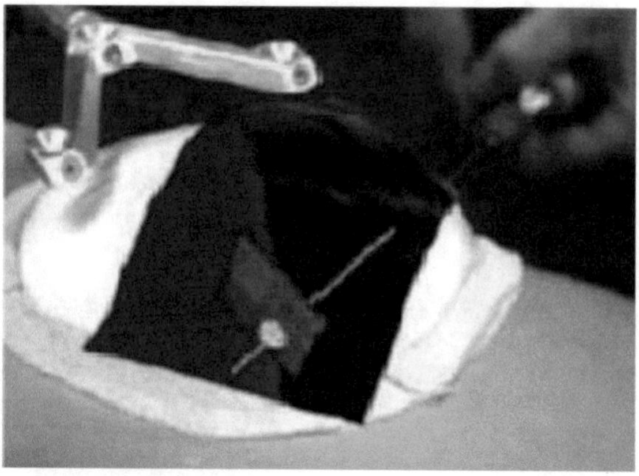

Fig. 4 Mockup of breast tumour biopsy. 3-D graphics guide needle insertion. *Source* (Courtesy UNC Chapel Hill Dept. of Computer Science)

need for larger incisions. AR might also be helpful for general medical visualization tasks in the surgical room. Surgeons can detect some features with the naked eye that they cannot see in MRI or CT scans, and vice-versa. AR would give surgeons access to both types of data simultaneously Fig. 4.

2.4 *Military Training*

AR can serve as a networked communication system that renders useful battle field data onto a soldier's goggles in real time. From the soldier's viewpoint, people and various objects can be marked with special indicators to warn of potential dangers. Virtual maps and 360 view camera imaging can also be rendered to aid a soldier's navigation and battle eld perspective, and this can be transmitted to military leaders at a remote command centre. And there is a lot of applications that can be made in military training like military aircraft and helicopter training where Head-Up Displays (HUDs) and Helmet-Mounted Sights (HMS) used to superimpose vector graphics upon the pilot's view of the real world Fig. 5 [2].

2.5 *Annotation and Visualization Manufacturing, Maintenance, and Repair*

Another category of Augmented Reality applications is the assembly, maintenance, and repair of complex machinery. Instructions might be easier to understand if they were available, not as manuals with text and pictures, but rather as 3-D drawings superimposed upon the actual equipment, showing step-by-step the tasks that need to be done and how to do them. These superimposed 3-D drawings can be animated, making the directions even more explicit Fig. 6.

2.6 Engineering Design

AR allowed industrial designers to experience a product's design and operation before completion. It also allows constructions engineers to test and visualize the building structure. AR interact with 3D models, animations, holograms and virtual environments in real time Fig. 7 [2].

Fig. 5 Augmented reality allows immersive experiences and analysis mixing real world and simulation

Fig. 6 Complex product development. *Source* http://www.advice-manufacturing.com/Virtual-and-Augmented-Reality.html)

Fig. 7 Engineering design

Fig. 8 Wikitude

2.7 *Navigation*

Navigation applications are possibly the most natural t of augmented reality with our everyday lives. Enhanced GPS systems are using augmented reality to make it easier to get from point A to point B. Wikitude Drive for the Android operating system which is currently in beta brings the GPS into the 21st century. Using the phone's camera in combination with the GPS, the users see the selected route over the live view of what is in front of the car. it provides information from Wikipedia on a smartphone and the information displayed as markers on the camera screen. Geo-tagged Wikipedia "Wikitude" provides information from Wikipedia on a smartphone. The Wikitude can add digital content in the smart phone. This digital content can include information about different places Fig. 8.

2.8 *Google Latitude and Bright Kite*

Location-based social networking and microblogging Latitude Friends can see each other's Locations shown on map (on phone) Pro le/ photos included for each friend depends on each users privacy settings Fig. 9.

2.9 *Alive Journals*

Alive application is one of the leading Augmented Reality applications that has empowered more than 200 brands and publications including the very popular TOI to enrich their print content with new and innovative Aug-mented Reality experiences. It enables users to connect and engage with your print content by scanning, interacting and sharing their digital experiences which includes watching videos, buying products, viewing 3D animations, photo gallery and much more. Alive is available on all platforms viz. iOS, Android, Windows, Blackberry, Symbian and Java see Fig. 10.

2.10 Alive Advertisement

Augmented reality can be used in advertisement to encourage customers to engage with their goods or services before purchase. This can form part of a companys marketing strategy where the aim is to identify and target a group of customers in order to satisfy their needs Fig. 11.

2.11 Smartphones Games and Application

The gaming industry embraced AR technology. A number of games were developed for prepared indoor environments, such as AR air hockey, Titans of Space,

Fig. 9 Google Latitude/Bright Kite

Fig. 10 Alive journals

Fig. 11 Alive advertisement

collaborative combat against virtual enemies, and AR-enhanced pool-table games. Augmented reality allowed video game players to experience digital game play in a real world environment. Companies and platforms like Niantic and LyteShot emerged as major augmented reality gaming creators. Niantic is notable for releasing the record-breaking Pokmon Go game [2].

Pokmon Go, a game that has quickly captured everyone's attention and given them a reason to go out into the world, walk around, and catch Pokmon. The game uses GPS to mark your location, and move your in-game avatar, while your smartphone camera is used to show Pokmon in the real world. For the most part, it works, provided the game hasn't Crashed or frozen [12]. Pokmon dates back to the mid-1990s with popular video games, trading cards, comics, and videos featuring the Pokmon creatures (pocket monsters). The mobile game Pokmon GO, created by Niantic, was introduced in July, 2016, and is available on both iOS and Android devices. It was released initially in a limited number of countries, presumably so that Niantic could build up its infrastructure to support increased server tra c. The object of the game is to catch as many Pokmon as possible; there are some 150 available to date. When starting up the game, a map view of the players location is displayed, showing the players avatar and any Pokmon in the area. The presence of Pokmon depends on the location and the time of day. Clicking on a Pokmon switches the display to an AR view, with the creature overlaid on the live scene as captured by the players camera, see Fig. 12 [13].

2.12 Face Recognition and Personal Information Systems Using AR

Augmented reality and image recognition playing a big part in our computing future. Now the same tourist saw someone walking in the street and he want to know some information about him. All he had to do is to put mobile cam-era and focus on his face. The software will recognizes a face and displays the information needed such as name, age, relationship status. See Fig. 15a, b, c. [14], AR could integrate phone and email communication with context-aware overlays, manage

Fig. 12 Using AR in smartphone games. *Source* (http://www.digitaltrends.com/mobile/best-augmented-reality-apps/)

personal information related to specific locations or people, provide navigational guidance, and provide a unified control interface for all kinds of appliances in and around the home Fig. 13 [15].

2.13 Read Any Language

By this technology we can put mobile in front of any language text and ask it to translate to our mother Tung language. In Fig. 14 a tourist from England is visiting Spain and he try to read a picture which was written in Spanish. All he had to do is to de ne the source and the target languages and put the mobile in front of the picture. The picture automatically will be turned to his target language as in Fig. 14 [14].

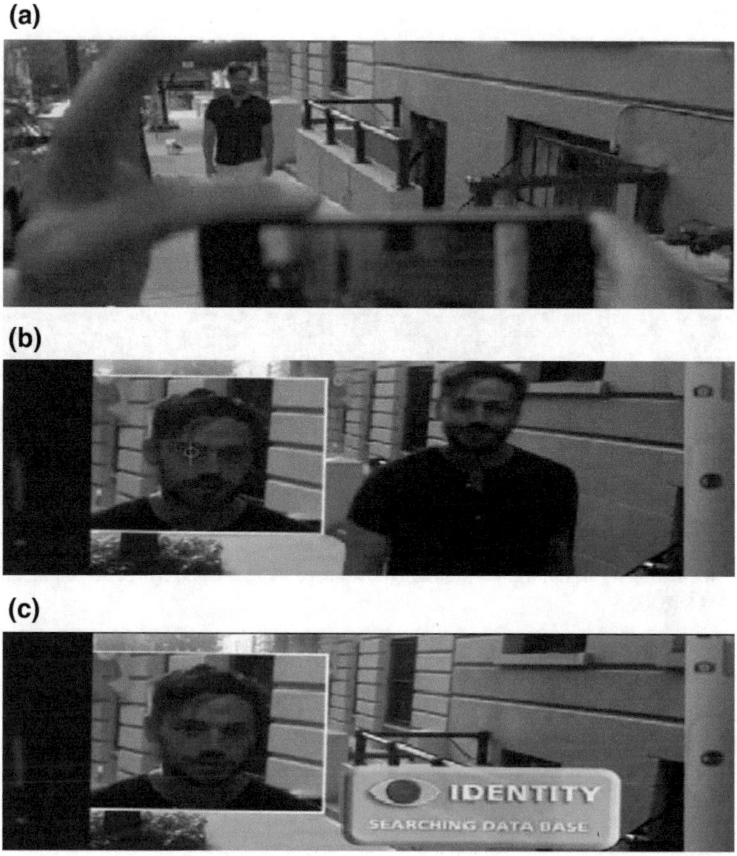

Fig. 13 **a** Putting the mobile camera in the right position. **b** Focusing on the face, **c** Recognizing the face

3 Components of AR

Major components needed in order to make an augmented-reality system work.

3.1 Display and Tracking System

Display enables user to view superimposed graphics and text created by the system. The tracking system pinpoints the users location in reference to their surroundings and additionally tracks the users head and eye movements or Marker movement also.

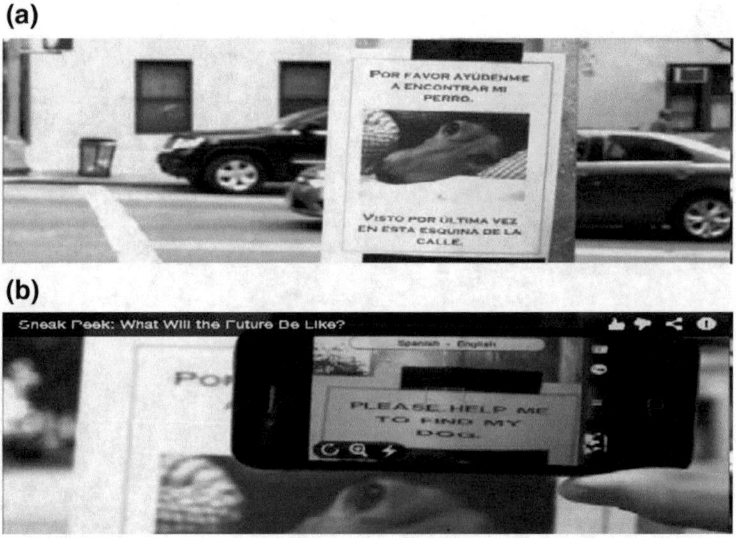

Fig. 14 **a** Tacking a snap shot from the foreign language. **b** Translated by the AR application into the needed language

- Head-Up Display (HUDs).
- Head Mounted.
- Projection Display (HMPDs).
- Occlusion Display Head Mounted Display (HMDs).

3.2 *Marker*

Some Markers Shapes

Different types of Augmented Reality (AR) markers are images that can be detected by a camera and used with software as the location for virtual assets placed in a scene. Most are black and white, though colours can be used as long as the contrast between them can be properly recognized by a camera. Simple augmented reality markers can consist of one or more basic shapes made up of black squares against a white background. More elaborate markers can be created using simple images that are still read properly by a camera A camera is used with AR software to detect augmented reality markers as the location for virtual objects. The result is that an image can be viewed, even live, on a screen and digital assets are placed into the scene at the location of the markers. The simplest types of augmented reality markers are black and white images that consist of two-dimensional (2D) barcodes [16]. They are black and white most of time but they may be coloured in companies applications Figs. 15, 16 and 17 [17].

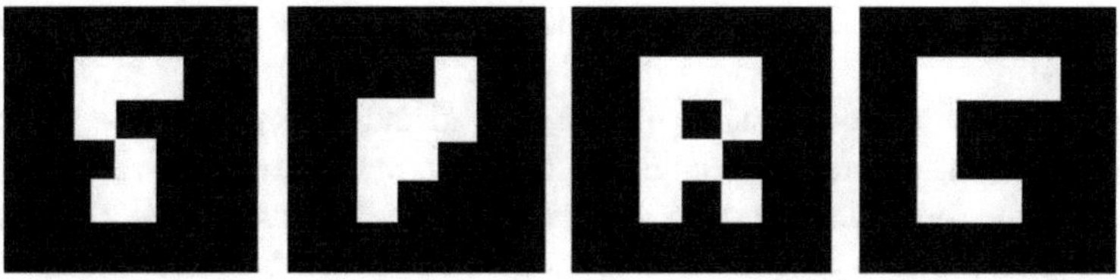

Fig. 15 Markers shapes

Fig. 16 Marker and how the system recognize it and display its related object on the mobile screen

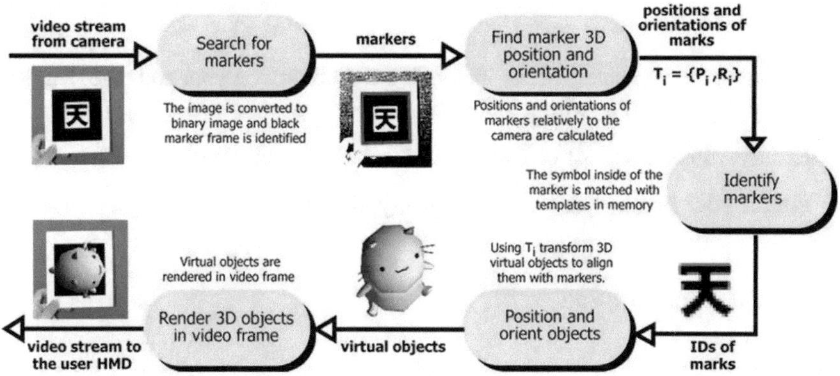

Fig. 17 ARToolKit process. *Source* http://hitl.washington.edu/people/tfurness/courses/inde543/READINGS-03/BILLINGHURST/MagicBook.pdf

3.3 Mobile Computing Power

Requires a highly mobile computer which superimposes the graphics and text created by the system over what is appears in reality Components of AR.

How does it work?

AR merges the three components (display, tracking system and mobile power) into a highly portable unit Research in the elds of computer vision, com-puter graphics, and user interfaces are actively contributing to advances in augmented reality systems [4].

A special code is incorporated within a print ad and users place this ad in front of their webcam. The software recognizes the code and activates a reaction which could be in the form of a 3-D modelling of a product. It turns out that a 3D baseball player is standing on a baseball card users hold up to their webcam [4]. AR allows superimposing 3D-content/products into the real world. AR allows interaction with 3D-content at real time also [3].

4 Augmented Reality and Mobiles

4.1 Main Features [1]

- Works on every mobile phone with a camera, average screen and Visio capability.
- All content is secured.
- Does not require an application to be installed on the mobile device. Compatible with both Windows PCs and Mac computers.
- Content is encrypted to protect your brand.
- Fully packaged for massive instantaneous deployment.

Benefits [1]:

- Extend the experience with an innovative technology on your web site. Runs on nearly any computer with a webcam.
- Reduces times and cost of deployment through a web interface.
- Protect your brand with a powerful encryption mechanism.

4.2 Software Requirement

Augmented reality is sti ed by limitations in software and hardware. Cell phones require superb battery life, computational power, cameras and tracking sensors. For software, augmented reality requires a much more sophisticated arti cial intelligence and 3-D modelling applications [1]. It needs also [18]

- Image processing like (Opens).
- Graphics like (OpenGL).
- Audio (Penal).
- Toolkit like (A Toolkit).

4.3 Hardware Requirements

In terms of hardware components we need the following requirements at any AR application: Marker: It is two dimensional black and white squares have a speci c pattern. It is used to display the virtual objects above it, when it moves; the virtual object should move with it and appear exactly aligned with the marker. Web Camera: to capture video of the real world and sends it to the computer. Plasma screen Display: is a type of at panel display to display the virtual objects on it. HMD Display: it is called Head Mounted.

Display and it is a display device, worn on the head or as part of a helmet, that has a small display optic in front of one (monocular HMD) or each eye (binocular HMD). HMD Display: it is called Head Mounted Display and it is a display device, worn on the head or as part of a helmet, that has a small display optic in front of one (monocular HMD) or each eye (binocular HMD). Input Hardware Devices: it is a device which helps us to input values to computer such as mouse and keyboard.

Types of Toolkits:

1. OSGART (OpenSceneGraph AR toolkit): is a C++ cross-platform de-velopment library that simpli es the development of Augmented Reality applications by combining computer vision based tracking libraries (e.g. ARToolKit,) with the 3D graphics library OpenSceneGraph.
   It provide it's user with
   (a) all the features of OpenSceneGraph (high quality rendering, multiple le Format loaders)
   (b) interaction between markers.
2. NyARToolkit: is an ARToolkit class library released for virtual machines particularly those which host Java, C and Android.
3. ATOMIC Authoring ToolKit: a Cross-platform Authoring Tool software, for Augmented Reality Applications, which is a Front end for the AR-Toolkit library. It was developed for non-programmers, to create small and simple, Augmented Reality applications.
4. FLARToolKit: (FlashARToolKit) it is AS3 ported version of ARToolKit. Actually, is based on NyARToolkit, Java ported version of ARToolKit. FLARToolKit recognize the marker from input image. And calculate its orientation and position in 3D world. You should draw 3D graphics by your own. But helper classes for major ash 3D engines (Papervision3D, Away3D, Sandy, and Alternativa3D) are included. Papervision3D is used in starter-kit.
5. ARToolKit: ARToolKit applications allow virtual imagery to be superim-posed over live video of the environment. Although this appears magical, the secret is in the black squares used as tracking markers. The ARToolKit tracking works as follows: [19]

(a) The camera captures video of the environment and sends it to the computer.
(b) Software on the computer searches through each video frame for any square shapes.
(c) If a square is found, the software uses some mathematics to calculate the position of the camera relative to the black square.
(d) Once the position of the camera is known a computer graphics model is drawn from that same position.
(e) This model is drawn on top of the video of the real world and so appears attached to the square marker.
(f) The nal output is shown back in the handheld display, so when the user looks through the display they see graphics overlaid on the real world.

6. Augment is a mobile application that lets you visualize your 3D models in Augmented Reality, integrated in real time in their actual size and environment. Augment is the perfect Augmented Reality app to boost your sales and bring your print to life.
7. OpenSpace3D is a free and open-source platform, designed to create virtual and augmented reality applications and games.

5 Conclusions

Augmented Reality is a relatively new eld, where most of the research ef-forts have occurred in the past few years. In this paper some applications have been presented like using augmented reality in education, medicine, mil-itary training, engineering, gaming and much more applications on AR. Some applications presented too like face recognition and language translating.

Augmented reality opens the door for interactive learning and education. So the aim of this research is to give an explanation about the AR and how it can be used in education. And in the future how to design educational Arabic books which will be helpful and interesting in education especially for kids by translating the letters, words and sentences to an observed seen in the reality.

References

1. Sullivan, P., Munoz, M., Repcinova, K., Brandenburg, K., Estabrook, J.: Augmented Reality, Artificial Intelligence, and Business Intelligence (2009). Presentation at http://www.slideshare.net/UAABA635/augmented-reality-artificial-intelligence-and-business-intelligence
2. [weki]. http://en.wikipedia.org/wiki/Augmentedreality
3. Yuen, S.C.: Augmented Reality (AR) in Education. Creating Future Through Technology Conference, Biloxi, Mississippi (2011). http://www.slideshare.net/scyuen/augmented-reality-ar-in-education

4. Cassie, A.C., Chris, D., Ryan, P., Tiliakos, S.N.: Augmented Reality. http://www.authorstream.com/Presentation/Saurabh0750-1053920-group-5-ar-presentation/
5. Milgram, P., Takemura, H., Utsumi, A., Kishino, F.: Augmented reality: a class of displays on the reality virtually continuum, ATR Communication Systems Research Laboratories, 2–2 Hikaridai, Seika-cho, Soraku-gun Kyoto 619-02, Japan
6. Yuen, S., Yaoyuneyong, G., Johnson, Johnson E.: Augmented reality: An overview and ve directions for AR in education. J. Educ. Technol. Deve. Exch. **4**(1), 119–140 (2011)
7. Azuma, R.T.: A survey of augmented reality. Teleoper. Virt. Environ. J. **6** (1997)
8. Kerawalla, Luckin, Selijefot, and Woolard. Pedagogical applications of smartphone integration in teaching lectures. students and pupils perspectives (2006)
9. Shelton and Hedley, Using Augmented Reality for Teaching Earth-Sun Relationships to Undergraduate Geography Students (2002)
10. Stanley, G.: Augmented reality language learning, virtual worlds meet m-Learning. Avatar Lang. Blog (2009). http://www.avatarlanguages.com/blog/arll/
11. Groundbreaking Augmented Reality-Based Reading Curriculum Launches, PRweb, 23 October (2011)
12. Kaufmann, H.: Collaborative Augmented Reality in Education, Institute of Software Technology and Interactive Systems, Vienna University of Technology. http://www.thgaftna.com/vb/showthread.php?t=18336page=1
13. Godwin-Jones, R.: "Augmented reality and language learning: form notated vocabulary to place-based Mobil games", Virginia Commonwealth University, (2016)
14. Peek, S.: What Will the Future Be Like. youtube (2012). http://www.youtube.com/watch?v=s4ZkOmbavgs
15. van Krevelen, D.W.F., Poelman, R.: A survey of augmented reality technologies, applications and limitations. Int. J. Virt. Real. (2010)
16. Shelton and Hedley: Using Augmented Reality for Teaching Earth-Sun Relationships to Undergraduate Geography Students, 200 Anuroop Katiyar1, Karan Kalra, and Chetan Garg, Marker Based Augmented Reality, Advances in Computer Science and Information Technology (ACSIT) (2015)
17. Larngear Technologies (2009). http://media.nmc.org/2010/03/augmented-books.mov
18. Klepper, S.: A survey: Augmented Reality—Display Systems. TU Muenchen (2007)
19. ARToolworks support library, "How ARToolKit works", last modification (2012). http://www.artoolworks.com/support/library/index.php?title=HowARToolKitworksandoldid=166

A Tutorial on Information Retrieval Using Query Expansion

Mohamed Yehia Dahab, Sara Alnofaie and Mahmoud Kamel

Abstract Most successful information retrieval techniques which has the ability to expand the original query with additional terms that best represent the actual user need. This tutorial gives an overview of information retrieval models which are based on query expansion along with practical details and description on methods of implementation. Toy examples with data are provided to assist the reader to grasp the main idea behind the query expansion (QE) techniques such as Kullback-Leibler Divergence (KLD) and the candidate expansion terms based on WordNet. The tutorial uses spectral analysis which one of the recent information retrieval techniques that considers the term proximity.

Keywords Information retrieval · Query expansion · Spectral analysis · KLD WordNet

1 Introduction

The most critical issue for information retrieval is the term mismatch problem. One well known method to overcome this limitation is the QE of the original query terms with additional terms that best retrieve the most related documents.

Most successful technique which has the ability to expand the original query with additional words that best capture the actual user goal. Most of the recent information retrieval models are based on the proximity between terms which is useful for

M.Y. Dahab (✉) · S. Alnofaie · M. Kamel
Faculty of Computing and Information Technology, Department of Computing Science, King Abdulaziz University, Jeddah 21589, Saudi Arabia
e-mail: mdahab@kau.edu.sa; mohamed.dahab@gmail.com
URL: http://www.kau.edu.sa/Home.aspx

S. Alnofaie
e-mail: salnefaie@kau.edu.sa

M. Kamel
e-mail: miali@kau.edu.sa

K. Shaalan et al. (eds.), *Intelligent Natural Language Processing: Trends and Applications*, Studies in Computational Intelligence 740,
https://doi.org/10.1007/978-3-319-67056-0_35

improving document retrieval performance (e.g., [1–8] and [9]), with consideration of the term positional information and different transformation algorithms to calculate the document score. The main idea of the document spectral analysis is extended to be used in document classification and clustering [6] and [9].

In the information retrieval community, Usually, the queries consist of two or three terms, which are sometimes not enough to understand the expectations of the end user and fail to express topic of search. The QE is a process of broadening the query terms using words that share statistical or semantic relationships with query terms in the collection or share meaning with query terms. To our knowledge, there is no technique uses the QE in information retrieval model and considers the proximity information except [7].

This tutorial provides more than a toy example with data to assist the reader to grasp the main idea behind the QE techniques such as KLD and the candidate expansion terms based on WordNet. The tutorial uses spectral analysis which one of the recent information retrieval techniques that considers the term proximity.

This research will proceed as follows, Sect. 2 presents the related background, Sect. 3 lists the preprocessing tasks, Sect. 4 introduces the a brief description about the synthetic dataset used in this research, Sect. 5 demonstrates the application of different techniques of the QE on the dataset and finally the conclusion is given in Sect. 6.

2 Background

Here we review some general aspects that are important for a full understanding of the following tutorials.

2.1 Automatic Query Expansion

The average length of the query is around 2–3 words where may the users and the document collection does not use the same words for the same concept that is known as the vocabulary or mismatch problem. Therefore, there is difficulty in retrieving the relevant documents set. To improve the performance of IR model, use the overcome mismatch problem approaches. One of the successful approaches is to automatically expand the original query with other terms that best capture the actual user intent that makes the query more useful [10] and [11].

Automatic QE process can divided into four steps: data source preprocessing, candidate expansion features generation and ranking, expansion features selection, the query reformulated [12, 13] and [14].

2.2 Data Source Preprocessing

In this step, the data source that is used for expanding the user query transforms into an effective format for the following steps. It consists of two phases. First, extract the intermediate features. Then, construct the appropriate data structures for access and manipulation this features in an easy way. Based on the source of the Candidate Expansion Terms (CET) the QE approach classifies to the external resources such as the WordNet and the target corpus. The WordNet approaches set some or all the synonyms terms of the synset the contain query term as candidate terms. The target corpus approaches are also divided into local and global. The global approaches set the whole corpus terms as candidate terms and analyze it while the local approaches set only the top relevant documents terms of the initial search results. The local approaches are known as pseudo relevance feedback. In the IR model, the documents collection or corpus is indexing to run the query. As seen in the above section, the documents store using inverted index file, which is useful in some QE approach such as the global approach while the local approach needs to the documents using direct index file.

2.3 Features Generation and Ranking

In this stage, the candidate expansion features generate and ranks by the model. The original query and the data source is the input to this stage while the candidate expansion features associated with the scores is the output. A small number of the candidate features add to the query. Therefore, the feature ranking is important. The relationship between the query terms and candidate features classify the generation and ranking approaches to:

A. One-to-one associations.
B. One-to-many associations.
C. Analysis of feature distribution in top-ranked documents.

2.4 Expansion Features Selection

After the candidate features ranking for some QE approach, the limited number of features is added to the query to process the new query rapidly.

2.5 Query Reformulation

This step usually involves assigning a weight to each expansion feature and re-weights each query term before submitting the new query to the IR model. The most popular query re-weighting scheme was proposed in [15].

3 Preprocessing Tasks

Preprocessing Tasks includes all steps to transform the query and the raw data source used for expanding the user query into a format that will be more effectively processed by subsequent steps. This includes the following steps:

- Text extraction from documents like HTML, PDF, MS Word, etc. (if the collection is made of such documents).
- Tokenization (i.e., extraction of individual words, ignoring punctuation and case).
- Stop word removal (i.e., removal of common words such as articles and prepositions).
- Word stemming (i.e., reduction of inflected or derivational words to their root form).
- Word weighting (i.e., assignment of a score that reflects the importance of the word, usually in each document).
- Some Automatic query expansion (AQE) techniques based on system that indexing the document use inverted index file which represent the document as a set of weighted terms. The indexing system may also store term positions, to provide proximity based search. Other AQE techniques, based on corpus analysis, require the extraction of particular features from the collection.
- Creating term signals for each document is an important preprocessing task but it will be shown in the processing phase for the purpose of giving the readers more details and explanation.

4 Synthetic Dataset

The tutorial is based on a single query q which is "*Suicide explosions in city*". Table 1 shows the synthetic dataset used in this research. It contains 20 rows, each row has three columns, the first column exhibits whether the document content is relevant or not (1 means relevant while 0 means not relevant). The query q is applied on the dataset shown in Table 1.

Table 1 Synthetic dataset

Relevant	Document content	Document No.
1	Explosions in Sanaa, city metropolis of Yamen, today. The terrorist group, Houthi organization, claimed the responsibility for the terrorist action. Three blasts in different places. The Downtown bomb left a number of victims	1
1	Burst Two explosions shake Sacramento city which is Located in California. The first one led to the death of five people. The second one resulted in wounding 65 people. Moreover, massive destruction in cars and buildings	2
1	A series of explosions in france. The terrorist Daesh organization, claimed responsibility for Terrorist action in paris city. one bomb burst in the Playground and another bomb in coffee. The detention of a number of hostages in the theater Pataklan	3
1	Burst Two explosions near the Yusufiyah mosque. The terrorist group, Daesh, claimed the responsibility for the accident. This action resulted in wounding two people and killing three. Upon the arrival of paramedics a malefactor bomb himself in the west part of Baghdad city	4
0	A new mall will be opened in Jeddah downtown on 23th of March	5
0	Large number of celebrity gathered in downtown Los Angeles	6
0	Syrians refugees flocking to Europe	7
0	Expo starting in Milano in April 2017	8
0	Snap chat has changed the Privacy Policy	9
0	Houthi organization Announces responsibility for security of Sanaa starting from Feb 2nd 2013	10
0	Arabic speaker passengers were denied boarding American plane	11
0	Terrorist word comes from the Latin word terrorism meaning great fear. Great fear is exactly what they want to create in order to achieve their goals. They are using violence, chaos, bomb, and destruction. By doing this, they are aiming to force people, and governments to take particular action especially for political Economic social and religious purposes. It is highly destructive phenomenon in recent years	12
0	Hundreds of people have demonstrated in Dublin to support anti-racism	13
0	A vigorous car blast in the east part of Beirut that has majority of Christian on Friday	14
1	Terrorist attack by nuclear bomb in Hiroshima Japan by United States of America in August 1945	15
0	Close Brussels Metro after New York state declared a state of emergency to the highest level	16

(continued)

Table 1 (continued)

Relevant	Document content	Document No.
0	More than 160 000 people were the victims of Tsunamis in 26 December 2004. Tsunami is a seismic sea wave. It is a series of waves in a water caused by the displacement of a large volume of water generally in an ocean or a large lake. The region was affected by Tsunami is the metropolis on Indian Ocean	17
1	Many victims in 3 big blasts today evening in Afghanistan metropolis	18
1	In 24th of Mar 2014 morning, Al-Qaeda organization claimed the responsibility for several bursts in Sanaa and Taiz	19
0	President Obama blasts Donald Trump's recent remarks	20

5 The Application of Different QE Techniques

In this section, statistical and semantic QE techniques will be applied using only one of the recent information retrieval techniques which is spectral based information retrieval ([5] and [8]).

5.1 *Spectral Based Information Retrieval with QE Using KLD*

As shown in Table 2, the matched query terms in document content are underlined. Document score increases as the query terms in documents content close to each other. Document score is computed by applying Haar discrete wavelet transform as explained in details in [2, 5] and [8]. Rows from 5 to 20 are empty because they do not have a matched query term. The notion of term signal introduced by [1–3, 8] is a vector representation of terms that describes frequencies of term occurrences in particular partitions within a document. In the example, 8 partitions or bins are used. Each line in the document content represents a bin, that is why there are 8 lines. The term signal of the term "*Suicide*" has been neglected because there is no term matched with it. The underline term means it matches with one of query terms. Some important preprocessing have been applied on document content. Documents number 15, 18 and 19 are relevant but they have zero document score so they have not been retrieved. By computing the Mean Average Precision (MAP) defined by the following equation, $MAP(q) = \frac{1}{7}(1 + 1 + 1 + 1 + 0 + 0 + 0) \approx 0.57$.

$$MAP(q) = \frac{1}{N}\Sigma_{j=1}^{N}\frac{1}{Q}\Sigma_{i=1}^{Q}P(doc_i) \quad (1)$$

Table 2 Document scores using spectral based information retrieval with the original query

No.	Document content	Term signal	Document score
1	– Explosions sanaa city – Metropolis yamen today – Terrorist group houthi – Organization claimed – Responsibility terrorist action – Blasts places downtown – Bomb left – Number victims	Explosions [1, 0, 0, 0, 0, 0, 0, 0] City [1, 0, 0, 0, 0, 0, 0, 0]	1.778
2	– Burst explosions shake – Sacramento city – Located california – Led death – People wounding – 65 people – Massive destruction – Cars buildings	Explosions [1, 0, 0, 0, 0, 0, 0, 0] City [0, 1, 0, 0, 0, 0, 0, 0]	0.889
3	– Series explosions france – Terrorist daesh organization – Claimed responsibility terrorist – Action paris city – Bomb burst playground – Bomb coffee detention – Number hostages – Theater pataklan	Explosions [1, 0, 0, 0, 0, 0, 0, 0] City [0, 0, 0, 1, 0, 0, 0, 0]	0.556
4	– Burst explosions yusufiyah – Mosque terrorist group – Daesh claimed responsibility – Accident action wounding – People killing arrival – Paramedics malefactor – Bomb west – Baghdad city	Explosions [1, 0, 0, 0, 0, 0, 0, 0] City [0, 0, 0, 0, 0, 0, 0, 1]	0.389
5			0.0
6			0.0
7			0.0
8			0.0
9			0.0
10			0.0
11			0.0
12			0.0
13			0.0
14			0.0

(continued)

Table 2 (continued)

No.	Document content	Term signal	Document score
15			0.0
16			0.0
17			0.0
18			0.0
19			0.0
20			0.0

Where:

N: is number of queries.
Q: is number of relevant documents for query q.
$P(doc_i)$: is the precision of ith relevant document.

To distinguish between useful candidate expansion term and unuseful expansion term by comparing the distribution of this term in the top relevant documents of the query with the distribution of this term in all documents. In other words, the score of the appropriate expansion term is high when the percentage of this term appearance in relevant documents more than in the collection.

Computing KLD. Carpineto proposed interesting query expansion approaches based on term distribution analysis [10]. The distributions variance between the terms in the top relevant documents and entire document collection where those terms obtain from the first pass retrieval using the query. The query expands with terms that have a high probability in the top related document compare with low probability in the whole set. The KLD score of term in the CET are compute using the following equation:

$$KLDScore(t) = P_R(t) log \frac{P_R(t)}{P_C(t)} \tag{2}$$

where $P_R(t)$ is the probability of the term t in the top ranked documents R, and $P_C(t)$ is the term t probability in the corpus C, given by the following equations:

$$P_R(t) = \frac{\Sigma_{d \in R} f(t,d)}{\Sigma_{d \in R} |d|} \tag{3}$$

$$P_C(t) = \frac{\Sigma_{d \in C} f(t,d)}{\Sigma_{d \in C} |d|} \tag{4}$$

After sorting the documents according to document scores to form the pseudo documents. Let the number of pseudo documents ($k = 4$) which includes the documents number 1, 2, 3 and 4 respectively. Apply the previous three equations on each term pseudo documents.

Let $NT_R = \Sigma_{d \in R}|d|$ and $NT_C = \Sigma_{d \in C}|d|$.

Now $NT_R = 80$ and $NT_C = 225$. $\Sigma_{d \in R} f(malefactor, d) = 1$. Also $\Sigma_{d \in C} f(malefactor, d) = 1$.

To compute the KLD score of the term *malefactor*, $P_R(malefactor) = \frac{1}{80} = 0.0125$ and $P_C(malefactor) = \frac{1}{225} = 0.0044$.

$KLDScore(malefactor) = P_R(malefactor) log \frac{P_R(malefactor)}{P_C(malefactor)} \approx 0.019$.

The KLD score for each term in pseudo documents show in Table 3.

Table 3 KLD score for each term in pseudo documents

Term	KLD score	Term	KLD score
City	0.075	Today	0.019
Explosions	0.075	Malefactor	0.019
Terrorist	0.063	Coffee	0.019
Bomb	0.045	Sacramento	0.019
Claimed	0.03	Detention	0.019
Burst	0.03	Baghdad	0.019
Action	0.03	California	0.019
Group	0.027	Hostages	0.019
Wounding	0.027	Buildings	0.019
Daesh	0.027	Killing	0.019
Responsibility	0.021	Accident	0.019
Number	0.017	Theater	0.019
People	0.015	Places	0.019
Arrival	0.019	Pataklan	0.019
Yusufiyah	0.019	Massive	0.019
65	0.019	West	0.019
Shake	0.019	Left	0.019
Led	0.019	Organization	0.01
Death	0.019	Series	0.005
Paris	0.019	Destruction	0.005
Paramedics	0.019	Houthi	0.005
Mosque	0.019	Blasts	0.005
France	0.019	Metropolis	0.0
Located	0.019	Victims	0.0
Playground	0.019	Downtown	0.0
Cars	0.019	Sanaa	0.0
Yamen	0.019		

Suppose the maximum length of candidate term list is four, so select the four terms that have maximum KLD score from Table 3. The candidate term list includes city, explosions, bomb and terrorist.

The candidate term list includes includes the original query terms in addition to extra terms that are supposed to be related to the original terms.

Apply the information retrieval technique, which was the spectral based information retrieval, again on the new query contains the new candidate term list.

Table 4 shows the document scores using spectral based information retrieval with expansion query using KLD. Documents, that do not have score, have been removed from table. The weight of original query terms has been increased by 100% to magnify their contribution in the document score more than the expanded query terms.

As a results from increasing the candidate term list to four, documents 12 and 15 are retrieved and it affects on MAP as following:

$MAP(q) = \frac{1}{7}(1 + 1 + 1 + 1 + 0.83 + 0 + 0) \approx 0.69.$

MAP as information retrieval measure has been improved when using statistical query expansion. Also, the order of documents according to document scores becomes 3, 1, 4, 2, 15 and 12 respectively.

5.2 *Spectral Based Information Retrieval with QE Using WordNet*

To apply spectral based information retrieval using semantic QE approach with the semantic lexicon WordNet, the following steps should be carried out:

1. **Determining the number of both pseudo documents and candidate term list**
   To continue using the same data described in the previous subsection, the number of pseudo documents is four, that is also include documents from $1 - 4$. The top two related terms will be added to the candidate term list.
2. **Computing the semantic similarity**
   For each term, t, in pseudo documents and term q_i in the query, compute related score using the following formulas:

   (a) Compute the semantic similarity between the term t and q_i using WordNet by considering the definitions of t and q_i as two sets of words, and the overlap between these two sets is taken as $Rel(t, q_i)$.

$$Rel(t, q_i) = \frac{2 * C_{t,q_i}}{C_t + C_{q_i}} \tag{5}$$

   where c_t, c_{q_i} is the number of words in t, q_i definitions respectively.
   C_{t,q_i} is the number of common words.
   To compute $Rel(t, q_i)$ when t = metropolis and q_i = city. The definition of the term "*metropolis*" in WordNet is :

Table 4 The document scores using spectral based information retrieval with the expansion query using KLD

No.	Document content	Term signal	Document score
1	– Explosions sanaa city – Metropolis yamen today – Terrorist group houthi – Organization claimed – Responsibility terrorist action – Blasts places downtown – Bomb left – Number victims	Explosions [1, 0, 0, 0, 0, 0, 0, 0] City [1, 0, 0, 0, 0, 0, 0, 0] Terrorist [0, 0, 1, 0, 1, 0, 0, 0] Bomb [0, 0, 0, 0, 0, 0, 1, 0]	5.635
2	– Burst explosions shake – Sacramento city – Located california – Led death – People wounding – 65 people – Massive destruction – Cars buildings	Explosions [1, 0, 0, 0, 0, 0, 0, 0] City [0, 1, 0, 0, 0, 0, 0, 0] Terrorist [0, 0, 0, 0, 0, 0, 0, 0] Bomb [0, 0, 0, 0, 0, 0, 0, 0]	1.28
3	– Series explosions france – Terrorist daesh organization – Claimed responsibility terrorist – Action paris city – Bomb burst playground – Bomb coffee detention – Number hostages – Theater pataklan	Explosions [1, 0, 0, 0, 0, 0, 0, 0] City [0, 0, 0, 1, 0, 0, 0, 0] Terrorist [0, 1, 1, 0, 0, 0, 0, 0] Bomb [0, 0, 0, 0, 1, 1, 0, 0]	6.44
4	– Burst explosions yusufiyah – Mosque terrorist group – Daesh claimed responsibility – Accident action wounding – People killing arrival – Paramedics malefactor – Bomb west – Baghdad city	Explosions [1, 0, 0, 0, 0, 0, 0, 0] City [0, 0, 0, 0, 0, 0, 0, 1] Terrorist [0, 1, 0, 0, 0, 0, 0, 0] Bomb [0, 0, 0, 0, 0, 0, 1, 0]	3.6
12	– Terrorist word latin word – Terrorism meaning great fear – Great fear create order – Achieve goals bomb violence – Chaos destruction aiming force – People governments action political – Economic social religious purposes – Highly destructive phenomenon years	Explosions [0, 0, 0, 0, 0, 0, 0, 0] City [0, 0, 0, 0, 0, 0, 0, 0] Terrorist [1, 0, 0, 0, 0, 0, 0, 0] Bomb [0, 0, 0, 1, 0, 0, 0, 0]	0.2
15	– Terrorist attack – Nuclear bomb – Hiroshima – Japan – United – America – August – 1945	Explosions [0, 0, 0, 0, 0, 0, 0, 0] City [0, 0, 0, 0, 0, 0, 0, 0] Terrorist [1, 0, 0, 0, 0, 0, 0, 0] Bomb [0, 1, 0, 0, 0, 0, 0, 0]	0.32

a large and densely populated urban area; may include several independent administrative districts people living in a large densely populated municipality metropolis

The definition of the term "*city*" in WordNet is:

a large and densely populated urban area; may include several independent administrative districts an incorporated administrative district established by state charter people living in a large densely populated municipality city

$C_t = C_{metropolis} = 22$, $C_{q_i} = C_{city} = 31$, $C_{t,q_i} = C_{metropolis,city} = 18$.
$Rel(metropolis, city) = \frac{2*18}{31+22} = 0.679$.

(b) Compute idf_t for each term in the collection.
$idf_{metropolis} = 0.673$.

(c) Select the most relevant document of the pseudo documents in which t occurs. This is intended to capture the intuition that terms coming from relevant document are better than the terms coming from non-relevant documents.

$$S(t, q_i) = Rel(t, q_i) * idf_t * \Sigma_{(d \in R)} \frac{sim(d, q)}{max_{(d' \in R)} sim(d', q)} \tag{6}$$

The term "metropolis" exist in only one pseudo documents 1 with score 1.778.
$Sim(metropolis) = \frac{1.778}{1.778} = 1$.

(d) The relatedness score of t with the whole documents is given by

$$S(t) = \Sigma_{q_i \in q} \frac{S(t, q_i)}{1 + S(t, q_i)} \tag{7}$$

$S(metropolis, city) = Rel(metropolis, city) * idf_{metropolis} * Sim(metropolis)$
$= 0.679 * 0.673 * 1 = 0.457$.
compute $S(metropolis, explosions)$ and $S(metropolis, Suicide)$ to compute $S(metropolis)$.

3. **Selecting the best related score**
The relatedness score of the remaining terms in pseudo documents shown in Table 5.

Table 5 The relatedness score of the terms in pseudo documents

Term	Relatedness score	Term	Relatedness score
Explosions	0.52	Detention	0.084
City	0.52	Malefactor	0.082
Metropolis	0.377	Playground	0.08
Victims	0.252	Mosque	0.067
Organization	0.23	Paramedics	0.065
Sanaa	0.229	Killing	0.065
Group	0.214	Series	0.061
Downtown	0.207	Accident	0.05
Action	0.203	West	0.041
Blasts	0.187	People	0.035
Left	0.182	Sacramento	0.027
Today	0.176	Located	0.027
Responsibility	0.165	Death	0.026
Claimed	0.164	Cars	0.024
Number	0.159	California	0.023
Terrorist	0.152	Massive	0.023
Bomb	0.148	Destruction	0.02
Places	0.139	Buildings	0.016
France	0.129	Shake	0.014
Arrival	0.123	Led	0.01
Paris	0.115	Yusufiyah	0.0
Theater	0.113	65	0.0
Burst	0.111	Daesh	0.0
Coffee	0.107	Yamen	0.0
Hostages	0.099	Houthi	0.0
Wounding	0.091	Pataklan	0.0
Baghdad	0.085		

4. **Applying spectral based information retrieval with new expansion query using WordNet**
   The top related terms that will be added to the candidate term list are Metropolis and victims. Table 6 shows the results of applying the spectral based information retrieval with new expansion query using WordNet.
   As shown in the Table 6, in addition to the original retrieved documents, i.e. the pseudo documents, documents 17 and 18 have been retrieved. Note that document number 17 is irrelevant while document number 18 is relevant.
   Addition weight, 100%, has been added to the original query terms because they contribute more in the document score. The underline term means it matches with one of query terms.

Table 6 The document scores using spectral based information retrieval with the expansion query using WordNet

No.	Document content	Term signal	Document score
1	– Explosions sanaa city – Metropolis yamen today – Terrorist group houthi – Organization claimed – Responsibility terrorist action – Blasts places downtown – Bomb left – Number victims	Explosions $[1, 0, 0, 0, 0, 0, 0, 0]$ City $[1, 0, 0, 0, 0, 0, 0, 0]$ Metropolis $[0, 1, 0, 0, 0, 0, 0, 0]$ Victims $[0, 0, 0, 0, 0, 0, 0, 1]$	6.38
2	– Burst explosions shake – Sacramento city – Located california – Led death – People wounding – 65 people – Massive destruction – Cars buildings	Explosions $[1, 0, 0, 0, 0, 0, 0, 0]$ City $[0, 1, 0, 0, 0, 0, 0, 0]$ Metropolis $[0, 0, 0, 0, 0, 0, 0, 0]$ Victims $[0, 0, 0, 0, 0, 0, 0, 0]$	1.28
3	– Series explosions france – Terrorist daesh organization – Claimed responsibility terrorist – Action paris city – Bomb burst playground – Bomb coffee detention – Number hostages – Theater pataklan	Explosions $[1, 0, 0, 0, 0, 0, 0, 0]$ City $[0, 0, 0, 1, 0, 0, 0, 0]$ Metropolis $[0, 0, 0, 0, 0, 0, 0, 0]$ Victims $[0, 0, 0, 0, 0, 0, 0, 0]$	0.8
4	– Burst explosions yusufiyah – Mosque terrorist group – Daesh claimed responsibility – Accident action wounding – People killing arrival – Paramedics malefactor – Bomb west – Baghdad city	Explosions $[1, 0, 0, 0, 0, 0, 0, 0]$ City $[0, 0, 0, 0, 0, 0, 0, 1]$ Metropolis $[0, 0, 0, 0, 0, 0, 0, 0]$ Victims $[0, 0, 0, 0, 0, 0, 0, 0]$	0.56
17	– 160000 people victims – Tsunamis 26 december 2004 – Tsunami seismic sea wave – Series waves water caused – Displacement large volume water – Generally ocean large – Lake region tsunami – Metropolis indian ocean	Explosions $[0, 0, 0, 0, 0, 0, 0, 0]$ City $[0, 0, 0, 0, 0, 0, 0, 0]$ Metropolis $[0, 0, 0, 0, 0, 0, 0, 1]$ Victims $[1, 0, 0, 0, 0, 0, 0, 0]$	0.14
18	– Victims – 3 – Big – Blasts – Today – Evening – Afghanistan – Metropolis	Explosions $[0, 0, 0, 0, 0, 0, 0, 0]$ City $[0, 0, 0, 0, 0, 0, 0, 0]$ Metropolis $[0, 0, 0, 0, 0, 0, 0, 1]$ Victims $[1, 0, 0, 0, 0, 0, 0, 0]$	0.14

Also in Table 6, 8 partitions or bins are used. Each line in the document content represents a bin, that is why there are 8 lines. The term signal of the term "*Suicide*" has been neglected because there is no term matched with it.

5. **Sorting documents based on score**
   After sorting the documents according to the document score, documents 1, 2, 3, 4, 17 and 18 have the following score 6.38, 1.28, 0.8, 0.56, 0.14 and 0.14 respectively.

6. **Evaluation**
   As a results from increasing the candidate term list to four, documents 17 and 18 are retrieved and it affects on MAP as following:
   $MAP(q) = \frac{1}{7}(1 + 1 + 1 + 1 + 0.83 + 0 + 0) \approx 0.69.$
   MAP as information retrieval measure has been improved when using WordNet query expansion.

6 Conclusion

This tutorial shows the impact of extending the query by adding statistical and semantic related terms to the original query terms over proximity based IR system. This is done by combining the spectral based information retrieval model with the best QE approaches such as the distribution approach (KLD) and WordNet. The toy examples results show that the spectral based information retrieval with QE using KLD and WordNet outperformed the spectral based information retrieval in precision at top documents and MAP metric. The toy examples provided in this research demonstrates that dividing the documents into a specific number of segments (8 bin).

References

1. Palaniswami, M., Ramamohanarao, K., Park, L.: Fourier domain scoring: a novel document ranking method. IEEE Trans. Knowl. Data Eng. **16**(5), 529539 (2004)
2. Park, L.A.F., Ramamohanarao, K., Palaniswami, M.: A novel document retrieval method using the discrete wavelet transform. ACM Trans. Inf. Syst. (TOIS). pp. 267–298 (2005)
3. Park, L.A.F., Palaniswami, M., Ramamohanarao, K.: Internet documentltering using fourier domain scoring. In: de Raedt, L., Siebes, A. (Eds.) Principles of Data Mining and Knowledge Discovery, September 2001, number 2168 in Lecture Notes in Articial Intelligence, pp. 362–373. Springer-Verlag (2001)
4. Park, L.A.F., Palaniswami, M., Ramamohanarao, K.: A novel document ranking method using the discrete cosine transform. IEEE Trans. Patt. Analys. Mach. Intell. pp. 130–135 (2005)
5. Aljaloud, H., Dahab, M., Kamal, M.: Stemmer impact on Guranic mobile information retrieval performance. Int. J. Adv. Comput. Sci. Appl. (IJACSA) **7**(12), 135–139 (2016). https://doi.org/10.14569/IJACSA.2016.071218

6. Al-Mofareji, H., Kamel, M., Dahab, M.Y.: WeDoCWT: a new method for web document clustering using discrete wavelet transforms. J. Inf. Knowl. Manage. **16**(1), 1–19 (2017). https://doi.org/10.1142/S0219649217500046
7. Alnofaie, S., Dahab, M., Kamal, M.: A novel information retrieval approach using query expansion and spectral-based. Int. J. Adv. Comput. Sci. Appl. **7**(9), 364–373 (2016). https://doi.org/10.14569/IJACSA.2016.070950
8. Dahab, M.Y., Alnofaie, S., Kamel, M.: Further investigations for documents information retrieval based on DWT. In: Hassanien, S.K.A. (Ed.), International Conference on Advanced Intelligent Systems and Informatics, vol. 533, pp. 3–11. Springer, Cairo (2016). https://doi.org/10.1007/978-3-319-48308-5_1
9. Diwali, A., Kamel, M., Dahab, M.: Arabic text-based chat topic classification using discrete wavelet transform. Int. J. Comput. Sci. **12**(2), 86–94 (2015). Retrieved from http://www.ijcsi.org/papers/IJCSI-12-2-86-94.pdf
10. Kakde, Y.: A Survey of Query Expansion Until. Indian Institute of Technology, Bombay (2012)
11. Singh, J., Sharan, A., Siddiqi, S.: A literature survey on automatic query expansion for effective retrieval task. Int. J. Adv. Comput. Res. **3**(3), 170–178 (2013)
12. Carpineto, C., Romano, G.: A survey of automatic query expansion in information retrieval. ACM Comput. Surv. (CSUR) **44**(1), 1–50 (2012)
13. Ooi, J., Ma, X., Qin, H., Liew, S.C.: A survey of query expansion, query suggestion and query refinement techniques. In: Proceedings of the International Conference on Software Engineering and Computer Systems, pp. 112–117. IEEE (2015)
14. Rocchio, J.J.: Relevance feedback in information retrieval. In: Proceedings of the SMART Retrieval System-Experiments in Automatic Document, pp. 313–323 (1971)
15. Carpineto, C., De Mori, R., Romano, G., Bigi, B.: An information-theoretic approach to automatic query expansion. ACM Trans. Inf. Syst. (TOIS) **19**(1), 1–27 (2001)

Printed in the United States
By Bookmasters